H. Ellard
Personnal copy
July 2010

Clay Materials Used in Construction

Geological Society Special Publications
Society Book Editors
R. J. PANKHURST (Chief Editor)
P. DOYLE
F. J. GREGORY
J. S. GRIFFITHS
A. J. HARTLEY
R. E. HOLDSWORTH
J. A. HOWE
P. T. LEAT
A. C. MORTON
N. S. ROBINS
J. P. TURNER

Special Publication reviewing procedures

The Society makes every effort to ensure that the scientific and production quality of its books matches that of its journals. Since 1997, all book proposals have been refereed by specialist reviewers as well as by the Society's Books Editorial Committee. If the referees identify weaknesses in the proposal, these must be addressed before the proposal is accepted.

Once the book is accepted, the Society has a team of Book Editors (listed above) who ensure that the volume editors follow strict guidelines on refereeing and quality control. We insist that individual papers can only be accepted after satisfactory review by two independent referees. The questions on the review forms are similar to those for *Journal of the Geological Society*. The referees' forms and comments must be available to the Society's Book Editors on request.

Although many of the books result from meetings, the editors are expected to commission papers that were not presented at the meeting to ensure that the book provides a balanced coverage of the subject. Being accepted for presentation at the meeting does not guarantee inclusion in the book.

Geological Society Special Publications are included in the ISI Index of Scientific Book Contents, but they do not have an impact factor, the latter being applicable only to journals.

More information about submitting a proposal and producing a Special Publication can be found on the Society's web site: www.geolsoc.org.uk.

It is recommended that reference to all or part of this book should be made in the following way:

REEVES, G. M, SIMS, I. & CRIPPS, J. C. (eds) 2006. *Clay Materials Used in Construction*. Geological Society, London, Engineering Geology Special Publication, **21**.

GEOLOGICAL SOCIETY ENGINEERING GEOLOGY SPECIAL PUBLICATION NO. 21

Clay Materials Used in Construction

EDITED BY

G.M. REEVES

UHI Millennium Institute, Thurso, UK

I. SIMS

STATS Limited, St Albans, UK

and

J.C. CRIPPS

University of Sheffield, UK

2006
Published by
The Geological Society
London

THE GEOLOGICAL SOCIETY

The Geological Society of London (GSL) was founded in 1807. It is the oldest national geological society in the world and the largest in Europe. It was incorporated under Royal Charter in 1825 and is Registered Charity 210161.

The Society is the UK national learned and professional society for geology with a worldwide Fellowship (FGS) of 9000. The Society has the power to confer Chartered status on suitably qualified Fellows, and about 2000 of the Fellowship carry the title (CGeol). Chartered Geologists may also obtain the equivalent European title, European Geologist (EurGeol). One fifth of the Society's fellowship resides outside the UK. To find out more about the Society, log on to www.geolsoc.org.uk.

The Geological Society Publishing House (Bath, UK) produces the Society's international journals and books, and acts as European distributor for selected publications of the American Association of Petroleum Geologists (AAPG), the American Geological Institute (AGI), the Indonesian Petroleum Association (IPA), the Geological Society of America (GSA), the Society for Sedimentary Geology (SEPM) and the Geologists' Association (GA). Joint marketing agreements ensure that GSL Fellows may purchase these societies' publications at a discount. The Society's online bookshop (accessible from www.geolsoc.org.uk) offers secure book purchasing with your credit or debit card.

To find out about joining the Society and benefiting from substantial discounts on publications of GSL and other societies worldwide, consult www.geolsoc.org.uk, or contact the Fellowship Department at: The Geological Society, Burlington House, Piccadilly, London W1J 0BG: Tel. +44 (0)20 7434 9944; Fax +44 (0)20 7439 8975; E-mail: enquiries@geolsoc.org.uk.

For information about the Society's meetings, consult *Events* on www.geolsoc.org.uk. To find out more about the Society's Corporate Affiliates Scheme, write to enquiries@geolsoc.org.uk.

Published by The Geological Society from:
The Geological Society Publishing House
Unit 7, Brassmill Enterprise Centre
Brassmill Lane
Bath BA1 3JN, UK

Orders: Tel. +44 (0)1225 445046
Fax +44 (0)1225 442836

Online bookshop: www.geolsoc.org.uk/bookshop

The publishers make no representation, express or implied, with regard to the accuracy of the information contained in this book and cannot accept any legal responsibility for any errors or omissions that may be made.

© The Geological Society of London 2006. All rights reserved. No reproduction, copy or transmission of this publication may be made without written permission. No paragraph of this publication may be reproduced, copied or transmitted save with the provisions of the Copyright Licensing Agency, 90 Tottenham Court Road, London W1P 9HE. Users registered with the Copyright Clearance Center, 27 Congress Street, Salem, MA 01970, USA: the item-fee code for this publication is 0267-9914/06/$15.00.

British Library Cataloguing in Publication Data
A catalogue record for this book is available from the British Library.

ISBN 10: 1-86239-184-X
ISBN 13: 978-186239-184-O

Typeset by The Charlesworth Group
Printed by Cromwell Press, Trowbridge, UK

Distributors
USA
AAPG Bookstore
PO Box 979
Tulsa
OK 74101-0979
USA
Orders: Tel. + 1 918 584-2555
Fax +1 918 560-2652
E-mail bookstore@aapg.org

India
Affiliated East-West Press Private Ltd
Marketing Division
G-1/16 Ansari Road, Darya Ganj
New Delhi 110 002
India
Orders: Tel. +91 11 2327-9113/2326-4180
Fax +91 11 2326-0538
E-mail affiliat@vsnl.com

Japan
Kanda Book Trading Company
Cityhouse Tama 204
Tsurumaki 1-3-10
Tama-shi, Tokyo 206-0034
Japan
Orders: Tel. +81 (0)423 57-7650
Fax +81 (0)423 57-7651
Email geokanda@ma.kcom.ne.jp

Dedicated to Professor Sir Alec Skempton, F.R.S., F.R.Eng.
1914–2001
Fellow of the Geological Society from 1944

Mangla Dam, West Pakistan, 1963
Alec Skempton (right) and Peter Fookes (left)

Contents

Foreword	xxi
Editors' note	xxii
Members of the Working Party	xxiii
Acknowledgements	xxiii
Preface	xxiv
List of symbols and units	xxv

1. Introduction 1
    1.1. Clay 1
    1.2. Definitions of clay 2
        1.2.1. Definitions of clay and clay minerals by the AIPEA Nomenclature and CMS Nomenclature Committees 2
        1.2.2. Civil engineering definitions of clay in British practice 4
        1.2.3. International Civil Engineering Soil Classification by particle size distribution (grading) 5
    1.3. Some British clay production statistics 5
    1.4. Geographical and stratigraphical distribution 5
    1.5. The Working Party 7
        1.5.1. The report 10

2. The composition of clay materials 13
    2.1. Clay minerals 13
        2.1.1. The kaolin and serpentine groups 15
        2.1.2. The illite–mica group 15
        2.1.3. Smectites 16
        2.1.4. Vermiculite 17
        2.1.5. Chlorite 17
        2.1.6. Mixed layer clay minerals 19
        2.1.7. Sepiolite and palygorskite 20
        2.1.8. Swelling properties of clay minerals 20
        2.1.9. Ion exchange properties of clay minerals 21
    2.2. Non-clay mineralogy 23
        2.2.1. Quartz and chert 23
        2.2.2. Feldspars 23
        2.2.3. Carbonates 23
        2.2.4. Iron sulphides 24
        2.2.5. Oxides and hydroxides 24
            2.2.5.1. Iron oxides and hydroxides 24
            2.2.5.2. Aluminium oxides and hydroxides 24
            2.2.5.3. Ion exchange in oxides and hydroxides 24
        2.2.6. Sulphates 24
            2.2.6.1. Ettringite group 24
        2.2.7. Zeolites 24
        2.2.8. Phosphates 25
        2.2.9. Halides 25
    2.3. Organic matter 25
        2.3.1. Organic-clay complexes and interactions 26

	2.4.	Water		26
	2.5.	Conclusions		27
	References			27
3.	**Formation and alteration of clay materials**			**29**
	3.1.	Introduction		29
		3.1.1.	The clay cycle	29
		3.1.2.	Clay materials and plate tectonics	29
		3.1.3.	Origins of clay minerals by neoformation inheritance and transformation	30
	3.2.	Weathering environment and soils		32
		3.2.1.	Introduction	32
		3.2.2.	Controls on clay mineral formation during weathering and soil formation	33
			3.2.2.1. Climate	33
			3.2.2.2. Drainage and topography	34
			3.2.2.3. Parent rock type	34
			3.2.2.4. Time	35
	3.3.	Sedimentary environment		36
		3.3.1.	Introduction	36
		3.3.2.	Origin of clay minerals in sediments	37
		3.3.3.	Source and supply of mud	38
		3.3.4.	Erosion, transport, and deposition of mud	38
		3.3.5.	Erosion, transport and deposition by water	39
			3.3.5.1. Transport by rivers	39
			3.3.5.2. Chemical composition of river suspended sediment	39
			3.3.5.3. Transport in the sea and ocean	39
			3.3.5.4. Deposition by settling and flocculation	41
			3.3.5.5. Fabric and microstructures	42
		3.3.6.	Erosion transport and deposition by ice	43
		3.3.7.	Erosion transport and deposition by wind	44
		3.3.8.	Sedimentary neoformed clays	45
	3.4.	Diagenetic to low grade metamorphic environment		48
		3.4.1.	Introduction	48
		3.4.2.	Early diagenetic processes	48
		3.4.3.	Main pathways and reactions of clay mineral diagenesis	49
			3.4.3.1. Smectite-to-illite reaction and illitization.	50
			3.4.3.2. Illite 'crystallinity'	51
			3.4.3.3. Formation of chlorite	53
			3.4.3.4. Diagenesis of kaolin	53
		3.4.4.	Burial diagenetic reactions of non-clay phases in mudrocks	53
		3.4.5.	Compaction, porosity and permeability	54
	3.5.	Hydrothermal environment		55
		3.5.1.	Introduction	55
		3.5.2.	Hydrothermal clay mineral formation	56
	3.6.	Clay materials: products of weathering, sedimentary, diagenetic/low grade metamorphic, and hydrothermal processes		57
		3.6.1.	Clay rich soils	58
		3.6.2.	Residual kaolin deposits	59
		3.6.3.	Vermiculite	59
		3.6.4.	Terrigenous clays, mudstones and shales	59
		3.6.5.	Sedimentary or secondary kaolins	63

	3.6.6.	Sedimentary bentonites and tonsteins	64
	3.6.7.	Chemical perimarine and lacustrine clay deposits	65
	3.6.8.	Deep sea clays	65
	3.6.9.	Hydrothermal kaolin deposits	66
	3.6.10.	Other hydrothermal clay deposits	67
References			67

4. Properties of clay materials, soils and mudrocks — 73

- 4.1. Composition, classification and description of clay materials — 74
  - 4.1.1. Mineralogy — 74
  - 4.1.2. Classification of clays — 75
    - 4.1.2.1. Geological classification — 75
    - 4.1.2.2. Engineering classification — 76
    - 4.1.2.3. Pedological classification — 78
  - 4.1.3. Particle size and grading of clays — 78
  - 4.1.4. Description of clays — 80
  - 4.1.5. Classification of mudrocks — 81
  - 4.1.6. Breakdown of mudrocks — 82
  - 4.1.7. Durability of mudrocks — 83
  - 4.1.8. Description of mudrocks — 84
  - 4.1.9. Classification of weathering in mudrocks — 85
- 4.2. Framework for clay material behaviour — 85
  - 4.2.1. Moisture content, density and saturation — 87
  - 4.2.2. Flow and rheology — 87
  - 4.2.3. Plasticity of clays: Atterberg limits — 89
  - 4.2.4. Permeability and diffusion — 94
  - 4.2.5. Stresses in the ground: pore pressures and effective stress — 95
  - 4.2.6. Unsaturated clays and suction — 96
  - 4.2.7. Clays for ceramic applications — 98
  - 4.2.8. Thermal properties of clays — 99
  - 4.2.9. Compaction of clays — 100
- 4.3. Geotechnical parameters: strength and stiffness of clay materials — 101
  - 4.3.1. Deformation — 102
  - 4.3.2. Strength parameters — 103
  - 4.3.3. Compression — 104
  - 4.3.4. One-dimensional compression and swelling — 107
  - 4.3.5. Critical state strength — 109
  - 4.3.6. Shear strength parameters — 109
  - 4.3.7. Undrained shear strength parameters — 110
  - 4.3.8. Effects of structure — 111
  - 4.3.9. Modulus of rupture — 111
  - 4.3.10. Strength and deformation of mudrocks — 111
- 4.4. Potentially troublesome clay materials — 112
  - 4.4.1. Shrink and swell — 112
  - 4.4.2. Dispersive soils — 114
  - 4.4.3. Collapsible soils — 114
  - 4.4.4. Quick clays — 115
  - 4.4.5. Tropical residual soils — 115
    - 4.4.5.1. Red tropical soils — 116
    - 4.4.5.2. Other tropical soils — 117

	4.5.	Properties of some British clays and mudrocks	117
		4.5.1. Recent and Quaternary deposits	119
		4.5.1.1. Tills	119
		4.5.1.2. Varved and laminated clays	120
		4.5.1.3. Alluvial and Marine clays	120
		4.5.2. Tertiary clays	121
		4.5.2.1. Ball Clay	121
		4.5.2.2. London Clay	121
		4.5.2.3. London Clay—density and moisture content	121
		4.5.2.4. London Clay—particle size and consistency limits	121
		4.5.2.5. London Clay—strength and stiffness	123
		4.5.2.6. London Clay—compaction, consolidation and permeability	124
		4.5.3. Cretaceous clays	124
		4.5.3.1. Gault Clay	124
		4.5.3.2. Gault Clay—density and moisture content	124
		4.5.3.3. Gault Clay—particle size and consistency limits	125
		4.5.3.4. Gault Clay—strength	125
		4.5.3.5. Gault Clay—compressibility and permeability	125
		4.5.4. Jurassic clays	125
		4.5.4.1. Lias Clay	126
		4.5.4.2. Lias Clay—moisture content and density	126
		4.5.4.3. Lias Clay—particle size and consistency limits	127
		4.5.4.4. Lias Clay—strength and stiffness	127
		4.5.5. Permo-Triassic mudrocks	127
		4.5.5.1. Mercia Mudstone	127
		4.5.5.2. Mercia Mudstone—density and moisture content	128
		4.5.5.3. Mercia Mudstone—particle size and consistency limits	128
		4.5.5.4. Mercia Mudstone—strength and deformation	128
		4.5.5.5. Mercia Mudstone—consolidation and permeability	129
		4.5.5.6. Mercia Mudstone—compaction	129
		4.5.6. Carboniferous mudrocks	129
		4.5.6.1. Coal Measures mudrocks	129
		4.5.6.2. Coal Measures mudrocks—moisture content and plasticity	130
		4.5.6.3. Coal Measures mudrocks—strength	130
		4.5.6.4. Coal Measures mudrocks—consolidation and deformation parameters	130
		4.5.6.5. Coal Measures mudrocks—durability	130
	4.6.	Conclusions	131
		4.6.1. Variation in UK clays and mudrocks with geological age	132
		4.6.2. Degradation of over-consolidated clays and indurated mudrocks	133
		4.6.3. Controls on properties	134
	References		134
5.	**World and European clay deposits**		**139**
	5.1.	The character of clay	139
		5.1.1. General	139
		5.1.2. Climate	139
		5.1.3. Time	139
	5.2.	Environments of clay formation	140
		5.2.1. Residual clays of tropical regions	140
		5.2.2. Hydrothermal clays	140

		5.2.3.	Transported clays	141
		5.2.4.	Clays of glacial origin	141
	5.3.	Palaeogeography and clay depositions		141
		5.3.1.	The Palaeozoic	141
			5.3.1.1. Cambrian	141
			5.3.1.2. Ordovician	142
			5.3.1.3. Silurian and Devonian	143
			5.3.1.4. Carboniferous and Permo-Triassic	143
		5.3.2.	The Mezozoic	144
			5.3.2.1. Jurassic	144
			5.3.2.2. Cretaceous	145
		5.3.3.	The Cenozoic	146
	5.4.	Climatic changes in the Quaternary		147
	5.5.	World clay deposits		147
		5.5.1.	'Common' clay	147
		5.5.2.	Kaolin clays	149
		5.5.3.	Smectite clays	150
	References			150

6. British clay stratigraphy . . . 153
6.1. Pre-Carboniferous . . . 156
6.1.1. Silurian shales . . . 157
6.2. Carboniferous . . . 157
6.2.1. Bowland Shale Formation . . . 158
6.2.2. Warwickshire Group . . . 158
6.3. Permian and Triassic . . . 159
6.3.1. Mercia Mudstone Group . . . 159
6.3.2. Penarth Group . . . 161
6.4. Jurassic . . . 161
6.4.1. Lias Group . . . 162
6.4.2. Upper and Lower Fuller's Earth and Frome Clay Formations . . . 164
6.4.3. Blisworth Clay & Forest Marble Formations . . . 164
6.4.4. Oxford Clay Formation . . . 164
6.4.5. West Walton Formation . . . 165
6.4.6. Ampthill Clay Formation . . . 165
6.4.7 Kimmeridge Clay Formation . . . 165
6.5. Cretaceous . . . 166
6.5.1. Wadhurst Clay Formation . . . 166
6.5.2. Weald Clay Formation . . . 166
6.5.3. Atherfield Clay Formation . . . 166
6.5.4. Speeton Clay Formation . . . 166
6.5.5. Gault Formation . . . 167
6.6. Tertiary . . . 167
6.6.1. Ormesby Clay Formation . . . 168
6.6.2. Lambeth Group . . . 168
6.6.3. London Clay Formation . . . 169
6.7. Quaternary deposits . . . 170
6.8. Summary . . . 173
References . . . 173

7. Exploration 177

- 7.1. Introduction 177
  - 7.1.1. The exploration cycle 178
  - 7.1.2. Standards relevant to exploration 178
  - 7.1.3. Personnel 178
- 7.2. Non-intrusive exploration 179
  - 7.2.1. Desk studies 179
  - 7.2.2. Walk-over surveys 181
  - 7.2.3. Field mapping 183
  - 7.2.4. Preliminary assessment report 185
- 7.3. Geophysical investigations 185
  - 7.3.1. Geophysical mapping 185
  - 7.3.2. Borehole geophysical logging 187
- 7.4. Sampling programmes 188
  - 7.4.1 Aims of sampling 188
  - 7.4.2 Sampling strategies and grids 188
  - 7.4.3 Sample numbers 189
- 7.5. Intrusive exploration methods 189
  - 7.5.1. Trial pitting and trenching 189
  - 7.5.2. Dynamic probing and window sampling 191
  - 7.5.3. Boring and drilling 191
  - 7.5.4. Cone penetration testing 192
  - 7.5.5. Logging and description 193
  - 7.5.6. Groundwater measurement 193
- 7.6. Soil samples 194
  - 7.6.1. Sample quality 194
  - 7.6.2. Sample types 195
  - 7.6.3. Sample preservation and storage 196
- References 196

8. Compositional and textural analysis of clay material 199

- 8.1. Mineralogical analysis 199
  - 8.1.1. X-ray powder diffraction analysis 199
    - 8.1.1.1. Mineral identification 201
    - 8.1.1.2. Crystallinity indices of illite/mica 203
    - 8.1.1.3. Peak decomposition techniques 204
    - 8.1.1.4. Quantitative mineralogical analysis 204
  - 8.1.2. Infra-red spectral analysis 206
  - 8.1.3. Thermal analysis 207
    - 8.1.3.1. Differential thermal analysis 207
    - 8.1.3.2. Thermal gravimetric analysis 208
    - 8.1.3.3. Evolved gas detection and analysis 209
- 8.2. Petrographic analysis 209
  - 8.2.1. Optical mineralogy 209
  - 8.2.2. Electron microscopy 209
    - 8.2.2.1. Scanning electron microscopy 209
    - 8.2.2.2. Transmission electron microscopy (TEM) 212
  - 8.2.3. Image analysis techniques 212
  - 8.2.4. X-ray goniometry 213
  - 8.2.5. Specific surface area 213

8.3.	Inorganic geochemical analysis		214
	8.3.1. Bulk compositions and selective leach techniques		214
		8.3.1.1. X-ray fluorescence spectrometry	214
		8.3.1.2. ICP-AES analysis	215
		8.3.1.3. Carbonate determination	215
		8.3.1.4. Sulphur species	216
		8.3.1.5. Selective leaching techniques	216
		8.3.1.6. Interstitial water analysis	216
	8.3.2. Electron microprobe analysis (EMPA)		216
	8.3.3. Ion exchange		217
		8.3.3.1 Cation exchange capacity	217
		8.3.3.2 Anion exchange capacity	218
8.4.	Organic matter		218
	8.4.1 Total organic carbon (TOC) analyses		218
	8.4.2 Rock-Eval pyrolysis		218
	8.4.3 Organic petrography		219
	8.4.4. Biological assays		220
References			220

9. Laboratory testing . . . 223

9.1.	Introduction		223
9.2.	Sample preparation and conduct of testing		223
	9.2.1. Sample preparation		224
		9.2.1.1. Disturbed specimens	224
		9.2.1.2. Undisturbed specimens	224
		9.2.1.3. Recompacted specimens	225
		9.2.1.4. Remoulded specimens	225
	9.2.2. Laboratory environment		225
	9.2.3. Calibration of equipment		225
	9.2.4. Precision of data		226
9.3.	Classification tests		227
	9.3.1. Scope		227
	9.3.2. Moisture content		227
	9.3.3. Liquid and plastic limits		228
		9.3.3.1. Liquid limit	228
		9.3.3.2. Plastic limit	229
	9.3.4. Shrinkage tests		230
		9.3.4.1. Shrinkage limit—definitive method	230
		9.3.4.2. Shrinkage limit—ASTM method	231
		9.3.4.3. Linear shrinkage	231
	9.3.5. Density tests		231
		9.3.5.1. Linear measurement	232
		9.3.5.2. Immersion in water	232
		9.3.5.3. Water displacement	232
	9.3.6. Particle density		233
		9.3.6.1. Small pyknometer method	233
		9.3.6.2. Methods for clays containing coarse particles	233
	9.3.7. Particle size		233
		9.3.7.1. Hydrometer test	234
	9.3.8. Clay suction		234

9.4.	Chemical tests		235
9.5.	Compaction related tests		236
	9.5.1.	Compaction tests (moisture content—density relationship)	236
		9.5.1.1. B.S. 'light compaction' test	239
		9.5.1.2. B.S. 'heavy compaction' test	239
		9.5.1.3. Compaction in CBR mould	240
	9.5.2.	Moisture condition value	240
	9.5.3.	California bearing ratio	242
9.6.	Dispersibility and durability tests		242
	9.6.1.	Pinhole method	243
	9.6.2.	Crumb and cylinder tests	243
	9.6.3.	Dispersion method	243
9.7.	Consolidation tests		243
	9.7.1.	One-dimensional consolidation in an oedometer	245
		9.7.1.1. Procedure	245
	9.7.2.	Constant, or controlled, rate of strain oedometer consolidation	247
	9.7.3.	Consolidation in a hydraulic consolidation cell	249
		9.7.3.1. Test procedures	252
	9.7.4.	Isotropic consolidation in a triaxial cell	253
		9.7.4.1. Equipment and test procedure	253
	9.7.5.	Continuous-loading consolidation tests	254
		9.7.5.1. Constant rate of loading	254
		9.7.5.2. Constant pore pressure gradient	254
		9.7.5.3. Constant pressure ratio	255
		9.7.5.4. Consolidation with restricted flow	255
		9.7.5.5. Back pressure control	255
9.8.	Permeability tests		255
	9.8.1.	Falling head test	255
	9.8.2.	Permeability in a hydraulic consolidation cell	256
		9.8.2.1. Scope	256
		9.8.2.2. Procedure and calculations	257
	9.8.3.	Permeability in a triaxial cell	257
		9.8.3.1. Procedure and calculations	257
9.9.	Shear strength tests: total stress		258
	9.9.1.	Vane shear test	258
	9.9.2.	Unconfined compression test	259
		9.9.2.1. Portable autographic apparatus	259
		9.9.2.2. Load frame	259
	9.9.3.	Undrained triaxial compression test	259
	9.9.4.	Direct shear test	260
		9.9.4.1. Consolidation stage	261
		9.9.4.2. Shearing stage	261
	9.9.5.	Direct shear test for residual strength	262
	9.9.6.	Ring shear test for residual shear strength	262
9.10.	Effective stress shear strength tests		263
	9.10.1.	Principles	264
	9.10.2.	Samples and equipment	264
	9.10.3.	Saturation and consolidation stages	265
	9.10.4.	Consolidated undrained compression (CU) test	266
	9.10.5.	Consolidated drained compression (CD) test	267

9.11.	'Advanced' tests			268
	9.11.1.	Measuring devices		270
	9.11.2.	Stress path testing		270
	9.11.3.	B-bar test		271
	9.11.4.	Other 'advanced' test procedures		271
		9.11.4.1.	Initial effective stress	272
		9.11.4.2.	Triaxial extension	272
		9.11.4.3.	Stress ratio K_o	272
		9.11.4.4.	Anisotropic consolidation	272
		9.11.4.5.	Resonant column	272
		9.11.4.6.	Simple shear	273
		9.11.4.7.	Cyclic triaxial	273
References				273

10. Earthworks 275

10.1.	Types of engineered fill			276
	10.1.1.	Infrastructure embankments		276
	10.1.2.	Earth dams		279
	10.1.3.	Environmental mitigation structures		284
	10.1.4.	Substructure fills		285
	10.1.5.	Quarry tips		285
10.2.	Planning requirements			286
	10.2.1.	Planning considerations		286
	10.2.2.	Environmental and safety considerations		287
		10.2.2.1.	Visual intrusion	287
		10.2.2.2.	Stability of slopes	287
		10.2.2.3.	Waste disposal	287
		10.2.2.4.	Vehicular traffic	288
		10.2.2.5.	Noise	288
		10.2.2.6.	Dust	288
		10.2.2.7.	Ground and surface waters	288
		10.2.2.8.	Flora and fauna	289
		10.2.2.9.	Amenity	289
		10.2.2.10.	Archaeology	289
		10.2.2.11.	Restoration	289
	10.2.3.	Parties to the development		290
10.3.	Material characteristics of new fills			290
	10.3.1.	Earthmoving in clay		290
	10.3.2.	Fill characteristics		291
	10.3.3.	Design		293
		10.3.3.1.	Shear strength	293
		10.3.3.2.	Compaction	293
		10.3.3.3.	End-product compaction specifications	293
		10.3.3.4.	Method compaction specifications	294
		10.3.3.5.	Slope stability	294
		10.3.3.6.	Setting moisture content limits for the construction of fills	294
		10.3.3.7.	Sub grade properties for infrastructure embankments and substructure fills	295
		10.3.3.8.	Making maximum use of clay in construction	296
	10.3.4.	Deleterious chemical effects		296

10.4.	Long-term material characteristics and maintenance		297
	10.4.1.	Clay degradation	297
	10.4.2.	Types and effect of maintenance	299
	10.4.3.	Long-term requirements of clay fills	299
		10.4.3.1. Operational safety	300
		10.4.3.2. Synergy with other assets	300
		10.4.3.3. Costs of repair	300
		10.4.3.4. Disruption and customer satisfaction	300
		10.4.3.5. Costs of failure	300
	10.4.4	Remedial and preventative measures	301
10.5.	Improved clay		301
	10.5.1.	Mechanical improvement	302
	10.5.2.	Chemical treatment	304
		10.5.2.1. Lime	304
		10.5.2.2. Lime improvement	305
		10.5.2.3. Lime stabilization	305
		10.5.2.4. Lime and cement stabilization	306
		10.5.2.5. Stabilizing material for use in highways and substructure capping	306
		10.5.2.6. Treating materials for use in bulk fill	306
		10.5.2.7. Materials characteristics	307
		10.5.2.8. Lime piles and lime-stabilized clay columns	307
		10.5.2.9. Safety and the environment	307
10.6.	Case histories		308
	10.6.1.	A34 Newbury Bypass	308
	10.6.2.	Orville Dam	310
References			311

11. Earthmoving — 315

11.1.	Earthmoving practice		315
	11.1.1.	Preparation works for earthworks in construction and for mineral deposits	316
	11.1.2.	Bulk earthmoving for construction	317
	11.1.3.	Bulk earthmoving in quarries and other mineral workings	317
11.2.	Methods of earthmoving		320
	11.2.1.	Manual work	320
	11.2.2.	Plant that can only excavate	320
	11.2.3.	Plant that can excavate and load materials	321
	11.2.4.	Plant that hauls and deposits	322
	11.2.5.	Plant that can excavate, load, haul and deposit materials	323
	11.2.6.	Plant that can compact materials	324
	11.2.7.	Specialist earthworks plant	326
	11.2.8.	Plant used in excavation and spoil removal in tunnelling	326
	11.2.9.	Plant teams	327
11.3.	Selection of earthmoving and compaction plant		328
	11.3.1.	Material factors	329
	11.3.2.	Quantities	334
	11.3.3.	Spatial factors	335
	11.3.4.	Surface water and groundwater	335
	11.3.5.	Plant and other factors	336

11.4.	The control of earthworks		336
	11.4.1. Acceptability criteria for earthworks materials		336
	11.4.2. Control during earthmoving		337
	11.4.3. Behaviour of clays during earthmoving		339
11.5.	Examples of earthmoving schemes		340
	11.5.1. Clay and sand quarry		340
	11.5.2. Leighton–Linslade Bypass		340
11.6.	Safety on earthmoving sites		343
	11.6.1. Safety of earthworks		343
	11.6.2. Safely related to earthmoving plant		343
References			344

12. Specialized applications — 347

- 12.1. Principles — 347
  - 12.1.1. Scope of chapter — 347
  - 12.1.2. Bentonite — 347
  - 12.1.3. Interaction of slurries and natural ground — 349
  - 12.1.4. Test procedures — 350
- 12.2. Clay slurries — 352
  - 12.2.1. Introduction — 352
  - 12.2.2. Applications — 352
    - 12.2.2.1. Diaphragm walls and piles — 352
    - 12.2.2.2. Slurry tunnelling — 355
    - 12.2.2.3. Pipe jacking and microtunnelling; shaft sinking — 358
    - 12.2.2.4. Soil conditioning in earth-pressure-balance tunnelling — 358
    - 12.2.2.5. Vertical and horizontal (directional) drilling — 358
  - 12.2.3. Re-use and disposal of slurries — 361
- 12.3. Plastic clay — 362
  - 12.3.1. Introduction — 362
  - 12.3.2. Waste disposal facilities — 362
    - 12.3.2.1. Landfill liners and covers — 362
    - 12.3.2.2. Geosynthetic clay liners — 367
  - 12.3.3. Radioactive waste storage and disposal — 368
    - 12.3.3.1. Introduction and background — 368
    - 12.3.3.2. Functions and materials for disposal vaults — 369
    - 12.3.3.3. Buffer materials — 370
    - 12.3.3.4. Room and tunnel backfills — 372
    - 12.3.3.5. Bulkheads, plugs and grout — 372
    - 12.3.3.6. Shaft seals — 373
    - 12.3.3.7. Exploration borehole completions — 373
  - 12.3.4. Bentonite-cement and grouts — 373
    - 12.3.4.1. Bentonite-cement soft piles and cut-off walls — 373
    - 12.3.4.2. Grouting — 375
    - 12.3.4.3. Treatment of contained land — 377
  - 12.3.5. Puddle clay — 377
    - 12.3.5.1. Introduction — 377
    - 12.3.5.2. Puddle clay and the 'Pennines' embankment dam — 377
    - 12.3.5.3. Puddle clay: nature and characteristics — 378
    - 12.3.5.4. Puddle clay cores: construction practice and specification — 379
    - 12.3.5.5. Puddle clay in cores: characteristics — 379

		12.3.5.6.	Consistency	379

                    12.3.5.6. Consistency 379
                    12.3.5.7. Shear strength 380
                    12.3.5.8. *In situ* permeability 380
                    12.3.5.9. Compressibility 381
                    12.3.5.10. Erodibility and dispersivity 381
                    12.3.5.11. Puddle clay service performance 381
           12.3.6. Other applications 381
                    12.3.6.1. Sealing of lagoons, ponds and reservoirs . 381
                    12.3.6.2. Tanking systems 381
                    12.3.6.3. Borehole sealing 381
   Appendix 12.1 Bentonite slurry for diaphragm walling 382
   Appendix 12.2 British Waterways: Technical Services—Standard Specification for Puddle Clay. . 382
   References 383

13. Earthen architecture **387**
   13.1. The use of earth as a global building material 387
   13.2. Applications of earth in building construction 389
   13.3. Characterization of earth for building 391
   13.4. The performance of earth in building 392
         13.4.1. Structural performance 392
         13.4.2. Durability 392
         13.4.3. Thermal performance 392
         13.4.4. Fire resistance 392
   13.5. Engineered earth 392
   13.6. Specification and practice 393
   13.7. The tradition of earth building in the UK 393
   13.8. New earth building in the UK 396
   13.9. Advantages of earth as a contemporary building material 398
   13.10. Conclusions 399
   References 399

14. Brick and other ceramic products **401**
   14.1. Historical utilization of clay in structural and fine ceramics 401
   14.2. Ceramic product range 402
         14.2.1. Ceramic bodies 402
   14.3. Clay minerals in ceramic products 404
         14.3.1. Kaolinite 404
         14.3.2. Illite 405
         14.3.3. Halloysite 405
         14.3.4. Montmorillonite (smectite) 405
         14.3.5. Chorite 405
   14.4. Structural or heavy clay products 405
         14.4.1. Bricks 406
         14.4.2. Pipes 406
         14.4.3. Tiles and other applications 406
   14.5 Clays utilize in structural ceramic products 407
         14.5.1. Common clay and shale 407
         14.5.2. Fireclay 407
         14.5.3. Ball clay 408

14.6.	Structural or clay extraction, processing and manufacture		408
	14.6.1.	Extraction	408
	14.6.2.	Pre-treatment and preparation	408
	14.6.3.	Forming	409
	14.6.4.	Drying	410
	14.6.5.	Firing	410
14.7.	Engineering properties of bricks		411
	14.7.1.	Porosity	411
	14.7.2.	Strength	412
14.8.	Fine ceramic or whiteware products		412
	14.8.1.	Wall tiles	412
	14.8.2.	Floor tiles	412
	14.8.3.	Tableware and electrical porcelain	413
	14.8.4.	Sanitaryware	413
14.9.	Clay raw materials for fine ceramic products		413
	14.9.1.	Ball clay	413
	14.9.2.	Kaolin (china clay)	417
	14.9.3.	Flint clay	418
14.10.	Fine ceramic clay production and refining		418
	14.10.1.	Ball clay	418
	14.10.2.	Kaolin	420
14.11.	Fine ceramic manufacturing processes		422
	14.11.1.	Ceramic body preparation	422
	14.11.2.	Ceramic article forming	422
	14.11.3.	Drying and firing	423
14.12.	Future trends in ceramics		424
	14.12.1.	General trends	424
	14.12.2.	Processes and materials	424
References			425

15. Cement and related products — 427

15.1.	Introduction		427
	15.1.1.	What is cement?	427
	15.1.2.	History and development of modern cement	428
15.2.	Portland cement		429
	15.2.1.	Types of clay used in the manufacture of Portland cement	429
	15.2.2.	Role of clays in the manufacture of Portland cement	430
	15.2.3.	Composition and hydration reactions of Portland cement	431
	15.2.4.	White Portland cement	432
	15.2.5.	Portland–Pozzolana cement	433
	15.2.6.	Calcined clays as Portland cement additions	434
15.3.	Lime and hydraulic lime cements		435
	15.3.1.	Natural hydraulic lime cements	435
	15.3.2.	Lime mortars	435
15.4.	Other types of cement		436
	15.4.1.	'Inorganic polymer cements' or chemical cements	436
		15.4.1.1. Geopolymer cement	436
		15.4.1.2. Pyrament concrete	436
		15.4.1.3. Tectoaluminate cement	437
		15.4.1.4. Earth ceramics and calcium silicate bricks	437

		15.4.2. Clays in calcium aluminate cements	437
		15.4.3. Gypsum–Portland–pozzolana cements	438
	15.5.	Use of clay in specialist applications	439
		15.5.1. Lightweight aggregates	439
		15.5.2. Rheology control	439
		15.5.2.1. Ready mixed concrete	439
		15.5.2.2. Cement grouts	439
		15.5.3. Immobilization of toxic waste	440
	15.6.	Recycled clays in concrete	441
		15.6.1. Introduction	441
		15.6.2. Ash from municipal solid waste (MSW)	441
		15.6.3. Ash from industrial processes	441
		15.6.4. Waste bricks	442
		15.6.5. Waste cracking catalyst	443
		15.6.6. Building rubble as recycled aggregate	443
		15.6.7. Dredging spoils	444
		15.6.8. Other waste materials	444
	15.7.	Deleterious effects of clays in concrete	444
	References		445

Appendix A Mineralogical and chemical data . . . 449

Appendix B Properties data . . . 461

Appendix C Some World, European and UK clay deposits . . . 475

Appendix D Test methods for clay materials . . . 481

Glossary . . . 485

Index . . . 513

CAUTION
Use of Formation-specific Data for Clay Deposits

In this Working Party report, various example data and 'typical values' of properties are included for particular types of material and named geological formations. These are given strictly for guidance purposes. As the performance of clay materials and clay deposits is likely to be highly variable and dependent not only on material properties, but on mass-specific and site- or source-specific conditions, great care must be taken in using these values. Assessments of particular clay materials should be based on property data obtained for representative samples from actual sites or sources and must not rely on the example data or 'typical values' given in this report.

Foreword

Lord Oxburgh of Liverpool, Chairman of Shell plc and former President of the Geological Society (2000–2002).

Clay is one of the most abundant constituents of the Earth's surface. It covers most of the floors of the oceans and is common on land. Since ancient times it has provided human beings with a very useful material from which they could fashion objects in an astonishing range of sizes and with an astonishing range of properties, from fine brooches to great dams, from paper to building bricks, not to mention its medicinal uses.

Yet for all its value and importance, clay has been an elusive material to understand. The reason, of course, is the very nature of clay—the fact that it is composed of grains too fine for the eye to see and only thousandths of a millimetre across. Furthermore, it has few characteristics to excite the imagination of traditional field geologists, except that it was the medium in which fossils could often be near-perfectly preserved.

It had been known for many centuries that there were different kinds of clay. Some were suitable for cement making, some for fine china, some for making bricks, etc. Although the optical microscope provided the first clues to the fine structure of clays, it was the development of the electron microscope and X-ray structural methods that revealed the fine details and secrets of the differences between different clays, and between the different clay minerals.

It is the interaction of the minute clay grains with water that gives clays some of their unique properties both in bulk and in detail. Water can lie along the boundaries of the grains in clays and its concentration can profoundly affect the bulk physical properties – from a creamy slurry to the heavy digging ground, unbeloved of gardeners, to the geologist's solid mudstone. Water molecules can also lie loosely within the structures of some of the clay minerals or they can be an integral and tightly bound part of the mineral structure. Paradoxically, perhaps, these properties mean that in bulk, clays are among the least permeable of rocks; their very low porosity makes them excellent containment for aquifers or hydrocarbon reservoirs.

The combination of fine grain size and water has another unique and important consequence—chemical reactivity. Water along grain boundaries provides an excellent medium for the diffusion of chemical species on the local scale, and the fine grain size means that the surface area/volume ratio is very favourable to chemical reactionss. These reactions may take place at ambient temperatures such as in the setting of cement, or at high temperatures in the potter's kiln, and result in a profound change in mineralogy and in bulk physical properties.

I hope that even the most sceptical readers will be convinced that clays are both important and interesting and that at the very least they should dip into this volume. If they do, I have no doubt that they will be amply rewarded. To my mind the book succeeds in meeting most, if not all, of the requirements of a professional reference work with a wide range of information important to clay professionals and to the civil engineer, general geologist, or for that matter archaeologists, alike. Even those with a lay interest in clays and their historical and practical importance will find a great deal to interest them.

The Geological Society is doing a great service by publishing this unique volume that represents a number of years of hard work on the part of the Clay Working Party of the Engineering Group. Each of the fifteen chapters was written by a small team of experts in their fields and the Working Party has brought together information from an enormous range and diversity of sources, reinforced and collated in the light of the experience of professionals in many areas. It will become a standard reference for those whose work involves dealing with clay and a source of pleasure and information for those who choose to dip into it at random. Both the Society and the Working Party are to be congratulated.

Ron Oxburgh
29 August 2004

Editors' note

During a distinguished career over more than 50 years, the Chairman of the Clay Working Party, Professor Peter G Fookes, F.R.Eng., has been a tireless supporter of the Engineering Group of the Geological Society, including the initiation and chairmanship of many working parties that have generated influential, often milestone, reports. This important work has included the unique trilogy of major reports on geological materials in construction of which this current volume forms the concluding part. Peter has said that, in retirement, he should make way for others to undertake the chairmanship of any future working parties. The editors of this valedictory report therefore wish to record their appreciation, on behalf of engineering geologists throughout the profession, for the monumental contribution to their discipline made by Peter Fookes, both personally and also by encouraging or helping others. Peter has always explained that 'his work is his hobby', so it must be hoped that his wisdom and experience will continue to be available, as a consultant and teacher, for many years to come, even as he enjoys a well-earned retirement.

The photograph was taken by Ken Head somewhere in the Salt Range in West Pakistan in March 1965, when a young Peter Fookes (looking at the camera) had characteristically organised a weekend geological field trip.

Members of the Working Party

Professor Peter G. Fookes	*Chairman*—Consultant, Winchester SC & 1998–2004
Dr Ian Sims	*Secretary & Editor (2)*—STATS Limited, St Albans SC & 1998–2004
Mr John H. Charman	Consultant, Dorchester SC & 1998–2004
Dr John C. Cripps	*Editor (3)*—University of Sheffield 1998–2004
Mr Kenneth H. Head	Retired soil engineers, Cobham (Surrey) 2003–2004
Dr Stephen Hillier	Macaulay Institute, Aberdeen 1999–2004
Mr Peter Hobbs	British Geological Survey, Nottingham 1999–2004
Mr Philip C. Horner	Consultant, Luton SC & 1998–2004
Dr Thomas R. Jones	Imerys Minerals Limited, formerly ECC International plc (retired) 1999–2004
Mr John R. Masters	Geolabs Limited, Watford 1999–2003
Dr George W. E. Milligan	Geotechnical Consulting Group, London SC & 1998–2004
Mr David R. Norbury	CL Associates, Wokingham 1999–2004
Dr John Perry	Mott MacDonald Group, Croydon SC & 1998–2004
Dr Alan Poole	University of London, Queen Mary College (retired) 2003–2004
Dr George M. Reeves	*Editor (1)*—University of Newcastle (now at the Decommissioning & Environmental Remediation Centre, UHI Millennium Institute, Thurso) 1998–2004
Dr Harry F. Shaw	University of London, Imperial College (Department of Earth Science & Engineering) SC & 1998–2004
Dr Michael R. Smith	*Advisor*—The Institute of Quarrying, Nottingham SC & 1998–2000
Ms Linda Watson	University of Plymouth (School of Architecture) 1999–2004
Dr Graham West	*Geological Society Engineering Group Representative*—Retired engineering geologist, Wokingham SC & 1998–2004
Mr Gordon J. Witte	WBB Minerals Limited, Sandbach SC & 1998–2004

The Working Party was inaugurated in 1998 and completed its work in 2004.
SC, Served on Steering Committee, 1998

Acknowledgements

The Working Party has been most grateful to receive considerable help from numerous individuals, organizations and companies who have contributed in various ways by providing advice, information and illustrations or by acting as reviewers for particular chapters. We give thanks to all these people and organisations. In addition, with apologies to anyone who might have been inadvertently omitted, those to whom specific appreciation is due include the following:

Prof. John M. Adams, Ms Ruth E. Allington (Geoffrey Walton Practice), Association of Geotechnical & Geoenvironmental Specialists (AGS: Ms Dianne Jennings), Prof. John H. Atkinson (City University, London), Dr Maxwell E. Barton (University of Southampton), Prof. Frederick G. Bell, Mr Andrew Bloodworth (British Geological Survey), Mr Michael Cambridge (Knight Piesold), Dr Andrew Charles (Building Research Establishment, retired), Mr Rodney Chartres (Bullen Consultants), Mr Adam Czarnecki (Cory Environmental (Central) Limited), Dr Maurice A. Czerewko (University of Sheffield), Mr William Darbyshire (Environment Agency), Mr Anthony B. Di Stefano (Knight Piesold), Dr Mark Dyer (University of Durham), Dr Eric Farrell (University of Dublin, Trinity College), Mr John Ferguson (Balfour Beatty Limited), Mrs Margaret Ford (University of Plymouth), Dr David A. Greenwood, Dr John Greenwood (Nottingham Trent University), Dr W. J. Rex Harries, Mr Arthur Harrisson (Rugby Cement plc), Mr Ian Higginbottom (Wimpey Laboratories, retired), Mr Graham Holland (British Waterways), Dr Jennifer Huggett (Petroclays), Prof. Richard Jardine (University of London, Imperial College), Prof. Stephan A. Jefferis (University of Surrey), Dr Ian Jefferson (Nottingham Trent University), Mr G. Peter Keeton (Soil Mechanics Limited), Miss Siân Kitchen (STATS Limited), Prof. David Manning (University of Newcastle), Dr Richard J Merriman (British Geological Survey), Mr A. Iain Moffat (University of Newcastle), Dr Dewey Moore (University of New Mexico, USA), Mr Timothy Murdoch (Allott & Lomax), Dr Paul Nathanail (University of Nottingham), Dr Philip J. Nixon (Building Research Establishment Limited), Dr Myriam Olivier, Dr Kevin Privett (SRK Consulting), Mr Nicholas Ramsey (Fugro Limited), Mr David Richardson (The Banks Group), Mr Peter Shotton (Cementation Foundations Skanska), Dr Alan Smallwood (W. S. Atkins Consultants Limited), Dr Andrew S. Smith (Hanson Brick Limited), Dr Douglas I. Stewart (University of Leeds), Mr Michael G. Sumbler (Rockvision Consulting), Dr Mark Tyrer (University of London, Imperial College), Dr Peter N. W. Verhoef (University of Delft, Netherlands), Dr Paul Wignall (University of Leeds), Dr J. Michael Woodfine.

The very significant contribution made by private individuals and employers whose staff were afforded the time and permission to be members of the Working Party cannot be overstated and we offer grateful thanks to those companies and institutions.

Preface

This book is about clay, as seen through the eyes of construction and production engineers and geoscientists. So, what is clay and why do we need a new book on it? The answer to the first question cannot be given here in the Preface, since the word has different meanings, depending on the situation and context, and its occurrence and uses are ubiquitous and diverse. Definitions of clay, from various viewpoints are given in Chapter 1, although these are by no means exhaustive, and further references for the interested reader are also to be found in that chapter.

In the following chapters, clay is generally used in the commonly understood sense of being a fine-grained, natural, earthy material which, at certain moisture contents, displays plasticity. Where, within a chapter, usage differs from this general meaning, the appropriate definition for the particular context is given.

This book, which is a Working Party Report of the Engineering Group of the Geological Society, considers most aspects of clay, including its identification and evaluation; occurrence and distribution; extraction; various forms of processing; use in man-made products; and engineering uses, for example in embankments, but excluding the engineering aspects of clay *in situ*. The latter topic was considered by the Engineering Group Committee to be such a large subject that inclusion would have distorted the intention and thrust of the book. With so many applications in various industries, it was considered a book was needed to present and explain, in a coherent format, the very wide range of geological information about clay. In this way information known within one discipline can be disseminated to the benefit of other applications.

The Report is an addition to the group of several important Working Party Reports and other documents published by the Engineering Group in the past four decades or so, which focus, from a geological viewpoint, on the help that applied geology can provide to practitioners in the diverse industries related to the subject matter. This particular Report is the final part of a trilogy of such books on geological materials in construction:

- *Aggregates*. First edition, 1985; second edition, 1993; third edition, 2001.
- *Stone: Building Stone, Rockfill and Armourstone in Construction*. First edition, 1999.

These first two members of the trilogy have proved most successful and it is hoped that *Clay Materials used in Construction* will also prove to be so.

The membership of this Working Party, like its predecessors, was drawn widely from industrial, academic and institutional concerns. Every member of the Working Party worked long and hard and members gave freely of their time or their employers' time, for which I am most grateful. Any credit should be attributed equally to every member of the Working Party.

This book, also like its predecessors, is the combined work of the whole Working Party membership. Each chapter was initially drafted by one or more members of the working party, combined with many day-long meetings. It was subsequently edited and cross-checked by all the members as an iterative process and commented upon by corresponding members. This book, like its predecessors, therefore carries the collective authority of the whole Working Party. Appendices were handled in a similar way and particular effort was made to make the book as comprehensive as possible and, again despite being drafted by a committee, to achieve reasonable continuity of style and uniformity in technical level and mode of presentation.

I conclude with two personal sentiments. Firstly, a very warm and special thank you to Dr Ian Sims who for the past twenty-five years of the forty or so years I have been Chairman of various Working Parties, has been the Secretary. As Secretary, he has done an absolutely superb job in the face of many difficulties, with consistently sound advice, immaculate meeting minutes, endless correspondence and telephone calls, with virtually no complaint about any chore with which he was saddled. It therefore gave me, and the Working Party, great pleasure to see that he was chosen to receive the Engineering Group Award for 2003. Congratulations, Ian.

My second is the dedication of this Working Party Report to the memory of Professor Sir Alec Skempton, F.R.S., F.R.Eng. I had the enormous privilege of working with him on the Mangla, Jari and Sukian Dams in West Pakistan in the 1960s, when I was a young man on site and he was the visiting consultant. His visits were inspirational and I learnt much from him that has stood me in good stead for the rest of my career. I came to regard him with tremendous respect and awe. These views I still hold today. 'Skem' worked in many engineering soils, even rock on occasion, but his principal medium was clay. He was always more than willing to pass on his knowledge to others, and this book has been written in the same spirit. The Working Party thus has no hesitation in dedicating our book to him. This has also given me much pleasure.

Peter Fookes
Chairman of the Working Party

Winchester
May 2002 and 2004

List of symbols and units

Below are listed units and symbols referred to in this book. Further symbols and explanation of those below are presented in the relevant chapters.

Basic units

Symbol	Quantity	Description
Å, μm, mm, m, km	distance	Angstrom unit (10^{-10} m), nanometre (10^{-9} m), microns (10^{-6} m), millimetre (10^{-3} m), metre, kilometre (10^3 m)
m^2	area	square metres
s, min, h, a, Ma	time	second, minute, hour, year (annum), millenia (10^6 a)
μg, mg, g, kg, Mg, T	mass	microgram (10^{-6} g), milligram (10^{-3} g), gram, kilogram (10^3 g), megagram (10^6 g), tonne (10^6 g)
g	acceleration due to gravity	9.81 m/s²
N, kN, MN, GN	weight, force	newton, kilonewton (10^3 N), meganewton (10^6 N), giganewton (10^9 N) [1 N = 1 kg × g = 1 kg × 9.81 m/s²]
Pa, kPa, MPa, GPa; N/m² kN/m² MN/m², GN/m²	pressure; stress	pascal, kilopascal, megapascal, gigapascal; newton, kilonewton, meganewton and giganewton per square metre [1 Pa = 1 N/m²]
m/s, km/h	velocity	metres per second, kilometres per hour
°C	temperature	degrees Celsius, centigrade
ml, m³, Mm³	volume	millilitres, cubic metres, mega cubic metres (10^3 m³)
kW, MW	power thermal capacity	kilowatt, megawatt

Soil characteristics

Symbol	Unit	Description
e		voids ratio [e = volume of voids/volume of soil grains]
v		specific volume [total volume of soil/volume of soil grains]
n	%	porosity [n = volume of voids/total volume of soil]
w	%	moisture content [mass of water × 100/mass of dry soil]
w_{opt}	%	optimum moisture content
w_L, w_p, w_s	%	liquid, plastic and shrinkage limits for soils
MCV		moisture condition value
I_p	%	plasticity index [$I_p = w_L - w_p$]
A		activity [$A = I_p$/weight % clay size]
G_s		density of soil particles [G_s = weight of soil particles/volume of soil particles]
γ	kN/m³	unit weight [γ = weight of soil/volume of soil]
γ_w	kN/m³	unit weight of water [γ_w = 9.81 kN/m³]
γ_b, γ_d	kN/m³	bulk (wet) and dry unit weight of soil
γ_{dmax}	kN/m³	maximum dry unit weight
ρ	Mg/m³ tonne/m³	density [ρ = mass of soil/volume of soil]
ρ_b, ρ_d	Mg/m³ T/m³	bulk (wet) and dry density of soil
ρ_{dmax}	Mg/m³	maximum dry density
S_r		Saturation ratio [S_r = volume of water/total volume of voids]
A	%	air voids percentage [A = volume of voids × 100/total volume]
A_r		aggregation ratio [A_r = % clay mineral/% clay size]
U		coefficient of uniformity (of soil grading)
C		coefficient of curvature (of soil grading)
K	m/s	hydraulic conductivity, coefficient of (field) permeability
k_i	m², darcy	intrinsic permeability

Soil and rock properties

Symbol	Unit	Description
σ	kPa, kN/m²	normal stress in terms of total and effective stress (compression positive)
σ_v, σ_v'	kPa, kN/m²	vertical or overburden total and effective stresses
σ_h, σ_h'	kPa, kN/m²	horizontal or lateral total and effective stresses

Symbol	Units	Description
$\sigma_1, \sigma_2, \sigma_3$	kPa, kN/m²	major, intermediate and minor principal total stresses
$\sigma_1', \sigma_2', \sigma_3'$	kPa, kN/m²	major, intermediate and minor principal effective stresses
τ	kPa, kN/m²	shear stress
P, p'	kPa, kN/m²	mean principal total and effective stresses respectively
u	kPa, kN/m²	pore (water) pressure
s	kPa, pF	pore (water) suction
ε_s		shear strain
$\varepsilon_1, \varepsilon_2, \varepsilon_3$		major, intermediate and minor strains
ε_{vol}		volumetric strain
γ		engineering shear strain
v_κ		specific volume of over-consolidated soil at $p' = 1.0$ kPa
N_p		specific volume of isotropically normally consolidated soil at $p' = 1.0$ kPa
λ		slope of isotropic normal compression line (v versus $\ln p'$)
κ		slope of isotropic normal swelling line (v versus $\ln p'$)
R_p		one dimensional over-consolidation ratio (ratio of maximum effective over burden stress to the present value)
K_0		coefficient of earth pressure at rest (σ_n'/σ_v')
$K_0 C_c$		compression index (slope of e-log effective consolidation pressure for compression)
C_s		swelling index (slope of e-log effective consolidation pressure for swelling)
m_v	m²/MN	one-dimensional coefficient of compressibility
c_v	m²/s, m²/yr	coefficient of consolidation
E'_0	kN/m²	one-dimensional stiffness modulus
c_u, c'	kPa, kN/m²	cohesion intercept in terms of total stresses and effective stresses respectively
ϕ_u, ϕ'	degrees	angle of friction in terms of total stresses and effective stresses respectively
c_{crit}', c_p', c_r'	kPa, kN/m²	cohesion intercept in terms of effective stresses respectively for critical, peak and residual conditions
$\phi_{crit}', \phi_p', \phi_r'$	degrees	angle of friction in terms of effective stress respectively for critical, peak and residual conditions
c_a, ϕ_a	kPa, degrees	apparent cohesion intercept and angle of friction
s_u	kPa, kN/m²	undrained shear strength
σ_c	MPa	uniaxial (unconfined) compressive strength (UCS)
m, s		empirical strength parameters for rocks
E, K, G	MPa, GPa	deformation moduli in terms of linear, volumetric and shear strain respectively for total stresses
E', K', G'	MPa, GPa	deformation moduli in terms of linear, volumetric and shear strain respectively for effective stresses
E_i, E_{sec}, E_{tan} K_i, K_{sec}, K_{tan} G_i, G_{sec}, G_{tan} E_i', E_{sec}', E_{tan}' K_i', K_{sec}', K_{tan}' G_i', G_{sec}', G_{tan}'	MPa, GPa	deformation moduli in terms of linear strain respectively in terms of total stresses measured respectively at the initial part of the stress–strain curve, total stress/total strain at a particular strain on stress–strain curve and as the slope of the stress–strain curve at a particular strain. May also be expressed in terms of effective stresses. Similar moduli in terms of volumetric and shear strain may also by defined, as shown
ν		Poisson's ratio
I_{dn}	%	slake durability index (n = number of cycles)

1. Introduction

Clay, noun. Old English <u>Claég</u>. A stiff viscous earth. (Blackies Compact Etymological Dictionary. Blackie & Son, London and Glasgow. 1946. War Economy Standard)

Clay: The original Indo-European word was 'gloi-', 'gli-' from which came 'glue' and 'gluten'. In Germanic this became 'klai-', and the Old English 'claeg' became Modern English 'clay'. From the same source came 'clammy' and the northern England dialect 'claggy' all of which describe a similar sticky consistency. (Oxford English Dictionary and Ayto's Dictionary of Word Origins, Bloomsbury, 1999)

Clay: from Old Greek γλἰα, γλοἰα "glue", γλἰνή "slime, mucus", γλοιός "anything sticky"' from I.-E. base *glei-, *gli- 'to glue, paste stick together. (Klein E. A comprehensive etymological dictionary of the English language. Elsevier, Amsterdam, 1967; Skeat W. An etymological dictionary of the English language. Oxford University Press, 1961; Mann S.E. An Indo-European comparative dictionary. Buske Verlag, Hamburg, 1987)

1.1. Clay

Definitions of clay are given in Section 1.2. The *uses* of clay are ubiquitous and diverse. On a world scale, clay is of major economic significance, touching virtually every aspect of our everyday lives, from medicines to cosmetics and from paper to cups and saucers. It is very difficult to over-estimate its use and importance. The treatment of clay in this book is therefore wide ranging to reflect this situation.

The *occurrence* of clay is also ubiquitous and diverse (see Text Box below) and, with its various mineral species, properties and behavioural characteristics, the industrial applications of clay are thus manifold and complex. As well as their traditional major uses for brickmaking, pottery and porcelain manufacture, refractories and the fulling of cloth, clays are now used for refining edible oils, fats and hydrocarbon oils, in oil well drilling and synthetic moulding sands, in the manufacture of emulsified products in paper and, as noted in Chapters 13, 14 and 15, many hundreds of other uses, including medicine, cosmetics and, on a larger scale, as fillers, as well as many uses in geotechnical engineering e.g. for grouts, membranes etc.

In foundation engineering, clay often provides poor foundation support, and can be responsible for slope instability. It finds extensive use as a construction material in embankments and in water-controlling structures as an impermeable barrier and in many other specialist ways (see Chapters 10, 11 and 12).

Given the worldwide distribution and variability of clay deposits, the production of an authoritative text, which is the aim of the Working Party, on all aspects of

---

Occurrences of clay

Sedimentary rocks occupy about three-quarters of the world's land surface and clays (including shales) form well over 50% of the sedimentary rocks. Clay sediments are collected by the agencies of water (e.g. marine clays, alluvial clays, lacustrine clays), wind (aeolian clays), or of ice (e.g. glacial clay, till or boulder clay). The majority of the common sedimentary clays, however, are the marine deposits typically comprising mixtures of coarser materials with clay in which the clay mineral, illite, usually predominates. See Text Box on p. 4 for descriptions of clay minerals.

Clay mineral-rich deposits can be formed in two other principal ways:

- by weathering of parent minerals in situ to form a clay rich *residual soil* in which the clay mineral kaolinite frequently predominates, especially common in landscapes undergoing tropical weathering, and
- by ascending fluids, i.e. *by hydrothermal alteration* of the host rock. Cornish china clay is a good example, the feldspar of the local granite having been converted mainly into clay minerals of the kaolinite group.

For fuller discussions see especially Chapters 3, 5 and 6.

clay, was considered a most daunting task. The original brief from the Engineering Group committee, the parent committee of the Working Party, was that the Working Party report, *viz.* this book, should *exclude* the engineering aspects of clay *in situ* (i.e. not consider it as a foundation material) but that other engineering aspects, or example use in embankments or in specialized engineering applications, should be considered.

In addition to this particular omission concerning the *in situ* use of clay, the Working Party decided that in order to retain depth of discussion it should concentrate only on the principal construction and industrial applications of clay with the omission of other specialist non-engineering uses, and in order to limit the scope of this current publication to a sensible size.

1.2. Definitions of clay

The term '*clay*' has no genetic significance. It is used for material that is the product of *in situ* alteration, e.g. by weathering, hydrothermal action or, alternatively, deposited as a sediment during an erosional cycle or developed *in situ* as an authigenic clay deposit. 'Clay' can be used as a rock term and also can be used as a particle size term in mechanical analysis of sedimentary rocks or soils. As a rock term, it is difficult to define precisely because of the wide variety of materials that have been called clays. A universal implication of the term 'clay' conveys that it is a natural, earthy, fine-grained material that develops plasticity when mixed with a limited amount of water. By 'plasticity', it is meant that within a certain range of moisture content the material will deform under the application of pressure, the deformed shape being retained when the deforming force is removed. Chemical analysis of clay minerals shows them essentially to comprise silica, alumina and water in variable combinations, frequently with appreciable quantities of iron, alkalis and alkaline earths.

A difficulty is that some material called 'clay' does not meet all the above descriptors. A glance at any comprehensive dictionary will show that clay has a plethora of definitions, scientific and colloquial, often steeped in history and clearly demonstrating that the definition of the word 'clay' depends on the context in which it is being used. The reasons for this situation undoubtedly lie in the many and diverse industries in which clay is used, each having developed, over the years, a definition appropriate to its requirements. A summary of the definitions of clay and clay minerals, presented in the joint report of the Association Internationale pour l'etude des Argiles (AIPEA) Nomenclature Committee and the Clay Minerals Society (CMS) Nomenclature Committee is given in Section 1.2.1; civil engineering definitions of clay in British practice are given in Section 1.2.2 and international civil engineering soil classification by particle size distribution (grading) is given in Section 1.2.3.

Examples of common, non-specific dictionary definitions are given in the Text Box on p. 3.

There are many *geological* dictionaries which seek to define **clay**. These are typically in general accord with the Working Party views. An example is given in the Text Box on p. 4, together with closely related terms defined in the same dictionary. These related terms are also in general accord with their use in this book.

It is necessary to state, however, that within the following chapters, where the term *clay* is used in a *general* sense as a material, it implies any fine-grained, natural, earthy, argillaceous material. This includes clays, shales and argillites of the geologists, and soils as defined by geologists, engineers and agronomists, provided such material is argillaceous (i.e. it contains clay minerals).

Any specific definition of the term *clay* that depends on the context, in which it is being used, is given within the chapter related to that definition. This is because, as is clear from the above dictionary definitions, the term *clay* is imprecise and variously defined. Description of clays is discussed in more detail in Chapters 4 and 8. The term *clay mineral* is described in the second Text Box on p. 4.

1.2.1. Definitions of clay and clay minerals by the AIPEA Nomenclature and CMS Nomenclature Committees

The definitions of clay and clay minerals in the Joint Report of the AIPEA Nomenclature and CMS Nomenclature Committees should be read in full for the complete expression of their views (Guggenheim & Martin 1995), together with the subsequent Discussions on this Report (Moore 1996; Guggenheim & Martin 1996). The following summarises some of the salient points and demonstrates the difficulties in making a precise definition of clay.

'Clay' Definition
The term 'clay' refers to a naturally occurring material composed primarily of fine-grained minerals, which is generally plastic at appropriate water contents and will harden when dried or fired. Although clay usually contains phyllosilicates, it may contain other materials that impart plasticity and harden when dried or fired. Associated phases in clay may include materials that do not impart plasticity and organic matter.'

In the Discussion to the Definition, the Committees make the point that, '*The 'naturally occurring' requirement of clay excludes synthetics and that based on the standard definition of mineral, clays are primarily inorganic materials excluding peat, muck, some soils, etc. that contain large quantities of organic materials. Associated phases, such as organic phases, may be present. 'Plasticity' refers to the ability of the material to be moulded to any shape. The plastic properties do not require quantification to apply the term 'clay' to a material. The 'fine-grained' aspect cannot be quantified, because a specific particle size is not a property that is universally accepted by all disciplines. They say that, for example, most geologists and soil scientists use particle size less than 2 μm, sedimentologists use 4 μm, and colloid chemists use 1 μm for clay-particle size.*'

Non-specific dictionary definitions of clay

These include, for example:
'*The Cassell Concise Dictionary*' (1998)
  '**clay** n. *1* heavy, sticky earth. *2 a hydrated silicate of aluminium with a mixture of other substances, used to make bricks, pottery etc. 3 (poet.) the human body. 4 (coll.) a clay pipe. ~v.t. 1 to cover or mix with clay. 2 to purify and whiten (sugar etc.) with clay. 3 to puddle with clay.* **Feet of clay** *FOOT.* **Clay court** n. *a tennis court with a hard surface covered with crushed shale or stone.* **Clayey** a. *clayish* a. **claylike** a. **clay-pan** n. *(Austral.) a hollow (often dry in summer) where water collects after rain . . .*'

and '*The Concise Oxford Dictionary*' (1980)
  '**clay** n. *Stiff tenacious earth, material of bricks, pottery, etc. (feet of clay, see FOOT; clay PIGEON[1]); (material of) human body;* ~ **(pipe)**, *tobacco-pipe made of clay;* ~-**pan**, *(Austral.) natural hollow in clay soil retaining water after rain; hence* ~**ey**[2] *a. . . . . . .*'

Scientific dictionaries take a more rigorous view, as might be expected. For example,

The '*Penguin Dictionary of Civil Engineering*' (1991)
  '**clay** [s.m.] *Very fine-grained soil of* colloid *size, consisting mainly of hydrated silicate of aluminium. It is a* plastic *cohesive soil which shrinks on drying, expands on wetting, and when compressed gives up water. Under the electron microscope clay crystals have been seen to have a platy shape in which for Wyoming* bentonite *the ratio of length to thickness is about 250 : 1 (like mica). For other clays it is about 10 : 1. Clays are described for engineering purposes by their* consistency limits. *A 'very soft clay' is one with an unconfined compressive strength of less than 35 kPa (5 psi); soft clay has 35–70 kPa; medium or firm has 70–140 kPa; stiff clay has 140–280 kPa. Clays are further described as organic, intact, etc.*'

The '*Chambers Dictionary of Science and Technology*' Volume One (1975) gives a general definition that is in accord with the Working Party views:
  '**clay** *(Geol., etc.). A fine-textured, sedimentary, or residual, deposit. It consists of hydrated silicates of aluminium mixed with various impurities. Clay for use in the manufacture of pottery and bricks must be fine-grained and sufficiently plastic to be moulded when wet; it must retain its shape when dried, and sinter together, forming a hard coherent mass without losing its original shape, when heated to a sufficiently high temperature. In the mechanical analysis of soil, according to international classification, clay has a grain-size less than 0.002 mm.*'

They also say that '*Plasticity is a property that is greatly affected by the chemical composition of the material. For example, some species of chlorite and mica can remain non-plastic upon grinding macroscopic flakes even where more than 70% of the material is less than 2 μm esd (equivalent spherical diameter). In contrast, other chlorites and micas become plastic upon grinding macroscopic flakes where 3% of the material is less than 2 μm esd*' and that, '*Plasticity may be affected also by the aggregate nature of the particles in the material.*'

In the Discussion on the Report, surface effects, grain-size effects, practical considerations and other issues are discussed in some detail, to which the reader is referred. In this discussion, Moore (1996) urges the Committees to state clearly that they are aware that clay is used in three different ways in their discipline: as a size term, as a rock term and as a mineral term. '*Users who do not clearly separate these meanings provide one of our most consistent sources of confusion. Each of these uses has utility or each would not have survived*'.

Clay mineral
Definition:
  '*The term "clay mineral" refers to phyllosilicate minerals and to minerals which impart plasticity to clay and which harden upon drying or firing.*'

In the Discussion, the Committees say that, '*Currently, minerals known to produce the property of plasticity are phyllosilicates. Because minerals are not defined based on their crystallite size, appropriate phyllosilicates of any grain size may be considered 'clay minerals'. Likewise, clay minerals are not restricted, by definition, to phyllosilicates.*'

They go on to say that minerals that are non-phyllosilicates, which impart plasticity to a clay and harden upon drying or firing, are also 'clay minerals'. They quote as example,' *if an oxy-hydroxide mineral in a clay shows plasticity and hardens upon drying or firing, it may be properly referred to as a 'clay mineral'. Thus a clay is not required to be predominantly composed of*

Geological dictionary definitions

An example of a family of geological dictionary definitions taken from '*Collins Dictionary of Geology*' (1990):

*'**clay**, n. **1.** A DETRITAL mineral particle of any composition having a diameter less than 0.004 mm. See ARGILLACEOUS. **2.** A smooth, earthy sediment or soft rock composed chiefly of clay-sized or colloidal particles and a significant content of CLAY MINERALS. Clays may be classified by colour, composition, origin, or use.'*

*'**argillaceous**, adj. **1.** Pertaining to rocks or SEDIMENTS composed of CLAY MINERALS or having a significant amount of clay in their composition. **2.** Pertaining to rocks or sediments composed of SILT or clay-sized DETRITAL particles.'*

*'**clay minerals**, n. a member of the aluminosilicate mineral group, which forms about 45% of all minerals in sedimentary materials. Most clay minerals belong to the kaolinite, montmorillonite, and illite groups. Their crystal structure is the same as that of mica, i.e. sheeted layer structures with strong intra- and intersheet bonding but weak interlayer bonding. The type of clay that is dominant in sedimentary material depends, during early stages of weathering, on the mineral composition of the parent rock; but in later stages, it depends completely upon the climate. Compare CLAY.'*

*'**sediment**, n. solid material, organic or inorganic in origin, that has settled out from a state of suspension in a fluid and has been transported and deposited by wind, water, or ice. It may consist of fragmented rock material, products derived from chemical action or from the secretions of organisms. Loose sediment such as sand, mud and till may become consolidated and/or cemented to form coherent sedimentary rock.'*

*'**detrital**, adj. (of mineral grains) transported and deposited as sediments that were derived from pre-existing rocks either within or outside the area of deposition.'*

*'**silt**, n. **1.** a detrital particle, finer than very fine sand and coarser than clay, in the range of 0.004 to 0.062 mm. **2.** a loose clastic sediment of silt-sized rock or mineral particles. **3.** fine earth material in suspension in water.'*

Clay minerals

The term ***clay mineral*** is used in the following chapters for clay materials that are essentially composed of extremely small crystalline particles of one or more members of a small group of minerals commonly known as '*the clay minerals*'. These minerals are essentially hydrous aluminium silicates, with magnesium or iron substituting wholly or in part for the aluminium in some minerals, and in some of them alkalis or alkaline earth are also present as essential constituents. Some clay materials are composed of a single clay mineral but in many there is a mixture of minerals. In addition to the clay minerals, some clays contain varying amounts of non-clay minerals such as calcite, feldspar, pyrite and quartz. Also, many types of clay contain organic material and water soluble salts. Some clay materials may contain clay material which is non-crystalline (e.g. allophane). There are examples in which allophane is a major component of the clay material. However, such examples are relatively rare and in general clay materials are composed wholly or predominantly of crystalline components and are therefore the components which largely determine the property of the clay. Clay minerals are discussed in detail in Chapter 2.

phyllosilicates. Minerals that do not impart plasticity to clay and non-crystalline phases (regardless if they impart plasticity or not) are either 'associated minerals' or 'associated phases' respectively'.

The Committees make the important point that their definition of clay mineral departs from previous definitions (e.g. Bailey 1980) where clay minerals were equated to phyllosilicates. They say that the current definition broadens the scope of possible minerals defined as clay minerals.

For an exhaustive account of the history of definition of clays up to 1963, the reader is referred to Mackenzie (1963), and for the more recent developments, Weaver (1989).

1.2.2. Civil engineering definitions of clay in British practice

When used in the mechanical analysis of soils following *British civil engineering practice*, clay is defined as material of particle size less than 0.002 mm (BS 5930:1999, *Code of Practice for Site Investigations*). However, some clay mineral particles may be larger than this. BS 5930 defines silt by mechanical analysis as between 0.002

and 0.06 mm. Clay is thus defined without reference to mineralogical composition.

A practical example of the way in which the term *clay* is used in context for civil engineering practice, and therefore of importance to this book, is the way that BS 5930:1999, handles the term. It recommends that soils are divided into two classes: fine soils and coarse soils. It suggests that as a first appraisal of the engineering properties, the soil's nature and composition are described visually, assisted by a few simple hand tests. *'Soils that stick together when wet and can be rolled into a thread that supports the soil's own weight (i.e. have cohesion and plasticity), contain sufficient silt and/or clay in them to be described as fine soils. Soils that do not exhibit these properties behave and are described as coarse soils'*.

The actual soil name *'is based on particle size distribution of the coarse fraction and/or the plasticity of the fine fraction as determined by the Atterberg Limits. These characteristics are used because they can be measured readily with reasonable precision and estimated with sufficient accuracy for descriptive purposes'*. It is suggested that where a soil (omitting any boulders or cobbles, i.e. particles greater than 60 mm in size) contains about 35% or more of fine material it is described as a fine soil ('clay' or 'silt', dependent on its plasticity). The Code goes further and says that 'the effects of clay mineralogy and organic content are significant. Fine soil should be described either as 'silt' or 'clay', depending on the plastic properties; these terms are to be mutually exclusive so that terms such as *'silty CLAY' are unnecessary and not to be used'*.

It should be noted that many geologists call all bedrock material 'rocks', even if their behaviour is 'soil-like' (cf. the Eocene 'London Clay'), whereas engineers tend to call bedrock clays 'engineering soils'. Clay rocks (e.g. a strong mudstone) which ring to the hammer would be called a rock by engineers. Current civil engineering thinking on the term 'cohesion' and the divisions between 'soil' and 'rock' are discussed in Chapter 4.

1.2.3. International Civil Engineering Soil Classification by particle size distribution (grading)

Figure 1.1 shows some of the most used soil classifications by particle size distribution ('grading') by various countries, determined by their national standardization institutes. An important point in this figure is that the boundary between sand and silt is 0.06 mm in European countries and 0.075 mm in the USA and Japan. These numbers have been used since the beginning of the standards in each country and both values are in use around the world to identify the boundary between sand and silt, which can be defined differently in different countries.

The current ISO Standard 14688:1996 placed the boundary at 0.06 mm. Japan has recently proposed that ISO should reflect the two values 0.06 and 0.075 mm.

1.3. Some British clay production statistics

In 2000 it was reported that clays represent only 4% (approximately 15 million tonnes, *Quarry Managers Journal* 2000) of the minerals quarried annually in the UK, but this greatly understates the quantity of clay materials actually excavated per annum. Statistics of production (see the Text Box on p. 7) are generally collected for clay materials extracted and transported away from a quarry for a specific purpose, say, manufacture of paper, ceramics, cement, bricks, tiles, pipes, sanitary ware and other clayware and take no account of large volumes of clay materials excavated during the course of civil engineering projects and reused within the construction site.

For example, also in 2000, a comprehensive survey of waste generated by construction and demolition showed that this activity creates 24 million tonnes of soil (unconsolidated material) and 15 million tonnes of mixed soils and demolition waste requiring disposal (Environment Agency 2000).

The quantities of clay materials excavated and reused internally during construction must be many times greater but no national statistics are recorded. It has proved very elusive to obtain the actual quantities of clay worked in heavy civil engineering; no statistics have been found but from the Working Party members' experience, it was considered, as a rough approximation, that some 15 km of new major highway in the UK required the handling of some 3 million cubic metres of clay material, i.e. for the approximately 100 km of new major highway in the UK per year, in very round figures, some 20 million cubic metres of clay material is handled, the highest total being in England. Similarly, the Working Party estimated some 10 million cubic metres of clay material was handled in all other engineering construction per year.

1.4. Geographical and stratigraphical distribution

Clays are widely distributed both geographically (i.e. across the surface of the Earth) and stratigraphically (i.e. over geological time). The outcrops of the *main* formations containing clay material in Great Britain are shown in Figure 1.2; similar maps could be prepared for all countries that have been geologically surveyed. Maps of this kind show both the areal extent and the outcrop of the clay formations at the present time. Chapters 5 and 6 contain maps from elsewhere.

The stratigraphical positions of the *main* clay formations in Great Britain are shown in Figure 1.3; again, similar stratigraphical columns could be prepared for other countries. The column shows the geological periods during which the main clay formations were deposited. See also discussion in Chapters 5 and 6.

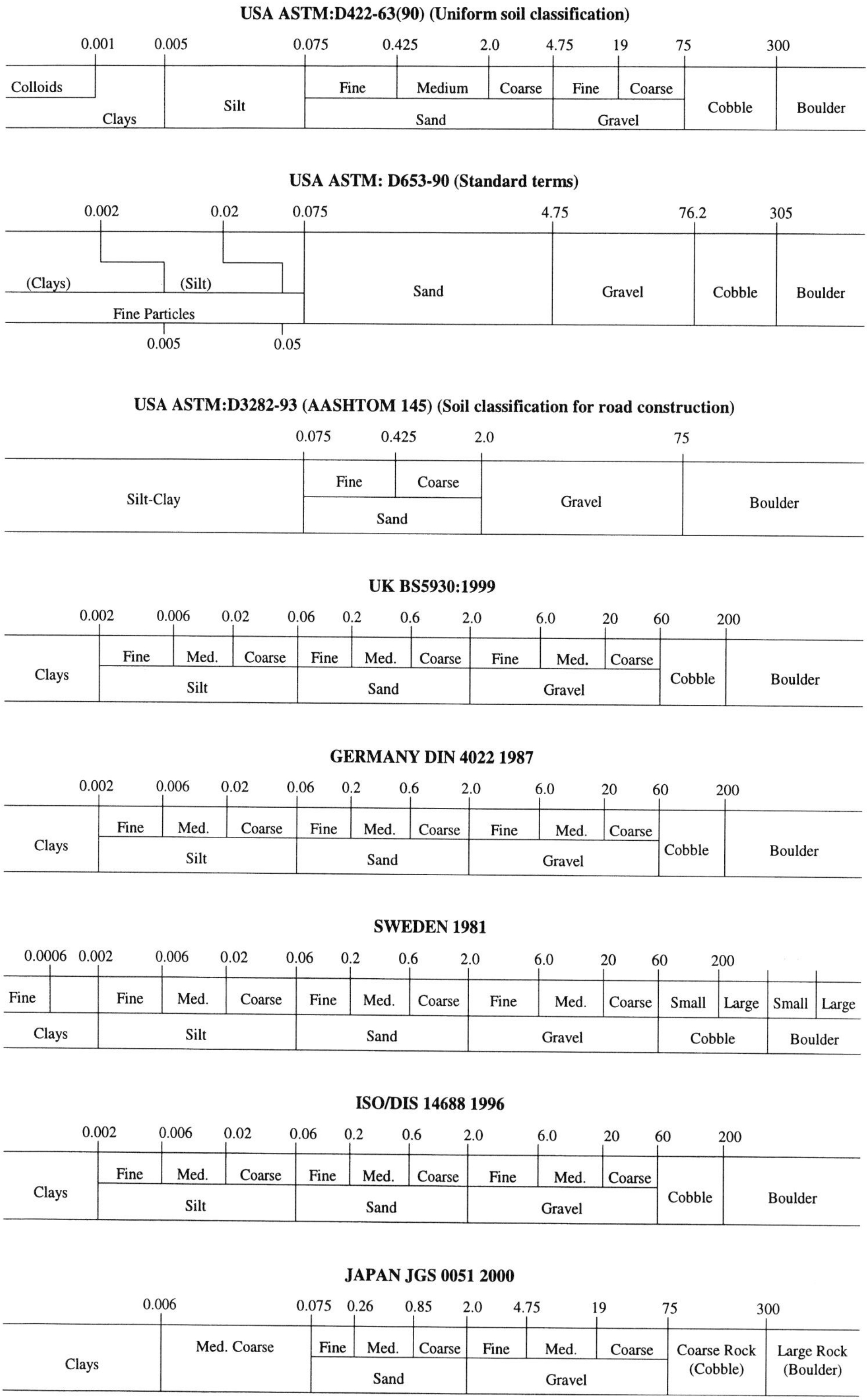

INTRODUCTION

USA ASTM:D422-63(90) (Uniform soil classification)

	0.001	0.005		0.075	0.425	2.0	4.75	19	75	300	
Colloids			Silt	Fine	Medium	Coarse	Fine	Coarse	Cobble	Boulder
	Clays				Sand			Gravel			

USA ASTM: D653-90 (Standard terms)

	0.002	0.02	0.075		4.75	76.2	305	
(Clays)		(Silt)		Sand		Gravel	Cobble	Boulder
	Fine Particles							
	0.005	0.05						

USA ASTM:D3282-93 (AASHTOM 145) (Soil classification for road construction)

	0.075	0.425	2.0	75
Silt-Clay	Fine	Coarse	Gravel	Boulder
		Sand		

UK BS5930:1999

	0.002	0.006	0.02	0.06	0.2	0.6	2.0	6.0	20	60	200	
Clays	Fine	Med.	Coarse	Fine	Med.	Coarse	Fine	Med.	Coarse	Cobble	Boulder
		Silt			Sand			Gravel				

GERMANY DIN 4022 1987

	0.002	0.006	0.02	0.06	0.2	0.6	2.0	6.0	20	60	200	
Clays	Fine	Med.	Coarse	Fine	Med.	Coarse	Fine	Med.	Coarse	Cobble	Boulder
		Silt			Sand			Gravel				

SWEDEN 1981

0.0006	0.002	0.006	0.02	0.06	0.2	0.6	2.0	6.0	20	60	200	
Fine		Fine	Med.	Coarse	Fine	Med.	Coarse	Fine	Med.	Coarse	Small	Large	Small	Large
Clays			Silt			Sand			Gravel		Cobble	Boulder

ISO/DIS 14688 1996

	0.002	0.006	0.02	0.06	0.2	0.6	2.0	6.0	20	60	200	
Clays	Fine	Med.	Coarse	Fine	Med.	Coarse	Fine	Med.	Coarse	Cobble	Boulder
		Silt			Sand			Gravel				

JAPAN JGS 0051 2000

	0.006	0.075	0.26	0.85	2.0	4.75	19	75	300	
Clays	Med. Coarse	Fine	Med.	Coarse	Fine	Med.	Coarse	Coarse Rock (Cobble)	Large Rock (Boulder)
			Sand			Gravel				

FIG. 1.1. Soil classification by particle size (mm) in various countries.

British clay production statistics

The following table is compiled from information published by the British Geological Survey (BGS 1988, 1986, 2001; Office for National Statistics, 2005)

Mineral statistic		1985	1994	1999	2004
Common clay and shale	Quarries	178	176	193	180
	Production (million tonnes)	19	12	11	11
Cement	Production (million tonnes)	13.3	12.3	12.7	11.4
Bricks	Production (million)	3.7	2.8	2.8	2.7
China clay	Production (million tonnes)	2.9	2.5	2.3	1.9
Ball clay	Production (million tonnes)	0.59	0.83	0.98	0.97

The following points should be noted:

1. The figures exclude large quantities of clay materials excavated and reused within construction sites.
2. The figures for the year 1985 may be anomalously high owing to the boom in construction that reached a peak in 1989.
3. Although the production of cement has remained fairly constant, the consumption of clay will be declining as the less efficient 'wet-process', based upon chalk and clay, is progressively replaced by the 'dry-process' based on limestone and shale. Indeed, several cement manufacturers use argillaceous limestones and marls, actually known by the name 'cement rock' in the U.S.A., to obviate the need to separately extract shales.
4. Most china clay (kaolin) is used in papermaking, not construction nor china-ware, and almost 90% is exported.
5. Similarly, most ball clay is used to manufacture 'white' ceramic ware such as tiles and more than 80% is exported.
6. Production of fireclay is very low, about 0.5 million tonnes per annum, and most is used for bricks and tiles, not refractories, owing to the decline of the metal smelting and refining industries.

1.5. The Working Party

The Clay Working Party was convened by the Engineering Group of the Geological Society to produce a Report to be published as a book. The published book is now the third of a trilogy produced by the Working Parties on Construction Materials. Together the three books cover the complete range of particle sizes of natural geological materials (geomaterials) used in civil engineering construction. The two previous books are:

Aggregates: sand, gravel and rock aggregates for construction purposes Collis & Fox (eds), 1st edn (1985); Smith & Collis (eds), 2nd edn (1993), reprinted in 1998; and Smith, M.R. & Collis, L. (eds) 3rd edn. revised by Fookes, Lay, Sims, Smith & West (2001). *Aggregates: sand, gravel and rock aggregates for construction purposes*. Geological Society, London, Special Publications, **17**.
'Stone: building stone, rock fill and armourstone in construction' Smith, M.R. (ed.) 1999. *Stone: building stone, rock fill and armourstone in construction*. Geological Society, London, Special Publications, **16**.

This trilogy is part of a series of Working Party Reports on engineering geology topics produced over the last four decades by the Engineering Group of the Geological Society.

The clay book is intended to be practical, authoritative and informative and to be of use to a wide spectrum of readers in the UK, Europe and around the world, from a diversity of backgrounds and employments.

Each member of the Working Party is knowledgeable in some aspects of clay and it is hoped that their combined expertise covers the full spectrum of clay from its geology and mineralogy to geological, chemical and engineering properties, investigation, testing, its practical use in construction and as commercial products.

Members were largely from industry, together with some from academia, and included geologists, engineering geologists, industrial geologists, geotechnical engineers, laboratory specialists and industrial chemists.

To achieve the maximum possible breadth and balance in the report, advice and constructive criticism were canvassed from a range of individual specialists, professional institutions, learned societies, industrial associations and research bodies.

8 INTRODUCTION

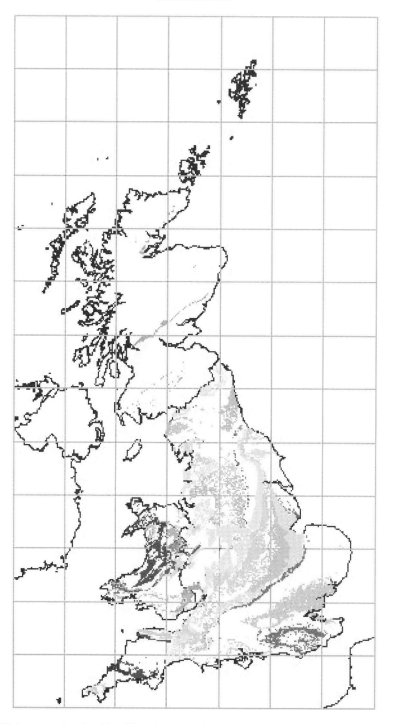

FIG. 1.2. Map of the UK showing major clay-rich solid-geology formations (more detailed UK maps are included in Chapter 6).

Thus it should be clear that the concepts and techniques described in the report are drawn from a wide scientific spectrum in which the Working Party sought advice from beyond the confines of the committee. The report links diverse fields in a common theme and as such will inevitably be used by persons of widely different vocational backgrounds. Consequently, basic material which is included in one section might be considered by an expert

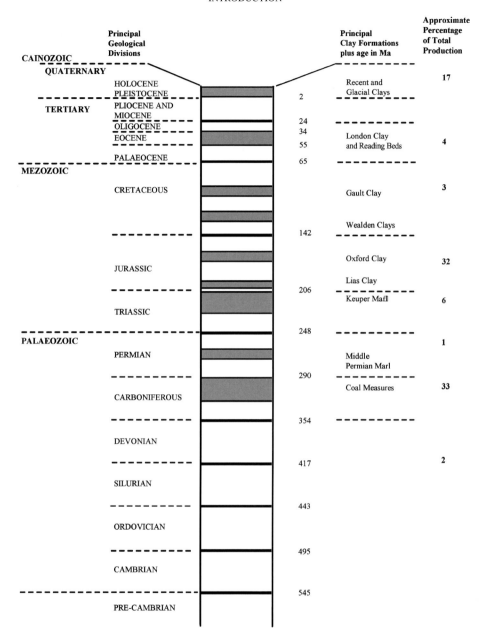

FIG. 1.3. Stratigraphical distribution of the principal British structural clays (ages after Gradstein & Ogg 1996).

in that particular field to be oversimplified. However, it is thought that this approach is justified in order to serve a wide readership and it is hoped that elementary facts might assume a new significance when presented in relation to the central theme. To help this, text boxes have been used to discuss a topic or part of a topic in specialist detail, or to give a simple basic background on some aspects of the subject where the text has assumed some prior knowledge.

As part of the extension of the work of the Working Party and its specialist corresponding members, a consultative seminar, '*Clay materials used in construction*', was held on 19 April 2000, in Manchester, at *Geoscience 2000*, the major biennial conference of the Geological Society. In this, the preliminary drafts of chapters in summary form were made available as preprints. The chapters were then presented during the session and wide-ranging discussion, for each chapter, held with the audience. This

> **Terms of Reference of the Working Party**
>
> The objectives, formulated into Terms of Reference, decided upon in the first meetings of the Working Party, on which this booked are based are:
>
> 1. These Terms of Reference have been agreed by the Clay Working Party (CWP).
> 2. The CWP has been established by the Engineering Group of the Geological Society and comprises officers and specialist participating members who will act as chapter authors or co-authors. The participating members may be assisted by any number of corresponding members based in the UK and occasionally overseas.
> 3. The CWP will produce a report, in book format, to complement the two earlier reports on geomaterials for use in construction: Aggregates and Stone & Rock. The report will be a comprehensive, state-of-the-art review on clay and clay-like materials in construction. The engineering behavior of clay *in situ* will not be included.
> 4. It is the aim of the CWP to produce a report that will act as an essential reference handbook for professionals, as well as a valuable textbook for students. The style will be concise and digestible by the non-specialist, yet be authoritative, up-to-date and extensively supported by data and collations of technical information. The use of jargon will be minimised and necessary specialist terms will be defined in an extensive glossary. There will be copious illustrations, many of which will be original, and many good quality photographs.
> 5. There will be collective responsibility for the whole report. Although each participating member will be the drafting author or co-author of one or more chapters, all members will be expected to review and contribute to the chapters drafted by all the other members. It is intended that the report will be completed within five years.
> 6. The content of the report will embrace a full range of topics from the latest research findings to the most practical of applications. There will be an endeavour to identify likely directions of future research and to predict future developments. Based principally on UK experience and practice, the CWP will aim to include a thorough coverage of the subject throughout Europe and worldwide.
> 7. A geological and mineralogical background to the subject will be fully described and explained. Exploration for clay materials, plus their characterisation, testing and properties will be reviewed. The uses of clay materials in construction will be fully assessed, including a range of direct applications, plus other processes in which clay is an essential raw material.

procedure produced valuable feedback for the Working Party and many ideas and suggestions developed at this meeting were included in the Terms of Reference (see the above Text Box).

1.5.1. The report

Based on the objectives outlined above, it was decided to structure the book round its sequence of chapters. The chapters have been subdivided into three main thematic groups: *Geology*, 2 to 6; *Investigation*, 7 to 9, and *Application*, 10 to 15 (see Text Box on p. 11).

The ***Geology*** chapters begin with fundamental scientific aspects, composition, formation and alteration of clays, and continues with discussions of their basic properties and reviews of clay deposits around the world as well as in Britain. This approach was designed to set the scene for the following ***Investigation*** chapters evaluating clays, their exploration, composition, textural analysis and their testing. This book is completed by the practical ***Application*** chapters concerning clays in earthworks and engineering situations, in buildings and in industrial use, principally bricks, ceramics and cement and related products.

It was considered that extensive *appendices* would be required to provide mineralogical and property data, detailed information on deposits around the world, test methods and a glossary. The intention was that, with the appendices, repetitive information, factual information and details which would hinder the flow of the text but were otherwise necessary in an authoritative book, would be included.

Each chapter is thoroughly cross-referenced to the other chapters, as appropriate, and contains detailed references for further reading.

Attention is drawn to the implications of the use of such terms as 'weathering' and 'weathered clay' throughout the text. Near surface clay is usually in a weathered condition. This weathering changes the characteristics of the material and most sources of clay around the world are weathered to a lesser or great extent. Therefore the form and influence of the weathering on the characteristics of the material must be taken into consideration in its evaluation and use. Often, reported test properties are those of the fresh material, whereas the material as delivered may be weathered to some extent and have different properties. Users and specifiers must be aware of this. See further discussion on weathering in Chapter 4.

The four **Appendices** (plus a **glossary**), together with parts of the text, have been assembled mainly from data of which the Working Party have no first-hand knowledge of the provenance or the reliability, although every effort has

Structure of the book/report

Chapter	Brief notes on scope of chapter
1. Introduction	Terms of reference, scope of report, definitions, statistics, distribution of clays in time and space, outline of chapters.

Geology

2.	Composition	Clay mineralogy: chemical composition and atomic structure of clay minerals. Non clay mineral phases. Swelling and ion-exchange, fluids and gases present, etc.
3.	Formation and alteration	Mode of formation of clay minerals and their alteration in relation to their environment. Diagenesis, lithification and low-grade metamorphism.
4.	Properties	Fundamental properties of clays, effects of clay mineralogy on properties. Classification schemes for clays. Clay material behaviour in flow, plasticity and brittle mechanics.
5.	World and European clay deposits	Survey of distribution of clays, world-wide and in Europe. Plate tectonics, palaeogeographic and climatic factors affecting clays.
6.	British clay deposits	Occurrence and stratigraphy, distribution, nature and properties of main British clay deposits. Maps of major clay formations.

Investigation

7.	Exploration	Procedures and techniques used in site investigation for clays: both for clays as a resource and for civil engineering. Specific investigation for industrial product clays is dealt with in the appropriate chapter.
8.	Compositional and textural analysis	Methods of identifying clay minerals, e.g. X-ray diffraction, thermal methods, chemical analysis, electron microscopy.
9.	Testing	Methods of testing clays: both for clays as a resource and for civil engineering, classification of clays. Standards.

Application

10.	Earthworks	Classification, specification, properties and use of clays as earthworks in civil engineering. Standards. Case histories.
11.	Earthmoving	Behaviour, performance and use of mechanical plant in earthmoving: both for clays as a resource and for civil engineering. Case histories.
12.	Specialized applications	Use of clay in civil engineering: clay slurries, diaphragm walls, piles, slurry tunnelling, pipe jacking, liners, borehole sealers, etc. Case histories.
13.	Earthen building	Use of earth (mainly clayey) in the construction of houses and walls, both traditional and modern, UK and world-wide.
14.	Brick and other ceramic products	Bricks, tiles, pipes and other ceramics used in building and civil engineering. Refractories and electrical insulators. Clay evaluation.
15.	Cement and related products	Composition and properties of various types of cement, metakaolin, clay-based cement additives, etc.

Appendices

A. Mineralogical data
B. Property data
C. Deposits: World, Europe, UK
D. Test methods

Glossary

been made to check the authenticity of the data sources. The quality of analyses and reliability of the information will vary considerably around the world and therefore data assembled in the appendices (as well as chapters) should only be used with considerable caution and it must not be assumed that similar clays from different locations will have similar properties: they may or may not. Instead, reliance should be placed on comprehensive sampling, testing and quality control at the original location.

Appendix A calls for a special comment. It includes new and valuable clay mineral analyses of over one hundred clay samples spanning the stratigraphic succession of the British Isles, collected, or provided, by members of the Working Party, during the period of its work. The samples were divided into two sub-samples: the whole clay, and the fraction less than 0.002 mm particle size. These were analysed by automated X-ray powder diffraction methods at the laboratories of the Macaulay Land Use Research Institute, Scotland, all at the same time by the same method and the results analysed by the same person (Dr Stephen Hillier). These studies provided a self-consistent set of results for the mineralogical composition of the clays, both the clay minerals present and non-clay-minerals. The samples were also analysed by X-ray fluorescence at the laboratories of Watts Blake Bearne and Company in Germany. This study provided information on the major elements and supplemented the results obtained by the X-ray diffraction method.

This appendix is considered a most important piece of work: it is the first comprehensive analytical study of this kind carried out in Britain. It follows the pioneering work of Perrin (1971) in the 1960s, and gives an authoritative conspectus of the mineralogical composition of British clays using modern methods of analysis. The results should be valuable in both the academic study of the origin and nature of clays and in the application of clay mineralogy to practical uses.

It is hoped that this book provides a fitting third member to the geomaterials trilogy. Improving and/or updating comments for possible further editions are welcome.

References

ASTM : D422-63 (90), STM for particle size analysis of soils.

ASTM : D653-96, Standard Terminology relating to soil, rock and contained fluids.

ASTM : D2487-93, Classification of Soils for Engineering Purposes (Unified Soil Classification System).

ASTM : D3282-93, Standard Classification of soils and soil aggregate mixtures for highway construction purposes.

BAILEY, S. W. 1980. Summary of recommendations of AIPEA Nomenclature Committee. *Clays & Clay Minerals*, **28**, 73–78.

BRITISH GEOLOGICAL SURVEY 1996. *United Kingdom Minerals Yearbook 1995*, Keyworth.

BRITISH GEOLOGICAL SURVEY 1988. *United Kingdom Minerals Yearbook 1988*, Keyworth.

BRITISH GEOLOGICAL SURVEY 2001. *United Kingdom Minerals Yearbook 2000*, Keyworth.

BRITISH STANDARD BS 5930, 1999. *Code of Practice for Site Investigations*. British Standardst Institution, London.

CASSELL CONCISE DICTIONARY, Revised. 1998. Cassell.

COLLOCOTT, T. C. & DOBSON, A. B. (eds) 1975. *Chambers Dictionary of Science and Technology*. Vol. 2. Chambers, Edinburgh.

DIN 1987. Classification and Description of soil and rock. DIN 4022, Part 1 Swedish Geotechnical Society : 1981 : Soil Classification and Identification. Laboratory Committee of Swedish Geotechnical Society.

ENVIRONMENT AGENCY 2000. *Construction and Demolition Waste Survey*, R&D Report P402, Swindon.

GUGGENHEIM, S. & MARTIN, R. T. 1995. Definition of clay and clay mineral: Joint Report of the AIPEA Nomenclature and CMS Nomenclature Committees. *Clay and Clay Minerals*, **43** (2), 255–256.

GUGGENHEIM, S. & MARTIN, R. T. 1996. Reply to the comment by D.M. Moore on: Definition of clay and clay mineral: Joint Report of the AIPEA Nomenclature and CMS Nomenclature Committees. *Clay and Clay Minerals*, **44** (5), 713–715.

GRADSTEIN, F. M. & OGG, J. G. 1996. A Phanerozoic time scale. *Episodes*, **19** (1&2).

ISO 2002. Geotechnical Engineering—Identification and classification of soil—IS 14688 Part 1 Identification and Description ISO/TC 182/SC 1.

JGS 2000. Method of classification of geomaterials for engineering purposes. Japanese Geotechnical Society 0051–2000.

KLEIN, E. 1966 & 1967. A comprehensive etymological dictionary of the English language. 2 Vols. Elsevier, Amsterdam, London, New York, 177 p.

LAPIDUS, D. F. 1990. *Collins Dictionary of Geology*. Collins, London and Glasgow.

MACKENZIE, R. C. 1963. *De Natura Lutorum. Clays and Clay Minerals*. v. XI. Proceedings of the Eleventh National Conference on Clays and Clay Minerals. Pergamon Press, 1–28.

MANN, S. E. An Indo-European comparative dictionary. Helmut Buske, Hamburg, 1984–1987. XIV, 1682 S.

MOORE, D. M. 1996. Comment on: Definition of clay and clay mineral: Joint Report of the AIPEA Nomenclature and CMS Nomenclature Committees. *Clay and Clay Minerals*, **44** (5), 710–712.

OFFICE FOR NATIONAL STATISTICS 2005. *Mineral extraction in Great Britain*, Business Monitor PA1007.

PERRIN, R. M. S. 1971. *The Clay Mineralogy of British Sediments*. Mineralogical Society, London.

QMJ PUBLISHING, 2000. *Directory of Quarries, Pits and Quarry Equipment 2001/2002*, Nottingham.

SCOTT, J. S. 1991. *The Penguin Dictionary of Civil Engineering*, 4th Edn. Penguin.

SKEAT, W. 1961. *An etymological dictionary of the English language*. Oxford University Press.

SMITH, M. R. (ed.) 1999. *Stone: Building stone, rock fill and armourstone in construction*. Geological Society, London, Engineering Geology Special Publications, **16**.

SMITH, M. R. & COLLIS, L. (eds) 2001 (3rd edn). Aggregates: sand, gravel and rock aggregates for construction purposes. Geological Society, London, Engineering Geology Special Publications, **17**.

SYKES, J. B. (ed.) 1980. *The Concise Oxford Dictionary of Current English*. Based on The Oxford English Dictionary and its Supplements, first edited by H.W. Fowler and F.G. Fowler. Book Club Associates, London.

WEAVER, C. E. 1989. *Clays, Muds and Shales*. Elsevier, Amsterdam.

2. The composition of clay materials

Clay materials are composed of solid, liquid and vapour phases. The solid phases are of mineral and organic phases that make up the framework of the clay materials. The mineralogy can be broadly subdivided into the clay and non-clay minerals, including poorly crystalline, so-called 'amorphous' inorganic phases. By definition, minerals are crystalline solids with well-ordered crystal structures but clay minerals and other inorganic phases in clay materials are often poorly crystalline compared to minerals such as quartz and feldspar.

Some clay materials may be dominated by one mineral phase, e.g. smectite in bentonites, opal in diatomaceous earths. However, most clay materials are composed of heterogeneous mineral mixtures. Based on the bulk mineral analysis of over 400 samples, Shaw & Weaver (1965) reported the modal mineralogical composition of siliciclastic mudrocks to be:

    60% clay minerals
    30% quartz and chert
    5% feldspar
    4% carbonates
    1% organic matter
    1% iron oxides

There is a general increase in the predominance of clay minerals in sedimentary rocks with decreasing grain size (Fig. 2.1) (Blatt *et al.* 1972). However, it needs to be stressed that, whilst clay minerals are usually significant, if not predominant, phases in clay materials, other mineral phases are usually present in varying amounts and can significantly affect the properties and behaviour of the materials.

In soils, mineral and organic compositional variations reflect the weathered parent rocks and the physical, chemical and biological factors controlling the soil forming processes (see Chapter 3).

The liquid and vapour phases, of which water is usually the most important, occur either as 'bound phases', adsorbed onto the surfaces of the solid particles, or as 'free phases' within pore spaces.

In this chapter the nature of the mineralogical, organic and aqueous phases present in clay materials are described plus the nature of swelling and ionic exchange properties in solid phases.

2.1. Clay minerals

The clay minerals are a group of hydrous aluminosilicates that are characteristically found in the clay ($<2\ \mu m$) fractions of sediments and soils. The majority of clay minerals have sheet silicate structures (see Text Box on p. 14).

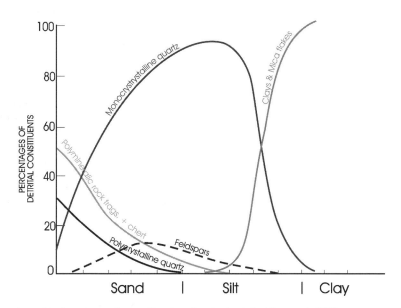

FIG. 2.1. Detrital mineralogy of sedimentary rocks as a function of grain size (after Blatt *et al.* 1972).

The sheet silicates

The sheet silicates consist of 'composite layers'—sheets of tetrahedrally co-ordinated Si, Al and octahedrally co-ordinated cations (principally Fe^{3+}, Fe^{2+}, Al, Mg) (Fig. 2.2).

The octahedral sheets can be defined as being dioctahedral, 2/3 of the octahedral sites occupied with Al^{3+} or other trivalent ions, and trioctahedral, all the octahedral sites occupied by Mg^{2+}, Fe^{2+} or other divalent cations.

The composite layers are stacked together and linked by cations and/or water molecules in the interlayer sites.

All sheet silicates, including the clay minerals, exhibit polymorphism arising from different modes of stacking of the composite layers, known as polytypism.

Polymorphism—minerals have the same chemical composition but different crystal structure.

Identification of the different polytypes of individual clay minerals is possible by X-ray powder diffraction analysis (see Chapter 8). Different polytypes can characterize different environments of formation and are thus of use in defining formation conditions.

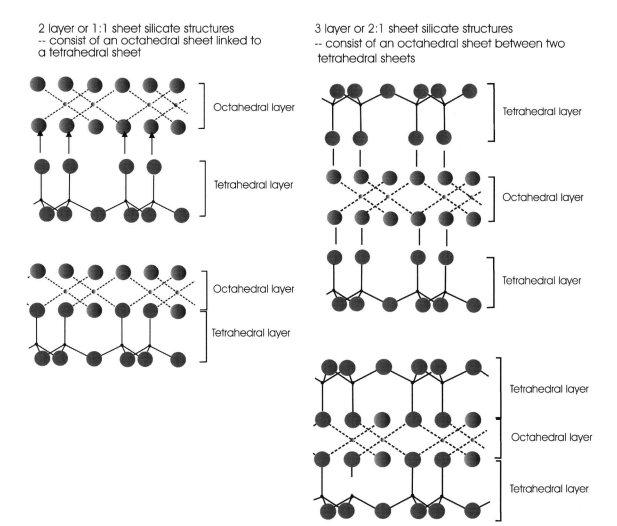

Fig. 2.2. Composite layers in clay mineral structures.

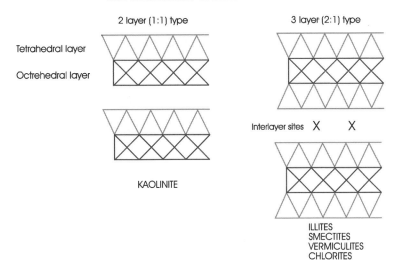

FIG. 2.3. General structures of clay minerals.

TABLE 2.1. *Classification of clay minerals*

Sheet silicate type	Property
1:1 type	
Kaolin and serpentine	Non-swelling
2:1 type	
Illites	Non-swelling
Chlorites	Non-swelling
Smectites (Montmorillonites)	Swelling
Vermiculites	Swelling
Mixed layer clays with smectite/vermiculite	Swelling
Mixed layer clays without smectite/vermiculite	Non-swelling
Palygorskite and sepiolite	Non-swelling

There are two types of composite layer structures in the clay minerals:

- the two layer or 1:1 type represented by the kaolin and serpentine groups;
- the three layer or 2:1 type represented by the illite–mica, smectite, vermiculite and chlorite groups (Fig. 2.3 and Table 2.1).

2.1.1. The kaolin and serpentine groups

The kaolin group of minerals is the most common of the clay minerals with the two layer 1:1 type of structure. The term *kandite* has also been used to describe kaolin group minerals but is not the preferred name.

Kaolins have dioctahedral structures with, ideally, no net negative charge on the composite layers and consequently no compensating interlayer cations (Fig. 2.4) or water layers. There are three principal polymorphs (or polytypes), kaolinite, dickite and nacrite of general formula $Al_4Si_4O_{10}(OH)_8$. Specific names have also been given to kaolin-rich deposits that occur in different formation environments (see Text Box on p. 16)

Kaolinite can range from well crystallized varieties to poorly crystalline forms that can be differentiated by X-ray powder diffraction analysis (see Section 8.2.2). Halloysite is a kaolin group mineral that contains one water layer in the interlayer sites, producing a 10Å interlayer spacing rather than the 7Å spacing of the kaolinite structure.

Its general formula is $Al_4Si_4O_{10}(OH)_8 \cdot H_2O$. Removal of the water layer causes the halloysite structure to collapse to a 7Å interlayer spacing. The collapsed form is referred to as 7Å halloysite (or metahalloysite) as opposed to the hydrated 10Å halloysite. Halloysites usually have a characteristic tubular morphology though other spheroidal habits have been reported (Fig. 2.5). The tubular morphology arises from the distortion and curving of the 1:1 layer structure from which the tubular habit develops.

The serpentine group of minerals is the trioctahedral equivalent of the kaolin group. Berthierine $(Fe^{2+}, Mg)_{6-x}(Fe^{3+}, Al)_x Si_{4-x} Al_x O_{10}(OH)_8$ is the most important member of the serpentine group in studies of sedimentary rocks. It commonly occurs in oolitic ironstones and in pedogenic environments but has often been misnamed chamosite. Chamosite is not a 7Å serpentine mineral but a 14Å chlorite mineral (see Section 2.1.5).

2.1.2. The illlite–mica group

Illite is the term generally used for all clay grade micas and most commonly has a dioctahedral 2:1 or three-layer composite layer structure (Fig. 2.6).

In the illite structure substitution of $[Si^{4+}]^{IV}$ by $[Al^{3+}]^{IV}$ and $[R^{3+}]^{VI}$ by $[R^{2+}]^{VI}$ produces a net negative charge of about 0.7–1.0 per $O_{10}(OH)_2$ formula unit. Potassium is the principal interlayer cation and lenses of water may also be present in the interlayer sites (Fig. 2.6).

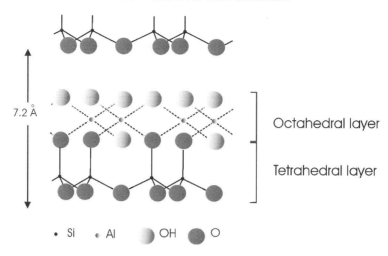

FIG. 2.4. Crystal structure of kaolinite ($Al_4 Sc_4 O_{10} (OH)_8$).

Specific names for kaolin-rich clay materials

Ball clays, fireclays, flint clays, seatearths or underclays and tonsteins.

Ball clays are impure sedimentary kaolin deposits that have plasticity, green and dry strengths suitable for making ceramic products that yield a white product when fired.

In the UK excellent examples of ball clays occur in the Tertiary fluvio-lacustrine basins of SW England, e.g. the Bovey Tracey Basin (see Chapter 6).

Fireclays are kaolin-rich underclays similar in composition to ball clays that can occur beneath coals in Coal Measure sequences.

Underclays or seatearths are palaeosols below coals on which grew the vegetation from which the coals was derived

Flint clays are very fine grained kaolin-rich clays that break with a conchoidal fracture and develop no plasticity when wet.

Some flint clays are considered to be derived from alteration of igneous or volcanoclastic rocks but the term tonstein is generally used for kaolin-rich clays formed by the alteration of volcanic material by acidic formation waters, e.g. in coal swamps.

Illite $K_{0.8}(R^{3+}_{1.65}R^{2+}_{0.35})(Si_{3.55}Al_{0.45})O_{10}(OH)_2$
Muscovite $KAl_2(Si_3Al)O_{10}(OH)_2$

Sericite is also used to generally signify fine-grained micacaeous material of an indeterminate nature that may include illite but also other clay and sheet silicate minerals.

Glauconite (*sensu stricto*) is a dioctahedral Fe-rich illite of general formula

$$K(R^{2+}_{0.67}R^{3+}_{1.33})(Si_{3.67}Al_{0.33})O_{10}(OH)_2 \text{ with}$$
$$Fe^{3+} \gg Al, Mg > Fe^{2+} \text{ and } Fe^{3+} > Fe^{2+}$$

However, 'glauconite' is also used to describe any green (pelletal) clay material irrespective of composition (see Text Box on p. 18). The term 'glaucony' has been proposed as a general term when the specific mineral composition is not known (Millot 1970).

Celadonite is the more Mg-rich equivalent of glauconite of ideal formula $K(Mg, Fe^{2+}, Fe^{3+})Si_4O_{10}(OH)_2$.

Unlike glauconites that are most often found in modern and ancient sediments, celadonites are most commonly formed in association with the alteration of volcanic, usually basaltic, rocks.

2.1.3. Smectites

Smectites are three-layer or 2:1 clay minerals (Fig. 2.7) that are most commonly dioctahedral but trioctahedral varieties do exist. They have a layer charge of 0.2–0.6 per $O_{10}(OH)_2$ unit of structure, which is offset by hydrated interlayer cations, principally Ca and Na.

The hydration of the interlayer cations causes the interlayer crystalline swelling that characterizes the smectites. Water is sorbed into the interlayer sites

THE COMPOSITION OF CLAY MATERIALS

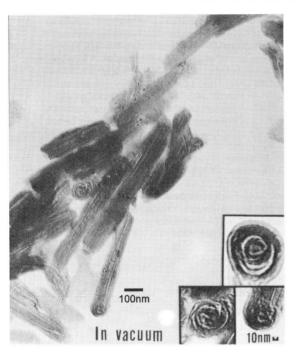

FIG. 2.5. Photomicrograph of tubular habit of halloysite (from Kohyama et al., 1978, reproduced by permission of the authors and publishers).

composed predominantly of montmorillonite and/or to indicate genesis of montmorillonites from alteration of volcanic ash. In the UK Fullers Earth is used in a similar way to bentonite though in North America Fullers Earth can be composed of palygorskite rather than smectite (montmorillonite).

<u>Dioctahedral smectites:</u>
Montmorillonites $\quad R^+_y(Al_{2-y}R_{2+y})Si_4O_{10}(OH)_2 \cdot nH_2O$
Beidellite $\quad R^+_y Al_2(Si_{4-y}Al_y)O_{10}(OH)_2 \cdot nH_2O$
Nontronite $\quad R^+_y Fe^{3+}_2(Si_{4-y}Al_y)O_{10}(OH)_2 \cdot nH_2O$

<u>Trioctahedral smectites:</u>
Saponite $\quad R^+_{x-y}(Mg_{3-y}(Al, Fe^{3+})_y)(Si_{4-x}Al_x)$
$\qquad O_{10}(OH)_2 \cdot nH_2O$
Hectorite $\quad R^+_{x-y}(Mg_{3-y}Li_y)Si_4O_{10} \cdot (OH)_2 \cdot nH_2O$

2.1.4. Vermiculite

Vermiculites are similar in structure to the smectites (Fig. 2.9) but have a larger net negative charge on the composite layer of 0.6–0.8 per $O_{10}(OH)_2$. The principal interlayer cations are hydrated magnesium. Vermiculites exhibit swelling properties similar to the smectites but to a lesser extent due to the higher layer charge. Vermiculites are trioctahedral with the general formula:

$$Mg_{(x-y)/2}(Mg, Fe^{2+})_{3-y}(Al, Fe3^+)_y$$
$$(Si_{4-x}Al_x)O_{10}(OH)_2 \cdot nH_2O$$

2.1.5. Chlorite

The chlorites have a 2:1 type structure with a second octahedral layer in the interlayer sites having a net positive charge to offset the net negative charge on the 2:1 layers (Fig. 2.10).

in monomolecular sheets (Fig. 2.8). Na and Ca-montmorillonites can absorb up to three layers at 90% relative humidity but in some instances at 100% humidity, Na-montmorillonites can absorb more and the 2:1 layers disperse.

Bentonite is often used synonymously with montmorillonite but is a rock term rather than a mineral name. As a rock term, it used to signify a commercial clay deposit

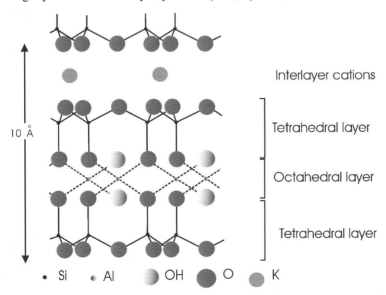

FIG. 2.6. Crystal structure of muscovite.

Variations in the mineralogy of 'glauconite' (after Burst 1958)

(1) Well ordered illitic structure
Non-swelling clay mineral
8% K_2O, $Fe^{3+}:Fe^{2+} = 4-7:1$
Glauconite *sensu stricto*

(2) Poorly ordered illitic structure
Non-swelling clay mineral
8% K_2O; $Fe^{3+} > Fe^{2+}$

(3) Poorly ordered illite–smectite structure
Swelling clay mineral
Low K_2O content; $Fe^{2+} > Fe^{3+}$

(4) Berthierine, chlorite and other green minerals

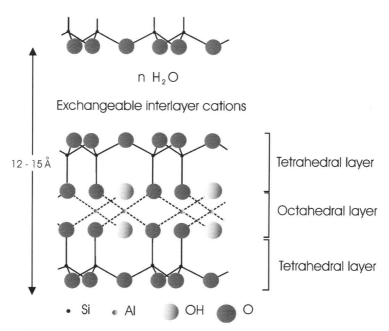

FIG. 2.7. Crystal structure of dioctahedral smecite.

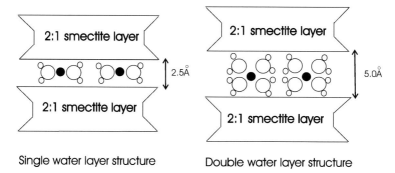

FIG. 2.8. One layer and two layer water structures in smectite (from Velde 1992).

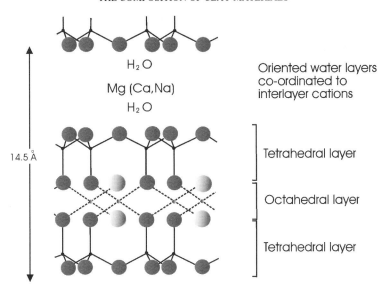

FIG. 2.9. Crystal structure of vermiculite.

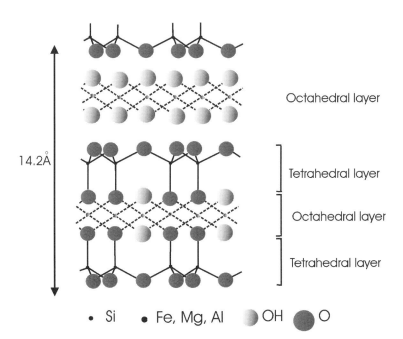

FIG. 2.10. Crystal structure of chlorite.

The general formula of the chlorites is

$$(Mg, Fe^{2+})_{6-x}(Al, Fe^{3+})_x Si_{4-x} Al_x O_{10}(OH)_8$$

The majority of chlorites are trioctahedral but some dioctahedral and mixed dioctahedral (2:1 layers) – trioctahedral (interlayers) forms are known and can be identified by X-ray diffraction analysis (see Chapter 8).

2.1.6. Mixed layer clay minerals

Mixed layer clay minerals describe those mineral phases in which different sheet silicate units occur in the stacking sequence. Generally the 2:1 clay minerals are most commonly involved in forming mixed layer clay minerals, though kaolinite–smectites and serpentine–chlorites have been reported. The most commonly reported mixed layer

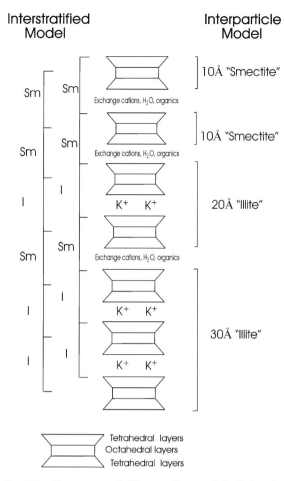

FIG. 2.11. The nature of illite–smectite as defined by the interstratified and interparticle models (after Nadeau & Bain 1986).

clays are the illite–smectites and to a much lesser extent the chlorite-smectites. The interstratification may be random, with no discernible pattern in the stacking sequence, or ordered, e.g. ABABAB, AABAABAA or AAABAAB. Specific names are also given to certain types of ordered mixed layer clay minerals.

The mixed layer phases may also show segregations into AB sequences with separate domains of A or B rather than continuously alternating sequences.

Generally mixed layer illite–smectites with more than 45% smectite are randomly interstratified, those with 30–45% smectite can be random or ordered and with less than 30% smectite are ordered (Bethke & Altaner 1986).

2.1.7. Sepiolite and palygorskite

Although sepiolite and palygorskite are often described as 2:1 sheet silicates, the sheets are not continuous but form alternate ribbons of 2:1 sheet silicate structures (Fig. 2.12). Consequently they have fibrous rather than the typically platy habits of the true sheet silicates. Compositionally they are Mg-rich, Mg being the principal cation in the octahedral sites but in some palygorskites aluminium can be the more common octahedral cation, i.e. $[Al^{3+}]^{VI} > [Mg^{2+}]^{VI}$.

The name attapulgite is often used for commercial deposits of palygorskite, but palygorkite is the preferred mineral name.

2.1.8. Swelling properties of clay minerals

The clay minerals can be classified into swelling and non-swelling varieties (Table 2.1).

Water and organic molecules may be sorbed onto the surface of and into the interlayer sites of clay minerals. The sorption of water and organic molecules into the interlayer sites is responsible for the characteristic intraparticle swelling properties of smectites and vermiculites (Table 2.1). A 'dry' smectite absorbs water into the interlayer sites in discrete layers with the water forming hydration sheaths around the interlayer cations. At very high humidities Na-smectites when immersed in water exhibit osmotic swelling and may completely dissociate. Complete dissociation of the clay layers occurs when the repulsive forces of the negatively charged clay surfaces exceed the forces of attraction between the hydrated interlayer cations and the clay particles.

Vermiculites, having a higher interlayer charge than smectites, show less swelling and will not completely dissociate in water.

Fundamental particle hypothesis for mixed layer clays

Nadeau *et al.* (1984) proposed a new hypothesis regarding the nature of mixed layer clays based on XRD and TEM data—the interparticle model (Fig. 2.11). They indicated that random mixed clays could be aggregates of very thin 'fundamental particles' of the two-component minerals (e.g. illite and smectite) whilst the regular mixed layer minerals could be composed of aggregates of thin 'fundamental particles' of only one phase (e.g. illite in the case of regular mixed/layer illite–smectite). The mixed layer X-ray diffraction characteristics (see Chapter 8) are explained as the result of interparticle X-ray diffraction effects.

The data and interpretations of Nadeau *et al.* have been questioned by several authors notably McKinnon (1987) and Sawnhey & Reynolds (1985).

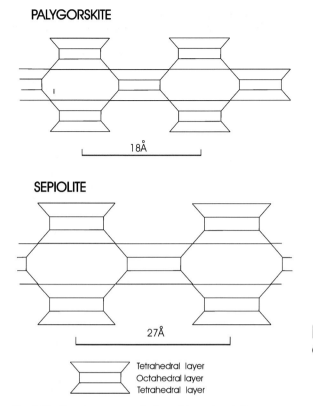

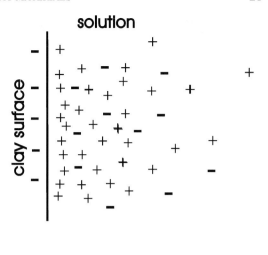

FIG. 2.12. Crystal structures of palygorskite and sepiolite.

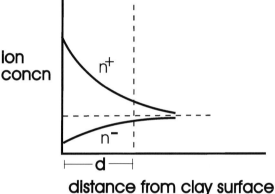

FIG. 2.13. Distribution of ions in the clay-water layer of thickness d and formation water.

Illites may have lenses of water in the interlayer sites, but not complete layers and do not show intraparticle swelling.

Chlorites are non-swelling clay minerals but there have been reported instances of 'swelling chlorites'. Moore & Reynolds (1997) commented that swelling chlorites are such a poorly described phase they were unable to calculate their diffraction characteristics. It seems probable that 'swelling chlorites' are most likely to be mixed layer chlorite/smectites or chlorite/vermiculites rather than a swelling chlorite *sensu stricto*.

Halloysites in the hydrated state have a water layer between the 7Å kaolin layers producing their characteristic 10Å interlayer spacing. Kaolinite in contrast to halloysite does not swell in the presence of water though increasingly disordered kaolinites contain lenses of water.

Sepiolite and palygorskite have open channel sites similar to zeolites (see Section 2.2.7) into which water and organic compounds can be absorbed or desorbed without significantly affecting the unit cell dimensions and thus are not regarded as swelling clays.

In addition to intraparticle swelling, all clay minerals can show interparticle swelling which is governed by similar factors that control intraparticle swelling, that is the nature of the clay minerals and the nature of and concentration of cations adsorbed onto the clay surface in the diffuse double layer and hydration of surface cations (Guven 1993; Guven & Pallastro 1993; Low 1992).

Interparticle associations of clays control their flocculation and dispersion in natural waters.

2.1.9. Ion exchange properties of clay minerals

The 2:1 clay minerals have a net negative charge on their composite layers due to cation substitutions (e.g. Al^{3+} for Si^{4+}; Fe^{2+}; Mg^{2+} for Al^{3+}, Fe^{3+}). In minerals such as illites and smectites the net negative charge is offset by interlayer cations and in the chlorites by the interlayer octahedral sheet. Kaolins ideally have neutral composite layer structures, but in reality limited cation substitution may occur producing a very small net negative charge on the composite layer offset by a small number of interlayer cations. When suspended in water some of the interlayer cations on the surface of the clay particles go into solution producing a negatively charged clay surface surrounded by a diffuse 'double layer' of hydrated cations (Fig. 2.13) (van Olphen 1977).

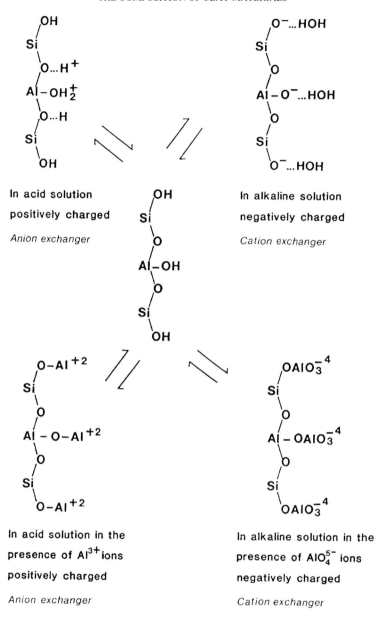

FIG. 2.14. Edge site anion exchange as a function of pH (from Yariv and Cross, 1979).

On the edges of the clay particles the disruption of the clay structure produces broken bond edge sites which may be negatively or positively charged. The nature of the charge is determined by the presence of certain ions, notably H^+, OH^-, Al^{3+} and AlO_4^{5-}, whose presence is pH dependent, being negatively charged in alkaline solutions and positively charged in acidic solutions (Fig. 2.14). This pH dependent charge accounts for only a small percentage of the total charge in illites and smectites, but is more significant for kaolins and chlorites.

In addition to cation exchange it is also possible to have anion exchange but usually to a lesser extent than cation exchange.

The ions principally adsorbed onto the clay surfaces and, to a lesser extent, on the edges of the clay particles, can be exchanged with other available ions. This gives rise to the phenomenon of cation/anion exchange. The degree to which exchange reactions take place depends on the nature of the clay minerals and the concentrations of the cations/anions involved and of other ionic species. Oxides and organic matter commonly coat clay particles and also have ion exchange properties, which are strongly

TABLE 2.2. *Cation exchange capacities of some clay minerals (meq/l00g) at pH = 7 (after Drever 1982)*

Smectite	80–150
Illite	10–40
Kaolin	1–10
Chlorite	< 10

dependent on pH, and can alter the cation/anion exchange capacities of associated clays in natural systems.

Ion exchange is a dynamic process governed by the law of mass action. Generally, the greater the concentration of a particular cation in solution, the more readily it will replace equivalent cations adsorbed onto the clay surfaces and interlayers. Cations of higher valence will usually replace those of lower valence and, if of the same valence, smaller sized cations will replace larger sized cations. However, it is important to note that in the diffuse layer, the cations will be hydrated and it is the size of the hydrated cation that needs to be considered.

In swelling clay minerals, the interlayer cation sites will also provide exchange sites, whereas for other clays only the outer surface sites will be involved. This explains the higher cation exchange capacities of the smectites compared to the illites (Table 2.2).

Because the edge sites are also involved in ion exchange and as the nature of the charge on the edge sites is pH dependent the cation exchange capacity of individual clay minerals are also pH dependent.

2.2. Non-clay mineralogy

2.2.1. Quartz and chert

Quartz is commonly present in clay materials, predominantly in the silt grade material (Fig. 2.1). It is generally detrital, but some may be biogenic or authigenic in origin.

Silicification may occur during early diagenesis with the dissolution of biogenic silica, a possible source for the later inorganic precipitation of quartz (Millot 1970). Diagenetic modification of clay minerals may also cause the mobilization of SiO_2 which may reprecipitate as a quartz cement in mudrocks or adjacent sandstone beds.

Chert is a general description for fine-grained siliceous sediments of chemical, biochemical or biogenic origin. It is a dense, relatively hard material composed of finely crystalline quartz crystals or fibrous chalcedonic quartz, usually derived from the diagenetic alteration of volcanogenic or biogenic opaline silica (see Text Box below).

Cherts usually occur as either bedded or nodular forms. The bedded forms are commonly associated with volcanic sequences whereas nodular cherts are mainly hosted in chalks or mudrocks. The nodular forms hosted in chalks and other carbonates are specifically referred to as flints. Jasper is a red coloured chert that contains hematite impurities.

2.2.2. Feldspars

As with quartz, feldspars in clay materials are predominantly detrital in origin, but there is petrological evidence to suggest they may also form diagenetic cements around existing feldspar grains.

2.2.3. Carbonates

According to Shaw & Weaver (1965) there is an average of about 4% by weight of carbonate minerals in siliciclastic mudrocks but this can be highly variable. Calcite ($CaCO_3$) and dolomite ($CaMg(CO_3)_2$) are the predominant carbonate minerals in clay materials, though siderite ($FeCO_3$) may also be locally important. Metastable polymorphs of $CaCO_3$, aragonite (orthorhombic) and, less commonly, vaterite (hexagonal) also occur in Recent clay materials. The carbonate may be present as clasts derived from organic skeletal remains and the weathering of pre-existing carbonate rocks or as diagenetic cements and nodules.

The source of carbonate for the formation of diagenetic carbonate cements or nodules may be from the recrystallization of primary (skeletal) carbonate material and/or direct precipitation from alkaline bicarbonate-rich pore waters. These processes are considered to be often related to the maturation of organic matter during diagenesis (Curtis 1983).

Dolomite and siderite formation are favoured in sulphate-depleted environments (Baker & Kastner 1981; Berner 1981). The presence of authigenic siderite is indicative of anoxic sulphate-depleted conditions at the time of its formation (Table 2.3).

Biogenic opaline silica

Biogenic opaline silica is an amorphous form of opal (opal-A) that mainly occurs in microorganisms such as radiolarians (marine zooplankton), diatoms (marine and non-marine phytoplankton) and siliceous sponges that form oozes in modern oceans. Opal-A is a hydrous form of silica, $SiO_2 \cdot nH_2O$, which transforms to chert via opal-CT. Opal-CT has a disordered interlayered cristobalite-trydimite framework silicate structure, more open than that of quartz.

TABLE 2.3. *Classification of oxic-anoxic environments using authigenic Fe-mineral indicators (Berner 1981)*

Environment	Authigenic Fe-minerals
Oxic	Hematite, goethite (no organic matter preserved)
Anoxic (sulphidic)	Pyrite (organic matter preserved)
Anoxic (non-sulphidic)	
(a) post-oxic	Glauconite, Fe^{2+}/Fe^{3+} silicates siderite (no sulphides some organic matter preserved)
(b) methanic	Siderite *after* pyrite (organic matter preserved)

2.2.4. Iron sulphides

Pyrite (FeS_2) is the most common of the sulphides found in clay materials occurring as veins, nodular aggregates, scattered crystals and infilling/replacing/pseudomorphing skeletal fragments. Sedimentary authigenic pyrite forms in anoxic sulphidic environments (Table 2.3). Weathering of pyrite produces iron hydroxides/oxides and mobilizes sulphate which may be incorporated into sulphate minerals such as gypsum and jarosite. Marcasite is an orthorhombic dimorph of pyrite that has occasionally been reported in clay materials. There is some evidence to suggest that marcasite forms under more acidic conditions than pyrite. Metastable monosulphide (e.g. mackinawite and griegite) are found in modern muds as precursors to later formation of pyrite. Pyrrhotite ($Fe_{1-x}S$) is a stable monosulphide mineral phase that occcasionally has been reported in clay materials but stoichiometric FeS, the mineral troilite, is only found in meteorites.

2.2.5. Oxides and hydroxides

2.2.5.1. Iron oxides and hydroxides. Iron oxides and hydroxides are present in various mineral forms in clay materials, notably hematite (α-Fe_2O_3), maghemite (γ-Fe_2O_3), goethite (α-FeO·OH), lepidocrosite (γ-FeO·OH). Limonite is a general term for a mixture of iron oxides and hydroxides of a poorly crystalline nature. The presence of these iron oxides and hydroxides produces the characteristic red/brown/yellow colours of many clay materials depending on which mineral is present. Goethite is yellow-brown, lepidocrosite orange, maghemite brown-black and hematite red.

2.2.5.2. Aluminium oxides and hydroxides. Gibbsite ($Al(OH)_3$), boehmite (γ-AlO·OH) and diaspore (α-AlO·OH) are the most common of the aluminium oxides and hydroxides and are typically significant constituents of laterites and bauxites. Boehmite is isomorphous with lepidocrosite and diaspore with goethite.

In addition there are aluminium – silicon oxyhydroxides, imogolite and allophane, that characteristically occur in soils derived from the weathering of volcanic rocks. Their nominal formula is $SiAl_2O_3(OH)_4$ but both have poorly ordered crystal structures and are often described as amorphous. Generally allophane forms spherules and appears to be less well ordered than imogolite with its typically cylindrical habit. The $SiO_2:Al_2O_3$ ratio varies from 1–1.2 for imogolite to 1.3–2.0 for allophane (Moore & Reynolds 1997).

2.2.5.3. Ion exchange in oxides and hydroxides. In the oxides and hydroxides the charge on the surfaces arises largely from broken surface bonds producing unshared surface and edge O and OH rather than the cation substitutions that are largely responsible for imparting charge to the clay mineral structures. As described above, charge arising from broken bonds is pH dependent (Fig. 2.14). The pH at which the charge is zero is referred to as the zero point of charge or ZPC; when conditions are more acidic than the ZPC the oxide/hydroxide will be positively charged and negatively charged for conditions more alkaline than the ZPC. For example, the ZPC for boehmite is 8.2 and 6 to 7 for goethite (Eslinger & Pevear 1988).

2.2.6. Sulphates

Sulphates may form nodules or veins in clay materials, with gypsum ($CaSO_4 \cdot 2H_2O$), followed by anhydrite ($CaSO_4$) being the most common, but celestite ($SrSO_4$), barite ($BaSO_4$) and the more water-soluble sodium and magnesium sulphates may also occur. Their occurrence may suggest hypersaline conditions of formation, though gypsum and various iron sulphates may also form during weathering of shales containing pyrite, the oxidation of the sulphide being the source of the sulphate.

Jarosite ($KFe_3(SO_4)_2(OH)_6$) is a distinctively yellow mineral that characteristically forms as surface coatings and aggregates from the oxidation of pyrite to iron sulphate, followed by further reaction with illite/mica or K-feldspars:

$$12FeSO_4 + 4KAl_2Si_3O_8(OH)_2 + 48H_2O + O_2 \\ = 4KFe_3(SO_4)_2(OH)_6 + 8Al(OH)_3 + \\ 12Si(OH)_4 + 4H_2SO_4.$$

2.2.6.1. Ettringite group. The ettringite group of minerals, mixed calcium carbonate sulphate silicates, and especially thaumasite ($Ca_3(SO_4)(CO_3)(Si(OH)_6) \cdot 12H_2O$), has attracted attention recently because of the observation of the growth of such minerals at concrete-clay boundaries, which can cause significant damage to concrete structures (Thaumasite Expert Group 1999).

2.2.7. Zeolites

Zeolites are hydrous framework silicates and those that occur in clay materials are predominantly alkali varieties (Table 2.4), They are commonly associated with aluminous smectites formed by the weathering and alteration of

TABLE 2.4. *Common zeolites occurring in clay materials*

Natrolite	$Na_2Al_2Si_3O_{10} \cdot 2H_2O$
Analcite	$NaAlSi_2O_6 \cdot H_2O$
Phillipsite	$(Ca,Na,K)_3Al_3Si_5O_{16} \cdot 6H_2O$
Erionite	$(Na_2K_2CaMg)_{4-5}Al_9Si_{27}O_{72} \cdot 27H_2O$
Heulandite	$(Na,K)Ca_4Al_9Si_{27}O_{72} \cdot 24H_2O$
Clinoptilolite	$(K,Na)_6 Al_6Si_{30}O_{72} \cdot 20H_2O$
Mordenite	$(Na,K,Ca)Al_2Si_{10}O_{24} \cdot 7H_2O$

volcanics. Analcite is also characteristically found as an authigenic mineral in hypersaline muddy sediments.

Zeolites can be used as indicators of diagenesis and low grade metamorphism of argillaceous sediments. The mixed alkali (K, Na and some Ca) zeolites are progressively replaced by more sodic, less siliceous varieties which with increasing metamorphism are in turn replaced by albite and eventually K-feldspars. The calcic forms persist to higher temperatures.

Zeolites have open framework silicate structures with the capability of acting as 'molecular sieves' by absorbing cations, anions and polar organic molecules into the open channels within their structures.

2.2.8. Phosphates

The most common phosphate minerals occurring in clay materials are the carbonate apatite group $(Ca_5(PO_4, CO_3)_3(OH, F, Cl)$. Phosphates may occur in a cryptocrystalline form referred to *sensu lato* as 'collophane'. The iron-rich phosphate mineral, vivianite, which has a very distinctive blue colour, may form in lacustrine environments and soils.

Phosphates may be bioclastic (e.g. accumulations of vertebrate skeletal fragments), faecal or diagenetic in origin. Authigenic phosphates can precipitate in mudrocks and sandstones as nodules, ooliths, pisoliths, cements or replacements of calcareous skeletal clasts in environments with low clastic sedimentation and high organic productivity enriched in phosphate.

In addition to being present in phosphate minerals, phosphorus is sorbed onto clay and other colloidal particles and this mechanism can be a major factor in controlling phophorus mobility in soil and aqueous environments.

2.2.9. Halides

Halite (NaCl) and, to a much less extent, other halides such as sylvite (KCl) and carnallite $(KMgCl_3 \cdot 6H_2O)$ are characteristically found in clay materials associated with evaporitic sedimentary and soil environments. Evaporation of sea water and chloride-rich brines can create a sequence of evaporitic minerals that are often cyclic due to replenishment of the evaporating brines or interbedded with fine-grained siliciclastic or carbonate sediments.

2.3. Organic matter

Organic matter is present in a variety of forms in clay materials including discrete organic particles, absorbed onto clay and other associated colloidal particles (e.g. iron oxides) and micro-organisms. The nature, abundance and distribution of organic matter can influence the physico-chemical behaviour of clay materials.

Generally the organic materials can be subdivided into labile and refractory organic fractions. Labile organic matter generally does not survive burial whereas refractory organics are non-metabolizable, generally survive burial and can be preserved in ancient sediments as kerogen. Kerogen makes up 95% or more of the organic matter preserved in sedimentary rocks. Kerogen has been defined as the organic constituents that are insoluble in aqueous alkaline and common organic solvents (Tissot & Welte 1984) which is the equivalent of 'humin' in soil science (Tyson 1995).

The labile organic components degrade via leaching and decomposition processes. The leaching process involves the breakdown of cellular material by intracellular enzymes producing soluble compounds. The decomposition process relates to the bacterial colonization and the fermenting activity of the bacteria which is greatest under warm, oxic conditions.

In addition to bacteria, other micro-organisms may also be present such as protozoa and small algae, fungi and viruses. In soils there are 10^7 to 10^8 bacteria per gram of dry mass, 10^5 to 10^6 fungi and 10^4 protozoa (Campbell 1992). These micro-organisms show a wide range of sizes and morphologies and can also exist in various dormant states such as spores and cysts that are resistant to dessication and temperature changes. Micro-organisms can flourish even under apparently very adverse conditions such as hydrothermal vents on mid-ocean ridges with temperatures above 200°C, hypersaline soda lakes with a pH > 10 or in acid mine drainage waters with a pH of <2.

Some micro-organisms are aerobic, acquiring their energy from reactions with oxygen, which is consequently consumed, leading to oxygen-deficient conditions. Others are anaerobic, generating their energy by reducing inorganic chemical species (sulphate, nitrate, carbon dioxide) or organic species (fermentation reactions) and/or oxidizing other inorganic species (sulphide, ferrous ions) in the absence of oxygen. The bacteria *Thiobacillus denatrificans* uses sulphide, carbon dioxide and nitrate as energy sources and produces sulphate, sugars and reduced nitrogen compounds (Campbell 1992). This is one of many thiobacillus bacteria that exist in clay materials which derive energy from oxidizing inorganic sulphur usually under aerobic conditions. One variety, *Thiobacillus ferrooxidans* also oxidizes ferrous ions to ferric in acidic environments and is a key influence in the natural degradation of pyrite (Hawkins & Pinches 1992).

The nature and abundance of micro-organisms in clay materials and the inorganic compounds available for oxidation and/or reduction will have a major influence in determining their chemical behaviour.

2.3.1. Organic-clay complexes and interactions

Organic-clay complexes represent part of the humin or kerogen organic fraction. Organic molecules are absorbed onto the charged clay mineral surfaces and interlayer sites by:

- competing with and replacing the water molecules and forming complexes with interlayer cations;
- being bonded to the cations by bridging water molecules;
- bonding of organic cations, molecules with a strong tendency to form hydrogen bonds and polar organic molecules onto the charged clay surfaces.

Most organic compounds of interest as environmental pollutants are small non-ionic molecules and their sorption onto clay minerals involves weak interactions via van der Waal forces and hydrogen bonding (Sawhney 1996).

Such complexes and organo-clay reactions can have significant influences on chemical mobility and catalysis of chemical reactions. These can affect, for example, the chemical behaviour of pollutants, fertilizers, pesticides in soils and the physical properties of flocculation and dispersion in aqueous environments (Johnston 1996).

The catalytic behaviour is related to the large surface areas of clays and other colloids, which is a function of mineralogy, crystal size and habit. Clay surfaces also have acidic properties due to the presence of Al substituting for Si in the tetrahedral sites. The resultant net negative charge can be offset by the presence of charge balancing H_3O^+ corresponding to a proton-donating Bronsted acid site. In addition, Al^{3+} ions on the edges of clay particle are capable of being electron pair acceptors and thus can act as Lewis acid sites (Rupert et al. 1987). The degree to which this surface acidity is produced is an essential factor in creating the conditions for clay minerals to act as effective catalysts of various organic reactions.

2.4. Water

In clay materials water is present in the inter-particle pore spaces, adsorbed onto or absorbed into clay minerals, other minerals and organic matter. Guven (1993) defined three forms of hydration in clay materials:

- interlamellar hydration involving adsorption of water onto the internal surfaces of clay mineral particles;
- continuous osmotic hydration involving unlimited adsorption of water onto the internal and external surfaces of primary mineral and organic particles;
- capillary condensation of water into micro-pores.

It needs to be emphasized that all mineral surfaces are capable of adsorbing water onto their surface. However, this is most significant for the clay minerals because their fine grain size and platy habit produce very large surface areas and thus give a relatively greater capacity for water adsorption.

The presence of adsorbed water layers covering the clay particles produces the characteristic cohesive plastic behaviour of clay materials. The water adsorbed onto the clay mineral surfaces has properties midway between bulk liquid water and ice and forms a structured water layer and a more diffuse less structured water layer. The thickness of the adsorbed surface water layer varies depending on the clay surface charge, the exchange cations and cation hydration and the salinity of the aqueous solution. With increasing ionic strength there is a lesser tendency for cations to diffuse away from the negatively charged clay particles and the absorbed water layer becomes compressed and less diffuse than at lower ionic strengths.

The nature and behaviour of the water absorbed into the interlayer sites of clay minerals is a function of :

- the polar nature of the water molecules;
- the size and charge of the interlayer cations and their hydration state;
- the location and value of the charge on the silicate layers of the clay mineral structures.

In the interlayer sites of expandable clay minerals the water forms coordinated shells around interlayer cations. In the smectites the water molecules form single, double or triple water layers, depending on the humidity of the surrounding environment. The vermiculites can exhibit single or double water layers depending on the humidity conditions.

There are significant differences in the interlayer water present in halloysites compared with the smectites and vermiculites that help explain why rewetting of dehydrated halloysites does not cause the structure to rehydrate. In the hydrated state there appears to be no preferential ordering of the water molecules and only weak interactions with the halloysite surfaces. There is also evidence that dehydration of the halloysite is accompanied by delamination of the halloysite layers thus inhibiting halloysite rehydration (Newman 1987).

The absorption of water into sepiolites and palygorskites involves adsorption onto the external surfaces and absorption into the channel sites within the crystal structure, which is analogous to the absoption of water into the channel sites of zeolite structures. The packing of the water molecules absorbed onto the outer surfaces of the sepiolites and palygorskites is much less dense than than that of a surface water monolayer on other sheet silicates (Newman 1987).

The density of the water in the interlayer sites of swelling clay minerals is considered to be greater than that of liquid water and to increase with increasing pressure (Skipper et al. 1993). However, Moore & Reynolds (1997) argued that with increasing pressure the interlayer water should have a similar density to that of liquid pore waters.

2.5. Conclusions

Clay materials are composed of a varying mix of solid, liquid and vapour phases, with the clay minerals usually the most abundant mineral phases. The varying properties of the individual clay minerals and their often complex interactions with fluid phases under different physical and chemical conditions can play a major role in controlling the behaviour of clay materials. However, it is simplistic and misleading to equate clay materials with clay minerals. The nature and abundance of other minerals and also organic matter, and their varying interactions with fluid phases, can also have a significant effect on the behaviour of clay materials.

Clay materials are used in a variety of applications as construction materials and a detailed knowledge of their compositions is of fundamental importance if we are to maximize the efficient use of such materials for a given application.

References

BAKER, P. A. & KASTNER, M. 1981. Constraints on the formation of sedimentary dolomite. *Science*, **213**, 214–216.

BERNER, R. A. 1981. A new geochemical classification of sedimentary environments. *Journal of Sedimentary Petrology*, **51**, 359–366.

BETHKE, C. M. & ALTANER, S. P. 1986. Layer-by-layer mechanism of smectite illitization and application to a new rate law. *Clays and Clay Minerals*, **34**, 135–145.

BLATT, H., MIDDLETON, G. & MURRAY, R. 1972. *Origin of Sedimentary Rocks*. Prentice-Hall, New York.

BURST, J. F. 1958. Glauconite pellets: their mineral nature and applications to stratigraphic interpretations. *AAPG Bulletin*, **42** (2), 310–327.

CAMPBELL, R. 1992. Microbiology of soils. *In*: HAWKINS, A. B. (ed.) *Proc. IGCC 92 Conference on The implications of ground chemistry and microbiology for construction*, Bristol, 1–17.

CURTIS, C. D. 1983. Geochemistry of porosity enhancement and reduction in clastic sediments. *In*: BROOKS, J. (ed.) *Petroleum Geochemistry and Exploration of Europe*. Geological Society, London, 113–126.

DREVER, J. I. 1982. *The Geochemistry of Natural Waters*. Prentice-Hall, New York.

ESLINGER, E. & PEVEAR, D. 1988. *Clay Minerals for Petroleum Geologists and Engineers*. SEPM Short Course, **22**.

GUVEN, N. 1993. Molecular aspects of clay-water interactions. *In*: GUVEN, N. & PALLASTRO, R. M. (eds) *Clay Water Interface and its Rheological Properties*. CMS Workshop Lectures, **4**, Clay Minerals Society, 2–79.

GUVEN, N. & PALLASTRO, R. M. (eds) 1993. *Clay Water Interface and its Rheological Properties*. CMS Workshop Lectures, **4**. Clay Minerals Society.

HAWKINS, A. B. & PINCHES, G. M. 1992. Understanding sulphate generated heave from pyrite degradation. *In*: HAWKINS, A. B. (ed.) *Proc. IGCC 92 Conference on The Implications of Ground Chemistry and Microbiology for Construction*. Bristol.

JOHNSTON, C. T. 1996. Sorption of organic compounds in clay minerals: a surface functional group approach. *In*: SAWNHEY, B. L. (ed.) *Organic Pollutants in the Environment. CMS Workshop Lectures*, **8**. Clay Minerals Society, 2–44.

KOHYAMA, N., FUKUSHIMA, K. & FUKAMI, A. 1978. Observations of the hydrated form of tubular halloysite by an electron microscope fitted with an environmental cell. *Clays and Clay Minerals*, **26**, 25–40.

LOW, P. F. 1993. Interparticle forces in clay suspensions: flocculation, viscous flow and swelling. *In*: GUVEN, N. & PALLASTRO, R. M. (eds) *Clay Water Interface and its Rheological Properties*. CMS workshop lectures, **4**, Clay Minerals Society, 157–190.

MCKINNON, I. D. R. 1987. The fundamental nature of illite-smectite mixed layer clay particles: a comment on the papers by P.H.Nadeau and coworkers. *Clays & Clay Minerals*, **35**, 74–76.

MILLOT, G. 1970. *Geology of Clay*. Chapman-Hall, London.

MOORE, D. M. & REYNOLDS, R. C. 1997. X-ray difraction and the identification and analysis of clay minerals. Oxford University Press.

NADEAU, P. H. & BAIN, D. C. 1986. Composition of some smectites and diagenetic illitic clays and implications for their origin. *Clays and Clay Materials*, **34**, 455–464.

NADEAU, P. H., WILSON, M. J., MCHARDY, W. J. & TAIT, J. 1987. Interparticle diffraction: a new concept for interstratified clays. *Clay Minerals*, **19**, 757–769.

NEWMAN, A. C. D. 1987. The interaction of water with clay mineral surfaces. *In*: NEWMAN, A. C. D. (ed.) *Chemistry of Clays and Clay Minerals*. Mineralogical Society Monograph, **6**, 237–274.

RUPERT, J. P., GRANQUIST, W. T. & PINNAVAIA, T. J. 1987. Catalytic properties of clay minerals. *In*: NEWMAN, A. C. D. (ed.) *Chemistry of Clays and Clay Minerals*, Mineralogical Society Monograph, **6**, 275–318.

SAWNHEY, B. L. 1996. Sorption and desorption of organic compounds by soils and clays. *In*: SAWNHEY, B. L. (ed.) *Organic Pollutants in the Environment. CMS Workshop Lectures*, **8**. Clay Minerals Society, 45–68.

SAWNHEY, B. L. & REYNOLDS, R. C. 1985. Interstratified clay as fundamental particles: a discussion. *Clays & Clay Minerals*, **33**, 559.

SHAW, D. B. & WEAVER, C. E. 1965. The mineralogical composition of shale. *Journal of Sedimentary Petrology*, **35**, 213–222.

SKIPPER, N. T., REFSON, K. & MCCONNELL, J. D. C. 1993. Monte Carlo simulations of Mg-and Na-smectites. *In*: MANNING, D. A. C. HALL, P. L. & HUGHES, C. R. (eds) *Geochemistry of Clay-pore Fluid Interactions*, Chapman & Hall, London, 40–61.

THAUMASITE EXPERT GROUP 1999. *The Thaumasite Form of Sulphate Attack: Risks, Diagnosis, Remedial Works and Guidance on New Construction*. Department of Environment, Transport & Regions, HMSO, London.

TISSOT, B. P. & WELTE, D. H. 1984. *Petroleum Formation and Occurrence*. Springer, Berlin.

TUCKER, M. E. 1981. *Sedimentary Petrology: an Introduction*. Blackwell, Oxford.

TYSON, R. V. 1995. *Sedimentary Organic Matter*. Chapman & Hall, London.

VAN OLPHEN, H. 1977. *An Introduction to Clay Colloid Chemistry*. Wiley Chichester.

VELDE, B. 1992. *Introduction to Clay Minerals*. Chapman & Hall, London.

YARIV, S. & CROSS, H. 1979. *Geochemistry of Colloidal Systems for Earth Scientists*. Springer, Berlin.

3. Formation and alteration of clay materials

3.1. Introduction

The formation and alteration of clay minerals and their accumulation as clay materials can occur by a very wide range of processes. In one way or another, however, most of these processes and the environments in which they operate involve the chemical actions and physical movement of water. As such, clay minerals can be considered the characteristic minerals of the Earth's near surface hydrous environments, including that of weathering, sedimentation, diagenesis/low-grade metamorphism and hydrothermal alteration (Fig. 3.1). Simply defined, the weathering environment is that in which rocks and the minerals they contain are altered by processes determined by the atmosphere, hydrosphere and the biosphere. Soil formation, also known as pedogenesis, occurs in the weathering environment. The sedimentary environment is the zone in which, soil, weathered rock and mineral (and biogenic) materials are eroded, mixed and deposited as sediments by water, wind and ice. Diagenesis involves all those physical and chemical processes that occur between sedimentation and metamorphism, whilst hydrothermal alteration encompasses the interactions between heated water and rock.

In this chapter, the origins of the various clay minerals that may occur in each of these environments are reviewed along with the processes that may lead to their accumulation and alteration, usually together with other components, to form clay materials. In many instances, clay materials are formed in one environment by the accumulation or alteration of clay minerals formed in others. Thus the geological history of a clay material, and consequently its properties and behaviour, may depend on many processes separated in both time and space.

In order to deal with this complexity in a logical manner the first part of this account of clay material formation and alteration deals sequentially with processes and features that characterize the four environments of clay material formation; namely the weathering, sedimentary, diagenetic/low-grade metamorphic, and hydrothermal environments. A second part deals with the products of these processes; that is the various kinds of common clay materials that may be encountered in the field. This structure is deemed necessary since the formation of most clay materials can only be understood in terms of the action of processes characteristic of more than one environment. Indeed, the specific circumstances that link key processes together to form particular clay materials are often unique. Nonetheless, general linkages also exist between the four environments where clay materials are formed and these can be most easily understood in terms of the concept of the clay cycle.

3.1.1. The clay cycle

Ultimately, the clay cycle (Fig. 3.2) begins with the formation of clay minerals and their accumulation in soils by the weathering of primary rock forming minerals and glassy volcanic ash. Erosion and transport of clay minerals from soils and weathering profiles followed by their selective sorting, segregation and deposition by physical processes of sedimentation lead to their further accumulation as the main components of muds in sedimentary basins. Burial of mud by further mud, or other sediments, may follow and in response to the physical and chemical changes that accompany burial (known as diagenesis) a progressive transformation occurs to change muds into mudstones and shales. Eventually, if tectonic forces are involved, mudstones and shales are transformed to slates. Inevitably, over geological time, tectonic uplift brings buried mudrocks (mudstone, shale and slate) back to the Earth's surface where the agents of weathering begin the cycle once again.

Although the clay cycle is a part of the larger geotectonic cycle, when attention is focused on the surface and shallow crustal portions of the Earth the global importance of clays is very clear. This is because mudrocks, of one kind or another, are by far the most abundant kind of sedimentary rock (Pettijohn 1975; Blatt *et al.* 1980; Potter *et al.* 1980). In fact most estimates indicate that more than two thirds of the Phanerozoic stratigraphic record is 'written' in clay. Thus many geotectonic processes are recycling clays and clay materials. In addition, but largely as a consequence of their abundance in the stratigraphic sedimentary veneer that covers much of the Earth's surface, mudrocks form the bedrock over more than one third of the World's land area (Meybeck 1987). The result is that clay materials are themselves the most common parent materials for weathering and soil formation. Thus many of the component clay minerals of these materials may have already been through one or more previous cycles of weathering, erosion, deposition, burial and uplift.

3.1.2. Clay materials and plate tectonics

The dynamic flux of clay from one setting to another and the cycles of formation of clay materials are driven by geotectonic forces. At a global scale the Earth's crust is divided into a series of plates, large coherent regions of

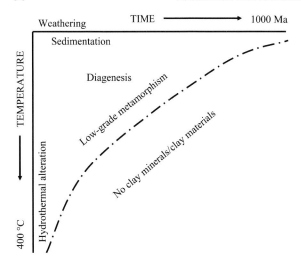

FIG. 3.1. The geological time–temperature realm of clay minerals/clay materials in the weathering, sedimentation, diagenesis low temperature metamorphism and hydrothermal environments.

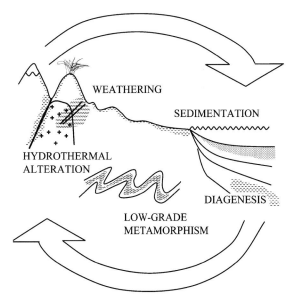

FIG. 3.2. Schematic clay cycle, whereby clay minerals formed in one environment are frequently recycled into others.

continental or oceanic crust moving with respect to each other, and interacting at their boundaries (Cox & Hart 1986). The relative movements of these plates are known as plate tectonics and it is this movement, especially at their boundaries, that ultimately determines the development of sedimentary basins, the distribution and elevation of mountain ranges, and the location of hydrothermal alterations (Fig. 3.3).

Sedimentary basins, the potential long-term stores of all kinds of sediments, are located where regions of the Earth's surface undergo prolonged subsidence. In a very simple fashion most of these regions are associated with movements at plate boundaries which may be either extensional, convergent or strike-slip. Furthermore, patterns of subsidence and burial of sediment, and thereby the character and intensity of diagenesis/metamorphism, are determined by the plate tectonic setting of the sedimentary basin and associated factors such as heat flow regime.

Similarly, mountain ranges, elevated by plate tectonics processes, provide the relief necessary to generate high volumes of muddy detritus with which sedimentary basins may be filled. Indeed, the location of hydrothermal clay formations is also intimately linked to plate tectonic processes.

For example, hydrothermal alteration and the weathering of volcanic rocks are particularly widespread sources of clay in Japan (Sudo & Shimoda 1978). This is a direct result of Japan's tectonic setting at a destructive plate boundary with its consequences for active volcanism. Thus the location and weathering of primary volcanic rocks, including lava and ash, as a potential source of clay minerals to the sedimentary environment is particularly significant along active margins of the Earth's crust (Merriman 2002a). In contrast to tectonic activity at plate margins, tectonic stability may be an equally important factor for the development of some clay materials. For example, the stability of the cratonic interiors of large continental plates may be an essential prerequisite for the development and preservation of deep (metres) clay-rich weathering profiles, provided of course that other factors combine to develop and preserve them.

Palaeogeographic reconstructions such as those figured in Chapter 5 show how the configuration of tectonic plates and their boundaries have changed through geological time. The distribution and character of clay materials preserved in the geological record depend on many factors but overall these factors all operate within the context of plate tectonics. In fact, it may still be possible to discern the influence of geotectonic setting on the character of clay materials, even after long histories which have progressed as far as low grade metamorphism (Merriman 2002b).

3.1.3. Origins of clay minerals by neoformation, inheritance and transformation

Within any of the four environments (weathering, sedimentary, diagenetic/low-grade metamorphic, hydrothermal) in which clay materials may be formed there are three possible origins for the constituent clay minerals of which they are composed (Eberl 1984). Firstly, clay minerals may be formed by direct precipitation from solution or via the ageing of amorphous materials, such as glassy volcanic ash, an origin known as neoformation. Secondly, clay minerals may be inherited, i.e. not actually formed in the environment in which they are now found. And thirdly, clay minerals may be *transformed*, i.e. have some combination of inherited and neoformed parts. For many clay minerals, one particular origin in one particular

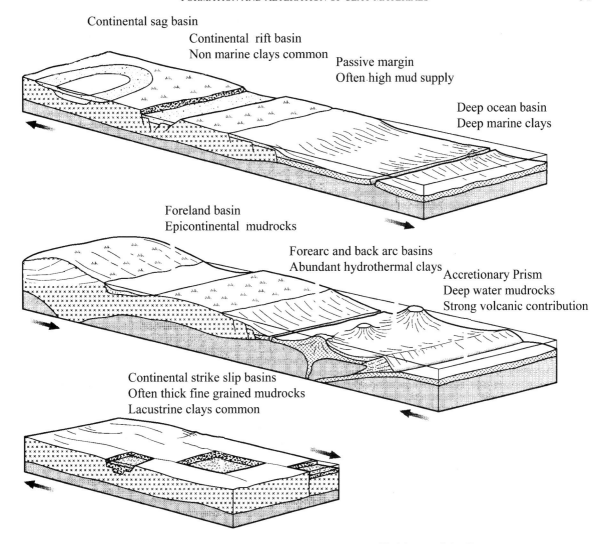

FIG. 3.3. Plate tectonic settings and this influence on the geresis of clay materials modified from Duff (1993).

environment may dominate, though none is exclusive. For example, most kaolinite in contemporary world soils is neoformed, whereas most kaolinite in marine muds is inherited (from soils). Furthermore, the three different origins also give different kinds of environmental and geological information. Thus neoformed clay minerals reflect the geochemical conditions at the time of their formation, such as solution chemistry. In contrast, inherited clays must be stable/metastable enough, either through equilibrium or slow reaction, to be inert under the conditions of their new environment. Inherited clays thereby give information about the environment from which they have been derived, i.e. their provenance. Transformed clays may hold both geochemical and provenance information.

Generally, the importance of the different origins varies considerably from one environment to another. The clay minerals found in the weathering environment and in soils are probably the most variable since, as explained in terms of the clay cycle, many are often inherited directly from whatever rocks are being weathered in the source area. Nonetheless, depending on the intensity of weathering and soil forming processes, inheritance, transformation or neoformation may be the dominant origin. In the sedimentary environment and at a global scale, inheritance is undoubtedly the most important origin for clay minerals since the components of most muds are detrital. That is, they are derived from other sources (soils, older sediments, hydrothermal alterations) by the various agents and processes of erosion, transport and deposition. Under certain sedimentary conditions, however, transformation and neoformation may be locally important. Indeed, neoformation of certain clay minerals in the sedimentary environment, for example smectites, may be almost as widespread as it is in the weathering environment, but inherently more difficult to recognize because of massive dilution of neoformed clays by detrital clays. As sediments are buried and move into the realm of diagenesis and subsequent metamorphism, the largely inherited clay mineral assemblages are increasingly subject to processes involving transformation, and neoformation. These processes proceed via geochemical reactions involving dissolution of some components and precipitation and growth of others. Clay minerals formed in the hydrothermal environment are mostly neoformed.

3.2. Weathering environment and soils

3.2.1. Introduction

Of all the environments in which clay minerals may form, none is more central to their origin than the weathering environment. Stated in very simple terms, clay minerals are formed by the interaction of rocks and water; and this interaction is a fundamental aspect of most weathering processes (Drever 1985; Lerman & Meybeck 1988; White & Brantley 1995; Cotter Howells & Paterson 2000). For example, when fresh rocks containing minerals such as feldspars are exposed to the weathering environment, clay minerals may form as a by-product of the weathering of the feldspars. Volumetrically, feldspars and volcanic ash are the most prolific sources of clay minerals formed by weathering, but the weathering of other rock forming minerals, especially olivines, pyroxenes, amphiboles, micas and chlorites, are also important. The weathering environment is therefore the birthplace for many clay minerals, which once formed may be re-cycled numerous times through weathering, sedimentary, diagenetic and metamorphic environments. Of course clay minerals do form abundantly in other environments and may enter the clay cycle at other points. However, the importance of the weathering environment as the ultimate source is underlined by the fact that most clay materials found in the sedimentary environment are, by and large, the redistributed products of weathering.

In certain circumstances, weathering processes alone may lead directly to the formation of clay materials. However, because of the importance of weathering as a primary source of clay minerals, an understanding of the clay mineralogy of weathering is also a starting point for understanding the origin of very many clay materials whose components were ultimately formed elsewhere. The clay mineralogy of soils and weathered rocks (saprolites) is extremely varied because many factors may influence it. Nonetheless, various generalizations concerning the origin of clay minerals in soils have been established since the mid-1960s (Millot 1970; Jackson 1964) and the review by Wilson (1999) summarizes progress since then.

An origin by inheritance depends only on the availability of a mineral in the parent or source material and the physical processes by which clay minerals may be transported or accumulate. There are therefore an infinite number of pathways by which clay minerals may be inherited by the weathering environment or soil in which they are observed. Origins by transformation and neoformation are much more restrictive since they reflect chemical processes occurring in soils. It is, therefore, possible to identify a variety of general and common pathways for the in situ formation of clay minerals in soils, to which may be added the ever present origin by inheritance (Fig. 3.4). Thus mixed-layer clay minerals formed in soils tend to form by the transformation of micas and, to a lesser extent, chlorite. Further developed transformations may eventually lead to the formation of vermiculite and smectite. Smectite may also form by direct neoformation and kaolinite and halloysite are almost exclusively formed in this way.

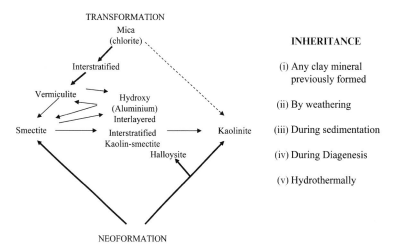

FIG. 3.4. Common pathways for the formation of clay minerals in soils. Thickness of transformation and neoformation arrows indicates relative importance of the process for the origin of the respective mineral. Modified from Wilson (1999).

3.2.2. Controls on clay mineral formation during weathering and soil formation

As a result of the common pathways of clay mineral formation there are also patterns to the distribution of clay minerals in soils at various scales, from the soil profile, to the landscape (catenas) and more generally at a global scale. Soil profiles develop in response to various chemical, biological and physical processes. Accumulations of organic matter and the redistribution and alteration of mineral matter produce changes in colour, texture and structure. Collectively such changes often lead to the development of distinct soil horizons, formerly named A, B, C with increasing depth. Together the various soil horizons define the soil profile and are the basis for many soil classification schemes. Frequently, clay minerals in one part of the profile are quite distinct from those in another part. Such patterns to clay mineral distribution within profiles and also at other larger scales can be understood in terms of the environmental factors that control the processes of inheritance, transformation and neoformation of clay minerals in the weathering and soil environment. Some of the principal factors are climate, drainage, topography, parent rock type and time, i.e. the very same factors that control the formation and development of soils themselves (Cotter-Howells & Paterson 2000).

3.2.2.1. Climate. Very broadly the global zonation of chemical and physical weathering, and of the intensity of these processes, are related to climate. As a result the clay mineral assemblages found in the World's soils are also broadly distributed in relation to climatic zones (Fig. 3.5). At the most general level, inheritance prevails where weathering is mainly physical, neoformation under conditions of intense chemical weathering, and transformation where conditions are intermediate. The depth of weathering and soil and clay formation is also similarly related to climate with the deepest weathering and consequently the most clay-rich soils generally characteristic of the tropical regions of the World.

Soils in tundra, polar and hot desert climates, where physical weathering dominates, may contain virtually any clay mineral inherited from the parent material. Notably, this includes those that would otherwise be very easily destroyed by any form of chemical weathering such as trioctahedral chlorite. As far as weathering is concerned, the mechanism of inheritance is probably the easiest to understand. However, it also tends to produces the most

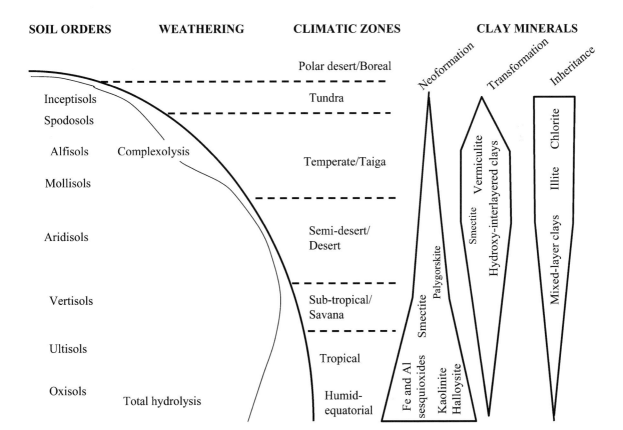

FIG. 3.5. Generalized global relationships among soil orders, weathering, climatic zones and common modes of formation of various clay minerals in soils.

diverse clay mineral assemblages because of the inherent variability of parent materials, as well as the potential for clays to be inherited from previous weathering episodes under different climatic conditions. Indeed many of the World's oldest soils are polygenetic, and so may contain clay minerals that are not related to present-day climatic conditions. Thus all clay minerals may be inherited. Nonetheless, for some clay minerals in soils, the most important of which are illite and chlorite, inheritance is undoubtedly their most common origin. In fact, there is no evidence that illite or chlorite *sensu stricto* are ever neoformed in soil or weathering environments (Wilson 1999).

More temperate climates are characterized by clay mineral assemblages containing notable quantities of vermiculites, mixed-layer clay minerals, and aluminium interlayered clays (sometimes called 'soil chlorites'). All are formed principally by transformation processes. Many of the processes of transformation in temperate soils are related to the presence of increased quantities of organic matter and organic acids that promote the process of 'acidolysis' (Pedro 1982; Duchaufour 1982). Indeed it has been suggested that the formation of mixed-layer clay minerals in soils is invariably by the process of transformation involving the break down and comminution of larger sand and silt size particles of mica and chlorite (Wilson & Nadeau 1985). This can be contrasted with the mechanisms of neoformation and crystal growth which are undoubtedly involved in the formation of mixed-layer clay minerals in the diagenetic and hydrothermal environments (see Sections 3.4 & 3.5). In soils the most widely recognized transformation processes involve the formation of vermiculites and smectite from mica and chlorite. In both cases the process proceeds via the formation of various intermediate mixed-layer clays, such as mica-vermiculite and chlorite-vermiculite. However, many mixed-layer clay minerals found in soils are directly inherited from diagenetic and hydrothermally altered parent materials. Thus, although transformation processes are common place, inheritance is typically the dominant origin for most clay minerals in soils developed under temperate climates. It should also be noted that mixed-layer clays formed in soils by transformation processes are likely to have substantially different properties to those that are inherited from diagenetic or hydrothermal sources (Wilson & Nadeau 1985).

Tropical climates are most favourable for the processes of neoformation, which may be promoted either as a result of the accumulation or of the depletion of base cations (Ca^{2+}, Mg^{2+}, Na^+) in solution. These contrasting conditions have been termed partial and complete 'hydrolysis' (Pedro 1982). In semi-arid tropical areas neoformation of smectite may be common place, particularly if the parent materials are enriched in base cations and conditions tend to be conducive to their accumulation in the soil. In contrast, under humid tropical conditions clay mineral assemblages are characteristically dominated by neoformed kaolinite, halloysite, gibbsite and oxides and hydroxides of iron. Such mineral assemblages result from the leaching of all but the most insoluble elements (Fe, Al, Si) from the soil profile. They are, therefore, often described as 'residual' clays. Finally, although the clay mineral assemblages of soils may reflect the overriding control of climate, the clay mineral assemblages found in deeply weathered rock (saprolite) are usually more strongly controlled by rock type.

3.2.2.2. Drainage and topography. On a smaller scale the precise distribution of clay minerals often vary in response to local changes in landscape and topography. Such distributions are related to the way in which landscapes control leaching. This is generally most evident within tropical climatic regions (Fig. 3.6) for example where so called 'red black toposequences' reflect the development of kaolinite and oxide-rich soils in uplands and smectite-rich soils in (less well drained) lowlands. Differences under other climatic regimes may be more subtle, but do exist, for example in Great Britain changes in mineralogy are commonly observed between soils differing only in drainage status (Loveland 1984; Wilson *et al.* 1984).

In tropical climates the clay mineralogy of soils may be strongly correlated with leaching indices. The greater the amount of leaching, the less base rich, the less siliceous, and the more aluminous the mineral species which develop. This correlation is simply a direct reflection of the fact that sequences such as smectite > kaolinite > gibbsite form from progressively more dilute aqueous solutions, which in a general way equates with a sequence from the driest to the wettest conditions. Such facts can be represented quantitatively by means of mineral stability diagrams such as that shown in Figure 3.7. In relation to the neoformation of clay minerals in soils, such diagrams serve simply to emphasise that the factors of climate, drainage, topography, vegetation and parent rock type are only important in so far as they may affect solution chemistry.

3.2.2.3. Parent rock type. The influence of rock type on clay mineral formation in soils is clear from the manner in which it obviously dictates which minerals are available to be inherited or transformed. Similarly, it also dictates which elements may be dissolved and removed or recombined in neoformations. Nonetheless, very intense chemical weathering produces more or less the same products irrespective of original rock type. For example, in humid climates rocks as different as basalt and granite weather to assemblages of kaolinite, halloysite, gibbsite and Fe-oxides. The classic study by Barshad (1966) of the clay mineralogy of surface (0–15 cm) soils developed on both felsic and mafic igneous rocks as a function of precipitation along the Sierra Nevada in California is instructive (Fig. 3.8). In general, smectite only occurs in soils on both rock types where precipitation is less than 100 cm per year, whereas gibbsite only occurs where precipitation is more than 100 cm per year. Kaolin minerals occupy a much wider precipitation range and dominate above 50 cm per year. Notably, illite is only present in the

Equatorial to Humid Tropical

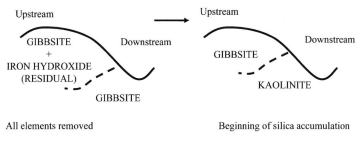

Humid Tropical with Alternate Seasons

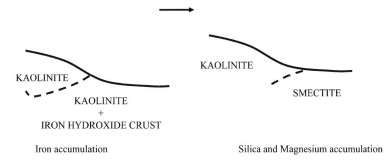

Tropical Arid

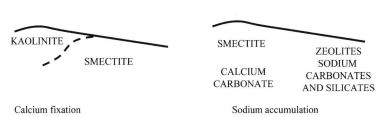

FIG. 3.6. Patterns of clay mineral development and distribution in soil catenas (topographic profiles) under various tropical climates (modified from Tardy *et al.* 1973).

soils developed on felsic rocks, indicating an origin by inheritance and transformation from dioctahedral micas which are absent in the mafic rocks. More subtle differences are also apparent in the relative proportions of clay minerals present within a given range of rainfall. This is most apparent at the lower values for precipitation, where the abundance of the various minerals varies somewhat between the felsic and mafic parent materials. Barshad (1966) suggested that this shift may be related to the generally higher cation content of the soil solutions from mafic compared to felsic rocks. The end points for both rock types at high values of annual precipitation are nonetheless very similar. In the earliest stages of weathering, or deeper within the soil profile, or where weathering occurs under less intense conditions such as in temperate climates, then original rock exherts a much greater influence, and may persist indefinitely. It is also worth noting that as a result of the disturbance of soils by a wide variety of sedimentary and mass movement processes, for example on slopes or due to glaciation, many soils are developed from materials that are quite different from the bedrock upon which they now rest.

3.2.2.4. Time. Time is another important factor in weathering and soil formation. (Righi & Meunier 1995). The formation of soils and clay minerals are typically very slow processes requiring from hundreds to millions of years. In the extensive areas that were glaciated in the Northern Hemisphere most soils have only been developing since the last glaciation about 10 000 years

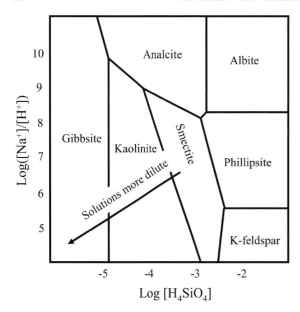

FIG. 3.7. Example of a mineral stability diagram in terms of the activities of various chemical species in solution, redrawn from Eberl (1984).

soils become, the more likely that they will have developed through significant changes in climate. Then clay minerals formed in one climatic regime become inherited when the soils adapt to another. In fact, the dynamic nature of weathering and soil formation and hence clay mineral formation in soils should be emphasized. It is doubtful that any soils ever reach a true equilibrium state, but instead are constantly adjusting to changing environmental conditions. This also leads to the cautionary note that the present global distribution of soils may be very different from episodes in the geological past when soils have had either more or less time to develop as a result of tectonic and/or climatic stability or instability. For example, Mesozoic weathering profiles rich in kaolin, typical of present day tropical conditions, appear to have formed at much more northerly latitudes under much colder climatic conditions than their present-day counterparts (Chivas & Bird 1995). Thus the present-day picture of clay mineral distribution in soils in relation to climatic zones may be one that is largely a reflection of differences in the rate-determining factors of weathering such as temperature, vegetation and atmospheric CO_2. When geological stability persists for long periods, the influence of such rate determining factors become progressively less important as soil profiles progress towards essentially the same set of end products.

3.3. Sedimentary environment

3.3.1. Introduction

Fine-grained sediments with a high proportion of clay minerals are by far the most abundant kind of sedimentary

ago. In areas that were not glaciated, however, soils may have had much longer times to develop. Indeed, in the stable cratonic areas of some continents such as Africa and Australia, much of the landscape is veneered with soils originally developed under tropical conditions in some cases dating back to Miocene (5.3–23.8 Ma) even Eocene (33.7–55.5 Ma) times (Thiry 2000). The older

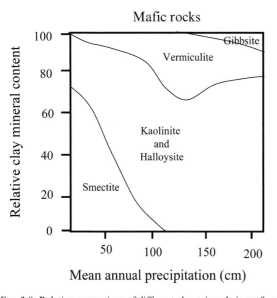
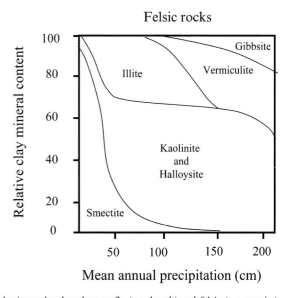

FIG. 3.8. Relative proportions of different clay minerals in surface soil horizons developed on mafic (e.g. basalt) and felsic (e.g. granite) rocks as a function of rainfall, redrawn from Barshad (1966).

rock. They account for about 65% of the stratigraphic record and they are three times more abundant than sandstones. Not surprisingly then, most clay materials are sedimentary in origin, although post-sedimentary processes have altered and contribute much to the character of many. Clay materials formed in the sedimentary environment are generally the products of physical sedimentary processes that erode, transport and deposit clay. Their formation and preservation in the geological record, however, depends on several factors such as the supply of clay to the sedimentary system and the right conditions for its accumulation and accommodation. In certain sedimentary environments chemical processes can also play a dominant role in the formation of clay minerals. The materials so formed are often so highly enriched in one particular clay mineral that their properties are very specific, and many are prized as industrial raw materials. Their unusual nature may also render them problematic from an engineering point of view, especially if their presence is unrecognized.

3.3.2. Origin of clay minerals in sediments

Most clay minerals in sediments are detrital, that is they are not formed in the sedimentary environment but are derived from external sources, principal amongst which are weathered rocks and soils. However, many of the clay minerals derived from soils may readjust to the new geochemical conditions of the sedimentary environment, giving rise to so called transformations. According to Millot (1970) the majority of transformation reactions in the sedimentary environment are aggradative, in contrast to the degradative transformations in soil environments. However, although there is some evidence for transformation reactions such as the aggradation of soil vermiculites back to more 'mica-like' minerals in the marine environment (Weaver 1989), evidence for transformation processes that involve anything more substantial than the alteration of interlayer complexes and/or ion exchange and fixation is generally lacking. Overall, the neoformation of clay minerals in sediments is also minor compared to the amounts that are detrital, although in certain sedimentary environments such as alkaline saline lakes or perimarine settings neoformation is commonplace.

The dominantly detrital origin of clay minerals in sediments is attested by the pattern of clay mineral distribution in the modern ocean basins (Biscaye 1965; Griffin et al. 1968; Rateev et al. 1969; Windom 1976). The pattern is broadly a latitudinal zoned distribution (Fig. 3.9) mirroring the distribution of clay minerals in the World's soils, which are thereby identified as the principal source of clays to the marine environment. The two minerals that illustrate this distribution best are kaolinite and chlorite. Kaolinite is concentrated in humid tropical latitudes where it is derived from kaolinite-rich soils formed under the intense chemical weathering regime via thorough leaching characteristic of humid tropical climates. Kaolinite cannot form in the oceans themselves since leaching does not occur in this environment. In contrast, chlorite is a very easily weathered mineral that is

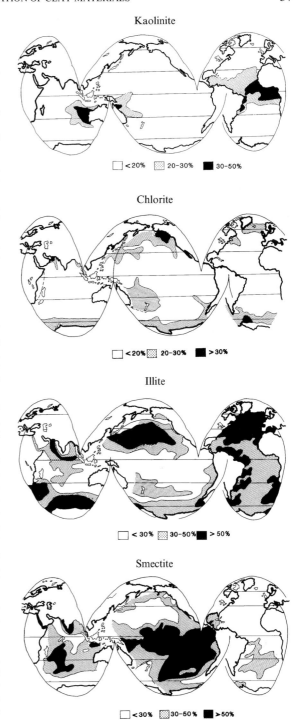

FIG. 3.9. Relative distribution of kaolinite, chlorite, illite and smectite in the clay fraction of fine-grained sediments from the World Ocean, modified after Windom (1976).

usually absent from low-latitude sediments but may be abundant at high latitudes where physical weathering, which does not destroy it, prevails. The distribution of illite and smectite also shows broad latitudinal zonations but these are not as well defined as those of kaolinite and chlorite.

Thus illite abundance tends to increase towards Polar regions but is more abundant in the northern than in the southern oceans. This possibly reflects the concentration of continental landmasses and hence terrigenous dust sources (Section 3.7) in the Northern Hemisphere. Notably, K-Ar dating of illitic clays from the ocean basins also demonstrates the dominantly detrital, terrigenous origin of most material since the ages obtained are generally older than the ages of the ocean basins themselves (see Weaver 1989; and Chamley 1989 for references). Smectite is much more abundant in the southern Oceans. Again this may be related to the distribution of terrigenous sources. With much less terrigenous input, fine particles of volcanic material, derived from both subaerial and submarine sources, which may readily alter to smectite, as well as smectites formed by other authigenic processes at the sea bed, may be proportionally more important in the southern oceans. Finally, it should be emphasized that the latitudinal patterns of clay mineral distribution are only very general patterns. There are many exceptions where local patterns are in total contradiction to the general climatically related global picture.

3.3.3. Source and supply of mud

Given that most clay minerals are detrital it is important to understand the factors that control their source and supply to sedimentary environments. In terms of the global supply to the sedimentary environment, areas of both high relief and high rainfall produce the most mud (Fig. 3.10). High relief gives high potential energy and high rainfall provides ample kinetic energy for erosion and transport (Potter 1998). Conversely low-lying arid regions produce the least clay. Modern examples where high relief (2000–4000 m) and high rainfall (2–4 m y^{-1}) are combined include the Ganges-Bramaputra system and the Amazon. In the Amazon system about 80% of the sediment load is derived from the Andes, even though this region represents only 12% of the drainage basin area (Gibbs 1967). Many islands in the Western Pacific also have very high sediment yields due to the combination of high relief and high rainfall. Bedrock type is another very important factor and, according to Potter *et al.* (1980), most mud comes from the erosion of pre-existing mudrocks.

3.3.4. Erosion, transport, and deposition of mud

All three agents of erosion, transport and deposition, that is water, wind, and ice, may erode, transport and deposit mud. Collectively, the formation of a mud deposit may be viewed in terms of how these processes lead to the

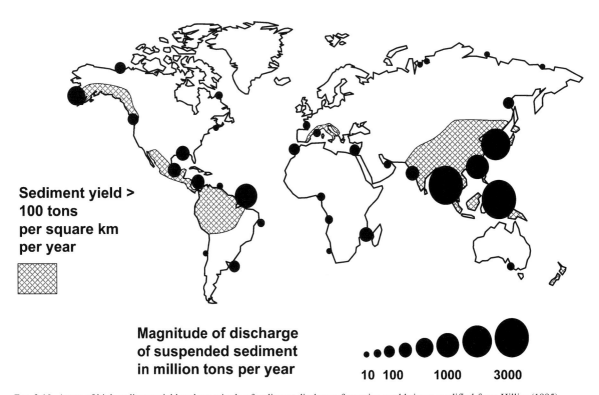

FIG. 3.10. Areas of high sediment yield and magnitude of sediment discharge for major world rivers modified from Hillier (1995).

segregation of mud from coarser sedimentary material, principally sand. Of the three agents, processes involving water are by far the most important, but there are also circumstances where ice and wind are responsible or contribute to the formation of clay materials.

3.3.5. Erosion, transport and deposition by water

3.3.5.1. Transport by rivers. On the continents, rivers are the principal means of transport of mud, mostly as part of the so-called suspended sediment load. The grain-size of sediment that can be carried in suspension obviously depends on flow conditions (energy) but typically suspended sediment is mostly <30 μm in size. In fact, on a global scale, suspended sediment accounts for most of the sediment transported by rivers. For example, at the mouths of large rivers such as the Amazon and the Mississippi, more than 90% of the total sediment load transported is fine-grained suspended sediment. Generally, suspended sediment concentration increase with increasing water discharge. Consequently, floods are especially important for the transport of large amounts of suspended sediment. Indeed, many rivers transport more than half of their annual sediment load in only 5–10 days of the year, when they are in flood (Meade & Parker 1985). However, many factors affect the supply of suspended sediment to rivers, including antecedent conditions, such as the amount of time since the last flood. Hence relationships to discharge are complex. Indeed, most systems are supply limited so that much more sediment is carried in suspension at points on the rising limbs of flood hydrographs compared to equivalent points on the falling limb.

In addition to loads, the sediment source area also dictates the characteristics of suspended sediment in terms of its grain-size and mineralogical composition. This is superbly illustrated by studies of the tributaries of the Amazon River (Gibbs 1967; Meade 1988). Tributaries which rise in tropical low-lands carry relatively little material, most of which is very fine-grained (<2 μm) and is compositionally dominated by kaolin. In contrast, those tributaries that rise in the Andes are turbid with sediment of a much greater range in both size and composition. As a result of the dominance of the Andean regions in terms of sediment yield, the composition of the suspended sediment at the mouth of the Amazon is also dominated by minerals derived from the Andes (Fig. 3.11).

Fluvial transport always provides potential for the segregation of mud and silt from sand, and the larger the river the more cumulative potential there is. As a result there is a general trend for the distal and deltaic deposits of large rivers to be proportionally more muddy than rivers draining smaller basins (Fig. 3.12) (Orton & Reading 1993).

3.3.5.2. Chemical composition of river suspended sediment. The composition of suspended sediment is determined by its mineralogy, which is in turn determined by weathering and other geological characteristics of the source areas. An estimate of the World average composition of river suspended sediment (Table 3.1), weighed for sediment discharge, was made by Martin & Meybeck (1979). Notably the composition is very similar to those reported for the 'average shale' (Section 3.6.4). Calculation of average compositions (un-weighted for discharge) for those rivers draining tropical lowlands (Congo, Niger, Orinoco, Parana), compared to those from other climatic zones or draining mountainous regions, shows that tropical suspended sediment is enriched in aluminium, iron and titanium, and depleted in magnesium, calcium, sodium and potassium.

Globally the present-day sediment yield of rivers that drain tropical lowlands is minor compared to the sediment yields of rivers draining regions of high relief and high rainfall (Section 3.3.5). Nonetheless, local supply and geological preservation of mud from tropical lowlands should be identifiable by its clear chemical (and mineralogical) signature.

3.3.5.3. Transport in the sea and ocean. Most mud that reaches the ocean is supplied by rivers and most of this is initially transported no further than the shallow shelves of the continental margins. Broadly speaking, factors such as climate and geotectonics influence the amount of mud present in shelf sedimentary environments (Hayes 1967). Shelves in tropical climatic zones tend to have the highest proportion of muddy environments and polar environments the least. Similarly, shelves located off passive margins are muddier than those associated with active margins and shields (Table 3.2). This may in part be related to the fact that most large low gradient rivers (i.e. the muddiest) characteristically discharge at passive margins.

Locally, fluvial supplies of mud to the shelf often result in the formation of mud belts whose location is controlled by the dynamics of supply and dispersal. In addition to fluvial supply, organic matter from marine organisms and re-suspension of fine-grained sediment by wave and current activity can also represent very significant sources of mud on continental shelves (Meade 1972). According to McCave (1972) the transport of mud across shelves is achieved by a combination of advective and diffusive transport.

Advective plumes such as those associated with the fluvial input from major deltas are often substantial enough to blanket the shelf completely with mud, especially in tropical regions. However, there is also a general diffusive transport of suspended sediment towards the shelf edge and beyond. The concentration of suspended sediment in shelf waters typically decreases exponentially from values of 100–10 mg l^{-1} in coastal regions to 1–0.1 mg l^{-1} at the shelf edge. McCave described the location and type of mud belts on shelves in terms of the interplay of this exponential decrease and a notional parameter related to wave and current activity at the sea bed (Fig. 3.13). Muddy coasts and blanket deposits result from very high supplies of mud to the shelf and often form

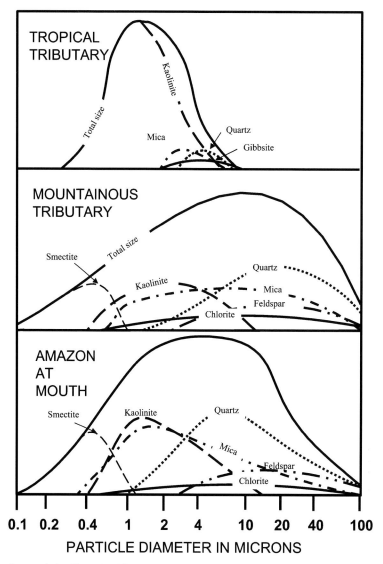

FIG. 3.11. Composition of suspended sediment in different parts of the Amazon River system modified from Gibbs (1967).

in regions of high wave and current activity. Thus the often-made assumption that muds are only characteristic of deposition in low-energy environments may sometimes be very far from the truth. Indeed, mud is often deposited under very high-energy conditions, simply because there is a lot of it about. In this respect it is also likely that many ancient mudrocks and shales, especially thick, widespread, marine shales, owe their origin as much to abundant supplies of mud to the sedimentary system as to the establishment of protected low-energy environments of deposition.

Further transport of mud from the shelves to the ocean basins depends on sea level and shelf width. High stands of sea level and wide shelves promote deposition on the shelf. In contrast low stands and narrow shelves promote deposition of mud in the deep ocean basins. In the oceans as a whole transport and deposition of fine sediment occurs by a wide variety of processes (Fig. 3.14).

In the deep ocean sedimentation occurs mainly by pelagic settling. This refers to settling out of material suspended in the water column, much of which is of biological origin, originating from calcareous and siliceous organisms that inhabit the trophic zone. Far from land mineral components are mainly supplied from aeolian inputs. Closer to continental sources so-called hemipelagic muds are deposited from diffuse plumes of fine sediment extending for hundreds of kilometres out into the oceans at depths of 2–3 km. These plumes are

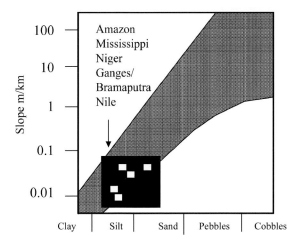 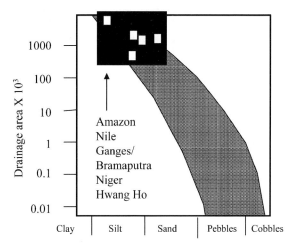

FIG. 3.12. Correlation amongst slope and drainage basin size with sediment grain-size, redrawn from Orton & Reading (1993).

TABLE 3.1. *Chemical compositions (wt%) of river suspended sediment. Source Martin & Meybeck (1979)*

	World average	Tropical lowland	Others
SiO_2	60.95	56.14	61.64
Al_2O_3	17.76	23.23	16.15
TiO_2	0.93	1.44	0.84
Fe_2O_3	6.86	9.61	6.59
MgO	1.96	1.32	2.37
MnO	0.14	0.10	0.11
CaO	3.01	0.72	3.60
Na_2O	0.96	0.44	1.33
K_2O	2.41	2.04	2.91
P_2O_5	0.26	0.36	0.26

TABLE 3.2. *Percentage of mud on World's inner shelves, after Hayes (1967)*

Weathering/climatic regime	Active	Shield	Passive	Average
Strong chemical/tropical and subtropical	40	34	61	**45**
Moderate to slight chemical/ arid and semi-arid	19	19	12	**17**
Moderate chemical/temperate	20	18	31	**23**
Moderate mechanical/polar and sub-polar	12	6	22	**13**
Average	**23**	**19**	**32**	

derived from both advective and diffusive transport and winnowing of shelf deposits. A large amount of mud is also subjected to re-sedimentation processes, most important amongst which are turbidity currents and contour currents that may transport sediments over thousands of kilometres (Stow 1994).

3.3.5.4. Deposition by settling and flocculation. Most deposition of clay minerals from water bodies occurs by settling from suspension. For clay particles <2 μm in size, settling times are extremely slow. For example, it takes of the order of 1 hour for a particle 2 μm in diameter to settle 1 cm in still water, whereas a particle of sand 125 μm in diameter would only take about 1 second to settle over the same distance. Such contrasts in fall velocity are the key hydraulic control that inevitably results in the segregation of mud from sand in the sedimentary environment. In fact, once in suspension the settling of clay particles from turbulent flows would be an extremely slow affair indeed were it not for various processes of aggregation which effectively increase particle sizes by orders of magnitude (McCave 1984). A variety of processes may operate including electrochemical or salt flocculation and biogenic or organic flocculation. The process of salt flocculation occurs between particles, such as clay minerals, that are electrically charged. This charge must be balanced by counter ions in solution. In weak solutions, such as fresh water, the cloud of counter ions that balance this charge extends further from the particle than in stronger solutions, such as salt water (Fig. 3.15). As a result the charged particles can approach each other much more closely in salt water, often close enough that attractive van de Waals forces may stick them together.

However, it is not certain that salt flocculation is a particularly important process in nature since there is increasing evidence that most flocculation/aggregation is the result of the activities of organisms (Meade 1972; Eisma 1986). Thus biological processes may act both to bring particles together and to stick them together with organic glues such as polysaccharides. It is also doubtful that differential flocculation is an important process in nature, although particle size segregation does appear to have the potential to selectively sort clay minerals in natural environments (Gibbs 1977). Notably smectites, which

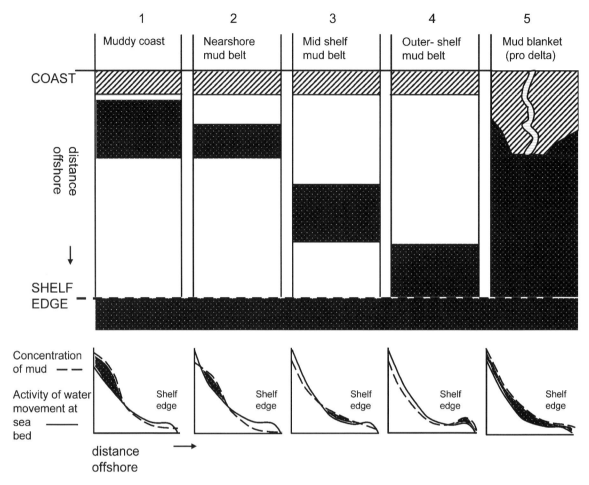

FIG. 3.13 Conceptual model of types of mud deposits on continental shelves in terms of mud supply and a notional parameter describing wave and current activity at the sea bed (modified from figures in McCave 1972).

tend to occur in the very smallest particle size fractions often appear to be held in suspension longer than other clay minerals such as illite or kaolinite that tend to be larger in size (Chamley 1989).

3.3.5.5. Fabric and microstructures. Many of the processes that occur during or shortly after deposition of muds determine the fabric and microstructures of mudrocks. Other processes related to deep burial and clay mineral diagenesis act over much longer periods of geological time. Bennett *et al.* (1991) summarized the processes that determine fabric and microstructure. They grouped the various processes into physico-chemical, bio-organic and burial diagenetic (Section 3.4) but stressed that there is a continuum of processes and mechanisms (Fig. 3.16).

In terms of the fabric of the clay material, it used to be believed that fabrics were composed of individually arranged clay plates, but many studies using electron microscopy have now demonstrated that this is rarely the case. Instead, the basic unit is a so-called domain or tactoid composed of several or more clay plates, often in a so-called stepped face-to face domain. Several such domains are commonly linked together to form a chain or aggregate, which may in many instances represent the depositional entity.

The two most common fabrics observed in mudrocks are laminated and random (Fig. 3.17). Each may be produced by a variety of processes and mechanisms. Laminated fabrics are commonly observed in organic-rich mudrocks, such as black shales. The laminar fabric can result from deposition of dispersed particles and unaggregated face to face domains and by the shearing affect of bottom currents which may break up and re-disperse sedimented aggregates. A laminar fabric may also be developed during compaction.

The preservation of millimetre-scale variations in sediment composition, largely due to the absence of bioturbation by organisms, is another major factor that may result in a laminar fabric. The development of anoxic

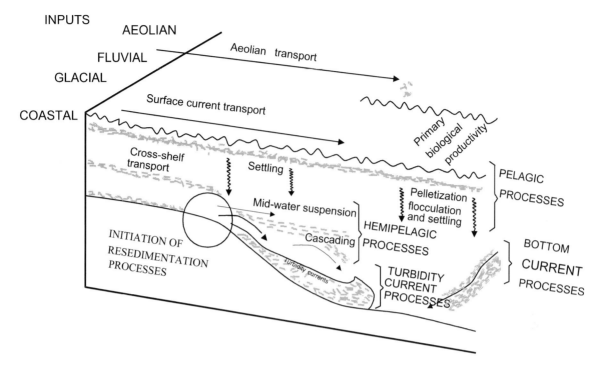

FIG. 3.14. Processes of sedimentation of fine-grained sediment in the sea and ocean (Modified after Stow 1994).

conditions at and above the sediment–water interface is a common way in which bioturbation may be limited. Such conditions are also favourable for the preservation of sedimentary organic matter so that black shales (Leventhal 1998) are often laminated. Additionally, grain-size range and sorting of the sediment is important. For instance if clay is well mixed with silt at the time of deposition, it is likely to show a more random fabric as a result of the differences in the size and shape of the silt and clay particles. Laminar fabric is the main cause of the fissility of some shales. Although fissility may also be developed and enhanced during burial diagenesis by both physical rearrangement of particles and by mineral dissolution and growth. However, the degree of fissility is by no means related simply to the degree of burial diagenesis.

Random fabrics can also be developed by a variety of processes. In most depositional settings flocculation and aggregation are the normal mode by which clay particles are deposited and the preservation of these entities results in random fabric. Indeed, dispersions of single clay plates that can be readily achieved in laboratory experiments may only be achieved in nature in extremely dilute suspensions. The structure of flocs is typically one of larger particles linked by smaller particles. As far as platy clays are concerned, flocs are often composed of stepped face-to-face domains linked edge-to-face with other domains, in a sort of 'cardhouse' structure. If they are not destroyed during deposition, or subsequently, the resulting deposit will have a random fabric, inherited from the processes of flocculation and aggregation. Clays deposited from turbidity currents also exhibit a random fabric, as do pro-delta clays in this case due to flocculation, whilst faecal pellets represent another form and origin of random fabric. Bioturbation also produces random clay fabrics. According to Bennett et al. (1991) it can sometimes be distinguished from random fabrics due to flocculation by the presence of more individual particles rather than domains and by the presence of thoroughly mixed silt and/or sand grains.

Studies by O'Brien & Slatt (1990) suggest that most mudrock fabrics are the result of early burial. After arriving at the depositional surface, cardhouse flocs begin to compact. As pore water is expelled it facilitates movement to a more stable parallel arrangement. The fabric of muds and shales is important not only from the point of view that it holds many clues to the depositional and geological history of the deposit but also because fabric is a primary control on many physical properties such as porosity, permeability (Section 3.4.4) and stress/strain behaviour. In general, random fabrics are more permeable than laminated fabrics (Davies et al. 1991) whose permeability is usually anisotropic, being greatest in the plain of the lamination.

3.3.6. Erosion transport and deposition by ice

The erosion, transport and deposition of clay materials by ice may occur in two settings; glacial environments and high-latitude oceans where sea ice develops. Glacial

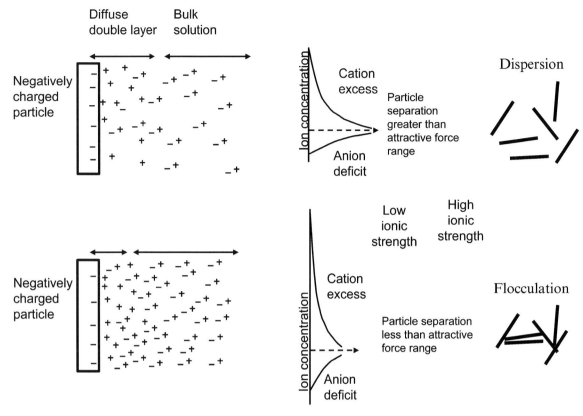

FIG. 3.15. Mechanism of salt flocculation due to compression of electric double layer in solutions of high ionic strength. Modified from Arnold (1978).

deposits, known as tills, cover large areas of northern Europe and North America and many of them contain large amounts of clay-sized material. Water deposited glacio-lacustrine and glacio-marine clays are also common and often record seasonal changes in the sediment load of glacial meltwater as couplets of silt and clay rich layers known as varves. In the oceans much fine-grained sediment is also transported by a process known as ice-rafting (see Hillier 1995). Sea ice in particular is an important vehicle for ice rafted transport of fine-grained sediment. By this process, sediment suspended in shallow shelf waters by storms is entrained and concentrated in sea ice by cycles of freezing and thawing. Subsequently this sediment may be transported great distances from the source of the ice, eventually being released back into the ocean as the ice melts.

3.3.7. Erosion transport and deposition by wind

The erosion transport and deposition of fine-grained material by wind is nothing like as important globally as that by water, but nevertheless there are instances where the role of wind is an important factor. For example, vast areas of the World are covered by wind-blown deposits of silt and clay known as loess (Catt 1988). Also wind-blown dust from both terrigenous and volcanic sources is an important component of pelagic sediments in the deep ocean basins. The major source areas and trajectories for wind-blown dust are shown in Figure 3.18. Windom (1975) estimated that the typical contribution of aeolian dust to deep-sea clays was 10–30% and in some instances contributions of more than 75% are apparent.

Erosion and transport of particles by wind is most active in deserts and semi-arid regions due to the deficiency of soil moisture and lack of plant cover. Such areas act principally as sources of dust. As with fluvial transport of suspended sediment, dust transport tends to be dominated by events with one or two large spring storms transporting most material in any one year (Rae 1994). Present-day deposition of loess is not extensive and most deposits were formed during the Pliestocene glacial maxima associated with increased aridity and dust production at these times (Catt 1988). According to Tsoar & Pye (1987), loess is typically 80% silt with a modal grain size between 20 and 40 µm, but weathered loess may contain up to 70% clay. Aeolian transport of grains of > 20 µm tends to occur by short-term suspension no more than a few metres above ground level. Hence loess deposits are located relatively close to their source areas

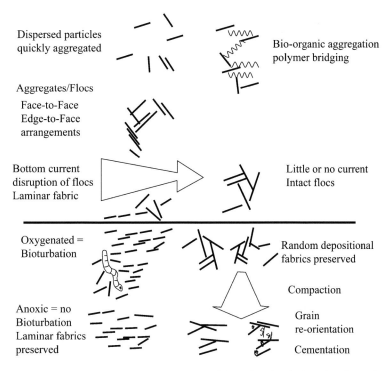

FIG. 3.16. Sedimentary and diagenetic processes that determine the microfabric of mudrocks, based on figures in Bennett et al. (1991).

and the trapping effect of vegetation may be essential for their effective accumulation (Tsoar & Pye 1987). Dust particles in the fraction finer than 20 μm tend to be transported to much greater heights up to 3000 m. It is mostly components of this fraction that may travel thousand of miles around the globe and that may contribute substantially to deep-sea sedimentation (Prospero 1981).

3.3.8. Sedimentary neoformed clays

Neoformed clay minerals are found in a very wide variety of sedimentary settings (Fig. 3.19) and although they are often not a major component of mudrocks they can in some special cases be the dominant component. As a general rule, most neoformed sedimentary clay minerals are either Fe or Mg-rich. In continental environments, the most important clay minerals are varieties of Mg-rich smectites including stevensite and saponite and the Mg-rich fibrous clay minerals sepiolite and palygorskite. More rarely a variety of Fe-rich smectite called nontronite, an Fe-rich illite known as ferric illite and a talc-related mineral known as kerolite are reported. All of these clays usually are found in association with mixed carbonate, evaporite, clastic sediments deposited in various kinds of alkaline lakes. As such they are all rather rare, but may form economically important deposits. Similarly, the Mg-rich clays, stevensite, palygorskite and sepiolite are also found in association with perimarine lagoonal environments such as the Miocene of the Southeastern United States (Weaver & Beck 1977). Many ancient continental sediments associated with evaporites, for example the UK Triassic Mercia Mudstone, contain suites of Mg-rich clay minerals including chlorite and corrensite. Often these minerals have been interpreted as syn-sedimentary neoformed, but it is much more probable that they are the neoformed products of clay mineral burial diagenesis (Hillier 1993).

In marine environments, probably the best known neoformed clays are the Fe-rich minerals of the glauconite family. Detailed mineralogical analysis of these minerals have shown that there are end members of Fe-rich smectite and Fe-rich mica (glauconite) and a complete sequence of mixed-layer glauconite–smectite minerals in between. The sequence is analogous in many respects to the mixed-layer illite–smectite series. However, whereas the illite–smectite sequence is associated with the elevated temperatures of diagenesis or hydrothermal alteration, Fe-smectite may be completely altered to glauconite (*sensu stricto*) at the sea floor. Petrographic obervation of glauconite pellets in sediments gives no information about the mineralogical stage of the glauconitization process. This prompted Odin & Matter (1981) to suggest the term glaucony for this facies, since this term has no precise mineralogical connotations. Most current theories regarding the origin of glaucony favour a neoformation mechanism. Much of the evidence for this origin has been gained from observations with the electron microscope.

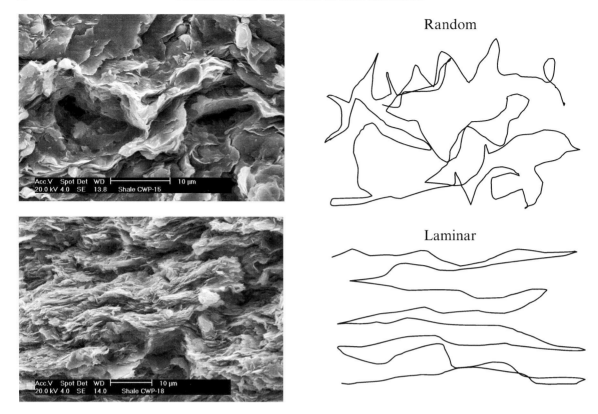

FIG. 3.17. Examples of random and laminar fabrics of mudrocks as revealed by scanning electron microscopy.

Studies of the glaucony facies on modern continental shelves indicate that it is largely concentrated at water depths between 100 and 300 m. Tropical latitudes also seem to be particularly favourable for glaucony development, perhaps because of the abundance of poorly crystalline sources of Fe in sediment derived from tropical landscapes, but it is not restricted to these latitudes. Generally, glaucony is characteristic of fully marine conditions, especially where sedimentation rates are very slow. One factor that may be very important for glaucony formation is cycles of changing redox conditions which may naturally result from repeated periods of reworking under sediment starved conditions.

Another group of Fe-rich clays usually associated with shallow marine conditions and, in contrast to glaucony, exclusively confined to tropical latitudes, are the green clays belonging to the Verdine facies. The nature of the minerals in the Verdine facies is still debated but essentially they appear to be mainly comprised of a form of 7Å mineral belonging to the trioctahedral serpentine group of minerals, but with a dioctahedral component. The Verdine mineral odinite may be the precursor of the serpentine group mineral berthierine, and later diagenetic alteration may transform berthierine to the Fe-rich chlorite chamosite. Note that berthierine and chamosite do not appear to occur in modern sediments.

Smectite is another common clay mineral that can be neoformed in marine sediments. The most obvious examples of smectite formation are those associated with the alteration of deposits of volcanic glass and ash, and those formed as the plumes issuing from deep ocean hydrothermal vents mix with sea water. However, much smectite also forms diffusely by processes that in some sense may be thought of as 'submarine weathering' or halmyrolysis. Since the processes involved are slow compared to most sedimentation rates the evidence for the neoformation of smectites is most striking in deep sea deposits such as large parts of the Pacific ocean where sedimentation rates are extremely slow. Nonetheless, in areas of higher sedimentation rate, smectite neoformation may be more common than is generally appreciated, but so diluted by other materials including allochonous smectite that it may only be identified by very careful mineralogical investigation (Belzunce-Segarra et al. 2002). Aplin (2000) gives a useful summary of clay mineral neoformation from a geochemical perspective in modern marine sediments.

When volcanic ash is deposited by air-fall on the seabed it often alters almost completely to a smectite and, if preserved, the resulting deposit is known as bentonite (Grim & Güven 1978). However, as summarized by Weaver (1989) there are many occurrences of volcanic

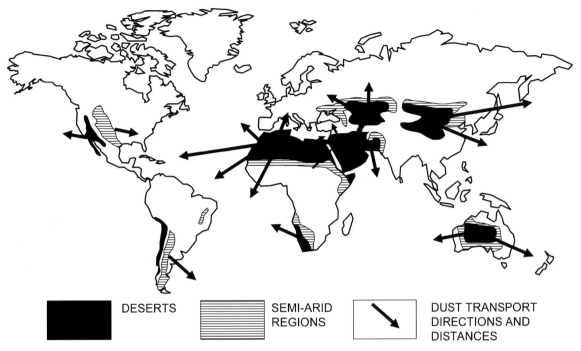

Fig. 3.18. Distribution of arid regions and major trajectories of aeolian dust over the continents and oceans, based on figures in Péwé (1981) and Pye (1987).

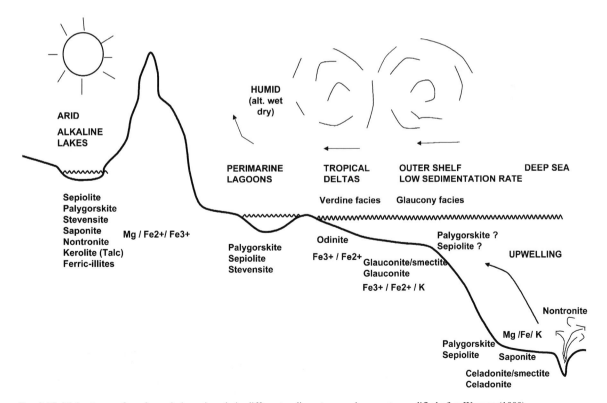

Fig. 3.19. Major types of neoformed clay minerals in different sedimentary environments, modified after Weaver (1989).

ash layers buried in Tertiary sediments, which have not yet altered completely to smectite. This suggests that the formation of bentonite may be, at least in part, a diagenetic process since the alteration of the ash to smectite is not necessarily completed at or near the seabed. Most sedimentary bentonites have formed from the submarine alteration of volcanic ash of acid to intermediate igneous composition. The process of submarine alteration of wet volcanic glass to form bentonite is also dependent on the nature of the depositional environment. Generally, ash deposited in a marine environment alters to smectite, whilst ash deposited in a non-marine fresh water environment is altered to kaolinite. This process has been especially well-documented in Coal Measure sequences, where altered volcanic ash deposits consisting primarily of kaolinite are known as tonsteins (Spears & O'Brien 1994).

3.4. Diagenetic to low grade metamorphic environment

3.4.1. Introduction

Before long the muds and clays deposited in sedimentary basins are buried under more sediment and enter the realm of diagenesis. The definition of diagenesis is not universally agreed, but very generally it can be defined as the set of processes that occur between the end of sedimentation and the beginning of metamorphism. Some definitions even include processes that operate during weathering and sedimentation but such a definition is regarded as too broad by many. Diagenesis involves both physical and chemical processes and the changes that they bring about result in the transformation of unconsolidated sediments into rock. For mudrocks, two aspects of their diagenesis have been studied in most detail.

The first is the diagenesis that occurs at a very early stage over a fairly restricted time–temperature–depth interval. Such early or shallow burial diagenesis is strongly influenced by the chemical and physical characteristics of the depositional environment and the time that sediment is resident in this shallow zone determined by the rate of sedimentation. In terms of depth of burial so called early diagenesis occurs over a scale of metres to hundreds of metres. However, the upper depth–temperature boundary is both arbitrary, as well as variable because of the strong links to depositional processes (Curtis 1977). The suggestion of Hesse (1986) that the upper limit be placed at the point where biologically mediated reactions of organic matter are superseded by thermally driven reactions is a useful reference because biological processes are of immense importance in early diagenesis. Hesse (1986) suggested an upper temperature limit of about 75°C corresponding to burial depths of 2–3 km.

The second realm of diagenesis is that which occurs during deeper burial in a sedimentary basin over an often protracted and varied time-temperature-depth interval. This realm is often termed deep burial diagenesis. Of course, diagenetic processes are part of a continuum but the distinction is useful because of a general change in the sediment components that participate in diagenetic reactions. From the point of view of clay materials, the earliest diagenetic processes tend to involve the most reactive solids in the sediment (organic matter, amorphous silica, volcanic glass, iron and aluminium oxides etc), although such substances may only represent a small proportion of the mass of the sediment. The products are also of such variety, that they may be of considerable importance in relation to the chemical properties of the resulting clay material. On the other hand, processes that occur largely in response to the increase in temperature and pressure during deeper burial are progressively dominated by reactions involving the clay minerals and other silicates. Along with the clay minerals, these are principally quartz and feldspars, which together make up the bulk of the solid framework of the sediment. During early diagenesis most of these minerals act as a comparatively inert matrix.

Eventually diagenesis gives way to metamorphism. Again there are problems of adequately defining precise boundary conditions (Frey & Robinson 1999). Most low-grade metamorphic mudrocks, such as slates, usually have a clay mineral assemblage consisting essentially of well-crystallized illite (white mica) and chlorite. Thus the diagenetic and subsequent metamorphic processes which convert mud into slate is essentially the story by which diverse detrital clay mineral assemblages are progressively transformed to an assemblage of white-mica and chlorite.

3.4.2. Early diagenetic processes

Freshly deposited fine-grained sediments are compositionally diverse consisting not only of clay minerals but of many other minerals and biogenic materials. Among others, some of the more common components include silicates in variously weathered states, Fe, Mn and Al oxy-hydroxides, carbonates, and various kinds of sedimentary organic matter. Together these components make up a loosely packed solid fraction with pore space of 90–30% by volume occupied by water and instantly home to all sorts of sediment biota. The large porosity of newly deposited sediments allows zones of open exchange of dissolved components with the overlying water column. Consequently the nature of the depositional milieu, for example marine versus non-marine, exerts an overriding control on many early diagenetic processes. At the same time mechanical compaction (Section 3.4.4), which is particularly rapid in the early stages of burial, reduces porosity and slowly isolates the mud from the influence of the depositional waters. Many of the components present in the complex mixture that is freshly deposited mud will be far from equilibrium with the conditions of the sedimentary environment and it is little wonder that diagenetic modification of mud begins as soon as deposition is done. However, the main driver of the varied reactions and mineral products that result is the oxidation of organic matter by the sediment biota, mainly bacteria, as a source of energy for metabolism. Biological processes are therefore of immense importance in early diagenesis. In detail, a sequence of early diagenetic redox reactions has been recognized in relation to the availability of potential oxidants as exploited by a depth wise layered community of micro-organisms. Ideally, this sequence is oxygen, nitrate, manganese-oxide, iron-oxide, sulphate and carbonate. The details are complex and varied but generally zones of oxic respiration and sulphate reduction are the most important zones in marine environments, whereas in fresh water settings sulphate reduction is unimportant. Additionally, sedimentation and burial rates control the residence time of sediment in any zones that are developed, thus establishing a further link between depositional and diagenetic environments (Curtis 1977). By combination with reduced iron, the eventual product of early diagenetic sulphate reduction in marine settings is pyrite, although it does not form directly but via a sequence of iron monosulphides (see Aplin (2002) for a recent summary of pyrite formation). In contrast, in muds deposited from fresh waters, the activity of sulphide is generally too low for iron sulphide formation and carbonate reduction and fermentation zones extend to shallower depths such that iron carbonate (siderite) is commonly formed instead. The zone of sulphate reduction in marine sediments is generally limited to a depth of no more than 1–2 m below the sediment–seawater interface. Below this depth and in non-marine environments also, is a zone of carbonate reduction followed by a further zone of fermentation. Both processes often result in the formation of early diagenetic carbonates. Some of the important environmental controls, processes, and sediment components involved in early diagenetic reactions, along with some of the more distinctive clay material products that may result are summarized in Figure 3.20.

3.4.3. Main pathways and reactions of clay mineral diagenesis

Following the phase of early diagenetic reactions, further burial of mudrocks and consequent exposure to progressively higher temperatures and pressures initiates more and more diagenetic reactions that involve the most important and abundant component of virtually all mudrocks, the clay minerals. The result of such processes is a progressive reorganization of the mineralogical components present, via a host of processes and chemical reactions involving transformation, dissolution, precipitation and recrystallization. Since the clay minerals tend to be the most abundant as well as the most reactive silicate component present most of what happens is termed clay mineral diagenesis. Nonetheless, it must not be forgotten that the other major mineralogical components of mudrocks namely quartz, feldspars and carbonates are also commonly involved. Taken to its most advanced stages, clay mineral diagenesis will transform most mudrocks to an assemblage of illite (white-mica), chlorite, feldspar and quartz. In the process, average crystal size will be increased, and the mudrock becomes denser, more compact, less porous and less permeable until by any definition or practical classification scheme it has all the characteristics of a 'hard rock'. The important thing to understand is that the processes of clay mineral diagenesis which effect these changes are progressive. Therefore the properties, and hence behaviour, of a clay material may be at any intermediate state between a semi-lithified

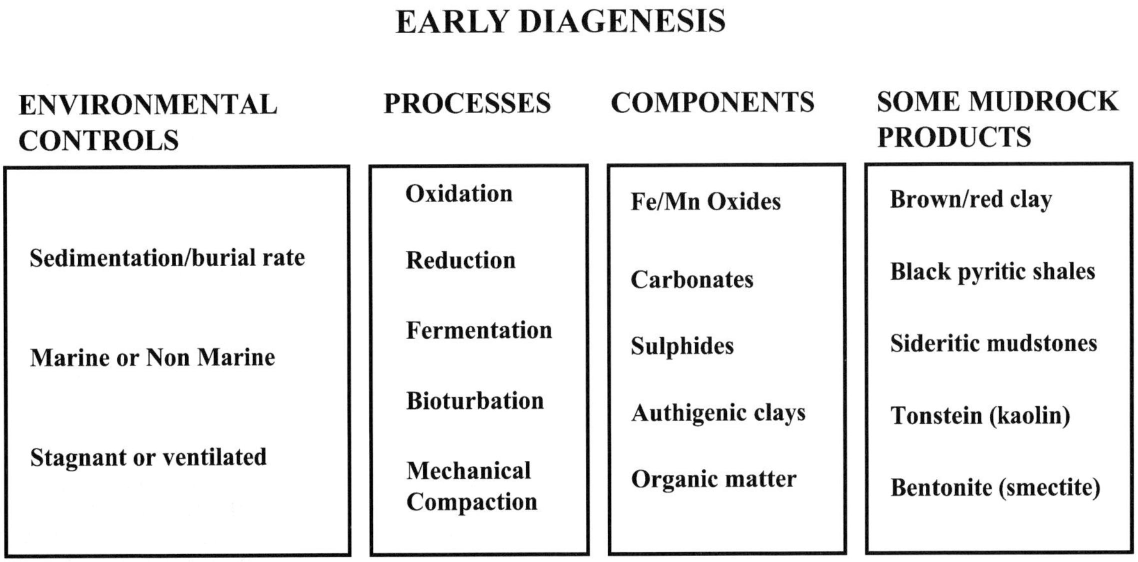

EARLY DIAGENESIS

ENVIRONMENTAL CONTROLS	PROCESSES	COMPONENTS	SOME MUDROCK PRODUCTS
	Oxidation	Fe/Mn Oxides	Brown/red clay
Sedimentation/burial rate	Reduction	Carbonates	Black pyritic shales
	Fermentation	Sulphides	Sideritic mudstones
Marine or Non Marine	Bioturbation	Authigenic clays	Tonstein (kaolin)
Stagnant or ventilated	Mechanical Compaction	Organic matter	Bentonite (smectite)

FIG. 3.20 Summary of early diagenesis in terms of the main environmental controls, important processes, solid components affected, and examples of particular products that may result.

mudstone and a slate. Furthermore, the same clay material found in different locations may have experienced quite different degrees of clay mineral diagenesis, for instance due to greater or lesser amounts of burial in the geological past. As a consequence, from an engineering perspective, the same clay material may behave in radically different ways from one place to another. In other words, clay minerals diagenesis depends on geological history, including the specific burial and time-temperature history of a deposit and this history is almost always variable from place to place. As an example, Cambrian mudrocks from the Russian Platform behave much like the London Clay, and bear few similarities with Cambrian mudrocks in the UK. This is a consequence of diagenetic histories, although the Russian clays are 600 Ma years old they have been little altered by burial or tectonic processes, such that they still exhibit properties that resemble those of London Clay. In contrast, the Cambrian rocks of Wales have experienced low-grade metamorphism and bear virtually no resemblance to London Clay. Generally speaking, older clay formations have had more chances to be altered by diagenetic processes, but there are always exceptions to a generalization such as this. The following sections describe the main pathways and reactions of clay mineral diagenesis, i.e. the chemical reactions, and mineralogical aspects of the conversion of mud into rock.

3.4.3.1. Smectite-to-illite reaction and illitization. The reaction of smectite-to-illite and the subsequent increase in illite crystallinity is probably the best known and certainly the most widely studied diagenetic reaction in mudrocks. It is also a common reaction in most other diagenetic and hydrothermal settings. In essence, many sediments when initially deposited contain some proportion of aluminous dioctahedral smectite. Invariably, this smectitic component is derived from a variety of sources (soils, older sediments, volcanic ash), and in detail it may be more or less smectitic, but upon burial it is universally and progressively converted into the clay mineral illite. The reaction is often described as progressive because it occurs via a series of intermediate mixed-layer illite–smectite minerals. The general sequence of this reaction has been documented in sedimentary basins worldwide. Indeed, considering that the clay mineral composition of mudrocks can vary enormously, the apparent ubiquity of the smectite-to-illite reaction in sedimentary basins of all types and ages suggests that a large array of mudrocks must have broadly similar bulk compositions. Or at least that most mudrocks must span a range of compositions that are conducive for this reaction to occur.

The main features of the smectite-to-illite conversion in mudrocks were first described by Burst (1959, 1969), Shutov *et al.* (1969), Perry & Hower (1970), and Hower *et al.* (1976). This early period of work has been reviewed by Hower (1981) and by Środoń & Eberl (1984). The main conclusions reached by many studies over this period and which remain valid today are as follows. Firstly, as mentioned above, the conversion of dioctahedral smectite into illite is a virtually ubiquitous diagenetic reaction in mudrocks from sedimentary basins of all ages worldwide. Secondly, the main controls on reaction progress are temperature, chemical variables and, more controversially, time. Thirdly, in terms of the intermediate mixed-layer clay illite–smectite, the conversion involves a progressive change from random mixed-layering to ordered mixed-layering (Chapter 2). For Tertiary and Mesozoic mudrocks the transition from smectite to random mixed-layer illite-smectite begins at burial diagenetic temperatures of about 50°C. Thereafter the transition from random to ordered mixed-layering occurs at about 100°C and the complete transition to end member illite (or some form of white-mica) occurs typically somewhere in the range of 200 to 300°C (Hoffman & Hower 1979). Fourthly, the conversion involves substantial changes in the composition of the component silicate layers, which gain aluminium and lose silicon, iron and magnesium. Additionally, exchangeable hydrated interlayer cations, principally sodium and calcium which compensate the net negative layer charge are replaced by a larger amount (charge equivalent basis) of non-exchangeable potassium related to a concomitant increase in layer charge.

Other aspects of the smectite-to-illite reaction remain controversial. From a purely chemical point of view the compositional changes that occur may come about in more than one way. In the classic study by Hower *et al.* (1976) the reaction was written as:

$$K + Al + Smectite \rightarrow Illite + Na + Ca + Fe + Mg + Si + H_2O.$$

This reaction is essentially iso-volumetric, it results in a decrease in pore fluid pH, and aluminium is mobile with respect to the two clay components. An alternative reaction proposed by Boles & Franks (1979) assumes that aluminium is immobile. As such this alternative reaction requires that the mass of clay is reduced by about 30% over the course of the reaction and the pH of the pore fluid is increased.

$$K + Smectite \rightarrow Illite + Na + Ca + Mg + Si + O + OH + H_2O.$$

Written in this form the reaction also releases substantially more silicon with obvious potential for quartz cementation both in adjacent sandstones and of course in the mudrocks themselves. The only way to address this question about the precise chemical nature of the reaction is to find ways to obtain accurate data on the amounts of clay minerals and other components in diagenetic sequences of mudrocks. Apart from analytical difficulties this problem is exacerbated by the potential for the presence of non-diagenetic illite–smectite, and well as the likelihood of initial differences in mineral and chemical composition of different sediments at the time of deposition.

During this early period of work it was generally assumed without question that the smectite-to-illite conversion involved a solid-state transformation mechanism. However, following the development of the concept of fundamental particles (Nadeau *et al.* 1984) it has become clear that the smectite-to-illite conversion also involves aspects of dissolution and crystal growth. Nevertheless, despite the enormous amount of work generated by the resulting controversy over the physical nature of mixed-layer illite–smectite there is still no generally accepted mechanism for the smectite-to-illite reaction (Środoń 1999). Indeed it is very likely that somewhat different mechanisms are applicable to the smectite-to-illite conversion in mudrocks compared to bentonites, sandstones and hydrothermally altered rocks (Altaner & Ylagan 1997). Additionally, it is probably also prudent to question the assumption implicit in many studies that diagenetic mixed-layer illite–smectite, and illite always form from a smectite precursor, especially since we know this not to be the case in many sandstones. It seems more likely that all that is known about sandstone diagenesis is just as likely to be happening somewhere in the heterogeneous microcosms that comprise a mudrock. Thus there are likely to be a variety of reactions and pathways by which diagenetic illite-smectite and illite can be formed. Obviously some, such as from a smectite precursor will tend to be volumetrically more important than others. It may be better therefore simply to refer in a general way to the process of illitization, defined as reaction(s) progress towards illite.

The degree of illitization can be used as an indicator of the degree of diagenetic alteration and has been correlated with many other such indicators. A recent compilation is that figured by Merriman & Frey (1999). Traditionally, illitization is measured in terms of the percentage of smectite layers, ranging from pure end member smectite (100% smectite layers) to pure end member illite (100% illite layers) via intermediate mixed-layer illite-smectite. Continuing debate over the very nature of the clay minerals involved suggests that the more generic measure, termed expandability, is better. Many studies have shown, however, that the process(es) of illitization is a function of both temperature and time (Fig. 3.21) as well as various chemical variables, so that caution must be applied when attempting to use it as an indicator of palaeotemperatures or diagenetic grade, especially without reference to corroborative indicators such as vitrinite reflectance. It remains a useful indicator of the degree of diagenetic alteration, but is best used in a comparative way for a group of related samples, for example the same mudrock with a wide regional distribution. It is generally not wise to use it to estimate the diagenetic grade of a single sample, except in the most general terms.

3.4.3.2. Illite 'crystallinity'. Once the illization process is more or less complete, i.e. there is little evidence for the presence of expandable layers (<10%), then further reaction progress towards true micas due to still further diagenesis and low-grade metamorphism is often measured in terms of illite 'crystallinity'. Thus the use of illite crystallinity is mainly confined to studies of mudrocks which have already been highly altered by diagenetic processes. There are a number of ways by which illite 'crystallinity' can be measured (see Chapter 8).

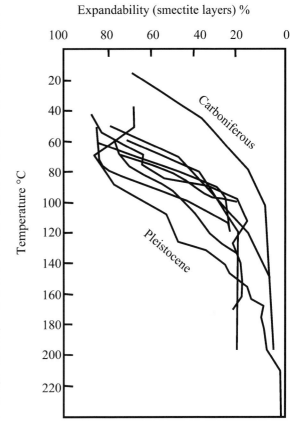

FIG. 3.21. Pathways of smectite illitization in terms of decreasing expandability of mixed-layer illite–smectite with increasing temperature due to burial in a range of sedimentary basin of different ages. After Środoń & Eberl (1984), with additional data from Jennings & Thompson (1986). Profiles from Pleistocene and Carboniferous rocks bracket profiles from Tertiary and Mesozoic rocks, suggestive of a kinetically controlled process, although in many profiles there are large uncertainties concerning the thermal history.

The most commonly used technique is to measure the width at half peak height of the 001 illite peak on an X-ray powder diffraction pattern, a measure now formally called the Kubler index (Guggenheim et al. 2002). With increasing diagenetic grade the width of this peak (and other basal peaks) becomes increasing narrower, by and large, in response to an increase in the average thickness of the illite crystals in the sample, i.e. peak width and crystal size are inversely proportional. If measurements of illite 'crystallinity' are to be interpreted in a wider context there are many factors to consider and standards to which they can be calibrated.

When used in this way specific values of illite crystallinity have been used to define a series of zones in terms of the degree of diagnosis/low-grade metamorphism. In order of increasing intensity of alteration these zones are called the diagenetic zone, the anchizone and the epizone. Illite 'crystallinity' values corresponding to these zones

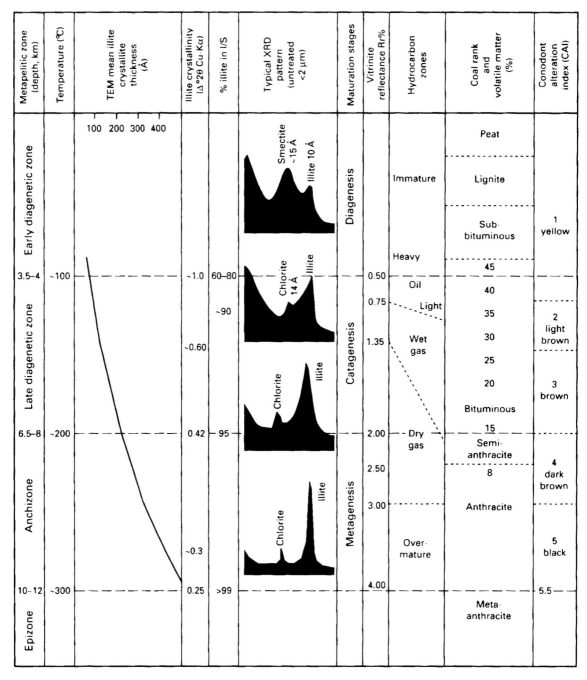

FIG. 3.22. Generalised correlations amongst different indicators of 'maturity' including % illitization and illite 'cystallinity', reproduced with permission from Merriman & Kemp (1996) as modified by Merriman & Frey (1999).

along with the general correlation amongst other indicators of diagenetic grade or 'maturity' are indicated in Figure 3.22. More detail about the clay mineralogy and microfabrics that characterise these zones can be found in Merriman & Peacor (1999). The progressive increase in crystal size and perfection which underlies the concept of illite 'crystallinty' can be viewed most simply as an extension of the processes of illitization, but under the more extreme conditions (temperature, pressure etc) of low-grade metamorphism.

3.4.3.3. Formation of chlorite. Just as with illite, there are numerous pathways by which chlorite may be formed in mudrocks. However, bulk rock compositional controls are much more apparent, although they doubtless exist for illites as well. For example, in mudrocks rich in Al and Fe the chlorites formed in such rocks will be similarly enriched. Likewise, Mg-rich chlorites tend to be found in Mg-rich mudrocks. Examples of the latter are quite well known, for instance in the UK the Triassic Mercia Mudstone group is often characterized by the presence of Mg-rich chlorite. Most work on chlorite formation in sedimentary rocks has been done on sandstones, but it is very likely that all the mechanisms and reactions identified in sandstones also occur in the microcosmic environment of mudrocks. The main routes for chlorite formation are: by the reaction of early formed Fe-rich 7Å phases such as odinite and berthierine which react upon burial to form the iron-rich chlorite chamosite; by the alteration of mafic minerals, such as the chloritization of biotite grains; by the progressive reaction of trioctahedral smectites via mixed-layer chlorite–smectites and/or corrensite; as a by product of the smectite to illite reaction which results in a net release of Fe and Mg to solution; by the reaction of the breakdown products of kaolinite in Fe-bearing mudrocks; and as a result of reactions between dioctahedral clay minerals and carbonates (Hillier 2003).

3.4.3.4. Diagenesis of kaolin. Again the diagenesis of kaolin minerals has been mostly studied in sandstones but by analogy it is likely that similar patterns occur in mudrocks. Thus halloysite from soils would be expected to alter to kaolinite, and kaolinite eventually to its high-temperature polytype dickite. This polytypic transformation is well known in sandstones and it has also been documented in mudstones (Ruiz Cruz & Reyes 1998). The direct precipitation of kaolinite has also frequently been observed in pore space inside mudrocks. For instance, the observation of kaolinite filling small pores inside the tests of calcareous microfossils is a common one (e.g. Primmer & Shaw 1987; Bloch 1998). Examination of shales from the US Gulf Coast has also revealed that diagenetic kaolinite may form within and from smectite (Ahn & Peacor 1987). Ahn & Peacor (1987) suggested that this was a reflection of the heterogeneous availability of K^+ within the shales, kaolinite forming in K^+-deficient regions. Advanced diagenesis of mudrocks that are rich in kaolinite, and hence chemically 'aluminous' shales, may eventually lead to the formation of pyrophyllite.

3.4.4. Burial diagenetic reactions of non-clay phases in mudrocks

Very few sedimentary clay materials are composed entirely of clay minerals. Most contain other minerals, principally quartz, feldspars and carbonates. Many of these minerals may be involved in clay mineral diagenetic reactions either as reactants or products (Totten & Blatt 1993). Additionally, they may also undergo a variety of diagenetic reactions of their own (Primmer & Shaw 1987). Such reactions will undoubtedly affect the character of a clay material in much the same way as clay mineral diagenetic reactions do. However, with the sole exception of the diagenesis of silica, comparatively little is known about the burial diagenesis of non-clay minerals in mudrocks (Bloch 1998).

Silica diagenesis involves the transformation of amorphous and para-crystalline forms of silica, known as opal-A and opal CT, eventually to α-quartz (the most common form of quartz) (Fig. 3.23). The opal-A form of silica is the form of which most siliceous microfossils, such as diatoms and radiolaria, are comprised. These organisms may make a substantial contribution to the mass of some fine-grained sediments and when they are particularly abundant beds of chert may result. Upon burial the amorphous opal-A is transformed to Opal CT and eventually to quartz. These transformations have been studied in some detail especially in deep-sea sediments (Kastner 1981; Knauth 1994). The temperature at which the conversion is complete may vary due to a host of geochemical factors but generally it is completed by about 40°C. Thereafter opal-CT may persist to temperatures up to 80°C. Opaline silica, particularly opal-CT, may also be a common component of many bentonites. This will similarly be transformed to quartz if the bentonites are subjected to burial diagenesis.

Details of other diagenetic reactions that may occur in mudrocks are less well known. As far as carbonates are concerned, calcite, dolomite and siderite are common in mudrocks, as a result of both biogenic contributions (fossils and microfossils) and early diagenetic processes (Section 3.4.2). Upon deep burial, however, almost universally mudrocks appear to lose carbonates as a result of diagenetic processes (e.g., Hower *et al.* 1976). This may be a result of straightforward solubilisation or as

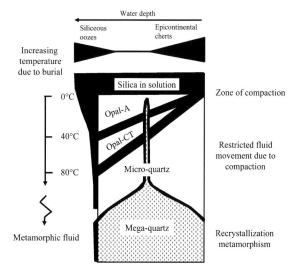

FIG. 3.23. Scheme of silica diagenesis in sediments, Knauth (1994).

a result of diagenetic reactions with other phases (e.g. Hillier 1993), but details are sketchy. As for feldspars, even less is known. As in sandstones feldspars may commonly undergo albitization processes during deep burial (Milliken 1992; Totten & Blatt 1993). In addition, there is some evidence to suggest a general augmentation of the albite content of many mudrocks by the stage of low-grade metamorphism. Nonetheless, details of the diagenesis of feldspars in mudrocks are not well understood.

3.4.5. Compaction, porosity and permeability

Freshly deposited clays and muds typically have large porosities of 60–80%. Geologists call the reduction of this porosity by burial and diagenetic processes compaction. The term is more or less synonymous with the term consolidation as used in soil mechanics. Compaction involves a wide variety of processes (Fig. 3.24) but most conceptual models of compaction recognize an early mechanical stage and a later stage where chemical diagenetic reactions involving dissolution and crystal growth processes become predominant (Hedberg 1936; Hinch 1980; Bjørlykke 1999).

Mechanical compaction is driven by effective stress (σ_e), which is the difference between overburden stress (σ_v) due to gravitational pressure and the pressure (P_p) of the fluid (usually water) in the pores ($\sigma_e = \sigma_v - P_p$).

Chemical compaction results from a number of diagenetic reactions foremost amongst which is that known as the smectite to illite reaction or more generally illitization (Section 3.4.3). Many such diagenetic reactions are strongly dependent on temperature and consequently rates of chemical compaction are also largely temperature dependent (Bjørlykke 1999).

The initial stages of compaction are rapid due to the dominance of mechanical processes such as grain rearrangement, and deformation. Thus porosity versus depth curves show a characteristically rapid decline down to depths of about 2.5 km. At deeper depths the rate of compaction decreases considerably, as chemical diagenetic processes become the dominant driving force. There is, however, no such thing as a universal compaction curve (Chilingarian 1983) and mudstones from different basins often show very different compaction versus depth trends (Fig. 3.25). Intrinsically, such differences are related to many factors including differences in sedimentation rate, depositional fabrics, grain-size distributions, and permeability characteristics. In some instances the apparent difference between some curves may also be related to the uplift and erosion of section, resulting in the displacement of some curves to shallower depths.

Compaction also tends to reduce permeability. Kasube & Williamson (1998) indicate that during mudrock burial and compaction permeability decreases by 6–10 orders of magnitude. Thus at the sediment water interface typical values range from 10^{-14} to 10^{-11} m^2 reducing to 10^{-21} to 10^{-20} m^2 at 2–4 km burial depth. They also indicate that the pore-size distribution in compacting mudrocks typically changes from modal values of around 200 nm at around 1 km depth (30% porosity) to values of 10–40 nm

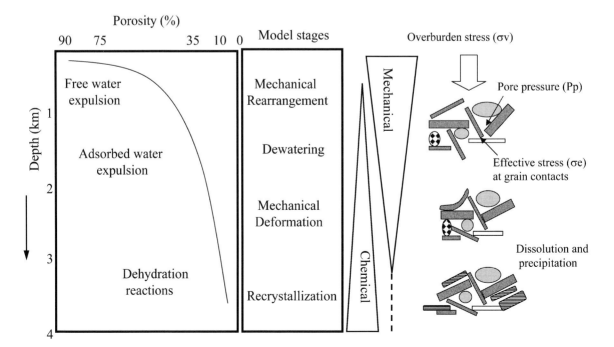

FIG. 3.24. Conceptual model of compaction and consequent loss of porosity of mudrocks with increasing depth of burial.

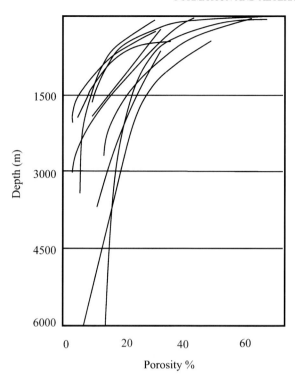

FIG. 3.25. Compaction curves for mudrocks from various different sedimentary basins modified from Chilingarian (1983).

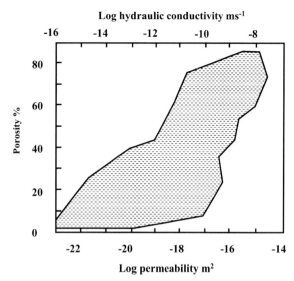

FIG. 3.26. Range of laboratory measured porosity versus permeability data for mudrocks, as compiled by Neuzil (1994).

at greater depths. Compilations of porosity versus permeability data for different mudstones (Neuzil 1994) show that for any given porosity permeability may range widely over several orders of magnitude (Fig. 3.26). There may be several reasons for the observed range of values but differences in grain-size and grain-size distributions between different mudrocks are probably one of the most important factors (Dewhurst et al. 1999). Thus at a specified porosity a silty mudstone is likely to be more permeable than a clay-rich mudstone. Permeability measurements on mudrocks are notoriously difficult to make, so data should always be critically evaluated (Dewhurst et al. 1999). Furthermore, it may be necessary to consider questions of scale, since fractures or the presence of other more permeable pathways may be present, depending upon the nature of the mudrock (Neuzil 1994).

The effect of mechanical compaction and consequent porosity loss on permeability is widely accepted, although still poorly understood. As far as chemical compaction is concerned, however, there are disparate views on its likely effect on permeability. As discussed above, the smectite-to-illite reaction is certain to have a major influence on the texture and fabric of mudstones, and this in turn may affect permeability. In this respect Ho et al. (1999) using transmission electron microscopy demonstrated that bedding parallel fabrics develop during diagenesis as a result of illitization processes. Nonetheless, Dewhurst et al. (1999), have suggested that illitization is unlikely to reduce permeability and may even enhance it. More recently, Nadeau et al. (2002) have presented evidence that the precipitation of tiny crystals of illite in porespace in mudstones, as expected from the perspective of neoformational models of illitization, will have a dramatic affect on permeability reduction, although only a minor effect on porosity. An analogy may be drawn with the well-known effect of the precipitation of fibrous illite on permeability in sandstones. Nadeau et al. (2002) further propose that this mechanism may be responsible for the development of overpressure (fluid pressure in excess of hydrostatic) in some mudrocks. Previously, Magara (1986) has claimed that the smectite to illite conversion is insignificant in relation to overpressure development. This claim, however, was made from the perspective of an illitization model based on the collapse of clay layers and release of interlayer water, not a model based on neoformation and its effect on permeability.

For the most part the engineer must deal with mudstones and shales that are at or near the surface of the Earth. However, it should not be forgotten that all of these materials have a generally irreversible burial and diagenetic history that exerts a major control on both their porosity and permeability characteristics and thereby their ability to transmit or retard fluids.

3.5. Hydrothermal environment

3.5.1. Introduction

Hydrothermal alteration may be simply defined as local rock alteration by fluids higher in temperature than those expected based on the regional geothermal gradient

(Utada 1980). The formation of clay minerals by hydrothermal alteration is very variable. In terms of scale, it may range from the localized crystallization of clays in fault and mineral vein systems, to the replacement of huge volumes of igneous rock, such as entire plutons. However, irrespective of scale, hydrothermal processes almost invariably involve the formation of clay minerals. Most present-day hydrothermal alteration is associated with active magmatic volcanic or plutonic igneous systems, both continental and oceanic. Fossil hydrothermal alteration also is often closely linked to igneous activity. In most instances this reflects the fact that the igneous activity is the source of heat. Such activity may also be a source of magmatic fluids, but for the most part hydrothermal fluids are usually derived from meteoric or seawater drawn into the hydrothermal system from the surrounding rocks. In young orogenic belts, hydrothermal alteration is typically widespread and may thereby represent a significant contribution to the clay cycle as for example in the Japanese Islands. Furthermore, many clay materials of economic value are hydrothermal in origin.

3.5.2. Hydrothermal clay mineral formation

Two features are notable concerning clay mineral formation by hydrothermal alteration. Firstly, hydrothermal clay mineral assemblages are usually quite simple, sometimes consisting of a single clay mineral. Secondly, clay formation usually occurs in well-defined zones. Inoue (1995) noted that the first of these observations might be related to the way in which the Gibbs phase rule is modified in metasomatic systems that encompass intense hydrothermal alteration. Since more chemical components are effectively mobile, the maximum numbers of phases is reduced. The second is related to factors such as changes in mass transfer between minerals and solution, changes in temperature of alteration, and changes in fluid to rock ratio. Thus hydrothermal alterations are characterized by steep gradients in both physical and chemical properties.

The types of clay minerals formed during hydrothermal alteration allows the nature of the alteration to be broadly classified as acidic, neutral or alkaline (Fig. 3.27) (Utada

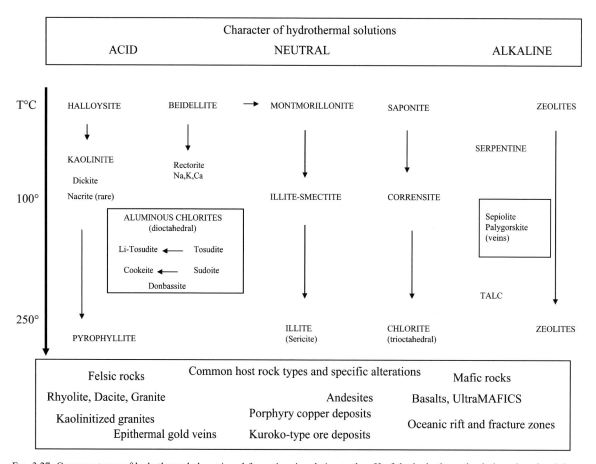

FIG. 3.27. Common types of hydrothermal clay mineral formation, in relation to the pH of the hydrothermal solutions, based mainly on data in Utada (1980) and Inoue (1995).

1980; Inoue 1995). The most acid hydrothermal solutions are capable of completely leaching most chemical components other than Al_2O_3, and SiO_2 from the host rock with the result that the clay minerals formed in order of increasing temperature of alteration are halloysite, kaolinite and pyrophyllite. Acid-type alteration may also lead to the formation of a variety of aluminous dioctahedral chlorites, such as sudoite and donbassite, as well as the mineral tosudite an interstratified mineral consisting of a perfectly regular alternation of dioctahedral chlorite layers and dioctahedral smectite layers. In some cases dioctahedral chlorites are rich in lithium such as the mineral cookeite, and tosudite may also occur as a Li-rich variety. The occurrence of dioctahedral chlorites usually indicates fairly high temperatures of alteration. In slightly less acidic conditions, the mineral rectorite, a regular interstratification of paragonite (Na-mica), and beidellite may occur. Varieties of rectorite in which the mica layer is a potassium or calcium variety are also known. The main conditions under which dioctahedral smectites are formed are neutral to mildly alkaline and, along with halloysite, smectites both di- and trioctahedral indicate the lowest temperature alterations. The most acidic conditions tend to be characterized by smectites of increasing beidellitic nature whereas more alkaline conditions favour montmorillonite. A smectite-to-illite reaction series characterized by the presence of intermediate mixed-layer illite–smectite is also common in many neutral to alkaline hydrothermal alterations of felsic rocks. Inoue (1995) has argued that the formation of mixed-layer illite–smectite in hydrothermal alterations is analogous to that in burial diagenesis in that it is progressive prograde formation from smectite to illite. However the direct crystallization of illite–smectites of any particular composition from solution is another possible pathway and is perhaps suggested by sequences in which only ordered illite–smectites occur with compositions ranging from R1 illite(50)–smectite to illite.

Under more alkaline conditions sequences of trioctahedral clays minerals are common. The sequence saponite–corrensite–chlorite is often recorded in hydrothermal alterations of andesites and basalts. The exact nature of the transformation of saponite to chlorite is controversial and in particular there is controversy over whether or not random interstratifications of chlorite and smectite occur as intermediate states. Several workers have suggested that randomly interstratified chlorite–smectite is characteristic of disequilibrium and high water–rock ratios whereas pure phase alterations, i.e. where saponite, corrensite and chlorite occur as discrete phases are characteristic of low water–rock ratios. Most alterations can be interpreted in terms of the scale and heterogeneity of intergrowths amongst saponite, corrensite and chlorite. Ultramafic rocks are often characterized by alteration to serpentine and talc. Additionally the minerals sepiolite and palygorskite may be found in hydrothermal veins in ultramafic rocks. The most alkaline forms of hydrothermal alteration are usually characterized by the formation of abundant zeolites as opposed to clay minerals. Hydrothermal alterations which involve the trioctahedral clay minerals are very common in submarine alteration of basaltic rocks at active spreading centres where the fluids involved are essentially derived from sea water circulating through the oceanic crust. Paragonite is also formed by this type of alteration in modern mid-oceanic hydrothermal systems (Alt & Teagle 1998).

In general, the various types of hydrothermal clay mineral formation described above occur most commonly in specific rock types. Thus acid and neutral types of alteration are most common in acid igneous and felsic rocks, whilst more alkaline alterations are most common in mafic rocks. In addition to the more acidic nature of magmatic fluids, Inoue (1995) suggests that the common occurrence of acid type hydrothermal alteration of felsic rocks may be due the more explosive nature of the acid volcanism forming large calderas which in turn act as effective catchments for meteoric water. Despite the general relationships of hydrothermal alteration to rock type, it should also be pointed out that acid hydrothermal alteration can also occur in mafic rocks where the acidity may be derived from meteoric water or the oxidation of sulphides. Indeed, juxtaposition or gradations from one type of alteration to another are commonplace. A striking example of the potential role of sulphide oxidation as a source of acidity was recently described by Marumo & Hattori (1999) at Jade in the Okinawa Trough, Japan. Here hydrogen sulphide appears to be too abundant to be completely consumed in the formation of metal sulphides and its oxidation provides the necessary acidity to form both abundant kaolinite and halloysite by hydrothermally driven circulation of a mixture of low pH sea water and hydrothermal fluids through basaltic pyroclastics.

Based on this summary of hydrothermal alteration it is evident that there are very close parallels with the formation of clay minerals by burial diagenesis and low-grade metamorphism. There are, however, important differences. One is the more pervasive nature of hydrothermal alteration, which often results in great volumes of clay formation per unit volume of original rock. Another is the fact that the composition of the solution phase i.e. the hydrothermal solution is typically determined by factors external to the rock so that it is a major variable in the alteration process. In diagenesis and low-grade metamorphism, the main variables are temperature and pressure, and fluid compositions that tend not to be as far from equilibrium with the rock, unlike the fluids in hydrothermal alterations. Lastly, the temperature-related mineral sequences observed in both settings tend to be displaced towards higher temperatures in hydrothermal alterations because of the shorter geological times over which the alterations occur and the kinetic nature of many reactions.

3.6. Clay materials: products of weathering, sedimentary, diagenetic/low grade metamorphic and hydrothermal processes

Clay materials may be soils, sediments, sedimentary rocks, metamorphic rocks or hydrothermal deposits. The factor common to all of them is a high proportion of clay and the properties that this may impart. Beyond this common factor, the diversity of clay materials is striking. Indeed, many are known by specific names derived from their properties, occurrence and, most particularly, their uses and history as industrial minerals (Carr 1994; Kendall 1996). Many others fall into more general categories, although generalization is a dangerous practice where clay materials are concerned. The previous sections of this chapter represent an attempt to lay a foundation for the understanding of clay materials as the products of processes that operate in the weathering, sedimentary, diagenetic, low-grade metamorphic and hydrothermal environments. However, the formation of many clay materials will represent unique combinations of circumstances and processes, often acting over protracted periods of geological time. For example, the development of stable tropical climatic conditions may produce a clay-rich weathered mantle, which later, during a time of tectonic instability, is stripped from the landscape to be deposited as a clay-rich sediment over a wide area, only to be preserved as such where it has completely escaped later contact metamorphism to hornfels (a hard rock), and all gradations in between, by the intense heat from a suite of granite igneous intrusives. The example is fictitious, but not unrealistic. It serves simply to emphasize that the origin of many clay materials will be complex, and depend upon processes from several environments. Furthermore, identifying the degree or even inaction of some processes may be just as important as an identification of those where positive action has occurred.

3.6.1. Clay rich soils

Most soils contain an important component of clay minerals but in some they may be particularly dominant. In terms of the textural classification used by soils scientists three types of clayey or clay-rich soils are recognized, namely sandy clay, silty clay and clay. All three classes contain at least 40% clay (Bridges 1970; Brady 1974). Clay-rich soils are widespread and in many instances, they are formed upon parent materials that are themselves rich in clay, or that readily weather to clay, such as shales and slate. In these instances most of the clay in the soil is simply inherited from the parent material. Nevertheless, soil clays are often considerably transformed and modified, so that their properties may be distinctly different from the parent clay material, particularly in the topsoil, and particularly if the extent of soil horizonation is well developed.

Clay soils formed directly from non-clay parent materials are generally restricted to areas of intense chemical weathering in the tropics. Two types deserve specific mention, namely vertisols and oxisols. Vertisols are clay soils that contain abundant smectite, or more rarely other swelling, i.e. water sensitive clays. The properties of smectite are imparted to the soil so that it shrinks and develops cracks when dried and swells inducing the development of pressure faces with slickensides when wet (Fig. 3.28).

When dry, vertisols are extremely hard, and when wet they become extremely sticky. As a result of repeated wetting and drying cycles and the opening and closing of cracks, the soil material is turned; hence the name vertisol derived from the Latin *vertere*, to turn. Vertisols occur mainly in the tropics where the climate is characterized by alternating wet and dry seasons. They tend to develop from rocks such as basalts, and are typically found in the lower parts of landscapes, as a result of the concentration of base cations in such areas which promotes the formation and stability of smectites. In most vertisols the species of smectite present is an Fe-rich beidellite (Wilson 1987).

Oxisols are characterized by abundant kaolinite and accumulations of oxides of iron and aluminium in an oxic horizon at depth. They are usually very deep soils (metres) and uniform in terms of their lack of horizonation

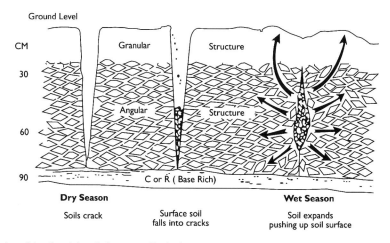

FIG. 3.28. Characteristics of the clay rich soils known as Vertisols.

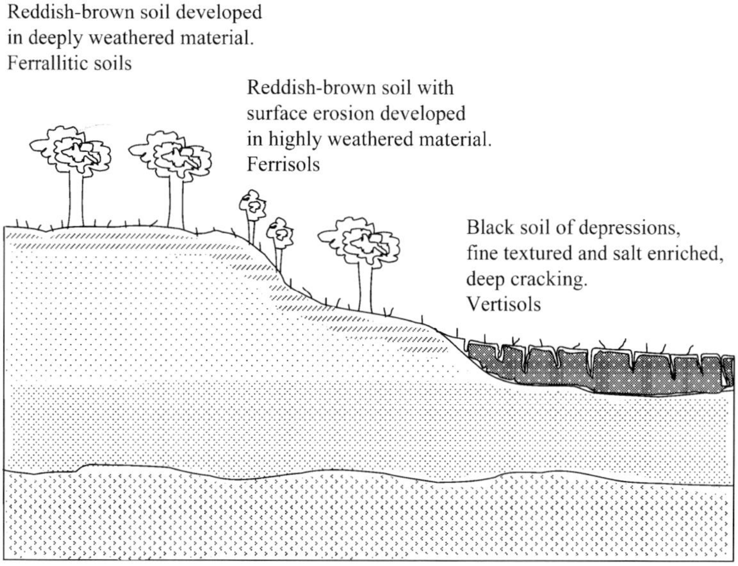

FIG. 3.29. Schematic illustration of so-called 'red black toposequence'. Modified from Bridges (1970).

and with a high content of kaolinite throughout. These soils occur essentially in the humid tropics, in areas that have been stable geologically for long periods (tens of thousands to millions of years). Iron pans inherited from previous land surfaces are a common feature. Oxisols are thus the products of extreme weathering promoted both by high temperatures and high rainfall acting over long periods of time. In many tropical landscapes both oxisols and vertisols may be juxtaposed in a so-called 'red and black' complex or toposequence, the development of which is usually related to topographically controlled differences in the degree of leaching (Fig. 3.29).

3.6.2. Residual kaolin deposits

Many of the most important kaolin deposits in the world are classed as residual deposits, i.e. they are formed mainly by weathering processes. Since they are formed *in situ* they are also classified as primary kaolins. Along with primary kaolins of hydrothermal origin and mixed residual hydrothermal origin, such deposits account for most of the world's commercially exploitable kaolin. Examples include the kaolin deposits of the Czech Republic, East Germany and the Ukraine, reflecting intense weathering during the Mesozoic and Tertiary periods which formed deposits often as much as 50 m thick (Murray & Keller 1993). The dominant kaolin mineral is almost invariably kaolinite formed by the alteration of feldspars and micas in granites or in acid volcanic rocks such as rhyolites or ignimbrites. The necessary conditions for the formation of residual kaolins include high rainfall in temperate or tropical climates, with low water table, rapid drainage and intense leaching.

3.6.3. Vermiculite

Although commercially exploited vermiculite is typically much larger than clay size, it is a natural material sought essentially for its clay-like properties. According to Hindmann (1994), most vermiculite deposits have been formed by the supergene alteration (weathering) of Fe-bearing biotite or phlogopite in rocks such as coarse-grained ultramafic intrusives, biotite schist and gneiss. The alteration of biotite into vermiculite requires a volume increase of 10–40%. This places a limit on the maximum depth of vermiculitisation in weathered saprolite which empirical evidence from vermiculite deposits suggests is about 250 m (Hindmann 1994). Below this depth no vermiculite formation may take place because lithostatic pressure exceeds the swelling pressure that results from oxidation, ion exchange and decrease of layer charge.

3.6.4. Terrigenous clays, mudstones and shales

Most clay materials formed in the sedimentary environment are shales or mudrocks of one kind or another. Both terms are used as generic general group names for the entire class of fine-grained sedimentary rocks; that is those rocks that contain at least 50% of grains of silt and clay size, i.e. finer than 62 microns (Blatt *et al.* 1980; Potter *et al.* 1980). Another term, in common use by the extractive minerals industry, which encompasses this class of clay materials is 'common clay and shale' (Ridgway 1982; Murray 1994). In the geological literature, the all-encompassing term mudrock is preferred by many since the term shale is commonly used to imply a

TABLE 3.3. *Classification of mudrocks proposed by Potter et al. (1980)*

		Percentage clay sized constituents	0–32	33–65	66–100
		Field adjective	Gritty	Loamy	Fat or slick
Nonindurated	Beds	>10 mm	Bedded silt	Bedded mud	Bedded claymud
	Laminae	<10 mm	Laminated silt	Laminated mud	Laminated claymud
Indurated	Beds	>10 mm	Bedded siltstone	Mudstone	Claystone
	Laminae	<10 mm	Laminated silstone	Mudshale	Clayshale
Metamorphosed	Degree of metamorphism	Low	Quartz argillite		Argillite
			Quartz slate		Slate
		High		Phyllite and/or mica schist	

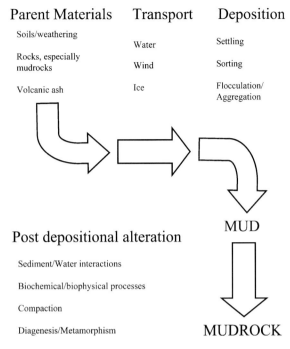

FIG. 3.30. Provenance, sedimentary and post depositional factors that determine the nature of mudrocks.

fissile mudstone. As yet, there is no universally accepted classification scheme for mudrocks but that proposed by Potter *et al.* (1980) is fairly comprehensive and easily understood (Table 3.3).

Potter's classification scheme is based on the relative amounts of clay and silt sized mineral particles, the presence of bedding or lamination, and the state of induration. In terms of particle size, only rarely are clay-sized minerals the only ones present in mudrocks. For instance, Picard (1971) determined that modern muds average 15% sand, 45% silt and 40% clay. Aplin *et al.* (1999) also point out that muds are characterized by a much wider range of grain sizes than is usual in any other clastic sediments. Features such as lamination and bedding are often closely related to subtle or distinct changes in the amount of clay and organic matter on the one hand and the amount of sand, silt and carbonate on the other. However, the use of fissility as a parameter for classification has been criticised since weathering can enhance it (Aplin *et al.* 1999). The degree of induration is largely a reflection of the diagenetic to metamorphic grade (Section 3.4). Increased grade is typically associated with an increase in the proportion of mineral grains which have undergone processes such as mechanical rearrangement, recrystallization, dissolution, reprecipitation, and crystal growth, all of which combine to produce a more indurated material. Many of the other classification schemes proposed for mudrocks are illustrated in O'Brien & Slatt (1990).

Additionally, Spears (1980) proposed an interesting scheme based on quartz content.

The mineralogical composition and hence the chemical composition of mudrocks are determined by some combination of provenance, sedimentary processes and post-depositional and diagenetic history (Fig. 3.30).

Thus most mudrocks are usually unique in some respect, although common mineralogical and chemical compositions are observed and it is useful to have an indication of average compositions, as well as compositions that are unusually enriched or depleted in various components. Details of the chemical composition of various mudrocks and of 'average shale' as calculated by Clarke can be found in Pettijohn (1975). In Table 3.4 the composition of the average shale is compared with aluminous, siliceous, carbonaceous and calcareous examples. An average calculated from the XRF analyses reported in Appendix A of this report (labelled 'CWP' in Table 3.4) is also included for comparison. Compositions of some British mudrocks as reported in Ridgway (1982) are also listed in Table 3.5 for comparison. Frequently some kinds of shale are named on the basis of non-clay components, for example black shales, which are generally rich in organic matter. Other examples include marls (calcareous shales), ferruginous shales and phosphatic shales. In most instances such shales reflect their formation under specific environmental and early diagenetic conditions.

The underlying control of the chemical composition of mudrocks is their mineralogical composition, which varies widely. Indeed, mudrocks generally have a much

TABLE 3.4. *Chemical composition (wt%) of the 'Average shale' and of some shales enriched in certain elements. For original sources of data see Pettijohn (1975), except for 'CWP' data which is derived from Appendix A of this report*

	Average shale	CWP	Aluminous	Siliceous	Carbonaceous	Calcareous
SiO_2	58.10	54.81	51.38	84.14	51.03	25.05
TiO_2	0.65	0.87	1.22	–	–	–
Al_2O_3	15.40	17.46	23.89	5.79	13.47	8.28
Fe_2O_3	4.02	5.21*	2.05	1.21	8.06	0.27
FeO	2.45		5.01			2.41
MnO			0.02			4.11
MgO	2.44	1.84	2.71	0.41	1.15	2.61
CaO	3.11	4.78	0.24	0.13	0.78	27.87
Na_2O	1.30	0.5	0.59	0.99	0.41	–
K_2O	3.24	2.95	7.08	0.5	3.16	–
H_2O	5.00		4.87	5.56	0.81	4.3
P_2O_5	0.17	0.13	0.01		0.031	0.08
CO_2	2.63		0.14	–		24.2
S					7.29	
C			0.16	1.21	13.11	
LOI		11.26				

* Total iron expressed as Fe_2O_3.

TABLE 3.5. *Chemical composition (wt%) of some British mudrocks. Source and identifiers from Ridgway (1982)*

	Ordovician No.1	Devonian No. 2	Coal Measure Shales No. 7	Etruria Marl No. 4	Mercia Mudstone No.2	Middle Lias No. 1	Oxford Clay No. 2	Upper Estuarine Clay No. 1	Weald Clay No. 3	London Clay No. 4
SiO_2	56.27	54.26	62.50	60.62	46.2	57.8	43.68	79.03	54.98	57.18
TiO_2	1.05	0.32	1.15	1.22	0.65	1.2	0.85	2.11	1.01	1.08
Al_2O_3	22.29	16.82	21.53	20.62	12.80	23.2	17.10	10.97	18.43	17.18
Fe_2O_3	0.64	7.54	7.40	7.4	4.95	9.3	2.08	1.95	10.37	7.98
FeO	7.12									
MnO	0.23		0.09	0.03						
MgO	2.11	2.94	2.03	0.73	8.77	2.5	1.57	0.35	0.91	2.82
CaO	0.39	2.46	1.00	0.28	7.32	1.0	10.55	0.52	2.66	2.41
Na_2O	0.79	0.68	0.77	0.12	0.08	0.9	0.17	0.20	0.46	0.27
K_2O	3.16	3.50	2.90	1.63	5.00	2.9		0.26	3.25	3.27
H_2O	5.60									
P_2O_5	0.18									
CO_2										2.10
S										
C	0.27									
LOI		11.46	8.20	7.59	13.50	1.0	18.10	4.56	7.71	8.00

wider range of composition than either sandstones or limestones because many more minerals may occur as major components. On average, however, the major mineral suite of most mudrocks is restricted to quartz, feldspars, carbonates and clay minerals. Average mineralogical compositions (Table 3.6) have been computed by Yaalon (1962) and measured by Shaw & Weaver (1965). Shaw & Weaver's (1965) average mudrock is based on an analysis of over 400 shales from North America. Also listed in Table 3.6 is the average bulk composition derived from detailed mineralogical analyses of a suite of mudrocks analysed for this report (Appendix A). These data emphasize that apart from clay minerals the main components of most mudrocks are quartz, feldspars and carbonates, with the last two varying most widely in abundance.

Shaw & Weaver (1965) noted that shales with greater than 50% quartz were commonly what would be described as siliceous shales, usually found in association with cherty formations. They further suggested that much

TABLE 3.6. *Average mineralogical composition of shale after Yaalon (1962), Shaw & Weaver (1965), and Appendix A of this report*

	Yaalon (1962)	Shaw & Weaver (1965)			CWP
		Mean	SD	RSD	
Quartz	20 (& chert)	30.8	15.2	49%	23.9
Feldspar	8	4.5	4.8	106%	3.7 (K-spar)
					2.4 (Plag.)
Carbonate	7	3.6	8.4	233%	7.5 (Calcite)
					1.3 (Dolomite)
					0.5 (Siderite)
Fe-oxides	3	0.5			0.8
Clay minerals	59	60.9	9.4	15%	48.7 (Di. clay)
					7.5 (Tri. clay)
Other minerals	2	2			0.5 (Pyrite)
Organic matter	1	1			

Note standard deviations reported by Shaw & Weaver (1965) are derived from a smaller subset (300) than used to calculate the means. SD, standard deviation; RSD, relative standard deviation.

of the quartz in such shales was authigenic in origin and that it was doubtful that detrital quartz in shales ever exceeded 50%. Indeed, in shales that are not obviously silty, quartz contents in excess of 35% are potentially an indication of authigenic formation of quartz from amorphous silica. In this context it is possible that authigenic quartz or other forms of authigenic silica may explain the siliceous compositions of the Jurassic 'Upper Estuarine Clays' and several Cretaceous clays as reported by Ridgway (1982), an example of which is given in Table 3.5.

Shaw & Weaver (1965) also noted that quartz/feldspar ratios in shales tended to be related to those in associated sandstones. It is also likely that the types of feldspars present in genetically related sandstones and mudrocks will also be related. Carbonate in shales is frequently a mixture of detrital, biogenic and authigenic material. Calcite is the most common carbonate in shales, but siderite is also common especially in non-marine shales, and dolomite is common in shales associated with limestones or evaporites. A UK example of the latter is the Mercia Mudstone, which often contains abundant dolomite. Other non-clay minerals frequently present in minor to moderate amounts include zeolites, iron oxides, titanium oxides, sulphates, sulphides and phosphates.

Volcanic ash and glass is also a common component of many mudrocks, although in many instances it may be so altered that its former presence can only be deduced by careful petrographic and geochemical analysis. Indeed, because the components of volcanic ash alter so readily to clay minerals it may have made a much greater contribution to many mudrocks than is generally acknowledged (Zimmerle 1998; Haynes *et al.* 1998; Jeans *et al.* 2000). Based on K-Ar dating and on the distribution of mixed-layer illite-smectite, Nadeau & Reynolds (1981) suggested that the contribution of volcanic components to mudrocks may also have fluctuated through geological time (Fig. 3.31) and that this variation can be correlated with first-order eustatic sea level rise. Such a correlation

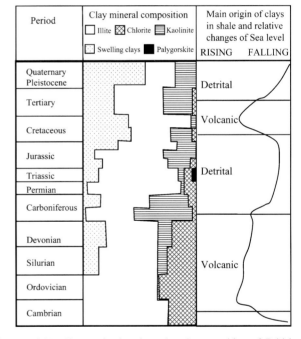

FIG. 3.31. Changes in the clay mineral composition of British mudrocks through time after Shaw (1981) and suggested major changes in provenance of mudrocks through geological time after Nadeau & Reynolds (1981).

is based on the fact that first-order sea level cycles are related to rates of sea floor spreading which has consequences for the volume of mid-ocean ridges and subaerial volcanism as a result of increased subduction rates at destructive plate boundaries. There are essentially two periods in earth history when sea levels were very high, one during the middle Palaeozoic and one during the late Mesozoic. These high stands have been related to the break-up of super-continents and are paralleled by peaks in the records of igneous activity. Peaks are also evident in the number of bentonites recorded from Ordovician and from Cretaceous rocks. Thus the middle-Palaeozoic and late Mesozoic may have been times when volcanic ash and glass made a much bigger contribution to mudrocks than at other times during Earth history, although at present the evidence for such trends is rather speculative.

Other trends in the clay mineralogy of mudrocks through time have been documented by Weaver (1967, 1989) for the United States and for the UK by Shaw (1981) whose data are also reproduced in Figure 3.31. Essentially, smectite, mixed-layer clays and kaolinite, are progressively replaced by assemblages dominated by illite and chlorite in older rocks. These trends have mostly been interpreted as reflecting the greater chance that diagenetic processes will have altered clay minerals in older mudrocks. However, the possibility of secular variations in the clay minerals supplied from the continents cannot be ignored. Thus some differences could also be related to changing distributions of source materials such as the proportions of crystalline, volcanic and sedimentary rocks exposed at the surface and/or changes in the nature of weathering process and/or soil forming factors, including the evolution of soil biota and vegetation. Another interesting trend that has captured the attention of many authors is the tendency for the potassium content of shales to increase with geological age. Some have argued that this trend is simply related to changes in the relative contribution of detrital sources such as cratonic and volcanic components (Hutcheon *et al.* 1998). Others have evoked open system behaviour with potassium enrichment of shales sourced from adjacent sandstones or from regional phases of potassium metasomatism (Awwiller 1993; Melson *et al.* 1998).

3.6.5. Sedimentary or secondary kaolins

Clay materials that are especially enriched in minerals of the kaolin group are common in many parts of the stratigraphic column. Indeed in many parts of the World such sedimentary deposits are important industrial sources of kaolin. Since they are largely formed by erosion, transport and deposition of kaolin formed elsewhere, they are also termed secondary kaolins. There are many kinds of sedimentary kaolins but the most common factor that unites them is that they are invariably associated with non-marine sedimentary environments of deposition, especially those that are also characterized by the presence of coal or lignite-bearing strata.

Many are known by specific names. Amongst these 'ball clay', 'underclay' and 'fireclay', are in frequent use. Generally, these names are a reflection of the specific properties and uses of these clay materials, but collectively such features also reflect some differences in their occurrence and origin.

Ball clays are plastic sedimentary clays with a high kaolinite content whose main distinguishing feature is the fact that they fire to white colour when used to make pottery or ceramics (Highley 1975). The name is thought to be derived from the early practice of cutting the clay out of the ground in cubes weighing about 15 kg, which rounded rapidly into a ball like shape due to handling. The term underclay refers to a type of clay commonly found beneath beds of coal. Underclays are usually massive mudstones often with extensive evidence that plants were rooted in them. Along with other indications of soil forming processes this may account for their general lack of bedding. Underclay is in fact an American term (Patterson & Murray 1984), the British equivalent of which is seatearth, a direct reference to the probable palaeosoil origin of many. Fireclay is a name used to refer to clay materials that exhibit refractory properties and so can be used to make furnace linings and in other high temperature ceramic applications (Highley 1982). The refractory properties arise because of the low content of alkalis (typically $< 1\%$ expressed as oxides), which if present in larger amounts would act as fluxing agents. In fact most fireclays occur as underclays. Indeed, most ball clays are also varieties of underclay (Patterson & Murray 1984). Although the detailed origins of ball clays, fireclays and underclays vary the common association of these kaolinite-rich clays with coal-bearing strata reflects a set of common factors in their origin. Essentially, the association with coals identifies a humid tropical climatic regime where weathered detritus is characteristically enriched in kaolinite. Additionally, this non-marine depositional environment has the greatest potential for the concentration and segregation of the products of intense tropical weathering, without dilution from other sources. The association with coals may also indicate that leaching by organic acids may play a role in the formation and purity of some sedimentary kaolins.

In the United Kingdom, as well as in many other parts of the World, most underclays and fireclays are found in association with Carboniferous coal bearing strata. Ball clay deposits tend always to be much younger. In part this probably reflects the fact that younger deposits have experienced less diagenesis and so tend to be less indurated and naturally more plastic in nature. The plasticity of ball clays is related among other things to their very fine particle size. The most highly prized may have more than 90% $< 1\,\mu m$ in size. The presence of colloidal organic matter, and also of other clay minerals beside kaolinite, such as mixed-layered clays, may also play important roles in determining plasticity.

Often these types of clay materials occur in lens-like deposits and it is well known that ceramic properties can vary widely between one bed and another. Such properties depend ultimately on the mix of minerals and organic components and engineering properties, likewise, will vary. In the United Kingdom most ball clay deposits are restricted to Southwest England in the Bovey and Petrockstowe Basins in Devon and across parts of Dorset, such as around Wareham, although similar clays are also known from Lough Neagh in Northern Ireland. All of these deposits are Eocene–Oligocene in age. The Bovey and Petrockstowe basins formed largely in response to dextral strike slip movements (Section 3. 1.2) on the Sticklepath fault system (Vincent 1983).

As discussed above, the term ball clay has no precise mineralogical significance but generally such clays are composed of three main components: kaolinite, quartz and micaceous/illitic clays. According to data in Highley (1975) for ball clays from Devon and Dorset, kaolinite ranges from 20–90%, quartz from 0–60% and micaceous clay from 0–40. Finely disseminated organic matter can also be a significant component up to 16% in some cases. An average composition derived form a suite of ball clays analysed for this report (Appendix A) indicates that they contain about 70% clay minerals, divided into around 40% kaolin, and 30% discrete micaceous and micaceous mixed-layer clay minerals. Ball clays are also typically very fine grained, many having more than 60% <0.5 μm in size (Vincent 1983). In the deposits of Devon and Dorset, most often the kaolinite is a highly disordered variety. However, there are stratigraphic intervals and beds where more ordered varieties predominate. Such differences in the type of kaolinite are related to provenance (Vincent 1983). Kaolinite derived from the hydrothermally altered Cornubian granites (Section 3.6.9) is characteristically well ordered. Disordered kaolinite predominates in the ball clays and this is believed to identify the source as a tropical weathered mantle that was developed on the Carboniferous and Devonian low grade metamorphic shales and slates as the main source of kaolinite for the Devon ball clays. Elsewhere in Europe important deposits of ball clay occur in the Cheb Basin in the Czech republic, the Don Basin in the Ukraine, and in various Tertiary basins in Germany and in France. In the United States important deposits occur in Tennessee and Kentucky (Watson 1982).

Outside of the United Kingdom, probably the best-known secondary kaolins are those from the State of Georgia in the USA deposited along the Cretaceous and Tertiary coastal plain and derived from weathered mantles on the piedmont to the northwest. Deposits are mainly located in Georgia, but extend into the neighbouring states of Alabama and South Carolina. In some cases, the kaolin is present in amounts up to 20–30 wt% in a sandy matrix and is thought to have formed by the alteration of arkosic (feldspathic) sediments by groundwater movement. In others direct deposition of fine-grained kaolin may form beds or lenses of clay with a kaolin content of more than 90%. The Cretaceous clays are coarser grained than the Tertiary clays and contain prominent vermicular stacks of kaolinite, differences which are thought to be related to different parent materials. The Cretaceous clays are believed to be derived from granite whereas the Tertiary clays are believed to be derived from finer-grained phyllites. Associated sedimentary facies and other indicators of depositional environment seem to consistently indicate that deposition occurred under non-marine conditions. Indeed as stated previously, as a general rule sedimentary kaolins are invariably found in non-marine sedimentary sequences. Another point worthy of mention is that many sedimentary and residual kaolins often have higher TiO_2 contents than hydrothermal kaolins. The TiO_2 content is mainly present in the form of anatase, and may be related to the presence of minerals such as biotite in the parent materials (Weaver & Pollard 1973; Weaver 1976).

A final kind of kaolin to which reference is sometimes made is a variety known as flint clay. Flint clay derives it name from the fact that the raw material, although predominately clay, does not slake when immersed in water, weathers into jagged fragments and fractures with a subconchoidal (flint-like) aspect when struck with a hammer.

3.6.6. Sedimentary bentonites and tonsteins

The term bentonite has no genetic significance and simply refers to a rock whose major component is a mineral of the smectite group (Grim & Güven 1978). As such, bentonites may be formed by a variety of processes. Mostly the term is used to refer to deposits of smectite (montmorillonite) formed by the subaqueous (wet) alteration of air fall volcanic ash. This is the origin of the Cretaceous Clay Spur deposit originally discovered in the Benton Shale, named after Fort Benton, in Wyoming, USA (Elzea & Murray 1990). In detail the composition of the parent volcanic ash from which bentonites are formed has little influence on the alteration process, such that bentonites may be formed from either mafic or silicic volcanic ash, although most major sedimentary bentonites seem to be formed from evolved silicic volcanic ash. This is probably because silica-rich magmas produce the most explosive volcanic eruptions and thereby distribute more volcanic ash over wider areas. A further point to note is that the subaqueous alteration of wet volcanic ash to bentonite is normally restricted to marine environments. Equivalent volcanic ash falling into a non-marine environment is instead altered to kaolinite forming a deposit known as a tonstein, a German term, literally meaning claystone. The equivalence of tonsteins and bentonites in terms of their common origin by the alteration of volcanic ash has been proved by tracing and correlating individual beds from one environment to the other (Spears & O'Brien 1994; Spears et al. 1999). It is important to stress that bentonites and tonsteins result from the subaqueous alteration of air-fall volcanic ash in a sedimentary environment. Although a background component of volcanic ash may be continuously contributed to many sedimentary environments, bentonites and tonsteins record major ash falls. Exceptionally, they may form beds up to a metre or more in thickness, for example the Cretaceous Clay Spur bentonite from Wyoming (Elzea & Murray 1990). Most commonly, however, they are only of the order of a few to a few tens of centimetres in thickness. Geographically, bentonites can often be traced over large areas up to tens of thousands of km^2 reflecting the original widespread distribution of ash from an explosive volcanic eruption. As a result they are also frequently used as important event stratigraphic markers, with the added advantage that many contain suitable minerals for geochronological studies. Economically, a distinction is sometimes made between 'swelling' and 'non-swelling' bentonites. This distinction reflects differences in the exchangeable cation population of the smectite component of the bentonite. The term 'non-swelling' is really a misnomer since all bentonites have the capacity to swell but smectites whose exchangeable cations are mainly sodium have a much greater capacity for swelling compared to those saturated with other cations, for example Ca^{2+}. Other exchangeable cations are usually present to some degree e.g., Mg^{2+}, but are not normally dominant. Because of their extraordinary capacity to swell, sodium saturated bentonites are highly prized commercially, but most bentonites are naturally calcium saturated. Treatment with various sodium salts, for example sodium carbonate, is often used to convert naturally calcium-saturated bentonite into a sodium-saturated form for industrial purposes. The volcanic origin of most bentonites in sedimentary successions has not always been accepted (Grim & Güven 1978), but studies of associated minerals and trace elements have proved decisive. Notable examples of bentonites in the United Kingdom stratigraphic succession include the Jurassic and Cretaceous Fuller's Earth (Jeans et al. 1977, 2000). Elsewhere in Europe the most important occurrences are in the Upper Miocene, Bavarian Molasse (Grim & Güven 1978). Unfortunately, the term Fuller's Earth has been, and is still, used in a variety of ways. In the United Kingdom, Highley (1972) used the term specifically to designate deposits of calcium saturated smectite (bentonite) and limited his use of the term 'bentonite' to natural sodium-saturated smectite. In industry the term is also sometimes used for clay materials whose principal component is palygorskite (Heivilin & Murray 1994). Thus Fuller's Earth is really a term based entirely on use, although most materials so described are bentonites in the sense of Grim & Güven (1978).

Most bentonites worked for industrial purposes are Cenozoic or Mesozoic in age. This is because older bentonites have generally been diagenetically altered, the effect of which is to decrease the exchange and swelling properties of the material as smectite is transformed to mixed-layer illite–smectite. As such, bentonites subjected to moderate or advanced diagenesis are often termed metabentonites or K-bentonites. The process of illitization requires potassium and metabentonites are therefore enriched in potassium relative to bentonite, hence the term K-bentonites. Thus metabentonites tend to be found in geologically older formations. For example, most Palaeozoic bentonites are metabentonites. In the United Kingdom they are common in Ordovician and Silurian rocks, such as those exposed in the Southern Uplands of Scotland. Clearly the extent of alteration from smectite to illite will have a direct bearing on the geotechnical behaviour of a metabentonite. Bentonites have very many industrial uses. Additionally, they may prove troublesome from an engineering point of view, especially when they occur as thin beds that may not be recognized for what they are. This is because they tend to be much more clay-rich and much richer in swelling clays (smectite or illite-smectite) compared to the surrounding beds in which they occur.

3.6.7. Chemical perimarine and lacustrine clay deposits

In certain lacustrine sedimentary environments and in some evaporitic peri-marine environments geochemical conditions may be suitable for the formation of beds of clay that are dominated by authigenic clay minerals. A well-known and well-documented example is the Miocene of the Southwestern United States, where thick beds of clays rich in palygorskite and sepiolite have developed (Weaver & Beck 1977).

3.6.8. Deep sea clays

The clay materials formed in large parts of the deep sea and oceanic basins are, generally, quite distinct from terrigenous clays (Fig. 3.32). This is because many such

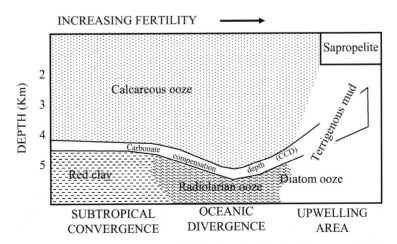

FIG. 3.32. Characteristics of deep-sea clays in relation to water depth and productivity of surface waters after Berger (1974).

areas are so far removed from land that detrital terrigenous material becomes a minor, even insignificant, source of sediment. As a result, the products of other processes make a more important contribution to the fine-grained sediments that accumulate in these environments (Berger 1974). Globally, the most important of these are the minute skeletal components of microfossils which form a continuous pelagic rain (Section 3.3.5.3) from surface to deeper waters. Their contribution to the fine-grained sediment accumulating at the ocean floor depends upon the dynamic balance between the processes of their production in surface waters and their destruction by dissolution on their journey down through the water column following death of the organisms. The two most important biogenic components are calcareous and siliceous microfossils. The calcareous microfossils include foraminifera and coccoliths composed of calcium carbonate ($CaCO_3$) mainly in the form of calcite, whilst the silceous microfossils include diatoms and radiolaria composed of opaline amorphous silica (SiO_2), in the form known as opal-A. The rate of production of these organisms in surface ocean waters depends on biological fertility. Diatoms dominate in more fertile nutrient rich water whereas coccoliths dominate in less fertile regions. Since seawater is universally undersaturated with respect to amorphous silica, most silica is dissolved and recycled as the skeletons of opaline microfossils descend through the water column. A further fraction arrives at the sediment water interface where more is dissolved but in regions of high productivity some is preserved and may accumulate. Thus the distribution of siliceous pelagic sediments mirrors the patterns of the most highly productive ocean waters such as in regions of oceanic divergence and upwelling where nutrient-rich deep ocean waters rise to the surface. The fate of calcareous pelagic sediment is similar except that the degree of undersaturation of seawater with respect to carbonates increases with depth. This gives rise to a depth in the oceans, known as the Calcite Compensation Depth (CCD), below which calcite does not accumulate. In the deepest parts of the Ocean basins (>3500 m) below the CCD, sedimentation rates may be extremely slow and hydrogenous processes involving iron and manganese oxides take on an important role. Such areas accumulate deposits know as deep sea red clays (Glasby 1991).

Red clays are extremely fine grained with often more than 80% <2 μm in size. They cover about 30% of the ocean basins and are most prevalent in the Pacific Ocean. Most of the components of Pacific red clays are allogenic, the most important being aeolian dust. Red clays accumulate very slowly with the highest rates of sedimentation coeval with Pleistocene glacial periods when aeolian dust production was at a maximum (Glasby 1991). Because of their fine grain-size and long term stability serious consideration has been given to using red clays as sites for radioactive waste disposal (Burkett et al. 1991).

3.6.9. Hydrothermal kaolin deposits

Along with residual kaolin deposits formed by weathering hydrothermal kaolin deposits are the other major so-called primary kaolins, i.e. kaolins formed *in situ* (Murray & Keller 1993). Many of these kaolin deposits are also known as 'china clay' and often the terms are used interchangeably, the latter a reference to the first use of this type of clay in China some 3000 years ago (Pickering & Murray 1994). The most common rock types that are altered to kaolin are silicic igneous rocks such as rhyolite and granite and silicic pyroclastic rocks. However, almost any igneous rock type may be altered along with any sedimentary or metamorphic rock, although examples of the last two kinds are rarer and tend to be on a smaller scale. Alteration of granites is typically in the form of funnel shaped zones of alteration, whereas volcanic rocks often show dome like alteration zones. Although the major kaolin mineral is nearly always kaolinite, major deposits of halloysite have also been formed through hydrothermal alteration such as in Utah in the United States (Patterson & Murray 1984) and in New Zealand (Harvey & Murray 1993). Dickite and nacrite are also often reported from hydrothermal kaolins, particularly those associated with fault zones. Alunite, along with some other sulphates, is a common secondary mineral. In Japan, deposits known as Roseki often contain substantial zones with abundant pyrophyllite and diaspore (Sudo & Shimoda 1978). This mineral assemblage indicates a higher temperature intense alteration.

Several major primary kaolins, such as those formed in the Hercynian granites of Southwest England are now thought to be of mixed residual and hydrothermal origin (Bristow 1977, 1993). The deposits occur in funnel or trough like forms in some cases over 1000 m in length and 250 m depth. The kaolin content of the altered granite is typically 10–20%, formed by the alteration of feldspar, mainly the plagioclase feldspar albite, and also, according to Bristow (1993), by the alteration of an earlier formed hydrothermal argillic alteration consisting of smectite and illite. The long-running debate over the origin of these deposits has been controversial and essentially polarized between advocates of hydrothermal and residual origins. As stated by Bristow (1977), however, the uniqueness of these deposits in terms of quantity and quality is more easily understood if they have arisen by a combination of processes, rather than by one single process in isolation. In essence this combination of processes involves an early hydrothermal part which 'softens' up zones of the granite making them susceptible to later supergene alteration. Bristow (1993) recognizes six stages in the formation of the china clay deposits, which are summarized in Figure 3.33. The main phase of kaolinitization is related to the reversal of convective circulation of meteoric fluids through the granite, which likely began in Jurassic times. This was probably augmented by deep weathering processes especially during the early Tertiary when the climate of the area was tropical. The high contents of radiogenic elements present in the granites means that they are unusually high heat producers. This may have been a key factor in the kaolinitization process since it provides a source of energy, which continues to this day to drive the circulation of water in the granites. Kaolin deposits have been worked in all the Hercynian plutons but most are located in the St Austell pluton. According

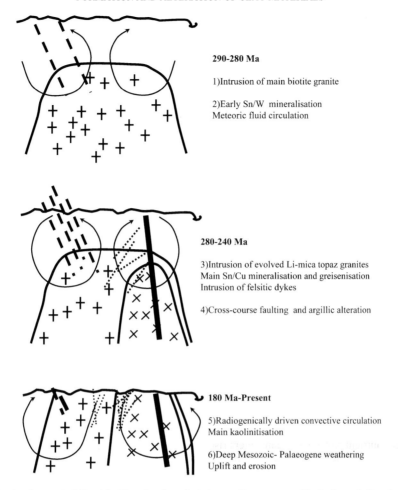

FIG. 3.33. Stages in the development of Cornish china clay deposits in terms of a sequence of hydrothermal alterations and deep weathering. Note reversal of convective fluid flow between Stages 4 and 5. Based on Bristow (1993).

to Bristow (1977) similarly Sn/W mineralized granites in other areas of the World show similar patterns of kaolinitization to the Hercynian granites of Southwest England, suggesting that certain kinds of granite may be prone to kaolinitization.

3.6.10. Other hydrothermal clay deposits

Clay materials formed by hydrothermal alteration may be encountered in many other geological settings, although most commonly they occur as veins or masses in association with igneous rocks or various kinds of metaliferous ore deposits. From an engineering standpoint extensive clay materials along veins or as a stock work through an otherwise competent rock could clearly prove troublesome. Vein and stockwork type alterations have been studied in detail around many ore deposits. One particularly well-studied type is that which occurs in association with so-called porphyry copper deposits. Here various alteration zones are recognized and called propylitic, potassic, phyllic and argillic, the latter, as their names imply being the most clay rich. Examples of more massive hydrothermal clay materials include deposits of halloysite (Patterson & Murray 1984; Harvey & Murray 1993) and hydrothermal bentonites (Grim & Güven 1978). The latter are masses of rock altered to a smectite dominated material by hydrothermal processes, an example being the well known deposit on Ponza Island, Italy (Grim & Güven 1978). Both hydrothermal halloysite deposits and hydrothermal bentonites are usually formed by the alteration of volcanic rocks.

References

AHN, J. H. & PEACOR, D. R. 1987. Transmission electron microscopic study of the diagenesis of kaolinite in Gulf Coast argillaceous sediments. *In*: SHULTZ, L. G., VAN OLPHEN, H. & MUMPTON, F. A. (eds) *Proceedings of the International Clay Conference, Denver 1985*. The Clay Minerals Society, Bloomington, Indiana, 151–157.

ALT, J. C. & TEAGLE, D. A. H. 1998. Probing the TAG hydrothermal mound and stockwork: oxygen-isotope profiles from deep ocean drilling. *In*: HERZIG, P. M., HUMPHRIS, S. A., MILLER, D. J. & ZIERENBERG, R. A. (eds) *Proceedings of the Ocean Drilling Program, Scientific Results*, **158**, 285–295.

ALTANER, S. P. & YLAGAN, R. F. 1997. Comparison of structural models of mixed-layer illite-smectite and reaction mechanisms of smectite illitization. *Clay and Clay Minerals*, **45**, 517–533.

APLIN, A. C. 2000. Mineralogy of modern marine sediments: a geochemical framework. *In*: VAUGHAN, D. J. & WOGELIUS, R. A. (eds) *Environmental Mineralogy*. EMU Notes in Mineralogy, Eötovös University Press, Budapest, 125–172.

APLIN, A. C., FLEET, A. J. & MACQUAKER, J. H. S. 1999. Muds amd mudstones: physical and fluid flow properties. *In*: APLIN, A. C., FLEET, A. C. & MACQUAKER, J. H. S. (eds) *Muds and Mudstones: Physical and fluid flow properties*. Geological Society, London, Special Publications, **158**, 1–8.

ARNOLD, P. W. 1978. Surface-electrolyte interactions. *In*: GREENLAND, D. J. & HAYES, M. H. B. (eds) *The Chemistry of Soil Constituents*. Wiley, New York, 355–404.

AWWILLER, D. N. 1993. Illite-smectite formation and potassium mass transfer during burial diagensis of mudrocks: a study from the Texas Gulf Coast Paleocene-Eocene. *Journal of Sedimentary Petrology*, **63**, 501–512.

BARSHAD, I. 1966. The effect of variation in precipitation on the nature of clay mineral formation in soils from acid and basic igneous rocks. *Proceedings International Clay Conf. (Jerusalem)*, **1**, 167–173.

BELZUNCE-SEGARRA, M. J., WILSON, M. J., FRASER, A. R., LACHOWSKI, E. & DUTHIE, D. M. L. 2002. Clay mineralogy of Galician coastal and oceanic surface sediments: contributions form terrigenous and authigenic sources. *Clay Minerals*, **37**, 23–37.

BENNETT, R. H., O'BRIEN, N. R & HULBERT, M. H. 1991. Determinants of clay and shale microfabric signatures: processes and mechanisms. *In*: BENNETT, R. H., BRYANT, W. R. & HULBERT, M. H. (eds) *Microstructure of Fine Grained Sediments from Mud to Shale*. Springer, New York, 5–32.

BERGER, W. H. 1974. Deep sea sedimentation. *In*: BURK, C. A. & DRAKE, C. L. (eds) *The Geology of the Continental Margins*. Springer, Berlin, 213–241.

BISCAYE, P. E. 1965. Mineralogy and sedimentation of recent deep-sea clay in the Atlantic Ocean and adjacent seas and oceans. *Geological Society of America Bulletin*, **76**, 803–831.

BJORLYKKE, K. 1999. Principal aspects of compaction and fluid flow in mudstones. *In*: APLIN, A. C., FLEET, A. J. & MACQUAKER, J. H. S. (eds) *Muds and Mudstones: Physical and Fluid Flow Properties*. Geological Society, London, Special Publications, **158**, 73–78.

BLATT, H., MIDDLETON, G. & MURRAY, R. 1980. *Origin of Sedimentary Rocks*. Prentice-Hall, Englewood Cliffs, N.J.

BLOCH, J. 1998. Shale diagenesis: a currently muddied view. *In*: SCHIEBER, J., ZIMMERLE, W. & SETHI, P. S. (eds) *Shales and Mudstones II. Petrography, Petrophysics, Geochemistry, and Economic Geology*. E. Schweizerbart'sche Verlagsbuchhandlung, Stuttgart, 95–106.

BOLES, J. R. & FRANKS, S. G. 1979. Clay diagenesis in Wilcox sandstones of south west Texas: implications of smectite diagenesis on sandstone cementation. *Journal Sedimentary Petrology*, **49**, 55–70.

BRADY, N. C. 1974. *The Nature and Properties of Soils*. Macmillian, New York.

BRIDGES, E. M. 1970. *World Soils*. Cambridge University Press, London.

BRISTOW, C. M. 1977. A review of the evidence for the origin of the kaolin deposits in S.W. England. *Proceedings of the 8th International Symposium and Meeting on Alunite*. Madrid-Rome, Sept. 7–16, 1977. 1–19.

BRISTOW, C. M. 1993. The genesis of the China clays of south-west England—A multistage story. *In*: MURRAY, H., BUNDY, W. & HARVEY, C. (eds) *Kaolin Genesis and Utilisation*. Clay Minerals Society Boulder, Colorado, Special Publications, **1**, 171–203.

BURKETT, P. J., BENNETT, R. H. & BRYANT, W. R. 1991. The role of the microstructure of Pacific red clays in radioactive waste disposal. *In*: BENNETT, R. H., BRYANT W. R. & HULBERT, M. H. (eds) *Microstructure of Fine Grained Sediments from Mud to Shale*. Springer, New York, 489–507.

BURST, J. F. 1959. Post diagenetic clay mineral-environmental relationships in the Gulf Coast Eocene in clays and clay minerals. *Clays and Clay Minerals*, **6**, 327–341.

BURST, J. F. 1969. Diagenesis of Gulf Coast clayey sediments and its possible relation to petroleum migration. *AAPG Bulletin*, **53**, 73–93.

CARR, D. D. 1994. *Industrial Minerals and Rocks*. 6th edn. Society for Mining, Metallurgy and Exploration, Inc, Littletin, Colorado.

CATT, J. A. 1988. Loess—its formation, trnasport and economic significance. *In*: LERMAN, A. & MEYBECK, M. *Physical and Chemical Weathering in Geochemical Cycles*. NATO ASI series C Mathematical and Physical Sciences, **251**, 113–142.

CHAMLEY, H. 1989. *Clay Sedimentology*. Springer, Berlin.

CHILINGARIAN, G. V. 1983. Compactional diagenesis *In*: PARKER, A. & SELLWOOD, B. (eds) *Sediment diagenesis*. NATO, ASI series C: Mathematical and physical sciences, **115**. Reidel, Dordrecht, 57–168.

CHIVAS, A. R. & BIRD, M. I. 1995. Palaeoclimate from Gondwanaland clays. *In*: CHURCHMAN, G. J., FITZPATRICK, R. W. & EGGLETON, R. A. *Clays Controlling the Environment*. Proceedings of the 10th International Clay Conference, Adelaide. CSIRO publishing, Melbourne, Australia.

COTTER-HOWELLS, J. D. & PATERSON, E. 2000. Minerals and soil development. *In*: VAUGHAN, D. J. & WOGELIUS, R. A. (eds) *Environmental Mineralogy*. European Mineralogical Union, Notes in Mineralogy, Eötvös University Press, Budapest, 91–124.

COX, A. & HART, R. B. 1986. *Plate Tectonics: How it Works*. Blackwell, Oxford.

CURTIS, C. D. 1977. Sedimentary geochemistry: environments and processes dominated by involvement of an aqueous phase. *Philosophical Transactions of the Royal Society of London*, **286**, 353–372.

DAVIES, D. K., BRYANT, W. R., VESSELL, R. K. & BURKETT, P. J. 1991. Porosities, permeabilities, and microfabric of Devonian shales. *In*: BENNETT, R. H., BRYANT, W. R. & HULBERT, M. H. (eds) *Microstructure of Fine Grained Sediments from Mud to Shale*. Springer, New York, 109–122.

DEWHURST, D. N., YAND, Y. & APLIN, A. C. 1999. Permeability and fluid flow in natural mudstones. *In*: APLIN, A. C., FLEET, A. C. & MACQUAKER, J. H. S. (eds) *Muds and Mudstones: Physical and Fluid Flow Properties*. Geological Society, London, Special Publications, **158**, 23–43.

DREVER, J. I. 1985. *The Chemistry of Weathering*. Advanced research workshop, NATO ASI series. Series C, Mathematical and physical sciences, **149**. Reidel, Dordrecht.

DUCHAUFOUR, P. 1982. *Pedology*. Allen and Unwin, London.

DUFF, P. McL. D. 1993. *Holmes' Principles of Physical Geology*. 4th edn. Chapman and Hall, London.

EBERL, D. D. 1984. Clay mineral formation and transformation in rocks and soils. *Philosophical Transactions of the Royal Society of London*, **311**, 241–257.

EISMA, D. 1986. Flocculation and deflocculation of suspended matter in estuaries. *Netherlands Journal of Sea Research*, **20**, 183–199.

ELZEA, J. M. & MURRAY, H. H. 1990. Variation in the mineralogical, chemical and physical properties of the Cretaceous Clay Spur bentonite in Wyoming and Montana (U.S.A). *Applied Clay Science*, **5**, 229–248.

FREY, M. & ROBINSON, D. 1999. *Low-grade Metamorphism*. Blackwell, Oxford.

GIBBS, R. J. 1967. The geochemistry of the Amazon River system: Part I. The factors which control the salinity and the composition and concentration of the suspended solids. *Geological Society of America Bulletin*, **78**, 1203–1232.

GIBBS, R. J. 1977. Clay mineral segregation in the marine environment. *Journal of Sedimentary Petrology*, **47**, 237–243.

GLASBY, G. P. 1991. Mineralogy, geochemistry, and origin of Pacific redclays: a review. *New Zealand J. of Geology and Geophysics*, **34**, 167–176.

GRIFFIN, J. J., WINDOM, H. & GOLDBERG, E. D. 1968. The distribution of clay minerals in the world ocean. *Deep Sea Research*, **15**, 433–459.

GRIM, R. E. & GUVEN, N. 1978. *Bentonites: Geology, Mineralogy, Properties and Uses*. Developments in Sedimentology **24**. Elsevier, Amsterdam.

GUGGENHEIM, S., BAIN, D. C. *et al.* 2002. Report of the Association Internationale pour l'Etude des Argiles (AIPEA) Nomenclature Committee for 2001: Order, disorder and crystallinity in phyllosilicates and the use of the 'Crystallinity Index'. *Clay Minerals*, **37**, 389–393.

HARVEY, C. C. & MURRAY, H. H. 1993. The geology mineralogy and exploitation of halloysite clays of Northland, New Zealand. *In*: MURRAY, H., BUNDY, W. & HARVEY, C. *Kaolin Genesis and Utilisation*, Special Publications, **1**, The Clay Minerals Society, Boulder, Colorado, 233–248.

HAYNES, J. T., MELSON, W. G., O'HEARN, T., GOGGIN, K. E. & HUBBEL, R. A. 1998. High potassium Mid-Ordovician Shale of the Central Appalachian foredeep: Implications for reconstructing Taconian explosive volcanism. *In*: SCHIEBER, J., ZIMMERLE, W. & SETHI P. S. (eds) *Shales and Mudstones II. Petrography, Petrophysics, Geochemistry and Economic Geology*. E. Schweizerbart'sche Verlagsbuchhandlung, Stuttgart, 129–141.

HAYES, M. O. 1967. Relationship between coastal climate and bottom sediment type on the inner continental shelf. *Marine Geology*, **5**, 111–132.

HEDBERG, H. D. 1936. Gravitational compaction of clays and shales. *American Journal of Science*, **31**, 241–287.

HEIVILIN, F. G. & MURRAY, H. H. 1994. Hormites: palygorskite (attapulgite) and sepiolite. *In*: CARR, D. D. (ed.) *Industrial Minerals and Rocks*. 6th edn. Society for Mining, Metallurgy and Exploration, Inc,. Littleton, Cororado.

HESSE, R. 1986. Diagenesis #11. Early diagenetic pore water/sediment interaction: Modern offshore basins. *Geoscience Canada*, **13**, 165–196.

HIGHLEY, D. E. 1972. Fullers Earth. Mineral Resources Consultative Committee. Institute of Geological Sciences. HMSO, London. Mineral Dosier No. 3.

HIGHLEY, D. E. 1975. Ball Clay. Mineral Resources Consultative Committee. Institute of Geological Sciences. HMSO, London. Mineral Dossier No. 11.

HIGHLEY, D. E. 1982. Fire Clay. Mineral Resources Consultative Committee. Institute of Geological Sciences. HMSO, London. Mineral Dosier No. 24.

HILLIER, S. 1993. Origin, diagenesis, and mineralogy of chlorite minerals in Devonian lacustrine mudrocks, Orcadian Basin, Scotland. *Clays and Clay Minerals*, **41**, 240–259.

HILLIER, S. 1995. Erosion, sedimentation, and sedimentary origin of clays. *In*: VELDE, B. (ed.) *Clays and the Environment*. Springer, Berlin, 162–219.

HILLIER, S. 2003. Chlorite in sediments. *In*: MIDDLETON, G. V., CHURCH, M. J., CONIGLIO, M., HARDIE, L. A. & LONGSTAFFE F. J. *Encyclopedia of Sediments and Sedimentary Rocks*. Kluwer, Dordrecht.

HINCH, H. H. 1980. The nature of shales and the dynamics of hydrocarbon expulsion in the Gulf Coast Tertiary section. *In*: ROBERTS, W. H. III & CORDELL, R. J. (eds) *Problems of Petroleum Migration*. AAPG Studies in Geology, **10**. American Association of Petroleum Geologists, 1–18.

HINDMAN, J. R. 1994. Vermiculite. *In*: CARR, D. D. (ed.) *Industrial Minerals and Rocks*. 6th edn. Society for Mining, Metallurgy and Exploration, Inc,. Littleton, Cororado.

HO, N. C., PEACOR, D. R. & VAN DER PLUIJM, B. A. 1999. Preferred orientation of phyllosilicates in Gulf Coast mudstones and relation to the smectite-illite transition. *Clay and Clay Minerals*, **47**, 495–504.

HOFFMAN, J. & HOWER, J. 1979. Clay mineral assemblages as low grade metmorphic geothermometers: application to the thrust faulted Disturbed Belt of Montana, U.S.A. *In*: SCHOLLE, P. A. & SCHULGER, P. R. (eds) *Aspects of Diagenesis*. Society of Economic Paeontology and Mineralogy Special Publication, **26**.

HOWER, J. 1981. Shale diagenesis. *In*: LONGSTAFFE, F. J. (ed.) *Short Course in Clays and the Resource Geologist*. V7, Mineralogical Association of Canada, 60–77.

HOWER, J., ESLINGER, E. V., HOWER, M. E. & PERRY, E. A. 1976. Mechanism of burial metamorphism of argillaceous sediments: 1. Mineralogical and chemical evidence. *Bulletin of the Geological Society of Amercia*, **87**, 725–737.

HUTCHEON, I., BLOCH, J., DE CARITAT, P., SHEVALIER, M., ABERCROMBIE, H. & LONGSTAFFE, F. 1998. What is the cause of potassium enrichment in shales? *In*: SCHIEBER, J., ZIMMERLE, W. & SETHI, P. S. (eds) *Shales and Mudstones II. Petrography, petrophysics, geochemistry, and Economic Geology*. E. Schweizerbart'sche Verlagsbuchhandlung, Stuttgart, 107–128.

INOUE, A. 1995. Formation of clay minerals in hydrothermal environments. *In*: VELDE, B. (ed.) *Origin and Mineralogy of Clays*. Springer, Berlin.

JACKSON, M. L. 1964. Chemical composition of soils. *In*: BEAR, F. E. (ed.) *Chemistry of the Soil*. Reinhold, New York, 71–141.

JEANS, C. V., MERRIMAN, R. J. & MITCHELL, J. G. 1977. Origin of Middle Jurassic and Lower Cretaceous Fuller's Earths in England. *Clay Minerals*, **12**, 11–44.

JEANS, C. V., WRAY, D. S., MERRIMAN, R. J. & FISHER, M. J. 2000. Volcanogenic clays in the Jurassic and Cretaceous strata of England and the North Sea Basin. *Clay Minerals*, **35**, 25–55.

JENNINGS, S. & THOMPSON, G. R. 1986. Diagenisis of Plio-Pleistocene sediments of the Colorado River Delta, southern California. *Journal of Sedimentary Petrology*, **56**, 89–98.

KASTNER, M. 1981. Authigenic silicates in deep-sea sediments: Formation and diagenesis. *In*: EMILIANI, C. (ed.) *The Sea*. Wiley, New York, 915–980.

KASUBE, T. J. & WILLIAMSON, M. A. 1998. Shale petrophysical characteristics: Permeability history of subsiding shales. *In*: SCHIEBER, J., ZIMMERLE, W. & SETHI, P. S. (eds) *Shales and Mudstones II. Petrography, Petrophysics, Geochemistry, and Economic Geology*. E. Schweizerbart'sche Verlagsbuchhandlung, Stuttgart, 69–91.

KENDALL, T. 1996. Industrial clays. 2nd edn. An *Industrial Minerals Special Review*. Industrial Minerals, London.

KNAUTH, P. L. 1994. Petrogenesis of chert. *In*: HEANEY, P. J., PREWITT, C. T. & GIBBS, G. V. (eds) *Silica Physical Behaviour, Geochemstry, and Materials Applications*. Mineralogical Society of America, Reviews in Mineralogy, **29**, 233–258.

LERMAN, A. & MEYBECK, M. 1988. Physical and chemical weathering in geochemical cycles. NATO ASI series C mathematical and physical sciences, **251**.

LEVENTHAL, J. S. 1998. Metal-rich black shales: Formation, economic geology and environmental considerations. *In*: SCHIEBER, J., ZIMMERLE, W. & SETHI, P. S. (eds) *Shales and Mudstones II. Petrography, Petrophysics, Geochemistry, and Economic Geology*. E. Schweizerbart'sche Verlagsbuchhandlung, Stuttgart, 255–282.

LOVELAND, P. J. 1984. The soil clays of Great Britain: I. England and Wales. *Clay Minerals*, **19**, 681–707.

MCCAVE, I. N. 1972. Transport and escape of fine-grained sediment from shelf areas. *In*: SWIFT, D. J. P., DUANE, D. B. & PILKEY, O. H. (eds) *Shelf Sediment Transport*. Dowden, Hutchinson & Ross, Stroudsboug, 225–248.

MCCAVE, I. N. 1984. Erosion transport and deposition of fine grained marine sediments. *In*: STOW, D. A. V. & PIPER, D. J. W. (eds) *Fine-Grained Sediments Deep Water Processes and Facies*. Geological Society London, Special Publications, **15**, 69.

MARGARA, K. 1986. Porosity depth relationships during compaction in hydrostatic and non–hydrostatic cases. *In*: BURUS J. (ed.) *Thermal Modelling in Sedimentary Basins*. Editions Technip, Paris, 129–148.

MARTIN, J. M. & MEYBECK, M. 1979. Elemental mass balance of material carried by major World Rivers. *Marine Chem.*, **7**, 173–206.

MARUMO, K. & HATTORI, K. H. 1999. Seafloor hydrothermal clay alteration at Jade in the back-arc Okinawa Trough: Mineralogy, geochemistry and isotope characteristics. *Geochemica et Cosmochemica Acta*, **63**, 2785–2804.

MEADE, R. H. 1972. *Transport and Deposition of Sediments in Estuaries*. Geological Society of America, memoirs, **133**, 91–120.

MEADE, R. H. 1988. Movement and storage of sediment in river systems. *In*: LERMAN, A. & MEYBECK, M. (eds) *Physical and Chemical Weathering in Geochemical Cycles*. Nato ASI series C. Mathematical and physical sciences. **251**, Kluwer, Pandraht, 165–180.

MEADE, R. H. & PARKER, R. S. 1985. Sediment in rivers of the United States. *US Geological Survey Water Supply Paper*, **2275**, 49–60.

MELSON, W. G., HAYNES, J. T., O'HEARN, T., HUBBELL, R., GOGGIN, K. E., LOCKE, D. & ROSS, D. 1998. K-shales of the central Appalachian Paleozoic: Properties and origin. *In*: SCHIEBER, J., ZIMMERLE, W. & SETHI, P. S. (eds) *Shales and Mudstones II. Petrography, Petrophysics, Geochemistry, and Economic Geology*. E. Schweizerbart'sche Verlagsbuchhandlung, Stuttgart.

MERRIMAN, R. J. 2002*a*. The magma-to-mud cycle. *Geology Today*, **18**, 67–71.

MERRIMAN, R. J. 2002*b*. Contrasting clay mineral assemblages in British Lower Palaeozoic slate belts: the influence of geotectonic setting. *Clay Mineral Society*, **37**, 207–209.

MERRIMAN, R. J. & FREY, M. 1999. Patterns of very low grade metamorphism in metapelitic rocks. *In*: FREY, M. & ROBINSON, D. (eds) *Low-grade Metamorphism*. Blackwell, Oxford, 161–107.

MERRIMAN, R. J. & PEACOR, D. R. 1999. Very low-grade metapelites; mineralogy, microtextures and measuring reaction progress. *In*: FREY, M. & ROBINSON, D. (eds) *Low-Grade Metamorphism*. Blackwell, Oxford, 10–60.

MEYBECK, M. 1987. Global chemical weathering of surficial rocks estimated from river dissolved loads. *American Journal of Science*, **287**, 401–428.

MILLIKEN, K. L. 1992. Chemical behaviour of detrital feldspars in mudrocks versus sandstones, Frio Formation (Oligocene) South Texas. *Journal of Sedimentary Pettology*, **62**, 790–801.

MILLOT, G. 1970. *The Geology of Clays*. Masson, Paris.

MURRAY, H. H. 1994. Common clay. *In*: CARR, D. D. (ed.) *Industrial Minerals and Rocks*. 6th edn. Society for Mining, Metallurgy and Exploration, Inc,. Littleton, Cororado.

MURRAY, H. H. & KELLER, W. D. 1993. Kaolins, kaolins and kaolins. *In*: MURRAY, H., BUNDY, W. & HARVEY, C. (eds) *Kaolin Genesis and Utilisation*. The Clay Minerals Society, Boulder, Colorado, Special Publications, **1**, 1–24.

NADEAU, P. H. & REYNOLDS, R. C. Jr. 1981. Volcanic components in pelitic sediments. *Nature*, **294**, No 5836, 72–74.

NADEAU, P. H., WILSON, M. J., MCHARDY, W. J. & TIAT, J. M. 1984. Interstratified clays as fundamental particles. *Science*, **225**, 923–935.

NADEAU, P. H., PEACOR, D. R. YAN, J. & HILLIER, S. 2002. I-S precipitaion in pore space as the cause of geopressuring in Mesozoic mudstones, Egersund Basin, Norwegian continetal shelf. *America Mineralogist*, **87**, 1580–1589.

NEUZIL, C. E. 1994. How permeable are clays and shales? *Water Reservoir Research*, **30**, 145–150.

O'BRIEN, N. R. & SLATT, R. M. 1990. *Argillaceous Rock Atlas*. Springer, New York.

ODIN, G. S. & MATTER, A. 1981. De glauconarium origine. *Sedimentology*, **28**, 611–641.

ORTON, G. J. & READING, H. G. 1993. Variability of deltaic processes in terms of sediment supply, with particular emphasis on grain size. *Sedimentology*, **40**, 475–512.

PATTERSON, S. H. & MURRAY, H. H. 1984. Kaolin, Refractory Clay, Ball Clay and Halloysite in North America, Hawaii, and the Caribbean. *USGS Prof. Paper*. 1306.

PEDRO, G. 1982. The conditions of formation of secondary constituents. *In*: BONNEAU, M. & SOUCHIER, B. (eds) *Constituents and Properties of Soils*. Academic Press, London, 63–81.

PERRY, E. A. Jr. & HOWER, J. 1970. Burial diagenesis of Gulf Coast pelitic sediments. *Clays and Clay Minerals*, **18**, 165–177.

PETTIJOHN, E. J. 1975. *Sedimentary Rocks*. 3rd edn. Harper and Row, New York.

PEWE, T. L. 1981. Desert dust: an overview. *Geological Society of America, Special Paper*, **186**, 1–10.

PICARD, M. D. 1971. Classification of fine-grained sedimentary rocks. *Journal of Sedimentary Petrology*, **41**, 179–195.

PICKERING, S. M. Jr. & MURRAY, H. H. 1994. Kaolin: *In*: CARR, D. D. (ed.) *Industrial Minerals and Rocks*. 6th edn. Society for Mining, Metallurgy and Exploration, Inc,. Littleton, Cororado.

POTTER, P. E. 1998. Shale-rich basins: controls and origin. *In*: SCHIEBER, J., ZIMMERLE, W. & SETHI, P. S. (eds) *Shales and Mudstones 1. Basin Studies, Sedimentology and Palaeontology*. E. Schweizerbart'sche Verlagsbuchhandlung, Stuttgart, 21–32.

POTTER, P. E. MAYNARD J. B. & PRYOR, W. A. 1980. *The Sedimentology of Shale*. Springer, New York.

PRIMMER, T. J. & SHAW, H. F. 1987. Diagenesis in shales: Evidence form backscattered electron microscopy and electron microprobe analyses. *In*: SHULTZ, L. G., VAN OLPHEN, H. & MUMPTON, F. A. (eds) *Proceedings of the International Clay Conference, Denver 1985*. The Clay Minerals Society, Bloomington, Indiana, 135–143.

PROSPERO, J. M. 1981. Eolian transport to the world ocean. *In*: EMILIANI, C. (ed.) *The Sea*. V7, 801–874.

PYE, K. 1987. *Aeolian dust and dust deposits*. Academic, London.

RAE, D. K. 1994. The paleoclimatic record provided by eolian deposition in the deep sea: the geologic history of wind. *Reviews in Geophysics*, **32**, 159–195.

RATEEV, M. A., GORBUNOVA, Z. N., LISITZYN, A. P. & NOSOV, G. L. 1969. The distribution of clay minerals in the oceans. *Sedimentology*, **13**, 21–43.

RIDGWAY, J. M. 1982. *Common Clay and Shale*. Mineral Resources Consultative Committee. HMSO. London. Mineral Dossier 22.

RIGHI, D. & MEUNIER, A. 1995. Origin of clays by rock weathering and soil formation. *In*: VELDE, B. (ed.) *Clays and the Environment*. Springer, Berlin, 43–157.

RUIZ CRUZ, M. D. & REYES, E. 1998. Kaolinite and dickite formation during shale diagenesis: isotopic data. *Applied Geochemistry*, **13**, 95.

SHAW, D. B. & WEAVER, C. E. 1965. The mineralogical composition of shales. *Journal of Sedimentary Petrology*, **35**, 213–222.

SHAW, H. F. 1981. Mineralogy and petrology of the argillaceous sedimentary rocks of the U.K. *Quarterly Journal of Engineering Geology, London*, **14**, 277–290.

SHUTOV, V. D., DRITS, V. A. & SAKHAROV, B. A. 1969. On the mechanism of a post sedimentary transformation of montmorillonite into hydromica. *Proceedings International Clay Conference, Tokyo, 1969*, **1**, 523–532.

SPEARS, D. A. 1980. Towards a classification of shales. *Journal of Geological Society, London*, **137**, 125–129.

SPEARS, D. A. & O'BRIEN, P. J. 1994. Origin of British Tonsteins and tectonomagmatic implications. *Zbl. Geol Paläont. Teil* I, **5/6**, 491–497.

SPEARS, D. A. KANARIS-SOTIRIOU, R., RILEY, N. & KRAUSE, P. 1999. Namurian bentonites in the Pennine Basin, UK—origin and magmatic affinities. *Sedimentology*, **46**, 385–401.

ŚRODOŃ, J. 1999. Nature of mixed-layer clays and mechanisms of their formation and alteration. *Annual Review Of Earth And Planetary Sciences*, **27**, 19–53.

ŚRODOŃ, J. & EBERL, D. D. 1984. Illite. *In*: BAILEY, S. W. (ed.) *Micas*. Reviews in Mineralogy 13, Mineralogical Society of America, Washington, DC, 495–544.

STOW D. A. V. 1994. Deep sea sediment transport. *In*: PYE, K. *Sediment Transport and Depositional Processes*. Blackwell, Oxford, 257–291.

SUDO, T. & SHIMODA, S. 1978. *Clays and Clay Minerals of Japan*. Developments in Sedimentology **26**, Elsevier, Amsterdam.

TARDY, Y., BOCQUIER, G., PAQUET, H. & MILLOT, G. 1973. Formation of clay from granite and its distribution in relation to climate and topography. *Geoderma*, **10**, 271–284.

THIR, Y. M. 2000. Palaeoclimatic interpretation of clay minerals in marine deposits: an outolook form the continental origin. *Earth Science Reviews*, **49**, 201–221.

TOTTEN, M. W. & BLATT, H. 1993. Alterations in the non-clay-mineral fraction of pelitic rocks across the diagenetic to low-grade metamorphic transition, Ouachita mountains, Oklahoma and Arkansas. *Journal of Sedimentary Petrology*, **63**, 899–908.

TSOAR, H. & PYE, K. 1987. Dust transport and the question of desert loess formation. *Sedimentology*, **34**, 139–153.

UTADA, M. 1980. Hydrothermal alteration related to igneous activity in Cretaceous and Neogene formations of Japan. *Mining Geology of Japan, Special Issue*. **8**, 67–83.

VINCENT, A. 1983. The origin and occurrence of Devon ball clays. *In*: WILSON, R. C. L. (ed.) *Residual Deposits: Surface Related Weathering Processes and Materials*. Geological Society, London, Special Publications **11**, 39–46.

WATSON, I. 1982. Ball and plastic clays –shaping up to the ceramic doldrums. *Industrial Minerals*, August 1982, 23–45.

WEAVER, C. E. 1967. The significance of clay minerals in sediments. *In*: NAGY, B. & COLOMBO, U. (eds) *Fundamental Aspects of Petroleum Geochemistry*. Elsevier, Amsterdam, 37–76.

WEAVER, C. E. 1976. The nature of TiO_2 in kaolinite. *Clays and Clay Minerals*, **24**, 215–218

WEAVER, C. E. 1989. *Clays Muds and Shales*. Developments in sedimentology, **44**. Elsevier, Amsterdam.

WEAVER, C. E. & BECK, K. C. 1977. Miocene of the S. E. United States: a model for chemical sedimentation in a peri-marine environment *Sedimentary Geology*, **17**.

WEAVER, C. E. & POLLARD, L. D. 1973. *The Chemistry of Clay Minerals*. Developments in Sedimentology **15**. Elsevier, Amsterdam.

WILSON, M. J. 1987. Soil smectites and related interstratified minerals: recent developments. *In*: SHULTZ, L.G., VAN OLPHEN, H. & MUMPTON, F. A. (eds) *Proceedings of the International Clay Conference, Denver 1985*. The Clay Minerals Society, Bloomington, Indiana 167–173.

WILSON, M. J. 1999. The origin and formation of clay minerals in soils: past, present and future perspectives. *Clay Minerals*, **34**, 7–25.

WILSON, M. J. & NADEAU, P. H. 1985. Interstratified clay minerals and weathering processes. *In*: DREVER, J. I. (ed.) *The Chemistry of Weathering*. NATO ASI series. Series C, Mathematical and physical sciences, **149**, Reidel, Dordrecht, 97–118.

WILSON, M. J., BAIN, D. C. & DUTHIE, D. M. L. 1984. The soil clays of Great Britain: II. Scotland. *Clay Minerals*, **19**, 709–735.

WINDOM, H. L. 1975. Eolian contribution to marine sediments. *Journal Sedimentary Petrology*, **45**, 520–529.

WINDOM, H. L. 1976. Lithogenous material in marine sediments. *In*: RILEY, J. P. & CHESTER, R. (eds) *Chemical Oceanography*, V5, Academic, New York, 103–135.

WHITE, A. F. & BRANTLEY, S. L. 1995. Chemical weathering rates of silicate minerals. Reviews in Mineralogy, **31**, Mineralogical Society of America.

YAALON, D. H. 1962. Mineral composition of the average shale. *Clay Minerology Bulltin*, **5**, 31–36.

ZIMMERLE, W. 1998. The petrography of the Boom Clay from the Rupelian type locality. Northern Belgium. *In*: SCHIEBER, J., ZIMMERLE, W. & SETHI, P. S. (eds) *Shales and Mudstones II. Petrography, Petrophysics, Geochemistry, and Economic Geology* E. Schweizerbart'sche Verlagsbuchhandlung, Stuttgart, 13–33.

4. Properties of clay materials, soils and mudrocks

The purpose of this chapter is to provide a background to the properties of natural clay materials used in the construction industry and other applications (see Table 4.1). There is a frequent need to evaluate clay materials for civil engineering projects, which has led to a bias in this chapter towards a consideration of their geotechnical properties. The intention is to present this information, and also some data applicable to other applications, in a manner suitable for scientists and engineers with an interest in clay materials. Both fundamental material behaviour and derived parameters are described.

This chapter considers the engineering behaviour of rock and soil materials that consist largely of clay mineral grains, together with minor amounts of other minerals. As discussed in Chapter 1, the term clay may mean a material made of clay-sized grains (smaller than 2 μm or 0.002 mm) or of grains consisting of clay minerals (see Chapter 2). Grains smaller than 2 μm may be clay minerals or other materials such as finely ground quartz or rock flour. Clay mineral grains may be larger than 2 μm and they are often bound into silt-sized (0.002–0.06 mm) aggregates. The Chapter concentrates on the properties of the material relevant to the exploitation and uses of excavated or extracted clays and mudrocks, rather than those of soil and rock masses in the ground, which are outside the scope of this report.

The behaviour of clays depends on their particle size distribution, mineralogy and moisture content. In terms of moisture content, it is convenient to identify three distinct ranges of liquid and plastic limits, as defined in Section 4.2.3:

- high water content (greater than liquid limit): slurry or liquid;
- water content between liquid and plastic limits: plastic solid (the usual state of engineering soils);
- low water content (less than plastic limit): non-plastic solid or solid.

These states are shown in Figure 4.4.

The principal feature of slurries is viscous flow. The behaviour of engineering soils is described by the theories of soil mechanics in which the material undergoes mainly plastic deformation. The behaviour of mudrocks, where the word 'mud' implies a mixture of clay and silt sized grains (<0.06 mm), is described by the theories of rock mechanics.

Due to the very small size of clay mineral grains and their properties, clay materials display certain basic features:

- clay minerals, in common with other constituents, will form colloidal suspensions in water and will flow;
- clay rich materials deform plastically at certain moisture contents such that a deformed shape can be achieved without rupturing and it is sustained once the stress causing the deformation is removed;
- clay soils tend to have very low permeability and they can sustain very large negative pore water pressures (suctions);
- certain varieties of clay mineral grains can expand and contract with changes in moisture content;
- increases in moisture content cause swelling and reductions in moisture content result in shrinkage and these changes are greater than for other soils;
- mudrocks tend to undergo relatively rapid breakdown when subjected to degradation or weathering processes.

TABLE 4.1. *Applications and relevant properties of clays*

Application	Properties
Earthworks, fills	Compaction, moisture condition value, plasticity, consolidation, in situ and remoulded strength, suction, permeability
Earth moving	Shear strength parameters, trafficability, bulking, durability
Ceramics	Plasticity, modules of rupture, rheology, shrinkage, colour, vitrification properties, thermal properties
Fillers	Colour, rheology, colloidal properties
Low permeability barriers, drilling muds	Colloidal properties, permeability, compaction properties, viscosity, thermal properties
Absorbents	Colloidal properties, cation exchange capacity
Agronomics	Bulk density, pore size
All	Description, density, porosity, particle size distribution, plasticity, activity

Definitions — clay, soils and rocks

The first part of Section 4.1 discusses compositional features of clay materials that affect their properties. Clay materials may occur in the form of soil or rock. Soil is used here to denote a particulate substance in which the particles consist of solid organic and inorganic particles that are either unbonded or only weakly bonded together. The pore space may or may not contain fluids and dissolved matter. In contrast to soils, rocks are usually defined as indurated (hard) materials consisting of solid organic or inorganic mineral grains that are firmly attached together by cementation, ionic forces or interlocking action.

As Table 4.1 shows, clays are used in many different situations and a wide range of parameters are used to characterize their behaviour. Some of these are restricted to one area of application. There is scope for misunderstandings about the properties of clays as some parameters are defined differently in different disciplines. For instance, moisture content may not be defined in the same way in geotechnical and ceramics applications.

There are many sources of information on the properties of clays. The behaviour of soils in engineering is covered by various books on soil mechanics, including Atkinson (1993), Berry & Reid (1988), Powrie (2004), Smith & Smith (1998), Scott (1994), Barnes (2000) and others. Clay behaviour is also considered in books that deal with ceramics, such as Worrall (1982), Dinsdale (1996) and Ryan & Radford (1987), and rheology, including Barnes et al. (1991) and Suklje (1969). In addition, texts on soil science, including Hillel (1980) and Marshall & Holmes (1979) provide relevant information. Gillott (1987) describes the behaviour of clay in engineering geology and Bell (2000) provides a reference source and discussion covering the engineering properties of soils and rocks, including clay-rich ones.

The chapter begins with a discussion of the composition of clay materials. The classification and description of clay materials are then explained, followed by a consideration of the parameters used in engineering and ceramics situations. The properties of clays and mudrocks are then explained, followed by a description of the general behaviour of different types of clay deposits. Appendix B of the report provides a compilation of the values of properties for a selection of UK clay deposits. These values are discussed with reference to composition and geological origins. It should be appreciated that the data, which are for guidance only, should **not** be used as a substitute for specific investigations.

4.1. Composition, classification and description of clay materials

In practice there is a graduation from stronger soils to weaker rocks and for engineering applications the distinction between soil and rock is usually defined arbitrarily in terms of strength, where a weak rock has an undrained shear strength in the range 300–625 kPa. (This is equivalent to a uniaxial compressive strength between 0.6–1.25 MPa, see Sections 4.3.7 and 4.3.10.)

4.1.1. Mineralogy

The mineralogy of clay materials is discussed in Chapter 2. As Figure 2.1 shows, the major components of clay materials are clay minerals, with mica and quartz. Features of commonly occurring clay minerals, are presented in Table 4.2. In comparison with other rock and soil forming minerals, clay minerals possess relatively large surface area and, as explained in Chapter 2, due to ionic substitutions, the surfaces of individual clay particles are usually negatively charged. This leads to cations becoming sorbed onto the surface and water molecules are also attracted to the mineral surface. The edges of the particles may be positively or negatively charged. In addition to electrostatic charges between clay particles, due to their small size, they also are also attracted to each other by Van der Waal forces.

TABLE 4.2. *Approximate values of properties for certain clay minerals (from Hillel 1980)*

Property	Clay Mineral				
	Kaolinite	Illite	Montmorillonite	Chlorite	Allophane
Planar diameter (μm)	0.1–4	0.1–2	0.0–2	0.1–2	
Basal layer thickness (Å)	7.2	10	10	14	
Particle thickness (Å)	500	50–300	10–100	100–1000	
Specific surface (m^2/g)	5–20	80–120	700–800	80	
Cation exchange capacity (meq/100g)	3–15	15–40	80–100	20–40	40–70
Area per charge ($Å^2$)	25	50	100	50	120

When clay particles are suspended in water, the surface charges result in the formation of a strongly bonded layer of water molecules together with other polar molecules, if present, attached to the surface. On account of the polar nature of water molecules, further water molecules are attracted and the surface cations may diffuse away from the surface. This leads to the formation of a double diffuse layer of water surrounding the clay particle (Van Olfen 1963). The thickness of this layer depends on the size, valence and concentration of ions in the water. A high concentration of high valence cations will tend to give rise to a relatively thin diffuse layer and the clay particles are attracted together into flocs. This contrasts with the more dispersed structure that occurs when the pore fluid has low cation concentration and/or contains low-valence cations and the particles repel each other. Further details of these phenomena are explained in Chapter 2.

The higher cation concentration of seawater compared with fluvial environments results in the flocculation of clay suspended in the water. This results in the deposition of clay flocs on entry to the sea. Worrall (1982) explains that deflocculation of a clay with adsorbed Ca can be achieved by replacing this cation with Na. In practice, it is also necessary for the Ca to be removed from the system.

In geotechnical situations, dispersal with deflocculating agents is used to separate clay particles for particle size analysis determinations, whereas flocculants are added to slurries to cause them to settle out of suspension. In ceramics applications dispersal agents and defloccul-ants may be used to provide pastes with the correct properties for casting and moulding.

Ions adsorbed onto the surfaces and edges of clay minerals may be exchanged for other ions. In addition, in the case of swelling clays, interlayer cations may be exchanged, as well as the surface cations. This gives rise to the higher cation exchange capacity of montmorillonite recorded in Table 4.2.

The presence of water attached to the surfaces of clay particles also gives rise to the plasticity properties displayed by clays and, as this water does not behave in an entirely liquid fashion, it is the reason that soils with a similar grain size but consisting of quartz or feldspar, do not display plasticity. As the amount of water increases, so there is more free water between the clay particles and the material acts in a more liquid manner. The addition or removal of water from the pore space also changes the ionic concentration and, because of changes to the electrostatic forces between particles, the properties are changed.

Swelling and shrinkage in response to changes in moisture content is an important feature of clays. Other soils are subject to this behaviour, but in clays the magnitude is greater. The amount of volume change may be affected by the loading conditions. If swelling is restricted by an applied load, then a swelling pressure will be generated. On the other hand, shrinkage generates tensile forces in the material that can result in it becoming cracked.

Grim (1962) explains that this volume change occurs by two processes: inter- and intra-crystalline swelling.

Inter-crystalline swelling occurs in all types of fine-grained materials but intra-crystalline swelling is a characteristic of smectite and some other varieties of clay minerals (see Chapter 2). Such clays are called swelling clays. Expansion arises when water enters interlayer sites within the clay crystal. The amount of swelling and the swelling pressure generated by the process are functions of the composition of the pore water. Monovalent cations, for example Na, in low concentrations tend to increase swelling, whereas a higher concentration of a divalent cation such as Ca will have the opposite effect. This forms the basis for the use of lime in the improvement in the geotechnical behaviour of clay soils.

4.1.2. Classification of clays

Clay materials may occur as slurries, soils or rocks. It is important to appreciate that these terms are not well defined and have diverse uses within different disciplines. Soils are of interest in three basic situations (Fang 1997):

- in engineering uses soil is any earthy material that may be excavated using simple hand tools;
- from a geological standpoint soil includes superficial, unconsolidated, disintegrated and decomposed rock material and also unconsolidated Pleistocene and Recent deposits;
- pedologists regard soils as the generally unconsolidated products of weathering and biological processes on surface materials.

Clays are also important raw materials in the manufacture of ceramics, cement and other products.

The classification of clays is of relevance in agricultural, engineering and mineral exploitation applications. It needs to be recognised that classifications have been devised by workers in each of these disciplines to suit their individual requirements. Clays are of importance in all applications of soils since, even if only minor constituents, they can significantly modify the physical and chemical properties of the material. For instance, in agricultural soils, clays impart soils with the capacity to retain moisture and nutrients. However, too much clay results in soils that are difficult to work, poorly drained and lacking good aeration. The classification of mudrocks is described in Section 4.1.5.

4.1.2.1. Geological classification. Gillott (1987) points out that in geological classifications the composition and texture of the soil are related to origin, particularly whether the material has formed *in situ* as a residual deposit or has resulted from the deposition of transported materials. As is explained later (Section 4.4.5), this seemingly esoteric distinction is important, as the properties of these two classes of soils are different.

Classification is based upon the average particle size. Unfortunately, the size ranges vary between different schemes and standards, as is shown by the examples in Table 4.3.

TABLE 4.3. *Different size classifications for clay, silt and sand*

Classification scheme	Maximum particle size (mm)		
	Clay	Silt	Sand
Wentworth (1922)	0.004	0.063	2.0
BS 5930:1999	0.002	0.060	2.0
ASTM D422-63 (2003)	0.005	0.074	4.75

Using the grain size categories specified, Wentworth (1922) proposed the following classification:

sandy silt	silt > sand > 10%; others < 10%
silt	silt > 80%
clayey silt	silt > clay > 10%; others < 10%
silty clay	clay > silt > 10%; others < 10%
clay	clay > 80%

Classification follows from considering the percentages of sand, silt and clay present as shown in Figure 4.1. As explained below, BS 5930:1999 recommends that for engineering applications clay and silt are distinguished from each other on the basis of hand or laboratory plasticity tests and descriptions such as silty clay and clayey silt based on particle size determinations should not be used.

In the UK, it has been common practice in the geological literature to separate materials into *solid* and *drift* deposits. The terms 'bedrock' and 'superficial deposits' are now preferred (see Chapter 6). Bedrock geological formations are the more ancient (pre-Quaternary) rock-like materials that form the solid crust of the Earth. On the other hand, superficial deposits are mainly unconsolidated Pleistocene and Recent materials that occur at or near the ground surface and which form a covering, absent in places, to the bedrock formations. Such materials may contain rocks. The use of the term *drift* in the UK is a reflection of the fact that much of the material involved is glacially derived. Relatively recently formed unconsolidated materials and superficial deposits at or near the land surface may also be referred to as regolith, alluvium, talluvium, colluvium, head and residual soils. It should be appreciated that from an engineering point of view some ancient bedrock geological formations behave as soils and conversely some superficial deposits are rock-like. Some formations, for example London Clay, contain both soil- and rock-like materials and are sometimes regarded by engineers to be 'engineering soils'.

4.1.2.2. Engineering classification. In engineering applications the main interest centres on the mechanical properties, such as strength and compressibility. Constituents, including the amount and type of clay minerals present exert a strong control on these properties. In engineering classifications both particle size and, in the case of soils, plasticity properties are considered.

Various soil classification systems are used around the world. Of those available, the UK and USA systems for classifying soils for engineering purposes are both widely used and they are briefly described below. The part relevant to fine-grained soils from BS 5930:1999, is summarized in Table 4.4, and the tests shown in Table 4.5 are used to assist with distinguishing between silt and clay. In fact, the Standard stipulates that it is these tests and laboratory determinations of plasticity (see Section 4.2.3) that should be used to distinguish between silt and clay, rather than an assessment of the particle size.

As shown in Table 4.4, fine-grained soils are defined as soils consisting of more than 35% particles smaller in size than 0.06 mm. Naturally, most clays belong to the latter

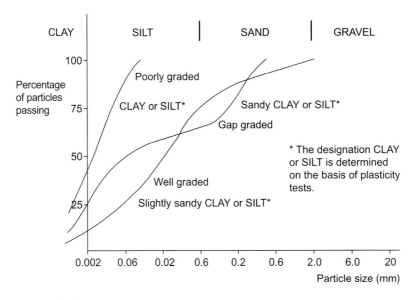

FIG. 4.1. Typical grading curves for clays.

TABLE 4.4. *Classification and description of clays (from BS 5930:1999, p. 114)*

Soil group	Density/compactness/strength		Scale of spacing of discontinuities			Bedding			Colour	Composite soil types (mixtures of basic soil types)			Particle shape	Particle size (mm)	PRINCIPAL SOIL TYPE
	Term	Field test	Term	Mean spacing (mm)		Term	Mean thickness (mm)			Term	Approx % secondary				
Fine soils (over about 35% silt and clay sizes)	Uncompact	Easily moulded or crushed in the fingers	Very widely	Over 2000		Very thickly bedded	Over 2000		Red Orange Yellow Brown	Slightly (sandy &/or gravelly)	<35		Angular Sub-angular Sub-rounded Rounded	0.06 Coarse	
	Compact	Can be moulded or crushed by strong pressure with the fingers	Widely Medium	2000–600 600–200		Thickly bedded Medium bedded	2000–600 600–200		Green Blue White Cream, Grey, Black etc	(sandy &/or gravelly)	35–65†		Flat Tabular Elongated	0.02 Medium	SILT
	Very soft (0–20)+	Finger easily pushed in up to 25 mm	Closely Very closely	200–60 60–20		Thinly Very thinly	200–60 60–20			Very (gravelly or sandy)	>65†		Minor constituent type	0.006 Fine	CLAY /SILT
	Soft (20–40)	Finger pushed in up to 10 mm	Extremely closely	Under 20		Thickly laminated	20–6						Calcareous, shelly, glauconitic, micaceous, etc using terms such as: slightly calcareous calcareous very calcareous	0.002	CLAY
	Firm (40–75)	Thumb makes an impression easily	Fissured	Breaks into blocks along unpolished discontinuities		Thinly laminated	Under 6		Light Dark Mottled						
	Stiff (75–150)	Can be indented slightly by the thumb	Sheared	Breaks into blocks along polished discontinuities		Inter-bedded	Alternating layers of different types, pre-qualified by thickness term if in equal proportions.								
	Very stiff (150–300)	Can be indented by thumb nail													
	Hard (or very weak mudstone) c_u > 300 kPa	Can be scratched by thumb nail (see Table 4.9)	Spacing terms also used for distance between partings, isolated beds, or laminae, desiccation cracks, rootlets etc			Inter-laminated	Otherwise thickness of and spacing between sub-ordinate layers defined						% derived on a site or material specific basis or subjective		

+ Value of undrained shear stength.
† Or describe as coarse soil depending on mass behaviour.

TABLE 4.5. *Simple tests to distinguish clay soils from silts (after BS 5930:1999)*

Visual identification
Silt: Only coarse silt particles are visible with a hand lens, exhibits little plasticity and marked dilatency, slightly granular or silky to touch, disintegrates in water, lumps dry quickly, possesses cohesion but can be powdered easily between the fingers.
Silt/clay: Intermediate in behaviour between clay and silt, slightly dilatent.
Clay: Dry lumps can be broken but not powdered between the fingers, disintegrates slowly in water, smooth to touch, exhibits plasticity but not dilatency, possesses cohesion, sticks to the fingers and dries slowly, shrinks appreciably on drying and usually shows cracks.

Dilatency
A volume of about 1 cm^3 of the soil from which particles larger than 0.4 mm removed, is remoulded with water until the soil is soft but not sticky. Shake this horizontally in the hand. Silt (or fine sand) displays dilatency, in which case water will appear at the surface of the soil and disappear when the soil surface is depressed and the soil stiffens and crumbles. Clay does not display dilatency.

Dry Strength
A volume of the soil from which particles larger than 0.4 mm have been removed, is remoulded with water until it has the consistency of putty. It is then dried. Clays will remain strong and more plastic clays are stronger, whereas silts and fine sands possess only low strength.

Toughness
A volume of about 1 cm^3 of the soil from which particles larger than 0.4 mm removed, is remoulded with water until it has the consistency of putty. The soil is then rolled into a thread about 3mm diameter. The thread is then folded and re-rolled repeatedly. The reduction in the moisture content causes the soil to stiffen and finally lose its plasticity at the plastic limit, when it crumbles. The separate pieces are then reformed into a lump, which is then gently kneaded until it crumbles. The tougher the thread and the stiffer the lump the greater is the effect of clay colloids. Inorganic low plasticity clays, organic and kaolinite-rich clays have low toughness.

Wet/dry behaviour
A dry silt can be disaggregated by gentle rubbing or dusting, whereas clays require washing to remove particles.

category. In some literature and older British Standards, coarse-grained soils are referred to as granular or frictional materials and fine-grained ones as cohesive materials. Although electro-static and other forces that exist between clay particles impart cohesion to clay soils, the effect is usually small in comparison with frictional forces. Accordingly, in geotechnical engineering, the term 'fine-grained' rather than 'cohesive' is now preferred to denote clay soils. However, because the difference between cohesive and granular behaviour is easy to recognize in the field, this distinction is used as the basis of the field classification and description of engineering soils (West 1991).

The part of the Unified Soil Classification System used in the USA, relevant to clay soils is displayed in Table 4.6. In this a two-letter code that depends on the plasticity is used to distinguish soils of different types.

4.1.2.3. Pedological classification. In pedological classifications of soils, the soil profile overlying bedrock is divided into a series of horizons depending upon colour, texture, structure and composition. In the United States, a Soil Taxonomy system is used to classify soils (Anon 1975). Brady (1990) explains that the soil is divided into a series of horizons, named as follows working down from the ground surface:

O Organic
A Top-most mineral horizon with humus
E Zone of maximum eluviation (or leaching)
B Zone containing accumulations of materials from the overlying horizons
C Unconsolidated horizon which may or may not be the same as the parent material of the overlying layers
R Underlying consolidated rock

Not all these horizons may be present and some profiles may contain additional transitional horizons.

The US Comprehensive Soil Classification System divides soils into six main categories, of which the order is one, where the order is determined by the main soil forming process. There are 11 orders. Further subdivision is achieved by consideration of common genetic factors, type and arrangement of horizons present, wetness, effects of animal activity, particle size and profile characteristics.

4.1.3. Particle size and grading of clays

The grading, or particle size distribution, of clay materials exerts a crucial influence over mechanical and chemical behaviour and it is an important aspect of their identification and description.

Particle size analysis of soils is carried out by sieving, sedimentation and other methods (see Chapter 9). Care must be exercised over the preparation of samples, as some clay materials are difficult to disaggregate. This is a particular problem with mudrocks, in which mechanical methods of disaggregation can result in the breakage of grains. On the other hand, aggressive chemical treatments can result in the removal of some minerals in solution.

Soils may display different grading types and the terms used by *engineers* are as follows:

- Well graded: a range of particle sizes without a bias towards presence or absence of a particular size;
- Poorly graded: a distribution with a predominance of one size;
- Gap graded: a distribution with an absence of or bias towards a particular size.

TABLE 4.6. *Unified Soil Classification System — fine-grained soils. More than half of material is smaller than No 200 sieve size (from Bell 2000)*

Identification procedures on fraction smaller than No 40 sieve size				Group symbols	Typical names	Information required for describing soils
Dry strength (crushing characteristics)	Dilatancy (reaction to shaking)	Toughness (consistency near plastic limit)				
Silts and clays: Liquid limit less than 50						
None to slight	Quick to slow	None		ML	Inorganic silts and very fine sands, rock flour, silty or clayey fine sands with slight plasticity	Give typical name: indicate degree and character of plasticity, amount and maximum size of coarse grains; colour in wet condition, odour if any, local or geologic name, and other pertinent descriptive information, and symbol in parenthesis.
Medium to high	None to very slow	Medium		CL	Inorganic clays of low to medium plasticity, gravelly clays, sandy clays, silty clays, lean clays	For undisturbed soils add information on structure, stratification, consistency in undisturbed and remoulded states, moisture and drainage conditions
Slight to medium	Slow	Slight		OL	Organic silts and organic silt-clays of low plasticity	
Silts and clays: Liquid limit greater than 50						
Slight to medium	Slow to none	Slight to medium		MH	Inorganic silts and micaceous or diatomaceous fine sandy or silty soils, elastic silts	Example *Clayey silt*, brown: slightly plastic: small percentage of fine sand, numerous vertical root holes: firm and dry in places; loess: (ML)
High to very high	None	High		CH	Inorganic clays of high plasticity, fat clays	
Medium to high	None to very slow	Slight to medium		OH	Organic clays of medium to high plasticity	
Highly organic soils						
Ready identified by colour, odour, spongy feel and frequent fibrous texture				Pt	Peat and other highly organic soils	

Use grain size curve in identifying the functions as given under field indentification

Plasticity chart for laboratory determination of fine-grained soils

These terms are in distinct contrast to those adopted in *geological* and *sedimentological* usages in which a well sorted or well graded sediment has a predominance of one size and a poorly sorted or poorly graded sediment has a wide range of sizes.

Grading curves for a range of typical clays are shown in Figure 4.1 as accumulative plots. The coefficient of uniformity U, and the coefficient of curvature, C, respectively defined in Equations 4.1.and 4.2, can be used to quantify the grading:

$$U = \frac{D_{60}}{D_{10}} \quad (4.1)$$

and

$$C = \frac{D_{30}}{D_{60}D_{10}} \quad (4.2)$$

where D_{10} is the size of particles at which 10% of the particles are smaller than this size. Similarly, D_{30} and D_{60} are the particle sizes at which 30% and 60% of particles are smaller than these sizes.

A well-graded soil has a uniformity of 10 or more, or a coefficient curvature between 1 and 3. A uniformly graded soil has a uniformity of less than 2.

In agricultural terminology soils are described as *monodisperse* if they contain only one size whereas *polydisperse* contain a variety of sizes.

4.1.4. Description of clays

Full and accurate descriptions are essential for effective communication of the type and properties of the material concerned. Assessment of the performance of a clay material usually depends on determinations of the behaviour of small samples. Description enables data to be extrapolated to a larger volume of material. In addition, it provides the means for taking into account any effects of variations in composition and structure on the mass properties. The importance of accurate description of clays cannot be over-emphasized.

As with classification, many systems are described in national standards and textbooks. The scheme for the description of soils for engineering purposes provided by BS 5930:1999, and the part relevant to clays is summarized in Table 4.4. This and other description schemes use a series of specific terms that are presented in a particular order, to build up a full description, as follows:

- Density/compactness/strength (Different terms are used depending on grain size). For fine gained soils the terms very soft, soft, firm, stiff, very stiff and hard are used. These terms are defined below.
- Discontinuities. The mean spacing and types of discontinuities present. The terms fissured or sheared and spacings of extremely closely, very closely, closely, medium, widely and very widely are defined for discontinuities or other features.
- Bedding. The spacing of bedding planes. The terms inter-laminated or inter-bedded and thinly laminated, thickly laminated, very thinly bedded, thinly bedded, medium bedded, thickly bedded, very thickly bedded are used.
- Colour. The predominant colour and any mixture judged according to a standard reference chart (e.g. Munsell Soil Color Charts). The colours red, orange, yellow, brown, green, blue, white, cream, grey, black may be supplemented by light, dark, mottled or reddish, orangish, yellowish, brownish etc.
- Composite composition. Term indicating the presence and approximate quantity of soil components other than the main one. In the case of fine-grained soils the term 'sandy' is used and this may be qualified by 'very' where the amount is more than 65% or 'slightly' if it is less than 35%.
- Minor constituent type. Presence and type of minor constituents, for instance organic matter. The terms 'slightly organic', 'organic' and 'very organic' are used to describe soils containing respectively 2 to 5%, 5 to 10% and more than 10% organic matter. The quantity of organic matter is usually reflected by the colour of the clay, where this will be respectively grey, dark grey and black.
- Principal soil type. Defined according to average grain size (e.g. clay, silt, sand, gravel, etc.) and written in capitals. In the case of fine-grained soils such as clays and silts the particles are smaller than can be discerned with the unaided eye. BS 5930:1999 stipulates that the distinction between clays and silts should be made on the plasticity behaviour of the material. In other words, either hand or laboratory tests are employed to determine whether or not the material possesses plasticity, and it is then named accordingly. This means that a fine-grained soil is described as either a silt or a clay, so that terms such as 'silty clay' should not be used. However, in cases where the tests are inconclusive, the term 'silt/clay' should be used as an indication that ambiguity exists.
- In composite soils containing mixtures of soil types, for instance clay with cobbles, it is necessary to describe the components separately, and to indicate the relative proportions of the different materials. Where boulders or cobbles constitute up to 5% of a soil, the terms 'with occasional' should be used. 'With some' and 'with many' are used to indicate amounts of 5 to 20% and 20 to 50%.
- Minor constituents. Presence and type of additional constituents within the soil. May be qualified by 'with rare', 'with occasional', 'with abundant/frequent/numerous' or as % of named materials such as shells, brick fragments, gypsum crystals, pockets of peat, rootlets etc.
- Stratum name. The geological or other name for the material where, ideally, this should be the currently defined name for the formations (e.g. London Clay), see Chapter 6.

As mentioned above, in fine-grained soils, predominant soil type is determined according to the plasticity properties, determined either using laboratory tests (see Section 4.2.3) or according to the hand tests suggested in Table 4.5.

Although used widely, the BS 5930:1999 scheme is not without faults. A particular difficulty is that the features are not described in the order of typical importance. The following scheme would be more logical for engineering applications:

- grains: grading mineralogy, Atterberg limits;
- state or consistency: liquidity index, effective stress conditions and value of suction;
- fabric and structural features.

Even though some of these features may require laboratory determinations, methods of visual and manual assessment also provide helpful data.

Soil scientists such as Hillel (1980) use the following terms to describe soils in terms of their structure:

- single grained: loose assemblage of individual grains;
- massive: tight packing into *cohesive* blocks;
- aggregated: soil grains formed into small quasi-stable aggregations or peds.

Aggregated soils are intermediate between single grained and massive soils and are the desirable structure for plant growth. The presence of aggregations depends both on the flocculation of clays and also on the binding effect of organic matter.

4.1.5. Classification of mudrocks

Many different schemes for the classification of mudrocks have been suggested. Like soils, particle size is one of the main factors used for classification but, although appearing simple and logical, such classifications are difficult to apply in mudrocks.

Due to their small grain size, mudrocks can be difficult to identify in hand specimens. Important features that assist identification are:

- at least half the grains being smaller than can be discerned with the naked eye;
- easy to scratch the surface with a knife;
- when combined with water the powdered rock can be formed into a plastic paste.

All these features are indicative of the presence of clay minerals.

Usually it is difficult to obtain an accurate measure of grain size. This is because in indurated mudrocks the particles are cemented together so the material is difficult to disaggregate by standard methods. For instance in Mercia Mudstone clay mineral grains are cemented together in clusters so the measured clay size fraction is less than the actual amount of clay present (see Davis 1967).

Various authors (see Spears 1980) have suggested that, in view of the difficulty of determining the grain size of indurated mudrocks, mineralogical data, rather than size, should be used as the basis for classification. One disadvantage of this is the reliance on relatively expensive X-ray diffraction or SEM methods for the determination of the mineralogy of fine-grained rocks.

TABLE 4.7. *Classification of mudrocks (from BS 5930:1999)*

Grain size mm	Bedded rocks (mostly sedimentary)		
Grain size boundaries approximate	Grain size description		Lithology
— 0.06 —	ARGILACEOUS	MUDSTONE	SILTSTONE Mostly silt
— 0.002 —			CLAYSTONE*

* Claystone is not included in BS5930:1999.

A useful standard classification scheme devised primarily for engineering applications is provided in BS 5930:1999. Part of this is shown in Table 4.7. Rocks that contain clay minerals include claystones, mudstones and siltstones in which the average grain size is less than 0.06 mm. A number of other types of rocks, for example chalk and some rocks of volcanic origin, contain grains of clay size (<0.002 mm) but these grains are not predominantly clay minerals. Other mudrock types that contain clay minerals include shale, clay-shale and argillite.

As mentioned in Section 4.1.8, the term *shale* can cause confusion because to geologists it implies a sedimentary rock that will easily split into thin layers, whereas others tend to use the word to describe any mudrock.

The approach is similar to that taken by IAEG (1979) where silt (0.002–0.06 mm) and clay (<0.002 mm) rich rocks are defined as either siltstones (>50% fine-grained particles) or claystones (>50% very fine-grained particles).

A review of the terminology and definitions used to describe argillaceous (clay rich) materials is provided by Czerewko (1997) who lists a plethora of definitions for the terms used above. It is therefore necessary to use the standard terms and interpretations and to be prepared to describe the material, rather than to arrive at an all-encompassing name. Various other schemes for classification have been proposed by Ingram (1953), Pettijohn (1975), Potter *et al.* (1980), Blatt *et al.* (1980) and Stow (1981) that are capable of reflecting the geomechanical differences between the rock categories (see Hawkins & Pinches 1992).

Whether the material behaves in a rock-like or soil-like fashion is highly relevant in many applications. Mead (1936) suggested the use of the terms:

- compaction shale: mudrock formed just by compaction due to burial;
- cementation shale: mudrock in which the grains are attached together by a significant amount of cement, or diagenetic processes have resulted in recrystallisation of clay minerals and cementation by quartz and/or carbonate cements.

Morgenstern & Eigenbrod (1974) suggested that disintegration in water could be used to distinguish between compaction and cementation shales. They introduced the term liquidity index, not to be confused with another use of this term defined in Section 4.2.3. Here, it is used to indicate the degree of softening suffered by samples on immersion in water. Wood & Deo (1975) suggested that a series of static and dynamic slaking tests plus a modified sulphate soundness test should be used for this. Hopkins & Dean (1984) also suggest a static and dynamic slaking test method that enables soil-like and rock-like materials to be distinguished from each other.

Most of the schemes described above were developed to meet the needs for material classification on particular projects. Grainger (1984) addressed this problem and devised the scheme shown in Figure 4.2, in which mudrocks are classified in terms of grain size and uniaxial compressive strength, σ_c (for definition see Equation (4.59)) and second cycle slake durability test (I_{d_2}), defined in Section 4.1.7:

Soil: $\sigma_c < 0.60$ MPa
Non-durable mudrock: $0.60 < \sigma_c < 3.6$ MPa; $I_{d_2} < 90\%$
Durable mudrock: $3.6 < \sigma_c < 100$ MPa; $I_{d_2} > 90\%$.
Metamudrock: $\sigma_c > 100$ MPa.

A uniaxial compressive strength of 0.60 MPa is equivalent to an undrained shear strength (see Section 4.3.7) of 300 kPa. The measurement is made perpendicular to any fissility or cleavage in the rock.

The quartz content is used to distinguish between claystone, shale, silty shale, mudstone and siltstone as follows:

Claystone and shale $<20\%$ quartz
Silty shale and mudstone 20–40% quartz
Siltstone $>40\%$ quartz.

The latter figure conforms with the suggestion made by Spears (1980). Note that here the mineralogical composition forms the basis of classification, rather than grain size as in BS 5930(1999) and other schemes.

The Grainger (1984) classification also separates fissile from non-fissile material, where fissile rocks will split into thin (<3 mm) laminae. Rocks which break into fragments with a flakiness (ratio of shortest to intermediate dimension) greater than 2/3 are regarded as fissile, while those with a ratio less than 1/3 are massive. In slaty materials a value of 2 for the ratio of the highest to the lowest strength determined by point load test is taken as the criterion.

4.1.6. Breakdown of mudrocks

The depth to which observable changes due to weathering breakdown occur is very variable, extending from a few millimetres to tens of metres, depending on the composition and structure of the rock mass, the climatic conditions and the length of time over which the weathering has been taken place. Not all profiles show regular progressive changes in properties. Also, as explained in Section 4.3.3, stress relief due to the erosion of the overburden leads to fissuring of the rock mass, which change the properties of the material.

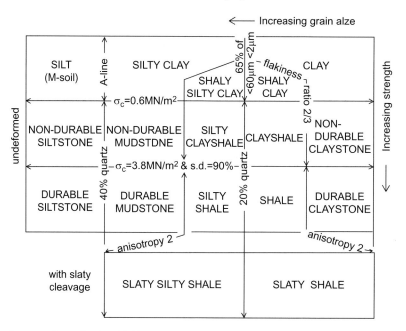

σ_c = Uniaxial compressive strength, see Section 4.3.10.
s.d. = second cycle slake durability test (Id$_2$), see Section 4.1.7.
Flakiness = ratio of shortest to intermediate dimension of fragments where $>2/3$ = fissile, $<1/3$ = non-fissile.
Anisotropy = ratio of the highest to the lowest point load strength, where >2 = slaty.
FIG. 4.2. Classification scheme for mudrocks (Grainger 1984).

In the case of mudrocks belonging to the Pietermaritzburg Formation in South Africa, Bell & Maud (1996) note that breakdown occurs preferentially along certain laminae within the rock mass, leading to the development of 3 to 5 mm thick clay seams within the rock mass. These horizons are weaker and more compressible than the underlying and overlying beds in the sequence and they are implicated in serious slope stability problems in this formation.

Mead (1936) divided mudrocks into compaction and cementation shales, to distinguish between stronger, more durable varieties and the weaker, less durable ones. While it is clear that durability is a function of porosity, cementation, the amount and type of clay minerals, pore water composition, fracturing, the presence of laminations and particle orientation, it is also affected by the degradation process. As explained by Czerewko & Cripps (2001), clay rich materials are particularly susceptible to breakdown by slaking action, which is a form of physical degradation brought about by cyclical wetting and drying.

For example, apparently well indurated mudrocks of Carboniferous age described by Kennard et al. (1967) slaked almost immediately to sand and gravel sized fragments on excavation exposure to high humidity conditions. Morgenstern & Eigenbrod (1974) found that there is a direct relationship between the strength loss during weathering and changes in initial porosity and moisture content. Thus physical factors as well as mineralogy exert an important influence over slaking. A high clay content, the presence of swelling clay minerals, laminated structure and incidence of microfracturing all increase the propensity of mudrocks to high rates of breakdown.

One of the processes apparently contributing to breakdown by slaking is by the generation of high pore air pressures as water is drawn by capillary forces into narrow pores (Taylor 1988). Pore water suctions during drying also result in tensile failure of weak inter-crystalline bonds. Thus the process of wetting and drying stresses the cements forming the skeletal framework of the material and acts to enlarge and extend existing pores and cracks. This has the effect of giving greater access for water and it also reduces the strength of the material.

In cases where cements strengthen the rock fabric, breakdown of the material may not be immediate, but repeated cycles of wetting and drying eventually induce failure, especially if cementing material is also removed by the process (Hudec 1982).

4.1.7. Durability of mudrocks

The durability of mudrocks is important in situations in which inappropriate handling of the material can cause it to degrade in an undesirable manner. This can make it difficult to transport and may adversely affect the trafficability of haul roads during extraction operations. In some applications, for example if the clay is to be processed, low durability may be a desirable feature. Low durability of mudrocks can also lead to problems with slope stability and the excavation of tunnels (Varley 1990). The durability can be assessed on the basis of specific tests or on the basis of strength, mineralogical composition and other characteristics.

Methods of quantitative assessment of durability are described by Czerewko & Cripps (2001), who also review the information about the controls of assessment of durability of mudrocks. The international standard slaking test (Brown 1981) involves measuring the breakdown that occurs when 40 to 60 g oven-died fragments of the mudrock are rotated in a drum formed out of 2 mm mesh and which is partly submerged in water. This test was originally devised by Franklin & Chandra (1972) who developed it from the end-over-end slaking test used by Badger et al. (1957). The Slake Durability Index, I_d is determined as follows:

$$I_d = \frac{\text{Dry weight of rock retained in drum}}{\text{Dry weight of sample}} \times 100\% \quad (4.3)$$

Gamble (1971) noted that results based on just one cycle of slaking could be unrepresentative and misleading so he suggested that a two-cycle test should be used. In this the dried material retained in the drum is subject to a second cycle of slaking. The material can then be classified according to the scale in Table 4.8.

The standard (Brown 1981) specifies the three-cycle test in rocks of higher durability and also indicates that if the I_d value is less than 10%, the material should be subject to soil classification using plasticity limits, where a plasticity index of 0 to 10% is low, 10 to 25 is medium and more than 25 is high. However, Taylor (1988) argued that a three-cycle test is required to provide reliable results across the range of durabilities, in which case the index is expressed as an I_{d_3} value.

One disadvantage of the test is that the mesh size for the degraded particles is fixed at 2 mm. This makes it insensitive to breakdown in both low- and high-durability materials. In low-durability rocks all the particles tend to break down to pass through the mesh whereas in high-durability materials breakdown of the original fragments results in particles larger than 2 mm in size, and is thus unrecorded. The standard suggests that the grading of the material in the drum should be determined, but this is seldom done in practice.

To overcome these problems, Czerewko & Cripps (2001) advocate the use of a static jar slake test for durability determination. In this the style and rate of breakdown are observed in a sample of the dry rock when it is placed in a beaker of distilled water. The test is relatively simple to apply and provides comparable results with

TABLE 4.8. *Classification of durability (Brown 1981)*

Slake durability, I_{d_2} %	Durability class
0–30	Very low
30–60	Low
60–85	Medium
85–95	Medium high
95–98	High
98–100	Very high

standard slake durability tests and natural weathering for a range of mudrock specimens. The results also showed good comparability with assessments based of saturation moisture content, which is an indication of the porosity, and methylene blue value, which indicates the presence of swelling clay minerals.

Dumbleton & West (1967) have described a simple test in which the sample is successively wetted and air dried, and the breakdown of the material is measured by sieve analyses carried out between each cycle of wetting and drying.

4.1.8. Description of mudrocks

Description is aimed not just at identifying the rock type, but also to enable the behaviour of the material to be predicted. As such, it is useful if information about the strength and durability of the materials are included in the description. Accordingly, Gamble (1971) proposed a classification scheme similar to that of Pettijohn (1949, 1975) in which the primary factors are grain size and breaking characteristics or durability. The description of rocks is carried out in much the same manner as for soils and attention is given both to features of the rock mass and of the rock material. The description includes information about the rock material, appropriate rock name (see Section 4.1.5) together with information about the discontinuities and other features of the rock mass.

In the UK, rock description for civil engineering purposes is usually carried out according to the scheme described in BS 5930:1999, part of which is shown here in Table 4.7. The terms used describe the material properties are listed below. This scheme is similar to that described by Fookes et al. (1971) and ISRM (1981). The sections and tables referred to below in square brackets are in BS 5930:1999.

- Strength [section 44.2.1]: very strong, strong, moderately strong, moderately weak, weak, very weak as defined in Table 4.9, which shows a range of values for the different strength tests.
- Structure [section 44.2.1]: thick, medium, thin, very thin, thickly laminated, thinly laminated, as defined here in Table 4.10.
- Colour [table 13]: general colour, whether light or dark and nature of mixtures and distribution e.g. banding or mottling.

TABLE 4.10. *Thicknesses of sedimentary bedding structures (BS 5930:1999)*

Term	Thickness
Very thick	Greater than 2 m
Thick	600mm to 2 m
Medium	200mm to 600 mm
Thin	60 mm to 200 mm
Very thin	20 mm to 60 mm
Thickly laminated	6 mm to 20 mm
Thinly laminated	Less than 6 mm

- Texture: arrangement of grains, crystalline, granular, amorphous etc.
- Grain size [table 14]: relative size, ground mass/matrix, clasts, inclusions
- Rock name [table 14]: in capitals, may be compounded e.g. silty MUDSTONE
- Other: minor constituents (e.g. fossil remains, pyrite), voids, inter-granular cement etc. Common cements (sedimentary rocks only) include quartz, iron oxides, calcite and clay.
- Geological formation (where known) e.g. (Mercia Mudstone)
e.g. Moderately weak, thickly bedded, dark grey MUDSTONE.

The description of the rock mass includes the following:

- Weathering: mass and material scales [table 19]: strength loss, colour change, description of weathered products
- Discontinuities in mass [section 44.4.3, inc table 15]
- Fracture state of mass [section 44.4.4]

Information about weathering is given in Section 4.1.9.

Some description schemes, for example ASTM D5434-03 (2003), include consideration of unit weight, measured by weighing the rock specimen in air and water using a spring balance. Such a procedure can cause the breakdown of some types of mudrock. This scheme also provides guidance on the estimation of the degree of weathering, rock strength, and the planar and linear elements contained in the rock.

Hawkins & Pinches (1992) discuss the application of the British Standard scheme (1981 edition of BS 5930) to

TABLE 4.9. *Strength classes for rocks (BS 5930:1999)*

Term	Field description	Unconfined compressive strength (MPa)
Very weak	Gravel sized lumps can be crushed between the fingers	<1.25
Weak	Gravel sized lumps can be broken in half by heavy hand pressure	1.25 to 5
Moderately weak	Only thin slabs, corners or edges can be broken off with heavy hand pressure	5 to 12.5
Moderately strong	When held in hand, rock can be broken by hammer blows	12.5 to 50
Strong	When resting on a solid surface, rock can be broken by hammer blows	50 to 100
Very strong	Rock chipped by heavy hammer blows	100 to 200
Extremely strong	Rock rings on hammer blows. Only broken by sledgehammer	>200

mudrocks, noting that the term fissility, the property of some mudrocks to split easily along depositional surfaces, is not defined. Weaver (1989) had previously argued that as fissility is a function of particle orientation, bedding, mineral composition, moisture content and weathering, the term should be used only as a descriptor rather than as a means of classification. Certain mudrocks rapidly develop fissility on exposure at the surface, whereas in others the process may be protracted or the material may retain its massive structure.

McKnee & Weir (1953) suggest the following terminology for describing the separation between the splitting surfaces in rocks:

Massive: blocky	<2 mm	papery
Flaky: wedge and chip shaped fragments	2–10 mm	platy
Flaggy: parallel sided fragments	10–50 mm	flaggy
Slabs: (usually siltstones and sandstones)	>50 mm	slabby

As noted in Section 4.1.5, the term shale is a frequent cause of confusion. In the UK it is usually taken to mean a fissile mudrock, or in other words one that will split along closely spaced planar surfaces. However, as fissility is a feature that develops at a certain stage of weathering, this term should not be used to define a particular rock type. The term is commonly used in geological literature to denote a sedimentary rock consisting predominantly of clay or silt particles of unspecified composition (Allaby & Allaby 1990).

4.1.9. Classification of weathering in mudrocks

A distinction needs to be made between the description and classification of weathered soils and rocks. The purpose of a description is to record important features of the material and to enable similar materials to be recognized. The description may also include references to engineering properties. Classification, on the other hand is carried out in order to group materials into a series of classes according to a set of pre-determined criteria. Anon (1995) notes that classifications of weathered rock tend to define classes in terms of either (1) the amount of change of particular features or properties, for example the chemical or mineralogical composition or physical features, or (2) on simple observed criteria, such as the proportion of degraded material present. These approaches are not mutually exclusive but the latter method presupposes that the state of the unweathered material is known, while the first approach usually requires complex laboratory work. The Working Party on the classification and description of weathered rocks (Anon 1995) concluded that the second approach provides a more practical method of classification.

An ideal classification should take cognisance of weathering effects that affect both the rock material and also the soil or rock mass. Depending on the type of material, topography, climatic conditions and other factors, these effects may occur at the same time, although their relative importance may differ. The approaches, shown in Table 4.11, advocated by Anon (1995) are as follows:

Approach 1. General description. Observations about changes in colour, strength and fracture state and the presence of the products of weathering are recorded according to the procedures used in BS 5930:1999. This may apply either at the material- or mass-scale, depending on the volume of rock or soil that can be observed and its heterogeneity. In variable materials, the different parts need to be described separately, and their relative proportions estimated. Classification of the material can then proceed using Approaches 2, 3, 4 or 5.

Approach 2. Classification for uniform materials. For materials which are moderately strong or stronger in the fresh state and which undergo a gradational style of weathering from a uniformly fresh rock to a soil (i.e. without the development of core stones).

Approach 3. Classification for heterogeneous masses. For materials which are moderately strong or stronger in the fresh state and which form a heterogeneous mixture of materials of different properties at certain stages of the weathering process (i.e. with the development of core stones that are little changed from the original material).

Approach 4. Classification incorporating both material and mass effects. For materials which are moderately weak or weaker in the fresh state and which undergo a gradational style of weathering that involves simultaneous changes to the material and the mass properties (i.e. with the development of lithorelicts that are weaker than the original material). Mudrocks often undergo weathering of this type. The scheme is shown in Table 4.11.

Approach 5. Special cases. For rocks that do not follow the patterns of weathering assumed in the other approaches. For example, the development of karst features in limestones.

4.2. Framework for clay material behaviour

Clay materials include all natural deposits in which clay minerals are an important component. They may also occur as powders, suspensions, slurries and pastes that contain clay minerals. The properties displayed by these materials relate, in particular to the small size, platy shape and the presence of surface charges of clay minerals. The presence and types of non-clay components, preferred orientation of minerals, porosity, state of saturation, stress conditions and pore water pressures (or suctions) are also important.

The plasticity, flow and rheology displayed by clay materials are introduced in this section, together with the compaction, permeability and thermal properties of clays. The strength and deformation properties of soils and rocks are described in Section 4.3.

TABLE 4.11. *Classification of weathering for mudrocks (Anon 1995)*

APPROACH 1: FACTUAL DESCRIPTION OF WEATHERING (MANDATORY)

Standard descriptions should always include comments on the degree and nature of any weathering effects at material or mass scales. This may allow subsequent classification and provide information for separating rock into zones of like character. Typical indications of weathering include

- Changes in colour
- Reduction in strength
- Changes in fracture state
- Presence, character and extent of weathering products

These features should be described using standard terminology, quantified as appropriate, together with non-standard English descriptions as necessary to describe the results of weathering. At the mass scale the distribution and proportions of the variously weathered materials (e.g. corestones vs matrix) should be recorded

Can classification be applied unambiguously? — No — Do not classify
Yes

Rock is moderately strong or stronger in fresh state | Rock is moderately weak or weaker in fresh state

APPROACH 2: CLASSIFICATION FOR UNIFORM MATERIALS

Grade	Classifier	Typical characteristics
I	Fresh	Unchanged from original state
II	Slightly weathered	Slight discoloration, slight weakening
III	Moderately weathered	Considerably weakened, penetrative discoloration. Large pieces cannot be broken by hand
IV	Highly weathered	Large pieces cannot be broken by hand. Does not readily disaggregate (slake) when dry sample immersed in water
V	Completely weathered	Considerably weakened. Slakes. Original texture apparent
VI	Residual soil	Soil derived by in situ weathering but retaining none of the original texture or fabric

Is a zonal classification appropriate and is there enough information available? — No — Use Rock Mass Classification if appropriate
Yes

APPROACH 3: CLASSIFICATION FOR HETEROGENOUS MASSES

Zone	Proportions of material grades*	Typical characteristics
1	100% G I - III (not necessarily all fresh rock)	Behaves as rock: apply rock mechanics principles to mass assessment and design
2	>90 % G I - III <10% G IV - VI	Weak materials along discontinuities. Shear strength stiffness and permeability affected
3	50 % to 90 % G I - III; 10 % to 50 % G IV - VI	Rock framework still locked and controls strength and stiffness: matrix controls permeability
4	30 % to 50 % G I - III; 50 % to 70 % G IV - VI	Rock framework contributes to strength; matrix or weathering products control stiffness and permeability
5	<30 % G I - III; >70 % G IV - VI	Weak grades will control behaviour. Corestones may be significant for investigation and construction.
6	100% G IV - VI (not necessarily residual soil)	May behave as soil although relict fabric may still be significant

APPROACH 4: CLASSIFICATION INCORPORATING MATERIAL AND MASS FEATURES

Class	Classifier	Typical characteristics
A	Unweathered	Original strength, colour, fracture spacing
B	Partially weathered	Slightly reduced strength, slightly closer fracture spacing, weathering penetrating in from fractures, brown oxidation
C	Distinctly weathered	Further weakened, much closer fracture spacing, grey reduction
D	Destructured	Greatly weakened, mottled, ordered lithorelics in matrix becoming weakened and disordered, bedding disturbed
E	Residual or reworked	Matrix with occasional altered random or 'apparent' lithorelics, bedding destroyed. Classed as reworked when foreign inclusions are present as a result of transportation

APPROACH 5: SPECIAL CASES

For rocks whose weathering state does not follow the other patterns indicated here, such as karst in carbonates and the particular effect of arid climates

*Grades GI to GVI defined in Approach 2

TABLE 4.12. *Parameters for moisture content, density and saturation of soils*

Property	Definition	Relationships
Bulk or wet density ρ_b (Mg/m³)	$\dfrac{\text{Mass of wet soil}}{\text{Total volume of soil}}$	$\dfrac{\rho_d(100+w)}{100}$
Dry density ρ_d (Mg/m³)	$\dfrac{\text{Mass of dry soil}}{\text{Volume of soil}}$	$\dfrac{100\,\rho_b}{100+w}$
Bulk or wet unit weight γ_b (kN/m³)	$\dfrac{\text{Weight of wet soil}}{\text{Volume of soil}}$	$\dfrac{\gamma_d(100+w)}{100}$
Dry unit weight γ_d (kN/m³)	$\dfrac{\text{Weight of dry soil}}{\text{Volume of soil}}$	$\dfrac{100\cdot\gamma_b}{100+w}$
Moisture content w (%)	$\dfrac{\text{Mass of water in soil} \times 100}{\text{Mass of dry soil}}$	$\dfrac{100\cdot w_w}{1+w_w}$
Moisture content w_w (%)	$\dfrac{\text{Mass of water in soil} \times 100}{\text{Mass of wet soil}}$	$\dfrac{100\cdot w}{1+w}$
Void ratio e	$\dfrac{\text{Total volume of voids in soil}}{\text{Volume of dry solids in soil}}$	$\dfrac{w\,G_s}{100\,S_r}$
Porosity n	$\dfrac{\text{Total volume of voids in soil}}{\text{Total volume of soil}}$	$\dfrac{e}{1+e}$
Specific gravity of soil particles, G_s	$\dfrac{\text{Mass of dry solids in soil}}{\text{Mass of an equal volume of water}}$	$\dfrac{100\cdot(1+e)}{100+w}$
Saturation ratio S_r	$\dfrac{\text{Volume of water}}{\text{Total volume of pores}}$	$\dfrac{w\,G_s}{100\,e}$
Saturation percentage s (%)	$\dfrac{\text{Volume of water} \times 100}{\text{Total volume of pores}}$	$\dfrac{w\,G_s}{e}$
Air voids percentage A (%)	$\dfrac{\text{Volume of air in soil} \times 100}{\text{Total volume soil}}$	$n(1-S_r)$

4.2.1. Moisture content, density and saturation

Methods of measuring the moisture content, density and degree of saturation of clays are given in Chapter 9. Commonly used parameters are listed in Table 4.12, together with their definitions and inter-relationships. It should be noted that in ceramics applications moisture content can be defined in terms of either the dry or the wet mass of soil, whereas in geotechnical applications it is always defined as the percentage of moisture in terms of the dry soil. Unless otherwise stated, moisture contents referred to here are in terms of dry soil mass.

4.2.2. Flow and rheology

Clays may undergo flow, deform in a plastic manner or undergo rupture in response to an applied stress. Moisture content exerts a strong control on the form of behaviour, but the mineralogy, grain size, system chemistry and environmental conditions are also important. Table 4.13 shows approximate ranges of water contents for different types of behaviour. It should be noted that the ranges are approximate and there will be a large variation in these values for different clays. Also the descriptions of the behaviour in terms of liquid and plastic are not meant to

TABLE 4.13. *The behaviour of clay suspensions at different approximate moisture contents (after Worrall 1982)*

Ceramic clay	Mixing	Casting	Moulding	Pressing
Geotechnical applications	Slurry	Liquid	Plastic soil	Brittle solid
Moisture content % dry mass	85–110	38–52	24–30	5–12
Moisture content % wet mass	47–54	28–35	19–25	3–10
Density (Mg/m³)	1.2–1.3	1.45–1.6	1.65–1.75	2.0–2.2

correspond with values of Atterberg Limits, as described in Section 4.2.3.

Very fine-grained clays (0.001 to 1 μm) may form colloidal suspensions in which gravitational forces will not cause the particles to settle out of the suspension. In dispersive clays (Section 4.4.2) such suspensions may form without the necessity for mechanical agitation or water movement, whereas in non-dispersive clays, water movement or mixing are required to form suspensions. As Gillott (1987) points out colloidal suspensions will pass through filter paper and the suspension may form a gel when the particles are caused by changes in the chemical composition of the suspension to form flocs. Such gels may absorb large volume of water, expanding in volume as they do so.

Weak suspensions may behave as Newtonian fluids in which viscous flow occurs in response to stress and the rate of flow is proportional to the applied stress. The constant of proportionality is called the coefficient of viscosity. Newtonian flow applies to conditions of laminar flow, and not to turbulent flow conditions.

Clay suspensions and slurries may display the characteristics of Bingham plastics, which display a yield stress. Once this is exceeded and flow starts, the rate of shear is proportional to the shear stress. They may also display thixotropy in that resistance to flow decreases with rate of stress increase. Materials possess 'apparent' viscosity if when left stationary the stress required to cause movement increases. Dinsdale (1996) explains that the thixotropy of liquids is determined by measuring the viscosity immediately after stirring and then again after a particular time lapse (for example 1 minute).

Baver et al. (1972) explain that the viscosity of colloidal clay suspensions depends on the structure of the adsorbed water associated with the surfaces of clay minerals. The property arises because of frictional forces between the adsorbed and free water molecules and between them and the clay particles. The value of viscosity depends on the particle shape, degree of aggregation and concentration of particles, the viscosity of the fluid as well as the surface density of exchangeable cations, their bonding energy and hydration state.

Both viscosity (or the fluidity, which is the inverse of viscosity) and thixotropy are very important parameters for clay suspensions used for casting and moulding articles in the ceramics industry (see Chapter 14). These properties are also relevant in grouting and other geotechnical applications described in Chapter 12.

Reducing the moisture content of a clay suspension produces a paste that will deform in a ductile manner. Further reducing the moisture content by a sufficient amount converts this to a brittle material. Brittle materials fail by fracture, whereas ductile materials undergo non-recoverable strain but may eventually rupture. Whether clay materials behave in a brittle or ductile fashion also depends upon the stress conditions, moisture content, rate of strain, temperature and other factors.

The behaviour of ideal plastic and elastic materials may be represented by stress–strain graphs shown in Figure 4.3. Definitions for stress and strain are given in the Text Box (p. 90). Gillott (1987) explains that an ideal plastic material deforms elastically up to its yield point at which stress it will continue to deform under constant applied stress. The strain is non-recoverable if the stress is removed and reloading results in elastic and plastic deformation until the former stress–strain relationship is followed. As Figure 4.3 shows, typical clays do not display this ideal behaviour and the response to stress also depends upon the moisture content of the material. Clays with a high moisture content possess a low yield stress so they undergo only a small amount of elastic deformation before yield. Drier clays undergo combined elastic and plastic deformation with little true elastic deformation. Indurated rocks tend to behave in an elastic manner up to the yield point and then suffer brittle failure with little or no plastic deformation.

Clays will behave plastically over a particular range of moisture contents. If the moisture content is very high then they will flow without a yield stress, whereas if it is low they fail by brittle fracture. This moisture content range depends on composition, grain size, pore fluid chemistry and environmental factors. Plasticity in clays is described in Section 4.2.3.

In an ideal elastic material, strain is proportional to stress and all strain is recoverable. Such a material is said to be a Hooke solid and the constant of proportionality is Young's modulus. This and other moduli of deformation, including modulus of rigidity, shear modulus and Poisson's ratio, are defined in Section 4.3.1. According to Gillott (1987), the yield stress is variously defined as the point at which there is permanent strain, or the point at which the stress–strain graph is no longer linear, or the minimum stress for continuous (plastic) flow, but all these are at similar values of stress in clays.

The strength of a clay material defines the amount of applied stress it can sustain without failing by brittle fracture or plastic yielding. As Figure 4.3 illustrates, elastic and plastic deformation occurs before the strength is exceeded but failure marks a pronounced change in the condition of the material and removal of the applied stress will not restore the material to its former state. Discussion of the post-failure behaviour of clays is given in Section 4.3.6.

The relationship between stress and strain in materials is affected by the rate of stress increase as well as the temperature and confining pressure conditions. A mudrock that behaves in a brittle manner in laboratory tests carried out with quickly increasing stress at room temperature and zero confining pressure ($\sigma_3 = 0$ and only atmospheric pressure constrains the specimen from undergoing lateral deformation), may behave in a ductile fashion in a geological situation in which the stress is increased very slowly, and due to the material being at a depth of some kilometres below the ground surface, the confining pressure and temperature are both elevated.

Gillott (1987) explains that rheological models built up of elements that represent different aspects of the response of materials to stress can be used to represent the deformation behaviour of materials. For instance the visco-elastic behaviour of clays may be represented by a spring and a dashpot connected in series. Immediate, recoverable strain is symbolized by the spring and delayed non-recoverable strain is symbolized by the displacement of the dashpot piston, which requires the

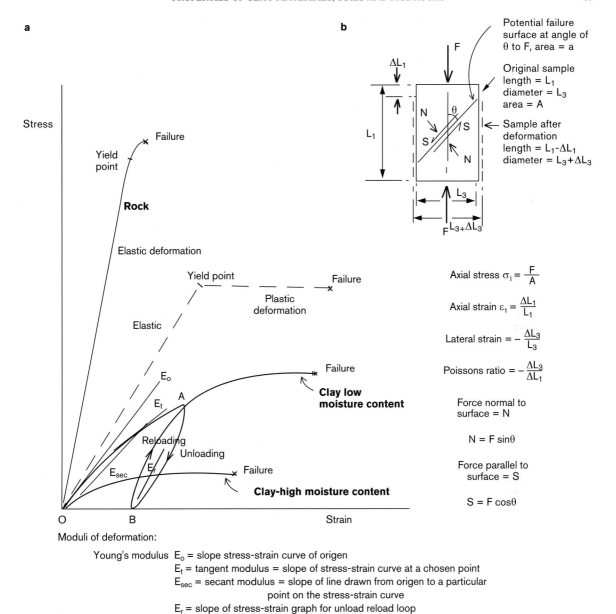

FIG. 4.3. Ideal and typical forms of stress–strain relationships in elastic and plastic solids.

leakage of a viscous fluid past the piston. Suklje (1969) uses these models to expound his linear visco-elastic continuum theory for soil deformation.

4.2.3. Plasticity of clays: Atterberg limits

As mentioned in Section 4.2.2, clays behave plastically over a particular range of moisture contents. This range is expressed in terms of consistency or Atterberg limits. The change in soil volume and consistency in terms of moisture content, w (defined in Table 4.12) is shown in Figure 4.4 and Table 4.14. The ranges of moisture contents for different consistencies are respectively defined as the shrinkage, plastic and liquid limits. At these, soil behaviour changes with progressive decreases in moisture content, as follows:

w_s (shrinkage limit) limiting moisture content for no change in volume;
w_p (plastic limit)   limiting moisture content for plastic behaviour;
w_L (liquid limit)   limiting moisture content for liquid behaviour.

Shear stress and normal stress

Stress is defined as force per unit area over which it acts, and is synonymous with pressure. In soil and rock mechanics it is usual to distinguish between normal and shear stresses, where a normal stress arises from a force acting normally (or perpendicularly) to a surface, whereas in the case of shear stress, the force is parallel to the surface. This point is illustrated in Figure 4.3(b), in which a force, F acts obliquely (at an angle of $\theta°$) to a plane surface of area a. The force F can be resolved into forces N and S which are respectively normal and parallel to the surface and in a plane that contains F and is orientated at right angles to surface a. The normal and shear stresses acting on the surface are respectively σ and τ, as given in equations (4.4) and (4.5):

$$\sigma = N/a \tag{4.4}$$

$$\tau = S/a \tag{4.5}$$

where $N = F\sin\theta$, $S = F\cos\theta$ and $a = A/\sin\theta$. Note that if $\theta = 0$, then $\tau = F/a$ and $\sigma = 0$, whereas if $\theta = 90°$, then $\tau = 0$ and $\sigma = F/A$.

Normal forces can be compressive or tensile, where compressive forces are conventionally regarded as positive and tensile forces as negative. Shear stresses acting in an anticlockwise direction are positive. Measuring forces in Newtons (N) and area in m^2 gives a force with units of N/m^2, or Pascals (Pa). Larger stresses are measured in kPa, MPa or GPa, which are respectively 10^3, 10^6 and 10^9 Pa.

Principal stresses

Any system of normal and shear stresses acting in a body of soil or rock can be expressed in terms of an equivalent system of three principal stresses called σ_1, σ_2, and σ_3. As shown in Figure 4.11, these are the normal stresses acting along three mutually perpendicular axes and, as by definition $\sigma_1 > \sigma_2 > \sigma_3$, they are respectively called the major, intermediate and minor principal stresses. In some situations σ_2 may equal σ_1 or σ_2 may equal σ_3, and in an isotropic stress situation $\sigma_1 = \sigma_2 = \sigma_3$. The latter condition applies where a body is immersed in a fluid as the fluid pressure acts equally in all directions. As in many ground engineering situations $\sigma_2 = \sigma_3$, it is common to consider problems only in terms of σ_1 and σ_3. The principal stresses defined above are in terms of total stresses. There is an equivalent set of effective stresses, where $\sigma_1' = \sigma_1 - u$, $\sigma_2' = \sigma_2 - u$, and $\sigma_3' = \sigma_3 - u$. Note that since the pore pressure, u acts equally in all directions it is the same in all these equations.

Strain

A force acting on a body will cause it to deform by a lesser or larger amount, where stiffer materials undergo less deformation. The deformation may entail displacement together with linear, volumetric or shear strain.

The linear strain is the change in length in a particular direction expressed as a ratio or percentage of the original length in that direction. In common with the three principal stress axes, linear strain can be measured in three mutually perpendicular directions, ε_1, ε_2, and ε_3. Thus, as Figure 4.3(b) shows, if the original size along the ε_1 axis was l_1 and the application of stress causes a reduction in length of Δl_1, then the linear strain is:

$$\varepsilon_1 = \frac{\Delta l_1}{l_1} \quad \text{or} \quad \frac{\Delta l_1 \times 100}{l_1} \quad \text{per cent.} \tag{4.6}$$

Conventionally, compressional or shortening strains are regarded as positive and extensional or lengthening strains as negative. Strains measured along the other two axes, ε_2 and ε_3, are respectively defined as $\varepsilon_2 = \Delta l_2/l_2$ or $\Delta l_2 \times 100/l_2$ per cent and $\varepsilon_3 = \Delta l_3/l_3$ or $\Delta l_3 \times 100/l_3$ per cent.

Compression or stretching results in lateral deformation as well as deformation along the axis of the imposed stress. In the case of a compressive stress being imposed, the ratio of the change in the amount of lateral expansion ($\Delta\varepsilon_3$) to the change in the shortening strain ($\Delta\varepsilon_1$) along the compression axis is defined as the Poisson's ratio for the material, v where

$$v = -\frac{\Delta\varepsilon_3}{\Delta\varepsilon_1} \tag{4.7}$$

The minus sign arises as the strain on the ε_3 axis is opposite in sign to the strain on the ε_1 axis. Because with undrained conditions (see Section 4.3.7) there is no volume change, the undrained value of Poisson's ratio $v_u = \frac{1}{2}$.

Shear stress and normal stress (continued)

Volumetric or bulk strain ε_v is a measure of the change in volume due to the imposition of stresses. Thus,

$$\varepsilon_v = \frac{\Delta v}{v} \qquad (4.8)$$

where the original volume is v and the change in volume, which can be positive or negative, is Δv.

The shear strain γ, is a measure of the distortion a body suffers as a result of the imposition of stress. The so called 'engineer's strain' is given by

$$\gamma = d/L \qquad (4.9)$$

where d is the linear distance a point in the body moves compared with its original position and L is the thickness of the body undergoing distortion measured at right angles to the plane of the surface over which the shear stress acts (see Figure 4.10a).

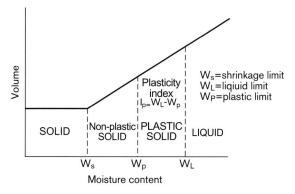

FIG. 4.4. Relationship between volume and moisture content in clays.

The limits listed above relate well to other properties of soils, such as strength, compressibility, and volume change. Other related parameters include the sticky point, plasticity index, liquidity index and activity. The sticky point is a parameter used in agricultural applications that is defined as the moisture content at which the soil will adhere to metal tools. This depends upon the clay mineralogy, surface chemistry and the separation between of clay minerals in the soil.

The *Plasticity Index* is the difference between the liquid and plastic limits ($w_L - w_p$). The other parameters are defined below.

As defined above, the shrinkage limit is the moisture content at which further drying causes little or no change in soil volume. This implies that the soil particles are no longer separated by water and at this point inter-particle shearing resistance is equal to the maximum capillary force that retains soil moisture in the pore space (Sridharan & Prakash 1998).

Sridharan & Prakash (1998) argue that the free-swell limit, which is the maximum water content for a free swelling remoulded sample, represents more truly the change from plastic to liquid behaviour than the BS 1377:1990 liquid limit test. Yong & Warkentin (1975) note that while plastic limit varies little between different soil types, both liquid limit and shrinkage limit are sensitive to the mineralogy.

The *swelling limit* and *air entry point* are parameters that are used in soil science. According to Braudeau *et al.* (1999), the air entry point is the minimum water content at which the soil remains saturated under atmospheric

TABLE 4.14. *Limiting water contents of a soil–water system (after Sridharan & Prakash 1998 and ELE 1998)*

Phase	SOLID	SEMI-SOLID	PLASTIC	LIQUID-PLASTIC	LIQUID	SUSPENSION
Water	← Water content decreasing ──────────────────────────────→					
Atterberg Limits	W_S	W_P	STL	W_L	WSL	WFS
Condition	Stiff - v. stiff	Stiff - workable	Sticky		Slurry	Suspension
Strength (kPa)	← Shear strength increasing 150			1.5	(~0–1.5)	
Shrinkage*	Residual	Normal	AE SWL Structural			

AE, air entry point; STL, sticky limit (%); SWL, swelling limit; WFS, free-swell limit (%); WSL, settling limit (%); * see Braudeau *et al.* (1999).

> ## Standard terms for liquid, plastic and shrinkage limits
>
> Standardized tests used to determine liquid, plastic and shrinkage limits are described in BS 1377:1990 (see Chapter 9). In this, the plastic limit is defined as the moisture content (w_p, %) at which rolling the soil out into a thin thread causes it to crumble rather than decrease in diameter. The liquid limit (w_L, %) is determined by carrying out a standard penetrometer test on the soil. For the shrinkage limit (w_s, %) the volume of a sample of the clay is measured as it is dried. As these consistency tests are carried out on the less than 425 μm fraction of remoulded soil samples, any aspects of behaviour connected with the presence of structure or larger grains are ignored.

conditions, whereas the swelling limit is the boundary between structured and normal shrinkage. Within the *normal* shrinkage zone volume change is proportional to the water loss, whereas in the *structural* zones the moisture content change is higher or lower than this.

Determination of the liquid and plastic limits of a soil enables it to be classified in terms of its plasticity behaviour as defined by the Casagrande plasticity chart shown in Figure 4.5, where *Plasticity Index* is defined as the difference between the Liquid and the Plastic Limits. Different degrees of plasticity are defined by liquid and plastic limit values as follows:

Low plasticity: $w_L < 35\%$
Intermediate plasticity: $w_L = 35–50\%$
High plasticity: $w_L = 50–70\%$
Very high plasticity: $w_L = 70–90\%$
Extra high plasticity: $w_L = >90\%$

Non-plastic: $w_p < 1\%$
Slightly plastic: w_p 1–7%
Moderately plastic: w_p 7–17%
Highly plastic: w_p 17–35%
Extremely plastic: $w_p > 35\%$

The A-line on the Chart in Figure 4.5 separates normal inorganic clays, which plot above it from organic clays and micaceous silty clays, which plot below it. Tropical soils containing hydrated halloysite also plot below the A-line (Dumbleton & West 1966a). The effect of increasing clay content is to move points up the A-line and parallel to it. It has been found that soils having similar engineering properties plot in similar areas on the plasticity chart, which makes the chart useful when assessing the likely properties of a new soil. Also, as mentioned in Section 4.1.2, the plasticity chart is used as the basis for classifying fine-grained soils.

Typical values of liquid and plastic limits for pure clays are given in Table 4.15.

Atkinson (1993) pointed out that the consistency limits relate to a number of other soil properties. The undrained shear strengths of a reconstituted sample (see Section 4.3.7) at the liquid and plastic limits are approximately:

at liquid limit $s_u = 1.5$ kPa (4.10)

at plastic limit $s_u = 150$ kPa. (4.11)

Strengths at intermediate moisture contents can be obtained by interpolation on a plot of log s_u versus moisture content.

Atkinson (1993) also shows that as a consequence of these relationships the compression index, C_c (see Section 4.3.4) under one-dimensional loading conditions, of a reconstituted soil can also be expressed in terms of the plasticity index, I_p:

$$C_c = (I_p \times G_s)/200 \quad (4.12)$$

where G_s is the specific gravity of the soil particles.

Figure 4.5 shows plasticity data for the three clay minerals montmorillonite, kaolinite and illite in mixtures with sand, where the numbers beside the points indicate the clay size percentages. It can be seen that the kaolinite and illite mixtures plot in a broadly similar part of the chart, fairly low down the A-line, while the montmorillonite mixtures plot much higher up the A-line. Thus kaolinite and illite have broadly similar properties associated with moderate plasticity, whilst montmorillonitic soils have high plasticity.

Liquidity index is used to indicate the state of a soil in situ. It is defined as follows:

$$\text{Liquidity index} = \frac{W - W_p}{W_L - W_p} \quad (4.13)$$

where a liquidity index of 1.0 corresponds with a low density or loose condition at which the natural moisture content ($w\%$) is equal to the liquid limit, whereas a liquidity index of zero signifies a high density or dense state at which the moisture content is equal to the plastic limit. Negative values are possible in highly compressed soils. The consistency index, defined in Equation (4.14), indicates the condition of the soil, as shown in Table 4.16. However, this parameter is not now in common usage.

$$\text{Consistency index} = \frac{W - W_L}{W_L - W_p}. \quad (4.14)$$

The Atterberg Limits depend on the grading and mineralogy, where the latter also affects the surface texture and shape of the grains. Activity quantifies the amount clay mineralogy controls soil behaviour:

$$\text{Activity}, A = (I_p/\text{clay content}) \quad (4.15)$$

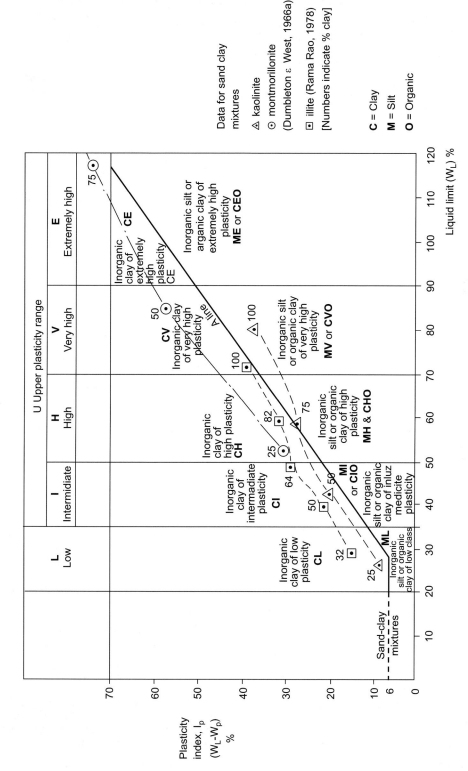

FIG. 4.5. Plasticity chart for clays.

TABLE 4.15. *Values of plasticity parameters for pure clays (Mitchell 1993)*

Mineral	Liquid limit %	Plastic limit %	Shrinkage limit %	Activity
Montmorillonite[a]	100–900	50–100	8.5–15	1.5–7
Illite[a]	60–120	35–60	15–17	0.5–1
Kaolinite	30–110	25–40	25–29	0.5
Chlorite[b]	44–47	36–40		
Attapulgite	160–230	100–120		0.5–1.2
Allophane[c]	200–250	130–140		0.5–1.2

a – Values at the higher ends of the ranges apply where monovalent cations are present, and the lower ones are for divalent cations.
b – Not all types of chlorite are plastic
c – Samples were not dried, which would be normal practice for liquid and plastic limit tests but, in the case of allophane, it would have caused irreversible changes to the clay.

TABLE 4.16. *Consistency of fine-grained soils*

Description	Consistency index	Approximate undrained shear strength (kPa)	Soil condition
Hard		>300	Simple tests to determine soil condition are shown in Tables 4.4 and 4.5
Very stiff	>1	150–300	
Stiff	0.75–1	75–150	
Firm	0.5–0.75	40–75	
Soft	<0.5	20–40	
Very soft		<20	

TABLE 4.17. *Activity classes (Skempton 1953)*

Description	Activity
Inactive	<0.75
Normal	0.75–1.25
Active	>1.25

where the clay content is the dry mass of clay sized particles divided by the total dry mass of the soil. Skempton (1953) defined the soil activity classes shown in Table 4.17.

Typical activity values for several clay minerals are shown in Table 4.15. Higher values indicate that the mineral causes the clay fraction to exert a relatively greater influence on the properties of the soil. The activity chart plotted in Figure 4.6 shows activity values for kaolinite, illite and montmorillonite mixtures with sand. As this plot illustrates, the activity value is a useful guide as to the types of clay mineral present in a soil. While kaolinite and illite plot close together, montmorillonite plots separately, exhibiting higher plasticity and activity, which emphasizes its particularly different properties from the other two clay minerals.

4.2.4. Permeability and diffusion

The permeability of a substance is a measure of the ease with which fluids can flow through it. One of the important features of clays is their very low permeability. This arises because of the small size of field-granular pores.

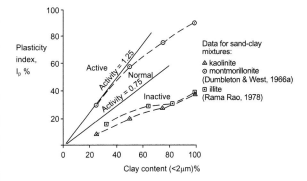

FIG. 4.6. Activity chart for clays.

The flow of water through a saturated soil is governed by Darcy's Law, which is expressed mathematically in Equation (4.16).

$$q = Kai \qquad (4.16)$$

where q is the quantity of water flowing in unit time (so the total amount of flow, $Q = qt$), i is the hydraulic gradient which is the ratio of the hydraulic or head difference between the inlet and outlets to the vertical height difference between these points, a is the area over which flow takes place measured at right angles to the direction of flow, K is the hydraulic conductivity or (Field) coefficient of permeability of the material and t is the time over which the flow is measured. Due to the fact that in clays some water is adsorbed onto clay particles and therefore is not free to flow, Darcy's Law does not fully explain the flow of water through clays (see Swarzendruber 1961). Fang (1997) reviews the results of several investigations into the variation of the permeability due to the types of clay minerals present, temperature and fluid composition.

Methods of determining the coefficient of permeability are described in Chapter 9. The value of this coefficient depends both on the properties of soil and the fluid and only applies to the flow of water at 20°C. For other liquids the viscosity, μ and unit weight, γ of the liquid must also be taken into account (Fetter 1988). This leads to the use of the term coefficient of intrinsic permeability, k_i that depends only upon the character of the pore network, where

$$k = \frac{k_i \gamma}{\mu} \qquad (4.17)$$

The units for coefficient of permeability are m/s, whereas for intrinsic permeability they are m² or darcy, where 1 darcy $= 9.87 \times 10^{-13}$ m². For water at 20°C a field coefficient of permeability of 1 m/s is equivalent to an intrinsic permeability value of 1.045×10^5 darcy or 1.031×10^{-7} m².

Clays usually have low permeability, typically between $k = 1 \times 10^{-10}$ m/s and 1×10^{-7} m/s. The high end of this range implies a rate of movement of fluid through the material of the order of 1 m in 0.3 years whereas at the low end the rate is of the order of 1 m in 300 years, where the difference in head between the inlet and out-flow is 1m. Fissures in clays may increase the value of field permeability by several orders of magnitude compared with the intact material. Table 4.18 shows typical permeability values for soils and rocks.

Water and ions in solution are transported through materials due to differences in energy between two points. Flow of fluids through the pore space in materials due to differences in fluid pressure is called advection. Water and dissolved ions may also diffuse through flow media as they move from zones of higher concentration to zones of lower concentration or in response to electrical potential difference. Such a movement need not be in the same direction as advective flow. In very low permeability clays, such as those used in low-permeability barriers, movement by diffusion can be the major means by which ions move. Fang (1997) indicates that the mass of ions passing through a given cross-section per unit time is proportional to the concentration gradient, dC/dx (mass/volume/distance), as shown in equation (4.18).

$$F = -D dC/dx \qquad (4.18)$$

where F is the mass flux of solute per unit area per unit time, D is the diffusion coefficient (area/time) and C is the solute concentration (mass/volume). The concentration gradient (dC/dx) is a negative quantity in the direction of diffusion. The value of the diffusion coefficient depends on the temperature and the chemistry of the soil–water system. Fetter (1988) indicates that for the majority of cations and anions in water, D ranges from 1×10^{-9} to 2×10^{-9} m²/s. Because in porous media diffusion only takes place through the pores and the pathways are longer than they would be in a liquid, the effective diffusivity is 0.01 to 0.5 of the diffusivity in a liquid. Mitchell (1991) provides further details of diffusion and also describes other mechanisms of solute movement. The movement of water in the vapour phase is also important in clays.

4.2.5. Stresses in the ground: pore pressures and effective stress

This section deals with the pressures in the ground and the theory of effective stress. Definitions are given for these terms and effective stress is important because it significantly influences the strength and deformation properties of soils, especially clays. (Stress is defined in the Text Box on p. 90.)

As indicated in Figure 4.7, the vertical stress, σ_v (kPa) at depth H (m) due to the self weight of the soil is given by:

$$\sigma_v = \gamma_s H \qquad (4.19)$$

where γ_s is the weighted average unit weight (kN/m³) of the soil over depth H (m). For reasons that are explained below, this stress is a total stress.

Some authorities prefer to use density rather than unit weights to calculate pressures. The unit weight of soil, $\gamma_s = \rho_s \times g$ where ρ_s is the density of the soil and g is the acceleration due to gravity (m/s²). As $g = 9.81$ m/s² and

TABLE 4.18. *Classification of permeability*

Field permeability range, k (m/s)	Description	Typical materials
1–10^{-2}	Very high	Unconsolidated coarse-grained sediments, heavily jointed rocks
10^{-2}–10^{-4}	High	
10^{-4}–10^{-6}	Medium	
10^{-6}–10^{-8}	Low	Silts, jointed rocks
10^{-8}–10^{-10}	Very low	Clays, silts, intact rocks
$< 10^{-10}$	Impermeable	Clays, mudrocks, intact rocks

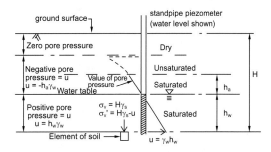

FIG. 4.7. Stress conditions and pore pressures below the ground surface.

$\rho_s \sim 2$ Mg/m^3, the unit weight of soil is approximately equal to 20 kN/m^3.

As soils, including clay materials are particulate substances, the inter-granular pore space contains fluids; typically air, water vapour and/or liquid water. In a typical soil profile such as that shown in Figure 4.7, in which the water table lies at some depth below the ground surface, the water in the ground below the water table will bear a pore water pressure. If the water is static, the value of the pore water pressure increases in proportion to the depth (h_w) below the water table as given by Equation (4.20).

$$u = \gamma_w h_w \text{ (kPa)} \quad (4.20)$$

where γ_w is the unit weight of water (kN/m^3).

As Equation (4.20) indicates, at the water table where $h_w = 0$, the pore water pressure is zero, which also defines the position of the phreatic surface. The soil may be saturated for some distance above the water table, within what is known as the closed capillary fringe, owing to capillary rise of water in the pore space. The value of pore pressure, which is given in Equation (4.21), is negative in this zone so the soil bears a soil suction

$$u = -\gamma_w h_a \text{ (kPa)} \quad (4.21)$$

where h_a is the height above the water table.

Above this zone soil is unsaturated in what is known as the open capillary fringe in which pores are occupied by both air and water. Surface tension forces in the menisci of water occupying the pore spaces in this zone results in suction pressures. In fact where water is removed from the capillary fringe, for example by evaporation from the ground surface, the flow of water through capillaries will be towards the ground surface, not necessarily vertically upwards. Further discussion of the properties of unsaturated soils is given in Section 4.2.6.

As the strength and deformation properties of soils are due to the forces between the particles, both the pore pressures or suctions and the forces due to the self-weight of the material as well as any externally applied forces, must be taken into account. The weight of a foundation structure resting on the soil mass would be an example of an externally applied force.

In a situation such as that represented in Figure 4.7, the effective vertical stress is given by the expression:

$$\sigma_v' = \sigma_v - u. \quad (4.22)$$

Note that $'$ is used to distinguish parameters such as σ_v', expressed in terms of effective stress from those like σ_v that are in terms of total stress. So, from Equations (4.19), (4.20) and (4.22), the vertical effective stress at depth H is:

$$\sigma_v' = \gamma_s H - \gamma_w h_w \quad (4.23)$$

where the water table is at depth $(H - h_w)$ m below the ground surface (see Fig. 4.7).

As the pore pressure can be positive or negative, the effective stress may be respectively lower or higher than the total stress. Also, as it is a fluid pressure it acts equally in all directions, hence the absence of a suffix to denote direction. Rearranging Equation (4.22) gives the formula for total vertical stress:

$$\sigma_v = \sigma_v' + u. \quad (4.24)$$

The significant effect pore pressure has on soil stress is demonstrated in Text Box on p. 97.

In the case of a flat area of ground on which no external stresses are applied, then the vertical stress, σ_v', acting in the ground induces a horizontal or lateral stress, σ_h'. The ratio of the horizontal stress to the vertical stress is known as the coefficient of earth pressure at rest, K_0, as defined in Equation 4.25.

$$K_0 = \frac{\sigma_h'}{\sigma_v'}. \quad (4.25)$$

Although in most cases the major principal stress (see Text Box on p. 90) is vertical so $\sigma_1' = \sigma_v'$ and the minor principal stress is horizontal so $\sigma_3' = \sigma_h'$, there are situations, for instance near to steep slopes or where stresses are imposed on the ground, in which this may not apply. Also, as explained in Section 4.3.3 in over-consolidated clays, due to a reduction in vertical loading, the major principal stress may be horizontal.

It is the effective stress that determines the strength and deformation properties of soils. Subjecting soils to stress or loading changes the pore pressure. The pore pressure may also change if the position of the water table or the pattern of water flow changes. From Equation (4.20), it can be seen that raising the water table in the soil profile increases the pore pressure, which, from Equation (4.22), causes a reduction in effective stress. Because, from Equation 4.37 this causes a reduction in strength, landslides are more likely to occur after a period of heavy rain.

Increases of pore pressure above the steady state value result in excess pore pressure. This excess pore pressure will dissipate as water flows from the region of high pressure into regions of lower pressure. During this process the soil undergoes consolidation (Section 4.3.3). Similarly, if the pore pressure is caused to decrease, water will be drawn into the zone of lower pressure. Due to the low permeability of clays these changes in pore pressures can take years or even tens of years to accomplish in typical engineering situations.

Failures in cuttings in London Clay some 40 to 50 years after construction were attributed by Skempton (1977) to the dissipation after that time of high pore water suctions that were previously responsible for maintaining the slopes in a stable condition. The high angle slope on London Clay cliffs on the Isle of Sheppey, which have not failed, have also been attributed to this (Dixon 1987).

4.2.6. Unsaturated clays and suction

The amounts of water and gas present within a soil or rock define its state of saturation. If all the pore space is occupied by water so there is no gas present, then the soil

Groundwater and pore pressures

The effect of the presence of a static groundwater table on effective stresses in the ground may be demonstrated by considering a simple example. If the water table is at the ground surface (so $h_w = H$) and a soil unit weight of $\gamma_s = 20$ kN/m³, which is typical for many clay soils, is assumed, and remembering that the unit weight of water is approximately 10 kN/m³, then the effective vertical stress is half the total vertical stress:

Pore pressure $u = \gamma_s \times H = 10 \times H$ (kPa)
Total vertical stress $\sigma_v = \gamma_s \times H = 20 H$ (kPa)
Effective vertical stress $\sigma_v' = (\sigma_v - u) = (20-10) \times H$

So for these special conditions

$$\sigma_v' = 10 \times H \text{ (kPa)}$$

And hence

$$\sigma_v' = \tfrac{1}{2}\sigma_v.$$

Thus the pore water significantly changes the magnitude of the resulting inter-particle stress.

In this case a water table at the ground surface was assumed. If it is lower than this, then the difference between the total and the effective stress is reduced. The pore pressure is zero if the water table lies below depth H. (If the water table is at or just below depth H, such that H lies within the capillary zone, the pore pressure may be negative.)

If the water level is above the ground level, as it would be in a body of water, then the effective stress in the ground below the water is independent of the depth of the water. This is because both the total stress and the pore water pressure increase by the same amount as the depth increases, so their difference (the effective stress) remains constant. If the depth of water is d then:

Pore pressure $u = \gamma_s \times (H+d) = 10H + 10d$ (kPa)
Total vertical stress $\sigma_v = \gamma_s \times H + \gamma_w \times d = 20H + 10d$ (kPa)
Effective vertical stress $\sigma_v = (\sigma_v - u) = 20H + 10d - 10H - 10d$ (kPa)

So for these conditions

$$\sigma_v' = 10 \times H \text{ (kPa)}.$$

And hence the effective stress depends only on the depth below the ground surface and not on the depth of the water above this.

---

or rock is saturated and, conversely, if there is no water it is dry. Where both water and gas are present, the soil or rock is said to be partially saturated or unsaturated.

As mentioned in Section 4.2.5, the soil above the water table contains water in the closed and open capillary fringes. The height to which water may rise above the water table due to capillary movement depends upon the size of pores in the soil and in clays it can be tens of metres (Lane & Washburn 1946), although the rate of flow is low. The suction pressure of the water within the closed capillary zone is given by Equation (4.21). However, in the open capillary fringe and in other situations in which the soil is partially saturated, suctions can be much higher than this and are greatest in finer-grained materials, like clays, as they contain smaller pores.

Total suction is made up of matric suction which is accounted for by surface tension forces in narrow capillaries together with the attraction of water and cations to mineral surfaces plus solute or osmotic suction. The latter is due to diffusion of water from regions of low concentration of solutes towards the regions of higher concentration.

Matric potential increases (i.e. suction or negative pore pressure increases) as the moisture content of a soil is decreased. Due to the resulting increase of effective stress, the strength of partially saturated clay is higher than that of otherwise identical saturated clay, an effect investigated by Croney & Coleman (1960).

In partially saturated soils pore air pressure will also be present and matric suction, s, is the difference between the pore air, u_a, and pore water suction u_w:

$$s = u_a - u_w \tag{4.26}$$

and as in most practical situations u_a is atmospheric pressure, and is small in comparison with u_w, matric suction is approximately equal to pore water suction, thus:

$$s \sim u_w \quad (4.27)$$

Fredlund & Rahardjo (1993) argue that any two out of the three stress state parameters: net normal stress $(\sigma - u_a)$, matric suction $(u_a - u_w)$ and total stress minus the pore water pressure $(\sigma - u_w)$ may be used to describe the stress conditions in an unsaturated soil. Of these the combination of net normal stress and matric suction is preferred as the change in total normal stress is separated from effects due to change in pore water pressure. Also, the pore air pressure is usually atmospheric and as indicated earlier, taken to equal zero. This leads to the modified form of the shear strength equation given in Equation (4.37) for saturated soils. Further discussion of the appropriate approaches to be taken when dealing with unsaturated soils is provided by Smith & Smith (1998).

Suction is often quoted using the pF scale of Schofield (1935), shown in Table 4.19 where the pF value is the power to base 10 of the suction measured in centimetres of water. As it is not an SI unit, it is preferable to use kPa.

Determination of the effective stress conditions in an unsaturated soil requires consideration of both pore water and pore air pressure. Smith & Smith (1998) refer to a model of soil behaviour proposed by Wheeler & Karube (1995). They point out that wetting a dry soil can cause either swelling or collapse (see Section 4.4.3). This is because the menisci of water between the soil particles tend to hold them together such that removal of this force causes collapse. On the other hand, the suction forces acting on the fabric of the soil cause the soil to be in a compressed state and reduction of this force allows the soil to swell as pore water is drawn into the pore space.

The relationship between suction value and moisture content takes the form shown in Figure 4.8. Note that greater suction is required to draw water out of narrow pores (drying) than is required to refill the larger pores (wetting) and that greater suction is required to reduce the moisture content of finer-grained deposits.

4.2.7. Clays for ceramics applications (see also Chapter 14)

Different types of clays, including ball clays, china clays, brick and tile clays, fireclays and refractory clays, are used for various industrial uses, including the manufac-

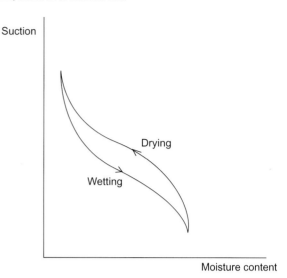

FIG. 4.8. Changes in clay suction with moisture content.

ture of ceramics and other fired clay products. These are distinguished mainly by their composition, which gives rise to different properties. Brick clays occur widely in the UK with suitable clays occurring as bedrock and superficial deposits. There has been a reduction in the number of brickworks and a concentration of production in fewer, but larger factories, which exploit various Carboniferous shales and clays, Mercia Mudstone and also Jurassic, Cretaceous and Tertiary clays. Blunden (1975) points out that Oxford Clay has highly desirable properties for brick making as it has a high content of carbonaceous matter, which reduces the fuel required for firing. Furthermore, Oxford Clay is sufficiently plastic in its natural state for moulding and it has favourable shrinkage properties. Bell (1992) assesses the suitability of various Carboniferous mudrocks as brick clays.

Gillott (1987) explains that china clays consist mainly of kaolinite and, if pure, can be used for the manufacture of china, pottery and other ceramics. Paper manufacture is another, if not *the* major use for kaolinite. Deposits of kaolinite are confined in the UK to Southwest England. Ball clays, which are used for the manufacture of earthenware, stoneware and other pottery products have high plasticity and usually contain illite as well as kaolinite and sometimes montmorillonite. The UK occurrences of such materials are in South and Southwest England. Refractory clays, including fireclays, occur as the seatearths of coal seams within the Coal Measures sequence. Besides being utilized in the manufacture of firebricks, they are also used in the production of sanitary ware, pipes, tiles and high-grade bricks.

Various non-clay minerals and impurities can have a significant influence on the performance of fine ceramic clays, as Table 4.20 shows.

In ceramics applications the moisture content may be defined either as percentage of water per unit mass of dry

TABLE 4.19. *pF scale for suction pressures*

	Soil suction pressure	
pF value	cm water	kPa
0	1	0.1
1	10	1
2	100	10
3	1000	100
4	10 000	1 000
5	100 000	10 000
6	1 000 000	100 000

TABLE 4.20. *Commonly used additives and impurities that influence the performance of fine ceramic clays*

Mineral or impurity	Effect in fired product
Quartz	Acts as a filler and reduces the fired shrinkage; can cause micro-cracks to form and reduces the fired strength
Tourmaline	Slight discolouration. Some HF generated in flue gasses
Haematite, pyrite, magnetite, siderite	Black spots. Pyrite will contribute to SO_2 in flue gasses
Goethite	Important controller of hue in final colour of structural clayware, e.g. strong red colour in roof tiles
Gibbsite, gypsum	Strong flocculants
Lignite	May improve green strength (strength after moulding) and rheological properties if very finely divided or in a colloidal form; may result in the formation of 'black holes' and black hearting in fast-fired products.
Rutile	Gives buff colour to fired product
Anatase	Gives slight yellow colour to fired product

soil or the percentage of water per unit mass of wet soil. The moisture content at which no further shrinkage occurs with drying is termed the critical moisture content. This is equivalent to the shrinkage limit defined in Section 4.2.3. It is an important parameter because the drying of clay can be much more rapid, without the fear of cracking, if the moisture content is below this value.

Ryan & Radfield (1987) explained that the shrinkage of pottery clays is determined by scribing marks 10 cm apart on prisms of prepared clay, allowing 24 hours drying at 110°C and then firing. The distances between the marks are noted before and after drying and after firing. The linear shrinkage values are then expressed as

- wet to dry: the shrinkage as a percentage of the wet length;
- dry to fired: the shrinkage as a percentage of the dry length;
- wet to fired: the shrinkage as a percentage of the wet length.

The definition of a parameter which provides a useful indication of the plasticity of clays used in the production of pottery has presented a challenge yet to be satisfactorily resolved. Although plasticity index described in Section 4.2.3 provides some useful information data about the range of moisture contents over which plastic behaviour occurs, it does not provide any indication of the amount of plastic deformation that can be achieved before rupture. Various approaches are available. Ryan & Radfield (1987) define the plasticity index as the amount of strain, which occurs between the yield point and rupture, this depends on the moisture content of the clay. Other indications of the plasticity include the value of dry strength of the clay, as plastic clays are also stronger than low-plasticity clays.

The moisture content can also be used to indicate the plasticity properties; as finer clays require higher moisture contents to achieve optimum plasticity. A further approach to the measurement of plasticity is provided by the Pfefferkorn test (see Ryan & Radfield 1987). In this cylindrical specimens of the remoulded clay at different moisture contents are deformed by dropping a standard weight on them. The Pfefforkorn number is the moisture content for a deformation ratio of 3 : 1. The BCR compression Plastometer test, in which a cylindrical sample is compressed between two parallel plates, may also be used.

It should be appreciated that the value of plasticity index obtained in these ways does not equate with the Atterberg Limits plasticity index defined in Section 4.2.3. A stiffness index can also be defined using this methodology: A further approach to plasticity assessment is to determine the amount of a non-clay material that needs to be added to the clay to make the mixture non-plastic. This is expressed as a percentage of the amount of clay.

4.2.8. Thermal properties of clays

Temperature changes in clays may arise from many sources, including hot buried pipes, radioactive wastes, buried electricity cables and both waste disposal and contaminated landfill sites (Mitchell 1993; Harbidge *et al.* 1996; Hueckel & Peano 1996). For instance, Collins (1993) notes that temperatures in excess of 60°C are typical in landfill sites. However, much greater temperatures can be experienced with high-level radioactive waste disposal. Besides heating, which occurs as a by-product of waste disposal, heating of soils would be deliberate in systems designed to store heat in the ground (see Adolfsson *et al.* 1985; Gabrielsson *et al.* 2000). Soils are also heated in the course of certain forms of contaminated ground treatment or to improve the geotechnical properties.

Youssef *et al.* (1961) and Mitchell (1993) noted that changes in temperature can occur during sampling and that the temperature during testing is usually higher than that in the ground. Mitchell (1993) showed that such a change might alter the composition of the pore solution so parameters measured in the laboratory may be unrepresentative of the true behaviour of a soil in the field. Much work, mainly in connection with radioactive waste disposal, has taken place on modelling temperature effects. These have utilized rate process ideas (see Campanella &

Mitchell 1968), double layer theory, rheological models (Virdi 1984) and stress equivalent proposals (Burrous 1973). Subsequently, modelling has adopted the critical state framework (see Section 4.3.5) for saturated soils and unsaturated soils (Romero *et al.* 2003).

As temperature affects the viscosity of fluids, the coefficient of permeability is sensitive to the temperature. As the intrinsic permeability depends only on the properties of the porous medium, its value does not change with viscosity. Barnes (2000) provides a correction factor for permeability values:

$$k_t = k_{20} \times \frac{\eta_{20}}{\eta_t} \qquad (4.28)$$

where k_t and k_{20} are the coefficients of permeability at temperatures of t and 20°C respectively and η_t and η_{20} are the dynamic viscosity of water at temperatures of t and 20°C.

Considering how useful Atterberg Limits are for clays, surprisingly little work has been carried out on the effects of temperature on soil plasticity. In most cases a reduction in liquid limit with temperature would be expected (Youssef *et al.* 1961; Laguros 1969; Ctori 1989).

Less consistency has been observed with plastic limit values by Youssef *et al.* (1961) and Ctori (1989), who respectively carried out tests between 15 and 35°C and between 6 and 35°C. All researchers observed reduction, whereas Laguros (1969), in tests between 2 and 41°C, observed an erratic trend overall. On the other hand, for temperatures between 8 and 22°C, Tippet (1976) detected no changes in plastic limit for three clays containing varying amounts of well-crystallized and disordered kaolinite and Reifer (1977) made similar observations. Previous authors have suggested that these results could be explained in terms of changes to the double layer, the viscosity of water, coagulation and/or the geometrical re-arrangement of the particles with temperature. Youssef *et al.* (1961) proposed correction factors for temperature, based on changes in the viscosity of water with temperature. However, these met with only a limited degree of success and it is likely that the results are produced by a combination of effects.

These conflicting results are further confused by questions concerning the reliability of plastic limit tests, which are more subjective and susceptible to human error than the liquid limit test. In the plastic limit test, transfer of heat from the hand may occur even in tests that are carried out at normal laboratory temperatures so it is likely that the temperature of the clay is somewhere between room and hand temperature. Attempts to overcome this difficulty with the use of gloves have not produced consistent results.

The effect of temperature on liquid limit also applies to other parameters. For instance, the decrease in strength with increase in temperature means that the optimum moisture content for maximum dry density in compaction tests (see section 4.2.8) is also reduced. Conversely there will be an increase in permeability and the coefficient of consolidation with increased temperature.

The effects of mineralogy on the response of soils to changes in temperature have received little attention in the literature. For example, montmorillonite may well become coagulated at a higher temperature and thus exhibit an increased liquid limit. This could account for the observations made by Reifer (1977), even though for the liquid limit test the sample should be in a completely remoulded state. This strength gain appears to be very rapid and therefore, the effect is significant even with relatively quickly conducted liquid limit tests. The heat capacity of clay soils may also be modified by changes in temperature, but unlike geotechnical properties, the change is either small or insignificant for kaolinite (Jefferson 1994; Button & Lawrence 1964) for temperatures between 0 and 60°C. By contrast, a marked temperature dependency occurs with montmorillonite, where the magnitude of zeta potential decreases linearly with temperature. Classical double layer theory indicates that temperature should have a negligible effect on the zeta potential as counteracting changes occur with temperature change. However, as indicated above, the properties of montmorillonite are susceptible to change in response to temperature. It is postulated that this is the result of changes in dissolved ions and the viscosity of water as well as the greater tendency for particle agglomeration to occur at higher temperatures.

4.2.9. Compaction of clays

Compaction and consolidation are terms that are used in different senses by geologists and engineers. To geologists, compaction is the process of increasing density and decreasing moisture content that occurs due to increases in overburden pressure in an accumulating series of sediments. Engineers regard this response to a static, but possibly steadily increasing pressure, as consolidation (see Section 4.3.3) and define compaction as the process by which the density is increased by the application of transient, or dynamic, energy (see Chapter 10). This energy may be supplied by rolling, vibration, impact or some combination of these processes. The word 'compaction' is used in this Chapter in its engineering sense. The methods of compaction are designed to reduce the volume of air voids present in the soil, without significantly changing the moisture content. The state of compaction of a soil is usually measured by the change in its dry density with change in moisture content (see Chapter 9). The increase in dry density achievable by compacting a soil is a function of the compactive effort during compaction and the moisture content of the soil, as well as depending on the soil grading and characteristics of the particles present.

The compaction properties of soils are usually found by measuring the change in dry density resulting from changes in moisture content. As shown in Figure 4.9, the maximum dry density is achieved at the optimum moisture content, this typically occurring at an air voids percentage (see Table 4.12) of approximately 5%. It is

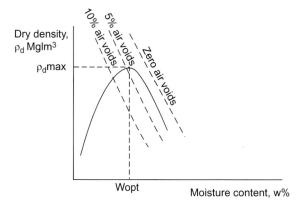

FIG. 4.9. Compaction behaviour of clays.

possible to vary the compactive effort used in testing so that the effects of using different compaction methods can be assessed.

Berry & Reid (1988) note that different changes occur to the soil structure at successively higher moisture contents. Higher moisture contents promote a greater degree of breakage and remoulding of soil lumps and also lead to greater dispersion and orientation of clay particles. Thus clays can be difficult to compact but increasing the moisture content results in an increase in density as it is then easier to remould soil lumps. The increase in density with increasing moisture content occurs until the optimum moisture content is reached. Increases in moisture content above this figure, result in lower density as excessive amounts of water occupy the pore space and it is impossible to remove this using compaction. Berry & Reid (1988) note that soil compacted at a moisture content that is higher than the optimum moisture content (wet of optimum) has higher strength, higher compressibility at low consolidation pressures and lower potential for swelling but greater potential for shrinkage than soil compacted dry of optimum.

The suitability of compacted clay used for road construction can be expressed in terms of its California Bearing Ratio (CBR). The test is described in Chapter 9. It entails measuring the resistance to a 2.54 mm penetration into the soil of a 49.6 mm diameter plunger. This is expressed as a percentage of the resistance measured when the plunger is pushed into a standard crushed stone.

The selection of soils for compaction can be aided by the use of the soil moisture condition value (MCV). The test is described by Parsons & Boden (1979). In it the penetration of a standard rammer into a compacted sample of the soil is measured. The difference in measured penetration for a particular blow count and four times that blow count is plotted against the number of blows. The MCV is determined from Equation (4.29) where B_5 is the number of blows for a 5 mm change in penetration. In effect this is a measure of strength of the soil in a condition of near full compaction.

$$MCV = 10 \times \log_{10} B_5 \qquad (4.29)$$

4.3. Geotechnical parameters: strength and stiffness of clay materials

The purpose of this section is to define and explain the parameters used to quantify strength and stiffness of clays in the manner usually adopted for geotechnical applications. Clays are particulate materials in which the particles interact with each other mainly via friction. Although the particles may undergo mutual attraction and repulsion due to electrostatic and Van der Waals forces, in most practical situations these effects are minor in comparison with friction. The stresses between particles are therefore a function of the characteristics of their surfaces and the confining pressure conditions. The latter may be modified by pore pressures or suctions (see Section 4.2.5). Some clays may contain some cemented clusters of particles or minor inter-granular cementation may be present; in rocks cementation and inter-particle bonds will be present.

This Section deals with clays as engineering soils where standard theories of soil mechanics apply so long as the soil is saturated and uniform. Such materials display the following basic features:

- their behaviour can be explained in terms of the principle of effective stress (see Section 4.2.5);
- behaviour may be drained or undrained depending on the rates of loading and drainage conditions (see Section 4.3.7);
- for isotropic compression volume varies with the logarithm of effective stress (see the Text Box on p. 90);
- strength is proportional to the mean effective stress (e.g. in effective stresses soil is frictional) and the logarithm of strength decreases with increasing water content;
- deformations are essentially plastic and irrecoverable and stiffnesses are highly non-linear (see Section 4.2.2).

Strength and stiffness are defined as follows (see the Text Box on p. 90):

Strength. The shear stress at failure. Failure may be taken to mean that the material ruptures or that it undergoes plastic flow at constant shear stress,
Stiffness is a measure of the resistance of the material to deformation. In other words, it is the amount of stress required to cause unit strain.

Clay materials derive their capacity to resist deformation by mobilizing shear resistance within the soil mass. In other words, deformation due to imposed stress is resisted by frictional forces between the particles as they move, or attempt to move, with respect each other. In some materials cements or electrostatic forces may also contribute to the strength. Hence strength and stiffness are influenced by mineralogy, grain size, structure, inter-granular cementation, moisture content and inter-particle pressures. Because pore pressures and structure exert

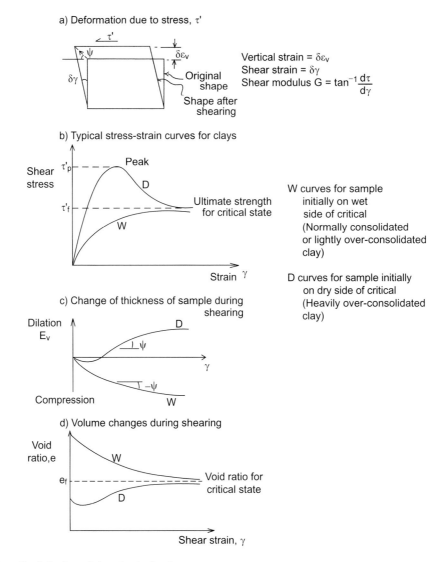

FIG. 4.10. Deformation behaviour of clays due to shearing.

significant controls over strength and stiffness, it is necessary to be precise about the pore pressure and state of disturbance of the soil.

4.3.1. Deformation

The typical strains caused to clay materials of medium or low moisture content by applied shearing stress are shown in Figure 4.10. The exact form of the shear stress-strain curve depends on the initial density and structure of the clay. If the density is high, the shear stress will rise to a peak value (Curve D in Fig. 4.10b) before dropping to a constant lower value, whereas for a low density, the shear stress rises to a constant value, as in curve W. The strength is the maximum shear stress that can be sustained by the material and, as shown, this may either be a 'peak' or an 'ultimate' value.

The stiffness of materials is expressed in terms of the ratio between the imposed stress and the resulting strain as a modulus of deformation. Different moduli are defined for the linear, volumetric and shear deformation. These are respectively the modulus for linear strain (Young's Modulus, E, for uniaxial stress), the Bulk Modulus, K and Shear Modulus, G, where a particular applied stress causes a linear strain ε, volumetric strain, ε_ω, or shear strain, γ. As the stress–strain relationship for soils is strongly non-linear, the value of the deformation modulus varies with the value of the stress. Thus it is necessary to define the value for a particular stress range. Tangent and secant values are defined in Figure 4.3.

These lead respectively to the modulus for linear strain (Young's Modulus, E), the Bulk Modulus, K, and Shear Modulus, G, as follows:

$$E_{\text{sec}} = \frac{\Delta\sigma}{\Delta\varepsilon} \tag{4.30}$$

$$K_{\text{sec}} = \frac{\Delta\sigma}{\Delta\varepsilon_v} \tag{4.31}$$

$$G_{\text{sec}} = \frac{\Delta\tau}{\Delta\gamma} \tag{4.32}$$

where Δ represents the change in the value of the parameter, σ is the applied stress, ε is the resulting linear strain, ε_v is the volumetric strain and γ is the shear strain (see Figure 4.3). The corresponding value of modulus is:

$$E_{\text{tan}} = \frac{d\sigma_1}{d\varepsilon_1}. \tag{4.33}$$

The initial tangent modulus, E_0 is defined as the slope of the stress stain curve at the origin. Values for G_0 and K_0 are similarly defined.

These moduli may be expressed in terms of either effective or total stresses, where E' and E_u refer to respectively to effective and total stress conditions and the same qualifiers are used for K and G, although $G_u = G'$ and $K_u = $ infinity (∞).

Useful relationships exist between the moduli, such that if two are known the third can be calculated:

$$G = \frac{E}{2(1+v)} \tag{4.34}$$

and

$$K = \frac{E}{3(1-2v)} \tag{4.35}$$

where v is Poisson's Ratio as defined in the Text Box on p. 90, Equation (4.7). As for undrained conditions, there is no volume change, $v_u = \frac{1}{2}$ and substituting this into Equation (4.35), it follows that $K_u = $ infinity (∞).

The deformation of soils may also be expressed in terms of compressibility, which is the inverse of bulk modulus. Compressibility is discussed in Sections 4.3.3 and 4.3.4.

4.3.2. Strength parameters

Strength is determined by carrying out tests in which a sample is subjected to an increasing shear stress until it fails. As further explained in Chapter 9, there are a number of ways of doing this depending on the application. By changing the arrangements for the test, compressive, tensile and shear strength can be determined (see Figure 4.11). All of these different types of strength are a function of the maximum shear stress the material can resist before failure occurs, where, as explained below failure may be defined by a number of different criteria.

a) Uniaxial tensile strength, σ_t'

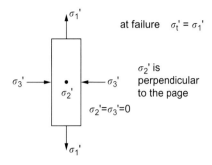

b) Uniaxial compressive strength, σ_c'

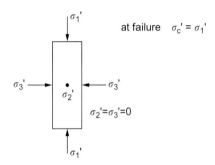

c) Shear strength, τ'

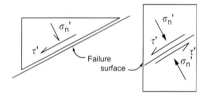

At failure
$\tau' = c' + \sigma_n' \tan\phi'$
where $c' = $ cohesion and
$\phi' = $ angle of friction for the surface

FIG. 4.11. Definitions of strength in clays and mudrocks (after Atkinson 1993).

Soils can be caused to fail by imposing a range of different stress combinations upon them. Combinations of stresses that cause failure are called a failure criterion. There are two main criteria of failure for soils, which define failure respectively caused by the undrained (or total stress) conditions and drained (or effective stress conditions). These are respectively the Tresca and the Mohr–Coulomb criteria. Figure 4.12 shows these failure criteria defined in terms of the shear stress–normal stress conditions present in the soil mass. Under the Tresca criterion, the shear stress to cause failure, τ_f, is a constant equal to the shear strength, s_u:

$$\tau_f = s_u. \tag{4.36}$$

d) Bending strength

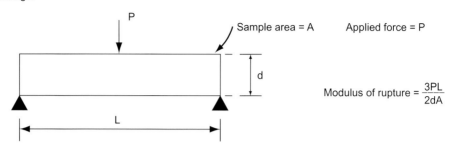

e) Diametral loading (Brazilian) test

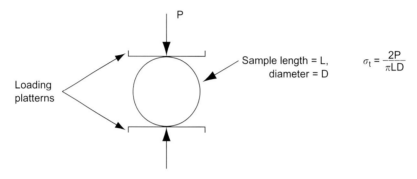

FIG. 4.11. (Continued) Definitions of strength in clays and mudrocks (after Atkinson 1993).

On the other hand, the Mohr–Coulomb criterion predicts that failure will occur when:

$$\tau' = c' + \sigma_n' \tan\phi' \qquad (4.37)$$

where σ_n' is the effective stress acting normal to the failure surface, ϕ' is the angle of internal friction of the soil and c' the cohesion of the soil. In the case of clays c' is very small and, as the Mohr envelope is curved, this linear equation applies only within a restricted range of σ_n' values. The shear strength parameters (c' and ϕ') and the stresses are all expressed in terms of effective stresses in this equation and it means that for a particular soil, its strength can be calculated if they and the effective stress (σ_n') are known. Figure 4.12 shows examples of Mohr circles appropriate to these criteria. The manner in which the values of the shear stresses and normal stresses that exist in the soil are represented by these Mohr's circles is explained in the Text Box on p. 106.

For disturbed soils and some undisturbed soils $c' = 0$. However, where soil particles are held together by cements, as occurs in some soils, the material will possess some strength even when $\sigma_n' = 0$ and thus c' does not equal zero in these cases. The effective stress conditions apply if any pore pressure generated as a consequence of volume change has dissipated so the stresses applied to the soil are resisted solely by inter-granular friction. As pore fluid cannot sustain a shear stress, $\tau' = \tau$ and consequently many authorities omit to distinguish between τ' and τ.

Deformation of soils results in the generation of pore pressures or suctions. With time, such pore pressures dissipate as water flows to equalise the pressure throughout the soil mass. In impermeable soils like clays, this may take a long time (tens of years in typical situations), whereas in sands and silts the process is much more rapid (minutes to weeks).

Discussion of the applicability of the various shear strength parameters in different situations is presented in Section 4.3.6.

4.3.3. Compression

The compression or swelling of soil is the change in volume response it makes to changes in effective stress. The related process of consolidation is the process by which the volume of a soil changes by the expulsion or ingress of pore water in response to the imposition of a higher constant total stress. In a natural soil consolidation may occur as a consequence of the deposition of overlying sediments or the weight of an ice sheet. In an engineering situation it may result from the loading imposed by a building or fill. If the deposit is saturated, the stress causes an increase in pore pressure. Dissipation of this

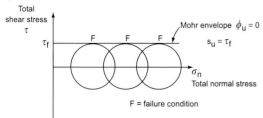
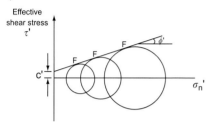
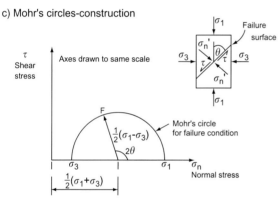

FIG. 4.12. Failure criteria and Mohr's circles.

pore water pressure by flow of pore water into neighbouring soil masses leads to an increase in effective stress, re-arrangement of soil particles and a reduction in volume as the water is squeezed out. The reduction in moisture content results in an increase in strength.

The process of compression is distinct from soil compaction described in Section 4.2.9 in which a reduction in pore air volume is brought about by the application of transient stresses due to, for example, rolling or vibrations.

Owing to the low permeability of clay soils, pore pressure change is slow. There are two important consequences of this:

- the increase in strength due to water content reduction can lag behind the increase in applied pressure and lead to failure of the soil mass;
- settlement of clay due to expulsion of pore water may continue after the construction period is completed and can result in damage to structures.

For a soil being compressed from a slurry by an isotropic (equal all-round) stress, as in the case of a freshly deposited sediment, empirical observations confirm that the volume (v) of the soil will decrease with the logarithm of the applied pressure (p').

The change in volume of a soil is usually expressed in terms of the specific volume (ratio of the total volume of the soil to the volume of solids in the soil) or it can be expressed in terms of the void ratio (e) (see Table 4.12). Figure 4.13 shows the behaviour of an ideal clay soil during loading from an initial condition O, where the volume decreases linearly with the logarithm of effective stress (curve OA). This is the normal compression line (NCL) for the soil.

The slope of this line depends mainly on particle size and is steepest for colloidal clays and lowest for silts. Reduction of the stress (unloading) causes some swelling (line AB) and the volume for any particular stress is less than during initial compression. With a subsequent increase in stress (reloading), the stress–volume relationship follows a new relationship (BC) until yield occurs (at the yield pressure p'_y) and the Normal Compression Line CD is resumed. The equations for the normal compression OD and swelling lines AB in terms of volume, e and the parameters λ, κ and N, that are constants for a particular soil, are as follows:

$$\text{loading} \quad v = N - \lambda \ln p' \tag{4.40}$$

$$\text{swelling} \quad v = v_k - \kappa \ln p' \tag{4.41}$$

where N is the value of v for $p' = 1.0$ kPa, $-\kappa$ is the slope of the swelling line AB, $-\lambda$ is the slope of the normal compression line OD and v_k is the value of v for $p' = 1.0$ kPa.

The bulk modulus, K' which is the ratio of the change in volumetric strain ($d\varepsilon_v$) to the stress change (dp') causing that strain, is given by:

$$K' = \frac{dp'}{d\varepsilon_v}. \tag{4.42}$$

Differentiating Equation 4.40 with respect to p' gives:

$$\frac{dv}{dp'} = \frac{-\lambda}{p'} \tag{4.43}$$

so

$$dp' = \frac{-p' dv}{\lambda}. \tag{4.44}$$

Since

$$d\varepsilon' = \frac{-dv}{v} \tag{4.45}$$

substituting Equations (4.44) and (4.45) into Equation (4.42) gives:

$$K' = vp'/\lambda \tag{4.46}$$

Construction of Mohr's circles

Mohr's circles are a convenient way of interpreting the results of imposing stresses on materials. As explained in the Text Box on p. 90, any system of stresses in a body can be represented by three mutually perpendicular principal stresses σ_1, σ_2, σ_3, where $\sigma_1 > \sigma_2 > \sigma_3$. As in many practical situations $\sigma_2 = \sigma_3$, it is convenient to consider the situation in terms of just σ_1 and σ_3. Applying stresses σ_1 and $\sigma_2 = \sigma_3$ to a body results in the generation of shear stresses and normal stresses within the body. The values and directions of these are predicted by the Mohr Circle.

The situation of applying σ_1 and $\sigma_2 = \sigma_3$ to samples of soils or rocks in triaxial tests is described in Chapter 9. In these tests the conditions for failure of a cylindrical sample are determined where σ_1 is the pressure applied along the axis and σ_3 is the confining pressure at right angle to this (see Figure 4.12). In the simplest type of triaxial test in which a saturated clay sample is tested in an undrained condition the total stresses imposed on the sample are equal to σ_1 and σ_3. Typically the sample will fail by the formation of a rupture surface at an angle of $45 - \frac{1}{2}\phi$ ° to an axial plane, where ϕ is the angle of internal friction for the soil. By resolving the forces within the sample it can be shown that the shear (τ) and normal (σ_n) stresses acting on the *failure surface* are:

$$\tau = \frac{1}{2}(\sigma_1 - \sigma_3)\sin 2\theta \quad (4.38)$$

$$\sigma_n = \frac{1}{2}(\sigma_1 + \sigma_3)\cos 2\theta - \frac{1}{2}(\sigma_1 - \sigma_3)\sin 2\theta \quad (4.39)$$

where θ is the angle between the σ_1 direction and the failure surface.

When τ is plotted against σ_n, for different values of 2θ, provided the same scale is used for both axes, as shown in Figure 4.12c, the graph is a circle that shows the relationship between τ and σ_n for different values of θ. The position and size of this circle depend on the values of σ_1 and σ_3. The radius is equal to $\frac{1}{2}(\sigma_1 - \sigma_3)$ and the centre lies at $\frac{1}{2}(\sigma_1 + \sigma_3)$ on the σ_n axis.

In a standard triaxial test, σ_3 is kept constant and σ_1 is increased until failure occurs along a surface orientated at θ to σ_1. The combination of τ and σ_n at an angle of 2θ that caused failure of the sample is represented by point F on Figure 4.12c. In order to determine the shear strength parameters (c and ϕ) for the soil, at least three tests are carried out to measure the axial stress (σ_1) that produces failure for three different σ_3 values and a Mohr circle is drawn for each test. The values of the shear strength parameters (c_u and ϕ_u or c' and ϕ') are found by drawing an envelope to these circles. This may be convex upwards but is usually approximated as a straight line.

For a test carried out on a saturated clay under undrained (total stress) conditions, the value of c_u is the value on the τ axis corresponding to $\sigma_n = 0$ and ϕ_u is the slope of the line (where both axes are plotted using the same scale). Generally speaking for undrained loading of a saturated clay, $\phi_u = 0$. This arises for undrained loading because an increase in σ_3 produces no change in moisture content of the clay and therefore the value for shear stress at failure (τ_f) does not change. Thus the Mohr envelope is a horizontal line and $\phi_u = 0$ and the undrained shear strength is $s_u = c_u$.

---

for compression and by a similar process

$$K' = vp'/\kappa \quad (4.47)$$

for swelling.

Where the volume–stress relation for a soil lies on the normal compression line the soil is said to be in a *normally consolidated* condition. At any particular point the effective stress value is the maximum borne by the soil since deposition. As the soil does not attain the pre-consolidation volume when it is unloaded, it is left in a denser state than it possessed at the same pressure on the virgin compression line. Such a soil is said to be in an *over-consolidated* state.

The amount by which a soil is over-consolidated is expressed in terms of the over-consolidation ratio (OCR), R_0:

$$R_0 = p_y'/p' \quad (4.48)$$

where p_y' is the maximum overburden pressure on the deposit and p' is the present overburden pressure. The previous maximum overburden pressure may be obtained from the $v - \ln p'$ plot in Figure 4.13b. The method of doing this is described by Berry & Reid (1988) and other books on soil mechanics.

Recent lacustrine clays and ablation tills (see Section 4.5.1) tend to be in a normally consolidated state because they have never been loaded by more than their present overburden. However, there are many older clays that are in an over-consolidated state because overlying sediments have been removed by erosion. Lodgement tills tend to be over-consolidated as they were deposited beneath an ice sheet that has subsequently melted. A similar, though less pronounced effect can also be produced by desiccation of clay deposits as pore water suctions increase the effective stress on the soil.

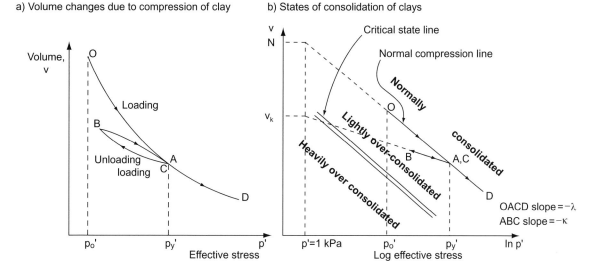

Fig. 4.13. Compression and consolidation condition for clays (after Atkinson 1993).

Before stress relief and weathering all the geologically aged (Tertiary and older) clay formations in the UK are over-consolidated (Section 4.5). As over-consolidated clays are denser and drier than normally consolidated clays at the same effective stress they are stronger and less compressible. The removal of overburden results in some expansion of the clay and the formation of structural discontinuities such as fissures and joints. A fissured or jointed clay or mudrock mass is weaker and more compressible than the material in a structurally continuous condition. Such a process may affect the clay at depths below those to which weathering processes usually penetrate. Nearer to the ground surface, degradation and weathering of the material commences at the surfaces of fissures leading to softening. This is accompanied by the formation of more discontinuities so that ultimately, the over-consolidated condition of the clay is destroyed.

Soils may be in a normally consolidated ($R_0 = 1$) lightly or heavily over-consolidated state. Most soils will be in a lightly over-consolidated state if the over-consolidation ratio is between 1 and 2 and in a heavily over-consolidated state if this ratio is greater than 3.

The critical state line (see Section 4.3.5) is the condition (in terms of volume for a particular effective stress value) the soil attains at shear failure where there is no change in volume for a particular value of normal stress. The critical state line lies parallel to the normal compression line. States between the critical state line and the normal compression line are said to be on the wet side of critical and those below the critical state line are on the dry side of critical. Note that all this discussion refers to soils in a saturated condition, and here soils are called dry because they have relatively low moisture content. They are denser, and therefore drier than equivalent soils on the wet side of critical.

The coefficient of earth pressure at rest, K_0 which is defined in Equation 4.25 as the ratio of the horizontal effective stress to the vertical effective stress, has a value of approximately $1 - \sin\phi_{crit}$ (see Section 4.3.5) in normally consolidated clays and sands. The value may increases to about 4.0 in over-consolidated clays. This is because although vertical stress is reduced lateral constraint prevents a reduction in horizontal stress from occurring. As the proportional change in vertical stress is larger towards the ground surface, K_0 increases towards the surface, but near the surface degradation of the clay by stress relief fissuring and softening in response to weathering action results in a reduction in K_0 value to less than unity.

4.3.4 One-dimensional compression and swelling

In many geological and engineering situations, soils are subjected to what approximates to one-dimensional loading. The amount of consolidation liable to occur due to an increase in applied pressure and the rate at which consolidation will occur is determined by testing samples of the materials concerned. As described in Chapter 9, the amount of compression due to a particular increase in applied pressure sustained by a cylindrical sample of soil held in an oedometer is measured against time.

The amount of compression can be quantified in terms of the one-dimensional compression modulus E_0' and the coefficient of compressibility m_v where:

$$m_v = d\sigma_z'/d\varepsilon_z \quad (4.49)$$

and

$$E_0' = 1/m_v. \quad (4.50)$$

Here $d\sigma_z'$ is the increase in effective stress in the z-direction and $d\varepsilon_z$ is the corresponding linear strain.

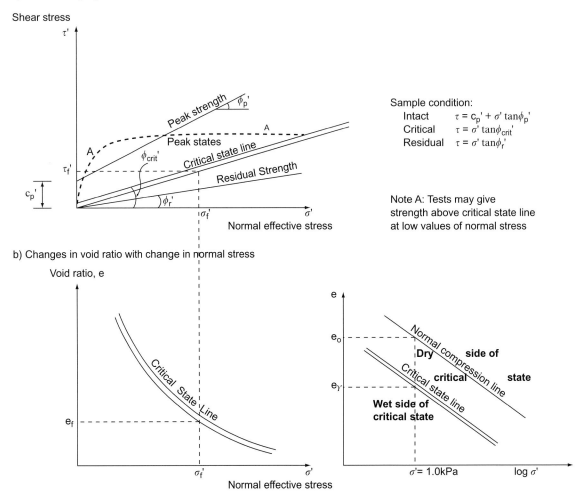

FIG. 4.14. One-dimensional compression and shear strength parameters for clays (after Atkinson 1993).

Figure 4.14 shows the typical form the volume change versus log pressure graph for one-dimensional consolidation and swelling. The parameters κ, λ and N in Figure 4.13(b) and Equations (4.40) and (4.41) are replaced respectively by the compression index, C_c, the swelling index, C_s and the initial void ratio, e_o:

$$e = e_o - C_c \log \sigma' \quad (4.51)$$

$$e = e_k - C_s \log \sigma' \quad (4.52)$$

where C_c and C_s are the slopes of the straight line portions of the e-log σ' graph respectively for compression and swelling. Both C_c and C_s are dimensionless parameters and they apply to normally consolidated clays. Typical values for the compression index for different clays are given in Table 4.21.

TABLE 4.21. *Typical values of compression index for different fine-grained soils* (C_c)

Compression index	Compressibility behaviour	Clay type
>3.0	Very high	Soft clay
0.3–0.15	High	Clay
0.15–0.075	Medium	Silt
<0.075	Low	Sandy clay

For overconsolidated clays, the over-consolidation ratio is:

$$R_0 = \sigma_{max}' / \sigma_0'. \quad (4.53)$$

where σ_{max}' is the maximum, and σ_0' the present, effective overburden pressure.

The rate of consolidation is expressed in terms of the coefficient of consolidation, c_v that depends on the permeability of the clay, as defined in Equation (4.54):

$$c_v = \frac{k}{\gamma_w m_v}. \quad (4.54)$$

As the permeability, k, of clay is usually very low, the rate of consolidation is also low.

As both m_v and c_v are sensitive to the stress range quoting a single figure for the parameter, does not accurately describe the performance of the material. BS 1377:1990 recommends that the value should be quoted for a pressure increment of 100 kPa in excess of the effective overburden pressure for the depth from which the sample was taken. It is useful to quote the compressibility behaviour in terms of the defined class ranges shown in Tables 4.22 and 4.23.

4.3.5. Critical state strength

Applying stress to a soil may cause it to fail by undergoing plastic strain, which may involve the development of a shear surface. Due to the process of turbulent flow of the soil particles, such shearing destroys the original structure of the soil. Once this process is complete the soil is in its Critical State in which the deformation occurs without change in shear stress, normal stress, or void ratio (and volume). All soils must ultimately achieve a critical state.

For soils in their critical state there is a unique relationship between the shear stress, normal stress and the void ratio or volume of the soil. In other words the void ratio at failure has a particular value that depends on the normal effective stress. This void ratio is a function of the grain size distribution (grading) and shape and surface texture of the grains (in turn this depends on mineralogy) but is independent of the initial density.

At the critical state the shear stress and void ratio are defined by:

$$\tau_f' = \sigma_f' \tan\phi_{crit}' \quad (4.55)$$

$$e_f' = e_\Gamma - C_c \log\sigma_f' \quad (4.56)$$

where f refers to the ultimate failure, C_c is the compression index, e_Γ is the value of void ratio for $\sigma_f' = 1$ kPa on the critical state line (equivalent to e_o for the normal compression line) and ϕ_{crit}' is the angle of friction of the soil for the critical state. These relationships are illustrated in Figure 4.14. As mentioned in Section 4.3.3, saturated soils may exist in a condition either wetter or drier than the critical state. In other words they can be denser or less dense than they would be in this condition. As illustrated in Figure 4.10 soils that are less dense than the critical state compress during shear until they reach a density commensurate with the critical state for the prevailing normal stress conditions. Such soils follow a stress–strain curve like curve W in Figure 4.10. On the other hand, denser than critical soils dilate during shear until the critical state is reached. The stress-strain relationship for these soils is like curve D with a peak shear stress. After the peak, the stress drops to the ultimate value, after which the volume and the shear strength remain constant.

These changes to the structure of the soil entail changes in volume that cause either dilation or compression of the soil (Fig. 4.10c) and changes in void ratio (Fig 4.10d) during shear. In the part of the soil undergoing shear the void ratio attains a constant value, e_f the value of which depends on the characteristics of the soil and the normal effective stress value, σ'. The typical form of the relationship between the void ratio and the normal effective stress for the critical state is shown as the Critical State Line in Figure 4.14b.

4.3.6. Shear strength parameters

As mentioned above and illustrated in Figure 4.10, dry of critical, dense soils display strengths higher than critical, as shown by curve A in Fig. 4.14a. This higher peak strength is traditionally represented by peak shear strength parameters c_p' and ϕ_p'. However, it should be appreciated that this implies a linear Mohr envelope, which may not be the case and it applies to a limited range of effective stressess. Furthermore, as cohesion, in terms of effective stress, of an uncemented soil is zero, or at least it is very small in comparison with friction, ascribing a c_p' value in such cases is potentially misleading.

Peak shear strength is displayed by intact, undisturbed soils. In practice the peak strength determined on small laboratory samples may be much greater than the mass

TABLE 4.22. *Typical values for the coefficient of volume compressibility (m_V) after Head (1982)*

Coefficient of compressibility (m²/MN)	Compressibility	Typical material
<0.05	Very low	Heavily over-consolidated clays and weathered rocks
0.05–0.1	Low	Till (Boulder clay)
0.1–0.3	Medium	Fluvio-glacial and lacustrine clays
0.3–1.5	High	Normally-consolidated alluvial clays
>1.5	Very high	Organic alluvial clays

TABLE 4.23. *Typical values of the coefficient of consolidation (c_V) after Lambe & Whitman (1979)*

Coefficient of consolidation, c_V (m²/year)	Rate of consolidation	Typical material
<0.01	Very low	
0.1–1.0	Low	>25% clay
1–10	Medium	15–25% clay
10–100	High	<15% silt
>100	Very high	

strength of the material. This is because fissures and joints are under represented in small laboratory test specimens.

With further shearing the soil attains the critical state and the corresponding shear strength parameters are: c_{crit}' and ϕ_{crit}'. This is the critical state line in Figure 4.14. It should be noted that this state may be localized just to the area of the soil undergoing shear failure, rather than throughout the mass of soil. The critical shear strength parameters are usually similar to those obtained for soils that have undergone natural disaggregation by weathering processes or, in the context of engineering operations involving clays, where they are in a disturbed condition, for example when compacted in an embankment structure. In fact, as the cohesion of a soil depends largely on structural features, as Figure 4.14 shows, c_{crit}' is usually zero, or close to this value.

When designing permanent structures founded on, or constructed out of soils, the effective shear strength parameters, should be used as it is the long-term strength of the material that is important. This implies that the effective stress conditions are known which might require determination of the pore pressure. As stress relief and weathering of a clay soil results in the reduction in strength from the peak to the ultimate value, the design of permanent structures should be on the basis of the effective critical shear strength parameters, c_{crit}', ϕ_{crit}'.

A third condition that needs to be described is a clay soil that has been previously sheared, as for example in a landslide or failed foundation. As Early & Skempton (1972) explain, shearing of clay soils leads to a reduction in strength below the critical value as flat, platy clay particles become arranged parallel to the direction of shearing. The establishment of such a particle arrangement may well be confined to a narrow zone. With sufficient shear displacement the strength reaches a minimum or the 'residual' value. This has been shown to be a constant that depends on the mineralogy and grain size of the soil. As depicted in Figure 4.14a, the cohesion in this state is zero so the shear strength is usually quoted as a value of ϕ_r'. In the case of clay soils ϕ_r' is usually approximately ½ϕ_{crit}'.

The different shear strength parameters are summarized in Table 4.24.

4.3.7. Undrained shear strength parameters

So far, attention has been devoted to the behaviour of saturated soils under drained conditions in which pore pressures due to deformation have dissipated or the pore pressure is known, so the shear strength parameters have been defined in terms of effective stress. However, when soils are subject to shearing in conditions in which pore pressures are not dissipated the shear strength parameters are expressed in terms of total stress. The cohesion and angle of internal friction are respectively denoted as c_u and ϕ_u and the shear strength parameter is s_u (see Section 4.3.2).

As in undrained loading the volume is constant, the void ratio is also constant and $\tau_f = s_u$ so, Equations (4.51) and (4.55) give:

$$\text{Log}[s_u/\tan\phi_{crit}'] = [e_r - e]/C_c \quad (4.57)$$

where e is the void ratio of the soil.

As explained in the Text Box on p. 106, for saturated soils $\phi_u = 0$. In the case of unsaturated soils, because loading results in an increase in density, ϕ_u is not zero. However such considerations do not lead to parameters that are useful in engineering situations.

In the discussion of effective stress conditions in Section 4.3.5 it was explained that the initial density of the soil gives rise to different responses during shearing. Thus a dense soil displays a peak strength, and with shearing this will fall to the critical value, whereas for a loose or reconstituted soil the maximum shear strength will be the critical value for the particular normal stress. In parallel with this, the undrained strength could be expressed in terms of either the peak or the critical values, however it is not usual to distinguish the parameters in this way. Thus the quoted s_u value for an undisturbed over-consolidated clay will be the peak value, whereas that for a normally consolidated or reconstituted clay will be similar to the critical value. It would also be possible to define an undrained residual shear strength but it would not be a useful parameter in engineering situations.

The undrained shear strength, s_u is equal to the undrained cohesion intercept, c_u only if the angle of friction, $\phi_u = 0$. Due to a lack of saturation of samples being tested or the breakdown of particles during tests, some

TABLE 4.24. *Shear strength parameters for clays (summary)*

Stress conditions	Applicability	Cohesion, shear strength	Angle of internal friction
Total stress Undrained shear—no dissipation of pore pressures during shearing	Short-term strength of clays.	c_u s_u	ϕ_u For saturated soil $\phi_u = 0$
Effective stress Drained shear—pore pressures dissipated such that $u = 0$	Strength of intact soil (peak strength)	c_p'	ϕ_p'
	Long-term strength of reconstituted soil at critical state (volume constant)	c_{crit}' Usually $c_{crit}' = 0$	ϕ_{crit}'
	Long-term strength for existing shear surface (residual)	c_r' Usually $c_r' = 0$	ϕ_r'

soils do display an undrained friction angle, leading to values of c_u and ϕ_u rather than s_u being quoted.

The different shear strength parameters for clay in different states of stress and different conditions are summarized in Table 4.24. As the undrained shear strength, s_u only applies in conditions in which pore pressures are not dissipated, it is only pertinent in conditions immediately after deformation, loading or unloading of low permeability materials such as clays. Hence this is the shear strength parameter that is used to predict the behaviour immediately after the change in total stress, for example, the stability of the sides of an excavation in clay immediate following excavation. Initially such an excavation may be stable but once pore pressures dissipate, failure may occur. The use of the undrained shear strength is only appropriate in the case of low-permeability materials such as clays, in which pore pressure changes occur slowly. However, permanent structures need to be designed so that they are also stable after dissipation of pore water pressures.

As mentioned in Section 4.1.4, the undrained shear strength needs to be quoted in a full description of a clay. This is done according to the strength classes given in Table 4.4, which also shows a series of tests that can be used to determine the appropriate strength range.

4.3.8. Effects of structure

Due to the disturbance to the structure of clay soils, the shear strength in the reconstituted state tends to be less than that in the undisturbed state. The ratio of the undrained strengths of the undisturbed to the reconstituted strength at the same moisture content is termed the sensitivity. It should be appreciated that, because of the presence of fissures, the mass strength of many clay soils is less than the peak strength determined for small intact specimens. The classification in Table 4.25 was proposed by Skempton & Northey (1952):

If disturbed, clays with very high sensitivity behave as viscous fluids and undergo flow. The cause of such behaviour is a metastable structure such that slight disturbance results in a large loss of strength (See Section 4.4.4). The liquidity index (See Section 4.2.3) of such materials is usually greater than unity. Typically the regain of strength once flow has ceased is slight. Such clays may display some degree of thixotropy.

Chandler (2000) used the concept of sensitivity to quantify the mechanical consequences of structural changes to clay sediments due to consolidation, excavation and weathering processes, including density and diagenetic changes. He defines the strength sensitivity as the ratio between the peak undrained shear strength and the undrained shear strength of a reconstituted sample at the same void ratio.

4.3.9. Modulus of rupture

The modulus of rupture is a strength determination used in the ceramics industry. The value is an indication of the tensile strength of the material and does not relate directly to the deformation properties. The test is carried out by subjecting square or circular prisms of prepared dried clay to three-point loading. The test is described by Ryan & Radford (1987) who indicate that the specimen should be prepared in the same manner as the production clay. The modulus of rupture (MOR) for a rectilinear prism is calculated using Equation (4.58):

$$MOR = 3PD/2dA \qquad (4.58)$$

where P is the force to cause failure, D is the distance between the supporting knife edges, d is the depth of the sample, A is the cross-sectional area of the specimen and the load is applied via a knife edge placed mid-way between the supports (see Figure 4.11d). This test gives a higher value of strength than that obtained using the direct tensile test.

4.3.10. Strength and deformation of mudrocks

There are several parameters used for expressing the strength of rock. These depend on the stresses applied to the rock and the confining stress conditions during the test. The simplest test is a compressive strength test in which a sample is subjected to an increasing compressive stress until it ceases to be able to carry stress (see Figure 4.11). In most cases this means the specimen has ruptured but in some cases failure may take the form of plastic deformation. The methods of testing are described by Brown (1981). The Unconfined Compressive Strength (or the Uniaxial Compressive Strength), σ_c is determined by loading the sample under conditions of zero confining stress, where

$$\sigma_c = \frac{\text{Force to cause failure}}{\text{Area of sample}}. \qquad (4.59)$$

The Uniaxial Tensile Strength, σ_t is determined by applying a tensile load to the sample, by carrying out three point loading tests on prisms (see Figure 4.11d and Section 4.3.9) or by applying a diametral load to a cylindrical sample (see Figure 4.11e). The latter is called a Brazilian test, where σ_t is given in equation (4.60):

$$\sigma_t = \frac{2P}{\pi LD} \qquad (4.60)$$

in which P is the force to cause failure and D and L are the diameter and length of the specimen, respectively.

TABLE 4.25. *Sensitivity classes*

Ratio (s_u undisturbed peak:reconstituted)	Class
<1	Insensitive
1–2	Low sensitivity
2–4	Medium sensitivity
4–8	Sensitive
8–16	Extra sensitive
>16	Quick

The Point Load Test is a variation of the latter test in which a length of rock core or irregular piece of rock is loaded using a special cone shaped platens. The force, P to cause failure is recorded and the Point Load Index, I_s determined from Equation (4.61), where D is the diameter of the core or, if the test is carried out on a sample of irregular shape, the distance between the platens at the start of the test. In the standard test the sample should be a 50 mm diameter by 75 mm long cylinder. Bell (1994a) provides a correction factor to be applied if the sample is otherwise dimensioned. The Uniaxial Compressive Strength, σ_c is given by Equation (4.62), where $C = 24$. Bell (1994a) points out that in tests the value of C commonly varies between 16 and 24, and thus it is prudent to carry out some uniaxial compressive strength tests to determine the appropriate value of C for the rock in question.

$$I_s = \frac{P}{D^2} \quad (4.61)$$

$$\sigma_c = C \times I_s. \quad (4.62)$$

As mentioned in Section 4.1.8, the Uniaxial Compressive Strength is quoted as part of the description of mudrocks. The categories defined in Table 4.9 are used for this, which also shows criteria that may be used to determine an approximate value of strength.

Applying stress to samples of mudrocks results in both elastic and plastic deformation. Figure 4.3 shows the form of a typical stress–strain plot for unconfined compression (OA), unloading (AB) and reloading to failure (BF). Because of the non-linearity of this relationship, there are various ways of defining the deformation modulus, E, as follows:

Initial tangent modulus	E_0
Secant modulus	E_{sec}
Modulus for 50% failure stress	E_{50}
Modulus for unloading and reloading cycles	E_r

Definitions for these parameters are shown in Figure 4.3, and reference should be made to Section 4.3.1, which includes definitions for, and relationships between, Bulk Modulus, Shear Modulus and Poisson's Ratio.

The hysteresis with loading and unloading, seen in Figure 4.3, is caused by energy loss during deformation. Because of the closure of cracks the rock becomes stiffer and the modulus, E_r measured for unloading and reloading cycles may be lower than the initial modulus, E_0.

The strength parameters and moduli may be expressed in terms of either effective or total stresses (see Section 4.2.5), depending on the water conditions and pore water pressures. However, with rocks it is not usual to make this distinction because in most situations the pore pressures are relatively small in comparison with the stresses required to cause deformation and failure of the rock. However, in weaker rocks, such as some mudrocks, the pore pressures may be significant. As in rocks failure usually occurs along existing discontinuities in the material it is the water pressure in these features that controls the behaviour. However, if the failure involves shearing through intact rock, then the pore pressure within the matrix of the rock needs to the taken into account. As with soils this pressure may be positive or negative.

Notwithstanding the effects of pore pressures, the presence of moisture in the pores of rocks can significantly reduce the strength of the material. Colbeck & Wiid (1965) carried out tests on quartzitic shale and other rocks which led them to report a 50% reduction in compressive strength when the rock was saturated, compared with when it was dry. These authors suggested that a reduction in the tensile strength of the material had occurred, with little effect on the angle of shearing resistance of the rock. Even a small amount of water makes a significant change. For the same reason, mudrocks are much stiffer when in an oven-dry condition than when damp.

As with soils, tests to determine the triaxial shear strength parameters, typically reveal a curved Mohr envelope, and it is common to quote the shear strength parameters as *apparent* values of cohesion, c_a and angle of internal friction, ϕ_a based on the assumption of a linear envelope. In the case of rocks the failure criterion can be expressed in terms of the principal stresses (σ_1, σ_3) for failure.

According to Hoek & Brown (1980), the equation for the Mohr envelope drawn in Figure 4.15a for a triaxially stressed, intact, isotropic rock is:

$$\sigma_1' = \sigma_3' + [m\sigma_c\sigma_3' + s\sigma_c^2]^{0.5} \quad (4.63)$$

where m and s are dimensionless constants and σ_c is the uniaxial compressive strength of the intact rock. These experimentally determined shear strength parameters are equivalent to the angle of internal friction and cohesive strength of a soil. The values of m and s vary with rock type and the state of intactness. The values given in Table 4.26 are provided by Hoek (1983).

4.4. Potentially troublesome clay materials

Certain types of clay materials can cause difficulties on construction sites and during exploitation if their behaviour is not anticipated and catered for at an early stage. This section contains a discussion of the typical behaviour of clays that display a propensity to shrink–swell, rapid dispersion and structural collapse. Quick clays that have high sensitivity such that they will flow if disturbed and tropical residual clays are also described. This is followed by Section 4.5, which contains discussion of the engineering properties of British clays and mudrocks, including tills and alluvial, laminated and over-consolidated clays and mudrocks, ranging in age from Carboniferous to Recent.

4.4.1. Shrink and swell

All clays are subject to shrinkage and swelling as a result of changes in moisture content. Such volume changes are experience alternating wet and dry seasons. The depth of the zone in which swelling and shrinkage occurs may extend to over 6 m in some semi-arid regions of South Africa, Australia and Israel. However, in temperate Southeast England seasonal volume changes are usually restricted to the upper 1.0 to 1.5 m of clay unless the zone of moisture change is extended to greater depth by the presence of tree roots (Driscoll 1983).

As a consequence of the presence of expandable clay minerals, many tropical residual clay soils of medium to high plasticity are subject to shrink-swell behaviour and, where constrained, high pressures are exerted during the wetting cycle. Pressures sufficient to break the bonded structure of the soil may be generated thereby giving rise to large heave movements. Such movements can occur when desiccated black clay soils are wetted. For example, the black cotton soil of Nigeria is typically a highly plastic silty clay in which the volume changes tend to be confined to an upper zone that is frequently less than 1.5 m thick. Typically swelling pressures up to about 240 kPa may be generated.

Expansive clay soils often contain fissures, which can lead to the short-term rapid ingress of water until closed by expansion of the clay adjacent to the fissures. Cemented and undisturbed, but potentially expansive, clay soils may have a high resistance to deformation, but disturbance and/or processing such clays may increase their tendency to expand. Clays with high porosity tend to undergo less expansion than low-porosity soils and flocculated clays tend to swell more than dispersed clays.

Swell-shrink behaviour of a clay soil under a given state of applied stress in the ground is controlled by changes in soil suction. The relationship between soil suction and water content depends on the proportion and type of clay minerals present, their micro-structural arrangement and the chemistry of the pore water. The extent to which vegetation is able to increase soil suction to the level associated with the shrinkage is important. In fact, the moisture content at the wilting point exceeds that of the shrinkage limit in soils with a high content of clay, and is less in those possessing low clay contents. This explains why settlement resulting from the desiccating effects of trees is more notable in low to moderately expansive soils than in clays of higher expansiveness. The removal of vegetation usually leads to expansion of expansive clay soils. In South Africa uplifts of 150 to over 350 mm have been recorded by Williams (1980).

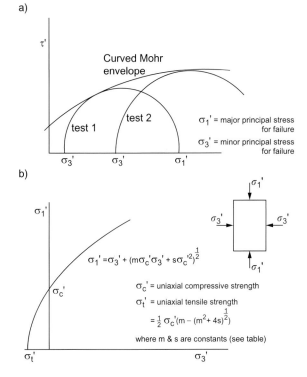

FIG. 4.15. Shear strength parameters for mudrocks.

independent of loading but loading can restrict the extent of swelling and, in the process, pressure is imposed on the constraint.

As mentioned in Section 4.1.1, Grim (1962) recognised that the volume change occurs by two processes: *inter-* and *intra-crystalline* swelling where inter-crystalline swelling occurs as a result of an increase in the amount of pore water present between the grains and intra-crystalline swelling occurs as a result of water entering the clay crystal, leading to expansion. Swelling clay minerals, such as montmorillonite, are subject this latter behaviour (See Chapter 2).

Most expansive clay materials are rich in montmorillonite. For instance Popescu (1979) found that expansive clays in Romania contained between 40 and 80% montmorillonite. Problems are more serious in regions that

TABLE 4.26. *Values of m and s for mudstone, siltstones, shale and slate tested normal to the cleavage (from Hoek 1983)*

Rock condition	s	m
Intact rock samples without fractures.	1	10
Very good quality rock mass – Undisturbed with rough unweathered joints spaced at 1–3m.	0.1	5
Good quality rock mass – Fresh to slightly weathered rock, slightly disturbed with joints spaced at 1–3m.	0.004	1
Fair quality rock mass – Several sets of moderately weathered joints spaced at 0.3–1m, disturbed.	0.0001	0.20
Poor quality rock mass – Numerous weathered joints spaced at 30–500mm with some gouge. Clean compact rock fill.	0.00001	0.05
Very poor quality rock mass – Numerous heavily weathered joints spaced at 50mm with gouge. Waste rock.	0	0.01

The volume changes that occur due to evapotranspiration from clay soils can be conservatively predicted by assuming the lower limit of the soil moisture content equals the shrinkage limit (see Section 4.2.3). Empirical and laboratory methods of predicting volume changes in soils are described by Bell (2000). Empirical methods relate the swelling potential to void ratio, natural moisture content, liquid and plastic limits, and activity. However, as such methods do not consider the influence of soil texture, soil suction or pore water chemistry, they should be used for guidance purposes only.

The change in soil suction from initial to final conditions may be used to obtain the degree of volume change. Suitable laboratory methods are described in Chapter 9, but unfortunately it is not easy to measure initial suction accurately by either *in situ* or laboratory methods. More direct methods of determining potential expansiveness are provided by various oedometer tests. However, as the larger fissures which are present in the clay *in situ* are not usually present in laboratory samples, the amount of heave and the expansion pressures are liable to be over-estimated.

The amount of heave due to the presence of expansive clay soils can be significantly reduced by compacting them to low densities at high moisture contents. In fact swell is usually reduced to a negligible amount if the soil is compacted at a moisture content higher than optimum but compaction below optimum is liable to result in excessive swell.

Bell (1993) explains that the expansiveness of clay soils may also be decreased by lime stabilization. This entails mixing 4 to 6% lime with the soil; however, this should only be done if less than about 5 g/kg SO_4 is present, as it is otherwise likely that the formation of ettringite $[Ca_6Al_2(OH)_{12}(SO_4)_3 \cdot 27H_2O]$ will result in expansion (Forster *et al.* 1995). The presence of a few percent of pyrite (FeS_2) will lead to similar problems as it may oxidise to produce sulphate rich pore water. Other treatment options include lime slurry pressure injection or cement stabilization, or treatments with a mixture of lime and cement (Anon 1982, 1986, 1990*a*).

4.4.2. Dispersive soils

Dispersion occurs in soils in which the repulsive forces between clay particles exceed the attractive forces. This occurs as a consequence of a reduction in the cation content of pore fluids that can result in deflocculation and dispersal of the clay particles. Thus, in the presence of relatively pure water the particles repel each other. In non-dispersive soil there is a definite threshold velocity below which flowing water causes no erosion and clay particles are not taken into suspension. However, with dispersive soils colloidal clay particles go into suspension even in still water and these materials are therefore highly susceptible to erosion by water (see Section 4.1.1). Serious piping damage to embankments, in which flow tubes have been formed, and failures of earth dams have occurred where dispersive soil has been used in their construction. Similarly, deep gullies can be formed on the unprotected slopes of earth embankments as a result of rainfall.

Aitchison & Wood (1965) indicated that presence of exchangeable sodium is the main chemical factor contributing towards dispersive behaviour in soil. The exchangeable sodium expressed as a percentage of the cation exchange capacity (ESP) can be used to identify susceptible clays. A threshold value of ESP of 10% has been recommended, above which soils that have their free salts leached by seepage of relatively pure water are liable to be prone to dispersion. Soils with ESP values above 15%, are highly dispersive (Bell & Maud 1994). Those with low cation exchange values (15 meq/100 g of clay) have been found to be completely non-dispersive at ESP values of 6% or below. However, soils with high cation exchange capacity values and plasticity indices greater than 35% swell to such an extent that dispersion is not significant.

The boundary between the flocculated and deflocculated states in clays depends on the value of the soil sodium adsorption ratio, salt concentration, pH value and mineralogy. Soils vulnerable to dispersion have more than 10% clay particles, together with a higher content of dissolved sodium (up to 12%) in their pore water than non-dispersive soils, and pH values in the range between 6 and 8.

Various procedures, including the crumb test and the pinhole test, are available to assist the identification of dispersive soils (see Chapter 9), although these tests have shortcomings. However, as no single test provides a definitive diagnosis of dispersivity, Bell & Walker (2000) proposed a rating system based on various parameters to assist with the recognition of vulnerable soils.

Sherard *et al.* (1977) explained that with careful construction control and the use of filters, then it is feasible to use dispersive clay in the construction of dams. Treatments with hydrated lime, pulverized-fuel ash (fly ash), gypsum and aluminium sulphate have also been used to improve dispersive clays used in earth dam construction (see McDaniel & Decker 1979). Dosing reservoir water with calcium sulphate or aluminium sulphate is another method of control. (see Bourdeaux & Imaizumi 1977).

4.4.3. Collapsible soils

Collapsible soils have high porosity and low density. Such soils have an open textured fabric that can withstand reasonably large stresses when partly saturated, but undergo a decrease in volume due to collapse of the soil structure on wetting, even at low stress. Particularly problematic soils react to wetting by collapsing under the influence of the overburden pressure. Many soils of this type consist mainly of silt-sized particles with clay minerals at inter-particle contacts, although the clay mineral content is usually less than 50%. Many tropical residual soils, including saprolites derived by the chemical weathering of granites, are subject to collapse behaviour. Most collapsible soils contain sufficient pore volume that when saturated their moisture content exceeds the liquid limit.

At their natural moisture content they have high apparent cohesion but this is lost when the moisture content is increased.

Loess is a Quaternary age collapsible soil that is formed as a wind blown fine granular, mainly silt-sized, deposit in continental arid, but not necessarily hot, areas. The extensive loess deposits of China, Australia, Central Europe and North America and Africa are commonly prone to this behaviour. The reader is referred to Fookes (1997) for further discussion of these materials. Older Quaternary loess deposits are less susceptible to collapse, because weathering processes and previous episodes of wetting have rendered the material more stable. Many loess deposits, for example the Brickearth deposits of Europe and southern England, occur in areas not now subject to arid climatic conditions. Some loess deposits are redeposited.

Various authors (see Bell 2000) have suggested diagnostic criteria to identify potential collapsible soils. These are based mainly upon the relationships between density or void ratio, moisture content and plasticity properties of the material. Laboratory oedometer and flooded oedometer tests on samples are used for confirmation. Various methods of stabilization are available, including increasing the density of the soil by the use of compaction with static or vibrating rollers, dropped weights and vibroflotation techniques (Bell 1993). The injection of grouts, including lime, cement or bitumen emulsions may also be used, but are generally less successful.

4.4.4. Quick clays

Quick clays are clay soils that exist in a meta-stable state such that they can liquify and flow as liquid if they are subjected to a relatively small disturbance. In other words they are extremely sensitive (Section 4.3.8). Most quick clays are composed predominantly of particles smaller than clay size (0.002 μm) but many deposits are very poor in clay minerals, instead containing a high proportion of fine quartz. Gillott (1979) indicated that quick clays from Canada, Alaska and Norway typically possess an open fabric and high moisture content. The grains, which consist of either aggregations of fine particles or primary mineral particles, are usually linked to each other by agglomerations of clay-sized quartz and feldspar particles with illite and chlorite and, in some cases, swelling clay minerals. Vulnerable soils have a high porosity, and moisture content.

Quick clays formed as a consequence of the Quaternary glaciation of the northern hemisphere occur in northern Europe, Siberia and Asia, where deposition of silt and clay particles occurred under marine conditions. High cationic strength pore waters brought about flocculation of clay particles that then formed links between silt particles. Subsequently, uplift of the deposits above sea level resulted in them now being subject to humid climatic conditions. The lack of consolidation coupled with leaching of the pore fluids has resulted in the dispersal of the linking clay particles such that a meta-stable structure has been created. In the absence of inter-particle bonding, a small disturbance to a deposit with a high moisture content causes it to become liquefied. Typically there is a small recovery of the original strength once flow stops.

Remoulding the clay will usually reduce the shear strength to near zero. Most quick clays have low plasticity indices of between 8 and 12%. The liquidity index normally exceeds 1, and the average liquid limit is usually less than 40%. Quick clays are usually inactive with $A < 0.5$ and high sensitivity (see Section 4.3.8), which increases with increasing liquidity index.

Quick clays are associated with several serious engineering problems, including low bearing capacity and high compressibility, but are best known for their dramatic flow slide failures, particularly in Alaska and Norway. It is not possible accurately to predict the magnitude of settlement of foundation structures by standard methods of testing and analysis.

4.4.5. Tropical residual soils

Weathering processes penetrate to greater depth and cause more rapid chemical change to rocks in tropical regions than in temperate areas. Thus in such regions, the bedrock tends to be mantled by perhaps tens of metres of residual soils, many of them clay-rich. To be a 'residual soil', weathered material must remain *in situ* rather than it being removed by erosional processes and deposited in another location.

The classification and properties of tropical soils with respect to engineering applications are reviewed by Fookes (1997), see the Text Boxes on pp. 116 and 118. Due to the presence of inhomogeneities within rock masses, complex weathering profiles can be formed. These may include highly variable mixtures of material degraded to lesser or greater degrees, with core stones of unweathered rock. The engineering properties, and performance of the mass can be subject to change over short distances and therefore the mass properties are very difficult to predict. Also it is usually difficult to fix the exact point of rock-head, which marks the maximum depth to which weathering processes have penetrated. Frequently, rock that shows no or only the slightest degree of weathering may overlie material of a much higher weathering grade.

In addition to the complex mixtures of materials present in weathering profiles the materials themselves can also display challenging forms of behaviour. The engineering behaviour depends on the mineralogy and structure of the soil. In turn these are functions of the original rock type as well as the climatic and drainage conditions during and subsequent to weathering. These factors result in wide variations in grain size, unit weight, strength and compressibility. Values of geotechnical parameters reported by Bell (2000) for a range of tropical residual soils are shown in Table 4.27. The engineering properties can be affected by the presence of structural features inherited from the parent rock. This can result in weaker zones being present within the mass and the material behaving in an anisotropic manner. The coarser grains in most residual soils are quartz, usually together with

TABLE 4.27. *Geotechnical parameters for a range of tropical residual soils (Bell 2000)*

	Laterite	Red clay	Latosol	Black clay	Andosol
Moisture content %	N 18–27 S 16–49		J 36–38		K 62 J 58-63
Liquid limit %	H 245	B 42, 28 G 48	B 47 G 52	C 75 K 103 I 132	K 107
Plastic limit %	H 135	B 26, 22 G 24	B 30 G 29	C 21 K 36 I 41	K 73
Clay content %	H 36	B 35, 27 G 31	J 80 B 35 G 23	I 48	J 29
Activity		B 0.46, 0.3 G 0.77	B 0.51 G 1.0		
Unit weight (kPa)	H 15.2–17.3			I 11.5	
Dry density (Mg/m^3)				C 1.27–1.75	
Specific gravity		G 2.73 B 2.8	J 2.74 B 2.86 G 2.83	I 2.72	
Effective cohesion (kPa)	H 48–345		J 47–58 G 26	I 25	K 46
Effective friction angle (°)	H 27–57		J 16–20 G 21	I 37	K 18
Compression index			G 0.03–0.06		K 0.01–0.04
Coeff. vol comp (m^2/kN)		B 0.47-0.08	G 1.0–0.01		K 0.01–0.04
Coeff consol (m^2/yr)		B 2.32–232			K 0.36–0.08
Permeability (m/s)	H 1.5 × 10^{-6}– 5.4 × 10^{-9}	B 5 × 10^{-6}– 2 × 10^{-6}	G 4.7 × 10^{-8}– 5 × 10^{-8}		K 4 × 10^{-8} J 2.8 × 10^{-8}– 5.6 × 10^{-9}

H, Hawaii; N, Nigeria; S, Sri Lanka; C, Cameroon; B, Brazil; G, Ghana; J, Java; K, Kenya; I, India.

some feldspar and hydrated oxides of iron and aluminium. The most commonly occurring clay minerals are kaolinite, illite and montmorillonite, plus allophane and halloysite, which occur in active volcanic terrains.

Changes in the properties of residual clay soils can be brought about by drying, which may either occur *in situ*, or as a result of sampling. Any reduction in moisture content may well result in the precipitation of sesquioxides, aggregation of clay particles and water loss from hydrated clay minerals. Some of these changes are irreversible, for instance halloysite becomes transformed into metahalloysite. The aggregation of clay particles increases the proportion of silt and sand-sized grains and results in a reduction or loss of plasticity. Because of this, special care needs to be taken when sampling and handling tropical clay soils (Fookes 1997).

Aggregation and inter-particle bonding are common features of tropical residual soils. The latter may be inherited from the parent material or may be the result of the formation of cements during weathering or the precipitation of minerals within inter-granular pores. Where replenishment of pore water occurs, for instance by capillary rise, a hard rock-like duricrust may be produced.

4.4.5.1. Red tropical soils. Lateritic soil is a general term, often misused, for red tropical soils. Buchanan's (1807) original 'laterite' from West India was a term used for red soils that harden on drying, often forming a natural red crust or latcrete (see Charman, 1988). These rocks consist typically of kaolinite with iron oxides and hydroxides. In the case of bauxite, aluminium minerals predominate. Other common soils types are the vertisols (black clays), which tend to consist of kaolinite and montmorillonite.

The dry unit weight of lateritic soils tends to span a larger range of values than is found in transported soils, with values varying between 9 kN/m^3 and 23 kN/m^3.

Tropical soils

Due to the structure of many tropical soils, it is not meaningful to classify them on the basis of grain size and consistency limits, as presented in Tables 4.4 or 4.6. Instead Fookes (1997) advocates the use of a generic pedogenic classification based upon the material produced under different climatic and environmental conditions. The soils produced are divided into mature soil types (ferrallitic and ferruginous, ferrisiallitic soils, ferrisiallitic andosols and vertisols) see the Text Box on p. 118, and duricusts. Fookes (1997) goes on to present a geomorphological model illustrating the formation and interrelationships of these soils in tropical climatic conditions in an area of varied geology. Fersiallitic soils are typical of Mediterranean climatic conditions, whereas ferruginous soils form under sub-tropical conditions and ferrallitic soils form under wet tropical conditions. Fersiallitic soils may also be the result of the incomplete development of ferralitic or ferruginous soils. Fersiallitic soils and vertisols usually display low to moderate permeability, moderate to high compressibility and low to high strength. They may also display a propensity for high shrink–swell behaviour and be dispersive. Ferruginous soils usually have moderate permeability, low compressibility and high strength, whereas ferrallitic soils tend to be variable in terms of permeability and compressibility but usually possess high strength.

Owing to the aggregation of soils particles, it can be difficult to determine a consistent value for the plasticity indices of tropical soils. Northmore *et al.* (1992) recommend carrying out remoulding trials so that the minimum amount of remoulding to provide reliable results may be ascertained.

Bell (2000) reports that the plasticity of lateritic soils tends to straddle the A-line and range from low to high plasticity. Most ferrallitic and ferrisol soils containing kaolinite and halloysite, plot in an area just below but parallel to the A-line, whereas andosols and halloysitic soils lie below the A-line and have high plasticity.

Clay particle aggregates may be dispersed as a result of chemical changes or leaching, which leads to an increase in liquid limit. Moreover, removal of cement by leaching gives a large increase in compressibility and reduction in strength. This can occur on a seasonal basis.

Lateritic soils may contain particles from clay size to gravel and larger. Due to the variation in texture of residual soils, the void ratio may be wide ranging with no particular relation to the soil type or conditions of formation. Because of this Vaughan *et al.* (1988) introduced the concept of relative void ratio (e_R) for residual clay soils, defining it as:

$$e_R = \frac{e - e_{opt}}{e_L - e_{opt}} \quad (4.63)$$

where e is the natural void ratio, e_L is the void ratio at the liquid limit and e_{opt} is the void ratio at the optimum moisture content. They suggested relating engineering properties to relative void ratio rather than to the *in situ* void ratio.

Many tropical clay soils behave as if they are overconsolidated in that they exhibit a yield stress at which there is a discontinuity in the stress–strain behaviour and a decrease in stiffness once this stress is exceeded. At low values of effective stress, saturation may cause structural collapse (see Vargas 1974).

Some properties of lateritic soils from Brazil are reported by Vargas (1974) who showed that the compression index correlates with the liquid limit w_L% as follows:

$$C_c = 0.005(w_L + 22). \quad (4.64)$$

The more porous residual soils tend to have high permeability values of up to 10^{-4} m/s. Remoulding or structural collapse of the material will cause this value to decrease by several orders of magnitude.

4.4.5.2. Other tropical soils. Problematic residual clay soils tend to have high void ratios and they may be in a metastable, stable-contractive or stable-dilatant state, such that at low stresses, the structure remains stable but a large contraction occurs once a particular yield stress is exceeded. Idealised consolidation data for these conditions are illustrated in Figure 4.16. Stable contractive and post-yield metastable soils will undergo contraction towards the critical state during shear, whereas stable

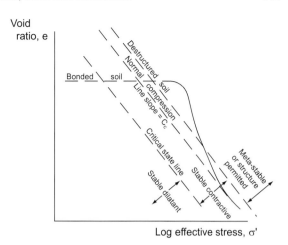

FIG. 4.16. Consolidation curves for collapsible laterite (Vargas 1974).

dilatant soils will dilate during shear until the critical state is achieved.

The shear strength of bonded soils may be underestimated if the bonds yield during testing. Hobbs *et al.* (1992) reported that yield usually occurs at stresses between 100 and 200 kPa. Mitchell & Sitar (1982) report the results of shear strength tests on a range of remoulded lateritic soils, for which the (presumably) effective shear strength parameters were an angle of friction of between 18 and 41°, with most values in the range 28 to 38°, and a cohesion of between 0 and over 48 kPa. Limited numbers of tests to measure the undisturbed strength of lateritic soils gave friction angles of between 23 and 33° and cohesion ranging from 0 to over 210 kPa.

Hobbs *et al.* (1992) report that the effective shear strength varied from 12 to 37° for red clays from Kenya, whereas the range was 19 to 41° for samples from Indonesia. The corresponding effective cohesion values ranged from ranged from 0 to 55 kPa and 2 to 39 kPa. As laterites may be hardened by the presence of precipitated minerals and amorphous phases near the ground surface, the strength may decrease with increasing depth. The strength will also decrease by perhaps 50% if the moisture content is increased from its natural value to 100% saturation.

4.5. Properties of some British clays and mudrocks

Caution: see Text Box on p. 471 of Appendix B.
In this section the properties of British clay soils and rocks related to their composition and mode of formation are reviewed (see Chapter 6 for details of formation). Typical values of various parameters for selected UK clay formations are presented in Appendix B of this report, but a commentary on these values is provided in this Section. In view of the scarcity of published data on some formations, in some cases the lack of geographical spread of

Types of tropical soils

Andosols are tropical soils that are rich in allophane and are mainly derived from mafic volcanic rocks in areas with a hot climate and high rainfall. The moisture content of such soils tends to be in the range 55 to 65% and they usually have high plasticity with liquid limits and plastic limits in the ranges of 80 to 300% and 20 to 100% respectively. The liquid limit decreases with increasing remoulding of the soil, a feature that can be used to identify these allophane-rich soils. Low dry density values with high void ratio of up to 6, are also commonly found. Mitchell & Sitar (1982) indicate that the angle of internal friction in (presumably) effective shear strength tests tends to be relatively high, with values in the range of 27 to 57° and cohesion of 22 to 345 kPa. Increasing the moisture content causes a large reduction in strength and the materials are highly sensitive.

Soils rich in *halloysite* also have high moisture contents of from 30 to 65% and liquid limits of 40 to 100%. The corresponding plastic limits range between 10 and 50%. These values increase with increased remoulding of the soil. Quoted clay contents are in the range of 50 to 80%. Some halloysitic soils are prone to collapse in that they will undergo a large volume reduction if flooded under constant load.

Black clay soils and *vertisols* are mainly formed in poorly drained areas with high seasonal rainfall and drought. Such soils tend to consist of up to 50% clay-size, which includes up to 70% montmorillonite, with sand. The soils are subject to large volume changes on wetting and drying, such that any inter-particle bonding and structure may be destroyed. Black cotton soils are highly plastic silty clays which tend to be formed from basaltic rocks, mudrocks or clays.

---

data and the high number of formations, this has been a challenging task.

The data are mainly those presented by Cripps & Taylor (1981, 1986 and 1987) and in British Geological Survey (BGS) reports on the engineering geology of particular areas and relating to particular deposits and from other published sources.

The data are intended to serve as a general guide as to the typical properties and variability of the materials concerned. The Appendix B tables also provide a reference source for further information about the formations. The values quoted are intended as guidance only and should not be used as a substitute for investigations into the properties of the materials themselves. The intention is that investigations should be more focussed and the data will also assist with the interpretation of test results. The properties covered in the tables are those that are commonly quoted in site investigation reports and publications. They include natural moisture content, density, clay-sized fraction and strength. Stiffness, consolidation, permeability and compaction parameters are also quoted for some formations. Unfortunately, the literature does not supply values for many of the parameters introduced in Section 4.3 of this Chapter.

The data presented by Cripps & Taylor (1981, 1986, 1987) were compiled from a great number of published sources available at that time. The sources were the result of an extensive literature review and the data are presented as a mean and range of values, where the ranges exclude what appear to be extreme values. Although extreme values may be an accurate representation of the material tested, they do not reflect typical behaviour. A less extensive literature review has been performed to complete the present data tables and, for these additional entries, data from several sources are presented for comparison.

The BGS reports, however, are the result of comprehensive studies involving the selection of data from what are judged to be the most reliable and comprehensive site investigations that, in some cases, have been supplemented with additional testing. Values taken from BGS Reports are usually quoted as a median or sometimes a range of medians, together with the first and third percentiles. This is so that the extreme values do not give rise to a biased view of the materials. In many cases, percentiles other than the first and third quartiles and the number of samples tested, together with the values for some other parameters such as the grading and amounts of sulphate present, are also given in some BGS Reports. Besides the numerical data presented here, many of the original data sources provided further information about the tests and materials tested. Salient points are included in this discussion of the values of the parameters, but where an important decision depends on the values of parameters, the reader is advised to refer to the original data sources.

Variation in the values of geotechnical parameters arises as a consequence of differences of grain size, mineralogy and structure of the materials. In turn these depend on the environment and processes of sedimentation, the loading history, including present depth of burial, and action of weathering processes. Disturbance to the material during sampling and the use of different methods and conditions of testing, particularly the actual moisture content and stress conditions during testing are further reasons for variation. Although data have been selected from sources judged to be reliable, unfortunately not all authors provide details of the methods and conditions of testing. For example, consolidation and deformation parameters are dependent on the stress range used for determinations, which may not be quoted. In view of the variation of consolidation parameters with stress level, most of the BGS Reports quote the coefficient of consolidation and volume compressibility in terms of defined classes. The values quoted in the tables are the upper and lower limits of these classes or, where appropriate, several classes.

Not all authors are explicit about the grade of weathering of the material tested. As most of the data have been derived in the course of site investigations, it can be expected that where not otherwise indicated, the conditions are those that would apply for standard site investigations. A range of weathered and unweathered materials derived from depths of less than about 20 m is likely to have been tested. Where possible, the weathering grade is specified and the data are assembled into those for the unweathered (zone I, fresh and zone II, slightly weathered) and weathered (zone III, moderately weathered and zone IV, completely weathered) zones.

Differences in the composition, loading history and weathering give rise to variation of geotechnical parameters with depth and location. In some cases, for instance where deposits are laid in thin laminae of differing composition, the properties will vary on a very small-scale and the material may also display anisotropy in that the properties depend on orientation. In other cases the deposit may be variable on a regional scale. For instance the amount of clay-sized material present in London Clay in the London Basin increases to the east. In the case of Lias Clay the materials differ depending whether deposition took place in rapidly subsiding basinal areas or the intervening shallow-water shelf areas. Gault Clay varies for a similar reason. Mercia Mudstone, which was deposited in a series of ephemeral water bodies, mudflats and alluvial plains, is a highly variable deposit.

The deposits are discussed in order of increasing age starting with the Recent and ending with the Carboniferous.

4.5.1. Recent and Quaternary deposits

Data about tills, laminated clays and alluvial clays, which are widely distributed in the UK, are presented in Appendix B, Tables B1, B2 and B3. The behaviour of these deposits is discussed below.

4.5.1.1. Tills. Tills are deposits formed by glacial processes that mostly consist of a variable assortment of rock debris ranging from fine rock flour to boulders. Their properties depend on composition, the position in the ice in which the debris was transported, mode of deposition and subsequent loading history and weathering processes. Mudstones and shales are easily abraded and produce fine-grained tills rich in clay minerals. In the UK, tills occur predominantly in the lowland areas of England north of an approximate line drawn from the Bristol Channel to the Wash. Many tills contain a low percentage of clay minerals, although they may contain significant quantities of silt- and clay-sized material. Till deposits can contain one or more layers of different materials with different structures and gradings.

A thorough review of engineering in glacial tills is provided by Trenter (1999), who considers the formation, classification and engineering implications of these materials. Most types are gap-graded, being deficient in sand sizes and containing about 20 to 40% rock fragments of various sizes with a clay fraction of between 15 and 40% (Bell 2002).

A distinction is made between lodgement tills derived from rock debris carried along at the base of a glacier, and ablation or melt-out tills, in which the material was transported within or on the ice. Lodgement tills are commonly stiff, dense, relatively incompressible, over-consolidated clays, whereas ablation tills tend to be weaker and more compressible, less over-consolidated or normally consolidated materials (see Section 4.3.3). Both sub-horizontal and sub-vertical fissures are frequently present in these deposits. Melt-out tills may become over-consolidated if they have been overlain by an ice sheet formed during a glacial period subsequent to their deposition.

The geotechnical properties of tills from various places are presented in Appendix B, Tables B1 and B2. Trenter (1999) points out that the plasticity of tills is usually low or intermediate ($w_L = 20\text{--}50\%$) and the liquid limit increases with decreasing grain size. The plasticity data plot on a plasticity chart above the A-line (along the T-line of Boulton & Paul 1976). In heavily over-consolidated, unweathered lodgement tills the natural moisture content is generally slightly below the plastic limit and the liquidity index typically lies within the range -0.1 to -0.35. Denness (1974) demonstrated that in melt-out tills from Milton Keynes, in southern England, the plasticity properties show large variations in value over short distances.

The undrained shear strength of glacial tills depends on their mineralogy, structure, grading, density and moisture content. As with other materials, the strength measured in tests also depends on the method of sampling, sample size and orientation and testing conditions. As samples tend to contain fissures and other heterogeneities, the results of strength tests typically show wide scatter. This is reduced if the sample size or the volume of material tested is increased. Trenter (1999) points out that foliated samples, in which the particles are orientated in a particular direction, tend to have lower strength than non-foliated tills and that the measured strength is also strongly influenced by the moisture content. Due to the effects of surface drying, the upper layers of clay may be stiffer and stronger than the underlying material. Vertical fissures are often present in the former materials.

Trenter (1999) reports extensive work carried out at the Building Research Establishment by Marsland and co-researchers. Using *in situ* plate loading and pressuremeter tests and 100 mm diameter laboratory triaxial tests, undrained cohesions for three different tills from different areas of the UK were measured. The values were in the range from about 80 kPa to almost 300 kPa. Similar values of shear strength are reported by Robertson *et al.* (1994) for tills from east Northumberland and by Bell (2002) for tills from north Norfolk, Holderness and Teesside. The latter three tills experienced small losses in strength when remoulded. Some tills included in Appendix B, Tables B1 and B2, have lower strength, where weathered clays have been tested. The average undrained cohesion is in the range of 100 to 150 kPa and the undrained angle of friction may be a few degrees. As would be expected, Robertson *et al.* (1994) found that, neglecting the influence of depth of overburden, lodgement till is stronger than the melt-out till.

Significantly higher values of undrained shear strength were measured by Radhakrishna & Klym (1974) for dense tills. Pressuremeter and plate loading tests gave values of about 1.6 MPa, while the values from triaxial tests ranged between 0.75 and 1.3 MPa. They attributed these differences between field and laboratory results to the effects of stress relief during sampling.

Research into the effects of weathering processes on the properties of tills has been performed by Sladen & Wrigley (1983), who noted that, as a consequence of particle breakdown, there is an increase in the proportion of clay- and silt-sized material in the more weathered material, which is usually up to approximately 5 m thick. This change is accompanied by reductions in density from 2.26 to 2.05 Mg/m³ and increases in mean moisture content from about 12% to 21%, and in plasticity from low (w_L <35%) to intermediate (w_L 35–50%) or high (w_L 50–70%) values. Such change may be accompanied by removal of calcite cements. Trenter (1999) gives further details of this work together with the findings of several other authors.

Data on the compressibility of tills, shows them to have low or medium compressibility and medium to fairly rapid rates of consolidation. These properties may be subject to change over short distances due to the presence of layers of silty and sandy materials.

4.5.1.2. Varved and laminated clays. Fine-grained sediments that have accumulated on the floors of glacial lakes may be composed of alternating laminae of units with coarser grained, usually silt-sized materials at the base with overlying finer material, usually consisting of clay sized particles. Each such unit, which may be between fractions of a millimetre to a few millimetres thick, is called a varve. According to Bell & Coulthard (1997), clay is the dominant particle size in the Tees Laminated Clay of Northeast England. In this, illite and kaolinite are present with lesser amounts of chlorite and between 4% and 26% quartz depending on the thickness of the silty layers. Traces of feldspar, muscovite mica and expandable mixed-layer clays are also present. The average silt content is 37% and fine sand accounts for up to about 10%. Not all varved deposits are predominantly clayey, and in some the finer part consists of clay-sized grains of quartz, feldspar and mica rather than clay minerals.

The compound nature of varved clays means that some standard soil mechanics tests may yield misleading results. For instance, the grading, moisture content and liquid and plastic limits of a bulk (mixed) sample will not be the same as those for the individual materials. Also difficulties may arise with obtaining undisturbed samples of varved clays. Spectacular foundation failures of structures have been reported, especially from Canada.

The properties of various laminated clays are shown in Appendix B, Table B3. Bell & Coulthard (1997) report that Tees Laminated Clay displays a wide range of plastic and liquid limit values, where the plastic limit is usually less than the natural moisture content and the liquidity index is usually positive. The wide scatter of plasticity values is attributed to variations in the proportions of clay and silt in the samples. However, tests on the individual clayey and silty components indicated that the clay fraction exerts the strongest control over the plasticity behaviour. These results are similar to those obtained by Taylor *et al.* (1976) for other laminated clays from the north of England and by Kazi & Knill (1969) for laminated clays from Norfolk. The Tees Laminated Clay is inactive, having an average value of $A = 0.54$.

The average undrained shear strength of Tees Laminated Clay tends to be about 60 kPa, although there is a wide variation, from the below 30 kPa to just over 100 kPa. These values are less than those obtained by Robertson *et al.* (1994) for laminated clays in Northumberland, England which ranges between 50 and 360 kPa with a mean of 100 kPa. The sensitivity of Tees Laminated Clay ranges from 3.5 and 4.2.

Varved clays tend to be normally consolidated or lightly over-consolidated, unless they have been overlain by ice sheets or later sediments. Horton (1975) reports that laminated glacial lake deposits of the Birmingham area are firm to stiff over-consolidated clays of low plasticity. The permeability ranges from high to very low and commonly is higher in a horizontal than vertical direction. This anisotropy is apparent in the consolidation properties in laminated clay from Teesside given in Appendix B, Table B3.

4.5.1.3. Alluvial and Marine clays. Data about alluvial clays that occur either in active river valleys or as terrace deposits from various parts of the UK are presented in Appendix B, Table B3. Information about sandy, silty and organic rich deposits with which such clays are frequently interbedded is specifically excluded in this compilation. The composition of the deposits tends to be very variable, with changes in lithology occurring both horizontally and vertically.

The deposits are usually very soft to stiff, compressible silty clay and clayey silt of low to intermediate plasticity. The strength ranges from less than 10 kPa in the wet, clay rich material, to over 100 kPa in the dryer and sandier deposits. The plasticity is usually intermediate, with most points plotting above the A-line and a few points plotting as high plasticity. Some deposits have low density, particularly when dry, and some organic-rich deposits plot below the A-line. The effect of increased organic material is to increase the natural moisture content, plasticity and compressibility and reduce the density. This is clearly demonstrated in Appendix B, Table B3 by comparing data for clays and silt with those for clays and peat given by Waine *et al.* (1990) for the Wrexham area.

Most alluvial clays have medium compressibility, with lower values corresponding with the denser and sandier materials. Organic-rich clays tend to have high compressibility. The rate of consolidation may be medium or high.

4.5.2. Tertiary clays

A number of clays of Tertiary age occur in the south-east and central southern areas of England. The oldest of these are the clays of the Lambeth Beds, including the Woolwich and Reading Beds, which are overlain by the London Clay formation. Above this lie various younger deposits, including the Claygate Beds, the Barton Clay and the Bracklesham Beds. Apart from London Clay, which consists predominantly of clay, all the other formations include other lithologies. Engineers tend to term these (and other similar clay deposits) 'engineering soils'.

All these clay formations are over-consolidated due to the erosion of later overlying deposits. Fookes (1966) and Cripps & Taylor (1986) report that between about 150 and 400 m of overburden has been removed, with an apparent westward increase in the degree of over-consolidation of London Clay, in the London area.

The ball clays of south and south-west England, which are much valued as pottery clays, are also of Tertiary age. The properties of these Tertiary formations are discussed below, where separate sections are devoted to London Clay and Ball Clay. This is because there is much information about London Clay, whereas the Ball Clays are rather different from the other clays of Tertiary age.

From the data presented in Appendix B, Table B4, the plasticity of Tertiary clays ranges from intermediate to extra high, with points plotting above the A-line on the plasticity chart. In most of the deposits the clay-size fraction is about 50%, but it is lower than this in the Claygate Beds and clays of the Lambeth Group. The activity of London Clay is in the range 0.75 to about 1.25, whereas in the latter two deposits, the average value tends to be near the lower end of this range.

Culshaw & Crummy (1991) report that the fine-grained parts of the Claygate Beds are similar in geotechnical properties to London Clay (see also Northmore *et al.* 1999). They are firm to very stiff when fresh and firm when weathered. They have low to very low compressibility with higher values than this in weathered clay. However, the drier clay near the ground surface tends to be stiffer and stronger that the material below.

The clays of the Lambeth Group are similar, with strength in the range of stiff and hard soils. The compressibility of these materials is liable to be low or medium. Further information about the Lambeth Group may be found in Hight *et al.* (2004) and Entwisle *et al.* (2005).

4.5.2.1. Ball Clay (see also Chapters 6 and 14). Ball Clays occur in a restricted area of south and southwest England. They are kaolinite rich, with illite, quartz and organic matter making up the remainder of the constituents. As the data in Appendix B, Table B4, show, they are high to extra high plasticity clays with undrained shear strength in the range from 79 kPa to over 1000 kPa (stiff to hard).

4.5.2.2. London Clay (see also Chapter 6). London Clay is typically a firm to stiff, becoming hard with depth, over-consolidated, fissured, inorganic, blue to grey silty clay (for example, see Forster 1997). It is more sandy at the base and top. Parts are laminated and it contains nodular claystones and rare sandy partings. The formation is commonly weathered to brown clay to depths of 5 to 10 m.

Owing to the large amount of construction that has taken place on the London Clay outcrop, there are relatively large numbers of site investigation reports that give details of its properties. Cripps & Taylor (1986) reviewed published data and gave the ranges and averages for a number of parameters for weathered and unweathered clay, as given in Appendix B, Table B4. Subsequently Culshaw & Crummy (1991) collated data from selected site investigation reports for the south- east of Essex and quoted these in terms of median values with various percentiles. A further collation of data for the London area, is given by Forster (1997) who lists the ranges given in a selection of reports. In addition to these sources, London Clay properties are given in many case histories, reviews and research papers, for example those published by Atkinson (2000), Chandler (2000) and others.

4.5.2.3. London Clay—density and moisture content. The weathered clay may have very high moisture content, but in dry periods, the material in the upper few metres may be desiccated resulting in high strength (see Driscoll 1983). The depth of desiccation may be much greater than this if trees are present.

The weathered material also becomes more fissured. Below the depth of season variation the moisture content of London Clay varies by only a few per cent with depth. The bulk density of London Clay varies between 1.70 and 2.04 Mg/m^3 depending on weathering grade and location.

4.5.2.4. London Clay—particle size and consistency limits. The plasticity of London Clay is generally high or very high, with some instances of extremely high plasticity. As the plasticity chart in Figure 4.17, which shows the trend lines for different clay mixes discussed in Section 4.2.3, demonstrates, this is a reflection of the presence of the swelling clay mineral montmorillonite with other clay minerals. The variation in index properties relates to grading and mineralogy and may occur on a very small scale, for example where varves are present, or on a much larger scale. For example, Burnett & Fookes (1974) attribute an eastward increase in plasticity of London Clay to an increase in clay fraction across the London basin area. Data given by Cripps & Taylor (1986) compares values of 64–72% in the east with values of 52–60% elsewhere. However, Burnett and Fookes (1974) and Forster (1997) point out that the deposit also shows cyclical vertical variation in clay fraction so it is possible that the clay fraction may vary by perhaps 20% at any particular locality. On average there is also an increase in clay size fraction with depth.

Liquid limit values in the east average about 80% and those in the western side of London about 70%, whilst

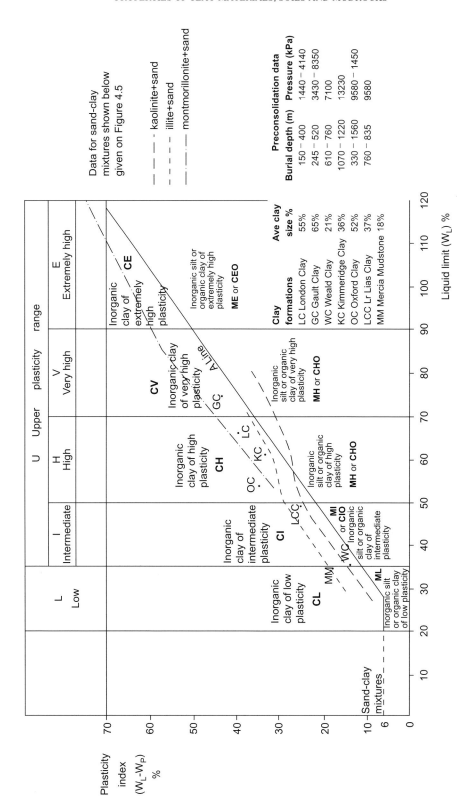

FIG. 4.17. Plasticity chart for selected British clays.

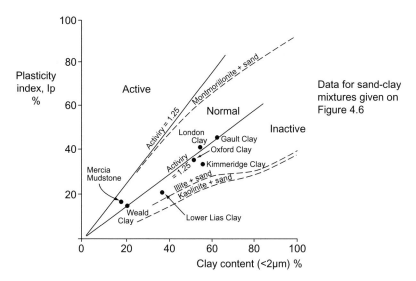

FIG. 4.18. Activity chart for selected British clays.

they are in the range 70–80% in central London. Apparently only minor changes in plasticity occur with depth and weathering grade. The average activity of London Clay is usually in the range 0.75 to 1.0. As Figure 4.18 shows, this implies the presence of a proportion of montmorillonite in this clay.

4.5.2.5. London Clay—strength and stiffness. The undrained shear strength ranges widely due to the variations in mineralogy, grading, moisture content, overburden pressure and density of the clay. From data compiled by Cripps & Taylor (1986) for London Clay from Hendon, desiccation in the top most few metres can mean that the near-surface clay can be stronger, with shear strength values in the range of 50 to 150 kPa (98 mm diameter triaxial tests on undisturbed samples), than the clay immediately beneath this. Below about 5 m, the strength usually increases from an average of about 80 kN/m^2 to about 120 kPa at a depth of about 20 m, where the scatter of data points is about +/− 50%. In part this scatter may be due to vertical changes in clay size fraction referred to above. Typical values range from 80 to 400 kPa (Atkinson 2000)

Comparing the average strength of the brown weathered clay with that of the blue or grey unweathered clay suggests that weathering action accounts for a strength reduction of about 50%. Lithological changes, particularly in clay size fraction, give rise to a decrease in strength from an average of 240 kPa for west and central London to 144 kPa in Hampshire and Essex and 96 kN/m^2 further east (Burnett & Fookes 1974).

The shear strength measured for larger specimens and large diameter plate bearing tests tends to be lower than that for small specimens, which usually contain fewer fissures. For instance, Marsland (1973) quotes the following values for the undrained shear strength, s_u, of London Clay from/at a depth of 15 m below the ground surface:

865 mm diameter in situ plate bearing tests 100 kPa
38 mm and 98 mm diameter triaxial tests 180 kPa
5.5 mm diameter in situ penetrometer tests 450 kPa.

The averages for the 38 mm and 98 mm diameter specimens were similar to each other but the results for the 98 mm diameter specimens showed less scatter.

Very much higher values of peak effective cohesion have been measured for London Clay from depths of 20 m than for shallower samples. Stress relief fissuring of the clay affects the clay below the levels penetrated by weathering action. At shallower depths the effective cohesion is in the range of 20 to 30 kPa and the angle of internal friction varies from about 23 to 26°. Concerns, expressed in Section 4.3.6, over the validity of ascribing an effective cohesion value to an unbonded clay should be borne in mind. Weathering causes a further progressive reduction in strength to near the critical value of $c_{crit}' = 0$, $\phi_{crit}' = 20°$ quoted by Skempton (1970) as the fully softened remoulded value of strength. As with plasticity and undrained strength, variation with depth and location due to changes in clay-size fraction will cause regional and vertical variation in these values. Due to curvature of the Mohr envelope, the values are also dependent on stress level.

The value of residual shear strength is about $c_r' = 0$, $\phi_r' = 9$ to 17° depending on lithology and normal pressure. A change in normal pressure from 50 to 100 kPa causes a drop of about 1° in the value of ϕ_r' and a change from 100 to 150 kPa causes a further drop of about 0.5°.

The value of stiffness depends on the methods of testing and the stress conditions. The results are also sensitive to sample disturbance, depth and difference in the amount of the clay size fraction present. The clay is much stiffer for small strains of less than 0.5%. Plots of modulus value against depth show an increase with depth from perhaps $E_v' = 7.5$ MPa near the surface to about 20 MPa at a depth of 20m. Typically E_o is 80 to 600 MPa. Typically the data also show a good deal of scatter and change with depth may not always be an increase.

4.5.2.6. London Clay—compaction, consolidation and permeability. The median value of the maximum dry density of unweathered London Clay is 1.60, with first and third quartiles of 1.58 and 1.81 Mg/m^3 respectively. The median value for weathered clay is 1.56 Mg/m^3. The median values for optimum moisture content are 24% for the weathered clay and 23% when it is fresh. The clay can be difficult to compact because this moisture content may be less than the natural moisture content.

The values of coefficients of volume compressibility and of consolidation are variable and depend on the stress range used in the testing. As the data in Appendix B, Table B4, are quoted without an indication of the relevant pressure range, the values give only a very rough idea of the amount and rate of consolidation settlement to be expected. However, the values are mostly in the range of very low or low compressibility with low to medium rates of consolidation. The values of compressibility for weathered samples are about 50% more than those for fresh samples of the same clay.

4.5.3. Cretaceous clays

Several important clay formations of Cretaceous age outcrop in southern, eastern and south-east England. Most of these deposits also contain lithologies other than clay, including sand, sandstone and ironstone. They are all heavily over-consolidated, with between 425 and 1370 m of overlying sediments reported by Smith (1978) to have been removed.

Special attention is paid below to Gault Clay of southern and eastern England. A brief discussion is given here of the properties of other Cretaceous Clays. Bell (1994*b*) discusses the engineering properties of Speeton Clay from East Yorkshire.

As indicated in Appendix B, Table B5, Cretaceous clays range in plasticity from intermediate to very high values, where the latter is caused by the presence of smectite in Fullers Earth Clay and Gault Clay. This is demonstrated in Figure 4.17, which shows the average plasticity for Gault Clay and Weald Clay plotted on a Casagrande chart on which the trend lines for clay mixtures are also plotted. Weald Clay plots in a position indicative of an illite and kaolinite mixture, whereas Gault Clay is a mixture including montmorillonite.

Most of the data about these clays comes from slope stability studies, which is an indication of an important aspect of their behaviour. Although the clay-size fraction is lower in part of the Weald Clay, the clay fraction usually accounts for more than 50% in these clays. The activity of such clays is high; Figure 4.18 shows an activity chart for Weald Clay and Gault Clay, together with comparative plots for other clays and clay mixtures.

Many of the formations in the south of England are mantled with solifluucted clay that has been subject to slope movements under periglacial conditions. This has resulted in the presence of shear surfaces within the disturbed clay. The moisture content is highest in these zones and is one of the reasons for the high mean moisture contents recorded for weathered clays.

The undrained shear strength of Cretaceous clays ranges from soft to firm, where the weaker materials may well have been disturbed by solifluction processes. The peak effective strength parameters for fresh clay show apparent cohesion but this is lost in the solifluucted material, which suggests that the disturbance to the clay has destroyed the effects of over-consolidation. In fact some reported effective shear strength values are between the critical and residual values, which may be because the samples tested contained existing shear surfaces.

Cretaceous clays tend to possess low or very low compressibility and slow or medium rates of consolidation.

4.5.3.1. Gault Clay (see also Chapter 6). A comprehensive review of the distribution, character and properties of Gault Clay is provided by Forster *et al.* (1995). Besides reviewing data from many slope stability investigations, they also consider the data obtained from several deep boreholes, tunnels and excavations. Gault Clay outcrops as a narrow strip that runs north eastwards from Dorset in southwest England to Norfolk, east England. It also outcrops in the North and South Downs in southeast England. The Lower Gault consists mainly of 1 to 2 m thick units of medium and dark grey, soft mudstone and silty mudstone with illite and kaolinite the predominant clay minerals. The Upper Gault, which outcrops in southeast England, is paler in colour and has a high calcium carbonate content with smectite the dominant clay mineral. Pyrite is also present. In weathered clay, this pyrite may have been oxidized, leading to an elevated value of sulphate being present and, because of the associated removal of calcite, reduced strength.

4.5.3.2. Gault Clay—density and moisture content. The median values of bulk density for different areas of the UK vary between about 1.89 and 2.01 Mg/m^3, with an overall median for all data of 1.94 Mg/m^3. The ranges are considerably greater and less than these values. The range of average values is 1.60 to 2.35 Mg/m^3, thus indicating appreciable data scatter. This may be a reflection of the effects of unloading induced stress relief and weathering. Hence down to 10 m, values in the range of 1.62 to 2.35 Mg/m^3 are quoted. Below this, the range is 1.88 to 2.10 Mg/m^3. Moisture content values range from about 5 to about 52%, where the highest values will correspond with smectite-rich weathered clay. The median values for different areas of the UK range from about 23 to about 30%, with an overall median of 28%. The lowest median value occurs for Gault Clay from the south-west area of outcrop.

4.5.3.3. Gault Clay—particle size and consistency limits. The liquid limit medians vary from 68 to 80%, with a lower value of 50% being recorded for the southwest area. The overall median value is 75%. Thus the clay is of high and very high plasticity. The values of plasticity show little constant variation with depth. This may be because of cyclic changes in the mineralogy, giving rise to a range of values being obtained for closely spaced samples in a vertical profile. Extremely high liquid limit values of up to 184% are reported by Forster *et al.* (1995) for smectite rich Upper Gault Clay from a depth of about 115m from a borehole at Harwell, Oxfordshire. Apart from this location, the overall range is from about 20 to about 125%, but the authors point out that, although unlikely, contamination of samples with bentonite drilling mud may have caused such high values. Lower values for both plastic limit and plasticity index are recorded for the southwest of England, which follows the trend set by the liquid limit values.

The clay size fraction is very variable with values lying between 15 and 65%, thus unless clay particles are aggregated into silt sized grains, some parts of the formation are low in particles of clay-size. The 15% lower limit which is for the south western (Dorset area) is much lower than the figure quoted by Cripps & Taylor (1987) that was derived from studies by Chinsman (1972) into shallow earthworks failures in Kent. This may be a reflection of the variable nature of the lithology, although incomplete disaggregation of the more carbonate-rich clay during sample preparation could also be a cause.

Values of activity vary from 0.5 to 3.0, with most values lying in the range 0.7 to 1.0. Data obtained for clay from boreholes to the north and south of Cambridge centre in the 'normal activity' area of the chart, however values for outcrops to the south of this show a range of activities from inactive to very active, which is a reflection of the variation in clay-size fraction and mineralogy. However the data need to be treated with some caution as, due to incomplete disaggregation of samples, some values of clay-size fraction may have been under-estimated.

4.5.3.4. Gault Clay—strength. Arithmetic means and percentile and median values for undrained cohesion are available for all areas of outcrop. Again the data span large ranges, with an overall range of undrained cohesion of from 1 to almost 600 kPa. The higher values were probably determined on samples of claystone. Higher values than this have been recorded for samples from deep boreholes and tunnels. The weathered smectite-rich clay would be expected to exhibit the lowest strength. Ignoring the extreme values gives a more consistent picture. Overall the median values for different areas of the UK span the relatively narrow range of 85 to 148 MPa, which means that most Gault Clay is stiff, with little variation from one area to another. The total stress (undrained) friction angle varies from 0 to 7°.

Consideration of the relatively few published data for effective shear strength parameters, show c_p' values of 0 to over 100 kPa with ϕ_p' of 2 to 36°. Presumably the materials tested range from highly weathered and sheared smectite rich clay to carbonate cemented claystone but Mohr envelope curvature rather than cementation bonding of particles may be responsible for reported cohesion values (see Section 4.3.6). Quoting just the 1st and 3rd quartile indicates values of $c' = 7$ to over 20 kPa, and $\phi' = 21$ to 26° respectively. The median values are 12 kPa and 12°. Plotting effective shear strength ($c' + \sigma\tan\phi'$, where σ = overburden pressure) against depth indicates a steady increase from near zero at the surface to about 100 kPa at a depth of 16 m. The range of values for the residual angle of friction is 5 to 25° with an average of about 10°.

4.5.3.5. Gault Clay—compressibility and permeability. Values of the coefficient of volume compressibility for Gault Clay from boreholes and deep foundations range from 0.04 to 0.1 m²/MN for a pressure equal to the estimated overburden stress at depths of 43.5 m and 23.5 m respectively. Values of the coefficient of consolidation range from 0.4 to 10.4 m²/yr. Thus, Gault Clay has low to very low compressibility and slow to medium rates of consolidation. Estimates of the previous overburden pressure for clay from boreholes to the north and south of Cambridge range from 330 to 502 kPa, which means that the over-consolidation ratio is between 1.6 and 5.2 at this location. These values are about an order of magnitude lower than those reported by Samuels (1975) and Hutchinson (1969) for locations in eastern and southern England and may indicate either a real change in the depth of overburden or that stress relief effects have led to a partial loss of over-consolidation effects in the latter tests. The permeability of Gault Clay tends to be very low with values quoted to be in the range of 1.6 to 3.5×10^{-11} m/s for samples from deep boreholes.

4.5.4. Jurassic clays

The Jurassic system in the UK contains several important clay formations. All of them are over-consolidated and, in some cases the deposits have become indurated into low-strength rocks. They outcrop in an area of England that stretches from the south coast in Dorset northwards and eastwards through the south midlands and eastern England to the coast in North Yorkshire. There are also outcrops on the Isle of Skye, Scotland. Appendix B, Table B6, lists the geotechnical properties of several Jurassic clay formations, whilst relevant lithological details are given in Table 4.28. The properties of these formations are described below. Data for Lias Clay, the properties of which are discussed in a subsequent specific section, are given in Appendix B, Table B7.

Most Jurassic clays are of intermediate to high plasticity, with many points plotting in the high area of the plasticity chart above the A-line. As indicated in Appendix B, Table B6, Upper Oxford Clay contains material with very high or extremely high plasticity. The natural moisture content is usually in the region of 20 to 30%, but tends to be higher in the Fullers Earth Clay, which has a high Ca-montmorillonite content and was sheared and Ampthill Clay in which the wetter samples were from a shallow,

TABLE 4.28. *Clays included in Table B6*

Formation*	Lithology
Kellaways Clay	Over-consolidated, fissured, silty, occasionally sandy, high plasticity clay
Lower Oxford Clay	Over-consolidated, fissured, silty, sometimes calcareous or organic, low to very high plasticity clay with at least four beds of cementstone nodules.
Middle Oxford Clay	Pale grey silty mudstone with thin beds of calcareous siltstone in the upper part
Upper Oxford Clay	Over-consolidated, fissured, silty, rarely sandy, intermediate to high and occasionally very high plasticity clay
Ampthill Clay	Over-consolidated, fissured, silty clay. Sometimes with sand grade shell fragments. The plasticity is either intermediate to high or very high to extremely high.
Kimmeridge Clay	Over-consolidated, fissured, silty, sometimes calcareous or sandy clay
West Walton Formation	Pale and dark grey, fossiliferous, silty mudstones with thin silt beds and fossiliferous horizons.

* See Chapter 6 for formation details.

weathered horizon. The average values of plasticity and activity respectively plotted in Figures 4.17 and 4.18, indicate that Oxford Clay and Kimmeridge Clay plot in the overlap region for illite, kaolinite and montmorillonite. The clay-sized fraction averages about 50%, except for lower values recorded for the Lower Oxford Clay.

The moisture content is very variable, with values of 10 to 60%, within the top 10 m. This changes to about 18 to 30% at approximately 5 m and 12 to 27% at between 7 and 25 m. Thus the effects of stress relief and weathering extend to depths of about 20 m. The wetter near-surface material has lower density with values of about 1.60 to 2.15 Mg/m^3 but this increases to 2.00 to 2.23 Mg/m^3 in the deeper fresh material.

The undrained shear strength recorded for Jurassic clays puts them in the stiff clay category, but some samples of Middle Oxford Clay and Kimmeridge Clay are stiff or hard clays. The strength values span large ranges due to differences in weathering grade and moisture content in the upper parts but depth of burial and lithological changes also cause the strength to vary. The wetter, near surface clays may be soft but, where the upper few metres are drier, they are stiffer and stronger than the underlying clay. Vertical profiles show an increase in undrained shear strength and reduction in moisture content below this. Work by Russell & Parker (1979) suggests that the undrained shear strength of moderately weathered Oxford Clay is about 50% that of the fresh clay.

The effective shear strength parameters are also sensitive to weathering processes, with effective cohesion as high as 480 kPa recorded by Jackson & Fookes (1974) for a cemented Oxford Clay. Weathering action, disturbance and loss of bonding reduces c' to zero or near zero. Average values of effective angle of friction are between 20 and 30°, and residual effective angle of friction averages about 15°.

The compressibility of Jurassic clays is from low to medium, depending on lithology and weathering grade. Most of these materials possess slow or medium rates of consolidation. The modulus of elasticity increases with depth with $E_{sec} = 50$ MPa at 17 m depth and 135 at 29 m.

The corresponding tangent modulus values were about 40 MPa higher than these values.

4.5.4.1. Lias Clay (see also Chapter 6). The Lias Clay is an important series of mudstones and clay formations that outcrop across England from Dorset on the southwest coast to Yorkshire and Teesside in the northeast. Data assembled by Hobbs *et al.* (2005) is summarised in Appendix B, Table B7, together with data collated by Cripps & Taylor (1987) and data reported by Coulthard & Bell (1993) for Lower Lias Clay from Gloucestershire. Data about Lower Lias Clay from Humberside is given by Bell (1994c). Chandler (1972) discusses the effects of weathering on the properties of Lias Clay.

Lias Clay is variable along its outcrop as it was deposited in a series of basins separated by platform areas of shallower water. Subsidence in the basinal areas has given rise to thick mudstones in these areas in which, due to being more deeply buried by later sediments, the clay is more highly over-consolidated. In certain parts of the sequence, burial diagenesis has given rise to mudstone rather over-consolidated clay. As shown in Appendix B, Table B7, the Lower Lias has been sub-divided into several formations and the Middle and Upper Lias have been re-named (see also Chapter 6).

4.5.4.2. Lias Clay - moisture content and density. The overall median value of moisture content is 21%, with first and third quartile values of 18 and 25%. The overall range of values is from 2 to 102%, which is probably a reflection of the range in the character of the material from an indurated mudstone to a highly weathered clay. As shown in Appendix B, Table B7, the median values for the individual formations vary between 17 and 18% for the Redcar and Scunthorpe Mudstones to 23% for the Blue Lias and the Dyrham Formation. This variation is a consequence of the differences in lithology between the clay-rich basin areas and the platform deposits. Cripps & Taylor (1987) point out that due to variation in lithology, the ranges are wide and reduced overburden and weathering action accounts for a 60 to 100% increase in moisture content in depth profiles.

Median bulk density values vary between 2.00 and 2.05 Mg/m^3, and thus the degree of variation is relatively small. The greatest differences between the lowest and highest values are recorded for the Blue Lias, in which the range of values is 1.18 to 2.72 Mg/m^3. It seems likely that the higher values are due to the presence of cemented clay ironstones.

4.5.4.3. Lias Clay—particle size and consistency limits. The liquid limit of Lias Clay varies between medians of 46 and 53% and the median plasticity indices are 23 to 29%, thus classifying the material as a clay of intermediate and high plasticity. The maximum liquid limits for the individual formations are in the region of 95 to 100%, indicating that some parts of all the formations can have high and extra high plasticity. However taking the average values for Lower Lias Clay, the plasticity and activity plots shown in Figures 4.17 and 4.18 respectively indicate that the predominant clay minerals are mixtures of illite and kaolinite. Data presented by Cripps & Taylor (1987) suggests that increased weathering causes only a minor increase in plasticity, but that in many failed cutting slopes the plasticity is very high or extra high.

Particle size data shown in Table B7 suggest that the Blue Lias formation contains the greatest percentage of clay. The range is from 30 to 60%, and the median is 42%. On the other hand the Dyrham Formation is more silty. In this the range is from 1 to 52% and the median is 22%. The particle size data do not correlate strongly with the plasticity data, suggesting that variation in clay minerals contributes to the variability in the latter data.

4.5.4.4. Lias Clay—strength and stiffness. Median and the first and third quartile strength values are given in Appendix B, Table B7, for undrained shear strength, and effective shear strength parameters. Undrained shear strength is highly variable which may be a reflection of the presence of fissures in some samples. Some samples, usually clay-rich and lower density clays have lower strength, but differences in moisture content and depth of sampling are further causes of variability. Most values catagorize the clay as firm or stiff but some strengths higher than this are reported by Haydon & Hobbs (1977) for Lower Lias Mudrock from deep excavations at Hinkley Point, Somerset. Here uniaxial compressive strengths in the range of between 25 and 90 MPa (moderately strong and strong rock) were recorded.

The shear strength parameters in terms of effective stresses show median cohesion values ranging from 12 to 34 kPa and angle of internal friction in the range 23 to 33°. Such values of cohesion may be due to cementation of the grains, or curvature of the Mohr envelope. Values of cohesion approaching zero are reported for degraded and brecciated Lias Clay in slopes (see Chandler & Skempton 1974). Average values of residual shear strength vary between about 10 and 15°, depending on the effective normal pressure. Typically the value drops steeply for pressures of between about 20 and 100 kPa.

Haydon & Hobbs (1977) record high values of Young's modulus of 10 to 35 GPa for borehole samples of Lias Mudrock. Such values are not typical of the material at outcrop in which values in the range 22 to 52 MPa have been recorded by Starsewski & Thomas (1977).

4.5.5. Permo-Triassic mudrocks

Mudrocks of this age in the UK include the Aylesbeare Mudstone, Whipton, Monkerton, Westbury and Blue Anchor Formations in southwest England, Mercia Mudstone, which occurs widely in England, the Edlington and Cadeby Formations of Yorkshire and the Eden Shales of Cumbria. The Blue Anchor Formation, Mercia Mudstone Group and the Cadeby and Edlington Formations were previously respectively called the Tea Green Marl, Keuper Marl and the Lower and Middle Permian Marls. Data about the properties of Mercia Mudstone are given in Appendix B, Tables B8 and B9. With the exception of the Mercia Mudstone, which is the subject of a separate section below, few data about the properties of these materials are reported in the literature.

Forster (1991*a*) indicates that the Whipton and Monkerton Formations consist mainly of sandstone, but mudstone is locally present. The Monkerton Formation mudrocks are described as heavily over-consolidated mudrocks of low and intermediate plasticity. Data about these deposits are given in Appendix B, Table B10. As Appendix B, Table B9, shows, the Aylesbeare Mudstone Group is similar to this but also contains some material of high plasticity. The strength values vary over a wide range, especially near the surface, of 10 to 300 kPa, which distinguishes it as a soft to very stiff clay. All these materials have low compressibility and undergo rapid consolidation.

The Edlington Formation is described by Forster (1992) as consisting of over-consolidated silty, sandy clay with clayey sandy and silty, gravelly sand. Most of the plasticity values indicate that it is mainly of intermediate and high plasticity. Very variable moisture contents occur in the upper 3 m of the formation, which marks the lower limit to the depth influenced by weathering. The compressibility is medium and the rate of consolidation ranges from low to high. The properties of the Cadeby Formation are similar to this but the bulk density and undrained cohesion are both a little higher. The properties are summarized in Appendix B, Table B10.

4.5.5.1. Mercia Mudstone (see also Chapter 6). The Mercia Mudstone Group includes a series of mudstones with sandstones and evaporite deposits of Triassic age that underlies large parts of the Midlands of England. There are also important outcrops in the eastern and western parts of northern England, southwest England and Northern Ireland. Hobbs *et al.* (2002) explain that the formation was deposited in a mudflat environment comprising a series of small terrestrial basins. Hot and arid climatic conditions gave rise to deposition of mud in temporary lakes, deposition of sheets of silt and fine sand by flash floods and the accumulation of wind-blown dust. Some of these deposits are finely laminated and others lack sedimentary structures. Fissures and the variable presence of evaporite deposits, the sub-surface dissolution of which has lead to complicated weathering effects, give rise to great variability in engineering properties.

The discussion below is based on recent reviews of the engineering properties of the mudstones of the Mercia Mudstone Group by Hobbs *et al.* (2002), Chandler & Forster (2001) and various BGS reports to which reference should be made for further details. Relevant data collating the engineering properties are presented in Appendix B, Tables B8 and B9. The data from Hobbs *et al.* (2002) and other BGS reports were selected from high quality site investigation reports and those compiled by Cripps & Taylor (1987) were the result of an extensive literature review.

4.5.5.2. Mercia Mudstone—density and moisture content. The data in Appendix B, Table B8, for various areas of England show that median values of bulk density vary between 1.96 and 2.27 Mg/m^3, with an overall median of 1.68 Mg/m^3. Hobbs *et al.* (2002) mention that the overall range is between 1.47 and 2.27 Mg/m^3. Although the lower value is for highly weathered material from Staffordshire, apparently weathering may lead to increased or decreased density. Moisture content values show a very large range with the possibility of the very high values being for poorly taken borehole samples. Overall median values are within the range of 18 to 20% but higher values were recorded for Humberside and East Devon and a lower value of 12% for Teesside.

4.5.5.3. Mercia Mudstone—particle size and consistency limits. Hobbs *et al.* (2002) make the comment that, as the particle sizes range from silty clay to silty sandy gravel, many samples are 'misidentified' (according to modern definitions) as mudstone, in which there should be more than 35% silt- and clay-sized particles. The clay content is reported to vary between 1 and 70%. Davis (1967) noted that although the measured clay size fraction was 10 to 40%, the clay mineral content determined using X-ray diffraction varied between 60 and 100%. This discrepancy arises due to the presence of silt-sized aggregations of clay particles, which do not break down using standard methods of sample pre-treatment for particle size analysis. Accordingly, Davis defined the Aggregation Ratio as the percentage of clay minerals divided by the percentage of clay-sized particles. Chandler *et al.* (1968) report aggregation ratio values of between 1.39 and 9.35, where a ratio of unity would indicate complete disaggregation of clay aggregates during sample preparation. Because of the difficulty of obtaining an accurate value for the clay-size fraction the quoted high values of activity (see Figure 4.18) are probably incorrect. However, the possible presence of swelling clay minerals, which would also give enhanced activity, could be the cause (see Atkinson *et al.* 2003). Plotting the average plasticity data on Figure 4.17 suggests that the average mineralogical affinities of Mercia Mudstone are commensurate with an illite and kaolinite mixture.

4.5.5.4. Mercia Mudstone—strength and deformation. Hobbs *et al.* (2002) report that it is difficult to give typical or average values of strength for Mercia Mudstone as variations in fabric, structure, cementation, composition and weathering grade all affect the values. The intact strength, moisture content and density are liable to undergo appreciable variation over a matter of mm or metres, with, in parts, weaker mudstone or clay underlying harder material. For instance plots of undrained cohesion against depth show considerable scatter and there is also poor correlation of strength with weathering grade. Chandler *et al.* (1968) noted that the value of sensitivity, which is the ratio of the undisturbed peak strength to ultimate or reconstituted value at the same moisture content, is about 4. (See Section 4.3.8).

The overall median values of undrained cohesion lie in the range 82 to 115 kPa. Within the data presented in Table B8, the medians for different areas range from 72 to 220 kPa. These values are liable to apply to the more soil-like weathered mudstone, which behaves as a stiff and very stiff clay. Also these values may not accurately represent the shear strength of the material as it is assumed that $\phi_u = 0$ which it may not be if samples contain fissures or are incompletely saturated.

Uniaxial compressive strength tests are liable to apply to rock-like mudstone. Unfortunately values are available for only a restricted range of areas. They vary between 0.3 and 8.1 MPa, corresponding with weak and moderately weak rock. Plots against depth show considerable scatter, where low values are liable to apply to fissured or weathered samples and higher ones to desiccated material or to cemented claystones or mudstones.

Chandler (1969) notes that the brittle behaviour of the unweathered mudstone is replaced by plastic deformation at a confining pressure of between 500 and 700 kPa. This transitional pressure reduces to 210–280 kPa in slightly weathered mudstone, 120 kPa in moderately weathered mudstone and 70 kPa in highly weathered mudstone. Chandler & Forster (2001) indicate that when fresh or slightly weathered, the effective cohesion is more than 25 kPa; presumably reflecting the cemented character of the material, and the effective angle of friction is greater than 40°. There is a reduction to effective cohesion values of less than 20 kPa in moderately weathered mudstone and this is accompanied by a reduction in the angle of friction to between 32 and 42°. Further weathering causes little change to the cohesion but the angle of friction may be reduced to between 32 and 25°. These values of effective cohesion for fresh and slightly weathered mudstone are lower than those quoted by Cripps & Taylor (1987). However caution needs to be exercised over all these values as, although cemented mudstones may possess high cohesive strength, apparent cohesion may also be due to curvature of the Mohr envelope, fissuring of samples and partial saturation of samples.

Hobbs *et al.* (2002) quote values of residual effective strength in the range from 22 to 30°.

Deformation properties quoted from Maddison *et al.* (1996) for *in situ* and laboratory measurements are presented in Appendix B, Table B8, for the second Severn

Crossing. The variations in value with stress levels indicate curvature of the stress-strain graph. Also, higher values are measured vertically than horizontally.

Chandler & Forster (2001) quote values of K_o, the coefficient of earth pressure at rest of between 1.52 and 1.8 for depths below the ground surface of 3.7 and 9.1 m.

4.5.5.5. Mercia Mudstone—consolidation and permeability. Median values of the coefficient of volume compressibility vary between 0.32 to 0.03 m²/MN over the stress range 50 to 1600 kPa. On the other hand, the coefficient of consolidation lies between 5.6 and 6.5 m²/year and does not vary greatly for the same stress range. This means that the mudstone is of low or medium compressibility and the rate of consolidation is medium.

Tellam & Lloyd (1981) quote permeability values of 1×10^{-9} to 1×10^{-11} m/s for intact laboratory samples of Mercia Mudstone tested perpendicular the bedding and this compares with horizontal field values of 1×10^{-6} to 1×10^{-8} m/s. Mercia mudstone compacted to its maximum dry density is reported by Chandler *et al.* (1968) to have a permeability of 1×10^{-8} to 1×10^{-10} m/s.

4.5.5.6. Mercia Mudstone—compaction. Median values of maximum dry density vary between 1.71 and 1.82 Mg/m³ with corresponding optimum moisture content of 19 and 15.5%. Chandler *et al.* (1968) found that for a low compactive effort a high moisture content gave the highest density, whereas for a high compactive effort a high density was obtained for a low moisture content. This implies that increasing the compactive effort will not improve the compaction of wet marls.

4.5.6. Carboniferous mudrocks

The Carboniferous system of the UK contains a great variety of rock types, among them diverse mudrocks. Limestone is the dominant rock type of the Lower Carboniferous (Mississippian of the USA) whereas the Upper Carboniferous (Pennsylvanian of the USA) contains sandstones, gritstones and shales of the Millstone Grit Formation, and sequences of sandstones, siltstones, mudstones and coals which form the Upper, Middle and Lower Coal Measures. The Coal Measures, in particular contain a great variety of mudrocks. Upper Carboniferous rocks are present in north and south Wales, midlands and north of England and central Scotland. The Carboniferous rocks of southwest England are folded and metamorphosed so they could only be considered as clay materials when weathered.

The engineering properties of Carboniferous mudrocks presented in Appendix B, Tables B11 and B12, have been compiled following an extensive literature review by Cripps & Taylor (1981) and also from BGS reports on the Engineering Geology of specific districts. Other sources of information are referred to in the discussion below. A separate section is devoted to a discussion of Coal Measures mudrocks.

Non-Coal Measures mudrocks include the Crackington Formation from the south-west of England and Etruria Marl of the West Midlands and Staffordshire. The plasticity of the Crackington Formation is intermediate and high, and values range up into the very high plasticity range if weathered and sheared samples are included. The material has an undrained shear strength of between 120 and 620 kPa, making it a stiff or hard clay. Values of the residual shear strength range between 11 and 15°, with a mean of 13° and a very small value of apparent cohesion.

Forster (1991*b*) explains that Etruria Marl is frequently found to comprise both soils and rock material. The data given in Appendix B, Table B11, are liable to relate to the clay and silt lithologies. The material consists of variegated or mottled red, green, yellow, grey and purple soft to stiff, and, where less weathered blocky and hard, generally highly to completely weathered clay with lithorelicts. Completely weathered material is reported to extend to a depth of about 12 m.

The median moisture content is 14% with a decrease from values of about 10 to 40% for the near-surface material and decreasing to between 5 and 22% at a depth of 10 m. Most plasticity data classify Etruria Marl as being of low or intermediate plasticity, but there is a significant presence of high and very high plasticity marl. Some of the siltier or sandier marls plot below the A-line. In terms of undrained strength the material ranges from soft to very stiff, with a trend of increasing strength with depth.

4.5.6.1. Coal Measures mudrocks (see also Chapter 6). The Coal Measures sequence consists of rhythmic sequences of seatearth, coal, mudstone, siltstone and sandstone beds. The mudrocks in these sequences may comprise mudstones, silty mudstones, argillaceous seatearths and shales. Some of these rocks contain appreciable amounts of organic material and the sequences are of variable lithology with rapid vertical and lateral changes. Typically the mudrocks may be described as medium and dark grey, weak to moderately strong, sometimes fissured and sometimes thinly laminated and fissile, mudstones and silty mudstones. Generally speaking the siltier, more quartz rich mudstones are lighter in colour. In many cases the silt may be concentrated into thin laminae, and with increasing the proportions of silt, the materials grade into muddy siltstones and siltstones. Mudstone units may contain strong or very strong ironstone nodules or bands, in which the rock has become cemented with inter-granular iron carbonate (siderite).

Lake *et al.* (1992) note that weathered mudrocks give rise to firm to stiff, orange-brown and pale grey mottled clay soils that extend from the ground surface down to a depth of between 2 and 6 m. Below this there may be a transition to less weathered mudstone which may include zones of highly weathered material consisting of softened mudstone in a silty clay matrix, extending to depths of 10 to 15 m (see Spears & Taylor 1972).The effects of weathering may extend to greater depths than this in faulted zones, or where the rock mass has been disturbed by mining subsidence, and, as noted below, stress relief results in fissuring (jointing) of the material below depths influenced by the other effects of weathering.

4.5.6.2. Coal Measures mudrocks—moisture content and plasticity. Fresh Coal Measures mudrocks from underground and freshly excavated spoil have values of moisture content between 5 and 10%. These materials tend to plot in the low plasticity part of the plasticity chart. Weathering results in increases of both moisture content and plasticity with typical values of moisture content in the range of 20 to 30%, and plasticity in the intermediate and high categories. Within this range, seatearths tend to have the highest and shales, or more fissile mudrocks, the lowest plasticity. Some individual unweathered seatearths possess high or very high plasticity, presumably this is a function of their clay-rich mineralogy.

4.5.6.3. Coal Measures mudrocks—strength. The strength parameters quoted in Appendix B, Table B11, from Cripps & Taylor (1981) were compiled from a large number of investigations, many of them in connection with the stability of slopes. Data from the testing of material from deep mine and tunnel excavations, as well as weathered and unweathered material from outcrops and boreholes, have also been considered. Appendix B, Table B12, contains data quoted from various BGS reports, which were originally from site investigation reports for various districts of the UK.

Depending upon weathering grade, lithology and structure, Coal Measures mudrocks may display properties ranging from brittle rocks to plastic soils. For material from the top 10 m, triaxial and in situ pressuremeter tests Cripps & Taylor (1981) quote undrained cohesion strength values ranging between about 15 kPa and over 300 kPa. Samples from shallow depths tend to contain fissures even if the rock is not weathered, and the strength values are lower than for intact samples discussed below. The data from the BGS reports suggest that typical values are liable to be in the range of 80 to 150 kPa which categorise the material as stiff and hard clay. The coarser grained and siltier, less well weathered and drier mudrocks would have the highest strength whereas the high plasticity and wet mudrocks, especially clay-rich seatearths, possess low strength. Owing to the lack of saturation of samples and/or curvature of the Mohr envelope, some undrained friction values, usually less than $\phi_u = 10°$, are recorded.

As shown in Appendix B, Table B11, the effective shear strength parameters range between $c' = 131$ kPa, $\phi' = 45.5$, which is for a series of tests on unweathered but fissured Coal Measures mudstone, shale, seatearth and argillaceous siltstone from depths of up to 18 m from South Yorkshire, to $c' = 15$ kPa, $\phi' = 21°$ for weathered mudstone from Walton's Wood, Staffordshire (Early & Skempton 1972). These values of cohesion may well relate to the presence of intergranular cements and curvature of the Mohr envelope. When remoulded, the strength of the latter mudstone dropped to $c' = 0$, $\phi' = 20°$. Thus, weathering causes a drop of about 90 or 100% in effective cohesion and 35 or 40% in effective angle of friction. The effects of fissuring and weathering on the Mohr envelope are shown in Figure 4.19. The less weathered material shows a curved Mohr envelope so these parameters do not accurately portray the strength of the material at low effective normal stresses.

The residual shear strength parameters are usually in the range of $c_r' = 0$ to 10 kPa, $\phi_r' = 12$ to 16°.

Strength values for compacted colliery spoil are also available in the literature. Here again, there is significant curvature of the Mohr envelope, and for effective normal stresses of up to 100 kPa, $c' = 0$, $\phi' = 50°$. At higher normal stresses, typically values are $c' = 0$ to 15 kPa, $\phi' = 35$ to 40°, depending in the mix of lithologies present and the weathering grade.

The uniaxial compressive strength (σ_c) of some UK fresh mudrocks of Carboniferous age can be as high as 200 MPa, which puts them into the category of very strong rock. Such high strength, which is probably accounted for by cementation in relatively silty samples, is not typical as most mudrocks tend to be weak ($\sigma_c = 1.25$–5 MPa) or moderately weak ($\sigma_c = 5$–12.5 MPa). Some types of mudrock are anisotropic on account of the fissility they possess. This is the result of a laminated sedimentary structure and particle alignment. Bell *et al.* (1997) measured ratios of the maximum to minimum uniaxial compressive strengths of 2.1–19.0.

Owing to the high strength of the rock skeleton, pore pressures generated during shearing are much smaller than the stresses being imposed on samples such that the total and effective stress parameters are approximately equal to each other. Data shown in Figure 4.19 indicate that for indurated and intact Carboniferous age UK mudrocks, values of apparent cohesion (c_a') of 5–30 MPa, and internal friction (ϕ_a') of 15 to 40° are typical, although Jackson & Lawrence (1990) quote values much higher than this for mudstone from northeast England. As the above figures illustrate, with weathering action the material loses its rock-like behaviour and, ultimately degrades to clay, the strength properties of which are a function of mineralogy, grain size, density and state of saturation.

4.5.6.4. Coal Measures mudrocks—consolidation and deformation parameters. Values of the modulus of elasticity span a very large range, depending upon the lithology, weathering grade and stress conditions during testing. Tests on small intact samples of mudrocks from underground locations give modulus values of about 2 to 50 GPa. Much lower values are obtained for fissured samples such that for samples from depths of less than about 20 m of the ground surface values between 10 and 200 MPa are more typical.

There are very few consolidation test results reported in the literature. Those quoted by Jackson & Lawrence (1990) indicate low compressibility and high rates of consolidation.

4.5.6.5. Coal Measures mudrocks—durability. As discussed in Section 4.1.7, the durability of mudrocks depends upon mineralogy, porosity, cementation and

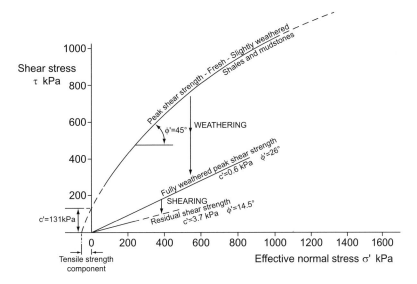

FIG. 4.19. Effects of weathering on the shear strength properties of mudrocks (after Taylor, 1988).

structure. Generally speaking, because geologically older sediments have been buried more deeply than younger ones, older mudrocks tend to contain a more stable assemblage of clay minerals, possess lower porosity and contain fewer micro-fractures; they are consequently more durable. This is consistent with the findings of Taylor & Spears (1970), who observed that an increase in durability of mudrocks interbedded with coal seams correlated with an increase in coal rank.

Notwithstanding the tendency for older mudrocks to possess higher durability, the slake durability of Carboniferous mudrocks can span large ranges. The results of tests on a selection of UK Carboniferous mudrocks, comprising claystones, silty mudstones and mudstones, reported by Czerekwo & Cripps (2001), indicate the durability of all three rock types span the complete range from non-durable to extra durable rocks. The relevant data are presented in Table 4.29 as the durability rating (see Table 4.8) for one, three and five cycles of slaking, the durability rating revealed by static jar slaking tests, saturation moisture content (an indication of porosity) and percentage of quartz present determined by XRD analysis. Also shown are durability data for seven samples exposed to 12 months of UK weathering conditions, where the assessment was made using the same rating criteria as was used in the static jar slaking tests.

The data indicate that, for the samples tested, there was a significant change in durability rating between first and third cycles of slaking, but that five cycles caused little further change. Thus three-cycle slaking tests provide a reliable indication of slaking performance. Non-durable samples tend to have higher saturation moisture contents (i.e. porosity of over 2%), but not all high porosity samples have low durability. This is because durability also depends upon structural features, such as laminations and fractures, as well as on mineralogical features such as the presence of swelling clay minerals and absence of cements.

It was evident in the tests that, although some samples had broken down, because the fragments were larger than the 2 mm size of the sieve mesh in the apparatus, they were classified as having high durability. Czerewko & Cripps (2001) suggest that the static jar slake test allows the true durability characteristics of such materials to be identified. In other cases the jar slake values are in good agreement with the three-cycle slake durability test results. The static jar slake test successfully predicted the durability of the seven samples that were also exposed to 12 months of UK weathering conditions.

4.6. Conclusions

The engineering properties of clays and mudrocks are controlled by the composition, structure and the moisture content of the materials. Variations in these controls arise as a result of differences in composition and structure of the original sediment and also as a consequence of processes associated with burial and subsequent exhumation of clay and mudrock formations. Further consideration is given below to the effects of burial and exhumation.

By definition clays and mudrocks have at least 35% of material less then 2 μm in size. This chapter has considered deposits in which clay minerals are an important component of this material, although quartz and other minerals will usually also be present. Clay and mudrock formations can be variable in composition and structure. The amount and type of clay minerals in the original deposit are a function of the source rocks from which they are being formed and sedimentological processes. This

TABLE 4.29. *Lithological features of Jurassic Clays included in Appendix B, Table B6*

Lithology	% quartz	w_{sat} %	I_{d1} %	I_{d3} %	I_{d5} %	I_j %	Rate	I_j * %
Organic claystone	3	0.2	E	E	E	D	F	D
Organic claystone	3	2.4	E	E	E	E	F	E
Fissile claystone	4	5.9	D	N	N	N	F	
Very fissile claystone	5	6.3	E	D	D	D	F	
Fissile claystone	5	13.1	D	N	N	N	F	
Organic claystone	5	6.8	E	D	D	N	S	
Claystone	7	5.7	D	D	D	N	F	
Fissile claystone	7	7.0	D	N	N	N	F	
Organic fissile mudstone	8	2.8	E	D	D	D	F	
Mudstone seatearth	10	4.2	N	N	N	N	S	N
Mudstone	11	4.0	D	ND	N	N	F	
Laminated mudstone	12	2.3	E	D	D	D	F	
Mudstone	15	5.7	E	D	D	D	F	
Mudstone	15	10.1	D	D	N	N	S	N
Mudstone	16	8.8	D	N	N	N	F	
Slightly fissile organic mudstone	17	9.0	D	N	N	N	F	N
Mudstone seatearth	17	4.1	D	D	D	D	F	
Slightly fissile mudstone	17	3.6	E	D	D	D	F	
Inter-laminated mudstone-siltstone	20	1.5	E	E	E	D	S	
Mudstone	21	5.9	D	D	N	N	S	
Inter-laminated mudstone-siltstone	22	2.9	E	D	D	D	F	
Mudstone	23	4.0	D	D	D	D	S	
Laminated mudstone-siltstone	24	3.6	E	D	D	E	F	E
Laminated mudstone-siltstone	27	2.7	E	D	D	D	S	
Siltstone	30	4.4	D	N	N	D	S	
Laminated slightly fissile mudstone-siltstone	34	4.2	D	D	D	D	S	
Laminated mudstone-siltstone	35	1.8	E	E	E	D	S	D

I_{d1}, I_{d3}, I_{d5} – Values of 1, 3 and 5 cycles slake durability determination (Brown 1981)
Slake durability rating: E, Extra high durability ($I_d > 95\%$); D, Durable ($65 < I_d < 95\%$), N, Non-durable ($I_d < 65\%$)
Jar slake test results: E, Extra high durability ($I_j' = 1$–2); Durable ($I_j' = 3$–5); Non-durable (($I_j' = 6$-8); F, fast, S, slow
12 months exposure to UK weathering condition—(I_j* values classified as above)

assemblage may be changed during burial and subsequent exhumation. The structure may be a product of the process of formation in residual and some glacial deposits, or in the case of sedimented deposits, laminations and particle alignment may be imparted by the processes of transportation and deposition. In residual deposits, structures may also be inherited from the original rock and all structures may be modified by burial, exhumation and weathering. The influence of these many factors can give rise to a variety of forms of behaviour in clays and mudrocks, and also to different patterns of lateral and vertical variation in their character and properties on all scales.

4.6.1. Variation in UK clays and mudrocks with geological age

As shown in Table 4.15 and Figures 4.17 and 4.18, the values of liquid and plastic limits and activity for different clay minerals lie within specific characteristic ranges. Thus variation in the type and amount of clay minerals account for differences in plasticity recorded by MacNeil & Steele (1999) for the seven British clays shown on Figures 4.17 and 4.18. Three of the clays (Weald Clay, Mercia Mudstone, Lower Lias Clay) plot in the general regions of the chart associated with illite and kaolinite mixtures, whilst two of the clays (Gault Clay, London Clay) plot in the lower part of the region characterized by montmorillonite mixtures. The remaining two clays (Oxford Clay and Kimmeridge Clay) plot in the overlap region of all three clay minerals. These observations are consistent with the clay mineral analyses given in Appendix A for samples of these clays. It can be noted that Mercia Mudstone appears to have high activity (0.94), but this may be because its clay content has been under-estimated because of difficulty with the dispersal of the material prior to particle size determination (Dumbleton & West 1966c) rather than as the result of the presence of montmorillonite, although Atkinson *et al.* (2003) do indicate that swelling clays may be present in parts of this formation.

The formations are listed on Figure 4.17 in order of age, with the youngest at the top. Thus, apart from the Gault Clay, a reduction in plasticity with geological age is apparent, reflecting a progressive conversion of active clay minerals to more stable forms with age. As mentioned previously, the high plasticity of Gault Clay is a result of it being relatively rich in montmorillonite.

Although related to age, it is the original clay mineralogy coupled with the maximum overburden pressure and temperature conditions that occur due to burial that are the key controls of the changes in plasticity.

According to Cripps & Taylor (1981), the moisture content of weathered and unweathered British clay formations varies from about 11% to 30% although much higher values may be encountered if sheared and highly weathered materials are included. There is a general trend of reducing moisture content with increasing geological age, although this is partly obscured by variations in the depth of sampling, lithology and weathering grade. Values of porosity and, as noted above, plasticity also show a general reduction with geological age.

Weathering processes account for a large increase in moisture content in over-consolidated clays to values mainly in the range from 35 to 45% regardless of age. The natural moisture content of unweathered mudrocks varies from around 5% to in excess of 20%, depending on density and composition. Again weathering accounts for a large increase in moisture content, with values of above 30% recorded for some weathered mudrocks.

Owing to variations in lithology, depth, stress conditions during testing and moisture content, undrained shear strength of clays varies over a very large range. For most over-consolidated clays, the mean undrained shear strength (s_u) lies in the range 100 to 200 kPa, but weathering processes typically reduce this to between 20 and 100 kPa. Some units display higher strength. For instance, Lower Lias Clay from a deep borehole has an undrained shear strength of up to 45 MPa (strong rock), implying a degree of cementation of the material.

Weathering action accounts for a reduction of undrained shear strength of approximately 50% in many over-consolidated clays and more than this in more indurated materials. Values of effective shear strength parameters span a wide range, with effective cohesion values of between about 10 and 250 kPa, and angle of internal friction between 20 and 35°. These data apply to confining pressures of up to about 400 kPa. Weathering reduces the strength to approximately $c' = 0$ to 55 kPa and $\phi' = 14$ to 30°. However, in uncemented clays, quoted values of effective cohesion are an artifact of curvature of the Mohr envelope. The residual shear strength (ϕ_r') of UK clays varies from around 6° to 22°, where $c_r' = 0$, depending upon the grading and clay mineral type. Figure 4.20 shows the relationship between residual strength and clay size fraction given by Skempton (1964). The residual shear strength data collated by Cripps & Taylor (1981) are in good agreement with these trends.

4.6.2. Degradation of over-consolidated clays and indurated mudrocks

As explained in Section 4.3.3, a reduction in loading brought about by erosion of the overlying sediments results in a deposit being in an over-consolidated state, and the recovery of void volume and increase in moisture

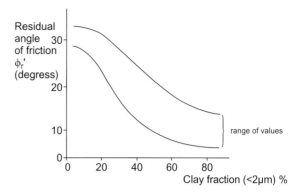

FIG. 4.20. Relationship between residual shear strength and clay-size fraction of clays (Skempton 1964).

content occur as a consequence of the reduced overburden pressures. The amount by which the moisture content increases due to unloading depends on the strength of the inter-particle bonds in the material. Such bonds are weakened, and may be completely destroyed, by weathering action. However, because this takes time, the loss of the effects of over-consolidation may occur over a protracted period of time. Where bonds are weak, the properties may change on timescales short enough to be of importance to engineers and those concerned with extraction and exploitation of the materials. Determinations of the slaking properties provide guidance as to the likely rate of deterioration (see Section 4.1.7).

Exhumation of a buried sediment leads to it experiencing the effects of stress relief. Initially an increase in porosity occurs by the development and opening of fissures and joints in the material. Subsequent changes due to weathering action are liable to include mineralogical changes accompanied by loss of cements and breakage of inter-particle bonds. Cripps & Taylor (1987) discuss the increase in coefficient of earth pressure at rest, K_o (indicative of relatively high horizontal stress) that occurs during the unloading process.

Thus in over-consolidated clays and indurated mudrocks fissuring accompanied by an increase in moisture content occurs before the other softening and mineralogical changes due to weathering processes have any influence on the material. The development of fissures brings about an increase in compressibility and loss of strength. In the more indurated mudrocks, such as those of Carboniferous age, the material displays a curved Mohr envelope with high values of apparent cohesion. Weathering processes result in increases in the number of discontinuities accompanied by degradation, increases in moisture content and softening, which penetrates from discontinuity surfaces. Thus the material becomes progressively weaker and more deformable as it becomes transformed from a rock into a soil. As mentioned above and illustrated in Figure 4.19, this results in a large reduction in apparent cohesion.

4.6.3. Controls on properties

The composition and structure, as well as the weathering grade, moisture content and stress conditions are all important controls over the engineering properties of clay materials. This Chapter has included descriptions of methods of classifying and describing clay materials and has explained the main parameters by which the mechanical behaviour can be elucidated. Attention has also been given to describing the behaviour typically exhibited by clays and mudrocks of different types.

Typical values for various parameters are also presented. As mentioned previously, care needs to be taken when applying these data as the values depend on many factors, including the method of sampling, variable features of the material concerned and the type and conditions of testing.

References

ADOLFSSON, K., RYDELL, B. G., SALLFORS, G. & TIDFORS, M. 1985. Heat storage in clay—geotechnical consequences and use of heat drains. *In*: *Proceedings 11th International Conference on Soil Mechanics and Foundation Engineering, San Francisco*, **3**, 1241–1244.

ALLABY, A. & ALLABY, M. 1990. *The Concise Oxford Dictionary of Earth Sciences*. Oxford University Press, Oxford.

AITCHISON, G. O. & WOOD, C. C. 1965. Some interactions of compaction, permeability and post-construction deflocculation affecting the probability of piping failures in small dams. *Proceedings Sixth International Conference Soil Mechanics Foundation Engineering*, Montreal, **2**, 444–446.

ANON 1975. *Soil Taxonomy*. Handbook 436, United States Department of Agriculture, US Government Printing Office.

ANON 1982. *Lime Stabilisation Construction Manual*, 7th edn, National Lime Association, Washington DC.

ANON 1986. *Lime Stabilisation Manual*. Imperial Chemical Industries, Buxton, UK.

ANON 1990a. State of the art report on soil-cement. *American Concrete Institute Materials Journal*, **87** (4), 395–417.

ANON 1990b. Tropical residual soils. Working Party Report. *Quarterly Journal Engineering Geology*, **23**, 1–101 (see also Fookes 1997).

ANON 1995. The description and classification of weathered rocks for engineering purposes. Working Party Report. *Quarterly Journal of Engineering Geology*, **28**, 207–242.

ASTM D422-63 2003. *Standard Methods for Particle Size-Analysis of Soils*. ASTM International.

ASTM D5434-03 2003. *Standard Guide for Field Logging of Subsurface Explorations of Soil and Rock*. ASTM International.

ATKINSON, J. H. 1993. *An Introduction to the Mechanics of Soils and Foundations*. McGraw-Hill, London.

ATKINSON, J. H. 2000. Non-linear soil stiffness in routine design. *Géotechnique*, **50**, 487–508.

ATKINSON, J. H., FOOKES, P. G., MIGLIO, B. F. & PETTIFER, G. S. 2003. Destructuring and disaggregation of Mercia Mudstone during full-face tunnelling. *Quarterly Journal of Engineering Geology and Hydrogeology*, **36**, 293–303.

BADGER, C. M., CUMMINGS, A. D. & WHITMORE, R. L. 1957. The disintegration of shale. *Journal of Institute of Fuel*, **29**, 417–423.

BARNES, G. E. 2000. *Soil Mechanics: Principles and Practice*. 2nd edn. Macmillan, Basingstoke.

BARNES, H. A., HUTTON, J. F. & WALTERS, K. 1991. *An Introduction to Rheology*. Elsevier, Amsterdam.

BAVER, L. D., GARDNER, W. H. & GARDNER, W. R. 1972. *Soil Physics*, 4th edn. Wiley, New York.

BELL, F. G. 1992. An investigation of a site in Coal Measures for brick-making materials: an illustration of procedures. *Engineering Geology*, **32**, 39–52.

BELL, F. G. 1993. *Engineering Treatment of Soils*. Spon, London.

BELL, F. G. 1994a. (ed.) *Engineering in Rock Masses*. Butterworth, Oxford.

BELL, F. G. 1994b. The Speeton Clay of North Yorkshire, England: an investigation of its geotechnical properties. *Engineering Geology*, **36**, 259–266.

BELL, F. G. 1994c. A survey of the geotechnical properties of some mudrocks of Lower Liassic age in the Scunthorpe area, Humberside, United Kingdom. *Bulletin International Association Engineering Geology*, **50**, 9–16.

BELL, F. G. 2000. *Engineering Properties of Soils and Rocks*, 4th edn. Blackwell, Oxford.

BELL, F. G. 2002. The geotechnical properties of some till deposits occurring along the coastal areas of eastern England. *Engineering Geology*, **63**, 49–68.

BELL, F. G. & COULTHARD, J. M. 1997. A survey of some geotechnical properties of the Tees Laminated Clay of central Middlesbrough, north-east England. *Engineering Geology*, **48**, 117–133.

BELL, F. G. & MAUD, R. R. 1994. Dispersive soils: a review from a South African perspective. *Quarterly Journal Engineering Geology*, **27**, 195–210.

BELL, F. G. & MAUD, R. R. 1996. Landslides associated with the Pietermaritzburg Formation in the greater Durban area, South Africa: some case histories. *Environmental and Engineering Geoscience*, **2**, 557–673.

BELL, F. G., ENTWISLE, D. & CULSHAW, M. G. 1997. A geotechnical survey of some British Coal Measures mudstones with particular reference to durability. *Engineering Geology*, **46**, 115–129.

BELL, F. G. & WALKER, D. J. H. 2000. A further examination of the nature of dispersive soils in Natal, South Africa. *Quarterly Journal of Engineering Geology and Hydrogeology*, **33**, 187–199.

BERRY, P. L. & REID, D. 1988. *An Introduction to Soil Mechanics*. McGraw-Hill, London.

BLATT, H. MIDDLETON, G. & MURRAY, R. 1980. *Origin of Sedimentary Rocks*. Prentice-Hall, New Jersey.

BOURDEAUX, G. & IMAIZUMI, H. 1977. Dispersive clay at Sobrandinho Dam. *In*: SHERARD, J. L. & DECKER, R. S. (eds) *Proceedings Symposium on Dispersive Clays, Related to Piping and Erosion in Geotechnical Projects*, Special Publication 623, American Society Testing Materials, Philadelphia, 13–24.

BLUNDEN, J. 1975. *The Mineral Resources of Britain*. Hutchinson, London.

BOULTON, G. S. & PAUL, M. A. 1976. The development of geotechnical properties of glacial tills. *Quarterly Journal of Engineering Geology*, **9**, 159–194.

BRADY, N. C. 1990. *The Nature and Properties of Soils*. Maxwell—Macmillan, New York.

BRAUDEAU, E., COSTANTINI, J. M., BELLIER, G. & COLLEUILLE, H. 1999. New device and method for soil shrinkage curve measurement and characterisation. *Soil Science Society of America Journal*, **63** (3), 525–535.

BROWN, E. T. (ed.) 1981. *Rock Characterization Testing and Monitoring: ISRM Suggested Methods*. Commission on Testing Methods, International Society for Rock Mechanics. Pergamon, Oxford.

BS 1377:1990 *British Standard Methods of Tests for Soils for Civil Engineering Purposes*. British Standards Institution, London.

BS 5930:1999 *Code of Practice for Site Investigations*. British Standards Institution, London.

BUCHANAN, F. 1807. *A journey from Madras through the countries of Mysore, Canara and Malabar*. Vol. 3. East India Company, London.

BURNETT, A. D. & FOOKES, P. G. 1974. A regional engineering geological study of the London Clay in the London and Hampshire Basins. *Quarterly Journal of Engineering Geology*, **7**, 257–296.

BURROUS, C. M. 1973. *The Effect of Temperature Change on Consolidation and Shear Strength of Saturated Cohesive Soils*. PhD thesis, University of Washington.

BUTTON, D. D. & LAWRENCE, W. G. 1964. Effect of temperature on the charge on kaolinite particles in water. *Journal of American Ceramic Society*, **47**, 503–509.

CAMPANELLA, R. G. & MITCHELL, J. K. 1968. Influence of temperature variation on soil behaviour. *Journal of Soil Mechanics and Foundation Division, Proc ASCE*, 94 no SM3: 709–734.

CHANDLER, R. J. 1969. The effect of weathering on the shear strength properties of Keuper Marl. *Géotechnique*, **19**, 557–673.

CHANDLER, R. J. 1972. Lias Clay: weathering processes and their effect on shear strength. *Geotechnique*, **22**, 403–431.

CHANDLER, R. J. 2000. Clay sediments in depositional basins: the geotechnical cycle. The third Glossop lecture. *Quarterly Journal of Engineering Geology and Hydrogeology*, **33**, 7–39.

CHANDLER, R. J. & FORSTER, A. 2001. *Engineering in Mercia Mudstone*. Report C570, Construction Industry Research and Information Association, London.

CHANDLER, R. J. & SKEMPTON, A. W. 1974. The design of permanent cutting slopes in stiff fissured clays. *Géotechnique*, **24**, 457–466.

CHANDLER, R. J., BIRCH, N, & DAVIS, A. G. 1968. *Engineering Properties of Keuper Marl*. CIRIA Research Report 13, Construction Industry Research and Information Association, London.

CHARMAN, J. M. 1988. *Laterite in Road Pavements*. Special Publication 47, Construction Industry Research and Information Association, London.

CHINSMAN, B. W. E. 1972. *Field and Laboratory Studies of Short-term Earthworks Failures Involving the Gault Clay in West Kent*. PhD thesis, University of Surrey.

COLBECK, R. S. B. & WIID, B. L. 1965. The influence of moisture content on the compressive strength of rock. *Proceeding of Symposium on Rock Mechanics, Canadian Department of Mineral Technology and Surveying*, Ottawa, 65–83.

COLLINS, H. 1993. Impact of the temperature inside the landfill on the behaviour of barrier systems. In: *Proceeding Sardinia '93, 4th International Landfill Symposium S.* Margheritta di Pula, Cagliari, Italy, 417–432.

COULTHARD, J. M. & BELL, F. G. 1993. The engineering geology of the Lower Lias Clay at Blockley, Gloucestershire, U.K. *Geotechnical and Geological Engineering*, **11**, 185–201.

CRIPPS, J. C. & TAYLOR, R. K. 1981. The engineering properties of mudrocks. *Quarterly Journal Engineering Geology*, **14**, 325–346.

CRIPPS, J. C. & TAYLOR, R. K. 1986 Engineering characteristics of British over-consolidated clays and mudrocks I Tertiary deposits. *Engineering Geology*, **22**, 349–376.

CRIPPS, J. C. & TAYLOR, R. K. 1987 Engineering characteristics of British over-consolidated clays and mudrocks II Mesozoic deposits. *Engineering Geology*, **23**, 213–253.

CRONEY, D. & COLEMAN, J. D. 1960. soil moisture suction properties and their bearing on the moisture distribution in soils. *Proceedings of a 3rd International Conference International Society Soil Mechanics and Foundation Engineering*, Zurich, **1**, 13–18.

CTORI, P. 1989. The effects of temperature on the physical properties of cohesive soils. *Ground Engineering*, **22** (5), 26–27.

CULSHAW, M. G. & CRUMMY, J. A. 1991. South West Essex—M25 Corridor: Engineering Geology. *Technical Report WON/90/2*. British Geological Survey, Keyworth, Nottingham.

CZEREWKO, M. A. 1997. *Diagenesis of Mudrocks: Illite Crystallinity and the Engineering Properties*. PhD thesis, University of Sheffield.

CZEREWKO, M. A. & CRIPPS, J. C. 2001. Assessing the durability of mudrocks using the modifies jar slake index test. *Quarterly Journal of Engineering Geology and Hydrogeology*, **34**, 153–163.

DAVIS, A. G. 1967. On the mineralogy and phase equilibrium of Keuper Marl. *Quarterly Journal of Engineering Geology*, **1**, 25–46.

DENNESS, B. 1974. Engineering aspects of the chalky boulder clay at the new town of Milton Keynes in Buckinghamshire. *Quarterly Journal of Engineering Geology*, **7**, 297–309.

DINSDALE, A. 1996. *Pottery Science: Materials, Process and Products*. Ellis Horwood, Chichester.

DIXON, N. 1987. *The Mechanics of Coastal Landslides in London Clay at Warden Point, Isle of Sheppey*. PhD thesis, Kingston Polytechnic.

DRISCOLL, R. 1983. The influence of vegetation on the swelling and shrinkage of clay soils in Britain. *Géotechnique*, **33**, 93–105.

DUMBLETON, M. J. & WEST, G. 1966a. Some factors affecting the relation between the clay minerals in soils and their plasticity. *Clay Minerals*, **6**, 179–193.

DUMBLETON, M. J. & WEST, G. 1966b. *The influence of the coarse fraction on the plastic properties of clay soils*. RRL Report, No 36, Road Research Laboratory, Crowthorne.

DUMBLETON, M. J. & WEST, G. 1966c. *Studies of the Keuper Marl: mineralogy*. RRL Report, No 40, Road Research Laboratory, Crowthorne.

DUMBLETON, M. J. & WEST, G. 1967. *Studies of the Keuper Marl: Stability of Aggregation under Weathering*. Report LR85. Road Research Laboratory, Crowthorne.

EARLY, K. R. & SKEMPTON, A. W. 1972. Investigations of the landslide at Waltons Wood, Staffordshire. *Quarterly Journal of Engineering Geology*, **5**, 19–41.

ELE 1998. *Construction Materials Testing Equipment*. 10th Catalogue, ELE International Ltd.

ENTWISLE, D. C., HOBBS, P. R. N., NORTHMORE, K. J., SELF, S., PRICE, S., SKIPPER, J., RAINES, M. G., BUTCHER, A. S., PEARCE, J. M., ELLISON, R. A., KEMP, S. J. & MEAKIN, J. L. 2005. *The Engineering Geology of UK Rocks and Soils: The Lambeth Group*. Technical Report No. IR/05/006. British Geological Survey, Keyworth, Nottingham.

FANG, HSAI-YANG 1997. *Introduction to Environmental Geotechnology*. CRC, Boca Raton.

FETTER, C. W. 1988. *Applied Hydrogeology*, Merrill, Columbus, Ohio.

FOOKES, P. G. 1966. London Basin Tertiary sediments. *Geotechnique*, **XVI** (3), 260–263.

FOOKES, P. G. (ed.) 1997. *Tropical Residual Soils: Engineering Group Working Party Report, Revised*. Geological Society, London.

FOOKES, P. G. 1997. Geology for engineers: the geological model, prediction and performance (The first Glossop Lecture). *Quarterly Journal of Engineering Geology*, **30** (4), 293–424.

FOOKES, P. G., DEARMAN, W. R. & FRANKLIN, J. A. 1971. Some engineering aspects of rock weathering with field examples from Dartmoor and elsewhere. *Quarterly Journal of Engineering Geology*, **4** (3), 133–185.

FORSTER, A. 1991a. *The Engineering Geology of the Exeter District, 1:50,000 Geological Map Sheet 325*. Technical Report WN/91/16. British Geological Survey, Keyworth, Nottingham.

FORSTER, A. 1991b. *The Engineering Geology of Birmingham West (The Black Country)*. Technical Report WN/91/15. British Geological Survey, Keyworth, Nottingham.

FORSTER, A. 1992. *The Engineering Geology of the Area around Nottingham, 1:50,000 Geological Map Sheet 126*. Technical Report WN/92/7. British Geological Survey, Keyworth, Nottingham.

FORSTER, A. 1997. *The Engineering Geology of the London Area, 1:50,000 Geological Map Sheet 256, 257, 270 and 271*. Technical Report WN/97/27. British Geological Survey, Keyworth, Nottingham.

FORSTER, A., HOBBS, P. R. N., CRIPPS, A. C., ENTWISLE, D. C., FENWICK, S. M. M., HALLAM, J. R., JONES, L. D., SELF, S. J. & MEAKIN, J. L. 1995. *The Engineering Geology of British Rocks and Soils: Gault Clay*. Technical Report WN/94/31. British Geological Survey, Keyworth, Nottingham.

FRANKLIN, J. A. & CHANDRA, A. 1972. The slake durability test. *International Journal of Rock Mechanics and Mineral Science*, **9**, 325–341.

FREDLUND, D. G. & RAHARDJO, H. 1993. *Soil Mechanics for Unsaturated Soils*. Wiley, New York.

GABRIELSSON, A., BERGDAHL, U. & MORITZ, L. 2000. Thermal energy storage in soils at temperatures reaching 90°C. *Journal of Solar Energy Engineering, Transactions ASME*, **122** (1), 3–8.

GAMBLE, J. C. 1971. *Durability-plasticity characteristics of shales and other argillaceous rocks*. PhD thesis, University of Illinois.

GILLOTT, J. E. 1979. Fabric, composition and properties of sensitive soils from Canada, Alaska and Norway. *Engineering Geology*, **14**, 149–172.

GILLOTT, J. E. 1987. *Clay in Engineering Geology*. Developments in Geotechnical Engineering, **41**, Elsevier, Amsterdam.

GRAINGER, P. 1984. The classification of mudrocks for engineering purposes. *Quarterly Journal of Engineering Geology*, **17**, 381–387.

GRIM, R. E. 1962. *Applied Clay Mineralogy*. McGraw-Hill, New York.

HARBIDGE, A. D. J., JEFFERSON, I., ROGERS, C. D. F. & COATES, M. 1996. A comparison of the heat dissipation performance of twin wall HDPE and earthenware cable duct. In: *Proc. 14th IEEE/PES Transmission and Distribution Conf.*, Los Angeles, 525–530.

HAWKINS, A. B. & PINCHES, G. M. 1992. Engineering description of mudrocks. *Quarterly Journal of Engineering Geology*, **25**, 17–30.

HAYDON, R. E. V. & HOBBS, N. B. 1977. The effect of uplift pressures on the performance of a heavy foundation on layered rock. *Proceedings Conference Rock Engineering*, Newcastle, 457–472.

HEAD, K. 1982. *Manual of Soil Laboratory Testing. Volume 2: Permeability, Shear Strength and Compressibility Tests*. Pentech, London.

HIGHT, D. W. ELLISON, R. A. & PAGE, D. P. 2004. *Engineering in the Lambeth Group*. Publication C583, Construction Industry Research and Information Association (CIRIA), London.

HILLEL, D. 1980. *Fundamentals of Soil Physics*. Academic, Orlando.

HOBBS, P. R. N., ENTWISLE, D. C., NORTHMORE, K. J. & CULSHAW, M. G. 1992. *Engineering Geology of Red Clay Soils: Geotechnical Characterisation, Mechanical Properties and Testing Procedures*. Technical Report WN/93/13, British Geological Survey, Keyworth.

HOBBS, P. R. N., HALLAM, J. R., FORSTER, A., ENTWISLE, D. C., JONES, L. D., CRIPPS, A. C., NORTHMORE, K. J., SELF, S. J. & MEAKIN, J. L. 2002. *The Engineering Geology of British Rocks and Soils: Mudstones of the Mercia Mudstone Group*. Research Report RR/01/02. British Geological Survey, Keyworth, Nottingham.

HOBBS, P. R. N., ENTWISLE, D. C., NORTHMORE, K. J., SUMBLER, M. G., JONES, L. D., KEMP, S. J., SELF, S., BUTCHER, A. S., RAINES, M. G., BARRON, M. & MEAKIN, J. L. 2005. *The Engineering Geology of UK Rocks and Soils: The Lias Group*. Technical Report No. IR/05/008. British Geological Survey, Keyworth, Nottingham.

HOEK, E. 1983. Strength of jointed rock masses. The Rankine Lecture. *Géotechnique*, **33**, 187–223.

HOEK, E. & BROWN, E. T. 1980. Empirical Strength Criterion for Rock Mass. Proc. ASCE Journal of the Geotechnical Engineering Div. 106 (GT9), 1013–1035.

HOPKINS, T. C. & DEAN, R. C. 1984. Identification of shales. *ASTM, Geotechnical Testing Journal*, GTJODJ, **7**, 10–18.

HORTON, A. 1975. *The Engineering Geology of the Pleistocene Deposits of the Birmingham District*. Report 75/4. Institute of Geological Sciences, HMSO, London.

HUDEC, P. P. 1982. Statistical analysis of shale durability factors. *Transport Research Record*, **873**, 28–35.

HUECKEL, T. & PEANO, A. 1996. Thermomechanics of clays and clay barriers, 3rd International Workshop on Clay Barriers, ISMES, Bergamo, Italy, October 22–24, 1993—Introduction. *Engineering Geology*, **41**, 1–4.

HUTCHINSON, J. N. 1969. A reconsideration of the coastal landslides at Folkstone Warren Kent. *Geotechnique*, **19**, 6–38.

IAEG 1979. Classification of rocks and soils for engineering geological mapping. Part 1: rock and soil materials. *Bulletin of the International Association of Engineering Geology*, **19**, 364–371.

INGRAM, R. L. 1953. Fissility of mudrocks. *Geological Society of America Bulletin*, **64**, 869–878.

ISRM 1981. Basic geotechnical description of rock masses. International Society of Rock Mechanics Commission on Classification of Rocks and Rock Masses. *International Journal Rock Mechanics Mining Science and Geomechanical Abstracts*, **18**, 85–110.

JACKSON, J. O. & FOOKES, P. G. 1974. The relationship of the estimated former burial depth of the Lower Oxford Clay to some soil properties. *Quarterly Journal Engineering Geology*, **7**, 137–180.

JACKSON, I. & LAWRENCE, D. J. D. 1990. *Geology and Land-use Planning: Morpeth-Bedlington-Ashington*. Technical Report WN/90/14. British Geological Survey, Keyworth, Nottingham.

JEFFERSON, I. 1994. *Temperature Effects on Clay Soils*, PhD Thesis, Loughborough University.

KAZI, A. & KNILL, J. L. 1969. The sedimentation and geotechnical properties of Cromer till between Happisburgh and Cromer, Norfolk. *Quarterly Journal Engineering Geology*, **2**, 63–86.

KENNARD, M. F., KNILL, J. L. & VAUGHAN, P. R. 1967. The geotechnical properties and behaviour of Carboniferous shale at the Balderhead Dam. *Quarterly Journal of Engineering Geology*, **1**, 3–24.

LAGUROS, J. G. 1969. Effect of temperature on some engineering properties of clay soils. *In*: MITCHELL, J. K. (ed.) *Proceedings of International Conference on Effects of Temperature and Heat on the Engineering Behaviour of Soils*, Washington D C, USA, Highway Research Board, Special Report, **103**, 186–193.

LAKE, R. D., NORTHMORE, K. J., DEAN, M. T. & TRAGHEIM, D. G. 1992. *Leeds: A Geological Background for Planning and Development.* Technical Report WN/91/1. British Geological Survey, Keyworth, Nottingham.

LAMBE, T. W. & WHITMAN, R. V. 1979. *Soil Mechanics, SI Version.* Wiley, New York.

LANE, K. S. & WASHBURN, D. E. 1946. Capillary tests by capillarimeter and soil filled tubes. *Proc. Highways Research Board,* **26,** 460–473.

MACNEIL, D. J. & STEELE, D. P. 1999. Swell test requirements for lime stabilised materials. TRL Project Report PR/CE/186/99. Transport Research Laboratory, Crowthorne.

MADDISON, J. D., CAMBERS, S. THOMAS, A. & JONES, D. B. 1996. The determination of deformation and shear strength characteristics of Trias and Carboniferous strata from in situ testing and laboratory testing for the Second Severn Crossing. *In*: CRAIG, C. (ed.) *Advances in Site Investigation Practice.* Thomas Telford, London, 598–609.

MARSHALL, T. J. & HOLMES, J. W. 1979. *Soil Physics.* Cambridge University Press, Cambridge.

MARSLAND, A. 1973. *Large in Situ Tests to Measure the Properties of Stiff Fissured Clays.* Current Paper 1/73. Building Research Establishment.

MCDANIEL, T. N. & DECKER, R. S. 1979. Dispersive soil problem at Los Esteros Dam. *Proceedings American Society Civil Engineers, Journal Geotechnical Engineering Division,* **105,** 1017–1030.

MCKEE, E. D. & WEIR, G. W. 1953. Terminology for stratification and cross-stratification in sedimentary rocks. *Geological Society of America Bulletin,* **64,** 381–389.

MEAD, W. J. 1936. Engineering geology of dam sites. *Transactions 2nd International Congress on Large Dams,* Washington DC, **4,** 183–198.

MITCHELL, J. K. 1991. Conduction phenomena: from theory to geotechnical practice. *Geotechnique,* **41,** 299–340.

MITCHELL, J. K. 1993. *Fundamentals of Soil Behavior,* 2nd edn. Wiley, New York.

MITCHELL, J. K. & SITAR, N. 1982. Engineering properties of tropical residual soils. *Proceedings of Speciality Conference on Engineering and Construction in Tropical and Residual Soils.* Geotechnical Engineering Division of American Society of Civil Engineers, Honolulu, 30–57.

MORGENSTERN, N. R. & EIGENBROD, K. D. 1974. Classification of argillaceous soils and rocks. *Journal of the Geotechnical Engineering Division, ASCE,* **100,** 1137–1156.

NORTHMORE, K. J., ENTWISLE, D. C., HOBBS, P. R. N., CULSHAW, M. G. & JONES, L. D. 1992. *Engineering Geology of Tropical Red Soils, Geotechnical Characterization: Index Properties and Testing Procedures.* Technical Report WN/93/12, British Geological Survey, Keyworth, Nottingham.

NORTHMORE, K. J., BELL, F. G. & CULSHAW, M. G. 1999. The engineering geology of the Claygate Beds and the Bagshot Beds of south Essex. *Quarterly Journal of Engineering Geology,* **32,** 215–231.

PARSONS, A. W. & BODEN, J. B. 1979. *The Moisture Condition Test and its Potential Application in Earthworks.* Report SR522, Transport and Road Research Laboratory, Crowthorne.

PETTIJOHN, F. J. 1949 (first edn) 1975 (third edn). *Sedimentary Rocks.* Harper and Row, New York.

POPESCU, M. E. 1979. Engineering problems associated with expansive clays from Romania. *Engineering Geology,* **14,** 43–53.

POPESCU, M. E. 1986. A comparison between the behaviour of swelling and collapsing soils. *Engineering Geology,* **23,** 145–163.

POTTER, P. E., MAYNARD, J. B. & PRYOR, W. A. 1980. *The Sedimentology of Shale: A Study Guide and Reference Source.* Springer, New York.

POWRIE, W. 2004. *Soil Mechanics: Concepts and Applications.* 2nd edn. Spon, London.

RADHAKRISHNA, H. S. & KLYM, T. W. 1974. Geotechnical properties of a very dense glacial till. *Canadian Geotechnical Journal,* **11,** 396–408.

RAMA RAO, R. 1978. *A Study of Swelling Clay.* PhD thesis, University of Glasgow.

REIFER, G. H. 1977. *The Effect of Temperature and Mineralogy upon the Atterberg Limits and Mechanical Properties of Cohesive Soils.* Undergraduate Project Report, Lanchester Polytechnic, Coventry.

ROBERTSON, T. L., CLARKE, B. G. & HUGHES, D. B. 1994. Classification and strength of Northumberland till. *Ground Engineering,* **27** (12), 29–34.

ROMERO, E., GENS, A. & LLORET, A. 2003. Suction effects on compacted clay under non-isothermal conditions. *Géotechnique,* **53** (1), 65–81.

RUSSELL, D. J. & PARKER, A. 1979. Geotechnical, chemical and mineralogical inter-relationships in weathering profiles in an over-consolidated clay. *Quarterly Journal of Engineering Geology,* **12,** 107–116.

RYAN, W. & RADFORD, C. 1987. *Whitewares—Production, Testing and Quality Control,* The Institute of Ceramics, Pergamon, Oxford.

SAMUELS, S. G. 1975. Some properties of the Gault Clay from the Ely-Ouse Essex water tunnel. *Géotechnique,* **25,** 239–264.

SCHOFIELD, R. K. 1935. The pF of the water in soil. *Trans. 3rd Int. Conf. Soil Science,* Oxford.

SCOTT, C. R. 1994. *An introduction to Soil Mechanics and Foundations.* 3rd edn. Spon, London.

SHERARD, J. L., DUNNIGAN, L. P. & DECKER, R. S. 1977. Some engineering problems with dispersive soils. *Proceedings of the American Society of Civil Engineers, Journal of the Geotechnical Engineering Division,* **102,** 287–301.

SMITH, T. 1978. *Consolidation and Other Geotechnical Properties of Shales with Respect to Age and Consolidation.* PhD thesis, University of Durham.

SMITH, G. N. & SMITH, I. G. N. 1998. *Elements of Soil Mechanics,* 7th edn, Blackwell, Oxford.

SKEMPTON, A. W. 1953. The colloidal activity of clays. *Proceedings 3rd Interational Conference Soil Mechanics and Foundation Engineering,* Zurich, **1,** 57.

SKEMPTON, A. W. 1964. Long-term stability of clay slopes. *Geotechnique,* **14,** 77–101.

SKEMPTON, A. W. 1977 Slope stability of cuttings in brown London Clay. *Proceedings 3rd International Conference Soil Mechanics Foundation Engineering,* Tokyo, **3,** 261–270.

SKEMPTON, A. W. & NORTHEY, R. D. 1952. The sensitivity of clays. *Géotechnique,* **2,** 30–53.

SLADEN, J. A. & WRIGLEY, W. 1983. Geotechnical properties of lodgement till—a review. *Glacial Geology: An Introduction for Engineers and Earth Scientists. In*: EYLES, N. (ed.) Pergamon, Oxford, 184–212.

SPEARS, D. A. 1980. Towards a classification of shales. *Journal of the Geological Society,* London, **137,** 125–129.

SPEARS, D. A. & TAYLOR, R. K. 1972. The influence of weathering on the composition and engineering properties of in situ Coal Measures rocks. *International Journal of Rock Mechanics and Mining Science,* **9,** 729–756.

SRIDHARAN, A. & PRAKASH, K. 1998. Characteristic water contents of a fine-grained soil-water system. *Géotechnique,* **48** (3), 337–346.

STARSEWSKI, K. & THOMAS, C. P. 1977. Anisotropic behaviour of an over-consolidated clay. *Proceedings 9th International*

Conference Soil Mechanics Foundation Engineering, Tokyo, **1**, 305–310.

STOW, D. A. V. 1981. Fine-grained sediments: terminology. *Quarterly Journal of Engineering Geology*, **14**, 243–244.

SUKLJE, L. 1969. *Rheological Aspects of Soil Mechanics*. Wiley, London.

SWARZENDRUBER, D. 1961. Modification of Darcy's law for the flow of water in soils. *Soil Science*, 93.

TAYLOR, R. K. 1988. Coal Measures mudrocks: composition, classification and weathering processes. *Quarterly Journal of Engineering Geology*, **21**, 85–99.

TAYLOR, R. K. & SPEARS, D. A. 1970. The breakdown of British Coal Measure rocks. *International Journal of Rock Mechanics and Mineral Science*, **7**, 481–501.

TAYLOR, R. K., BARTON, R., MITCHELL, J. E. & COBB, A. E. 1976. The engineering geology of Devensian deposits underlying P.F.A. lagoons at Gale Common, Yorkshire. *Quarterly Journal of Engineering Geology*, **9**, 195–218.

TELLAM, J. H. & LLOYD, J. W. 1981. Hydrogeology of British onshore non-carbonate rocks. *Quarterly Journal of Engineering Geology*, **14**, 347–355.

TIPPET, T., 1976. *An Investigation into the Effect of Temperature upon the Atterberg Limits and Mechanical Properties of Cohesive Soils*. Undergraduate Project Report, Lanchester Polytechnic, Coventry.

TRENTER, N. 1999. *Engineering in Glacial Tills*. Publication C504, Construction Industry Research and Information Association (CIRIA), London.

VAN OLFEN, H. 1963. *An Introduction to Clay Colloidal Chemistry*. Wiley, New York.

VARGAS 1974. Engineering properties of residual soils from south-central region of Brazil. *Proceedings 2nd Congress International Association Engineering Geology*, Sao Paulo, **1**, IVPC 5.1-5.26.

VARLEY, P. M. 1990. Susceptibility of Coal Measures mudstone to slurrying during tunnelling. *Quarterly Journal of Engineering Geology*, **23**, 147–160.

VAUGHAN, P. R., MACCARINI, M. & MOKHTAR, S. M. 1988. Indexing the properties of residual soil. *Quarterly Journal of Engineering Geology*, **21**, 69–84.

VIRDI, S. P. S. 1984. *Rheology of Soils, With Special Reference to Temperature Effects*. PhD Thesis, Coventry (Lanchester) Polytechnic.

WAINE, P. J., CULSHAW, M. G. & HALLAM, J. R. 1990. *The Engineering Geology of the Wrexham Area*. Technical Report WN/90/10. British Geological Survey, Keyworth, Nottingham.

WEAVER, C. E. 1989. *Clays, Muds and Shales*. Developments in Sedimentology, **44**, Elsevier, Amsterdam.

WENTWORTH, C. K. 1922. A scale grade and class terms for clastic sediments. *Journal of Geology*, **30**, 377–392.

WEST, G. 1991. *The Field Description of Engineering Soils and Rocks*. Open University Press, Milton Keynes.

WHEELER, S. J. & KARUBE, D. 1995. Constitutive modelling. State of the art report. *Proceedings 1st International Conference Unsaturated Soils*, Paris, **3**.

WILLIAMS, A. A. B. 1980. Severe heaving of a block of flats near Kimberley. *Proceedings 7th Regional Conference for Africa on Soil Mechanics Foundation Engineering*, Accra, **1**, 301–309.

WOOD, L. E. & DEO, P. 1975. A suggested method for classifying shales for embankments. *Bulletin of the Association of Engineering Geologists*, **XII**, 39–55.

WORRALL, W. E. 1982. *Ceramic Raw Materials*. The Institute of Ceramics, Pergamon, Oxford.

YONG, R. N. & WARKENTIN, B. P. 1975. *Properties and behaviour*. Developments in Geotechnical Engineering, **5**. Elsevier, Amsterdam.

YOUSSEF, M. S., SABRY, A. & RAMLI, A. H. EI. 1961. Temperature changes and their effects on some physical properties of soils. *Proceedings 5th International Conference on Soil Mechanics and Foundation Engineering*, Paris, **1**, 419–421.

5. World and European clay deposits

5.1. The character of clay

5.1.1. General

Clay can have a number of meanings in engineering geology. The term can relate to mineralogy, implying an assemblage of clay minerals; to size, implying an assemblage of clay sized particles (in British engineering terminology <0.002 mm); or to behaviour, implying a material which contains a sufficiently high clay size/mineral content to influence the material properties.

In traditional litho-stratigraphic nomenclature the term 'clay' is used loosely to refer to fine-grained deposits. These may comprise various constituents, including clay minerals, which have been derived from the weathering of the less stable components of the original rock, fine particles of quartz, feldspar and mica, and diagenetic minerals which have developed during the deposition, burial and uplift history of the deposit.

Broadly, the important clay mineral groupings are kaolinite, smectite and illite (Chapter 2) and one or more of these groupings tend to dominate a particular deposit. This is because the character of the clay mineral found in any particular soil depends on factors such as parent material, climate, topography, vegetation, and the length of time over which these factors have operated. In other words particular deposits form or are modified under specific environmental influences. Perhaps the most dominant of these are climate and time.

5.1.2. Climate

In tropical regions the relatively rapid chemical weathering associated with high annual temperature and high annual rainfall contributes to the formation of residual clay soils whose mineralogy is controlled by local environmental influences such as the local drainage regime. This mineralogy controls the engineering behaviour. In arid regions clay soils are relatively rare but tend to be saline and dominated by evaporite minerals. These conditions also seem to favour the formation of the clay mineral attapulgite.

In temperate regions most clay deposits are transported soils. Their engineering behaviour is dominated by the clay fraction and the particle size distribution is determined by fluvial, lacustrine, estuarine and marine surface processes. Finally in the arctic regions complex assemblages of clay till result from the actions of glaciers and ice sheets, perhaps modified by periglacial activity.

5.1.3. Time

Time is manifested in the effect of crustal plate movements and the major orogenies. What is now Britain, for example, as part of the Eurasian plate (Fig. 5.1) has drifted northwards from a latitude of about 30° south at the beginning of the Ordovician (~ 510 Ma) to its present position. During its travels it has suffered two major orogenies: the Caledonian (~ 400 Ma) and the Variscan or Armorican (~ 330 Ma), which largely obliterated the effects of any climatic influences on older rocks.

The effects of the Alpine orogeny (~ 30 Ma) were relatively minor in Britain and so climatic effects on rocks younger than the Carboniferous can still be observed. For example, the rocks of the Permian and Triassic periods reflect the tropical and arid climatic influences of the time as Britain drifted through the Equator and towards the Tropic of Cancer.

Although the Quaternary is a relatively short period in geological time its influence on shallow deposits has been immense. Climatic fluctuations have dominated the Quaternary and in the tectonically quiescent regions of the world these have been a major influence on the character of the surface soils that the engineer has to deal with today.

The character of each major clay deposit has, therefore, been the result of a complex history. Climate and palaeogeography largely determine its original mineralogy and its depositional environment. Modifications are caused as plate movements carry the deposit through different climatic regions or zones of tectonic activity and finally it has been affected by Quaternary climatic fluctuations.

In the following sections the major environments governing clay formation are outlined. The history of each of the major tectonic plates is summarised highlighting historic changes in climate and the influence of major orogenies and finally the modifying effects of the Quaternary are discussed. An important part of the history of a clay deposit is burial and uplift. The resulting overconsolidation of clay deposits is a significant influence on engineering behaviour. Finally, the major clay deposits are classified on the basis of initial environment of formation, pre-Quaternary modifications, and Quaternary effects.

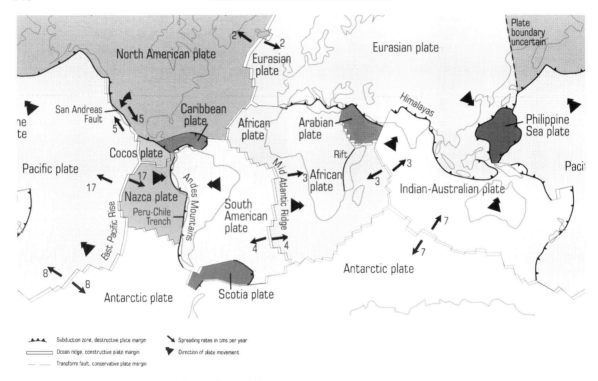

Fig. 5.1. Tectonic plates and their boundaries (after Toghill 2000).

5.2. Environments of clay formation

5.2.1. Residual clays of tropical regions

Tropical residual soils form *in situ* and are probably the largest group of soils in the tropical regions to influence engineering works. In these regions the combination of high mean annual temperature and high levels of humidity causes intense weathering of the primary minerals to great depth.

Several factors influence the mineralogical composition of the weathered product (Fookes 1997a) but the most important are the local environmental conditions, particularly climate and drainage. Dependent on local conditions, several mineral assemblages can be generated, and each assemblage is associated with engineering properties that are markedly different from the properties of the original material. Fookes (1997a) proposed an engineering classification of tropical residual soils that recognized five main groups: Vertisols, Ferruginous soils, Ferrisols, Ferrallitic Soils and Fersiallitic Andosols.

The behaviour of the Vertisols group is dominated by the presence of smectite which displays marked changes in volume when the moisture content changes. The ferruginous, ferrisol and ferrallitic soil groups contain increasing quantities of the hydrated oxides of iron and aluminium (sesquioxides) and these affect the results of laboratory tests in several ways.

- Clay-size particles can aggregate to silt-size or even fine sand-size particles by the cementing action of sesquioxide coatings. Such coatings also inhibit absorption of water. Intensive remoulding breaks down this coating so that plasticity is affected.
- Drying can cause irreversible changes in plasticity, and oven drying can drive off molecular water of hydration which normally takes no part in the engineering performance of the material.
- Some soils can collapse or show a significant decrease in volume on the addition of water.
- Fersiallitic andosols contain allophanes, halloysite and meta-halloysite which are also subject to loss of water of hydration and aggregation.

Therefore, tropical soils require a carefully planned laboratory testing programme to ensure that the results truly reflect the behaviour of these soils in the proposed engineering works.

5.2.2. Hydrothermal clays

Hydrothermal reactions are generally defined as taking place at pressures of greater than 1 atmosphere and temperatures of greater than 100°C (Gillott 1968). Hydrothermal clay deposits are associated with igneous masses, and are alteration products caused by the percolation of active solutions and vapours. Good examples are the kaolin china clay deposits associated with the granites of southwest England.

5.2.3. Transported clays

The action of surface water on the weathered mantle moves detrital particles into the fluvial regime. The original volume of clay size particles increases as coarser particles in the sediment load are abraded by river action. Sorting action as the energy of the river decreases transports the clay size material to the lower reaches of the river.

Minor clay deposits, on the global scale, are the result of overbank floods that build up alluvial clay deposits on flood plains. However, most of the sediment load moves on into the sea or into a lake. In delta areas the clay material forms the bottomset beds at the delta front. For example, the bottomset beds of the Nile Delta are found some 130 km offshore and extend to depths of 2000 m. Subsidence caused by the increasing weight of delta sediments can result in deposits of considerable thickness.

While the larger lakes may provide a similar depositional environment to the sea the deposits of smaller lakes reflect stiller conditions, the influence of seasonal deposition and the greater presence of organic matter. *Laminated clays* are the common result of lake deposition, often containing *varves* and reflecting annual changes in the depositional cycle.

In river estuaries there is a juxtaposition between river and tidal flows and also between fresh and sea water conditions. Sedimentation is controlled by the change between the situation where the two flows are opposite to the situation where they act together. As the tide rises the river flow slows and eventually reverses and the clay size material is dropped on the intertidal mud flats. As the tidal flow reverses, it supports the river flow and as levels drop the mud flats are drained by concentrated runoff channels. The mixing of sea and river water causes flocculation into larger silt size particles that sink more readily together with other silt size material. This slightly coarser grain size and the shallowness of the deposited material assists rapid drainage and drying before the next tidal cycle. Estuarine clays and silts are often laminated and contain organic matter.

The character of a transported clay is clearly largely determined by the depositional environment, but the mineralogy of the parent material may also be significant, for example when the river originates in a tropical region, or where the zone of deposition is influenced by local climate. One such example is provided where drainage flows into an interior basin. The soluble products of weathering remain in the system and as evaporation continues from the lake surface the water becomes more saline. In the arid desert regions of the world, ephemeral lakes dry out and recharge regularly and saline clay deposits are the result.

5.2.4. Clays of glacial origin

The majority of ice masses occur in ice sheets or smaller ice caps. While glacier ice is an important process, it is associated mostly with granular deposits and only localized clay deposits occur in moraines or lacustrine deposits. Most glacial clays are associated with ice sheets, and are the result of quarrying and abrasion of the 'bedrock' by the moving ice sheet above (Derbyshire & Love 1986).

Lodgement tills formed at the base of the ice sheet comprise crushed and abraded stones in a matrix of rock flour. The latter is often deficient in true clay minerals, being the result of abrasion and crushing of inactive bedrock minerals such as quartz and feldspar. The plasticity may therefore, be low. Exceptions are where the 'bedrock' comprises clays or mudrocks composed of clay minerals and the resulting till will display a greater plasticity. The release in high overburden pressures exerted on the till by the ice, when the ice front retreats, results in overconsolidation.

Lodgement tills may display a range of grain sizes from boulders down to clay and the local grading may change rapidly, but they are generally clay matrix dominated. They are typically stiff, dense soils of low compressibility.

5.3. Palaeogeography and clay deposition

In this section the relationship between environment of deposition and the location of clay deposits is described. Information on world and European clay deposits is provided in tabular form in Appendix C.

5.3.1. The Palaeozoic

5.3.1.1. Cambrian. In the Cambrian (600 Ma), a super-continent (Panotia) centred about the south pole began to break apart and several blocks drifted northward into the Panthalassic Ocean towards the Cambrian equator (Fig. 5.2). The most notable of these were the large continents of Laurentia (comprising ancestral North America, Greenland, Scotland), Baltica (comprising ancestral Northern Europe and Russia), and Siberia. A new ocean, the Iapetus Ocean, widened between Laurentia and Baltica. The remaining part of the original supercontinent, Gondwana (comprising ancestral South America, southern Europe, Middle East, Africa, India, southern China and Australia), stretched across the southeast of the globe from the Antarctic pole to the Equator.

In North America in the Cambrian the land mass of Laurentia comprised the ancestral Canadian Shield and Greenland and a transcontinental arch with associated metamorphic and volcanic rocks extending what is now southwest through the ancestral United States. The transcontinental arch was a source of clay sediments deposited both to what is now the east, in the Appalachian Basin, forming the shales of the Conasauga Formation, and to the west in Montana, Idaho and Nevada.

In Europe and Africa Cambrian marine shales are found in Israel. Middle Cambrian shales also occur in

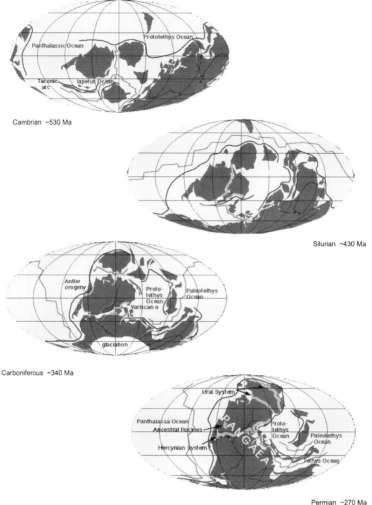

FIG. 5.2. Palaeozoic palaeogeography.

Norway and Sweden. These were transported from the eroding continent of Baltica.

These deposits were illite dominant with little kaolinite or chlorite. This is thought to be because there were no land plants at this time and rocks in highland regions were rapidly eroded and transported before chemical weathering could be initiated.

5.3.1.2. Ordovician. In the Ordovician and Silurian a general rotation caused Gondwana to move across the south pole and Laurentia and Baltica to close together (Fig. 5.2). During the Ordovician, warm water deposits, such as limestones and salt, formed in the equatorial regions of Gondwana (Australia, India, China and Antarctica), while glacial conditions and ice-rafted debris occurred in the south polar areas of Gondwana (Africa and South America). Numerous plates and continental blocks, including Avalonia (Ancestral England, Wales and New England), broke away from Gondwana and approached Laurentia from the south and east.

In the Ordovician, initial collision of Laurentia and Baltica (in what was the forerunner of the Caledonian Orogeny) marked the Taconic Orogeny. In North America as the Taconic Orogeny unfolded, deformation and metamorphism affected the eastern flank of Laurentia and mountains formed along the east coast of ancestral North America followed by the formation of a volcanic island arc as the Iapetus Ocean closed. There were two sources of clay sediment in ancestral North America. The acid rich rocks of the Canadian Shield and Transcontinental Arch continued to supply illite, with localized kaolinite rich sediments, to the Appalachian Basin, e.g. the Decorah and Glenwood Shales in Minnesota. The basic volcanic rocks of the island arc were uplifted and erosion in the Late Ordovician and deposited illite and chlorite sediments westward onto the former continental shelf, e.g. the Cincinnatian Shales of Ohio, Indiana and Northern Kentucky.

On the west coast of Baltica (ancestral Norway) shales were deposited from erosion of the volcanic island arc forming in the closing Iapetus Ocean to the west. The influence of the volcanic source materials is reflected in the increasing chlorite content in these shales. The Caledonian Orogeny is believed to have been active throughout the Ordovician and Silurian and the volcanic activity accompanying this contributed to the presence of potassium bentonite deposits in ancestral Norway, Sweden Estonia, Lithuania and Poland. In East Germany and in the Baltic Sea regions Ordovician sediments contain kaolinite, thought to reflect the development of tropically weathered crusts in the source areas of Baltica.

Clay materials were also deposited in the southern Ural geosyncline and on the Russian platform. Faulting had developed and the Russian and Siberian platforms were separating in the Ordovician and by the Silurian an open ocean geosyncline had developed.

Avalonia comprised Precambrian basement and volcanic ash and lava and was the source of Ordovician mudstones in Wales and in the Pyrenees, comprising illite and chlorite

5.3.1.3. Silurian and Devonian. By the Late Silurian and into the Devonian (approximately 400 Ma), the Iapetus Ocean had fully closed. The collision zone accompanied by the subduction of marginal island arcs was the focus of the Arcadian Orogeny forming the ancestral Appalachian Mountains, but had also moved northward and involved Siberia marking the Caledonian Orogeny affecting ancestral Britain and Northern Europe.

The Caledonian orogeny marked the assemblage of a new continent, comprising Laurentia, Baltica and Avalonia (ancestral North America and Europe), called Laurussia, or the 'Old Red Continent'. In the Silurian the orogenic activity spread northwards along the eastern flank of Laurentia and these highland areas were a continuing source of sediment to the Appalachian basin. Significant deposits include the Middle Silurian Clinton Group, extending from New York to Alabama, and Crab Orchard Shales of Kentucky. In the Upper Silurian the Wills Creek Shales of eastern West Virginia are also significant.

Whilst shallow seas had initially covered most of the old continental platforms, by the Devonian and continuing through to the Triassic, the continental plates began to emerge, reducing the shallow sea coverage and non-marine rocks became increasingly dominant. The Appalachian Mountains were the source of large volumes of continental sediment which spread both to the East (the Old Red Sandstone of Europe) and the West (the Catskill Formation of North America).

During the Devonian period fish evolved into the top predators. Plant life developed on land and became sufficiently abundant for coal deposits to form in the tropical swamps that covered much of what are now the Canadian Arctic Islands, northern Greenland, and Scandinavia.

By the late Devonian the eastern highlands had been denuded and there was limited continuing sedimentation in the Appalachian Basin. Orogenic activity was now centred on northeastern ancestral North America where highlands were developed in the Arcadian and Caledonian Orogenies as a result of continuing collision between Laurentia and Baltica. Finer-grained clastic sediments from this source were deposited in the Upper Devonian Catskill formations of Pennsylvania. Further west and southwest, Upper Devonian marine organic-rich black shales were deposited over much of ancestral eastern North America stretching from Texas in the south to as far north as Canada.

In northwest Africa Silurian and Devonian shales were deposited in the Polignac Basin. To the west the Tindouf Basin was also a focus of deposition of marine sediments. Silurian/Devonian slates also occur in South Africa.

In South America in the Upper Devonian dark marine shales were deposited in the Parano Basin of ancestral southern Brazil, Uruguay, Paraguay and Argentina.

5.3.1.4. Carboniferous and Permo-Triassic. In the Lower Carboniferous Gondwana continued its rotation and approached Laurussia from the south, causing ancestral South America and Africa to close with ancestral North America and Europe (Fig. 5.2). This collision formed an extensive and complex east–west orogenic belt, comprising (from west to east) the Ouachita, Appalachian (Alleghenian) and Hercynian (Variscan) orogenies. Laurussia had also collided with Siberia, resulting in the Ural orogeny. The Marathon orogeny was also developing as the forerunner of the Rocky Mountains in west–central North America. The result was a new supercontinent, Pangea, which straddled the Equator, but stretched from the South Pole to the North Pole, and separated the Palaeo-Tethys Ocean to the east, from the Panthalassic Ocean to the west. The western fringe of Pangaea was adjacent to a long subduction zone that formed the eastern margin of the Pacific.

In the Lower Carboniferous in ancestral North America the sedimentation patterns of the Devonian continued but depositional conditions changed from those favouring organic shales to those more favourable to carbonate formation. However, some Lower Carboniferous shales were deposited in Montana, the Besa River black shales were deposited in British Columbia and shales were extensively deposited across the southern United States. Further south in the Oachita–Marathon geosyncline the thick shales of the Tesnus and Stanley Formations were deposited.

By the Lower Pennsylvanian (mid-Namurian) volcanic activity had increased as part of the orogenic movements. This source of volcanic material led to the abrupt introduction of montmorillonite into mudstone deposits such as the Chesterian (mid-Namurian) shales of the ancestral southern and central United States, e.g. the Springer Shale of Oklahoma. The collision signalled the Hercynian Orogeny which continued through into the Permian.

In the Lower Pennsylvanian, in the south–central United States fine grained sediments were supplied from two main sources. The Appalachian Orogenic Belt to the southeast and south of the ancestral United States supplied illite/chlorite material to the north. However as sea level regressed from the craton, sediments were also supplied from residual highland areas to the north that had been subjected to weathering and kaolinite content increased to the north as these sources became more adjacent.

Moving into the Upper Pennsylvanian in the eastern and central United States, the Illinois Basin, sediments were deposited by low gradient rivers across extensive deltaic plains. The shoreline fluctuated repeatedly over hundreds of kilometres producing coal bearing cyclothems as muddy deltaic platforms divided heavily vegetated swamps.

During the Late Carboniferous and Early Permian the southern regions of Pangea (southern South America and southern Africa, Antarctica, India, southern India, and Australia) were glaciated. There is also evidence that a north polar ice cap existed in eastern Siberia during the Late Permian.

The term 'Pangea' means 'all land' and it included most of the landmasses that existed at that time. However, in the eastern hemisphere, to the east and south of the Palaeo-Tethys Ocean, North and South China and a long thin continent, Cimmeria (ancestral Turkey, Iran, Afghanistan, Tibet, Indochina and Malaya), were still separate. North and South China had drifted northward from Gondwana, into the Palaeo-Tethys Ocean, and were at the latitude of the northern tropic. Cimmeria is thought to have rifted away from the Indo-Australian margin of Gondwana during the Late Carboniferous – Early Permian as a new spreading centre formed and marked the opening of the Tethyan Ocean.

A broad Central Pangean mountain range formed an equatorial highland that during Late Carboniferous was the locus for the formation of coal deposits in an equatorial rain belt. By the mid-Permian, however, this mountain range had moved northward into drier climates and the interiors of North America and Northern Europe became arid as the continued uplift of the mountain range blocked moisture-laden equatorial winds.

The Chinese continents and Cimmeria continued to move northwards towards Pangea, ultimately colliding along the southern margin of Siberia during the late Triassic Period. After this collision all the world's landmasses were again joined.

5.3.2. The Mesozoic

Pangea had formed as a result of a series of collisions over an extended period beginning approximately in the Devonian and continuing through to the Triassic. Similarly, it broke apart again over an extended period through the Mesozoic in three main episodes (Fig. 5.3).

5.3.2.1. Jurassic. The first episode of rifting began in the early Jurassic (about 200 Ma). Pangea began to rotate and rifts began to form. Igneous activity, represented by up-welling basaltic magmas, occurred along what is now the east coast of North America and the northwest coast of Africa, and the Central Atlantic Ocean began to open as North America moved to the northwest. This movement also gave rise to the Gulf of Mexico as North America moved away from South America.

Initially the limited area of these new oceans favoured the formation of evaporites as water from the Tethyan and Pacific oceans began to ingress. However, from this time world temperatures began to decrease and this trend has continued to the present.

The Cordilleran arc initiated along the Pacific margin of North and South America. At the same time, on the other side of Africa, extensive volcanic eruptions along the adjacent margins of East Africa, Antarctica and Madagascar marked the beginning of the western Indian Ocean.

In North America, as the Central Atlantic Ocean opened, the continent to the North rotated clockwise, sending North America northward. In the central western United States the extensive thick quartzose sandstones of the Navajo were gradually submerged beneath a shallow sea advancing from the north and west, initial evaporite deposits were replaced by carbonates, and then fluvial shales and sandstones of the Morrison formation, sourced from granites and rhyolites to the west. In the northeast, closer to the developing rifts fluvial red and black mudstones and shales of the East Berlin formation developed with intercalated lava flows.

Asia moved southward as the Atlantic Ocean opened and coals, which were abundant in eastern Asia during the early Jurassic, were replaced by deserts and salt deposits during the Late Jurassic as Asia moved from the wet temperate belt to the dry subtropics. This led to the closure of the Tethys ocean.

In Europe during the Early Jurassic a transgression of the Tethys and Arctic seas covered much of Europe, but these began to regress again as sea floor spreading began from the mid-Atlantic and West-Tethys rift systems. The Jurassic of the Paris Basin contains significant clay deposits, analogous to the Jurassic clays of UK, and Fuller's Earth deposits of the Upper Jurassic reflect the alteration of volcanic layers into smectitic minerals. Jurassic Clays are also found in the Molasse basin, between the Jura and the Alps in Switzerland and in southwest France.

In northern and eastern Europe the warm and humid climate of the lower and middle Jurassic became more arid in the upper Jurassic. This caused chemical weathering on the Bohemian Massif, the Baltic Shield, northern Norway, the Urals, Romania and Ukraine with the development of a kaolinite rich weathering mantle.

In Africa lower Jurassic shales occur in an alternating sequence with limestone in the Guereif Basin, Morocco and upper Jurassic claystones occur in the Congo.

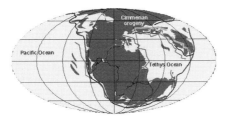

Jurassic ~200 Ma

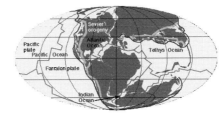

Cretaceous ~130 Ma

Cretaceous ~80 Ma

Tertiary ~20 Ma

FIG. 5.3. Mesozoic/Cenozoic palaeogeography.

5.3.2.2. Cretaceous. The second phase of rifting began in the early Cretaceous (about 140 Ma). The southern part of Pangea (the old Gondwana) continued to fragment as South America separated from Africa opening the South Atlantic progressively from south to north, and India together with Madagascar rifted away from Antarctica and the western margin of Australia opening the Eastern Indian Ocean.

Other plate tectonic events included, the counter-clockwise rotation of Iberia from France, the separation of India from Madagascar, the derivation of Cuba and Central America from the Pacific, and the uplift of the Rocky mountains in the Sevier Orogeny.

Globally, the climate during the Cretaceous Period, like the Jurassic and Triassic, was much warmer than today. Dinosaurs and palm trees were present north of the Arctic Circle as well as in Antarctica and southern Australia. There were no large ice caps during the Mesozoic Era, although there are thought to have been some remnants at the poles during the Early Cretaceous.

These mild climatic conditions were in part due to the fact that shallow seaways covered the continents during the Cretaceous. The Cretaceous was the time of rapid sea-floor spreading. It is thought that, because of their broad profile, rapidly spreading mid-ocean ridges displace more water than slow spreading mid-ocean ridges. Consequently, during times of rapid sea-floor spreading, sea level tends to rise and evidence suggests that it was some 100–200 metres higher than today. Warm water from the equatorial regions was also transported northward, warming the polar regions, and these currents also tended to make local climates milder, much like the modern Mediterranean Sea, which has an ameliorating effect on the climate of Europe.

In North America, the whole of the eastern edge of the continent, the Gulf coastal plain and the Great Plains area was flooded by Cretaceous seas. To the west uplift and mountain building associated with subduction of the Pacific plates produced the Rocky Mountains in the Laramide Orogeny of the Late Cretaceous. Even before this the uplifted area became a source of sediment flooding eastwards into the Cretaceous seas, as well as a focus for extensive volcanic activity. Sandstone and smectite rich shale sequences dominate the sedimentary sequence in North America and Europe. In North America the Lower Cretaceous deltaic and shallow marine Dakota Group mudstones are a source of commercial kaolinite. In Wyoming and Montana, the Mowry formation has a high smectite content, and 'Wyoming Bentonite' is sourced from the Cretaceous shales of the area. Kaolinite is present in all the Cretaceous deposits as a tropical weathering product. In the upper Cretaceous the Pierre Shale is found in Montana and North and South Dakota.

In Europe during the Cretaceous the Tethys Sea rose to cover almost all of the continent with the exception of the Baltic Shield, and in the Upper Cretaceous the Chalk was deposited. As Africa and Europe converged the seas regressed as the Alpine orogeny began to build.

Continental, fluviatile and lacustrine sediments of the Lower Cretaceous (the UK Wealden series) extended across Belgium, the Paris Basin, Spain, Switzerland and into North Africa. The emergent areas were heavily weathered and provided a source of kaolinite rich materials to the lower areas. Kaolinite is exploited from these deposits in Spain and France. During the Middle Cretaceous as sea levels rose volcanicity increased, associated with rifting in what is now the North Sea and in the Atlantic, and smectite rich deposits provide commercial deposits of Fuller's Earth.

This pattern of kaolinite rich clays derived from the weathered upland areas and later smectite rich deposits derived from increased volcanicity in the Upper Cretaceous deposits is also seen in Africa. Thick Cretaceous sediments (up to 4000 m) are noted in the Douala Basin, Cameroun, and continue southwestwards through the Benue Trough in Nigeria. These grade upwards into smectite predominant sediments in the Upper Cretaceous. On the other side of the Atlantic in the Campos Basin of Brazil marine Cretaceous shales are some 3500 m thick and contain both kaolinite and smectite.

5.3.3. The Cenozoic

The third, and final phase in the breakup of Pangea took place during the early Cenozoic. Though several new oceans opened during this period, it was a time of intense continental collisions. The most significant was the beginning of the collision between India and Eurasia, about 50 million years ago. During the Late Cretaceous, India was approaching Eurasia at a rate of 15–20 cm/year. After colliding with marginal island arcs in the Late Cretaceous, the northern part of India began to be subducted beneath Eurasia raising the Tibetan Plateau.

North America and Greenland split away from Europe, and Antarctica released Australia which like India 50 million years earlier, moved rapidly northward on a collision course with southeast Asia.

The most recent rifting events, all taking place within about the last 20 million years include: the rifting of Arabia away from Africa opening the Red Sea, the creation of the east African Rift System, the opening of the Sea of Japan as Japan moved eastward into the Pacific, and the northward motion of California and northern Mexico, opening the Gulf of California.

In addition other collisions have included: Spain with France forming the Pyrenean mountains, Italy with France and Switzerland forming the Alps, Greece and Turkey with the Balkan States forming the Hellenide and Dinaride mountains and Arabia with Iran forming the Zagros mountains.

This phase of continental collision has raised high mountains by horizontally compressing the continental lithosphere. Though the continents occupy the same volume, their area has decreased slightly. Consequently, on a global scale, the area of the ocean basins has increased slightly during the Cenozoic. Because the ocean basins are larger, they can hold more water. As a result, sea level has fallen during the last 66 million years, reflecting a pattern through history where, in general, sea level has been lower during times of continental collision (early Devonian, Late Carboniferous, Permo-Triassic).

During times of low sea level and emergent continents, land faunas flourish, migration routes between continents open up, the climate becomes more seasonal, and probably most importantly, the global climate tends to cool off. This is largely because land tends to reflect the Sun's energy back to space, while the oceans absorb the Sun's energy.

Also, landmasses at the poles permit the growth of permanent ice sheets, which because they are white, reflect even more energy back to space. The formation of ice on the continents also lowers sea level, which exposes more land, which cools the Earth, forming more ice, and so on. There is a tendency, therefore, that once the Earth begins to cool (or warm-up) positive feed-back mechanisms push the Earth's climatic system to greater and greater cooling (or warming).

During the last half of the Cenozoic, therefore, the Earth began to cool off. Ice sheets formed first on Antarctica and then spread in the northern hemisphere. For the last five million years or so the Earth has entered a major Ice Age.

In North America deposition in the Cenozoic was largely around the margins of the present continent. Palaeocene kaolinite occurs in the Santa Anna region of southern California, which is partially sedimentary and partially residual in origin. Extensive Eocene kaolinite deposits occur along the western margin of the Sierra Nevada (Ione Formation); these are sedimentary deposits derived from the weathered products of the igneous Sierra

Nevada mountains. Eocene kaolinite deposits also occur in Oregon, Washington and Idaho.

In the east, the Gulf of Mexico since the Miocene has become the focus for deposition of deltaic sediments from the Mississippi River. Periodic sea level changes have provided an alternating sequence of deltaic and marine sediments. Along the old pre-delta shoreline is a band of Eocene clay (Wilcox Formation) stretching through Texas, Illinois, Kentucky, Tennessee and into Georgia. This contains layers of kaolinite rich material. In the upper Eocene and Oligocene smectite becomes more dominant and bentonite deposits cropout along the coastal plain through Texas to Carolina and result from the volcanic activity of the Caribbean area.

In Europe, various mountain building movements relating to the collision of southern and northern land masses began to fragment the Tethys Sea. In the early Cenozoic, shallow seas and lagoons were located between the Baltic Shield and the emerging Alpine mountains. Large volumes of sediment were deposited in the early Cenozoic into the North Sea, Paris basin and the Hungarian Basin. This clay series contains dominantly smectite materials through the Palaeocene and Eocene. Further south the old Hercynian massifs were extensively weathered, producing thick kaolinite deposits. These filled the Aquitaine basin, flowing from the Armorican massif and the Massif Central. Many kaolinite quarries are located in the Charentes region. Kaolinite clays also extend eastwards into central and eastern Europe, reflecting weathering products from the Bohemian Massif. Further south again Spanish clay deposits were derived from the Pyrenees in rivers flowing towards the Atlantic.

In Africa, the coastal areas received sediment from the kaolinitic weathering products of the shield areas. The Niger delta contains a thick sequence of Cenozoic sediments very similar in composition to the Mississippi sequence.

5.4. Climatic changes in the Quaternary

The Quaternary period extends back from the present for some 2 million years and it is believed that there have been some 20 glacial cycles in that period (Fookes 1991). The reasons for these changes are beyond the scope of this summary but the interested reader is referred to Boulton *et al.* (1985) and Bennet & Glasser (1996). Each cycle comprises glacial episodes of ice advance and intervening interglacial episodes of ice retreat, but each main episode may contain several minor advances and retreats. The last main glaciation was the Devensian (British terminology) which reached its maximum some 18 000 years ago but it is now known that several minor advances have occurred in the last 10 000 years of the current interglacial period.

The effects of these changes are widespread and worldwide. Glacier ice and ice caps currently cover a little over a third of the area covered at the Devensian maximum (Fig. 5.4) when ice in the northern hemisphere reached to the northern outskirts of London. Permafrost and associated periglacial conditions with sparse tundra vegetation extended to the Mediterranean. Some 50% of the land area between the Tropics of Cancer and Capricorn was desert and tropical rain forest was considerably reduced to minor areas at the Equator.

Glacial till and fluvio-glacial sediments were deposited over much of northern North America, Europe and Asia. Extensive eolian loess deposits are located to the south of the glaciated regions in North America, Siberia and China.

In the last 18 000 years sea level has risen some 130 m and flooded the continental shelves, although the last 6000 years has seen a relatively consistent level.

5.5. World clay deposits

For convenience world deposits of clay can be grouped into three broad categories: 'common clay', largely comprising illite, kaolin clay and smectite clay. The former occurs widely as a sedimentary deposit from the Cambrian to the present. The latter two groups are commercially exploited as a source of industrial minerals.

Kaolin clays are derived from the hydrothermal alteration or *in situ* weathering of acid igneous and metamorphic rocks. They are therefore associated with the more stable shield areas of the world where these rocks have been exposed to a long period of sub-aerial weathering or have been affected by circulating meteoric fluids, producing 'primary' kaolin. Secondary kaolins are transported sediments originating from the primary mineral areas. Kaolin clays also occur as the product of tropical weathering in oxidising conditions of most rock types and are widespread throughout the tropical regions of the world since the Quaternary. Most of these processes have resulted in clays of Upper Cretaceous age or younger.

Smectite clays have either been derived by alteration of volcanoclastic rocks in a marine environment or by tropical weathering in reducing conditions. They are generally of Jurassic age or younger and associated with the volcanism that accompanied the rifting and drifting apart of Pangea, the beginning of the present continental framework. They are also associated with the tropical regions of the world through the Quaternary.

5.5.1. 'Common' clay

Clay is widely used as a low-grade construction material being a common source of fill for civil engineering purposes. It is also a source of low-grade material used in the production of bricks and tiles, and Portland cement.

The most important property as a low-grade material is its plasticity which is provided by the weathering of aluminosilicates that occur in most rock types, with the exception of pure sandstones and limestones. Clay therefore occurs widely as a weathering product in tropical and temperate environments. It also occurs as a mechanical

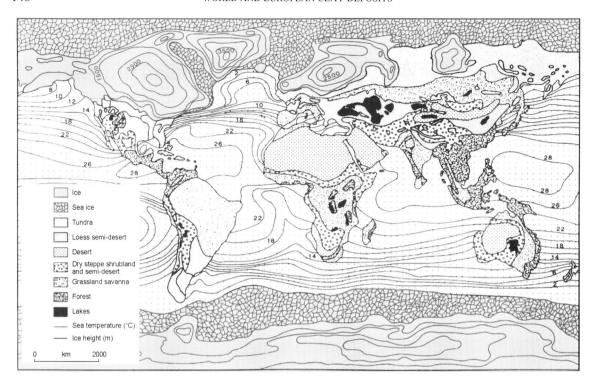

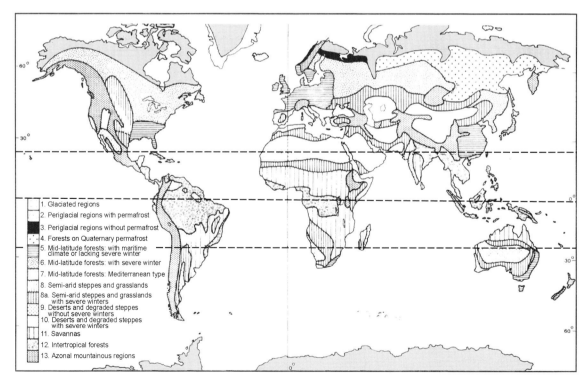

FIG. 5.4. Ice extent at Devensian maximum and present day (after Bowen 1979 and Tricart & Cailleux 1965).

product of ice action in glacial environments and as a low energy sedimentary deposit in fluvial and marine environments. Burial, diagenesis and metamorphism form mudstones, shales and slates. A major component of these deposits may be clay-sized clastic material such as quartz or mica. Authigenic minerals such as pyrite and siderite and biogenic minerals such as carbonates and hydrocarbons are also present.

As an undifferentiated clay-size material it is widely distributed and can be relatively easily located. In this usage transportation costs must be kept to a minimum and an economic radius of more than several tens of kilometres can rarely be justified.

No attempt has been made to provide a specific guide to the location of common clay materials. The principal processes governing the environments in which clays are formed are described in earlier sections and these control their global distribution.

When clay is used as a raw material for more refined industrial processes, specifications governing the selection and processing of clay become more rigid and particular mineral groupings are sought for commercial exploitation. The major locations of these more sought after clay deposits can be more easily defined and are described below.

5.5.2. Kaolin clays

Kaolin clays are found naturally as the weathering products or hydrothermal alteration products of feldspar rich rocks such as granites and gneisses. In tropical climates they are a natural weathering product of many parent rocks in wet, free-draining and oxidizing conditions and are a major mineral component of ferruginous and ferralitic soils and ferricretes. They also form large sedimentary accumulations in areas where the residual products are subject to fluvial erosion, particularly when a change in pH occurs, as the suspended sediment reaches brackish estuarine conditions. The major kaolin deposits are generally of Cretaceous and younger age.

In China Clay deposits the clay mineral component is almost entirely kaolinite. In Ball Clay deposits kaolinite is the most dominant clay mineral but other clay minerals such as illite, smectite, chlorite and mixed layer clays are present, together with some quartz and organic matter.

The major kaolin producing area of the world is on the upper coastal plain of Georgia and South Carolina in southeast USA. The deposits are sedimentary, and comprise the late Cretaceous Buffaloe Creek Formation and the middle Eocene (early Tertiary) Huber Formation, and are thought to be deltaic deposits derived from weathered igneous and metamorphic granites and gneisses forming a piedmont to the northwest. Other less important USA production areas occur in Florida, where kaolin is separated from the Middle Miocene Fort Preston sand formation; in Texas, where kaolin is separated from the Late Palaeocene Wilcox Group sands; and in Minnesota, where granites and gneisses underlying Cretaceous pyritic shales are thought to have been weathered by percolating organic and sulphur rich solutions.

The UK production of China Clay is centred on Cornwall and Devon and comes from residual deposits, representing hydrothermally altered zones of the Variscan granites of south-west England. The kaolinization is thought to have taken place from the Triassic and to be continuing to the Present.

In the Ukrainian Republic China Clay occurs in the Prosyanovski and Gluhovetsky areas as zones of hydrothermally altered granites and gneisses of the Ukrainian shield. The source rocks are Precambrian but the alteration must be pre-Neogene as Neogene deposits cover the kaolinized zone.

In Brazil major sedimentary deposits of China Clay of Pliocene age occur in the Amazon Basin. These are derived from weathered granitic source rocks of the Guyana Shield. They occur just below an extensive dissected plateau and it is possible that significant tropical weathering has also been influential in the kaolinization. Various other less significant deposits also occur worldwide (Table 5.1).

TABLE 5.1. *Less significant China Clay deposits*

Location	Deposit
Czechoslovakia: Karlov — Vary area.	Altered granites.
Germany: Hirscau and Schnaitenbach areas of Bavaria	Occurs within sedimentary Triassic sands derived from Granitic source rocks in the Naab basin mountains.
Spain: Guadalajara to Valencia	Occurs within Cretaceous arkosic sands derived from granitic Meseta Uberica, with further tropical weathering of the feldspars.
France: Brittany	Hydrothermally altered zones of the Variscan granites (as Cornwall)
India: Kerala, between Cochin and Trivandrum	Occurs in sedimentary sands, deposited in deltaic conditions from altered granites and gneisses to the east.
India: Jaipur	Altered zone within a granite pegmatite.
Indonesia: Belitung and Bangka islands.	Tropically weathered porhyritic biotite granite.
China	Numerous extraction areas, generally in zones of altered igneous rocks.
Australia: Weipa, Cape York peninsula	Within a sand layer underlying an extensive bauxite deposit.
New Zealand: Maungaparerua, North Island	Zones of hydrothermal alteration and tropical weathering in a rhyolite parent rock.

TABLE 5.2. *Other bentonite deposits*

Location	Deposit
USA: Hector, California, and Amargosa Valley, Nevada.	Hectorite, a magnesium/lithium bentonite, from Tertiary alkaline lake deposits.
USA: Taft, California.	Fuller's Earth within the Miocene Monterey Formation.
Peru and Chile:	Calcium bentonites occur in Tertiary strata in the coastal ranges.
Morocco: Trebia, Bouhoua and Providencia.	Interbedded volcanoclastic deposits within Cretaceous and Tertiary sediments.
South Africa: Orange Free State and Cape province.	

Ball Clay deposits are all sedimentary in origin. Large deposits occur in Tennessee and Kentucky in the USA. In the UK, thick sedimentary deposits of kaolin are present in the Bovey and Petrockstow basins of Devon. These Ball Clays are Oligocene in age and are thought to have been derived from the Dartmoor granite by fluvial deposition in a subsiding basin controlled by the Lustleigh-Sticklepath fault. Other significant deposits occur in the Westerwald area, northeast of Roblenz in Germany.

5.5.3. Smectite clays

Bentonite is the industrial mineral composed dominantly of smectite clay minerals. The majority of smectites either have calcium or sodium as the most abundant exchangeable cation. The low swelling calcium smectites are typified as the Southern bentonite of the USA and border the Gulf of Mexico. The higher swelling sodium smectites typify the Wyoming bentonite, produced in Wyoming and adjacent states in the USA. Other smectites such as lithium bentonite occur in California. Another smectite rich clay is Fuller's Earth, which has absorbent and bleaching properties.

The majority of the world's smectite rich clay deposits, or bentonites, occur as thin bedded sequences within sedimentary deposits and range in age from Jurassic to Pleistocene. Most commonly they originate from the *in situ* alteration of siliceous volcanic ash or tuff. This occurs as layered deposits of ash that have accumulated in marine conditions under water where a hydration reaction occurs which requires magnesium. The age and origin of these deposits coincides with the break-up of Pangea in the mid-Jurassic and the forming of rifts with accompanying igneous activity. Smectites are also the dominant constituent of residual fersiallitic vertisols which are the result of tropical weathering in poorly drained or reducing environments.

The location of the major bentonite deposits is indicated in Appendix C. In North America the Wyoming bentonite, a sodium bentonite, is mined in three areas in Wyoming, Montana and South Dakota. The bentonite occurs as several beds within Cretaceous shales, predominantly the Mowry Shale, an organic rich marine shale deposited in the Western Interior Seaway, and also the overlying Belle Fourche Shale and the Frontier Formation. Further north into Canada, bentonites of the same age occur in Alberta, Manitoba, British Columbia Saskatchewan and the Northwest Territories, but these are predominantly calcium bentonites.

The Southern bentonite, a calcium bentonite, is mined in Texas, Mississippi and Alabama generally bordering the Gulf of Mexico. It occurs in Mississippi and Alabama in the Upper Cretaceous Ewtaw and Ripley Formations and in the Tertiary Vicksburg Formation. In Texas it is mined from the Tertiary Jackson and Gueydan Formations. Fuller's Earth is exploited from the Palaeocene Porters Creek Formation around the Mississippi embayment. In Mexico many layers of volcanoclastic rocks occur in Tertiary marine deposits, and Fuller's Earth and calcium bentonite are exploited in Tlaxcala and Morelos in the Hildago region.

In Europe, Greece and Germany are second and third, respectively, to the United States in world bentonite production. In Greece extensive and thick deposits occur on the island of Milos, being the result of alteration of volcanic ash deposited in marine conditions in the Pliocene. In Germany bentonite is mined from Upper Miocene marine clays and sands, where it is thought to have formed from *in situ* tropical weathering. Extensive calcium bentonite deposits occur in Italy, on the island of Sardinia, formed from hydrothermal alteration of trachytic tuffs. In Almeria, Spain, calcium bentonite deposits are the result of hydrothermal alteration of rhyolites and dacites. In the UK Fuller's Earth, a calcium bentonite, is exploited from beds in the Jurassic and Upper Cretaceous of Oxfordshire, Bedfordshire and Surrey.

Japan is the fourth largest producer, mining sodium and hydrogen bentonite from volcanic ash beds in Miocene and Pliocene marine mudrock sequences. Various other deposits occur worldwide (Table 5.2).

References

BENNET, M. R. & GLASSER, N. F. 1996. *Glacial Geology: Ice Sheets and Landforms*. Wiley, London.

BOULTON, G. S. *et al.* 1985. Glacial geology and glaciology of the last mid-latitude ice sheets. *Journal of the Geological Society, London*, **142**, 447–474.

BOWEN, D. Q. 1978. *Quaternary Geology*. Pergamon, Oxford.

DERBYSHIRE, E. & LOVE, M. A. 1986. Glacial Environments. *In*: FOOKES, P. G. & VAUGHAN, P. R. (eds) *A Handbook of Engineering Geomorphology*. Surrey University Press.

FOOKES, P. G. (ed) 1997a. *Tropical Residual Soils*. Geological Society Professional Handbooks. Geological Society, London.

FOOKES, P. G. 1997b. Geology for engineers: the geological model, prediction and performance. *Quarterly Journal of Engineering Geology*, **30**, 293–424.

FOOKES, P. G. 1991. Quaternary Engineering Geology. Keynote address. *In*: FORSTER, A., CULSHAW, M. G., CRIPPS, J. C., LITTLE, J. A. & MOON, C. F. (eds), *Quaternary Engineering Geology*. Geological Society Engineering Geology, Special Publication **7**. Geological Society, London.

GILLOTT, J. E. 1968. *Clay in Engineering Geology*. Elsevier, New York.

TOGHILL, P. 2000. *Geology of Britain. An Introduction*. Stackpole Books.

TRICART, J. & CAILLEUX, A. 1965. *Introduction a la Geomorphologie Climatique*. SEDES, Paris.

6. British clay stratigraphy

There are three principal types of stratigraphy, each with its own terminology:

- *geochronology*—units of geological time (e.g. period, epoch), with a parallel and exactly corresponding stratigraphy;
- *chronostratigraphy*—time-rock units or time-stratigraphic units represent stratified rock successions (e.g. system, series, etc.) assigned to geological time units;
- *geochronometry*—the measurement of absolute time in years as numerical ages (e.g. 345 Ma), principally by means of radiometric dating, but increasingly, in Phanerozoic rocks, by the numerical dating of Milankovitch cycles.

This has led to some confusion. Current recommendations, led by the Geological Society (Zalasiewicz et al. 2004) are to adopt the geochronology scheme, which has been used in this chapter.

Part of the geological timescale, covering the last 545 million years of earth's history (Phanerozoic—from the greek meaning 'visible life'), is shown in Table 6.1

TABLE 6.1. *Broadest sub-divisions of the geological timescale*

Eon	Era	Sub-Era	Period
Phanerozoic	Cenozoic (Cainozoic)	Tertiary	Quaternary Neogene Palaeogene
	Mesozoic		Cretaceous Jurassic Triassic
	Palaeozoic	Upper	Permian Carboniferous Devonian
		Lower	Silurian Ordovician Cambrian

(Gradstein & Ogg 1996) with the youngest at the top. This shows the broadest sub-divisions of *era* and *period* (the Neogene and Palaeogene are often referred to as the Tertiary, though this name may fall into disuse). The names for the eras derive from the greek meaning 'ancient', 'middle', and 'new' life. Periods further sub-divide into early, middle, and late (or in some cases simply early and late), and thence into *epochs* and *ages* (refer to Appendix C). Time before the Phanerozoic is usually referred to as the Pre-Cambrian.

Clays are found in three principal geological environments: as primary deposits in the sedimentary environment, as alteration products of other materials in the weathering and diagenetic/hydrothermal environments, and as products of glaciation (Chapter 3). The last of these, though less important than the others, has particular significance in high latitude countries including Britain. Mudrock and clay formations in Britain are found throughout the stratigraphic column, from the Pre-Cambrian to the Quaternary, and of course some are still being formed at the present time. Important clay formations, in terms of thickness and areal extent, occur throughout the Mesozoic. Clays dominate the solid geological map of much of southeast England and the glacial and superficial map of Britain as a whole. Clays may also provide an important component of many predominantly 'non-clay' formations.

From an engineering and an economic perspective, important clay formations are found within the Carboniferous, Jurassic, Cretaceous and Palaeogene (early Tertiary), and as alteration products of granite in SW England. Clays are also found as infill material within joints, faults, and limestone caves. All of the Mesozoic, and many of the Tertiary, clay-rich formations are in a state of over-consolidation. Over-consolidation tends (Chapter 4) to impart greater strength and durability to a clay formation in its unweathered state, when compared with a normally consolidated equivalent. Compositional differences and burial history (depth/temperature/time)

William Smith

William Smith, blacksmith's son, canal surveyor, and self-taught geologist, known as the 'father of English geology', was one of the first to recognize the significance of fossils (and the evolution of their form) to the chronographic order of the strata, and the first to produce (single-handedly) a geological map of Bath as early as 1799, showing Jurassic strata sub-divided stratigraphically, and subsequently geology maps of English counties and of England and Wales (and parts of Scotland) as a whole in 1815 (Winchester 2001).

Over-consolidation of deposits

Over-consolidation is a natural process whereby the overburden stress acting on a deposit has been greater in the past than it is at present, or at some other time under consideration. Due to the hysteresis associated with clay volume change, this cycle of loading and unloading, at least on geological time scales, imparts greater density, strength, durability, and incompressibility to a clay deposit when compared with a deposit at the same depth which has undergone uninterrupted overburden stress increase up to the present value (normal consolidation). The amount of over-consolidation can be estimated either from geotechnical laboratory tests or from geological evidence. However, such estimates are often unreliable, due to the sometimes complex, and largely unknown, stress-history of a sediment. The nature of the clay mineral assemblages present in the clay may also be partly dependent on the extent of over-consolidation. Weathering, and reworking, tends to undo some of the effects of over-consolidation, bringing the geotechnical properties closer to those of normally consolidated clays (Chapter 4). This process of equalization becomes greatest near-surface. Of course, a clay deposit may have been subject to more than one cycle of over-consolidation in its stress history, but this is impossible to determine from conventional geotechnical data alone.

are likely to be the principal causes (Cripps & Taylor 1986, 1987, 1988), and the influence of this factor tends to increase with age.

Determining the location of clays in Britain by means of published geology maps is not necessarily straightforward, as the traditional geological map is subdivided into units based primarily on stratigraphy, and secondarily on other factors including lithology. Clays are found both in 'solid' (bedrock) and 'drift' (superficial) categories and are a major component in a variety of lithologies, for example: clay, mudstone, shale, and marl. The presence of clay may or may not be indicated in the 'group', 'formation', or 'member' names. For example, the London Clay Formation, Mercia Mudstone Group, Etruria Formation, and Gault Formation all contain clay as the major component. Thus a search for clay-rich deposits in a stratigraphic database using keywords such as 'clay' or 'mudstone' would be incomplete. Also, clay may form a minor component of a rock or soil formation or deposit. For example, a limestone may be described as 'argillaceous', meaning that clay is inextricably mixed in with its other materials but not as a major component. Alternatively, clay may occur in discrete layers within a formation dominated by another lithology; for example, as shale bands within a sandstone or limestone sequence, or as a clay layer within an otherwise coarse-grained glacial deposit. Such layers are unlikely to be indicated on a geological map unless they represent major, persistent, or otherwise readily mappable deposits, irrespective of their possible engineering geological significance. Minor clay layers within a largely non-clay formation may, nonetheless, be of considerable economic or engineering importance.

The physical properties of the wide range of clay materials are as varied as their mineralogies, depositional environments, and post-depositional histories, and depend on a cycle of deposition, consolidation, diagenesis, tectonism, weathering and erosion (Chapter 3; Chandler 2000). The distinction between mudrock and clay is, to some extent, an arbitrary one instigated mainly by the needs of classification, engineering and testing. The change from clay to mudrock, as in the process of induration, or from mudrock to clay, as in the process of weathering, is usually a gradual transition. General trends in the mineralogy and physical properties of clay formations can be discerned with increasing age (Chapter 4).

The wide variety of clay-forming and reworking processes characteristic of glaciation, periglaciation, and the Quaternary in Britain have resulted in significant deposits over much of the country. Stratigraphical classification and nomenclature of the Quaternary has long been a controversial issue in geological mapping. Recently, efforts have been directed at erecting a lithostratigraphical framework for the Quaternary of the UK (McMillan &

Solid and drift deposits

Solid and drift are two important divisions of geological materials, which are fundamental to the understanding of geological maps. Some geological maps show either one or the other, while others show both. The term 'drift' was first used by Charles Lyell (1797–1875), author of the 'Principles of Geology', to indicate only glacial deposits, thought to be derived from drifting sea-ice. More recently, the British Geological Survey (BGS) has used the term to signify all superficial deposits (i.e. all recent glacial, fluvio-glacial, alluvial, and marine sediments), such that the terms 'solid' and 'drift' are mutually exclusive and, at the same time, comprehensive. The term 'drift' has largely been replaced by the term 'superficial', but nevertheless appears on many existing geology maps.

Hamblin 2000). As with solid stratigraphy, the fundamental mapping unit is taken by the Geological Society Special Report (Bowen 1999) to be the 'formation'. In addition, the British Geological Survey (BGS) approach is to use groups and sub-groups irrespective of the presence of formations within them (McMillan & Hamblin 2000). Groups are units which contain materials having a common lithogenetic provenance. Two 'groups' containing glacigenic materials are proposed by the division of Britain into areas north and south of the southern limit of the last (Devensian) glaciation. North of the limit the deposits are divided into 15 'sub-groups', while to the south there is a single unit. Currently, the details of Quaternary stratigraphy are subject to debate.

Because of their physical characteristics, clay formations rapidly degrade at surface, and typically tend to form ground of low relief. Natural exposures of clay formations are usually confined to cliffs, beaches, riverbanks or landslides, but clay is perhaps best exposed in former and operational clay pits and quarries, temporary excavations for civil engineering works, foundations for buildings, and trenches for utilities. All cliffs and excavations can pose serious hazards. Extreme care should be taken if entering excavations, where up-to-date guidelines should be followed, and permission should be obtained from the landowner, site manager or operator. Clay exposed on actively eroding sea-cliffs should only be examined, on or below the cliff, where previously determined to be safe by a qualified person.

A summary of British clay-rich formations and deposits, having major engineering and economic importance, is given in Table 6.2.

A more comprehensive table of clay-rich formations is given in Appendix C. Near surface, clays and mudstones, in common with other rock types, undergo the weathering process, usually to a depth of several metres, dependent on the antiquity of the landscape, and the amount of time such processes have been operating (as well as other factors: chapter 3). Weathering is an ongoing process in most climatic and geological environments, but there may also be evidence of palaeo-weathering processes that have taken place under former climatic conditions. In Britain,

TABLE 6.2. *Stratigraphy of major clay formations*

Period	Age (million years BP)	Formation/deposit
Quaternary	1.8 to present	Till, Head, Brickearth, Lacustrine deposits, Alluvium
Tertiary	23.8 to 1.8	Ball clays, china clays
	65 to 23.8	Barton Clay, London Clay, Lambeth Group, Ormesby Clay
Cretaceous	142 to 65	Gault, Atherfield Clay, Weald Clay, Wadhurst Clay, Speeton Clay
Jurassic	206 to 142	Kimmeridge Clay, Ampthill Clay, West Walton Oxford Clay, Kellaways, Fuller's Earth Clay, Frome Clay Whitby Mudstone, Charmouth Mudstone
Permo-Triassic	290 to 206	Penarth Group, Mercia Mudstone Group
Carboniferous	354 to 290	Etruria Marl, Coal Measures, Bowland Shale

Weathering of rocks

Weathering is a natural process which has a major impact on engineering activity, particularly where erosion has been slow to remove evidence of it, or has been absent. This is because weathering alters near-surface rocks and their geotechnical and hydrogeological properties. Weathering is a chemical and mechanical process involving the actions of air, water, temperature, stress relief, flora and fauna on rocks. The precise combination of these factors involved in weathering depends on prevailing climatic conditions, and in fact some may be absent while others may dominate. In the case of clays, oxidation of certain minerals is usually accompanied by changes in fabric, colour, physical properties, and engineering behaviour. Typically, a grey clay weathers to a brown clay, due to the alteration of ferrous to ferric compounds (Chandler 2000). Gypsum (often as 'selenite' crystals) may be precipitated in the weathered zone due to pyrite breakdown, and acid formation reacting with calcite; though being moderately soluble is seldom found at the surface. Also, rock mass properties deteriorate compared with the fresh condition, as a result of stress-relief fissuring, swelling, slaking, and loss of strength and stiffness. Some near-surface clay deposits have been extensively altered by biological influences, and contain significant proportions of biogenic material, for example in the form of polysaccharides. These influences may be contemporaneous or relict, and may be sufficient to alter the geotechnical properties. Bacteria and fungi play an increasingly recognized role in weathering, sometimes to considerable depths. Palaeo-weathering processes, that is processes that have occurred under former climatic or geological environments, may have a significant influence on rock mass properties, bedrock profile, depth of mineral alteration, and other factors relevant to engineering geology.

unusually deep weathering is generally an indication of weathering under tropical or sub-tropical climates in the Tertiary. In areas that have undergone glaciation in recent geological time, such as Scotland, Wales, and northern and central England, much of the evidence of former weathering has been removed. This is not the case in southern England where old thick weathering profiles may be found.

The following account of Britain's geology and palaeogeography is taken from Duff & Smith (1992), Toghill (2000), Cope *et al.* (1992), Gradstein & Ogg (1996), and Chapter 5 unless referenced otherwise.

6.1. Pre-Carboniferous

As indicated in Chapter 5, during the pre-Carboniferous, the rocks of ancestral Britain were formed in a variety of palaeo-environments, both marine and continental. Many phases of plate tectonic movement and mountain building took place and, during the Devonian Period, which lasted 63 million years from 417 to 354 Ma, the essential fabric of today's Britain was formed by the joining of several formerly separate elements. During the Pre-Cambrian, substantial thicknesses of clay-rich sediments were deposited, but much of this material has subsequently been altered by metamorphic processes to more durable lithologies.

After a period of quiet sedimentation in the Cambrian, the Ordovician Period was characterised by volcanic activity associated with subduction at either side of the Iapetus Ocean, resulting in great thicknesses of lava and ash. During the Ordovician, southern Britain moved from tropical latitudes 3000 miles northward to meet Scotland which then formed part of North America. Warm seas were flanked by semi-desert landmasses. Muds were deposited on sinking sea-floors far from land. These were subsequently indurated to form mudstones and shales. Persistent volcanic activity took place, particularly in the Lake District and Snowdonia.

The Silurian Period saw the final episodes of marine sedimentation on either side of the narrowing Iapetus Ocean. Deposition of carbonate reefs took place and dark, clay-rich formations were laid down in deeper water before closure of the ocean towards the end of the period, and continental collision and uplift of the Caledonian mountains along the Iapetus Suture, running in Britain from Cumbria to Northumberland. By the end of the Silurian many deposits of sand and mud many kilometres thick had been formed.

Most of Britain now formed part of a new land area with mountains in the north and the northern edge of a huge alluvial plain in the south. Erosion of the newly-formed continent gave rise to the dominantly terrestrial deposition of the Devonian Period. In Devon and Cornwall thick muds, now shales and slates, were formed in offshore basins with limestones in the shallows. Throughout the Devonian sub-sea volcanic activity took place and the climate was hot and wet. By the end of the Devonian the Caledonian Mountains had been eroded sufficiently to allow the development of shallow seas in parts of Britain, in which fish flourished.

Mudrocks of the Cambrian, Ordovician, and Silurian occur as shales and turbidites (McKerrow 1981). The mudrocks are found principally in Scotland, Wales, and the Lake District, with limited outcrops in the west Midlands. The mudrocks are highly indurated and are generally found as shales and slates, the latter particularly in northwest Wales. Mudrocks of the Devonian occur in terrestrial environments in Scotland, south Wales and the Welsh Borders ('Old Red Sandstone'), and marine mudstones occur in southwest England, where great thicknesses occur in south Devon and Cornwall. The present outcrop of pre-Carboniferous rocks is shown in Figure 6.1.

FIG. 6.1. Map showing Pre-Cambrian to Devonian outcrop.

6.1.1. Silurian shales

Considerable thicknesses of shales were deposited originally as deep-water clays throughout the Silurian Period; notably the Hughley Shales, the Wenlock Shales, the Ludlow Shales, and the Temeside Shales of the West Midlands and Welsh Borders. These are seen forming the scarp and vale scenery of the Wenlock Edge and Severn Valley areas (Toghill 2000).

The Hughley (Purple Shales) form the uppermost part of the Llandovery Series and consist of over 100 m of grey, green-grey, maroon, and brown mudstones and silty mudstones, with thin beds of limestone and calcareous sandstone. The overlying Wenlock Shale Formation consists of over 300 m of grey and blue-grey mudstones with argillaceous limestones, forming most of the Wenlock Series of the Silurian Period; an upper limestone (Wenlock Limestone) overlaying it in the east. The Wenlock Shales occupy the low ground between Ludlow and Much Wenlock, Shropshire, and are divided into the Buildwas (lower) and Coalbrookdale (upper) Beds. The latter contain trilobite fossil remains, and several thin bands of bentonite, believed to be instrumental in the deep-seated Buildwas landslide (Hamblin & Coppack 1995). The Ludlow Shales, of the Ludlow Series, are divided, in the east, into upper and lower by the Aymestry Limestone. They consist of 170 m of grey and green-grey mudstones (lower) and siltstones (upper). The Temeside Shales of the Pridoli Series consist of up to 40 m of green and purple micaceous marls.

6.2. Carboniferous

The Carboniferous Period lasted for 64 million years from 354 to 290 Ma, and is named after the characteristic coal-bearing sequences within it. The environment was typified by warm shallow seas, deltas, and swamps. These covered Britain as far north as the Scottish Highlands, and extended far into present-day continental Europe, Ireland, and North America. Several large basins developed within which clays were deposited. In southwest England cherts were deposited. The climate was equatorial with abundant rain forest, subject to cycles of flooding and subsidence. The environment was similar to islands of the Caribbean today. Marine life and terrestrial plant life flourished. Insects, lizards, and amphibians flourished. Initially, tropical shallow water carbonates were formed, later being replaced with sands and clays entering the basins from river deltas. These then became swampy and the landmasses densely forested; repeated inundation, due to sea-level rises, resulting in the decay of this vegetation, thus producing the familiar coal seams of the Coal Measures during the later Carboniferous in central and northern England, Southern Scotland, and southern Wales. Inundation also deposited marine clays. Basin deposition resulted in considerable variations in thickness across Britain. The present outcrop of the Carboniferous is shown in Figure 6.2.

FIG. 6.2. Map showing Carboniferous outcrop.

The Carboniferous succession is divided, in ascending order, into the Carboniferous Limestone Group (Dinantian), the Millstone Grit Group (Namurian), and the Coal Measures Group (Westphalian). In Western Europe the Namurian and Westphalian are separated into the Silesian sub-group (Duff & Smith 1992). The Coal Measures crop out in England and Wales as a series of coalfields: Northumberland/Durham, Cumberland, Lancashire, North Wales, Yorkshire, Nottinghamshire/ North Derbyshire, North & South Staffordshire, Leicestershire/South Derbyshire, Warwickshire, South Wales, Bristol, and Somerset. It is thickest (1900 m) in the North Staffordshire and Lancashire Coalfields. Formations such as the Millstone Grit, and Lower and Middle Coal Measures, are mixed lithology formations, where major single lithology units (e.g. coal seams, sandstone beds) are indicated on the geology map by ornament and may also be named. The Carboniferous in West

Lothian, Scotland, supplied Europe's first oil industry in the mid-19th century, in addition to an established coal industry.

In the Lower Carboniferous, shelf and lagoonal sequence on the margins of land, received periodic influxes of muddy sediment due to uplift of land areas. Kaolinite and illite are the dominant clay minerals. Some volcanic input and contemporaneous weathering took place, forming bentonite layers in limestones and mudstones. Uplift and variable sediment supply formed the coarsening-upwards cyclic sequences of the 'Yordales' in northern England. In the Millstone Grit and Coal Measures, mudrocks are found as one component of cyclic sequences, where they form more than 50% of the deposit (Ramsbottom et al. 1981). The sequence is important for its mudrock formations, many of which are, or have been, exploited for clay products. Others have potential for use as engineering fills (e.g. colliery spoil) although high sulphur values have been a problem. Vast amounts of colliery spoil are available in some locations.

The Millstone Grit was mainly deposited under deltaic conditions with massive beds of delta-top and river-channel sandstones, alternating with shales deposited in the deep marine areas. Turbidite sequences that include shales were formed in the distal parts of the deltas. Within the Coal Measures the cyclic sequences include substantial thicknesses of mainly non-marine mudrocks. Contemporaneous weathering and soil formation produced seatearths and fireclays (see Text Box on p. 159); leaching giving rise to a typical mineralogy and structure (low durability and low strength). Some deeper water mudstones formed under marine conditions. Such marine bands are commonly found immediately above coal seams, at the base of the sedimentary cycle. They consist of dark grey, fossiliferous, pyritic shales which weather to form a relatively soft dark grey clay. These bands are thin but form important marker horizons to constrain the stratigraphy and help in identifying other important beds, such as coals. Most sequences were subject only to burial diagenesis, and 'gentle' tectonism. Some contact metamorphism, e.g. Central Valley of Scotland, and some intense tectonism and raised diagenetic rank, e.g. South Wales, took place. Typically, the major sandstones are named, but the mudstones are grouped under 'Millstone Grit (undifferentiated)'.

Events during the Pleistocene meant that some weathering profiles were largely removed but some slopes remained affected by peri-glaciation. An overall depth of weathering was retained in non-glaciated areas. This has important implications for engineering, in particular excavation. Carboniferous mudstones are particularly susceptible to reductions in strength and stiffness due to weathering (Cripps & Taylor 1981). Comparisons of mechanical properties may be misleading where samples are taken at different depths. A high incidence of landsliding is found within the Millstone Grit and Coal Measures, which includes some very large, deep-seated landslide complexes in the high Pennines, for example at Rishworth and Scammonden Moors.

6.2.1. Bowland Shale formation

Including the former Edale Shale Formation of Derbyshire, this formation is found in the Midlands, northwest England, and North Wales to a thickness of up to 350 m. It comprises dark grey organic and calcareous mudstones and bituminous and pyrite-rich shales, with thin siltstones and turbiditic sandstones. Limestone and minor coal bands are also present. The top of the formation occurs at the base of the Millstone Grit Group.

The Bowland Shale Formation is notable for its involvement in the fatal gas explosion at an underground pumping station at Abbeystead, Lancashire in 1984 in which 16 people were killed. The formation is a source rock for gas in commercial quantities in the East Midlands (Duff & Smith 1992). The formation is involved in the Mam Tor and Alport Castles landslides, amongst others, in the Derbyshire Pennines (Waltham & Dixon 2000). The Formation is generally too bituminous and pyritic to provide a useful source of brick-clay (Bloodworth et al. 2001).

6.2.2. Warwickshire Group

The red-brown and variegated mudstones and silty mudstones of the Etruria Formation, formerly known as the Etruria Marl, form part of the Warwickshire Group and have a limited outcrop principally within Staffordshire, but also in Warwickshire. Within the mudstones are mottled layers with root traces, resulting from the oxidation and leaching processes of sub-aerial soil formation. The formation contains the clay minerals kaolinite, illite, and chlorite, and the non-clay minerals quartz, haematite, and calcite. The high iron oxide content renders the clay suitable for the manufacture of premium quality bricks and tiles, including 'blue' engineering bricks. The formation is in demand but represents only 1% of brick clay resources (Bloodworth et al. 2001). In North Staffordshire the Etruria Formation overlies the productive Coal Measures and is up to 430 m thick, where it contains lenticular sandstone, conglomerate, and 'espleys' (pebbly sandstones). The sandstone layers increase in proportion in the upper part and also to the south of the outcrop. The Etruria Formation is the principal brick-clay resource in the West Midlands and, along with other clays of the Upper Coal Measures, is well known for its use in the manufacture of red and blue brick, and tiles in Birmingham, Shropshire, and the Welsh Borders (Chapter 14). The main area of industry has been established for over 180 years at Aldridge, north of Birmingham. This formation and others have also been used for the supply of marl. The Etruria Formation is also susceptible to landslides (Wilson 1987).

In addition to the above there are major thicknesses of mudrock in the Enville, Keele, and Halesowen Formations, which now form part of the Warwickshire Group (Smith et al. 1986; Hamblin & Coppack 1995). In the Coalbrookdale and the Flintshire/Denbighshire coalfields the correlative formations are named the

Marl

Marl is a term signifying a fine-grained, deep-sea sediment consisting of about two-thirds microfossils and one-third clay; the microfossil content is partly siliceous. In the past, the term has seen widespread use, in some cases incorrectly, to indicate simply calcareous clay. Classification schemes have hinged on the carbonate content. Marls have been used widely as a dressing in agriculture, horticulture, and also for sports pitches. The Chalk Marl of southeast England forms part of the lower Chalk (lowermost Upper Cretaceous). The Channel Tunnel was bored almost entirely within the Chalk Marl. It formed an ideal uniform, and relatively watertight, tunnelling medium (Mortimore & Pomerol 1996).

Fireclays

Fireclays are kaolinitic and quartzitic, clay-rich types of 'seatearth' or 'underclay', which are typically found underlying coal seams in the Westphalian of Britain, and forming part of the characteristic upward-shallowing succession cycle; the components of which are in ascending order: shale, mudstone/ironstone, siltstone, sandstone, seatearth, and coal. Seatearths represent the thin fossil *in situ* soil layers (palaeosols) which once hosted the coal-bearing plants. They played an important role in the industrial revolution due to their close association with coal and ironstone. Formerly, fireclays were used as refractories (hence the name) but are currently used, as a by-product from opencast coal mining, for brick and pipe making. Currently fireclay is exploited by mining in Yorkshire, Derbyshire, and Shropshire. Refer to Chapter 14.

Hadley Formation and the Ruabon Marl Formation. In the Wyre Forest area the Etruria Formation is included within the upper part of the Kinlet Group. In South Staffordshire a local name used is the Old Hill Marl.

In the Severn Valley, downstream of Ironbridge, the Halesowen and Etruria Formations are involved in complex landslides, in particular Jackfield and Lloyd's Coppice (Skempton 1964). These, together with the Buildwas landslide upstream of Ironbridge, in the Pre-Carboniferous Wenlock Shales, are associated with valley slopes in former glacial channels.

6.3. Permian and Triassic

The Permian Period (Smith, 1986, 1995) named after the Russian town of Perm, lasted 42 million years, from 290 to 248 Ma, and the Triassic Period 42 million years from 248 to 206 Ma. Traditionally, in Britain, the rocks of these periods are commonly grouped together under the term 'Permo-Trias', and their rocks have been known (somewhat misleadingly) as the 'New Red Sandstone'. Fossils are very rare in most of the British Permo-Trias. Britain was, at the time of deposition, north of the equator, and the environment was one of a huge intracontinental basin with a monsoonal to desert climate. A modern-day analogy for this climate might be parts of the Sahara or southwestern USA. The present outcrop of the Permian and Triassic is shown in Figure 6.3.

At the start of the Permian, vast fields of sand dunes ultimately became the red sandstones typical of the south Devon coast. Relic dune structures may be seen. Permian strata crop out mainly in a narrow area in northern and central England, but are extensive in the subsurface. Characteristic of the period are red mudstones and sandstones with beds of evaporites such as halite (salt) and gypsum. Stratigraphical subdivision is often difficult due to the severe lack of fossils. During the period the whole of Britain was uplifted, with high land to the northwest and, due to a position near the equator, desert type conditions prevailed. Some marine influences are evident in eastern Britain with Permian Marls and Magnesian Limestone deposited in marginal areas of the Zechstein Sea.

This was followed by the invasion of inland seas and the deposition of limestones and shales, and subsequently by deserts, river systems, and saltpans in the Triassic. The Triassic re-established marine environments over and beyond the Permian outcrop, with seas to the west and east of the Pennine ridge in the mid to late Triassic. The Variscan Mountains, in what is now southwestern England, provided a major source of sediment. The Mercia Mudstone Group is the major clay-rich sequence of the British Triassic, representing over 30 million years of deposition.

6.3.1. Mercia Mudstone Group

The Triassic Period derives its name from the three-fold lithostratigraphic division, recognized in Germany, of Buntsandstein (sandstone), Muschelkalk (marine carbonate) and Keuper (mudstone). However, in Britain this division does not apply (Forster & Warrington 1985). The

FIG. 6.3. Map showing Permo-Trias outcrop.

Tarporley Siltstone, Eldersfield Mudstone, Arden Sandstone, Twyning Mudstone and Blue Anchor Formations (Howard *et al.* 1998; Wilson 1993). However, these are given local formation names in the East Midlands, Cheshire and Lancashire. In particular, the Eldersfield Mudstone Formation has (in ascending order) four members in the East Midlands: Sneinton, Gunthorpe, Edwalton and Cropwell Bishop (Fig. 6.4); three in Cheshire: Bollin Mudstone, Byley Mudstone and Wych Mudstone; and three in Lancashire: Singleton Mudstone, Kirkham Mudstone, Breckells Mudstone.

A large part of British motorway construction during the 1960s took place across the outcrop of the Mercia Mudstone and the engineering properties and clay mineralogy of the deposit were extensively studied at that time (Dumbleton & West 1966). Recent re-assessments of the Mercia Mudstone as an engineering material have also been made (Chandler & Forster 2001; Hobbs *et al.* 2001). Engineering behaviour of the Mercia Mudstone is influenced by the presence at depth of gypsum and of voids produced by the dissolution of gypsum. In areas of salt mining (e.g. Cheshire) subsidence is a major problem.

Muschelkalk equivalent is not developed and Triassic rocks are represented by the mainly fluviatile Sherwood Sandstone Group (including the former Bunter Beds and Lower Keuper Sandstone), the Mercia Mudstone Group (formerly Keuper Marl), and the Penarth Group (formerly the Rhaetic).

The climate at the time of deposition was arid and monsoonal at a latitude equivalent to the present Sahara (Warrington & Ivimey-Cook 1992). The Mercia Mudstone Group was deposited in a sabkha mudflat environment of temporary lakes, and wind-blown dust. Salt (halite) deposits occur in Cheshire and Staffordshire. Evaporitic sulphate deposits (gypsum and anhydrite) and sandstones, many deposited in flash floods, are common. Because of a variety of local formational names cross-basin correlation, across the outcrop has been difficult to establish. Five main lithostratigraphical formations are recognized in most basins (in ascending order):

FIG. 6.4. Cropwell Bishop Formation, Mercia Mudstone Group, former gypsum quarry, Cropwell Bishop, Nott. Photo: A. Forster.

Sandstone bands, known as 'skerries' or 'doggers', are stronger, and more resistant to erosion and excavation, than the mudstones and may contain perched water tables. The Edwalton Formation in the East Midlands has unexpectedly been found to contain horizons containing significant amounts of mixed-layer swelling clays which, in common with similar formations, tend to become more plastic and markedly less workable with increased remoulding. This has been experienced in the clogging of full-face TBMs (tunnel boring machines).

Resources for the brick-making industry have traditionally been taken from beneath the Arden Sandstone; that is, the lower part of the Eldersfield Mudstone Formation and the Tarporley Siltstones Formation. The former is a structureless red-brown, mudstone with sandy horizons and gypsum, whilst the latter consists of red-brown micaceous mudstones and siltstones, usually lacking in gypsum, interbedded with impersistent micaceous sandstones. Both formations thicken considerably westward, the Eldersfield ranging from 120 m in the East Midlands to over 1000 m in Cheshire. The high carbonate content, usually in the form of dolomite, of the mudstones produces bricks with a distinctive pale body colour. The Group outcrop represents 39% of Britain's brick-clay resources. Currently, extraction is confined to the Tarporley Siltstones Formation and the lower part of the Eldersfield Mudstone (Bloodworth et al. 2001).

An example of a large-scale coastal landslide in which Mercia Mudstone hosted part of the main shear surface is that of Hooken which occurred in 1790 immediately to the west of Beer Head, Devon.

6.3.2. Penarth Group

The Penarth Group (formerly the Rhaetic Beds) overlies the Mercia Mudstone. This is co-extensive with the Lias, and represents a return to marine conditions at the end of the Triassic. Within this group are found the mudstones and shales of the Westbury Formation and Cotham Member. In the Midlands these tend to form escarpments between the Langport Member and the Blue Lias. The Westbury Formation has in part formed the shear plane of the famous Bindon, East Devon, coastal landslide of 1839 (Pitts & Brunsden 1987) located west of Lyme Regis, West Dorset, and landslides in the Axminster area (Croot & Griffiths 2001).

6.4. Jurassic

The Jurassic Period, named after the Jura Mountains of France and Switzerland, lasted for 60 million years, from 208 to 146 Ma. During the Jurassic the Atlantic Ocean opened out, and the North Sea subsided. It is most notable for the burgeoning period of dinosaurs and of invertebrate marine life. Its abundant fossils facilitated the early development of the science of stratigraphy in Britain and Europe during the early 19th century. The environment was dominantly marine, during which most of the clay-rich formations were deposited, but with deltaic and fluviatile sequences occurring during the Mid-Jurassic in Eastern England. A modern-day analogy for much of the marine Jurassic might be the West Indies. The present outcrop of the Jurassic, combined with the Cretaceous, is shown in Figure 6.5.

Some very important clay formations occur in this part of the geological column, especially in central and southern England and North Yorkshire (Taylor 1995). Deposits formed in areas of low relief and coast with high landmasses far to the northwest. Thin deposits were laid down in shallow seas, some preserved in basins, e.g. Skye and N. Ireland Chalks, that are isolated from the main outcrop. The climate was warm and humid and land areas well vegetated. Vulcanicity reached a peak in the Mid-Jurassic.

Dominantly marine conditions gave rise to mainly mudstone deposits in the deeper waters of the Early Jurassic. These Lower Jurassic calcareous mudrocks of the Lias Group are probably the most recent to be described today as 'strong rocks' (Cripps & Taylor 1981, 1987).

China clays (kaolin)

China clays (kaolin) are refined from the residual weathering (mainly Permian) and hydrothermal alteration product of granite, a kaolinised feldspar, and consist of kaolinite with trace amounts of mica and quartz. China clay, and other forms of kaolinite, are found in many igneous rocks and as secondary minerals produced from the alteration of aluminosilicates. China clays have been exploited mainly from the St. Austell granite, the west flank of Dartmoor, and on the west and south of Bodmin Moor, Cornwall, since the mid-18th century; with the St Austell granite accounting for about 75% of total production (BGS 2002). They are used in fine ceramics, paint, cement, papermaking, and the chemical industry, and are characterized by their fine texture, retention of whiteness on firing, and ease of moulding. China clay is also used in the manufacture of fibreglass. China clays are usually won by the application of high-pressure water jets (monitors). The low plasticity clay is usually mixed with other more plastic clays in manufacturing. The large quantities of waste produced by the industry are disposed of locally in engineered lagoons. China clay has recently been the UK's second most important mineral export after crude oil (Duff & Smith 1992). Refer to Chapters 14 and 15.

FIG. 6.5. Map showing Jurassic and Cretaceous outcrop.

the clay mineral illite, but contain significant proportions of mixed-layer illite–smectite and non-clay minerals (refer to Chapter 3).

Sedimentary cyclicity, in relation to orbital-climatic cyclicity, has been investigated within the Blue Lias, Belemnite Marl, and Kimmeridge Clay Formations by Weedon et al. (1999). Magnetic susceptibility and carbonate content were used to generate high-resolution time series. The presence of regular sedimentary cycles, that vary in wavelength according to sedimentation rate, and with patterns of orbital precession and eccentricity, was deduced.

6.4.1. Lias Group

The Lias Group principally comprises mudstones and limestones of Early Jurassic age, spanning 25 million years. The main outcrop extends from Dorset to Yorkshire and Cleveland, with many outliers. Within basins the thickness is several hundred metres. The Lias Group clays were deposited in a Mediterranean-like climate in warm, shallow seas. Carbonates were deposited as limestone beds and nodules, particularly in the lower part of the succession, to form the Blue Lias Formation. The Lias Group is very fossiliferous, and ammonites, being particularly abundant, have been used to sub-divide the group into twenty bio-zones. Traditionally, the group has been divided into the Lower, Middle, and Upper Lias, but the succession is better described in terms of the formations defined by Cox et al. (1999). Of these, the main clay-bearing formations are the Blue Lias (Fig. 6.6), Charmouth Mudstone (Fig. 6.7), and Whitby Mudstone Formations (Ambrose 2001). The sole major sand formation is the Bridport Sand, while the Dyrham Formation consists of siltstones and mudstones; both outcropping in the south-west of England. The Lias Group is best seen in coastal cliff sections in Dorset between Lyme Regis and Bridport, and in Yorkshire between Robin Hood's Bay and Redcar. Slopes within the Lias Group have been subject to

Some burial diagenesis has undoubtedly affected the Lias. The sequence includes impure limestones and marls. In Eastern England, much of the Middle Jurassic is composed of non-marine fluvial and deltaic facies, with the Ravenscar Group of Yorkshire featuring cyclic sediments, much like those of the Coal Measures. Some mixed beds with fluviatile clastics and some estuarine formations occur (e.g. in Yorkshire, the Alum Shale). Towards the southwest, shallow seas resulted in deposition of the Cotswold limestones.

Major clay formations of the Jurassic fall within the Lias Group (Lower Jurassic), Great and Inferior Oolite Groups (Middle Jurassic), and the Ancholme Group (Upper & Middle Jurassic). Examples are the Blue Lias, Charmouth, and Whitby Mudstones of the Lias Group, the Fuller's Earth and Frome Clay of the Great Oolite, and the Oxford, Ampthill and Kimmeridge Clays of the Ancholme Group. Jurassic clays tend to be dominated by

FIG. 6.6. Blue Lias Formation, Lias Group, Ware Cliff, Lyme Regis, Dorset. Photo: P. Hobbs.

extensive landsliding and cambering, particularly where overlain by Inferior Oolite limestones, for example in the Cotswolds where both processes may be seen within the same valley. Large coastal landslides occur within formations of the Lower Lias on the mid-Glamorgan coast. Shales within the Lias Group are highly bituminous and lignitic. Lias Group shales are believed to be the source rock for oil reserves discovered within the Great Oolite at Basingstoke, Hampshire in 1980 (Duff & Smith 1992). Ironstone bands are also found, and have been exploited for iron ore in the past.

The Lias formations from different basins have undergone different amounts of induration and overconsolidation. This is largely determined by the thicknesses of sediment (Jurassic to Quaternary) that may once have overlain a particular site and the amount of erosion that has taken place. A maximum previous burial depth of the order of 2 km is indicated for the Cleveland Basin (M. Sumbler, pers. com.). This has resulted in greater strength ('strong' to 'very strong') and durability overall in the case of the northern Lias mudstones, compared with 'weak' to 'moderately strong' in the south. However, this trend is reversed for plasticity. Fresh Lias mudstones tend to be strong and durable but, in common with other Jurassic clay-rich formations, undergo considerable deterioration of most engineering properties due to stress relief and weathering action (Cripps & Taylor 1981). Weathering by oxidation, resulting in a colour change from grey to brown, typically extends to depths of about 5 m below ground level. This is typically accompanied by a significant increase in water content, resulting in altered engineering properties (Chandler 2000). Shale lithologies are inherently weaker, less durable, and possess marked anisotropy.

FIG. 6.7. Charmouth Mudstone Formation, Lias Group, (Northcot Brickworks), Blockley, Gloucs. Photo: P. Hobbs.

Thaumasite attack

Sulphur occurs in a variety of forms, in mudrocks and other sediments, and is very common in British formations. The best known forms are gypsum and anhydrite. Sulphates, in particular pyrite (FeS_2), are also formed in anaerobic conditions through the action of sulphate-reducing bacteria and are found dispersed as microscopic minerals throughout the fabric of many British mudrocks (Czerewko *et al.* 2003). They may constitute up to 5% of the whole rock. The process of pyrite oxidation is complex and consists of three stages: (a) oxidation of pyrite to form ferric oxide (Fe_2O_3) and sulphuric acid (H_2SO_4), (b) the sulphuric acid reacts with calcium carbonate, if present, to form gypsum ($CaSO_4 \cdot 2H_2O$), (c) under certain conditions, particularly if there is insufficient calcium carbonate, the sulphuric acid reacts directly with clay minerals and leaches them of exchangeable cations (potassium, magnesium, and sodium). One of the end products of this process is the sulphate mineral thaumasite. Other sulphate mineral products are jarosite and ettringite (Floyd *et al.* 2002). Under natural circumstances the oxidation of pyrite associated with weathering may be a very slow process. However, buried concrete structures within the Lias Group mudrocks have been found to be susceptible to thaumasite attack, a particularly aggressive form of sulphate attack, particularly where saturated Lias-derived fill is in contact (Longworth 2002). The result is a transformation of the concrete fabric into a weak paste. The word thaumasite derives from the Greek 'thaumasion' meaning surprising. Indeed, the phenomenon has only been recognised in the last few years. The low temperature ($<15°C$) reaction involves water, hydrated calcium silicates in the cement and a source of carbonate (Floyd *et al.* 2002). Thaumasite attack (known as TSA) has been particularly marked on bridge substructures where concrete has been in contact with pyritic Lias Group clays and clay-fills on the M5 motorway in Gloucestershire. Recent research, including field trials, has led to a re-assessment of concrete standards (BRE 2001).

Weak horizons within the Whitby Mudstone have contributed to the processes of cambering, valley bulging, and landsliding, for example in Northamptonshire (e.g. Empingham) (Horswill & Horton 1976) and the Cotswold escarpment (e.g. Bredon Hill, Broadway Hill, Leckhampton Hill) where it is overlain by Inferior Oolite limestones. Lower Lias shale and limestone formations, underlying lavas, on the Trotternish escarpment in northeast Skye, form the Quiraing and Storr landslides, amongst the largest and most unusual in Europe. Reworking of Lias mudstones imparts considerable loss of strength and trafficability. This is seen in quarries, particularly within the Charmouth Mudstone, after periods of rain. A good negative correlation has been established between strength and water content (Cripps & Taylor 1987). Somerset's famous landmark, Glastonbury Tor, is made up of Bridport Sand Formation on Beacon Limestone Formation on Dyrham Formation; the latter also forming Wearyall Hill west of the Tor.

6.4.2. Upper and Lower Fuller's Earth and Frome Clay Formations

These clays overlie Inferior Oolite in the Cotswold to Dorset area (Penn & Wyatt 1979; Hawkins et al. 1988). The Fuller's Earth is divided into lower and upper divisions by the Fuller's Earth Rock limestone. Within the Upper Fullers Earth Clay is a band of commercially exploited 'fuller's earth' up to 1 m thick. This smectite-rich clay, derived from volcanic ash fall, has been exploited for various purposes, including as a de-greasing (or 'fulling') agent for wool and as a component of cosmetics and cat-litter. The engineering properties are dominated by the smectite content which, though variable, imparts low strength and extremely high plasticity and stickiness. For the same reasons, the Fuller's Earth Clays are implicated in slope instability (e.g. Bath). The Frome Clay, developed only in the southern part of the region is the lateral equivalent of much of the Great Oolite limestone succession of the Bath area. It was formerly included in the Fuller's Earth. In the East Midlands, the equivalent clays are found within the Rutland Formation, while in North Yorkshire the mudstones of the Ravenscar Group give rise to major landslides.

6.4.3. Blisworth Clay and Forest Marble Formations

Formerly known as the Great Oolite Clay, the Blisworth Clay Formation is found in the Midlands from Bedfordshire to Lincolnshire. It consists of mudstone, typically variegated (purplish-greenish-bluish grey), with very thin limestone and thin sandstone included within the clay. The Blisworth Clay Formation is typically a 5 to 10 m thick, 'firm' to 'stiff', grey-green, silty clay with a slightly crumbly texture. At depth it is described as a 'weak' mudstone with shell fragments. Bands of clay rich in iron pyrites and gypsum crystals are found at depth. Plasticity classification is typically 'high' to 'very high'. Clay contents are around 50 to 60%.

The Forest Marble Formation (the approximate equivalent of the Blisworth Clay Formation in the southwest) is dominated by mudstone with limestones developed particularly in its lower part, and thin sandstones. However, the formation is particularly heterogeneous. The limestones consist of lenticular beds of shell-fragmental, ooidal grainstone, and sandy limestone, are often argillaceous, and typically cross-bedded and forming banks and channel-fills.

6.4.4. Oxford Clay Formation

The Oxford Clay Formation is a fissured, heavily over-consolidated, bituminous clay and clay-shale, typically up to 70 m in thickness in Central England. In the past the succession was sub-divided into the Lower, Middle, and Upper Oxford Clay, but these are now known as the Peterborough, Stewartby, and Weymouth Members, respectively. The Peterborough Member is darker, more bituminous, and more shaly than the two younger members. Selenite (crystalline gypsum) occurs in the weathered zone. It weathers to form regular hard and soft bands. It was formed under marine conditions. It is up to 180 m thick, and extends from the Dorset coast (e.g. Weymouth

---

Fuller's Earth

Fuller's Earth is a sedimentary clay rich in calcium smectite (calcium montmorillonite), a highly plastic clay mineral derived from altered volcanic ash. Historically, it has been described as 'an unctuous yellow clay'. Its occurrence is intermittent, usually in the form of lenticular deposits up to 4 m thick, but typically less than 1 m thick. It has been exploited in both the Lower Cretaceous (Lower Greensand Group) of Surrey and the Middle Jurassic (Fuller's Earth Formation) near Bath. However, current production is centred on the Lower Greensand at Baulking (Oxfordshire) and Clophill (Bedfordshire). The material is quarried and then processed by sodium-exchange (BGS 2002). Traditionally, its use has been as a de-greasing (or 'fulling') agent for woollen cloth, but its present commercial functions are varied and include cosmetics, as well as industrial and domestic applications utilising its absorbent properties. An appraisal of Fuller's Earth resources is given in Moorlock & Highley (1991). The origins of the Fuller's Earth are described in Jeans et al. (1982).

Bay) to the Yorkshire coast near Scarborough. Former burial depth has been estimated (Jackson & Fookes 1974) at between 330 m (Hants.) and 1560 m (Dorset).

The Peterborough Member outcrop is up to 39 m thick in the Swindon area, reducing to 28 m in the Oxford area, 21 m at Bedford, but is absent in North Yorkshire, where the equivalent beds are in a sandstone facies. It is distinguished by its brown-grey, fissile organic-rich mudstones and bituminous shales, with subordinate beds of pale-grey, blocky mudstone. Argillaceous limestone (cementstone) nodules (up to 1 m diameter) often occur as bands throughout the member. The basal beds are typically silty and shelly. The Peterborough Member weathers to a pale grey and ochreous mottled clay containing dispersed selenite. The upper part of the Formation tends to be weaker and more fissured than the lower due to stress relief (Cripps & Taylor 1981) and mineralogy. The relationships between weathering, mineralogy, and geotechnical properties of the Oxford Clay have been described in Russell & Parker (1979). In particular, undrained strength has been found to correlate negatively with clay 'activity' (as defined by Skempton), and to correlate positively with cementing agents calcite and pyrite. Extensive cryo-turbation of the Peterborough Member, including relic shears, has been noted during construction of the M40 by Nicholls (1994). The relationships between strength and water content and the effects of stress-relief on excavation stability have been established for Middle and Lower Oxford Clays (Burland et al. 1978; Cripps & Taylor 1987). Fault zones, with throws of the order of 5 m, have been noted in the East Midlands allowing the Kellaways Formation to juxtapose the Oxford Clay Formation locally. A down-faulted section of Oxford Clay Formation has formed Castle Hill, Scarborough. On the coast the Oxford Clay is involved in the landslides between Bowleaze Cove and Red Cliff Point, to the northeast of Weymouth, Dorset.

The Peterborough Member is well known for its intensive exploitation as brick clay near Bedford and Peterborough where over 30% of Britain's brick-clay is currently produced, and where some of Britain's largest former brick pits are being used for landfill waste disposal (Chapter 14). The Peterborough Member differs from most other brick-clay resources in having a high carbon content, which supplements the fuel in the particular firing process used. The outcrop represents 18% of Britain's brick-clay resources (Bloodworth et al. 2001).

6.4.5. West Walton Formation

The West Walton Formation lies between the underlying Oxford Clay (Weymouth Member) and overlying Ampthill Clay Formation, particularly in Oxfordshire and North Lincolnshire. It has in the past been included within the Oxford Clay, and has been found to have generally similar geotechnical properties (Nicholls 1994). It consists of 10 to 20 m of calcareous mudstone, very stiff, silty, dark grey clays, mudstones and siltstones, locally with argillaceous limestone or siltstone nodules, sandstones, and ooidal and coralline marl at the top. The West Walton Formation features rhythmic alternations of dark grey clay, fossiliferous silty mudstone (Cox & Gallois 1979). The plasticity is described as typically 'high' (Cripps & Taylor 1981), but has a higher silt content than the Ampthill Clay Formation.

6.4.6. Ampthill Clay Formation

Until relatively recently, the Ampthill Clay Formation was not generally separated from the Kimmeridge Clay, and is shown as part of the latter on some geological maps. The Ampthill Clay Formation outcrops in Bedfordshire and consists of 10 to 20 m thickness of fossiliferous, pyritous, pale grey mudstone with occasional thin nodular cementstone bands. It is typically fissured, and may contain shear surfaces in the uppermost 1–2 m. Its plasticity classification is described as 'high' to 'very high' and the clay content high, at about 70% (Cripps & Taylor 1981). Unusual mud-springs are found, affecting the Ampthill Clay Formation, in the area of Wootton Bassett, Wiltshire (Bristow et al. 2000). These arise from artesian pressures within the underlying Corallian limestone, and bring fossils to the surface.

6.4.7. Kimmeridge Clay Formation

The Kimmeridge Clay Formation outcrops along the Dorset Coast between Weymouth and Swanage, where the type sections are found, and in Yorkshire (including the Vale of Pickering). It also outcrops in Lincolnshire, Cambridgeshire, Oxfordshire, and Wiltshire; including the conurbation of Swindon. The best outcrops are at the Isle of Purbeck. It consists of organic or calcareous, silty and sandy, mudstones, with thin siltstone and cementstone beds. It has traditionally been divided into 49 beds on the basis of lithology and macrofauna. The formation consists of marine, fossiliferous mudstones and calcareous mudstones showing both small and large-scale sedimentary rhythms, with bands of limestone and siltstone (Gallois 2000). As with many Jurassic clay formations, major planes of weakness parallel with bedding are formed by dense layers of whole and crushed fossils.

Horizons of pyritic oil shale also occur within the formation. The cycling of normal with bituminous shales relates to mildly oxygenated and totally anoxic conditions, respectively (Morris 1980). Bituminous shales have been exploited commercially at relatively shallow depth in Dorset. An area to the east of Osmington Mills, Dorset, is known as 'Burning Cliff' owing to spontaneous combustion of hydrocarbons at outcrop. More importantly, beneath the North Sea, oil released from the Kimmeridge Clay, and trapped against faulted-up Middle Jurassic sandstones, provides the large oil reserves currently exploited. The base of the formation is at the top of the smectic Ringstead Waxy Clay Member of the Ampthill Clay Formation (Upper Oxfordian). The Kimmeridge Clay has been involved in large coastal landslides at White Nothe point, to the east of Weymouth, Dorset.

6.5. Cretaceous

The Cretaceous Period is named after chalk (latin 'creta'), which dominates the upper part of the succession. It lasted from 146 to 65 Ma before present. However, the Lower Cretaceous contains some important clay-rich formations, notably the Gault, Atherfield, Weald, Wadhurst and Speeton Clays. Clays also form a substantial part of the Lower Cretaceous succession in Lincolnshire (e.g. Terlby Clay and Rooch Formations). These are generally clays and weak mudstones with 'high' or 'very high' plasticity, which present significant hazards in terms of volume change (swell/shrink) and landslides. The present outcrop is shown, combined with the Jurassic in Figure 6.5.

The Lower Cretaceous is dominated by terrigenous, clastic settlements including various mudrock formations formed in marine, fluviatile, and estuarine settings, with mixtures of poorly indurated silts, clays and sands. The Upper Cretaceous is almost entirely chalk, which nevertheless includes some thin mudrock layers that may have engineering significance (e.g. slope stability). With the exception of parts subjected to Alpine earth movements (in the Isle of Wight and along the south coast of England), most areas were little affected by tectonism. However, the influence of Pleistocene periglacial processes (e.g. solifluction, cambering etc.) was very important. Many of these features have remained intact in southern England.

The Cretaceous period was marked by active tectonism; as the north Atlantic began to open. Commercially exploited Fuller's earth deposits in Surrey (Lower Greensand) are believed to have been formed by volcanic ash (Jeans *et al.* 1982). Deltaic and river deposits produced the Wealden Beds, whilst a marine transgression resulted in the Gault and Lower & Upper Greensand Formations.

6.5.1. Wadhurst Clay Formation

The Wadhurst Clay Formation consists of a dark grey, heavily faulted and jointed, over-consolidated silt/clay and clay shale, of 'intermediate' to 'high' plasticity, cropping out irregularly in the Weald area of Sussex and Kent. Lenses of calcareous sandstone up to 3 m thick, and subordinate sandstones, siltstones, clay-ironstones, and lignites, are also found within it. It reaches a maximum thickness of 80 m in the Tunbridge Wells and Horsham area and forms the central unit of the Hastings Group, the basal part of the Lower Cretaceous Wealden Group. It was formed in brackish and freshwater lacustrine and fluvial environments. Following Tertiary uplift and exposure in the Wealden anticline, periglacial conditions resulted in cambering, valley bulging and solifluction, all of which have affected the present thickness of the deposit. It is typically highly weathered and subject to solifluction to a depth of about 5 m, and is particularly susceptible to shallow landsliding (Bromhead *et al.* 2000; Pugh *et al.* 1991). The mineralogy is dominated by kaolinite and illite, with some mixed-layer mica-vermiculite (Bloodworth *et al.* 2001). The Wadhurst Clay weathers to an ochreous green-grey colour.

Ironstone found as siderite nodules and tabular masses within the Wadhurst Clay were used as the basis of the Wealden iron industry in East Sussex and Kent, from Roman times and peaking in the 17th century (Duff & Smith 1992). Brick making was also carried out (Chapter 14). Remains of these industries are characterised by 'hammer ponds' dotting the outcrop (Bromhead *et al.* 2000). The Wadhurst Clay outcrop represents 2% of British brick clay resources (Bloodworth *et al.* 2001). The high density of discontinuities tends to impart high mass-permeability to this formation (Pugh *et al.* 1991).

6.5.2. Weald Clay Formation

The outcrop of the Weald Clay Formation is confined to low-lying areas of the Weald in southeast England, and to the cliffs on the southern coast of the Isle of Wight. It consists of up to 450 m of thinly bedded, grey and mottled red/green silty clay, shale, and mudstone, and clayey silts with silts, sands, with beds of ironstones, sandstone and shelly limestone. The crystalline limestones are thinly bedded, as are the uniform fine-grained calcareous sandstones. Red/green 'catsbrain' mottling is associated with some sandy horizons (Bloodworth *et al.* 2001). It was formed in brackish and freshwater lacustrine and fluvial environments. Dispersed gypsum and lignite are common. The Weald Clay weathers to a mottled orange-yellow, yellow-brown, light-grey, green, or red clay typically to a depth of 5 m, with secondary gypsum. The clay minerals illite and kaolinite predominate with some mixed-layer mica-vermiculite, giving the Formation 'intermediate' to 'high' plasticity. The Weald Clay Formation is a brick-making resource (Chapter 14) and its outcrop represents 6% of British brick clay resources (Bloodworth *et al.* 2001). The Weald Clay Formation has been involved in the complex landslides on the Lower Greensand escarpment at Sevenoaks, Kent.

6.5.3. Atherfield Clay Formation

The Lower Greensand Group is a marine sequence of Early Aptian to Early Albian age that overlies the fresh- and brackish-water Wealden Beds (Simpson 1985). The Atherfield Clay, reaching a thickness of 50 m, forms the lowermost part of the Lower Greensand Group, and has its type section at Chale Bay cliffs, Isle of Wight. It also crops out in the Weald and on the Dorset coast. It is a highly plastic clay or mudstone with a smectite content of around 10%, containing clay-ironstone nodules. Although less plastic than the Gault Clay, it is susceptible to landsliding (particularly where overlain by the Hythe Beds), and also to swelling and shrinkage.

6.5.4. Speeton Clay Formation

The Speeton Clay, present only in the Cleveland Basin and at Speeton, East Yorkshire, comprises around 100 m of dark grey, slightly shaly, marine mudstone spanning much of the Lower Cretaceous, and overlying non-sequentially the Kimmeridge Clay Formation. The Speeton Clay contains many ammonite zones. It is older than the Gault, its stratigraphic equivalents to the south being the Spilsby Sandstone, Sandringham Sands, and parts of the Purbeck. It consists of weak fossiliferous, pyritous mudstones with layers of bentonite-rich clay in the lower parts and layers of phosphatic nodules. The bentonite is derived from volcanic tephra. The Speeton Clay is typically of 'intermediate' to 'extremely high' plasticity. The Speeton Clay, like the Gault, is involved in coastal landslide activity (e.g. Filey Bay, Yorkshire) and to swell/shrink behaviour. Evidence of micro-tectonic structures can be found at Speeton.

6.5.5. Gault Formation

The Gault Formation is a sequence of grey and blue-grey fissured, over-consolidated clays, mudstones, and thin siltstones with bands of phosphatic nodules, laid down towards the end of the Lower Cretaceous in the Middle and Upper Albian age (Fig. 6.8). It is the youngest of the British Cretaceous over-consolidated clays (Marsh & Greenwood 1995). Small-scale (1 to 2 m) lithological rhythms have been identified (Gallois & Morter 1982) which, to a large degree, determine the geotechnical properties. The Gault Formation weathers to a lighter grey colour and forms a generally weaker material. The outcrop stretches southwestward from East Anglia through Wessex to West Dorset, and also surrounds the Weald in an arc; the only major conurbation on the Gault Formation being Cambridge. The Gault thickens southward to 100 m thick in the Weald, and is also less calcareous. The Gault is made up of the clay minerals illite/smectite, kaolinite, and illite/mica (refer to Chapter 3), and non-clay minerals quartz and calcite, and generally has a 'high' to 'very high' plasticity. However, some parts have 'extremely high' plasticity (liquid limit >90%), probably due to the high smectite content derived from volcanic ash (Jeans *et al.* 1982). The Gault Formation at outcrop poses serious engineering problems by way of subsidence due to swell/shrink, and slope instability. It exhibits marked intact anisotropy in terms of strength, deformability, and volume change due to shrink/swell. The range of strength encountered is considerable, depending on depth-related factors such as weathering, periglacial disturbance, structural features, and anisotropy. On reworking in a saturated state, the Gault becomes extremely sticky and untrafficable. The Gault Formation clays have been, and still are to a limited extent, used in the brick-making and cement industries. The engineering geology of the Gault is described in Forster *et al.* (1995) and Fookes & Denness (1969).

A large number of both shallow and deep-seated landslides involve the Gault Formation. Several well-known coastal landslides include those at Ventnor (Isle of Wight), Lyme Regis (Dorset), and Castle Hill and Folkestone Warren (Kent). Slope stability studies have been carried out at Selborne, Hants. (Cooper *et al.* 1998) and in Kent (Garrett & Wale 1985). The Gault Clay lies beneath the entire length of the Channel Tunnel which was driven through the overlying Chalk Marl (Gale *et al.* 1999) . The palaeoclimatology of the Gault Clay has been examined by Erba *et al.* (1992).

6.6. Tertiary

The Tertiary sub-Era lasted for 63 million years from 65 to 2 Ma before present. It encompasses the earlier and greater part of the Cainozoic Era (the second being the Quaternary), and comprises the Palaeogene and Neogene Periods. During the Tertiary what can be called modern life forms developed, while the landmass, on which Britain is situated, moved north and eastward by 1100 km (Curry 1992; Toghill 2000). During much of this period the climate was sub-tropical with creatures similar to those found in the sub-tropics today. The early Eocene London Clay represents the arrival of North Sea type marine deposits over southern England and the development of residual clays, used for pottery, in the west (Toghill 2000). Much of the earliest palaeontological

FIG. 6.8. Gault Formation, (Selborne Brickworks), Selborne, Hants. Photo: P. Hobbs.

and sedimentological work was carried out in the British Tertiary deposits (Curry 1992). The present outcrop of the Tertiary is shown in Figure 6.9.

The essentially soft sediments of the London and Hampshire basins include Lambeth Group, London Clay and Barton Clays formed as marine clays in areas remote from high landmasses. Some ingress of silty or sandy [+ carbonate] materials in some areas gave regional variations. Variations in burial depth have affected mechanical properties but the original sedimentary mineralogy has largely been preserved. Up to 200 m of London Clay was deposited in the London Basin. The Lambeth Group was deposited largely in a complex cyclical environment of coast, lagoon, and marsh, resulting in great variability in lithology and mechanical properties in this formation.

Ball Clays (see Box) were produced in the southwest of England as the sedimented weathering product of Devonian slates and granite. They are plastic, sedimentary kaolinitic clays developed from the sub-tropical Tertiary weathering of granite and other rock types, re-sedimented and preserved in basin environments. The influence of subsequent Pleistocene events was very important. China Clays (see Box) have been refined in Cornwall from the residual weathering product of granite, and consist of kaolinite with trace amounts of mica and quartz.

6.6.1. Ormesby Clay Formation

The Ormesby Clay Formation of Eastern Norfolk and Suffolk consists of grey, red-grey, and green-grey heavily over-consolidated clay or mudstone. It is poorly bedded, variably glauconitic and calcareous, highly bioturbated, and silty towards the base. Sporadic tephra layers are found in the lower part (Ellison et al. 1994). Stratigraphically it is equivalent to the Thanet Sand Formation of the London Basin and lies between the Lambeth Group and the Chalk (Hight & Jardine 1993).

6.6.2. Lambeth Group

The term Lambeth Group has recently been introduced to incorporate what were previously designated the Woolwich and Reading Beds (Ellison et al. 1994). The group features notable lateral and vertical lithological variations, which have contributed to a long history of engineering problems in London (Ellison 2000). The Lambeth Group deposits were laid down about 55 to 56 million years ago in embayments on the western margin of a deep-water marine basin centred on the present North Sea. Shallow marine sands dominate, but thin clays and silts were laid down in brackish water lagoons and alluvial plains. Much intersection and infilling of channels took place. Variation in vertical and lateral permeability is considerable. Gentle folding of the Lambeth Group strata occurred in the Miocene around 20 million years ago when the London and Hampshire basins became separated.

The Lambeth Group is divided into the Upnor, Reading, and Woolwich Formations. Of these, the Reading Formation contains the greatest proportion of clay-rich strata. The Reading Formation is sub-divided into the lower and upper mottled clays (Fig. 6.10); the former being smectite-rich and the latter illite-rich (Ellison 2000). The clays are distinctively multi-hued, due to pedogenic processes, bioturbation, and alternating oxidizing and reducing environments. The clays tend to fail along slickensided joints or shears. The sand layers of the Upnor Formation themselves contain clay minerals, some of which are smectitic, and have thin interlayered clays. The thickness of the Lambeth Group in the London Basin ranges from less than 10 m in the southeast to about 30 m in the central part. The deposit in the Hampshire Basin is between 25 and 50 m in thickness. The thickness of the Reading Formation is between 12 and 27 m. The Woolwich Formation contains two clay-rich layers: the Lower and Upper Shelly Clays, which have illite, smectite, and kaolinite clay minerals. Engineering experience in London has highlighted under-drainage due to historically depressed water table in the Thanet Beds.

FIG. 6.9. Map showing Tertiary outcrop.

Fig. 6.10. Reading Formation, Lambeth Group, (Knoll Manor Brickworks), Corfe Mullen, W. Sussex. Photo: L. Jones.

6.6.3. London Clay Formation

The London Clay Formation is a thick, relatively homogeneous, clay sequence forming the upper part of the Thames Group (King 1981), and accounts for the greater part of its thickness and area of outcrop. It underlies much of Greater London and also outcrops in Essex, Kent, Hampshire, Sussex and the Isle of Wight. In the London Basin the London Clay Formation is up to 200 m thick, but typically it is between 90 and 130 m. The London Clay Formation has been subject to tectonic and micro-tectonic displacements, notably on the Isle of Wight, where joints and pervasive slickensides are encountered. Regional variations have been described by Burnett & Fookes (1974), the principal of these being an overall increase in clay content and plasticity east of the London Basin. The litho-stratigraphy is currently under review by the British Geological Survey, and will form part of a new London memoir.

The London Clay Formation is an over-consolidated, fissured, silty clay deposited in a marine shelf environment subject to a tropical climate. It is dominantly illite/smectite (refer to Chapter 3) and weathers in the uppermost 10 m to form a characteristic brown clay. Thin smectitic layers are found, particularly in the eastern part of the outcrop and also beneath the North Sea. Sporadic beds of calcareous 'cementstone' nodules occur. Large isolated siliceous nodules resembling cannon balls are occasionally found within the London Clay. Individual and clustered crystals of selenite (gypsum) are found scattered through both weathered and unweathered material. Plant remains, including wood fragments, are also found. The base of the formation is characterized by glauconitic sandy silts.

Over the last two centuries the London Clay Formation has become a familiar tunnelling and excavation medium beneath the capital. Over 45 000 boreholes have been recorded within the London Clay outcrop. The London Clay Formation has a 'high' plasticity and is subject to shrink/swell behaviour resulting in foundation damage. Acid groundwater is also an engineering hazard. The formation has been extensively used for brick-making (refer to Chapter 14) and engineering fill. Detailed studies have been made of subsidence due to tunnelling and excavation, and the effects of groundwater rise due to diminishing industrialization (Brassington 1990; Craig & Simpson 1990). Deep localized depressions (scour hollows) in the London Clay subcrop have been reported (Lewis 1998). Landsliding within the London Clay is common, particularly in Kent and Essex and particularly on the coast.

Weathering of London Clay, seen as oxidation of grey to brown clay, typically extends to 10 m below ground level. This is believed to represent former periods of warmer or drier climate associated with a lower water table, rather than present weathering conditions (Chandler 2000), as the depth of weathering appears to increase with increasing elevation of the land surface. This also has implications for the near-surface structure of weathered London Clay, which contains a variety of discontinuities, some associated with former desiccation, in addition to features attributed to micro-tectonism. The effect of weathering on engineering properties has been investigated by Chandler & Apted (1988). Pre-existing sub-horizontal shear zones, due to flexural slip (folding and bedding slip), within the formation represent a significant hazard to engineering works (Chandler 2000). Sub-vertical shear zones associated with faulting are also found locally within the London Clay. Such planes exist at much lower strength than the surrounding material, and are thus prone to re-activation where exposed or loaded by engineering operations.

Landslides are found on slopes in excess of 10° throughout the London Clay outcrop in Kent, Essex, and North London (Hutchinson 1967). This includes built-up areas where they tend to be obscured. Coastal landslides in the London Clay are found at Herne Bay and the Isle of Sheppey in Kent (Fig. 6.11), and extensive landslides within former sea-cliffs are seen at Hadleigh Castle, Essex. Progressive failure is seen as a key process in London Clay slopes (Chandler 2000), and may result in failure occurring decades after construction of a slope. Temporary exposures around Bognor Regis and Chichester are described in Bone (1992). The exposure of London Clay (along with other Tertiary deposits) at Alum

170   BRITISH CLAY STRATIGRAPHY

Fig. 6.11. London Clay, Thames Group, Warden Point, Isle of Sheppey, Kent. Photo: P. Hobbs.

Bay, Isle of Wight, is notable for being inclined vertically due to folding.

6.7. Quaternary deposits

The Quaternary Period, comprising the Pleistocene Epoch (about 2.4 million years to 10 000 years B.P.) and

Fig. 6.12. Map showing Quaternary outcrop.

Ball clays

Ball clays are plastic, sedimentary kaolinitic clays developed (and transported) from sub-tropical Tertiary weathering of shales, granites, and other rock types, re-sedimented and preserved in basin environments. Ball clay deposits in Britain are of particularly high quality, and are used in the manufacture of white ceramic ware. Demand for the product is considerable worldwide, and in 2000 production peaked in the UK at over 1 million tonnes, almost 80% of which was supplied to the EU, in particular Italy and Spain (BGS 2002). It occurs in three basins; the Bovey (south Devon), Petrockstowe and Torrington (north Devon), and Isle of Purbeck, Wimbourne, and Wareham (Dorset); the Bovey Basin providing over 50% of the total. Ball clays are usually quarried or shallow-mined using power tools. The name is derived from the balls or blocks in which form the clay was previously recovered. Refer to Chapter 14.

the Holocene Epoch (last 10 000 years), has been marked by extreme climatic variations. Much of the present British landscape relates directly to Quaternary processes, and the resulting glacial deposits are widespread. Tertiary weathering profiles have been preserved in many places in England south of the limit of glaciation. North of the limit such features have generally been destroyed by glaciers, although some pockets remain. There have been a number of episodes of glaciation affecting Britain. Most workers believe there were two major glaciations, the Anglian (probably around 350 000 to 45 000 BP), which affected much of Britain, and the Devensian (which culminated around 18 000 BP) during which glacier ice affected only Northern Britain and Wales. Although there is some limited evidence for a glaciation between the Anglian and Devensian, deposits are very limited. The type-section for what was previously described as Wolstonian is now thought to be Anglian (Bowen 1999), but evidence for a pre-Anglian glaciation is found in north Norfolk. The glacial episodes were marked by considerable temperature fluctuations. Ice thicknesses probably reached a maximum of around 2 km in Britain. Water locked up within ice-sheets globally resulted in sea-level falls of 120 to 150 m compared with the present. The present outcrop of Quaternary deposits is shown in Figure 6.12.

During the temperate parts of the Quaternary the British coastline was close to that of today. The climate ranged from warm-temperate to glacial and humid to semi-arid. Successive episodes of glaciation occurred with the removal of cover and the development of tills and moraines. The main types of deposits are lodgement and ablation tills. (these blanket a large proportion of north and central Britain), fluvio-glacial deposits, lacustrine and laminated clays. The difference between environments north and south of the maximum glacial extent have greatly influenced pre-Quaternary rock-head and engineering properties to depths of up to tens of metres. Also, loess and brickearths were deposited in periglacial environments in south and eastern England. The latter are mainly represented by silt-size deposits, but have been used in brick manufacture (refer to Chapter 14). Landsliding and cambering, resulting largely from periglacial processes, are widespread within the British landscape. Associated with these slope processes are head and solifluction deposits. These deposits are

FIG. 6.13. Skipsea Till Formation, Holderness, S. Yorks. Photo: P. Hobbs.

Tills

Tills are deposits produced by the direct grinding and transporting actions of glacier ice. Many are composed essentially of stony clay (e.g. diamictons). Large quantities of rock were transported, sometimes from far afield. During this process the rocks beneath the ice-sheets were ground to rock flour, whilst boulders were transported within and on the ice. As the ice melted and retreated these rock materials were deposited in irregular mounds containing many lithologies and particle sizes. Ahead of the glaciers gravels, sands, silts, and clays were deposited within transitory, high energy alluvial and lake environments. Boulders of many lithologies are found within the fine-grained matrix of the tills, giving rise to the traditional name, 'boulder clay'. Those boulders not of local provenance and in excess of 0.6 m diameter are called 'erratics'. These are usually rounded and of high strength, and create problems for ground engineers. In contrast to the boulders, the matrix of tills is often relatively homogeneous. In some parts of Britain the tills are sufficiently mappable as to warrant formational names (Hughes *et al.* 1998). In some areas (including North Norfolk) massive 'rafts' of rock displaced by ice thrusting are found, with glaciotectonically altered till formations (Lee 2001). Geotechnical characteristics are discussed in McGown (1971), McGown *et al.* (1974), and Bell & Forster (1991). The hydrogeological properties of tills are described in Klinck *et al.* (1993) and engineering behaviour described in Trenter (1999). A typical till from Holderness is shown in Figure 6.13.

characterized by heterogeneity and the presence of shear surfaces within and beneath them.

The engineering geology of Quaternary sediments has been described by Funnell & Wilkes (1976), Higginbottom & Fookes (1970), Fookes (1991, 1997). Periglacial deposits, including brickearths and clay-with-flints, and features associated with them, are well preserved in the south of England, and particularly in the southwest, where subsequent glacial advances did not reach and remove them. Landslides developed on sea-cliffs formed in Quaternary deposits are found in various coasts, especially those of eastern England, notably North Norfolk, and North and East Yorkshire. Coastal recession rates tend to be very high (e.g. 1–2 m per year on average) where these materials constitute the whole sea cliff. The complex interrelations of Quaternary and Recent deposits and their

Head deposits

Head deposits are formed by the downslope creep movement of bedrock or superficial materials resulting from freeze–thaw effects (referred to as gelifluction/solifluction) on near-surface deposits. They occur frequently as sheets in valley floors, valley sides, and interfluves. Some deposits extend across the Quaternary deposits of coastal plains and merge with the alluvium (Aldiss 2002). They contain a wide range of clast sizes derived locally, both from solid and superficial formations. The Head deposit may retain some persistent features of the source material which have not been totally destroyed by the formation processes. They are frequently difficult to identify, and are inconsistently and incompletely mapped in Britain. They are generally heterogeneous, but may locally be markedly stratified and, most significantly, contain relict shear surfaces at, or near, the residual strength (see Chapter 4). The deposits tend to thicken downslope, and also within depressions in the underlying competent deposits. Typically, Head has a thickness of between 2 and 5 m, but locally may thicken to as much as 8 m. Head is usually a poor and unpredictable engineering material and may become unstable in excavations and slopes.

Brickearth

Brickearth deposits were formed in cold, dry climates, mainly during the Devensian (though in some cases earlier), by the wind erosion of rocks and soils, subsequent deposition, and in some cases secondary post-depositional processes. Brickearths occur in southern and eastern Britain, particularly in East Kent, South Essex, Sussex, Hampshire, the Thames Valley and the south Devonshire coast. These deposits are usually thin and are a form of loess. They are called 'Brickearths' because of their use in the brick-making industry (refer to Chapter 14). Whilst strictly well-sorted silts, brickearths may contain up to 30% clay fraction material, though not all of this consists of clay minerals, but includes quartz, feldspar, and secondary calcite (Jefferson *et al.* 2001). The fabric typically consists of silt-size particles bonded by clay-size materials to form a relatively open structure. Much of the brickearths in Britain are alluvial in origin, for example those in the mid-Thames area form part of the alluvial terraces of the River Thames. Due to their open, weakly bonded fabric (Northmore *et al.* 1996; Jefferson *et al.* 2001; Catt 1988), brickearths typically exhibit the mechanical property of 'metastability' (Fookes & Best 1969) and are prone to structural collapse upon wetting under constant stress.

Clay-with-flints

Clay-with-flints is a term used to describe heterogeneous superficial deposits consisting of stiff red-brown clay and sandy clay containing abundant unworn flints, quartz pebbles and nodules, and also sandstone fragments (Aldiss 2002; Harrington *et al.* 1995). They are found in southern England where they were preserved beyond the limit of the Anglian Glaciation, particularly on the Chilterns, Berkshire Downs, Marlborough Downs, North Downs, Hampshire Downs and in Dorset and Devon (Aldiss 2002; Catt 1984). It is found capping the Chalk, and frequently infills solution pipes within the chalk. As with Head, the mapped deposits are incomplete. It is believed to represent a residual deposit of the solution of the Chalk and reworking of the Chalk and also Lambeth Group deposits (Klinck *et al.* 1993). It may also include Quaternary head and fluvial deposits. North of the Hitchin area (Hertfordshire) the Anglian Glaciation removed any clay-with-flints which may once have overlain chalk. Small fissures, probably derived from permafrost (Klinck *et al.* 1998), impart enhanced permeability to the deposit. Some larger fissures have slickensided surfaces probably associated with subsidence movement.

influence on regional hydrogeological regimes has been described in a detailed study of the Sellafield and Drigg areas of northwestern England (McMillan et al. 2000). In such palaeo-environments local thickness reductions of till and clay deposits combine with glaciotectonic deformation to influence groundwater recharge or discharge.

A common feature of glacial deposits is that of re-working. Glaciers typically advance and retreat many times during their life, applying stress to both underlying strata and previously laid-down glacial material, and also supplying large volumes of sediment and melt-water to an area, often within a relatively short timescale. A primary glacial deposit (e.g. lodgement till) may have been partially or totally reworked by secondary glacial processes, such as mass movement and freeze–thaw, thus having its original structure and fabric altered, possibly many times. Some glacial deposits, which have been in direct contact with ice, may contain little clay mineral material despite being clay-sized. Such material is referred to as 'rock flour' and has different geotechnical properties, including low plasticity, from material composed of clay of the same particle size.

6.8. Summary

Clays and mudrocks constitute a significant proportion of British rocks and soils. There are few formations which do not contain some clay material, either as a primary sediment or as a secondary weathering product. Clays are widespread and variable in their nature, reflecting the wide variety of geological provenance and process which they have undergone. They are represented in both solid and superficial deposits, and may comprise an entire formation or only part of it, either as layers or as a constituent of the whole. Geological processes such as induration, glaciation and weathering affect their mechanical properties, and hence engineering behaviour. As such clays and mudrocks are foremost in ground engineering considerations. Clays have the unique property, amongst naturally occurring materials, of changing volume significantly when wetted and dried. They have been used by man for many millennia as a structural material, and are still used today, particularly in Britain where brick-faced dwellings are the norm. Clays act as effective barriers to fluids and are utilized as such, particularly in reworked form, in industry and engineering. The many forms of clay mineral, and the enormous variety of deposits in which they occur, provide continuing challenges and opportunities to engineering, industry and science.

References

ALDISS, D. T. 2002. Geology of the Chichester and Bognor district. Sheet description British Geological Survey 1:50,000 Series sheets 317 and 332 (England and Wales)

AMBROSE, K. 2001. The lithostratigraphy of the Blue Lias Formation (Late Rhaetian – Early Sinemurian) in the southern part of the English Midlands. *Proceedings of the Geologists' Association*, **112**, 97–110.

BELL, F. G. & FORSTER, A. 1991. The geotechnical characteristics of the till deposits of Holderness. In: FORSTER, A., CULSHAW, M. G., CRIPPS, J. C., LITTLE, J. A. & MOON, C. F. (eds) *Quaternary Engineering Geology*. Geological Society, London Engineering Geology Special Publications, **7**, 111–118.

BLOODWORTH, A. J., COWLEY, J. F., HIGHLEY, D. E. & BOWLER, G. K. 2001. *Brick Clay: Issues for Planning*, British Geological Survey, Technical Report No. WF/00/1R.

BONE, D. A. 1992. Temporary exposures in the London Clay around Bognor Regis and Chichester, West Sussex. *Tertiary Research*, **13**, 103–112.

BOWEN, D. Q. 1999. *A Revised Correlation of Quaternary Deposits in the British Isles*, Geological Society, London, Special Reports, **23**.

BRASSINGTON, F. C. 1990. Rising ground water levels in the United Kingdom. *Proceedings of the Institution of Civil Engineering*, **88** (1), 1037–1057.

BRE, 2001. *Special Digest 1: Concrete in aggressive ground* (SD1).

BRISTOW, C. R., GALE, I. N., FELLMAN, E., COX, B. M., with WILKINSON, I. P. & RIDING, J. B. 2000. The lithostratigraphy, biostratigraphy, and hydrological significance of mud springs at Templars Firs, Wootton Bassett, Wiltshire. *Proceedings of the Geologists' Association*, **111**, 231–245.

BRITISH GEOLOGICAL SURVEY, 2002. *United Kingdom Minerals Yearbook 2001*, British Geological Survey, Keyworth, Nottingham.

BROMHEAD, E. N., HUGGINS, M. & IBSEN, M. L. 2000. Shallow landslides in Wadhurst Clay at Robertsbridge, Sussex, UK. In: *Landslides in Theory, Research, and Practice*, Thomas Telford, London, 183–188.

BURLAND, J. B., LONGWORTH, T. I. & MOORE, J. F. A. 1978. A study of ground movement and progressive failure caused by a deep excavation in Oxford Clay. Current Paper 33/78. Building Research Establishment.

BURNETT, A. D. & FOOKES, P. G. 1974. A regional engineering geological study of the London Clay in the London and Hampshire Basins. *Quarterly Journal of Engineering Geology*, **7**, 257–296.

CATT, J. A. 1988. *Quaternary Geology for Scientists and Engineers*, Ellis Horwood Chichester.

CHANDLER, R. J. 2000. Clay sediments in depositional basins: the geotechnical cycle. *Quarterly Journal of Engineering Geology*, **33**, 7–39.

CHANDLER, R. J. & APTED, J. P. 1988. The effect of weathering on the strength of London Clay. *Quarterly Journal of Engineering Geology*, **21**, 59–68.

CHANDLER, R. J. & FORSTER, A. 2001. Engineering in Mercia Mudstone. Construction Industry Research and Information Association, CIRIA C570.

COOPER, M. R., BROMHEAD, E. N., PETLEY, D. J. & GRANT, D. I. 1998. The Selborne cutting stability experiment. *Géotechnique*, **48**, 83–102.

COPE, J. C. W., INGHAM, J. K. & RAWSON, P. F. 1992. *Atlas of Palaeogeography and Lithofacies*, Geological Society, London, Memoirs, **13**.

COX, B. M. & GALLOIS, R. W. 1979. Description of the standard stratigraphical sequences of the Upper Kimmeridge Clay, Ampthill Clay, and West Walton Beds. In: GALLOIS, R. W. *Geological Investigations for the Wash Water Storage Scheme*, Report of the Institute of Geological Sciences, **78/19**, 68–72.

COX, B. M., SUMBLER, M. G. & IVIMEY-COOK, H. C. 1999. *A Formational Framework for the Lower Jurassic of England*

and Wales. (Onshore Area), British Geological Survey, Research Report No. RR/99/01.
CRAIG, R. N. & SIMPSON, B. 1990. Potential problems associated with rising ground water levels in the deep aquifer beneath London. *Tunnel Construction '90. Inst. Mining & Metallurgy*, London, 1–8.
CRIPPS, J. C. & TAYLOR, R. K. 1981. The engineering properties of mudrocks. *Quarterly Journal of Engineering Geology*, **14**, 325–346.
CRIPPS, J. C. & TAYLOR, R. K. 1986. Engineering characteristics of British over-consolidated clays and mudrocks, I. Tertiary deposits. *Engineering Geology*, **22**, 349–376.
CRIPPS, J. C. & TAYLOR, R. K. 1987. Engineering characteristics of British over-consolidated clays and mudrocks, II. Mesozoic deposits. *Engineering Geology*, **23**, 213–253.
CROOT, D. & GRIFFITHS, J. S. 2001. Relict periglaciation: S & E Devon. *Quarterly Journal of Engineering Geology*, **34**, 269–281.
CURRY, D. 1992. Tertiary. *In*: DUFF, P. McL. D. & SMITH, A. J. (eds) *Geology of England and Wales*, Geological Society, London, 389–412.
CZEREWKO, M. A., CRIPPS, J. C., DUFFELL, C. G. & REID, J. M. 2002. The distribution and evaluation of sulphur species in geological materials and manmade fill. *Proceedings 1st International Conference on Thaumasite in Cementitious Materials*, BRE, Garston, UK. June 2002.
DUFF, P. McL. D. & SMITH, A. J. (eds) 1992. *Geology of England and Wales*, Geological Society, London.
DUMBLETON, M. J. & WEST, G. 1966. Studies of the Keuper Marl: Mineralogy. Road Research Laboratory. Ministry of Transport. RRL Report No. 40.
ELLISON, R. A., KNOX, R. W. O'B., JOLLEY, D. W. & KING, C. 1994. A revision of the lithostratigraphical classification of the early Palaeogene strata of the London Basin and East Anglia. *Proceedings of the Geologists' Association*, **105**, 187–197.
ELLISON, R. A. 2000. Geology of the Lambeth Group. *In*: Engineering properties of the Lambeth Group. CIRIA Report No. RP576 (in preparation).
ERBA, E., CASTRODORI, D., GUASTI, G. & RIPEPE, M. 1992. Calcareous nannofossils and Milankovitch cycles: the example of the Albian Gault Clay Formation (southern England) *Palaeogeography, Palaeoclimatology, Palaeoecology*, **93**, 47–69.
FLOYD, M., CZEREWKO, M. A., CRIPPS, J. C. & SPEARS, D. A. 2002. Pyrite oxidation in Lower Lias Clay at Concrete highway structures affected by thaumasite, Gloucestershire, UK. *Proceedings 1st Int. Conference on Thaumasite in Cementitious Materials*, BRE, Garston, UK. June 2002.
FOOKES, P. G. 1991. Quaternary engineering geology. *In*: FORSTER, A., CULSHAW, M. G., CRIPPS, J. C., LITTLE, J. A. & MOON, C. F. (eds) *Quaternary Engineering Geology*, Geological Society, London Engineering Geology Special Publications, **7**, 73–98.
FOOKES, P. G. 1997. Geology for engineers: the geological model, prediction and performance. *Quarterly Journal of Engineering Geology*, **30**, 293–424.
FOOKES, P. G. & BEST, P. 1969. Consolidation characteristics of some late Pleistocene periglacial metastable soils of East Kent. *Quarterly Journal of Engineering Geology*, **2**, 103–128.
FOOKES, P. G. & DENNESS, B. 1969. Observational studies on fissure patterns in Cretaceous sediments in southeast England. *Géotechnique*, **19**, 453–477.
FORSTER, S. C. & WARRINGTON, G. 1985. Geochronology of the Carboniferous, Permian & Triassic. *In*: SNELLING, N. J. (ed.) *The Chronology of the Geological Record*, Geological Society, London, Memoirs, **10**.
FORSTER, A., HOBBS, P. R. N. *et al.* 1995. Engineering geology of British rocks and soils: Gault clay, British Geological Survey, Technical Report No. WN/94/31.

FUNNELL, B. M. & WILKES, D. F. 1976. Engineering characteristics of East Anglian Quaternary Deposits. *Quarterly Journal of Engineering Geology*, **9**, 145–158.
GALE, A. S., YOUNG, J. R., SHACKLETON, N. J., CROWHURST, S. J. & WRAY, D. S. 1999. Orbital tuning of the Cenomanian marly chalk successions: towards a Milankovitch time-scale for the late Cretaceous: *Philosophical Transactions of the Royal Society of London*, **A357**, 1815–1829.
GALLOIS, R. 2000. The stratigraphy of the Kimmeridge Clay Formation (Upper Jurassic) in the RGGE Project boreholes at Swanworth Quarry and Metherhills, south Dorset. *Proceedings of the Geologists' Association*, **111**, 265–280.
GALLOIS, R. & MORTER, A. A. 1982. The stratigraphy of the Gault of East Anglia. *Proceedings of the Geologists' Association*, **93**, 351–368.
GARRETT, C. & WALE, J. H. 1985. Performance of embankments and cuttings in Gault Clay in Kent. *In*: *Failures in Earthworks*, Thomas Telford, London.
GRADSTEIN, F. M. & OGG, J. G. 1996. A Phanerozoic time scale. *Episodes*, **19**, Nos. 1 & 2.
HAMBLIN, R. J. O. & COPPACK, B. C. 1995. *Geology of Telford and the Coalbrookdale Coalfield*, Memoir of the British Geological Survey, parts of Sheets 152 & 153 (England & Wales).
HARRINGTON, J. F., WILLIAMS, L. A., KLINCK, B. A. & WEALTHALL, G. P. 1995. Characteristics of superficial Clay-with-Flint. British Geological Survey, Technical Report No. WE/94/24. March 1995.
HAWKINS, A. B., LAWRENCE, M. S. & PRIVETT, K. D. 1988. Implication of weathering on the engineering properties of the Fuller's Earth Formation. *Géotechnique*, **38**, 517–532.
HOBBS, P. R. N., HALLAM, J. R. *et al.* 2001. Engineering geology of British rocks and soils: Mudstones of the Mercia Mudstone Group, British Geological Survey, Research Report No. RR/01/02.
HIGGINBOTTOM, I. E. & FOOKES, P. G. 1970. Engineering aspects of periglacial features in Britain. *Quarterly Journal of Engineering Geology*, **3**, 85–118.
HIGHT, D. W. & JARDINE, R. J. 1993. Small strain stiffness and strength characteristics of hard London Tertiary clays. *Geotechnical Engineering of Hard Soils—Soft Rocks*, Athens, **1**, 533–552.
HORSWILL, P. & HORTON, A. 1976. Cambering and valley bulging in the Gwash Valley at Empingham, Rutland. *Philosophical Transections of the Royal Society, London*, **A283**, 427–462.
HOWARD, A. S., WARRINGTON, G., AMBROSE, K. & REES, J. G. 1998. Lithostratigraphy of the Mercia Mudstone Group of England and Wales. Discussion document for the BGS Stratigraphy Committee by the BGS Permo-Trias Stratigraphical Framework Committee, British Geological Survey.
HUGHES, D. B., CLARKE, B. G. & MONEY, M. S. 1998. The glacial succession in lowland Northern England. *Quarterly Journal of Engineering Geology*, **31**, 211–234.
HUTCHINSON, J. N. 1967. The free degradation of London Clay cliffs. *Proceedings of the Geotechnical Conference of Oslo*, **1**, 113–118.
JACKSON, J. O. & FOOKES, P. G. 1974. The relationship of the estimated former burial depth of the Lower Oxford Clay to some soil properties. *Quarterly Journal of Engineering Geology*, **7**, 137–179.
JEANS, C. V. 1978. Silicification and associated clay assemblages in the Cretaceous marine sediments of southern England. *Clay Minerals*, **13**, 101–126.
JEANS, C. V., MERRIMAN, R. J., MITCHELL, J. G. & BLAND, D. J. 1982. Volcanic clays in the Cretaceous of Southern England and Northern Ireland. *Clay Minerals*, **17**, 105–156.
JEFFERSON, I., TYE, C. & NORTHMORE, K. J. 2001. Behaviour of silt: the engineering characteristics of loess in the UK. *In*:

JEFFERSON, I., MURRAY, E. J., FARAGHER, E. & FLEMING, P. R. (eds) *Problematic Soils*, Thomas Telford, London.

KING, C. 1981. *The stratigraphy of the London Clay and associated deposits*, Tertiary Research Special Paper 6. Backhuys, Rotterdam.

KLINCK, B. A., HOPSON, P. M., SEN, M. A. & SHEPHERD-THORN, E. R. 1993. *A review of the hydrogeology of superficial clays with special reference to boulder clay and clay-with-flints*, British Geological Survey, Technical Report No. WE/93/21. Sep 1993.

LEE, J. R. 2001. Genesis and palaeogeographical significance of the Corton Diamicton (basal member of the North Sea Drift Formation), East Anglia, UK. *Proceedings of the Geologists' Association*, **112**, 29–43.

LEWIS, J. D. 1998. The engineering implications of a previously unidentified anomalous thickening of Quaternary deposits in the Thames Valley. Geoscience '98. Keele University, April 1998.

LONGWORTH, T. I. 2002. Contribution of construction activity to aggressive ground conditions causing the thaumasite form of sulphate attack to concrete in pyritic ground. *Proc. 1st Int. Conf. on Thaumasite in Cementitious Materials*, BRE, Garston, UK. June 2002.

MARSH, A. H. & GREENWOOD, N. R. 1995. Foundations in Gault Clay. In: EDDLESTON, M., WALTHALL, S., CRIPPS, J. C. & CULSHAW, M. G. (eds) *Engineering Geology of Construction*, Geological Society, London Engineering Geology Special Publications, **10**, 143–160.

McGOWN, A. 1971. The classification for engineering purposes of tills from moraines and associated landforms. *Quarterly Journal of Engineering Geology*, **4**, 115–130.

McGOWN, A., SALDIVAR-SALI, A. & RADWAN, A. M. 1974. Fissure patterns and slope failures in Till at Hurlford, Ayrshire. *Quarterly Journal of Engineering Geology*, **7**, 1–26.

McKERROW, W. S. 1981. Distribution of Lower Palaeozoic mudrocks. *Quarterly Journal of Engineering Geology*, **14**, 245–251.

McMILLAN, A. A. & HAMBLIN, R. J. O. 2000. A mapping-related lithostratigraphical framework for the Quaternary of the UK. Quaternary Science Reviews.

McMILLAN, A. A., HEATHCOTE, J. A., KLINCK, B. A., SHEPLEY, M. G., JACKSON, C. P. & DEGNAN, P. J. 2000. Hydrogeological characterisation of the onshore Quaternary sediments at Sellafield using the concepts of domains. *Quarterly Journal of Engineering Geology*, **33**, 301–323.

MOORLOCK, B. S. P. & HIGHLEY, D. E. 1991. An appraisal of fuller's earth resources in England and Wales. British Geological Survey Technical Report No. WA/91/75.

MORRIS, K. A. 1980. A comparison of major sequences of organic-rich mud deposition in the British Jurassic. *Journal of the Geological Society, London*, **137**, 157–170.

MORTIMORE, R. N. & POMEROL, B. 1996. Chalk marl: geoframeworks and engineering appraisal. In: HARRIS, C. S., HART, M. B., VARLEY, P. & WARREN, C. (eds) *Engineering Geology and the Channel Tunnel*, Thomas Telford, London, 455–466.

NICHOLLS, R. A. 1994. M40 Oxford-Birmingham: geology and geotechnics of the Waterstock-Banbury section. *Proc. Instn. Civ. Engrs. Transp.*, **105**, Nov., 283–295.

NORTHMORE, K. J., BELL, F. G. & CULSHAW, M. G. 1996. The engineering properties and behaviour of the brickearth of South Essex. *Quarterly Journal of Engineering Geology*, **29**, 147–162.

PENN, I. E. & WYATT, R. J. 1979. *The stratigraphy and correlation of the Bathonian strata in the Bath-Frome area*. Report of the Institute of Geological Science, London, **78** (22), 23–88.

PITTS, J. & BRUNSDEN, D. 1987. A reconsideration of the Bindon Landslide of 1839. *Proceedings of the Geologists' Association*, **98**, Part 1.

PUGH, R. S., WEEKS, A. G. & HUTCHINSON, D. E. 1991. Landslip and remedial works in Wadhurst clay. In: CHANDLER, R. J. (ed.) *Slope Stability Engineering: Developments and Applications*. Thomas Telford, London, 377–382.

RAMSBOTTOM, W. H. C., SABINE, P. A., DANGERFIELD, J. & SABINE, P. W. 1981. Mudrocks in the Carboniferous of Britain. *Quarterly Journal of Engineering Geology*, **14**, 257–262.

RUSSELL, D. J. & PARKER, A. 1979. Geotechnical, mineralogical, and chemical interrelationships in weathering profiles of an over-consolidated clay. *Quarterly Journal of Engineering Geology*, **12**, 107–116.

SIMPSON, M. I. 1985. The stratigraphy of the Atherfield Clay Formation (Lower Aptian, Lower Cretaceous) at the type and other localities in southern England. *Proceedings of the Geologists' Association*, **96** (1), 23–45.

SKEMPTON, A. W. 1964. Long term stability of clay slopes. Rankine Lecture. *Géotechnique*, **14**, 77–102.

SMITH, D. B. 1986. Permian. In: HARWOOD, G. M. & SMITH, D. B. 1986. *The English Zechstein and Related Topics*, Geological Society, London, Special Publications, **22**, 275–305.

SMITH, D. B. 1995. *Marine Permian of England*, Joint Nature Conservation Committee. Chapman & Hall, London.

SMITH, D. B., HARWOOD, G. M., PATTISON, J. & PETTIGREW, T. H. 1986. A revised nomenclature for Upper Permian strata in eastern England. In: HARWOOD, G. M. & SMITH, D. B. (eds) *The English Zechstein and Related Topics*, Geological Society, London, Special Publications, **22**, 9–17.

TAYLOR, P. D. (ed.) 1995. *Field Geology of the British Jurassic*, Geological Society, London.

TOGHILL, P. 2000. *The Geology of Britain*, Swan Hill Press.

TRENTER, N. A. 1999. *Engineering in Glacial Tills*, Construction Industry Research and Information, CIRIA Report No. C504.

WALTHAM, A. C. & DIXON, N. 2000. Movement of the Mam Tor landslide, Derbyshire, UK. *Quarterly Journal of Engineering Geology*, **33**, 105–123.

WARRINGTON, G. & IVIMEY-COOK, H. C. 1992. Triassic. In: COPE, J. C. W., INGHAM, J. K. & RAWSON, P. F. (eds) *Atlas of Palaeogeography and Lithofacies*, Geological Society, Memoir, **13**, 97–106.

WEEDON, G. P., JENKYNS, H. C., COE, A. L. & HESSELBO, S. P. 1999. Astronomical calibration of the Jurassic time-scale from cyclostratigraphy in British mudrock formations. *Philosophical Transactions of the Royal Society*, Ser. A, **357**, 1787–1813.

WILSON, A. A. 1987. Geological factors in land-use planning at Aldridge-Brownhills, West Midlands. In: CULSHAW, M., BELL, F., CRIPPS, J. & O'HARA, M. (eds) *Planning and Engineering Geology, Geological Society, London Engineering Geology Special Publications*, **4**, 87–94.

WILSON, A. A. 1993. The Mercia Mudstone Group (Trias) of the Cheshire Basin. *Proceedings of the Yorkshire Geological Society*, **49**, 171–188.

WINCHESTER, S. 2001. *The Map that Changed the World*, Penguin Viking.

ZALASIEWICZ, J., SMITH, A., BRENCHLEY, P., EVANS, J., KNOX, R. et al. 2004. Simplifying the stratigraphy of time. *Geology*, **32**, 1–4.

7. Exploration

7.1. Introduction

This chapter on the exploration for and of clay deposits for man's benefit covers a range of topics. It is essential to comprehend the difference between a clay deposit and a reserve. In order for an exploration programme to provide the information required in an efficient manner, the data collection needs careful planning. This process is best carried out in stages, with the plans for later stages depending on the results of earlier stages. The critical importance of the exploration being led by a qualified geologist who has relevant experience in the type of deposit under study is emphasized. Methods of non-intrusive investigation, used before significant expenditure on field work, are summarized, followed by a review of intrusive methods of investigation. This coverage is brief, not repeating the coverage of exploration techniques covered in most textbooks and Standards on field investigations, for instance Clayton *et al.* (1988) and BS 5930:1999. The use of geophysical methods to enhance the drilling programme is discussed, again emphasizing the need for competent persons, and sampling programmes and sample types are reviewed. The importance of iterative reporting of the phases of the investigation is threaded through the presentation, as it would be in the development of the assessment of the clay deposit.

The purpose of geological exploration is progressively to build a three-dimensional model of the ground, including the relevant location and characteristics of the mineral strata and the groundwater appropriate to the extraction of the clay mineral. In the context of mineral exploration, the requirement of the investigation process is to establish whether minerals of suitable quantity and quality are economically accessible. Clay strata are widely present as resources but, bearing in mind the generally low value of bulk clay, few of the resources investigated will turn out to be reserves. The combination of overburden removal and storage, and infrastructure factors such as transport logistics can combine to render the economics of a deposit unfavourable. The low unit value of most heavy or structural clays mean that they usually need to be worked and processed close to their point of sale. However, fine ceramic clays for whiteware applications have an intrinsically greater value and the international sales demand can be extensive. Similarly montmorillonite clays such as bentonite (including Fuller's Earth) are processed and exported to markets around the world.

The relationship between resources and reserves (see Text Box below) and the various international definitions are summarized in Smith (1999) and Anon (2001). Mineral resources can be estimated on the basis of geological information, with limited input from other disciplines. Mineral reserves are a subset of resources, definition of

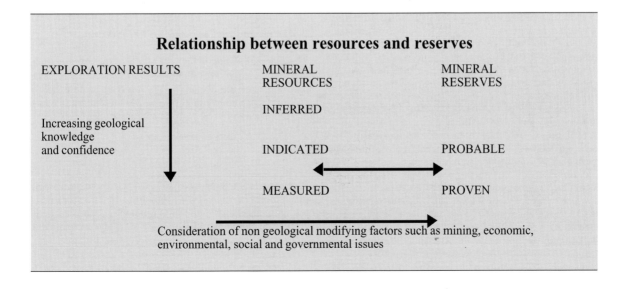

which requires consideration of modifying factors, mainly non-geological.

7.1.1. The exploration cycle

Exploration for a clay deposit should be divided into a number of steps:

- locate the clay deposit as a target for exploration;
- define the key properties which will control the utilisation and application of the raw material;
- investigate the deposit to determine its extent, quality and variability;
- evaluate the costs of material extraction, including the need to maintain a blended feedstock;
- cost the clay processing and transport of product to the market;
- carry out a market survey to establish predicted sales demand and growth;
- establish the environmental costs of extraction, including disposal of any waste, balanced against the costs and benefits of restoration.

These steps should be carried out in an iterative, structured and phased manner, based on the above sequence. The costs of the various studies and investigations necessary at each iteration can be substantial, and resource investigations often fail to confirm the presence of reserves for any of a number of reasons. The early stages of an investigation incorporate desk studies and data reviews, and so are relatively low cost. Once field investigation teams and drilling rigs are mobilized, the unit costs rise markedly, and it is important to make sure that the work of these teams is well targeted.

At all stages, and particularly in the early stages of investigation, close co-operation within a team of experts is essential to ensure that realistic cost estimates are available. These cost estimates will incorporate geological factors together with extraction, environmental, processing, plant and transport costs as well as finance costs balanced against expected market demand. An important element that needs to be included is consideration of the geotechnical aspects of extraction; too little attention is often paid to the costs that can arise when stable side slopes and efficient groundwater control are not achieved. An unstable pit is unlikely to be an economic pit (see Fig. 7.1). On the other hand, a conservatively designed pit will reduce profits.

At the completion of each stage of the investigation process, the continued viability of the site, and thus the investigation itself, needs to be considered carefully. This is conducted through a formal process of project management and control. Reports should be prepared by the team, and reviewed by managers and experts external to the team, in order to maintain the critical element of objectivity. The results of each stage of investigation should also revisit the conclusions from each previous stage, in order to update the ground and cost models. Any contradictions in the information and the model will need to be explained. In addition the assessed level of confidence in all aspects of the ground and economic models should be

FIG. 7.1. Pit instability affecting production.

defined. Identifying areas of uncertainty is one of the most important aspects of any geological investigation. The objective of each successive stage of investigation is to reduce such areas of uncertainty by providing more detailed models, built with progressively higher confidence levels. There should not be major surprises in the late stages of an investigation if this phased approach has been carried out successfully, a process that requires logical and thorough execution at all stages.

7.1.2. Standards relevant to exploration

The Standards under which the whole of the investigation is to be carried out should be defined at the outset. The scope of the Standards that need to be referenced includes drilling and field procedures and laboratory test specifications as well as financial assessment (see Text Box on p. 179).

7.1.3. Personnel

It is important that appropriately trained and experienced personnel are used on field investigations. The investigating team should as a minimum include a qualified geologist, for example a chartered or professional geologist from a professional body in UK (GSL), Ireland (IGI), Canada (CCPG), USA (AIPG) or Spain (ICOG). This qualified person will also need to be experienced in

> ### Standards used in clay exploration
>
> It is not appropriate to list those Standards which are normally considered relevant in matters relating to clay exploration, as the subject area is so wide, including investigation procedures, drilling, sampling, and laboratory testing. In addition there are Codes of Practice on reporting mineral exploration results, such as the European Code (Anon 2001).
>
> Publishing bodies include the International Standards Organisation (ISO), and for Europe the Comite Européen de Normalisation (CEN). These two bodies work closely together under the terms of the Vienna Agreement which provides mechanisms for avoiding duplication of work items. In addition each country has its own Standards organisation such as BSi in UK, DIN in Germany and American Society for Testing and Materials (ASTM) in the USA. National Standards are increasingly being required to fall in line with International Standards, a process which will take a good number of years to complete. In the meantime, individuals should be clear as to which Standards work apply to the work being carried out.

the particular clay product and geological setting under investigation. The definition of such a competent person is given in Anon (2001), which definition includes specific reference to the awarding body having enforceable rules of conduct (see Text Box below).

In addition to the team being led by a competent person, it is also important that all members of the team have sufficient and relevant experience to enable them to fulfil their roles. For instance, if geophysics is to play an significant role in the exploration programme, the importance of including a geophysical specialist to advise on the selection and deployment of appropriate techniques, as well as in the interpretation of the data, cannot be over-emphasized. This requirement should also extend to the geologist carrying out desk studies, walk over surveys and supervising drilling, as well as the driller and any field and laboratory technicians working in the team. Definitions of appropriate individual's qualifications to act as principal adviser and specialist are given in SISG (1993).

It is also important that those supervising the investigation work in the field should fully understand the purpose of the studies. This will often mean that the field supervisor will be a member of the design team, with sufficient knowledge of the ground model to be able to react to findings on site and maximize the benefit gained. Investigations carried out blindly to a scope written in advance are likely to waste significant amounts of money and time. More importantly, such shortfalls are likely to leave key questions unanswered, or at least not answered with sufficient confidence levels.

7.2. Non-intrusive exploration

Ground investigation consists of both non-intrusive and intrusive forms of exploration. The former is relatively inexpensive, and requires little attendance on and disruption to the site. The latter consists of the drilling of boreholes, sampling of materials and performance of in situ tests. Such activities are expensive to carry out and so need to be carefully designed and executed to ensure the benefit from the expenditure is realised. Such designs need to be based on the non-intrusive desk studies, walk-over visits and field mapping.

7.2.1. Desk studies

Field exploration by intrusive methods is expensive compared with the considered use of geological experience. The application of trained geoscientists in collecting available data is normally highly cost-effective. It is probable that there are relatively few mineral deposits, at least in the developed world, that have not already been investigated for some reason. Although this may have been for a completely different end use and may or may not have

> ### Definition of a competent person (Anon 2001)
>
> A competent person is a corporate member of a recognised professional body relevant to the activity being undertaken, and with enforceable rules of conduct. A competent person should have a minimum of five years experience relevant to the style of mineralization and type of deposit under consideration. If the competent person is estimating or supervising the estimation of mineral resources or mineral reserves, the relevant experience must be in the estimation, evaluation and assessment of mineral resources or mineral reserves respectively.

included intrusive investigations, there may nevertheless be some appropriate data arising. These data are normally available at low cost and can be very significant and useful, particularly in the early stages of the investigation process when site selection is determined. A useful summary of sources of information for such studies in the UK is given in Perry & West (1996).

The aim of a desk study is to access and evaluate all the readily available information (see Text Box below). Decisions need to be made as to which of the information available is relevant and to what constitutes the limit of something being readily available. Some potentially useful data may become available only after incurring higher than normal costs or after longer delays associated with overcoming confidentiality issues and data ownership. The investigator will need to make decisions on these matters on a case by case basis. As a rule of thumb, it is usually better to initiate collection of all available data, even at higher cost or delay, as crucially important items of information can arrive from the most unexpected quarters. A previous investigation report that arrives after some delay may still contribute significantly to the completeness of the overall study, even when the investigation has moved on to a subsequent stage. Negative information can also be very important; for instance to receive confirmation that no investigation had been carried out in a key area could be very helpful.

The key data sources that start most investigations are the topographic and geological map collections. Most of the earth's surface has now been mapped at some scale, although not necessarily at a large enough scale for mineral investigations. Map collections are housed in national collections, Geological Surveys, and universities throughout the world. In addition, there are usually geological reports to support these maps, as well as articles and papers in the technical literature and doctoral theses. Access to these is variable, depending on the location of the records, and their perceived value, often in security terms. The availability of such records through on line and CD-ROM based databases is increasing, but we are a long way from no longer having to make personal visits

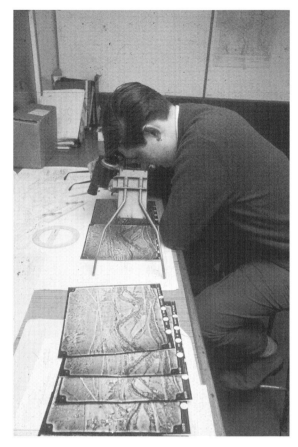

FIG. 7.2. Desk study activities, examining ariel photographs.

to libraries, departments and offices to obtain this basic information. It is however true to say that no geologist need go out into the field without at least some broad indication of the geology that they will encounter and with the first iteration of the ground model already established (see Fig. 7.2).

Generic sources of geological information

- Geological Survey publications, maps, memoirs;
- mining records;

- local and university libraries;
- doctoral theses and Master level dissertations, and supervisory academic staff;
- local knowledge, hearsay evidence;
- historical accounts of life and society, including building materials, water supply;

- historical hand held and professional photographs;
- aerial photographs;
- satellite images;
- airborne remote sensing surveys, such as gravity and magnetic contoured data.

The geological publications will usually indicate likely areas of clay material availability. However, the indigenous population will normally have already found and exploited minerals albeit at a small scale. The evidence of such activities could exist on maps, either as irregularities of surface form or in place names. Local libraries usually hold publications of local historical societies, which may have recorded previous mineral extraction activities; indeed, local library staff often have a personal interest in such subjects. County record offices and museums may also provide valuable data on historical extraction sites and activity.

Remote sensing techniques are increasingly being used in geological exploration. They include a wide range of products, several of which are potentially useful in the search for minerals (Gibson 2000). Due to the low permeability and relatively low resistance to erosion of clay materials, such deposits tend to be associated with areas of low relief and poor drainage. Land forms of this type may be delineated from satellite images at a regional scale, especially when using modern digital methods of image enhancement and interpretation tools. Further developments in such data processing are anticipated, which should result in satellite imagery providing useful data for site scale studies.

Remote sensing techniques may be applied in all parts of the world, even helping to identify sources near built-up areas (see Fig. 7.3). They are more particularly useful however in non-developed areas, especially those lacking topographical or geological maps at suitable scales or aerial photography cover (Rencz & Riverson 1999). The image in Figure 7.3 covers an area on the outskirts of Las Vegas and illustrates these points nicely; relict, buried fluvial channel systems can be seen, giving an indication of where to look for fine and coarse soils, and groundwater flows within the channel systems.

However, because of their low unit price, the demand for clays is normally centred around established areas of population. The need is therefore usually for large-scale information which can most readily be obtained from aerial photographs, rather than imagery from satellite based platforms. The interpretation of aerial photographs requires experience which needs to be developed, preferably by instruction and example with a skilled interpreter. A good quality stereoscope is a great help, although pocket stereoscopes are often adequate for initial views (see Fig 7.2). Advice and guidance on aerial photograph interpretation is given in Dumbleton & West (1970). Interpretation of the aerial photographs, in conjunction with large-scale maps, is to be recommended for a significant amount of relevant high-quality information that can be obtained relatively cheaply (Paine 2003). Topographic maps alone are not able to show the subtle landforms that, to the trained eye, indicate the underlying geology. Aerial photographs have the additional advantage that evidence of previous local extraction activities can usually be identified, contributing significantly to the knowledge of the area (see Fig 7.4). Transfer of this combined information onto tracing sheets or onto the actual map forms the base plan for the geologist to then carry out the walk-over survey and mapping discussed below.

The utilization of other aircraft-borne sensing systems, which are predominantly geophysical can also be beneficial (see also Section 7.3). Airborne multi spectral scanners and thematic mappers provide a good way of identifying different types of clay, relying on the varying spectral characteristics of different clay deposits. However, these methods are regional, or at least small scale, rather than site specific, and so their use in identifying clay deposits is often limited in the developed world. They provide markedly greater potential for location of clay deposits in non-developed parts of the world, particularly where vegetation cover is limited (Gibson 2000).

7.2.2. Walk-over surveys

The aim of the walk-over survey by the geologist is to establish the ground truth for information obtained to date and to achieve calibration with the local geology. The walk-over survey can also incorporate the planning of the next stage of the investigation, which can include knowledge on practical aspects such as access for drilling rigs and sensitivity of the local environment to disturbance. The disposal of mud slurries from a drilling programme can be considered, as can the potential disturbance to agricultural land and activities.

FIG. 7.3. Satellite platform image of clay deposits near Las Vegas.

Fig. 7.4. Air photo showing clay terrain.

In the desk study stage, background information will be gained on the geology. The walk-over enables the geologist to expand this knowledge, and identify any further gaps. The aim at this stage is to obtain a broad but accurate picture of the geology, without getting too involved in the small-scale aspects that could obscure the whole picture. A thorough understanding of the geology is required to establish the extent, location and quality of the clay deposit.

The overall character of the sequence needs to be clearly understood, as does the small-scale detail of the consistency or variability of the clay. This understanding needs to include appreciation of impurities within the sequence which may affect its value and application. The definition, importance and impact of impurities will depend on the market and application of the clay. Impurities within the clay material can include sand, lignite, carbonates, sulphides and soluble salts. In addition, it may include non-economic interbeds, which may complicate mineral extraction, or varying colours, which may enhance or hinder the sale price of the finished product. As blending of materials from different parts of the sequence is commonly required, this requires close liaison with the process specialists, and an appreciation of the market value of the various products. This consideration will include the means and cost of disposal of waste materials that are not suitable for use in the manufacturing process.

The walk-over survey (see Fig. 7.5) should visit all the previous extraction sites to identify what was extracted and (if possible) where it was used. This information will

Fig. 7.5. Walk-over survey in desert conditions.

provide useful background on matters such as clay quality and colours, durability of products and general suitability for the proposed application. This latter information will usually come from looking at buildings in the area, and from examining field boundaries and walls around estates or parks, which will all indicate a range of products obtained locally.

Information on environmental issues can also be obtained at the walk-over stage. The requirements for environmental protection increase every year, as evidenced by the position taken by the regulatory authorities at local and national levels. In developed areas, the suitable clay reserves are usually under farmland or recreational land and they may also occur within natural or water protection areas. There are liable to be objections to any extractive proposal, even when land remediation is part of the proposal, whenever extraction activities are proposed close to centres of habitation. This can occur even when the investigation team demonstrate from the outset that they understand and appreciate the existence of these issues, and are willing to resolve them in a sympathetic manner. The investigators have to be sure to compile comprehensive and high-quality scientific data to stand any chance of dealing with attempts to stop the development using less rigorous approaches.

Although the geologist doing the walk-over will not necessarily be a naturalist, he will be expected to highlight areas of potential concern, e.g. ancient woodland or wetland. The location of nearby residences, roads and public amenities needs to be established as these will influence the siting and extent of landscaping requirements, e.g. tree screens and bunds. The geologist is also usually one of the first to walk the site, and so will be in a position to gauge the local feeling for the proposed extractive activity. At the very least, the geologist must ensure that the situation is not inflamed by insensitivity, and that rumours are not generated. The landowner and/or occupant must have been informed why the geologist is there and have given permission before entering the land.

Hydrological information on surface-water courses and underground water will be important at this stage. There is always a real danger of polluting water courses during exploratory drilling activities and during exploitation of a deposit, and any implications need to be assessed in advance, not afterwards. There is often a need to divert streams during pit development which needs careful planning. If groundwater will need to be pumped out of the pit in any quantity, this may affect the final product cost and thus the viability of the whole operation.

Conservation areas can add to the cost of pit development or, at worst, sterilize certain parts of the clay reserve. Conservation areas in the UK can include National Parks, designated sites of special scientific interest (SSSI) or regional geological importance (RIGS). The status of any such sites needs to be assessed in order that the developer can be proactive in restoring any affected areas to a standard compatible with that existing before mining operations. The cost of such restoration, if involving say wetlands, may exceed the value of the mineral released.

Photographic records are advisable to establish a comprehensive record of the site and its environs. In typical explorations many sites may be investigated to some extent, but the investigation might not be concluded for all sites, concentrating on only one or two locations initially, with the preference for the most likely sites changing with time and the level of information available. Investigation of possible sites may also be suspended due to external factors such as changes in market conditions.

An album of annotated good quality photographs will be invaluable when investigation is resumed at a later date, saving significant time in revisiting old ground. Further, it is valuable to provide an archive record of the state of the site before commencing any exploration or extraction activities. These should include a record of the current state of farmland, other landscape features, and public roads or tracks, which may be affected by even the drilling rigs, and investigation plant. Consideration should also be given to recording the condition of houses or other buildings which may be affected by settlement, dewatering or blasting; in practice formal condition surveys would normally be carried out at a much later stage, but an early photographic record of the condition of structures can be valuable in avoiding arguments later.

Site infrastructure around the site, although not normally part of the geologist's remit, will be one of the significant factors in determining the viability of the mining or quarrying operation. The necessity to construct significant lengths of road will affect the economic case in the assessment of the reserve. In addition, the availability of a suitable work force and their ability to get to the site are important.

Although road access to the site is important, in many countries the availability of rail and port facilities are also a vital consideration in gauging the viability of the project. Other factors including the availability and reliability of electricity and water supplies can also have a significant impact on costs.

7.2.3. Field mapping

The basis for the geological model of the site, and the initial economic assessments that follow, spring from the relevance and quality of the geological map of the area. If available, published maps would have been used at the desk study stage, but there is now a need for the site geologist to compile a larger scale evolution of this, which should include all the information relevant to the particular clay being sought.

The base map should synthesize the relevant parts of a good topographic map, with the geological information and data transcribed from the aerial photographs. In areas less well served in such resources, aerial photographs or satellite images may serve as the base map. If these do not exist at a suitable scale, basic topographic surveying and geological mapping are needed to provide a map.

The use of modern surveying equipment assists greatly in these surveying tasks. The total stations now available enable rapid position fixing within sight lines without the need for a separate surveyor. The data are recorded electronically and can be downloaded into a drawing package. The map can therefore evolve whilst still on site.

On larger sites, the use of Global Positioning (GPS) to aid location of survey positions is a practical aid. Hand-held receivers, which are comparable in size to a mobile cellular phone, have a positional accuracy of about ten metres. The establishment of a local slave station enables the positional accuracy to be improved of the order of a metre. These degrees of accuracy are quite sufficient for geological surveying, and links can be established to portable, 'field-proof' computers for electronic capture of field notes and geological line work at the exposure. When using GPS receivers (see Fig. 7.6), the operator needs to be conversant with the latitude and longitude in his site area, as well as the details of the reference spheres used to provide the instrument readout.

Geological mapping principles will not be covered here, as they are still taught on many undergraduate geology courses around the world. The primary requirement for mapping a clay mineral resource is to have a very clear understanding as to what information is needed to facilitate the progress of the investigation steps described in Section 7.2.1. The potential importance of changes in the character of the clay have been mentioned above, but other aspects may need to be examined in the field.

A systematic approach to the recording of information is essential, to ensure that a full interrogation at any subsequent stage of the investigation is possible. It is important to record all information that may be of interest while at the exposure. It is easy to discard spare information, but it is difficult and expensive to go back to record something that has been missed. Where exposures are not available, they may need to be created using excavations (by hand or by using a hydraulic excavator), hand auger or lightweight drill. These aspects are covered in Section 7.5 below under intrusive investigations.

Many geological investigations involve more than one geologist. It is therefore imperative that all members of the team use the same terminology and present their information in an agreed format. The terminology used can either be in accordance with a national standard (e.g. BS 5930:1999 or ISO 14688 (Part 1, 2002, Part 2 in press) and 14689 (in press) which include detailed descriptive schemes for soils and rocks respectively), or locally determined, see Section 4.2.2. For instance the colour of clays for use in bricks is a key feature, and consistency of terminology is more important than is normal in say an engineering investigation. International colour charts such as those published by the Geological Society of America (Munsell) can be used. Alternatively, reference samples or colour charts from any convenient source can provide the basis of a site specific consistent scheme.

Discontinuities and geomorphological data recording is especially important in clay deposits, as these features can be critical in pit slope design. The discontinuities that need to be examined are of various geological origins. In any stiff (overconsolidated) clay, there will be fissures due to burial and exhumation of the strata, which are synonymous with joints. In many clay deposits in the northern hemisphere, periglacial processes have superimposed different types of discontinuity including desiccation cracks and solifluction shears. In addition, mass movements under periglacial conditions resulted in the clays being permanently reduced in mass strength. The particular features that may be observed, and the associated weathering effects are summarized in Spink & Norbury (1993); an example of the result of not assessing such features occurred at Carsington Dam on residual Carboniferous Mudstones (Skempton et al. 1991).

The data on discontinuities that need to be recorded is the same as that for rocks, and is given in BS 5930:1999 or in ISRM (1978) recommended procedures. Natural exposures of clay that allow discontinuity recording are rare, and therefore exposures need to be created. Particular care in logging such artificially created exposures is necessary as the information to be recorded will not be readily observable in the wet clay.

The effect of Quaternary processes on clay strata is unique, but allows many geomorphological aspects of the landscape to be interpreted by the trained eye (see Fig. 7.7). The presence of relict mass movements, which are common for instance in the southern part of Britain, firstly helps to identify the presence of clay strata at the surface, and secondly indicates that caution would need to be exercised in any excavation of such materials. The

FIG. 7.6. Field surveying with GPS receivers.

FIG. 7.7. Relict mass movements in a clay slope.

design of safe slopes in such a material would normally result in slopes too shallow to allow economic mineral extraction.

7.2.4. Preliminary assessment report

At one or more convenient stages during the course of these preliminary studies the findings of the investigation team should be formally reported. The discipline this requirement imposes is in properly collating all the information collected, in interpreting the data to provide estimates of the reserve and its accessibility, in identifying areas of uncertainty and in making formal proposals for the subsequent stages of the investigation.

This report will not only present the detailed strategies and costs for the next stages, but also outline programmes of investigation and budget costs for the whole of the remaining investigation exercise. In order to achieve this the report will normally include outline designs of the pit layout, processing plant and infrastructure.

This preliminary report would normally be a company confidential document, which makes its preparation all the more important, providing an opportunity for the investigating team to be subject to peer group review. In addition to being a report to the company board who are funding the investigation, the preliminary report can also be used for interim briefing of outside parties including financiers and shareholders. This means that the proper planning and thorough execution of the preliminary studies will be put fully under the spotlight.

It is also important that this preliminary report is prepared because, up to this stage, the investigation costs would have been modest. From this point, the unit rates for the individual elements of investigation are much higher and, in particular, significant numbers of sampling points will be required to fully assess the mineral deposit. Subsequent investigation stages involve significant expenditure.

7.3. Geophysical investigations

The importance of geophysics in the identification of a clay mineral deposit is high. In most circumstances clays give marked and distinctive response signatures, and so deposits are readily detected using geophysical mapping or logging. However, geophysics is best used in conjunction with direct observations of exposures or borehole cores made by a geologist and physical, chemical or mineralogical testing of the samples taken (BS 5930:1999). The inclusion of a properly planned programme of geophysics should provide the investigating team with greatly enhanced confidence that the direct observations made can be interpolated between the investigation points, and that the uniformity or variability of the deposit is quantified. Without the help of geophysics, understanding the three-dimensional form of the deposit is much more uncertain. The usefulness of geophysics is further enhanced in investigating a low unit value mineral like clay as its cost is relatively low, although the comments below and in BS 5930 (1999) on planning and supervising an investigation should be noted. An up-to-date review of the use of geophysics in the context of engineering investigations is given by McDowell *et al.* (2002).

7.3.1. Geophysical mapping

The use of geophysics in the early stages of an investigation as a mapping tool to delineate the extent of the clay deposit is recommended. However, it needs to be borne in mind that the depth of penetration of surface mapping arrays is limited, depending, *inter alia*, on the groundwater conditions. On the other hand, carrying out investigations by borehole geophysical methods considerably increases the costs.

There are a large number of variables involved in setting up and interpreting a geophysical survey. It is therefore essential that the survey is designed and supervised by a suitably experienced and qualified specialist. Unfortunately, due to a lack of good advice and forward planning, many geophysical surveys have not provided the information required.

There are a large number of geophysical surveying techniques available, which are based on different theoretical principles, such as seismic velocity or electrical resistivity. The interpretation is an iterative process that requires some initial knowledge of the geology. Detailed interpretation also requires direct geological control in the form of boreholes or trial pits. Carrying out a geophysical survey on its own without any such control would produce a result of limited reliability and thus of limited value.

It is normally advantageous to use geophysics in the early stages of field investigation (see Fig. 7.8). A reconnaissance survey can be used effectively to identify both the general conditions and any anomalous areas of the site, and the results used to site the relatively expensive boreholes. Subsequently, additional surveys can be carried out to provide more thorough exploration between the boreholes as an aid to borehole interpolation. The performance of all geophysical methods is influenced by four controls, namely depth penetration, vertical and lateral resolution, signal-to-noise ratio and the contrasts in physical properties, for example electrical resistivity or density. These controls are intimately linked, in that one improves at the expense of another.

The best results in geophysical mapping are obtained when the conditions are reasonably uniform, with clear contrast in relevant physical properties. The effect of clay deposits below the water table is to absorb much of the input energy, thus limiting the depth penetration and the ability to locate the lower boundary of the clay. Geophysical mapping techniques used for clay are best at defining weathering profiles, determining the thickness of any superficial deposits, and locating any erosional or structural interruptions to the clay continuity. They are not usually successful in identifying inhomogeneities within the deposit itself.

Electrical methods utilize an electrical current passed into the ground by a pair of electrodes which gives a potential difference between similar measuring electrodes from which the resistivity of the ground can be calculated. Changing the pattern of both the input and the measuring electrodes allows investigation of the variation of resistivity with depth. This variation can be mapped laterally and vertically and then related to the geology. Geophysical mapping can be used to measure the geo-electrical stratification of the ground (soundings or depth probes), lateral changes (constant separation traverses) or local anomalies (equipotential surveys). Computer controlled multiple electrode techniques are under development to produce

FIG. 7.8. Mapping survey using geophysical methods.

three-dimensional electrical image (tomogram) of the ground. The model can be adjusted to achieve a best fit with the field data by using iterative processing. Care is required however to take account of interactions; for instance depth probing utilises a progressively wider electrode spacing which could invoke lateral variation. Soundings are also preferably carried out away from the location of man-made linear features such as roads, water mains, metal fences and power lines. As with any geophysical method, sounding does not always produce a definite solution, at least in part because the theoretical models used for interpretation are idealized and may not mirror the actual geological conditions.

An alternative method for use in reconnaissance surveys is the measurement of ground electrical conductivity. A transmitter coil is placed on the ground surface and charged with alternating current. The resulting induced currents generate a secondary magnetic field that is sensed by the receiver coil. By varying the operating frequency, intercoil spacing and coil orientation, the ground conductivity, and thus resistivity, can be determined over a range of depths, and thus the geological layering can be deduced, but to a less refined extent than by depth probing sounding. Penetrations up to about 60 metres can be achieved but the great advantage of the system is that it can be operated by one person making it a rapid and cost effective reconnaissance tool.

Seismic refraction methods use a variety of tools to generate the required energy input, ranging from hammers to detonators or explosives; the source used depends on the locality and the amount of energy required. The waves generated travel through the surface layers and are refracted at material boundaries. Material boundaries in this context are usually contrasts in mass density. Interpretation provides layer thicknesses and velocities, to a reasonable level of confidence if the ground suits the idealized model and the geology is not unduly complex. However, the method is not well suited to investigating complex geology, and can also give misleading results if a lower velocity layer underlies a higher velocity layer. The measured seismic velocity can be usefully correlated with a number of geotechnical properties including the ease of excavation or rippability (see Chapter 11).

7.3.2. Borehole geophysical logging

It makes good sense to maximize the amount of information recovered once money has been spent on drilling boreholes. Downhole geophysical logging produces a profile down the borehole for each geophysical tool (sonde) deployed. The data produced against depth provide continuous records of the ground profile and particular combinations of sondes can yield enhanced results. Experience of the use of these methods is increasing rapidly and they are now used more or less routinely. There are a very wide range of tools available for a multiplicity of applications, and expert advice is necessary if sensible combinations of sondes are to be deployed for the problem in hand.

If properly used, downhole logging can avoid the use of expensive core drilling in some cases and significant time savings can also accrue. The first boreholes on any site should be drilled to recover high-quality cores. These cores are necessary to provide the ground truth for interpretation of geophysical mapping and logging. The cores also provide samples for laboratory testing and thus calibration of the geophysical sondes. However, once sufficient confidence has been obtained in the results of the geophysical logging, a large majority of the remaining boreholes could be put down by open hole drilling methods. These holes should be cheaper to drill, avoid the need to handle and store extensive numbers of core boxes and are usually significantly faster to drill. This is not standard practice in the evaluation of say most ceramic clay deposits. Coring is usually necessary to establish detailed variations in certain key properties, which (depending on application) include rheological properties, strength, subtle variations in fired colour, impurities etc.

The tools can provide information on the formation around the borehole, the fluid in the borehole, the physical properties of the ground mass around the borehole and details of the borehole construction. For formation studies, the natural gamma and electrical logs are most commonly used to provide stratigraphical correlation across a site, whilst formation porosity can be derived from neutron or sonic logs. Geotechnical parameters can also be derived from the logged parameters, but direct calibration against laboratory test results is always to be recommended.

It is usually advisable to include some borehole imaging within the suite of methods deployed. At its simplest level this could be a caliper survey using at least three arms; such measurements are usually and necessarily run with other sondes in any case. More sophisticated mapping tools using direct or indirect sensing are also available, as described below. If the fluid in the borehole can be flushed clean, optical imaging of the borehole walls can also provide useful information on the geological succession.

Radioactivity logging measures natural gamma radioactivity and is useful for lithological identification and correlation, and is particularly useful for strata rich in potash or uranium. The measurement of back-scatter from an in hole source, the gamma–gamma method, gives a reading of formation density, or, indirectly, formation porosity. The sonde can also be used for geological correlation within an area. Provided the sonde is appropriately calibrated, the measurement of density is repeatable and accurate. The neutron–neutron sonde is used to measure formation porosity or water content, although the accuracy is related to the care exercised in calibration of the sonde. Essentially these methods can provide distinctive signatures in clay strata, and deviations in these signatures are normally related to inhomogeneities in the deposit. These deviations can then be further investigated by an examination of borehole samples. These logs can all be run in a cased hole.

Electrical Downhole resistivity logs, with the second electrode at the ground surface or at the opposite end of the tool, provide shallow penetration into the borehole wall but a crudely reliable lithological log. The log also responds to fractures in the borehole wall and so can be used in rock mass fracturing assessments. In high-resistivity formations the use of a focused resistivity tool provides better resolution of stratum boundaries and variations in formation resistivity. These logs cannot be run in cased boreholes.

Seismic methods to better investigate the ground in between borehole positions include the use of vertical or cross-hole profiling. The seismic source is located on the ground surface, or progressively lowered down one borehole, and an array of receivers in another borehole record the wave arrivals. The data are then modelled to produce a synthetic image of the rock in terms of compressional or shear wave velocity. Apart from the usual limitation of the ground having to suit the available models, this sort of method is advancing rapidly in line with the increasing computing power available and so better results should be available in the future.

A range of dipmeter tools have been recently developed to map the walls of the borehole. The dipmeter tools comprise resistivity probes on 3-, 4- or 6-arm devices. The shift of characteristic patterns on the traces can be used to determine the attitude of a plane intersecting the borehole, normally a stratum boundary. The formation scanner has a dense array of micro-resistivity sensors; analysis of these closely spaced traces can produce a high-resolution image of the borehole wall. The penetration of this system is usually a few centimetres into the borehole wall and thus the better quality results can closely resemble the core recovered from the borehole (Reeves *et al.* 2001). The comments at the start of this section on the need for core recovery once reliable geophysical methods are in use on a site are therefore pertinent.

7.4. Sampling programmes

7.4.1. Aims of sampling

In designing a sampling programme it is necessary to define what is known and what is unknown about the clay deposit. Refinement of the ground model requires reduction in the uncertainty about the location, quantity and quality of the clay deposit. All of this information needs to be determined for a material that is usually not exposed at the surface.

This requires careful attention to the location, numbers, size and type of samples to be obtained. A successful sampling programme is one that gives the investigating team increasing confidence in their predictions. At each stage, the confidence in the ground model should be reassessed along with the potential costs associated with the remaining uncertainty; the programme can be considered complete when these expected costs are adequately constrained. It is not usually possible to design a sampling programme that will suit all phases of the investigation sequence. Indeed, it is important to sequentially re-evaluate the sampling strategy. As more information becomes available, the areas of predictive certainty and uncertainty will alter, affecting the requirements of the next phase of sampling. Inadequate attention to the design of the sampling patterns, which should incorporate geological expertise, can result in bias and lead to misguided optimism or pessimism. Either of these outcomes is unsatisfactory and a poor reflection on the science.

7.4.2. Sampling strategies and grids

The size of clay deposits, their unit value and the cost of establishing on-site processing facilities vary widely. A parameter commonly of great significance in the life of an extractive operation is whether to establish a process plant close to the source material or to transport the dug clay to a process operation that can serve several extraction sites. For instance the substantial capital cost of establishing kilns and associated infrastructure at a brick quarry or processing equipment at a kaolin quarry will necessitate the proving of substantial reserves. This gives a very different balance to the exploration compared with a shallow stripping operation to expose a high-value, low-process clay such as Fuller's Earth. Requirements such as these will significantly affect the scale of the sampling programme. As noted above, one important requirement is to establish what is necessary to achieve a satisfactory level of confidence in the clay deposit. This will be influenced by the financial investment required to commence the extraction and processing.

An important consideration is that the hole created by the extraction can itself become a valuable asset. This will require the inclusion of various non-geological variables in the investigation programme, such as local planning strategies, as these will influence the possible end use of the void. A common high value end use is as a landfill site. As the regulatory controls on commencing new landfills become more stringent every year, the exploration sampling programme may need to ascertain the technical viability of a range of end uses or restoration strategies as part of the overall financial evaluation of the site. Alternative uses increasingly encompass a range of leisure activities. It should also be borne in mind that the restoration phase may not begin until many years after the exploration programme, and the perception of what are acceptable end uses, in the opinion of the public and regulators, can change relatively rapidly.

A note of caution is required when considering the task faced in designing an adequate sampling programme. If investigatory boreholes were to be put down on a 100 m square grid, and the boreholes are of 100 mm diameter, the volume of ground sampled is one millionth (1×10^{-6}) of the total. Even at a grid spacing of 20 m, which is employed in the detailed production-planning stages of some ball clay and kaolin deposits, the ratio of the ground actually sampled is only 0.0025%. This ratio becomes even worse when, as is normal, only a selection of the borehole samples are subjected to testing or analysis.

The results of any sampling programme can at best only provide an estimate of the clay deposit and its variability. It will always be true that the best trial hole is the quarry itself. A successful sampling programme will reduce the uncertainty and thus the risk of the venture to within acceptable defined limits.

It is probably true to say that sampling densities are often more closely controlled by investigation costs than to the ongoing analysis of sample results. A formal risk analysis approach, where the areas of uncertainty are quantified, and the costs of possible outcomes priced, should enable the investigators to arrive at an objective and balanced judgement, and not an emotive view, of the adequacy of the investigation (Clayton 2001).

A great advantage of the initiation of this type of approach is the lateral thinking that can be introduced to the planning of the sampling. The questions that really need to be answered come to the fore and, more importantly, many of the questions that have traditionally formed the bulk of the investigation can be reduced in importance. At an early stage of any project, it is advisable to establish an analytical risk model. Initial versions of this model will necessitate the elicitation of expert views. These expert views can be obtained not only from the geologists within the investigation team, but from a range of professional disciplines, both within and outside the team. The latter often provide valuable reality checks on the overall sampling design. It should also be remembered that the initial criteria for defining the viability or markets for the clays in a deposit may change, both during the investigation and after production has commenced. This possibility needs to be considered when establishing the sampling density and is a cogent reason for retaining the investigation samples for possible future analysis.

Once the range of possible cost outcomes revealed by the sampling and testing programme is reduced to within acceptable bounds, the sampling is complete. The size of a sampling programme carried out in this way can be either larger or smaller than with a conventional approach, but the chances of surprises should have been much reduced. It is also notable that the instigation of a quantitative approach within the investigation framework allows greatly improved communication between the various disciplines within the investigation team. The traditional areas of difficult communication between say the geologist and the accountant are diminished as they are talking in a common language and objective terms.

7.4.3. Sample numbers

Accurate sampling of the mineral body requires a representative suite of samples to be obtained and numerous guidelines on sampling strategies have been published, see for example ISO (1996), BS EN 932-7 (1999) or Ferguson (1992). Consideration needs to be given to the form of the clay deposit and the variability within it. Sedimentary clay strata tend to be compositionally uniform in a lateral direction, but vertical variability in clay strata should be expected. On the other hand, primary kaolin deposits often exhibit marked variability and inhomogeneity with respect to colouring, oxide content or kaolin yield. Sampling programmes need to be designed to provide information on the expected geology and its variations (see Text Box below).

7.5. Intrusive exploration methods

A wide variety of field or intrusive investigation methods is available. The generic types of investigation are outlined below and are readily comprehended. However, within each investigation method there are many variations in practice which have arisen to solve particular problems. These are not usually recorded. It is therefore necessary to ensure that the appropriate expertise is consulted at a sufficiently early stage, and that the method(s) of investigation required to advance the investigation in hand are known and available, and their limitations understood. Many non-specialists think that they know sufficient about investigation methods to specify the methods to be used. It is usually far better to present the problem to potential contractors, and ask them to recommend the most technically appropriate and cost-effective methods to obtain the necessary information.

7.5.1. Trial pitting and trenching

Trial pits dug by hand or by hydraulic excavator are one of the most commonly used and cost effective methods of investigation. For the cost of the machine and a logging geologist, significant coverage of the site can be achieved

---

Sampling strategies

The selection of an appropriate sampling pattern will depend on the objectives of the investigation, the information already available, the site conditions and the prevailing geology. Sampling patterns need to be determined before field work starts otherwise the result will be over-dependent on the expertise of the field supervisor, and allows sampling results to be influenced by local factors such as ease of access, where the ground is easily penetratable. Less experienced, or locally knowledgeable, personnel tend to select locations which give very high or very low values, rather than representative results. However, experience and theoretical considerations show that in many cases systematic sampling on a regular grid is both practical and and will allow a sufficiently detailed picture of variations to be established. The number of sampling points can readily be increased in areas requiring more detail, and the application of additional judgemental sampling when appropriate.

Examples of different sampling patterns include:

- Irregular or non-systematic patterns used where the soil character is thought to be reasonably consistent, but is inadequate to asses lateral variability. Examples are S, W or X or irregular zig-zag patterns;
- systematic grid patterns with the grid size selected on the basis of the geology and likely rate of variation, practical to set up and follow in the field;
- random sampling appropriate where irregular variability is anticipated, selected using published tables, and should not be confused with sampling without predetermined planning of patterns;
- stratified random sampling avoids some of the disadvantages of the other methods, but can be difficult to interpolate between data points;
- systematic non rectangular grid, circular patterns and linear patterns are not usually advisable for clay exploration.

FIG. 7.9. Clay logging.

quickly and efficiently. The trial pit can be readily enlarged, vertically or horizontally to form a trench, to follow features of interest. Perhaps most importantly, a trial pit exposure allows detailed *in situ* logging of the materials present as well as identification and measurement of jointing and bedding orientations (see Fig. 7.9). The opportunity to take samples in a controlled manner of particular materials is also of value.

The merits of trial pits are limited by a number of practical aspects. Firstly, the reach of excavators available locally is usually limited. For this reason most trial pits are 3 m deep, the practical limit of wheeled excavators, but tracked machines can excavate to about 5 or 6 m depth. Larger machines with longer reach are available, but the high mobilization costs cannot often be justified for the short duration of a geological investigation. It is normally difficult to control groundwater inflows within trial pits; although the water can be physically pumped out, instability of side walls and smearing of the clay renders detailed logging very difficult.

A serious aspect to bear in mind is one of safety in that no unsupported trial pit more than 1.2 m deep should be entered. Current Health and Safety regulations in the UK forbid entry to any excavation unless a Risk Assessment has first been completed. The apparent stability of fresh clay faces is very deceptive, and collapse can occur without warning. The burial of workers in trench collapses is one of the most common causes of fatalities in the UK construction industry, and great care is always required. Mechanical or hydraulic shoring systems are available and should be used if entry to deeper excavations is required. Although requiring at least two people to carry around and install, they are not difficult to use and do provide a sufficiently safe working place for detailed logging to continue. Alternatively, one side of a trial pit can be battered back to a shallow angle to permit emergency egress; such an approach requires that other face is adequately stable to allow safe access for logging. Man entry to any pit should always be accompanied by gas monitoring; there are many causes of oxygen deficiency or hazardous atmospheres in confined spaces which are only detectable by appropriate gas monitors being in use at all times.

Deeper trial pits or trial shafts can also be excavated. These are commonly excavated by hand with traditional timber support, although auger or other piling rigs can be used. The comments above on groundwater are likely to apply even more strongly in deeper excavations. Working in deep shafts is very dangerous. Careful consideration must be given to the stability of the excavation; attention also needs to be given to the maintenance of a breathable air supply, and this includes ensuring that rig engines next to the shaft collar are not left running. Full-time banksman attendance should also be provided.

At all times that a pit or shaft excavation is open, clear barriers should be provided to avoid the possibility of someone falling in, even in remote locations. As soon as the logging and sampling exercises have been completed, backfilling should be carried out, with the arisings (excavated material) being compacted back into place to ensure that there is no remaining risk to third parties. It is very difficult to restore the location of trial excavations to their original conditions, and the level of reinstatement required is an important element of the investigation planning. Trial pits that are inadequately restored in the field will antagonise the farmer or landowner, whose co-operation will be needed in the next phase of the investigation.

In logging artificially created exposures of clay, care needs to be taken to clean the excavated smeared face back to undisturbed clay. The depth of material that needs to be cleaned off varies with the clay and with the method of formation of the face. In a normal trial pit, removal of up to 200 mm can be necessary. This cleaning off should be carried out over all or at least a large proportion of the pit face or shaft wall and is a time-consuming operation; the provision of local labour to carry out this task greatly enhances the productivity of the field geologist. The faces of the trial pit or shaft wall should be photographed after they have been logged. The clarity of the photographs can be greatly enhanced by marking key features such as bedding, boundaries or shear surfaces with spray paint on the face before taking photographs. The photographs should include small scale images of the whole exposure as well as larger-scale pictures of any features of particular interest. All photographs should include a scale and identifiers of pit reference, depth and orientation.

7.5.2. Dynamic probing and window sampling

Reconnaissance boring can be carried out using dynamic probing type rigs (see Fig. 7.10). These rigs are lightweight, fitting readily into the back of a medium size van. They also have the advantage of being easily manoeuvrable by hand and can thus be used in restricted positions not available for larger machines. These rigs are driven by a light hammer. In dynamic probing mode, a continuous blow count or strength record is obtained. In window sampling mode, a series of progressively smaller diameter tubes is driven into the ground; the disturbed core of soil can be examined through the tube's window and samples taken. Depending on the strength and plasticity of the clay, penetrations of up to 10 m are obtainable, although a lesser limit of about 6 m is normally achieved. The speed of operation of the sampling equipment renders this a highly cost-effective method, the limitations being the available depth and the small quantity of sample recovered.

7.5.3. Boring and drilling

Investigation of the clay deposit by boring or drilling is necessary where the clay lies beneath water bearing superficial deposits such as river gravels, or where investigation to depths of several tens of metres is required (see Fig. 7.11).

Boring is carried out by a variety of methods in different parts of the world. Cable tool boring (often erroneously called shell and auger) uses a mobile tripod rig which can penetrate soils and weak rocks to depths in excess of 50 m. The diameter of tools used is typically 150 to 250 mm diameter. Larger diameter casings may occasionally be used to support the borehole and exclude ground water, but at a rapidly rising unit cost. The drill tools are operated on the end of a wire rope and, although crude and damaging to the soil strata around the end of the borehole, have the advantages of relatively low cost, simplicity of operation and ability to penetrate most natural materials up to weak rock strength.

Mechanical augers can be used for investigations in clay deposits. The soils can either be sampled from the flights of the auger as they are cleared off, or a sampler can be deployed through the hollow stem and thus ahead of the auger. The depth records of stratum changes determined from auger holes are less accurate than by other drilling methods, and the samples recovered are disturbed. This may be an advantage rather than a disadvantage for different types of clay deposit. The stems of hollow augers are typically about 75 to 125 mm diameter. Continuous flight augering, or augering to any

FIG. 7.10. Light probing equipment with a hollow stem auger.

FIG. 7.11. Tripod boring rig.

great depth, requires high torque rig drives. These are often mounted on the base units of medium to large tracked hydraulic excavators, and come at a high hire rate. However, in the appropriate ground, their productivity and mobility can render their use cost-effective.

There are a number of other methods of boring developed and used in various parts of the world. Wash boring is widely used in North America and uses an even simpler rig than the cable tool rig in the UK. A rope held percussive bit is used to break up the ground, the cuttings being washed to the surface by circulating water. These cuttings are very disturbed and so it is often difficult to determine the *in situ* soil state. Rotary percussive boring and similar methods are not very efficient in drilling clay strata and so are not widely used.

Rotary core drilling (see Fig. 7.12) offers a technically effective but more costly method of investigation. Depending on the depth to which investigation is required and the strength of the clay deposit or interbedded horizons, core drilling is a valid method to be included in the range of available tools. A wide variety of core diameters, core barrels, drill bit designs and materials, and many flush media are available. The opportunities for the recovery of high-quality cores is greatly enhanced if experienced, accredited drilling contractors are employed.

Suitable advice should be taken on the best combination of tools and methods appropriate to the materials being investigated. In ball clays, it is common to utilise HQ diameter (80 mm) wireline to maximize core recovery. By contrast, in kaolin, an SF diameter core (110 mm) is obtained because of the friability of the material, and occasionally air-flush percussion rigs are used.

Core diameters usually fall in the range 60 to 140 mm, with the cost of core drilling increasing with diameter. Core drilling effectively has no depth limit, provided a large enough rig is available. In clay, it is advisable to keep the kerf (the annulus of ground that has to be removed from the borehole to produce the core) as small as possible. This requires the use of a thin walled core barrel. The choice of core liner in the barrel, the design of the bit and the flush employed contribute substantially to the quality of core that can be recovered. Given the adoption of the right combination of drilling variables for the particular clay deposit, there is no reason why 100% recovery of high-quality core is not achieved. Where the clay sequence contains interbedded, unconsolidated sands, there may be problems with washout and borehole collapse. This problem can be tackled by casing the hole, or by using specific drilling muds to penetrate and stabilise the walls of the borehole. However, if drilling muds are used, care must be taken to ensure that the core samples are cleaned and uncontaminated, especially in the case of clays being evaluated for ceramic applications.

7.5.4. Cone penetration testing

A number of types of static probe have been developed in various parts of the world, but typically a cylindrical probe is pushed into the ground on hollow rods at a uniform rate of penetration (Meigh 1987). The thrust required to achieve penetration and the friction on the side of the probe are measured (see Fig. 7.13). In most modern rigs, these measurements are standardised at 2 cm intervals throughout the process. Additional sensors can be used, separately or in combination, to measure pore pressures in the soil, to record the arrival of seismic waves and to measure parameters such as conductivity.

The records arising from the probe test can be used to determine a number of geotechnical parameters of the soil, such as its strength and stiffness (see Section 4). By comparing the end and side resistances, it is possible to interpret the soil types penetrated. The world is divided into those who do and those who do not have great faith in these interpreted soil profiles, but one great advantage of this method of investigation is its ability to identify thin bands of different materials (for instance a thin sand layer in a clay) at an accurately defined depth. Such observations are often difficult to obtain by other methods. However, as the method does not produce any samples for the geologist to look at, the reliability of these indirect soil descriptions will remain a matter of debate, unless used in conjunction with other methods, such as drilling, to confirm these conditions.

The rig used to carry out this test varies from a large enclosed lorry with all reaction systems in-built to small

FIG. 7.12. Rotary drilling rig recovering cores.

FIG. 7.13. Manually operated sounding rig.

tracked units which install ground anchorages to achieve the necessary thrust. The equipment used to carry out cone testing is expensive on a daily basis. However, about 40 to 60 m of probing per day is possible. As preliminary records should be produced immediately, close control of the findings and thus the scope of the work is feasible. The quality of the results obtained together with the rate of production render this method of investigation compatible with other methods.

7.5.5. Logging and description

Most of the methods of intrusive investigation described above produce some soil or rock in the form of an exposure, sample or core that needs to be described and recorded. Part of the purpose of an investigation is for this evidence to be examined by a geologist and incorporated in the reserve assessment. The high cost of intrusive investigations has been referred to above. In order to maximize the return on this expenditure, it is important that the geologist examines all the sample or cores and records the material descriptions in a systematic and comprehensive and appropriate manner. The information recorded by the logger needs to include factors pertinent to its use, as well as say geotechnical factors that will affect its exploitation such as the ease of excavation or pit stability. Too many reserve investigations do not follow this precept and it becomes necessary to drill additional holes to obtain this information.

Full coverage of the information that needs to be recorded for engineering purposes is given in BS 5930:1999. This includes information on the strength, structure, bedding, colour, and particle size grading (see Section 4). This Code recommends that detailed description of discontinuities within the soil or rock mass is also given wherever possible. The information to be recorded includes dip and dip direction of the discontinuities, their persistence, surface form and spacing. Other information on the mass is required to provide information on the weathering profile in the form of full and detailed description of all the changes of colour, fracture state and strength that are induced by weathering of the soil or rock.

More detailed logging of relevant aspects of colour, mineralogy or sand or impurity content may be necessary in certain clay deposits (particularly those destined for fine ceramic markets). It may be necessary to agree and clarify the basic terminology and highlight the most important features with the site geologist before logging commences. Although the use of standard terminologies is important, as it allows communication of information between professionals and between sites, the use of appropriate terminology consistently across any one site, and between all the loggers on that site, is more important. Where non-standard terminology is used, care should be exercised to ensure that already defined terms are not reused with a different meaning, otherwise great confusion will result.

7.5.6. Groundwater measurement

Groundwater occurs throughout the ground in which exploitation of most clay deposits is carried out. More problems in quarry design, excavation and operation are caused by lack of appreciation of the groundwater conditions than any other single reason. The problems that water can cause if not adequately understood include instability of side slopes, possible damage to adjoining infrastructure, base heave in the quarry, sterilization of reserve and excessive pumping costs.

In the investigation of clay it could be surmised that clay is largely impermeable, does not produce significant volumes of groundwater and therefore it is not a problem. Indeed, it has often been said that the clay does not hold any water. Groundwater problems manifest themselves either as excess inflows or as excessive water pressures. Unexpected and troublesome water inflows in clay pits can occur through even quite thin sand layers, for instance. The inflowing water could result in significant slope instability (see Fig. 7.1). The ability of clay to maintain negative pore water pressures, or suctions (see Section 4.2.6), is an intrinsic characteristic. The dissipation of these suctions with time can be very significant to the stability of cut faces and an understanding of these properties of a clay is important.

It is important for the exploration programme to achieve an understanding of the original groundwater conditions at the site, and to obtain the parameters necessary for designing any water control measures that will be necessary. As boreholes have been put down in the overall exploration programme, it is usually advisable to install groundwater monitoring instruments in as many holes as possible. The extra cost is minor once the hole has been formed. Ideally, these instruments should be located in areas where they will not be damaged by plant during exploitation of the reserve. This is often very difficult to achieve. Long-term water level and water quality records will prove valuable when the restoration options for the quarry are being discussed at a later stage. It is important that any instruments placed in boreholes are read on a regular basis and that damaged instruments are replaced.

It is probably good practice to install most instruments at a diameter of at least 50 mm to allow the recovery of water samples for quality analysis to be carried out at occasional intervals.

The cheapest, and most commonly ignored, information on groundwater is that available during the excavation of trial pits and execution of boreholes. It is unlikely that such records recorded from boreholes in a clay deposit will represent equilibrium levels. However, the presence of water at a level will strongly suggest that the water table is at least at that or a higher elevation. On completion of the boreholes, water level monitoring instruments should be installed (Dunnicliffe 1993).

The simplest form of installation is a standpipe comprising a tube with a slotted section; the filter material outside the pipe fills the annulus over most of the depth of the borehole. In a clay deposit this is, although cheap, not usually an appropriate instrument. Significant volumes of water need to move to record any change in water level, and so the response time is too slow to be of much use. Piezometers in which the length of filter is reduced to the short length of interest provide better response times, but even so equilibration times after installation will often be reckoned in months.

Pressure-measuring piezometers have better response sensitivity, as only very small water flows are necessary to influence the reading. Different types include hydraulic, pneumatic and electrical piezometers which function behind a diaphragm in the down hole cell. Use of such instruments in long-term monitoring requires care to ensure that air has not entered the system, and maintaining calibration (depending on instrument type), but they do have the advantage of being able to measure negative pore pressures. The piezometer cells are relatively expensive and read-out units are required. Of concern is whether these types of installation are appropriate for use in an operational environment. The connections to the surface and the read-out unit are reasonably robust but even so the life expectancy of such installations may be shorter than the exploitation period. Nevertheless good quality water level information is required and so some form of installation will be needed. Careful siting of the installation boreholes so that they are away from plant operating areas and haul roads, will help to improve the life expectancy of the instruments.

7.6. Soil samples

In exploration of clay, the volume of sample required for representative samples of the ground to be obtained is relatively small. For instance, the sample sizes required for typical geotechnical classification tests including particle size determination is normally less than 1 kg. Even for compaction tests (see Sections 9 and 10.4), where fresh material should be used for each point on the curve, a sample of about 20 kg should suffice. That sample would fit in a plastic bag and would be carried readily from the field to the laboratory. Although the amount of material for any particular test may be modest, it is always worthwhile taking more sample than may theoretically be required. It is easier to throw away excess material than have to revisit the site to collect additional materials.

Where core sampling of sedimentary ceramic clay deposits is being conducted, the entire length of core for each sampled interval should be thoroughly cleaned and retained. After logging, the core should be finely cut or crushed and thoroughly mixed, to ensure that the samples for analysis are representative.

7.6.1. Sample quality

A wide range of sampling techniques is available. The key characteristic of clay which simplifies recovery of representative samples is that it is self-supporting in excavations, at least in the short term. This means that samples of clay can be stood on the laboratory bench for long periods, apparently without the sample undergoing changes. In fact the soil is changing in response to the changed stress conditions imposed upon it since it was recovered from the ground, but these are visually imperceptible.

In deciding which sort of sample is required, it is important clearly to understand the objectives of and demands that will be made on each and every sample. If an examination of the nature of the soil is required any sample that is not contaminated with foreign material will suffice. The nature is the soil's intrinsic character, such as plasticity and reflects the clay mineralogy and the particle sizes present. Such samples are often recovered from the boring tools or the digger bucket in a disturbed condition. They are normally perfectly suitable for use in classification testing or particle size analysis, or indeed in any test that itself reconstitutes the specimen.

If on the other hand, it is the state of the soil that is of interest, more care is required in the sampling to ensure that the recovered sample state accurately reflects the ground state *in situ*. The state of a clay soil is essentially the water content and thus its strength and stiffness. The state of a clay soil can be changed by the sampling process, and will usually be changed by the excavation and handling processes within an excavation operation. The significance of any such changes on the soil state need to be investigated to understand the state of the clay that will be delivered to the process plant. Methods of preserving samples in a suitable condition for strength and stiffness testing are described in the Text Box on p. 195.

Preservation of samples

Samples cost a considerable sum of money to obtain and should be treated with great care. The usefulness of the test results depends on the quality of the sample when tested, and it is therefore important to establish a satisfactory procedure for handling, labelling and preserving the samples. This care extends to the transport and storage so that they do not deteriorate and can readily be drawn from storage for testing or examination (BS 5930:1999).

Samples selected for testing should be inspected to ensure they are free from defects such as incipient fractures or zones of weakness which could cause the specimen to break during preparation, which may include machining, or affect the test result. If recovered in a sample tube (pushed, driven or rotated) the sample ends should be sealed. If recovered as unconstrained core the sample should be wrapped in cling film, and coated with paraffin wax or resin to provide both increased rigidity and preservation of moisture. Alternatively the sample can be inserted into a tight fitting plastic tube.

Disturbed samples should be placed in robust bags, labelled inside and out. Light plastic bags, such as available from grocer's shops or hessian bags are not adequate to retain moisture and so the clay will be irreversibly changed when it dries out.

All samples should be cleaned of drilling fluid or debris, and labelled inside and outside the tube or wax before being packaged for storage or transport. Care should be taken that the samples are large enough for the expected tests, for instance if strength testing of cores is required, the samples should be at least three diameters long.

Higher quality of samples are required to investigate the structure of the clay deposit. The behaviour of a clay will be markedly affected by the presence and character of inhomogeneities such as interbedded coarser materials or discontinuities. The structure of a soil will usually be affected by all except the highest quality sampling procedures, and will almost certainly be destroyed by earthworks activities. However, the presence of, for example, thin beds of sand could crucially affect the processing parameters as well as affecting the stability characteristics of the side slopes to the excavations. Such aspects are everyday significant problems which can have a marked impact on the operating costs of the quarry, and are not the province of merely academic interest to which they are often consigned.

The quality of the sample as recovered from the borehole needs to be maintained in transport of the sample to the store, and then in storage. Samples should be preserved in their original containers and sealed, preferably at the sampling location, and then transported with a degree of protection appropriate to the intended sample quality. Bag samples of disturbed clay would not need any special protection, but undisturbed core samples intended for effective stress testing should be restrained and cushioned during transport. The requirement for such measures is particularly important if the clays are unusually sensitive, or if they are particularly soft or stiff.

Clay samples take some time to dry out, reflecting the suction properties and low permeability, but once dried cannot be restored to their *in situ* condition. This is critical for most examinations and testing purposes, as knowledge of the condition of the clay *in situ* is fundamental to assessing the excavation and processing of the product. Desiccation of the sample can be largely avoided by simple wrapping of the sample, within a sample tube, plastic bags or a film wrapping. Paraffin wax is also commonly used to seal the openings in sample containers.

Samples of expansive clay can usefully be sealed into an expansive grout, itself contained within a larger sample container, in order to restrain the sample. All samples should be at least double labelled. One label should be on the outside for ready visibility and retrieval, with a second label on the inside of the sample as a check, and in case the outer label is lost or becomes illegible.

7.6.2. Sample types

Sample types relevant in the exploration of clays fall into five main categories.

- Small disturbed samples recovered from exposures or from drilling tools used for identification and simple classification testing.
- Large disturbed samples recovered from the bulk of borehole arisings (and therefore usually spread over a relatively large depth increment) or recovered from the bucket of the excavator digging the trial hole. Large samples can be used to carry out a range of suitability tests, safe in the knowledge that the source material is directly comparable. For these reasons, such samples are usually best recovered from trial pits where the material being included in the sample bag can be closely controlled by the field geologist.
- Tube samples, normally of between about 75 and 100 mm diameter, can be obtained by pushing or driving the tubes into the clay exposed or through the base of a borehole. The cross sectional proportions of the sampling assembly and the method of driving influence the amount of disturbance to the sample. Such samples are widely termed undisturbed, but in reality the amount of disturbance can be very substantial, particularly in soft or in stiff soils. However, these samples can be obtained from depths below the water table, and are normally considered representative of the soil nature over the sampling depth.

Fig. 7.14. Block sampling.

- Rotary sampling, either by turning in a sample tube or by rotary core drilling (in its normal sense) can produce high-quality samples of clays. A sequence of consecutive samples can provide a complete profile through the clay deposit. The samples can be extruded, split and logged in detail. Modern wireline drilling techniques carried out with the aid of polymer mud flushes and semi-rigid core liners can produce very high-quality continuous samples of clay strata. The cores, about 90 to 120 mm diameter, can be obtained from significant depths below the ground and ground water and are generally reckoned to be as undisturbed as it is sensibly possible to recover. The recovered cores provide excellent opportunities to carry out detailed logging of the clay profile; and they are generally of sufficiently high quality to allow subsampling of identified materials for any testing that requires samples to be representative of the in situ state or structure of the ground.
- Block samples of clay can be usefully taken to provide large volumes of material from which undisturbed sub samples can be obtained for testing. Such sampling requires generous space for man access and so is limited to pits or shafts in reasonably dry conditions, and normally only from shallow depths (see Fig. 7.14).

7.6.3. Sample preservation and storage

Samples recovered represent a significant investment within the exploration programme. Although each sample may not appear to have been paid for at a high unit rate, replacement of a lost or damaged sample requires mobilization back to site to repeat the whole drilling or excavation exercise. It is important to establish an effective chain of custody system to ensure that the location of individual samples can be identified at all times. The laboratory is not a sample store, and so only those samples actually to be tested should be transported there, the remainder being deposited in a suitable designated store. The testing schedule is often amended, and so the store has to be revisited to recover additional samples. Without an efficient custody record, samples are commonly mislaid.

The quality of stored samples of clay will degrade with time (see Text Box on p. 195). This is the result of the changed stress conditions being applied to the clay sample, moisture content changes within the sample and chemical changes arising from exposure of the clay to oxidation (oxidation of sulphides gives rise to volume changes and acidic conditions). This happens even in temperature controlled sample stores, but at a much reduced rate provided, of course, that the samples have been effectively sealed to prevent moisture loss. The ideal is to store samples at an even temperature of about 4°C, but this is rarely achievable with the level of investment provided for a normal site store. Such economies result in the physical and chemical degradation of samples, thus potentially rendering them useless. This is particularly the case in the absence of effective temperature and humidity control. The sun beating on a locked container can quickly raise temperatures above 30°C in the UK, and higher in other parts of the world. The inside temperature will also fall rapidly on a frosty night. A frozen clay sample will have a disrupted fabric after it has thawed, and any test then carried out will not be on material representative of the in situ soil. If the condition of the clay samples is critical, for instance for use in testing to determine effective strength parameters, the samples should be removed at the first opportunity to a permanent store which will have an equable environment between about 5 and 25°C. It is recommended that clay samples required for chemical testing, and that need to be stored for more than one month, should be oven dried at 60°C, ground and stored in sealed containers at low temperatures (0 to 4°C) or in a desiccator (Czerewko et al. 2002).

References

Anon, 2001. *European Code for reporting of mineral exploration results, mineral resources and mineral reserves* (The European Code). Draft May 2001.

BS 5930:1999. *Code of Practice for Use in Investigations*, BSI, London.

BS EN 932-6:1999. *Testing Aggregates*, BSI, London.

CLAYTON, C. R., SIMONS, N. E. & MATTHEWS, M. C. 1988. *Site Investigation—A Handbook for Engineers*, Granada.

CLAYTON, C. R. 2001. *Managing Geotechnical Risk*, Thomas Telford, London.

CZEREWKO, M. A., CRIPPS, J. C., REID, J. M. & DUFFELL, C. 2002. An investigation into the effects of storage conditions on the sulfur speciation in geological materials. *Quarterly Journal of Engineering Geology and Hydrogeology*, **36**, 331–342.

DUMBLETON, M. J. & WEST, G. 1970. *Air-photograph Interpretation for Road Engineers in Britain*, RRL Report LR 369. Road Research Laboratory, Crowthorne.

DUNNICLIFFE, J. 1993. *Geotechnical Instrumentation for Monitoring Field Performance*, Wiley, Chichester.

FERGUSON, C. C. 1992. The statistical basis for spatial sampling of contaminated land. *Ground Engineering*, 34–38.

GIBSON, P. J. 2000. *Introductory Remote Sensing: Principles and Concepts*, Routledge, London.

ISO—DIS (Draft BS 7755:1996) *Soil Quality Part 2—Design of Sampling Programmes*.

ISO 2002. *The Identification and Description of Soils*, ISO 14688 Part 1.

ISO in press. *The Classification of soils*, ISO 14688 Part 2.

ISO in press. *The Identification, Description and Classification of Rocks*, ISO 14689.

ISRM 1978. Suggested methods for the quantitative description of discontinuities in rock masses. International Society of Rock Mechanics. *International Journal Rock Mechanics and Mining Sciences*, **15**, 319–368.

MCDOWELL, P. *et al.* 2002. *Geophysics in Engineering Investigations*, CIRIA publication C562, London.

MEIGH, A. C. 1987. *Cone Penetration Testing*, CIRIA Report 87.

PAINE, D. P. 2003. *Aerial Photography and Image Interpretation*, 2nd Edition. Wiley, New York.

PERRY, J. & WEST, G. 1996. *Sources of Information for Site Investigation in Britain*, TRL Report LR 192.

REEVES, G. M., REEVES, H. J. & COCCO, P. 2001. On the indentification and characterisation of discontinuities in igneous and sedimentary rocks. Seventh International Conference MGLS-SPWEA, Denver. Society of Professional Log Analysts, Minerals and Geotechnical Logging Society.

RENCZ, A. N. & RIYERSON, R. A. (eds.) 1999. *Manual of Remote Sensing, Volume 3, Remote Sensing for the Earth Sciences*, Wiley, New York.

SISG 1993. *Site Investigation in Construction*, Thomas Telford, London.

SKEMPTON, A. W., NORBURY, D. R., PETLEY, D. J. & SPINK, T. W. 1991. Solifluction shears at Carsington, Derbyshire. *In*: Quaternary Engineering Geology, Geological Society, London, Engineering Geology Special Publication, **7**.

SMITH, M. R. (ed.) 1999. *Stone: Building Stone, Rock Fill and Armourstone in Construction*, Geological Society, London, Engineering Geology Special Publications **16**.

SPINK, T. W. & NORBURY, D. R. 1993. The engineering geological description of weak rocks and overconsolidated soils. *In*: *The Engineering Geology of Weak Rock*, Engineering Geology Special Publication, **8**. Balkema, Rotterdam.

8. Compositional and textural analysis of clay materials

The composition and textural fabrics of clay materials control their applications for constructional and other purposes. Discussed below are the most commonly used techniques to characterize these aspects of clay materials. The analytical approaches, their advantages and their limitations are summarized without giving detailed discussions of each technique, but references to more detailed accounts are provided.

The overall sampling and analytical approach, and how it will help define and/or solve potential problems, needs to be carefully planned. A preliminary reconnaissance sampling and analysis programme, taking into account desk studies (see Chapter 7) and any related available information needs to be made before embarking on full scale detailed and comprehensive analyses.

Sampling and sample preservation methods are too often inadequately designed and this can limit or even negate the results from later sophisticated and costly analyses. Sampling needs to be planned to provide adequate representative samples taking into account the nature of the problem to be answered and the heterogeneity of the materials—clay materials are often very heterogeneous, even on a small scale. Once sampled, the material needs to be carefully preserved so as to minimize alteration of the sample from its original state prior to analysis. For example, drying of samples needs to be undertaken at temperatures which will not dehydrate minerals such as gypsum, i.e. at 40°C rather than the 105°C that is often required in standard procedures.

Also the sampling and preservation procedures need to be designed in the context of the objectives of the investigation and the planned analytical programme. The flow diagram in Figure 8.1 indicates generally the manner in which different sample preparation procedures are needed for the different types of analyses described below.

Different analytical approaches are often employed by different laboratories and these need to be carefully assessed to determine whether the approach used is satisfactory for the problem in hand. For example, one laboratory's method of quantitative mineral analysis may differ from that employed by another laboratory and yield significantly different results, even though both may accord with recognized procedures. In a recent comparison, the same clay sample was analysed by two reputable laboratories; one laboratory reported no presence of the swelling clay mineral smectite whilst the other laboratory reported 35% smectite. Before contracting a laboratory to undertake a large number of analyses it would be sensible to send test samples to several laboratories to assess the results provided. It is important that the analyst is also fully informed of the purpose of the investigation and is available to discuss the results in the context of the overall project.

In this chapter the principal analytical methods to determine the mineralogical and chemical characteristics of clay materials and their textural fabrics are described.

8.1. Mineralogical analysis

8.1.1. X-ray powder diffraction analysis

X-ray powder diffraction (XRD) analysis is the most widely used technique for the identification of minerals (see the Text Box on p. 200). It is particularly useful for the ready determination of the nature and relative abundance of the clay minerals and other fine-grained minerals that characterize different clay materials (see Chapter 2).

The theoretical basis for this type of analysis is that crystalline materials (i.e. minerals) will diffract a beam of incident X-rays (Moore & Reynolds 1997).

This effect produces a diffraction pattern that characterizes each mineral species. The diffraction patterns are monitored using X-ray powder diffractometers, modern versions of which allow them to be stored and interpreted using mineral identification and quantification programs.

It needs to be emphasized that XRD analysis can only be used to detect crystalline solids (i.e. minerals) present in abundances usually greater than 1–2% by weight. Very poorly crystalline or 'amorphous' phases cannot be detected using this method.

Samples of clay materials for XRD analysis are usually prepared in two forms; a finely ground (5–10 microns) randomly oriented powder and an oriented clay (<2 μm) fraction. The finely ground whole rock powder is needed for the whole rock analysis. Apart from the identification of non-clay minerals, X-ray powder diffraction analysis of the whole rock powder should indicate the general nature and approximate relative proportions of clay minerals. A typical diffractogram obtained from the analysis of a whole rock powder with its interpretation in terms of the minerals identified is shown in Figure 8.2. However, examination of the oriented clay fraction specimens is usually necessary for the detailed analysis of the clay minerals that are concentrated in the clay fraction. The majority of clay minerals have sheet silicate structures and hence platy habits; it is thus advantageous to use oriented specimens of the <2 μm fractions for their XRD analyses. A diffraction pattern from such specimens will emphasise the basal 00l reflections (see Text Box on p. 201), which are the most useful in differentiating the

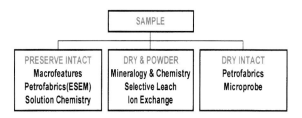

FIG. 8.1. Samples of clay materials required for different analytical techniques described in the chapter.

principal clay mineral species (Moore & Reynolds 1997). For specific problems it can also be useful to analyse finer sub-divisions of the clay fractions (i.e. < 0.5 or < 0.1 μm); for example, studies of expandable clay phases, which are often preferentially concentrated in the finest clay fractions.

Oriented specimens of the clay fractions are prepared either by (a) suction of the suspended material onto an unglazed ceramic disc (Shaw 1972) or a Millipore filter (Moore & Reynolds 1989) (b) smearing a clay paste onto a glass slide (Gibbs 1965).

It may be necessary to remove amorphous iron oxides and/or organic matter before preparing the oriented clay fractions as their presence can adversely affect the X-ray diffraction spectra (Moore & Reynolds 1997). However, such pre-treatments to remove these phases can sometimes affect the clay minerals and should be used with care or avoided unless absolutely necessary.

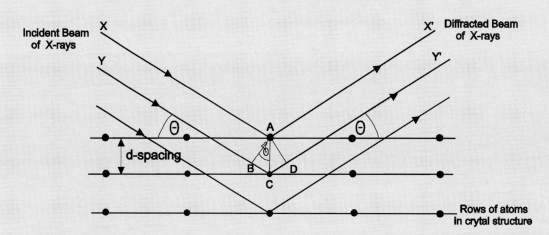

Principles of X-ray powder diffraction and Bragg's Law

X-ray powder diffraction is a method for the analysis of crystalline solids involving the diffraction of a monochromatic X-ray beam in which the regular array of atoms in the crystal structure act as planes of a diffraction grating to the incident X-rays.

This is schematically shown in the diagram above.

Positive X-ray diffraction effects will occur if the path difference between two parallel rays XAX' and YCY', BC + CD equals a whole number (n) of wavelengths (λ) of the incident X-rays–this is **Bragg's Law**. From the geometry of the above diagram the path difference BC + CD can be related to the spacing between the rows of atoms–the d-spacing. BC = CD = $d.\sin\theta$ i.e. BC + CD = $2d.\sin\theta$ where θ **is the angle of incidence of the X-ray beam to the row of atoms.**

Bragg's Law defines the conditions for X-ray diffraction to occur in terms of the wavelength of the incident X-rays (λ), the interatomic distance between the planes of atoms (d) and the angle of incidence of the X-ray beam to the planes of atoms (θ)

$$\text{i.e.} \quad n\lambda = 2d.\sin\theta.$$

Within a crystal structure of a mineral there will be numerous planes of atoms with different d-spacings capable of diffracting the incident beam of X-rays of a given wavelength. It is the X-ray diffraction effects from these different planes of atoms, with their characteristic d-spacings for a given mineral that produce its characteristic X-ray diffraction pattern and can be used for its identification and quantification. The d-spacings are most often measured in Angstroms (Å) where 1Å = 0.1 nm though nanometres may also be used.

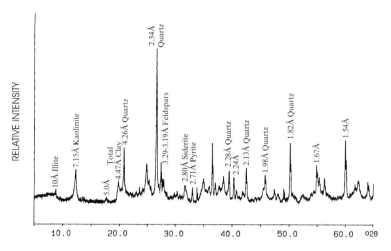

FIG. 8.2. XRD trace of whole rock material from a Nigerian mudrock.

X-ray reflections notation

The notation system used to describe X-ray reflections is an extension of the Miller indexing system used to describe crystal faces in morphological crystallography. As with the crystal faces, the planes of atoms in the crystal structure causing the diffraction of the X-rays are described in terms of whole number or zero indices e.g. 001, 002, 060, 111 etc. The indices are derived from the intercept ratios that the planes make with a set of axes in three-dimensional space (Whittaker 1981). The notation 00l denotes a general set of diffraction planes of the type 001, 002, 003, 004, etc.

8.1.1.1. Mineral identification
(a) Non-clay minerals. Mineral identification by XRD analysis involves comparison of the analysed spectra with the XRD spectra of reference standard minerals. The most comprehensive reference spectra are provided by the JCPDS (Joint Committee on Powder Diffraction Standards) mineral powder diffraction data bases and these are now usually incorporated within computer-based search-match programs for mineral identification. Compositional variations and preferred orientation effects can produce significant differences between analysed and standard XRD diffraction patterns of essentially the same mineral. Search-match programs use best fit critera and tests of significance to determine the reliability of each mineral identification. Such programs also incorporate procedures to 'strip-out' the spectra of individual standard minerals to allow analysis of the remaining spectra. This is often needed for the analysis of minor minerals in a mixture. The detection limits for identification of minor minerals is dependent on the nature of the individual minerals but generally it is about 1–2% of the total sample volume.

(b) Clay minerals. Each principal group of clay minerals has a characteristic sheet silicate structure except for the sepiolite-palygorskite group (see Chapter 2). The d-spacings of the basal 00l x-ray reflections, which characterize each clay mineral group, are summarized in Table 8.1 together with the principal low-angle reflections of sepiolite and palygorskite.

The JCPDS files that are widely used for mineralogical identifications from XRD data, as discussed above, are of more limited use in the identification of clay minerals. This is because they have been prepared from XRD analysis of randomly oriented powder specimens, not the oriented specimens usually used for clay mineral analysis.

There is increasing use of calculated clay mineral XRD patterns such as the NEWMOD program devised by Reynolds (1985) to aid the identification of clay minerals. The program simulates XRD patterns for individual clay minerals, including mixed layer clays and variable mixtures of clay minerals, which can then be compared with the XRD data from analysed samples.

It can be seen from Table 8.1 that there is considerable overlap of reflections from different groups. This can be overcome in the analysis of clay mineral mixtures by the use of various pre-treatments to differentiate individual clay minerals within a clay assemblage. These treatments principally include solvation with ethylene glycol (and/or glycerol) and heating between 300–550°C (Table 8.2) (Figs. 8.3 & 8.4).

TABLE 8.1. *Principal basal reflections of clay mineral groups and their 060 reflections plus the principal low-angle reflections of sepiolite-palygorskite group and halloysites in the 'air-dry' state*

	d 001 (Å)	d 002 (Å)	d 060 (Å)
Kaolin Group	7.15–7.20	3.57	1.49
Serpentine Group	7.25 7.35	3.65	1.536–1.540
Berthierine	7.05	3.52	1.55
Smectite Group			
- dioctahedral	12.5–15.0*		1.49–1.50
- trioctahedral	12.5–15.0*	7.18	1.54
Illite–mica Group	9.5–10.0	4.75–5.00	1.49–1.50
Glauconite	9.5–10.0	4.75–5.00	1.51
Chlorite Group			
- Mg rich	14.5–14.4	7.07–7.18	1.54–1.55
- Fe rich	14.1–14.3	7.06–7.12	1.56
	Principal low angle reflections (Å)		d060
Sepiolite	12.1	4.49	—
Palygorskite	10.4	4.47	—
Halloysite-10A	10.1	4.46	1.48
Halloysite-7A (* RH = 60%)	7.4	4.43	1.48

TABLE 8.2. *Summary of data for d 00l/00l values of randomly interstratified clay minerals*

Mixed layer phase	Air dry	After exposure to ethylene glycol (EG) or glycerol (G)		After heating at 300°C
Illite-smectite	10–15Å	EG	17Å*	10Å
Illite-chlorite	10–14Å	EG	10–14Å	10–14Å
Chlorite-smectite	12–15Å	G	14–17Å	10–14Å
Illite- Mg vermiculite	10–14Å	G	10–14Å	10Å
Chlorite – Mg vermiculite	14Å	G	14Å	10–14Å
Mg Smectite – Mg Vermiculite	14Å	G	14–18Å	10Å

* 17Å for Smectite100 to Illite70 Smectite30.

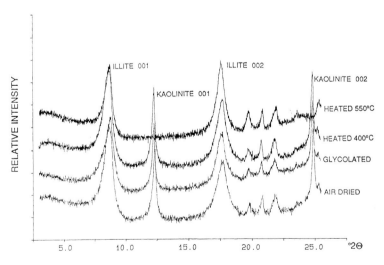

FIG. 8.3. XRD traces of a clay fraction containing illite and kaolinite.

The identification of smectites and vermiculites is based on their swelling behaviour with water, ethylene glycol or glycerol that is dependent on their interlayer cations (see Sections 2.1.3 and 2.1.4).

Other treatments that are occasionally used to differentiate between kaolinite, serpentine and chlorite minerals are solvation with DMS0 (dimethylsulphoxide) and boiling with 10% HCl. Solvation with DMSO causes kaolin

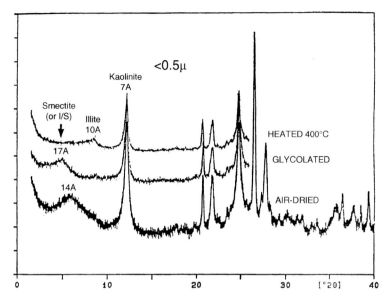

FIG. 8.4. XRD traces of a clay fraction containing kaolinite, illite-smectite and illite.

minerals to expand from 7Å to 11Å, whilst serpentine and chlorite minerals are unaffected.

Boiling with 10% HCl usually causes dissolution of chlorites (though Mg-chlorites may not dissolve) whereas kaolin and serpentine minerals do not dissolve. Fe and Mg chlorites can be differentiated from each other by differences in the relative intensities of their basal reflections.

(c) Mixed layer clay minerals. The composition and the random or regular nature of interstratified mixed layer clay minerals can be determined from the XRD data derived from analysis of the oriented clay fractions. Generally, random mixed layer clay minerals show variations in the basal reflections in the air dry state related to their changing compositions (Table 8.2).

Solvation with ethylene glycol causes expansion of the randomly mixed layer clay minerals with a smectite component (Table 8.2). On heating at 300°C, further changes occur (Table 8.2) and analysis of the XRD data on glycolation and after heating usually allows detailed identification of the mixed layer phases.

Ordered mixed layer clay minerals have characteristically one long d-spacing (>20Å) which is the sum of the component spacings. For a mixed layer 1:1 illite smectite with 50% illite and 50% smectite, $d001$ Illite50 Smectite50 = $d001$ Illite + $d001$ Smectite = 25Å. This long d-spacing is modified on solvation with ethylene glycol and/or glycerol and after heating depending on the composition of the mixed layer phase (Table 8.3).

8.1.1.2. *Crystallinity indices of illite/mica.* Well ordered highly crystalline minerals show sharp well defined X-ray diffraction peaks whereas less well ordered minerals have broad diffuse reflections. This phenomenon has been used to monitor variations in the crystallinity of illite/micas during diagenesis/ metamorphism and weathering cycles. The Weaver (1960) Crystallinity Index is based on the ratio of the intensity of the illite/mica 10Å–001 reflection to the background intensity at 10.5Å (Fig. 8.5); the peak becomes sharper and narrower in profile with increasing crystallinity. Kubler (1968) found the Weaver Crystallinity Index was satisfactory for poorly crystallized illites but produced large errors for well crystallized illites and devised his own index based on the width of the 10Å illite peak at half its height (Fig. 8.5 and Table 8.4).

In comparing the illite crystallinity indices it must be remembered that the Kubler and Weaver Index measurements critically depend on the X-ray diffractometer

TABLE 8.3. *Values of d001 (Å) for ordered mixed layer clay minerals (from Brindley 1981)*

Mineral	Air dry	Ethylene glycol	+ Glycerol	Heated at 300°C
1:1 illite-smectite	24.5–25Å	26.5Å	27.5Å	19.5Å
Illite-vermiculite	24.7Å	–	24.6Å	10Å
Chlorite-smectite	29.0Å	31.0Å	–	23.5Å
3:1 illite-smectite 75%illite/25%smectite	44.0Å	46.0Å	47.0Å	19.5Å

conditions used, i.e. radiation, width of divergence, scatter and receiving slits, monochromator and scan speed and these must always be defined when presenting measured data.

The factors that control illite crystallinity are not well understood but they include the thickness of the illite crystallites, crystal structure defects and their water and potassium contents. The main problem in the interpretation of illite crystallinity index values arises from the presence of illites from different sources in the same sample, i.e. detrital and diagenetic illites.

8.1.1.3. Peak decomposition techniques. Clay materials often contain poorly crystalline phases (clay minerals, oxides, etc) that produce broad diffuse X-ray diffraction reflections or peaks, which can be difficult to interpret. In modern powder diffractometry the use of peak stripping and peak decomposition techniques allow better definition and analysis of these reflections (Fig. 8.6) (Jones 1989; Lanson & Besson 1992). The broad diffuse X-ray 'peaks' are interpreted as being composed of a series of overlapping component 'peaks' and the decomposition and stripping techniques involve identifying and separating out these component 'peaks'.

8.1.1.4. Quantitative mineralogical analysis. The intensities of the X-ray reflections of a particular mineral are a function of the relative proportions of that mineral within a mixed mineral assemblage. This is the basis of all methods of quantitative mineralogical analysis by XRD techniques but is complicated by the intensity of the reflection being also a function of other factors. This can be represented by the equation:

$$Ip = Kp.Wp/\mu$$

where Ip is the integrated intensity of principal reflection of phase 'p', Kp = f (nature of phase 'p', the X-ray reflection measured and the experimental conditions), Wp is the weight fraction of phase 'p' and μ is the average mass attenuation coefficient of the total sample.

The quantitative analysis of minerals, and particularly the clay minerals, is complicated by the variability of their crystal structures and chemical compositions that affect their diffracted intensities in ways that can be difficult to predict.

Quantitative X-ray diffraction mineralogical analysis should be made using randomly oriented powder specimens. Because of their sheet silicate structures this is difficult to achieve with samples containing clay minerals but methods have been devised to try to overcome these difficulties.

The most common approach to solving the problem has been to use analysis of the bulk powder sample to estimate the non-clay mineral phases and the 'clay mineral' content. The relative proportions of the individual clay mineral are then estimated from analysis of the oriented clay fractions. Often the clay mineral proportions in the clay fraction are reported separately rather than incorporated into the total bulk mineralogy proportions. This can lead to misunderstanding and misinterpretation of the data. For example, if analysis of the clay fraction indicates the presence of 40% smectite, this alone might suggest a potentially damaging swelling clay problem. But if the clay material only contains 20% clay fraction there is therefore only 8% smectite in the bulk material. It is this value that is relevant to understanding the behaviour and properties of the bulk material.

The inherent bias in quantitative methods of clay mineral analysis that rely only on the analysis of the <2 μm fractions without due attention to the proportions of these

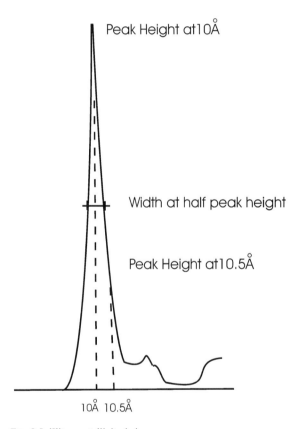

FIG. 8.5. Illite crystallinity index.

TABLE 8.4. *Definitions of diagenetic, anchizone and metamorphic zones based on the Weaver and Kubler Illite Crystallinity Index*

	Weaver Index	Kubler Index (CuKα)
Diagenetic		
	------------- 2.3 -------------	------------- 0.38°2θ -------------
Anchizone		
	------------- 12.1 -------------	------------- 0.21°2θ -------------
Metamorphic		

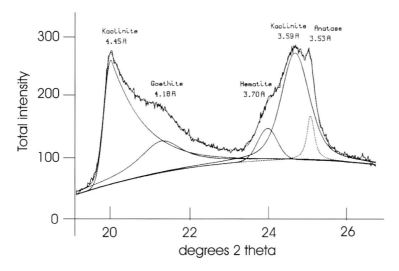

FIG. 8.6. Decomposition of complex overlapping X-ray reflections (from Jones 1989).

clay minerals in the total sediment was demonstrated by Towe (1974). It is still a major source of potential misinterpretation/misunderstanding in the application of mineralogical analysis.

Recent developments in the preparation of spray dried samples has offered the prospect that truly random powder samples including those containing clay minerals can be achieved rapidly, precisely and cheaply (Hillier 1998). Hillier has further shown that acceptable quantitative mineralogical analyses of spray dried samples of clay-bearing sandstones can be made using two different methods of quantification based on interpretations of their X-ray diffraction (Hillier 1999). However, Srodon et al. (2001) reported that similarly random powders could be produced by grinding samples under methanol in a McCrone rod mill and side loading the powder into a cavity mount for XRD analysis.

The most commonly used methods of quantification based on XRD analysis are the reference intensity ratio (RIR) and the Rietveld methods (e.g. Hillier 1999; Srodon et al. 2001)

In the RIR methodology the intensity of the chosen reflection of a mineral is measured relative to the intensity of the chosen reflection of an added internal standard. The measured intensities in the sample are then compared with the relative intensities of the same two phases in an artificial mixture with known amounts of the mineral and the internal standard to derive the proportion of the mineral in the sample. The use of the internal standard removes the need to make matrix absorption measurements and consequent corrections. For example, the method described by Srodon et al. (2001) used a single XRD pattern from an air dried random powder specimen with a zincite (ZnO) internal standard. The clay minerals are quantified from their diagnostic 060 reflections (after the application of peak decomposition procedures) that are much less affected by structural and compositional variations within individual clay mineral families than their basal 00 1 reflections.

The Rietveld method involves fitting the observed diffraction pattern with a synthetic diffraction pattern derived from combining patterns of the individual mineral phases present in the mixture. Using an iterative process, the relative proportions of each identified phase are systematically varied in the synthetic diffraction pattern to produce a 'best fit' with the diffraction pattern of the analysed sample. If the mineral phases in the analysed mixture have significantly different absorption coefficients a further correction has to be made to the quantitative estimates.

Batchelder & Cressey (1998) used a Rietveld method of quantification incorporating absorption coefficient corrections for the quantification of clay mineral-bearing samples using a position-sensitive detector diffractometer. This type of diffractometer differs from the more widely used scanning diffractometer in employing a curved array of fixed detectors over a 120°2θ arc rather than a scanning detector. This geometry minimizes the effects of preferred orientation without any special sample preparation but does not completely remove them. The geometry also means that the irradiated sample area is constant throughout the analysis which is of critical importance for quantitative analysis. This can also be achieved in modern scanning diffractometers but was a problem in earlier designs (Batchelder & Cressey 1998).

Recent developments in quantification of the mineral compositions of clay materials are encouraging and appear to offer methods capable of an accuracy of ±5% relative. In the analysis of clay material composed of a multi-mineral mixture mineral proportions should only be quoted to whole mineral percentage values—relative mineral % values quoted to 0.1 or 0.01% are spurious.

There are also significant problems to be overcome in dealing with complex mixtures especially those that contain mixed-layer clay phases. Differentiation of these phases and their full characterization is still likely to require detailed XRD analysis of the clay fraction to compliment the bulk XRD analyses, as outlined above.

Also it needs to be borne in mind that any quantification method based exclusively on XRD analysis will not include estimates of organic matter or poorly crystalline 'amorphous' phases. Estimates of organic matter and 'amorphous' phases require other methods of analyses, i.e. chemical, infra-red, thermal analysis, etc. Inorganic chemical analyses of both bulk powder and clay fractions have been incorporated into a combined XRD-chemical analysis approach to determine the compositions of clay materials (e.g. Calvert *et al.* 1989). The chemical data are used via computerized iterative procedures to modify the mineralogical estimates to provide 'best fits' with the chemical data. In the method described by Calvert *et al.* (1989) they also incorporate surface area and cation exchange analyses to further refine the quantitative mineralogical data. Thornley & Primmer (1995) described a combined XRD /thermal analysis approach to the analysis of clay-bearing sandstones and Johnson *et al.* (1985) incorporated thermal analysis into a complex multicomponent/matrix inversion analytical approach for the analysis of soils.

Although XRD analysis provides the most readily applicable means of quantifying the mineralogy of clay materials the use of supplementary chemical, infra-red, cation exchange and thermal analysis can often provide important additional information to improve the quantification methodology.

Having decided upon a method of mineral quantification it is imperative to retain that approach throughout as comparisons of data using different quantification methods are fraught with problems. Calvert *et al.* (1989) compared data from the analyses of four artificial 'rock-like' mixtures composed of varying proportions of clay and non-clay minerals made by five different laboratories. The significant variations in the results (Table 8.5) illustrate the difficulties when comparing mineralogical data obtained using different methods of quantitative analysis.

In deciding upon an appropriate method of quantitative analysis one also needs to bear in mind that for many applications only relative trends in mineralogical proportions are required and semi-quantitative estimates will suffice. There needs to be a judgement of how good the data have to be and what expenditure of time and money is justifiable for a given purpose.

8.1.2. Infra-red spectral analysis

Infra-red (IR) spectral analysis has been used to characterize minerals, including the clay minerals, and 'amorphous' phases. The basis of the method is the absorption of a multi-wavelength infra-red source and measuring the degree of absorption of the infra-red spectra. Absorption of the IR radiation is related to the molecular, stretching and bending vibration modes that are controlled by the overall crystal symmetry, the site symmetry of individual atoms in the structure, the overall atomic mass and the length and strength of interatomic bonds (Russell & Farmer 1994) (Figs. 8.7 & 8.8). The different absorption bands in the IR spectra can be used to identify minerals, variations in mineral structures and the causes of such variations that might be difficult to identify from other types of analysis.

Infra-red analysis has been promoted as an alternative means of identifying and making quantitative analysis of mineral mixtures. It has the advantage that it is capable of differentiating poorly crystalline 'amorphous' materials as readily as crystalline phases which is a limitation of XRD methods. However, in practice IR analysis of clays shows less sensitivity to differences in clay mineral structures and is not as versatile as XRD, for example, in differentiating mixed layer mineral species.

Generally IR analysis is used to supplement XRD data and more fully characterize individual minerals when specific information is needed on particular aspects of their structures and reactivity. IR analysis is widely used to monitor mineral-water and mineral-organic reactions where changes in bonding environment shift the interatomic vibration modes. Because organic-mineral interactions can significantly affect IR spectra it is often necessary to remove organic matter from clay materials before making IR analyses.

The application of infrared analysis has been greatly enhanced by the development of Fourier Transform infra-red (FTIR) spectrometers. The FTIR instruments with linked computerized control and data manipulation programs provide greater versatility, convenience and speed of use that have greatly enhanced the applicabilty of IR analysis. These instruments can now also be linked to an analytical optical microscope allowing the analysis of microscopic samples.

The FTIR instrument employs a detector that continuously monitors the full spectral range of the IR source. It is an inherently more sensitive instrument than the standard dispersive IR spectrometer. However, the power and versatility of such instruments can also be greatly improved by the use of computer programs to manipulate and interpret the data from multi-component mineral assemblages.

TABLE 8.5. *Comparison of the accuracy of estimates for clay and non-clay mineral proportions in four artificial 'rock-like' mixtures made by five different laboratories (Calvert et al. 1989)*

Laboratory	All minerals	Non-clay minerals	Clay minerals
A	92%	100%	78%
B	74%	79%	67%
C	46%	53%	33%
D	63%	67%	56%
E	57%	57%	56%

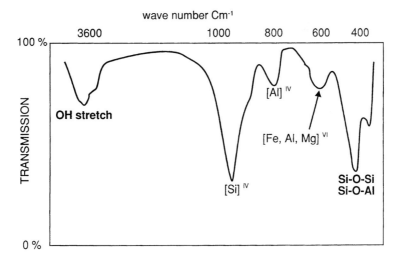

FIG. 8.7. Generalized interpretation of infra-red spectra (from Velde 1992).

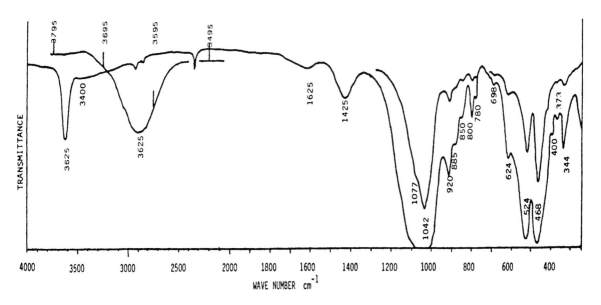

FIG. 8.8. Infra-red spectra for a montmorillonite (from van Olphen & Fripiat 1979).

8.1.3. Thermal analysis

Thermal analysis is used as a complementary technique to analytical methods such as XRD for the study of clay materials. Thermal analysis techniques monitor the changes in physical properties and/or reaction products when a substance is heated under controlled conditions. The principal types of thermal analysis are differential thermal analysis and thermal gravimetric analysis; other less commonly used techniques include evolved gas detection/analysis (Paterson & Swaffield 1987; Stucki et al. 1990). In addition to helping to characterize clay materials thermal analysis is also useful for the study of various reaction processes under varying temperature and pressure regimes, such as dehydration/hydration, interactions of clays with organic and inorganic compounds, oxidation and reduction processes (Stucki et al. 1990).

8.1.3.1. Differential thermal analysis. Differential thermal analysis (DTA) monitors changes in thermal properties by measuring differences in temperature between the sample and a thermally inert reference standard over a controlled temperature range up to about 1400°C. This type of analysis allows exothermic or endothermic thermal reactions that occur over this temperature range to be differentiated. An exothermic reaction in the sample produces heat and an increase in temperature of the sample relative to the standard and the reverse for an endothermic reaction in the sample in which heat is absorbed. Water loss reactions are generally endothermic whilst recrystallization reactions are usually exothermic (Fig. 8.9). An alternative means of monitoring changes in thermal properties is provided by differential scanning calorimetry (DSC). DSC involves measuring the different inputs of energy required to keep a sample and reference standard at the same temperature over a controlled heating range and thus allows monitoring of exothermic and endothermic reactions (Fig. 8.10). Paterson & Swaffield (1987) reported better sensitivity for the DSC method compared to DTA and DSC is far superior for determination of enthalpy values. In DTA and DSC analysis phase identification is made by comparing thermal 'events' as a function of temperature with those from known standards.

8.1.3.2. Thermal gravimetric analysis. Thermal gravimetric analysis (TGA) involves the monitoring of weight loss when a substance is heated and the temperatures at which such changes occur provide a useful identification characteristic. In the silicates this is principally a function of water content and how the water is present: surface adsorbed water, interlayer bound water, OH or crystalline water. Gypsum, hydrated oxides and hydroxides silicon, iron and aluminium also show weight loss due to dehydration. The carbonate minerals show a weight loss related to the loss of CO_2. Oxidation of

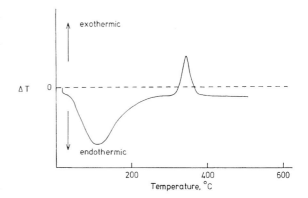

FIG. 8.10. Typical differential scanning calorimetry (DSC) output (from Paterson & Swaffield 1987).

organic matter results in a weight loss but some inorganic oxidation reactions involving absorption of atmospheric oxygen can be accompanied by an increase in weight.

Thermogravimetric analysis produces integral weight loss data curves that can be difficult to interpret. Consequently interpretations of the TGA data are better made using the first derivative of the weight loss curve. This is derivative thermogravimetric (DTG) analysis (Fig. 8.11).

As with DTA, qualitative TGA is achieved by comparing the weight loss characteristics with known standards. Quantitative thermogravimetric analysis is best attempted using DTG data but is often only used for the analysis of poorly crystalline material that cannot be adequately quantified by XRD analysis.

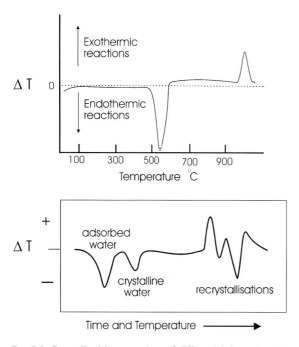

FIG. 8.9. Generalized interpretations of differential thermal analysis curves (from Velde 1992).

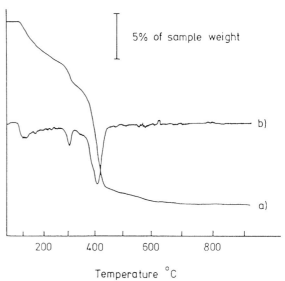

FIG. 8.11. Examples of (a) thermal gravimetric (TG) and (b) differential thermal gravimetric (DTG) curves in recommended format (from Paterson & Swaffield 1987).

8.1.3.3. Evolved gas detection and analysis. During heating of a substance, any evolved gases can be detected and analysed by gas chromatographs, FTIR or mass spectrometers attached to the thermal analysis equipment. In addition to characterizing the evolved gases such analyses can be used to interpret the thermal processes involved.

To generally summarise, thermal analysis allows identification of poorly crystalline phases that cannot be easily identified by XRD analysis. In mixed phase samples there can be problems in differentiating phases that have similar thermal characteristics in certain parts of the thermal spectra, resulting in overlapping peaks. The presence of organic matter and the strong exothermic reactions accompanying oxidation of the organic matter can obscure the thermal characteristics of minerals. Chemical pre-treatments of samples and/or carrying out the analyses in an inert atmosphere can help overcome some of these problems (Paterson & Swaffield 1987; Stucki *et al.* 1990).

Quantitative analysis is also possible using DTA and DSC based on standard calibration curves but there are problems in doing this satisfactorily for multiphase mixtures. However, it does offer a means of quantifying poorly crystalline phases that cannot be achieved by XRD analysis (Stucki *et al.* 1990).

8.2. Petrofabric analysis

In this section the applications of optical and electron microscopy techniques, image analysis methods, X-ray goniometry and specific surface area measurements to the study of clay materials are described.

8.2.1. Optical microscopy

The standard optical petrological microscope generally has limited use for the study of clay materials as individual particles are too small to be studied in a standard 30 micron thick thin section. The problem can be partly overcome by preparing thin sections that are less than the 30 micron thick but this is often difficult to achieve.

However, examination of standard thin sections of clay materials in a petrological microscope can be useful to determine the general distribution of clay and coarser grained material. The general orientation of platy clay particles, which can have an important influence on the development of anisotropic fabrics and properties in clay materials, can also be examined in the petrological microscope. The insertion of a Sensitive Tint Plate which increases the optical birefringence colours observed under crossed polars by one order can be a useful aid in the assessment of the preferred orientation of clay particles (Shelley 1985).

Tchalenko *et al.* (1971) quantified optical birefringence measurements and included them in a methodology for quantifying orientation in clays. Other methods for the quantification of rock fabrics based on optical microscopy observations have been described by O'Brien (1964), Brewer (1964), Smart (1965), Martin (1966), Morgenstern & Tchalenko (1967), La Feber (1967), Tchalenko *et al.* (1971). These methods have been largely superseded by image analysis of the higher magnification electron microscopy images (Bennett *et al.* 1991).

Optical microscopy can also be used to observe clay materials dispersed in water to determine non-clay minerals and rock fragments, micro-organisms and organic matter. To do this the preserved undried clay material is dispersed in water with the aid of a dispersion agent (e.g. sodium hexametaphosphate) and filtered through a 240 mesh (0.06 mm) sieve. The material retained is then dried and observed under reflected light using a binocular microscope.

8.2.2. Electron microscopy

There are essentially two basic types of electron beam instruments, the scanning electron microscope (SEM) and the transmission electron microscope (TEM), and all instruments and their modes of operation are essentially variations or combinations of these two instruments with their appropriate signal detectors (Mackinnon 1990).

The different modes of electron microscopy imaging and compositional analysis are outlined in Table 8.6 and these require different sample preparations. The purpose for which the analysis is required and the nature of the available samples needs to be carefully considered before deciding on which EM method is most appropriate.

Electron microscopy is used to examine the petrofabrics and composition of preserved samples in either the dried, or less commonly, the undried state. Drying of samples will modify the fabrics of clay materials and every effort needs to be made to minimize such modifications. Freeze drying and critical point drying have been found to be most effective in minimizing damage to soil fabrics prior to SEM examination of fracture surfaces (Gillott 1976; Shi *et al.* 1999).

8.2.2.1. Scanning electron microscopy. Scanning electron microscopy imaging involves scanning the sample surface with an electron beam in a TV-like raster and displaying the scattered electron signal from an electron detector on the TV screen where it can be captured photographically either directly or digitally (Mackinnon 1990).

The scattered electrons are of two types, low energy secondary electrons produced by inelastic collisions from the top 50–500 Å of the specimen and higher energy backscattered electrons produced by elastic collisions.

(a) Secondary electron imaging.
In secondary electron imaging both secondary and backscattered electrons are detected but the configuration of the detector is such as to maximise the detection of the lower energy secondary electrons compared to the higher energy backscattered electrons (Mackinnon 1990).

The number of the secondary electrons generated is a function of the angle of incidence of the beam onto the specimen. A tilted surface produces more secondary electrons than those normal to the beam and hence

TABLE 8.6. *Comparison of different electron optical modes of operation commonly used for the study of clay materials (Reed 1996; Huggett & Shaw 1997; Mackinnon 1990)*

Mode	Nature of sample	Analytical precision	Spatial resolution
CSEM	coated/uncoated fracture surfaces		5–10 nm
FESEM	coated/uncoated fracture surfaces		2–5 nm
ESEM	uncoated fracture surfaces		100 nm
BSEM	polished	$Z < 1$	100 nm
	C-coated surfaces	EDS < 1%	1–3 μm
		WDS <<1%	1–3 μm
Microprobe	polished	$Z < 1$	1 μm
	C-coated surfaces	EDS < 1%	1–3 μm
		WDS <<1%	1–3 μm
TEM	ion milled, microtomed, dispersed samples	0.3 nm 2%	30 nm

CSEM, Conventional SEM; FESEM, field emission SEM; ESEM, Environmental SEM; BSEM, Back scatter SEM; TEM, transmission EM; Z, atomic number; EDS, energy dispersive spectrometry; WDS, wavelength dispersive spectrometry.

topographic contrast is produced (Fig. 8.12A). Consequently secondary electron imaging is usually carried out on fractured surface specimens to enhance topographic contrast. This type of examination has made a major contribution to petrofabric analysis of geological materials. However, it has proved less successful in the study of finer grained rocks, such as clay materials, as they exhibit limited topographic contrast.

Usually the fracture surfaces of clay materials, which are non-conductors of electricity, need to be coated with a conducting material, usually gold or a gold-palladium alloy, to prevent charging under the electron beam.

The use of field emission crystal electron beam sources, compared to the usual thermionic sources, greatly increases the intensity of the electrons in the electron beam with a subsequent improvement in spatial resolution (Table 8.6). Field emission electron microscopy (FESEM) involves secondary electron imaging of fractured specimens. This type of electron microscopy has had little application in the study of clay materials but has major advantages for such analyses (Fig. 8.12B).

(b) Back-scattered imaging (Mackinnon 1990).
The number of higher-energy backscattered electrons which are produced by elastic collisions of the electron beam within the specimen vary as a function of:

(a) angle of tilt of the specimen producing topographic contrast as with secondary electrons;
(b) atomic number of the atoms in the specimen producing atomic number contrast.

The degree of back-scattering is measured in terms of the BSE coefficient (η).

$$\eta = \frac{\text{number of back-scattered electrons}}{\text{number of incident electrons}}.$$

The BSE coefficient is a function of the mean atomic number of the specimen and the higher the mean atomic number the brighter the image (Fig. 8.12 C,D; Fig. 8.13). It is possible to differentiate minerals with only a difference of 0.1 in their mean atomic number (Lloyd & Hall 1981). This would be equivalent to a 0.5 wt% change in the FeO contents or a 1 wt% change in the MgO contents of a chlorite; a change in K_2O content of 1 wt% will have the same effect for an illite. These changes in the atomic number contract make the use of BSE imagery very useful for analysis of small compositional changes in minerals.

Because the intensity of the back-scattered electron signal is also a function of the angle of tilt producing topographic contrast, polished thin sections or blocks, need to be used to examine the effects of backscatter contrast alone.

The samples are usually coated with carbon as a conductor to prevent surface charging. Minerals may be identified from their atomic number contrast but from Figure 8.13 it is apparent that different minerals can show similar atomic number contrasts. It is therefore necessary to couple microprobe analysis facilities with the BSE image to fully identify minerals.

(c) Environmental scanning electron microscopy.
The modes of scanning electron imaging discussed above require that the specimens be thinly coated with a conducting material to prevent charging and for the specimens to be observed under a high vacuum.

The environmental SEM (ESEM) allows the imaging of wet specimens with no coating which greatly increases the nature of the specimens and *in situ* processes that can be observed. This has been made possible by the development of a differential pumping system which maintains the electron gun under the required high vacuum but allows the specimen to be held in a low-pressure gaseous environment (Unwins 1994; Danilatos 1991).

Backscatter detectors can be successfully used in ESEM instruments with little loss of resolution. Conventional secondary electron detectors cannot be used in ESEM instruments but this problem has been overcome by development of an 'environmental SE' detector that can operate under low vacuum conditions (Unwins 1994).

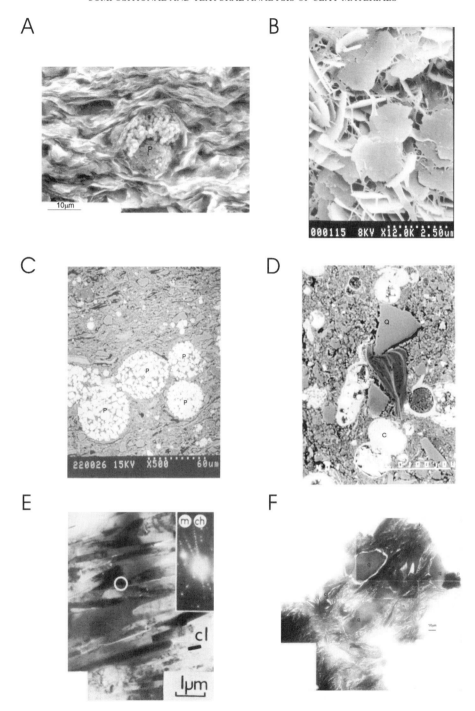

FIG. 8.12. Plate of electron microscope images. (A) Secondary electron image of a shale mudrock showing preferred orientation of platy clay particles and nodular framboidal pyrite (P) (from O'Brien & Slatt, 1990 reproduced with permission of the authors and publishers). (B) Field emission secondary electron image of a mudrock showing the growth of fibrous illite from illite plates. (C) Back scatter electron image of a mudrock showing sections through nodular framboidal pyrite (P). (D) Back scatter electron image of a calcareous sandy mudrock showing carbonate microfossils (C), quartz clasts (Q) and a deformed micaceous clast (M). (E) Transmission electron image of mica and chlorite intergrowths and the electron diffraction pattern from these intergrowths showing the different patterns exhibited by the mica and the chlorite. (F) Transmission electron image mosaic of a silty mudrock showing quartz grains (q) with small quartz overgrowth cements within the clay matrix.

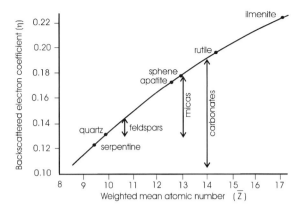

FIG. 8.13. Relationship between the back-scattered electron coefficient (η) and the weighted mean atomic number ($\bar{Z}$) for some common minerals (from Lloyd & Hall 1981).

The ESEM can also include X-ray micro-analysis facilites to monitor phase compositions. The resolution of the ESEM is generally less than that of a standard SEM depending on the conditions of the specimen chamber: the lower the relative pressure of the sample chamber, the higher the potential resolution (Table 8.6).

8.2.2.2. Transmission electron microscopy (TEM). Transmission electron microscopy (TEM) analysis is not routinely used for the examination of clay materials. The principal disadvantages in the use of TEM for the analysis of clay materials are: (a) specimen preparation, which is a lengthy process and it can be difficult to ensure that the appropriate material is being sampled; (b) the delicate clay specimens can be readily damaged in the electron beam and also during sample preparation methods. Consequently careful consideration needs to be given as to what useful additional information is likely to be provided by such analyses.

In the TEM the electron beam is transmitted through ultra-thin specimen sections and an image is produced by differential scattering and absorption that is a function of variations in electron density of the specimen. Higher electron density areas scatter and absorb more of the electron beam and thus appear darker than areas of lower electron density. Image resolutions down to tenths of nanometres can be routinely achieved compared to optimum resolutions of tens of nanometers for the SEM in secondary imaging mode.

Many mineralogical analyses involving TEM instruments have been on material extracted from the host rock and dispersed on a substrate to provide the required electron-optically transparent specimens. Consequently, this precludes any observations regarding the textural relationships between the mineral grains. However, techniques have been developed involving ion beam milling to allow TEM examination of specimens taken intact from standard thin sections (Huggett & White 1982) (Fig. 8.12F) or by microtoming sections down to 100–50 nm thickness from critically point dried clay samples embedded in epoxy resin (Mackinnon 1990). The TEM can also be used in diffraction mode (see the Text Box below).

8.2.3. Image analysis techniques

Observations of the petrofabrics of clay materials can be made using optical and electron microscopy techniques discussed above. A variety of image analysis techniques have been employed to quantitatively analyse or classify these fabrics. Quantification of the fabrics allows the data to be more readily tested and manipulated using statistical modelling techniques.

Examples of modern image analysis techniques applied to SEM by Sokolov & O'Brien (1990) and to TEM images by Smart *et al.* (1991) and Chiou *et al.* (1991). Sokolov & O'Brien (1990) described the application of a signal intensity gradient technique to construct rose diagrams to define grain orientation in clay fabrics (Fig. 8.14).

The system described by Smart *et al.* (1991) involves digitizing the gray scales to differentiate between particles and voids in negatives of TEM images of ultra-thin

Transmission electron microscopy—diffraction mode

When the electron beam is transmitted through a crystalline specimen, part of it is diffracted principally by the planes nearly parallel to the electron beam. The electron diffraction patterns so produced provide direct information on the crystal structure of the phase involved (Fig. 8.12E).

It is also possible to produce real-space images of crystal lattices using TEM techniques (lattice fringe imaging). This type of imaging is caused by interference of the diffracted electron beams as they pass through the crystalline material.

TEM instruments may also have microprobe analytical facilities that enable chemical analyses to be made of areas down to about 30 nm across. Unlike SEM microprobe analyses, TEM microprobe analyses determine elemental ratios with no direct measurement of water content. However, there is less likelihood of interference from adjacent mineral grains that can be encountered in SEM microprobe analysis because of the ultra-thin specimens being analysed in the TEM (Guven 1990).

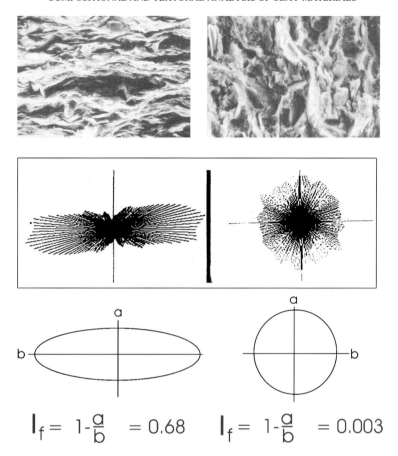

FIG. 8.14. (a) Scanning electron micrographs (b) fabric index diagrams of argillaceous rocks showing oriented and random microfabrics. Micrographs from O'Brien and Slatt (1990) published with permission of authors and publishers.

sections of clay materials. Digital image analysis allows mapping of domains and particle clusters and measurement of their dimensions and orientation.

Chiou et al. (1991) described a manual technique based on point counting and direct measurement of grain orientation in TEM micrographs.

8.2.4. X-ray goniometry

X-ray goniometry provides an alternative indirect method of quantifying clay fabrics and specifically clay particle orientation based on the method described by Gipson (1966). The method is based on comparing the intensities of the XRD reflections of illite from three mutually perpendicular surfaces of a sample of clay material. The orientation ratio (O/R) is calculated from the following ratio of the intensities of the illite reflections:

$$O/R = \frac{A_{002}/A_{020}}{B_{002} + C_{002}/B_{020} + C_{020}}$$

where A_{002} is the intensity of illite 002 reflection for the horizontal surface (parallel to bedding); A_{020} is the intensity of illite 020 reflection for the horizontal surface; B_{002}, C_{002} is the intensity of illite 002 reflection for the two vertical surfaces and B_{020}, C_{020} is the intensity of illite 020 reflection for the two vertical surfaces.

A ratio of one indicates random orientation and values greater than 1 represent an increasing degree of orientation. The method is independent of the amount of illite present but is influenced by the illite crystallinity. In the absence of illite another clay mineral can be substituted.

Tchalenko et al. (1971) also employed an XRD method involving measurement of kaolinite and illite reflection intensities together with optical analysis based on measurement of mineral birefringence to determine particle orientation in clays.

8.2.5. Specific surface area

Clay materials, being composed of fine-grained clay and other colloidal phases such as iron and aluminium oxides/hydroxides, have large and variable surface areas that can significantly affect their properties.

The clay minerals have variable surface areas (Table 8.7) that reflect their external surface areas, which are a function of their mean crystal size and their platy or

TABLE 8.7. *Surface areas of clay minerals (from van Olphen & Fripiat 1979)*

	Approximate Surface Area(m²/g)		
	Internal	External	Total
Smectite	750	50	800
Illite	5	25	30
Chlorite	0	15	15
Kaolinite	0	15	15

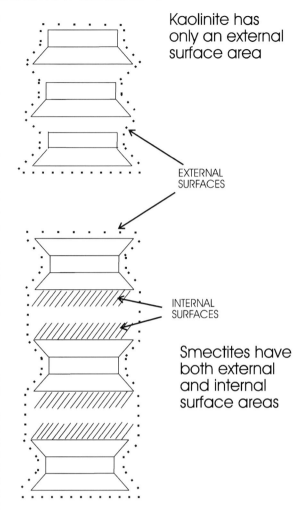

FIG. 8.15. Schematic illustration of the differences between external and internal surface areas of clay minerals (from Eslinger & Pevear 1988).

fibrous habits, and their internal interlayer surface areas, which are a function of the clay mineral type. For example, smectite has internal surface areas much greater than their external surface areas, whilst kaolinite has no measurable internal surface area (Fig. 8.15).

The external surface area can be measured using nitrogen adsorption having degassed the sample by evacuation and gentle heating to collapse any expandable clay interlayers prior to sorption of the nitrogen. From the amount of nitrogen adsorbed the surface area is determined by application of the Brunauer, Emmett and Teller (BET) equation

$$X/X_m = cR/\{(1 - R)[1 + (c - 1)R]\}$$

where X is the mass of nitrogen adsorbed; X_m is the mass of monolayer coverage; R is the relative vapour pressure and c is the constant related to the energy of sorption.

A problem with the use of nitrogen sorption is that the evacuation process can cause the platy clay particles to collapse together and thus reduce the external surfaces accesssible to sorption of the nitrogen. An alternative approach is to use glycerol or ethylene glycol rather than nitrogen. These polar molecules can be sorbed into the interlayer sites of expandable clays and thus give a measure of the internal and external surface areas. Water has the potential to be used to measure total surface areas but problems have been encountered in applying the BET equation to calculate the surface areas of swelling clays from water sorption (Newman 1987).

Adsorption of methylene blue has been used as a general measure of surface area. It has the advantage of being rapid and can be applied to wet materials (Hang & Brindley 1970). However, there is evidence that the mechanism of methylene blue absorption varies with different clay mineral compositions (Ramachandran *et al.* 1962) that can significantly affect surface area results from this method and significant discrepancies with surface areas measured using BET gas absorption methods.

8.3. Inorganic geochemical analysis

8.3.1. Bulk compositions and selective leach techniques

The most commonly employed techniques for bulk inorganic geochemical elemental analyses of clay materials are X-ray fluorescence spectrometry (XRF) and inductively coupled plasma atomic emission spectroscopy (ICP-AES)(Potts 1987; Fairchild *et al.* 1988).

8.3.1.1. X-ray fluorescence spectrometry. The basis of X-ray fluorescence (XRF) analysis is that when a material is irradiated with an X-ray or electron beam source the elements present are capable of acting as new sources of X-rays producing fluorescent X-rays. The energy of the emitted X-rays is a function of the atomic number and the abundance of that element and this forms the basis of quantitative elemental XRF analysis.

One limitation of XRF analysis is that only elements with an atomic number of that of sodium and above can be routinely analysed and that the higher the atomic number the lower the detection limit. For example, for routine XRF analyses the detection limit for sodium ($Z = 11$) is 1600 ppm and for strontium ($Z = 38$) it is 3 ppm (Potts

XRF Sample preparation

Samples for XRF analysis can be presented as pressed powder specimens requiring little preparation. However, such sample specimens do suffer from particle size, mineralogical and variable absorption–enhancement effects, arising from the presence of other elements (matrix effects). To help overcome these matrix effects the specimens for XRF analysis can be presented in the form of a fused glass disc. Inter-element absorption–enhancement effects in the fused glass disc are lessened by dilution with the borate glass and these effects are further reduced by the addition of lanthanum oxide as an absorber. One disadvantage of the fused glass specimens is that trace element constituents are diluted by the flux and thus more difficult to detect. For such trace element analyses the pressed powder specimens may, on balance, be better than the fused glass specimens (Potts 1987).

1987). The reason for this is that the fluorescence yield is inversely proportional to the square of the atomic number value—Moseley's Law. Detection of elements lighter than sodium is possible by XRF analysis but requires non-standard detection methods which are used in electron microprobe analysis (Section 8.4.2).

Fluorescent X-rays can be detected on the basis of their energy, using energy dispersive spectrometers (EDS), or their wavelengths, using wavelength dispersive spectrometers (WDS). Generally WDS instruments have better resolution than the EDS instruments but WDS systems are large and mechanically complex compared to EDS systems. Consequently, though the standard laboratory-based XRF instruments have WDS detection systems, effective portable XRF machines have been developed using EDS detectors.

8.3.1.2. ICP-AES analysis. Inductively coupled plasma atomic emission (ICP-AES) analysis is a destructive solution-analysis technique for inorganic elemental analysis.

The basis of ICP-AES analysis is the introduction of the sample solution into an argon gas plasma heated to about 10 000°C which causes complete atomization of the elements in the sample. The resulting emission of light is analysed in terms of its wavelengths and intensities and forms the basis for the quantitative inorganic elemental analysis. This type of analysis can be used to determine a wider range of elements than XRF techniques, including those with atomic number less than 11 and analyses can be carried out on small samples.

A further development of the ICP technique has been ICP-mass spectrometry (ICP-MS), which uses a mass spectrometer detection system and has much lower detection limits than ICP-AES. The ICP-MS is now a widely used routine method for rapid multi-element trace analysis.

8.3.1.3. Carbonate determination. The determination of total carbonate is based on the measurement of the CO_2 released by the acid dissolution of the carbonates or ignition of the sample at 1000°C. The released CO_2 is measured either directly as a volume measure or sample weight loss, by the amount absorbed onto an absorbent material or producing a measurable amount of a precipitate resulting from chemical reactions with the evolved CO_2.

In widely used C–H–N analysers, samples are first ignited to 550°C to decompose organic matter followed by ignition at 1000°C to decompose the carbonates. The measured weight losses thus provide measures of organic and carbonate carbon. There is evidence that some carbonates (e.g. siderite) can decompose below 550°C and also not all organic material decomposes below 550°C. Also other phases can be decomposed by high temperature ignition (e.g. sulphates and sulphides to evolve SO_2). This problem can be overcome by the use of infra-red detectors to differentiate and measure the CO_2.

The measurement of sample weight loss after dissolution in dilute hydrochloric acid is an inadequate measure of carbonate as other minerals can be dissolved by this acid treatment.

ICP–AES analysis—sample preparation

Decomposition of clay material prior to ICP-AES analysis is usually by acid dissolution in combined perchloric, nitric and hydrofluoric acids which allows rapid dissolution without loss of silica by volatilization.

An alternative method of sample digestion is to use the lithium metaborate fusion similar to that used for XRF analysis. The resultant cooled melt is then dissolved in nitric acid and diluted to produce the solutions for analysis (Potts 1987).

Sulphates in clay materials

In clay materials sulphate is present as calcium sulphates (gypsum, anhydrite), barite ($BaSO_4$), celestite ($SrSO_4$), sodium and magnesium sulphates. The sodium and magnesium sulphates are generally soluble in water, gypsum is soluble in dilute hydrochloric acid, anhydrite slightly soluble in acid, barite and celestite are insoluble in acid.

8.3.1.4. Sulphur species. The sulphur species in clay materials can be present as sulphate, sulphide and organic sulphur. Total sulphur content can be readily determined by XRF analysis or by ignition of samples at 1500°C causing decomposition of organic sulphur, sulphate and sulphide phases and measuring the evolved SO_2. The sulphate and sulphide sulphur can be chemically differentiated on the basis of their relative solubilities in water and acid.

A commmonly used procedure for determining water soluble sulphates employs a 2:1 water/solid extract and the sulphate concentration in the resultant solution is determined by ICP-AES. This method is more rapid and as reliable as the standard gravimetric method based on the addition of barium chloride resulting in the precipitation of barium sulphate (Reid *et al.* 2001; Bowley 1995).

Acid extraction results in the dissolution of sulphates, such as gypsum, and also monosulphides, such as pyrrhotite. Following acid extraction, using hot dilute hydrochloric acid, the resultant solution is analysed by ICP-AES. The ICP-AES method is more rapid than gravimetric analysis based on the addition of barium chloride to the acid extract solution resulting in the precipitation of barium sulphate (Reid *et al.* 2001; Bowley 1995).

Pyrite sulphur, or more strictly reduced inorganic sulphur, can be determined rapidly and reliably based on the decomposition of reduced sulphur to H_2S in hot acidic $CrCl_2$ solution and then precipitated as ZnS. The precipitated sulphur concentrations can be determined by iodometric titration or by ICP-AES (Canfield *et al.* 1986; Reid *et al.* 2001).

Lord (1982) has described a method based on the analysis of iron to determine pyrite sulphur that has lower detection limits and avoids the problems of overlap in differentiating pyrite, elemental and organic sulphur in the sulphur analysis methods.

Organic hosted sulphur occurs in various forms that make it difficult to analyse using a single coherent method. Consequently the organic sulphur is determined as the difference between the total sulphur and the separately determined different forms of inorganic sulphur. Alternatively the organic sulphur can be determined from analysis of the residue following sequential analysis of water-soluble, acid-soluble and reduced sulphur species (Tuttle & Goldhaber 1986; Reid *et al.* 2001).

8.3.1.5. Selective leaching techniques. Selective leaching techniques are used to try and determine the source of particular elements in the bulk material, e.g. is the Fe present hosted in clay minerals, Fe-carbonates, Fe-oxides, adsobed onto colloidal surfaces?

Bain *et al.* (1994) outlined a variety of leaching procedures to determine the source of elements in poorly ordered and organic phases in clay materials, which are summarized in Table 8.8.

A combination of these methods can be used to define the various sources of elements such as Fe and Al.

Crystalline Fe oxides = dithionate Fe—oxalate Fe
Poorly ordered Fe/Al = oxalate Fe/Al—pyrophosphate Fe/Al
Organically bound Fe/Al = pyrophosphate soluble Fe/Al

Leaching with hot dilute (10%) hydrochloric acid can be used to leach out carbonates and determine elements associated with carbonate phases.

8.3.1.6. Interstitial water analysis. Interstitial pore waters can be extracted by mechanical compaction of the preserved clay material in a low pressure, teflon-lined stainless steel squeezer (Manheim 1966; Presley *et al.* 1967). The extracted interstitial fluids are principally analysed using ICP-AES, specific ion electrodes, titriometric and microgravimetric techniques.

8.3.2. Electron microprobe analysis (EMPA)

As mentioned in Section 8.3.1, the irradiation of a specimen by an electron beam produces fluorescent X-rays that are characteristic and proportional to the concentration of the elements in the specimen. A focussed electron beam of variable size is used for this type of analysis and chemical analyses can be obtained for grains down to 1–3 μm

TABLE 8.8. *Characterization of poorly ordered materials (after Bain et al. 1994)*

Leaching method	Phases leached	Elements leached
Alkali dissolution	Poorly ordered Al silicates	Al,Si
Ammonium oxalate	Poorly ordered inorganics/organics	Al,Fe,Mn,Si
Pyrophosphate	Organics	Al,Fe,C
Na-dithionate	Poorly ordered inorganic, organic and crystalline Fe	Al,Fe,Si

Electron microprobe analysis

In EPMA fluorescent X-rays are detected by either wavelength dispersive (WDS) or energy dispersive (EDS or EDAX) spectrometers and usually both types of detectors are fitted to the instruments. Generally, the EDS detectors have less good resolution than the WDS detectors and the latter are usually used for more accurate/precise analyses of individual elements, particularly the lighter elements, and elemental mapping procedures. The EDS detectors can produce acceptable analyses with short count times and small beam current which reduce the chance of specimen damage or volatilization of labile elements. Velde (1984) has indicated the optimum conditions for microbeam analysis of clays to limit the degradation of the hydrous clay phases by the electron beam.

(Table 8.6). Accuracy of about $\pm 1\%$ and detection limits typically in the range of 50 ppm are readily obtainable (Reed 1996). Such methods now allow the microprobe analysis of the elements with atomic numbers between 5 and 9, i.e. boron to fluorine.

Quantitative EMPA is best carried out on flat polished specimens oriented normal to the incident electron beam. Analyses of fractured specimens can generally only be used for semi-quantitative 'finger-print' analyses of mineral phases. This type of analysis provides absolute element concentrations with the water content estimated by difference, which can then be converted to mineral formulae as required. One problem with this technique is that mineral grains beneath the grain being analysed can be inadvertently penetrated by the electron microbeam and included in the analysis.

Based on the microprobe analyses outlined above it is possible using on-line data interpretation programs to determine mineral formulae from point analyses of individual grains.

One potential limitation of EMPA is that, as mentioned for XRF analysis, the analysis of elements lighter than sodium require special detection methods. These include the use of thin window or windowless detectors to reduce absorption of the low-intensity long-wavelength fluorescent X-rays produced by these elements (Reed 1996). Another aspect of EMPA is elemental mapping which employs a scanning electron beam in a rectangular raster across the specimen and allows the distribution of an element to be mapped across a specimen. The method can be used with EDS or WDS detectors but there are advantages to using a WDS detector. This is a powerful tool that also allows compositional zonal variations in individual mineral grains to be monitored.

8.3.3. Ion exchange

The ion exchange capacity of clay materials is a measure of two fundamental properties, the external and internal surface areas and the charge on the surface area and is chiefly controlled by the presence of clay minerals, colloidal iron and aluminium oxides/hydroxides and organic matter. Ion exchange can involve both cations and anions but cation exchange processes predominate.

8.3.3.1. Cation exchange capacity. Cation exchange capacity can be defined as the sum of exchangeable cations that can be adsorbed at a specific pH. In addition to the cation exchange potential of clay minerals and other inorganic colloidal phases, organic matter can significantly affect cation exchange capacities. For example, the presence of 3–5% organic matter can increase the cation exchange capacity by 20–50% (Gillott 1987). The influence of organic matter arises by the formation of complexes with inorganic compounds and by large organic molecules blanketing potential exchange sites in mineral phases. There are two main methods of measuring cation exchange capacity:

(a) direct displacement of a saturating index cation;

Methylene Blue method (Hang & Brindley 1970)

The absorption of the organic dye, methylene blue, has been used as a general measure of cation exchange and of surface area. This is a rapid method and can be carried out on wet samples. However, there is evidence that the absorption of methylene blue is affected by clay mineral composition. For example, in the smectites the large methylene blue molecules 'cover up' potential exchange sites and thus produce underestimates of cation exchange capacity.

(b) a summation method based on the measurement, and summation of, displaced exchangeable cations i.e. Na + Ca + Mg + K (Rhoades 1982; Bain & Smith 1994).

The most commonly used index cation has been NH_4^+ but for some minerals this can produce erroneously low values. In 2:1 sheet silicates some of the NH_4^+ ion can become fixed and not exchangeable. For this reason and because they are easier to analyse after displacement, sodium and barium have been suggested as better index cations than NH_4^+.

The process involves first saturating the clay material with the index cation (e.g. NH_4^+), removing the excess and then displacing the index cation with an appropriate salt solution (e.g. acidified NaCl solution). The concentration of the displaced index cation in solution provides a measure of the cation exchange capacity. In the summation method this is done sequentially for each of the cations. The cation exchange capacity is usually expressed as milliequivalents per 100 grams of material.

8.3.3.2. Anion exchange capacity. Anion exchange capacity is the sum of the exchangeable anions that a clay material can adsorb at a given pH and is a measure of the overall positive charge on clay minerals and other colloidal phases. As with cation exchange the anion exchange is measured by saturating with an index anion (e.g. phosphate, chloride) and measurement of the amount adsorbed. Wada & Okamura (1977) described a combined method for cation and anion exchange capacity using ammonium chloride as the saturating salt. Anion exchange is also expressed as milliequivalents per 100 grams of material.

8.4. Organic matter

There are a wide variety of analytical techniques available for the characterization of organic matter in clay materials (Bordenave *et al.* 1993). However, in the context of the use of clay materials in construction, only a limited discussion of these techniques is included here.

8.4.1. Total organic carbon (TOC) analyses

The colour of a clay material can provide an approximate indication of the total organic carbon (TOC) (Fig. 8.16). The TOC content is measured by combustion of a sample after carbonate has been removed by acid dissolution in air or oxygen. In the widely used C-H-N Leco analyser method, after acid dissolution to remove any carbonate, the sample is mixed with CuO to convert the carbon monoxide released by combustion at 1500°C to CO_2, which is then detected by infra-red analysis. The results from the Leco method may suffer interference by decomposition of sulphate minerals that does not occur if lower combustion temperatures are used.

8.4.2. Rock-Eval pyrolysis

Organic matter can be characterised by elemental analysis of C,H,O, but requires separation of the organic matter and lengthy analytical procedures. Rock-Eval pyrolysis provides a relatively rapid screening tool of the organic matter and an indirect measure of the oxygen, hydrogen and carbon elemental compositions. It allows the total organic matter to be assessed in terms of its various components based on its behaviour on heating up to 600°C (see the Text Box on p. 219).

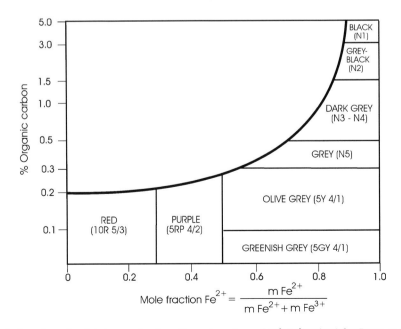

FIG. 8.16. Variations in the colour of wet shales as a function of organic matter and Fe^{2+}/Fe^{3+} ratios (after Potter *et al.* 1980).

Rock-Eval analysis

The Rock-Eval instrument consists of a pyrolysis and an oxidation module. The pyrolysis module involves analysis of the organic matter heated in a helium atmosphere. Free hydrocarbons, oil and gas, contained in the organic matter are vapourised by heating at 300°C and measured by a flame ionization detector- the S_1 peak (Fig. 8.17).

Heating between 300–600°C causes cracking of kerogen and heavy bitumens—the S_2 peak. Oxygen compounds decompose between 300–390°C and the resulting evolved CO_2 is measured as the S_3 peak (Fig. 8.17). The maximum temperature for the S_2 peak is the T_{max} value—the temperature of maximum hydrocarbon evolution.

The organic residue remaining after the recording of the S_2 peak is measured by combustion of the organic matter at 600°C in the oxidation module—the S_4 peak. The evolved CO/CO_2 is passed over CuO to convert the mixture to CO_2 for measurement. The low temperature of ignition can mean that the more mature organic matter e.g. coals will not be combusted and thus not measured. The summation of S_1, S_2, S_3 and S_4 provides a measure of the total organic carbon content. This TOC measure is often an underestimate of the TOC from Leco combustion because of the low combustion temperature used in the Rock-Eval. This underestimation depends on the composition of the organic matter, as discussed earlier.

From the Rock-Eval analyses the Hydrogen (HI) and Oxygen Index (OI) can be determined.; HI = S_2/TOC and OI = S_3/TOC. These can be used to classify kerogen types from the van Krevelen plot (Fig. 8.18).

Type-I kerogens are principally algal materials, Type-III terrigenous 'woody' materials and Type-II a mixture of Type-I degraded by bacteria and Type-III (Cornford 1984). This simplification of the true situation serves as a useful summary.

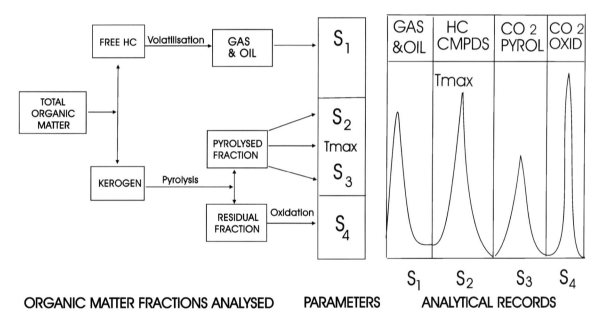

FIG. 8.17. Rock—Eval analysis (from Bordenave 1993).

8.4.3. Organic petrography

The study of preserved organic matter by optical microscopy in both reflected and transmitted light allows the identification of individual organic materials (macerals), their distribution and an assessment of organic maturity. The analysis of the material in transmitted light is often made on chemically separated organic matter rather than the *in situ* material.

Observations in reflected light are made on the *in situ* material using white light to examine images by reflection contrast and ultraviolet radiation to examine fluorescence effects. Such observations in reflected light have formed the basis of a classification of organic matter into macerals. Macerals are the organic equivalents of minerals that were first defined from studies of polished sections of coals in incident reflected light (Stopes 1935). There are three maceral groups, vitrinite/huminite,

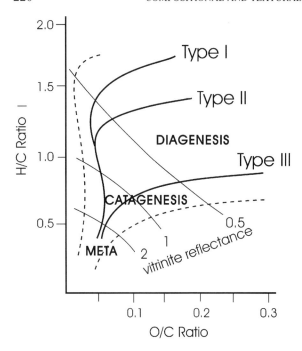

FIG. 8.18. *Van Krevelen diagram.*

TABLE 8.9. *Optical properties of macerals*

Maceral	Reflectance	Fluorescence	
Inertinite	High	$R_o > 2.5$	None
Vitrinite	Moderate	R_o 0.5–2.5	None
Huminite	Low	$R_o < 0.5$	
Liptinite	Low	$R_o < 0.5$	High

TABLE 8.10. *Correspondence between organic maturity, coal rank and vitrinite reflectance (R_o%)*

Hydrocarbon genesis	R_o %	% Org. carbon	Coal Rank
Immature	0.20		
	0.25		Peats
	0.30		-----------
	0.35	70	
-----------	0.40		Lignites
	0.50		
Oil Generation	0.60	77	-----------
	0.65		
	0.80		
-----------	1.00		Coals
Gas Generation	1.25	87	
	2.00		
	2.30	91	-----------
-----------	3.00		Anthracites
Over-mature	4.00	93	-----------
	>4.50		Meta - Anthracites

liptinite and inertinite, which can be differentiated optically, based on, standardized reflectance and fluorescence measurements (Table 8.9).

Inertinite macerals originate from woody tissue and fungi, vitrinite/huminite from plant debris especially wood and bark and liptinite from spores, cuticles and algae.

The standardized reflectance measurements, known as vitrinite reflectance (R_o), of the polished specimens are made using an optical microscope fitted with a microphotometer to measure the reflection of a monochromatic light source. In addition to helping to define macerals, standardized reflectance measurements have been used to monitor organic maturation processes; the standardized reflectance increases with increasing maturity (Table 8.10).

8.4.4. Biological assays

There has been increasing interest in the number and types of micro-organisms in soils and groundwaters. There are three main ways of estimating the micro-organism populations; examining them *in situ* microscopically, measuring consequences of their activity and/or growing them and monitoring activity (Campbell 1992).

References

BAIN, D. C. & SMITH, B. F. L. 1994. Chemical analysis. *In*: WILSON, M. J. (ed.) *Clay Mineralogy: Spectroscopic and Chemical Determinative Methods*, Chapman & Hall, London, 300–332.

BAIN, D. C., MCHARDY, W. J. & LACHOWSKI, E. E. 1994. X-ray fluorescence spectroscopy and microanalysis. *In*: WILSON, M. J. (ed.) *Clay Mineralogy: Spectroscopic and Chemical Determinative Methods*, Chapman & Hall, London, 260–299.

BATCHELDER, M. & CRESSEY, G. 1998. Rapid, accurate phase quantification of clay-bearing using a position sensitive X-ray detector. *Clays and Clay Minerals*, **46**, 183–194.

BENNETT, R. H., BRYANT, W. R. & HULBERT, M. H. (eds) 1991. *Microstructure of Fine-grained Sediments: from Mud to Shale. Frontiers in Sedimentary Geology*. Springer, Berlin.

BORDENAVE, M. L., ESPITALIE, J., LEPLAT, P., OUDIN, J. L. & VANDENBROUCKE, M. 1993. Screening techniques for source rock evaluation. *In*: BORDENAVE, M. L. *Applied Petroleum Geochemistry*, Editions Technip, 219–278.

BOWLEY, M. J. 1995. *Sulphate and Acid Attack on Concrete in the Ground: Recommended Procedures for Soil Analysis*. BRE Report 279.

BREWER, R. 1964. *Fabric and Mineral Analysis of Soils*. Wiley, New York.

BRINDLEY, G. W. 1981. X-ray identification (with their ancillary techniques) of clay minerals. *In*: LONGSTAFFE, F. J. (ed.) *Short Course in Clays and the Reservoir Geologist*, Mineralogical Association of Canada, 22–38.

CALVERT, C. S., PALKOWSKY, D. A. & PEVEAR, D. R. 1989. A combined X-ray powder diffraction and chemical method for the quantitative mineral anlaysis of geologic samples. *In*: PEVEAR, D. R. & MUMPTON, F. A. (eds) *Quantitative Mineral Analysis of Clays*, CMS Workshop Lectures v.1., Clay Minerals Society, 154–167.

CAMPBELL, R. 1992. Microbiology of soils. *In*: HAWKINS, A. B. (ed.) *Proceedings International Conference on Implications of Ground Chemistry and Microbiology for Construction*, University of Bristol, 1–16.

CANFIELD, D. E., RAISWELL, R., WESTRICH, J. T., REAVES, C. M. & BERNER, R. A. 1986. The use of chromium reduction in the analysis of reduced inorganic sulfur in sediments and shales. *Chemical Geology*, **54**, 149–155.

CHIOU, W. A., BRYANT, W. R. & BENNETT, R. H. 1992. Quantification of clay fabrics: a simple technique. *In*: BENNETT, R. H. *et al.* (eds) *Microstructure of Fine-grained Sediments: from Mud to Shale. Frontiers in Sedimentary Geology*, Springer, Berlin, 379–387.

CORNFORD, C. 1984. Source rocks and hydrocarbons of the North Sea. *In*: GLENNIE, K. W. (ed.) *Introduction to the Petroleum Geology of the North Sea*, Blackwell, Oxford, 171–204.

DANILATOS, G. D. 1991. Review and outline of environmental SEM at present. *Journal of Microscopy*, 391–402.

ESLINGER, E. & PEVEAR, D. 1988. *Clay mineralogy for petroleum geologists and engineers*. SEPM Short Course Notes 22.

FAIRCHILD, I., HENDRY, G., QUEST, M. & TUCKER, M. E. 1988. Chemical analysis of sedimentary rocks. *In*: TUCKER, M. E. (ed.) *Techniques in Sedimentology*, Blackwells, Oxford, 274–354.

GIBBS, R. J. 1965. Errors due to segregation in quantitative clay mineral X-ray diffraction mounting techniques. *American Mineralogist*, **50**, 741–751.

GILLOTT, J. E. 1976. Importance of Specimen Preparation in Microscopy: Soil Preparation for Laboratory Testing. ASTM Special Technical Publication, **599**, 289–307.

GILLOTT, J. E. 1987. *Clay in Engineering Geology: Developments in Geotechnical Engineering*, **41**, Elsevier, Austerdam.

GIPSON, M. 1966. A study of the relations of depth, porosity and clay mineral orientation in Pennsylvanian shales. *Journal of Sedimentary Petrology*, **36**, 888–903.

GUVEN, N. 1990. Electron diffraction of clay minerals. *In*: MACKINNON, I. D. R. & MUMPTON, F. A. (eds) *Electron Optical Methods in Clay Science*. CMS workshop lectures, **2**. Clay Minerals Society, 42–68.

HANG, P. T. & BRINDLEY, G. W. 1970. Methylene blue absorption by clay minerals. Determination of surface areas and cation exchange capacities. *Clays and Clay Minerals*, **18**, 203–212.

HILLIER, S. 1998. Use of an air brush to spray dry samples for X-ray powder diffraction. *Clay Minerals*, **34**, 127–136.

HILLIER, S. 1999. Accurate quantitative analysis of clay and other minerals in sandstones by XRD; comparison of a Rietveld and a reference intensity ratio (RIR) method and the importance of sample preparation. *Clay Minerals*, 35.

HUGGETT, J. M. & SHAW, H. F. 1997. Field emission scanning electron microscopy—a high resolution technique for the study of clay minerals in sediments. *Clay Minerals*, **32**, 197–203.

HUGGETT, J. M. & WHITE, S. H. 1982. High voltage electron microscopy of authigenic clay minerals in sandstones. *Clays and Clay Minerals*, **30**, 232–236.

JOHNSON, L. J., CHU, C. H. & HUSSEY, G. A. 1985. Quantitative clay mineral analysis using simultaneous linear equations. *Clays and Clay Minerals*, **33**, 107–117.

JONES, R. C. 1989. A computer technique for X-ray diffraction curve fitting/peak composition. *In*: PEVEAR, D. R. & MUMPTON, F. A. (eds) *Quantitative Mineral Analysis of Clays*. CMS workshop lectures v.1. Clay Minerals Society, 52–103.

KUBLER, B. 1968. Evolution quantitative du metamorphisme par la crystallinite de l'illite. *Bull. Centre Rech. Pau-SNPA*, **2**, 385–397.

LA FEBER, D. 1967. The optical determination of spatial (three dimensional) orientation of platy clay minerals in soil thin sections. *Geoderma*, **1**, 359–369.

LANSON, B. & BESSON, G. 1992. Characterisation of the end of illite-smectite transformation: decomposition of X-ray patterns. *Clays and Clay Minerals*, **40**, 40–52.

LLOYD, M. G. & HALL, G. E. 1981. The SEM examination of geological samples with a semi-conductor back-scattered electron detector. *American Mineralogist*, **66**, 362–368.

LORD, C. J. 1992. A selective and precise method for pyrite determination in sedimentary materials. *Journal of Sedimentary Petrology*, **52**, 664–666.

MACKINNON, I. D. R. 1990. Introduction to electron beam techniques. *In*: MACKINNON, I. D. R. & MUMPTON, F. A. (eds) *Electron Optical Methods in Clay Science*. CMS workshop lectures, **2**. Clay Minerals Society, 1–14.

MANHEIM, F. T. 1966. *A Hydraulic Squeezer for Obtaining Interstitial Water from Consolidated and Unconsolidated Sediments*. USGS Prof. Paper, **550-C**, 256–261.

MARTIN, R. T. 1966. Quantitative fabric of wet kaolinite. 14th National. Conference. *Clays & Clay Minerals*, 271–287.

MOORE, D. M. & REYNOLDS, R. C. 1989. *X-ray Diffraction and the Identification and Analysis of Clay Minerals* (1st edn) Oxford University Press.

MOORE, D. M. & REYNOLDS, R. C. 1997. *X-ray Diffraction and the Identification and Analysis of Clay Minerals* (2nd edn) Oxford University Press.

MORGANSTERN, N. R. & TCHALENKO, T. S. 1967. The optical determination of preferred orientation in clays and its application to the study of microstructure in consolidated kaolin, II. *Proceedings of the Royal Society London Series A*, **300**, 271–287.

NEWMAN, A. C. D. 1987. The interaction of water with clay mineral surfaces. *In*: NEWMAN, A. C. D. (ed.) *Chemistry of Clays and Clay Minerals*, Mineralogical Society Monograph, **6**, 237–274.

O'BRIEN, N. R. 1964. Origin of Pennsylvanian underclays in the Illinois Basin. *Bulletin Geological Society of America*, **75**.

O'BRIEN, N. R. & SLATT, R. M. 1990. *Argillaceous Rock Atlas*. Springer Verlag, New York.

VAN OLPHEN, H. & FRIPIAT, J. J. 1979. *Data Handbook for Clay Materials and Other Non-Metallic Minerals*. Pergamon, Oxford.

PATERSON, E. & SWAFFIELD, R. 1987. Thermal analysis. *In*: WILSON, M. J. (ed.) *A Handbook of Determinative Methods in Clay Mineralogy*, Blackie, Glasgow, 99–132.

POTTER, P. E., MAYNARD, J. B. & PRYOR, W. A. 1980. *Sedimentology of Shale*. Springer Verlag, New York.

POTTS, P. J. 1987. *A Handbook of Silicate Rock Analysis*. Blackie, Glasgow.

PRESLEY, R. J., BROOKS, R. R. & KAPPEL, H. M. 1967. A simple squeezer for removal of interstitial water from ocean sediments. *Journal of Marine Research*, **25**, 355–6.

RAMACHANDRAN, V. S., KACKER, K. P. & PATWARDHAN, N. K. 1962. Adsorption of dyes by clay minerals. *American Mineralogist*, **47**, 165–169.

REED, S. J. B. 1996. *Electron Microprobe Analysis and Scanning Electron Microscopy in Geology*. Cambridge University Press.

REID, J. M., Czerewko, M. A. & Cripps, J. C. 2001. *Sulphate Speciation for Structural Backfills*. TRL Report PR /IS/111/2000, 1–98.

REYNOLDS, R. C. 1985. NEWMOD © a computer program for the calculation of one dimensional diffraction patterns of mixed layer clays. R. C. Reynolds Jr, 8 Brook Road, Hanover, NH.

RHOADES, J. D. 1982. Cation exchange capacity. *In*: PAGE, A. L., MILLER, R. H. & KEENEY, D. R. (eds) *Methods of Soil Analysis*. Agronomy **9**, 149–157.

RUSSELL, J. D. & FRASER, A. R. 1994. Infrared methods. *In*: WILSON, M. J. (ed.) *Clay Mineralogy: Spectroscopic and Chemical Determinative Methods*. Chapman & Hall, 11–67.

SHAW, H. F. 1972. The preparation of oriented clay mineral specimens for X-ray analysis by a suction-onto-ceramic-tile method. *Clay Minerals*, **9**, 349–350.

SHI, B., WU, Z., INYANG, H., CHEN, J. & WANG, B. 1999. Preparation of specimens for SEM analysis using freeze-cut –drying. *Bulletin of Engineering Geology & Environment*, **58**, 1–7.

SHELLEY, D. 1985. *Optical Mineralogy* (2nd edn). Elsevier, New York.

SMART, P. 1965. Optical Microscopy and Soil Structure. *Nature*, **210**.

SMART, P., TOVEY, N. K., LENG, X., HOUNSLOW, M. W & MCCONNOCHIE, I. 1991. Automatic analysis of microstructure of cohesive sediments. *In*: BENNETT, R. H. *et al.* (eds) *Microstructure of Fine-grained Sediments: from Mud to Shale. Frontiers in Sedimentary Geology*. Springer, New York, 359–366.

SOKOLOV, V. N. & O'BRIEN, N. R. 1990. A fabric classification of argillaceous rocks, sediments, soils. *Applied Clay Science*, **5**, 353–360.

SRODON, J., DRITS, V. A., MCCARTY, D. K., HSIEH, J. C. C. & EBERL, D. D. 2001. Quantitative X-ray diffraction analysis of clay beraring rocks from random preparations. *Clays & Clay Minerals*, **49 (6)**, 514–528.

STOPES, M. 1935. On the petrology of banded bituminous coals. *Fuel*, **14**, 4–13.

STUCKI, J. W., BISH, D. L. & MUMPTON, F. A. (ed.) 1990. *Thermal Analysis in Clay Science: CMS Workshop Lectures*, **3**, Clay Minerals Society.

TCHALENKO, J. S., BURNETT, A. D. & HUNG, J. J. 1971. The correspondence between optical and X-ray measurements of particle orientation in clays. *Clay Minerals*, **9**, 47–70.

THORNLEY, D. M. & PRIMMER, T. J. 1995. Thermogravimetry/evolved water analysis (TG/EWA) combined with XRD for improved quantitative whole-rock analysis of clay minerals in sandstones. *Clay Minerals*, **30**, 27–38.

TOWE, K. M. 1974. Quantitative clay petrology: the trees but not the forest? *Clays and Clay Minerals*, **22**, 375–378.

TUTTLE, M. L. & GOLDHABER, M. B. 1986. An analytical scheme for determining forms of sulphur in oil shales and associated rocks. *Talanta*, **33**, 953–961.

UNWINS, P. J. R. 1994. Environmental scanning electron microscopy. *Materials Forum*, **18**, 51–75.

VELDE, B. 1992. *Introduction to Clay Minerals*. Chapman & Hall, London.

VELDE, B. 1984. Electron microprobe analysis of clay minerals. *Clay Minerals*, **19**, 243–247.

WADA, K. & OKAMURA, Y. 1977. Measurements of exchange capacities and hydrolysis as a means of characterising cation and anion retention by soils. *In: Proceedings of International Seminar on Soil Environment and Fertility Management in Intensive Agriculture, Tokyo*. Society of Scientific Soil Manure, Japan, 811–815.

WEAVER, C. E. 1960. Possible uses of clay minerals in the search for oil. *AAPG Bulletin*, **44**, 1505–1518.

WHITTAKER, E. J. W. 1981. *Crystallography*. Pergamon, Oxford.

9. Laboratory testing

9.1. Introduction

This chapter outlines the testing of clays for civil engineering purposes in the laboratory. The laboratory may range from a large permanent establishment to a small temporary facility set up on site. Field tests (or *in situ* tests), which are carried out on clay while it is still in the ground, are covered in Section 7.6.

Field tests and laboratory tests are not alternative options, but are complementary. Each has its advantages and limitations. Some of the advantages of laboratory tests can be summarized as follows:

- specific soil properties can be measured;
- the choice of material for testing can be controlled;
- control of test conditions can be exercised, and changes in conditions can be simulated;
- a relatively high degree of accuracy of measurements is possible;
- parameters can be derived within an acceptable timescale;
- tests can be performed on undisturbed, re-constituted or remoulded clay.

Data obtained from laboratory tests provide a closer understanding of the properties and behaviour of clays as engineering materials. This can lead to a reduction in uncertainties in the analysis of earthworks; more economic design; construction in difficult conditions which would otherwise not be feasible; and increased economy in the use of clay as a construction material.

The laboratory tests outlined in this document are those that are commonly recognized throughout the world as being appropriate for geotechnical purposes. The procedures can be conveniently divided into three categories:

- classification tests—to establish the type of clay and the engineering category to which it belongs;
- chemical tests—usually to identify deleterious or aggressive constituents that require protective measures;
- tests for the assessment of engineering properties—permeability, compressibility, shear strength.

Classification tests are covered in Section 9.3, chemical tests in Section 9.4, and the various types of test for determining different aspects of engineering properties are given in Sections 9.5 to 9.11. Details of procedures for all these tests are given in the relevant British and U.S. standards, and in 'Manual of Soil Laboratory Testing' (Head 1992, 1994, 1998). References to the relevant sections of these and other publications are listed in Appendix D.

In the UK, laboratory tests are generally carried out in accordance with BS 1377:1990, 'British Standard Methods of Test for Soils for civil engineering purposes', Parts 1 to 8. (Part 9 relates to *in situ* tests). This is the main reference quoted in this document. Mention is also made of practice in the USA, with reference to certain ASTM Standards. Test procedures described therein are generally very similar to those in British Standards, but there are often differences in details. Not all European countries have their own sets of Standards, and many current procedures are based on British or U.S. practice. Drafting of harmonised European Standards is now well advanced.

A laboratory entrusted with soil testing should be capable of carrying out the required tests in accordance with recognised Standards, in a competent, safe and efficient manner. Quality assurance accreditation provides evidence that a laboratory has these capabilities. Most reputable soil laboratories in the UK are accredited for specific test procedures under the United Kingdom Accreditation Service (UKAS). Accreditation by UKAS provides confidence in the laboratory's ability to carry out those procedures correctly. Some of the requirements for UKAS accreditation are as follows:

- all procedures are documented, and any departures from written standards are reported;
- test equipment is properly maintained, checked and calibrated;
- samples are suitably handled, protected and stored;
- staff are competent, experienced and fully trained;
- the laboratory provides an appropriate working environment;
- an audit system ensures that these requirements are maintained.

9.2. Sample preparation and conduct of testing

This Section provides information of a general nature, which is relevant to the testing of clays and to all types of test covered in this chapter. Reference is made to these details in the test procedures sections to avoid unnecessary repetition.

Topics covered are the preparation of specimens for testing; laboratory environmental requirements; calibration and checking of laboratory measuring devices and test equipment; and the recommended precision of reporting of test results.

9.2.1. Sample preparation

The essential requirement of the portion of clay used for a laboratory test is that it is as representative as possible of the stratum or material from which it was taken. In this context, the word *sample* normally applies to the material taken on site, and *specimen* refers to the portion of that sample which has been prepared for testing.

Retrieval of samples from site is covered in Chapter 7, Section 7.4. This section deals with the preparation of specimens on which the tests are carried out, from the samples received from site. For the purposes of preparation of test specimens in the laboratory, four categories of specimens are generally recognized:

- disturbed specimens—used mainly for classification and identification tests;
- undisturbed specimens—essential for determination of engineering properties such as compressibility, shear strength and permeability;
- re-compacted or re-constituted specimens—used for determining properties of clay that has been removed and re-deposited, often as construction fill;
- remoulded specimens—used for determination of the effect of remoulding on measured properties.

In the CEN document prENV 1997–2, soil samples for laboratory testing are classified in five Quality Classes (independent of the categories related to sampling methods given in Section 7.6.2). These range from Class 1, in which the clay has been subjected to the minimum of disturbance (there is no such thing as a completely undisturbed sample), to Class 5, which is useless for testing because of changes in water content and loss of some particles, but could be used to give a broad indication of the sequence of strata.

9.2.1.1. Disturbed specimens. Clay should not normally be dried before testing, but should be used in its natural state unless drying is specified. When drying is necessary, it should be done by one of the following recognized procedures.

1. Oven drying, usually at a temperature of $105 \pm 5°C$.
2. Partial drying in an oven at a specified temperature less than $100°C$.
3. Air drying—partial drying by exposure to air at room temperature, with or without the aid of a fan.

Not less than about 500 g of clay should be taken initially for classification and chemical tests. An initial clay sample is selected from a specific zone or layer, or to represent an identified feature. If the clay contains coarse particles, a larger initial sample is necessary. When very small test specimens (50 g or less) are required, these should be obtained from a larger portion of clay, which is thoroughly mixed and riffled down to provide test specimens of the required mass. Preparation procedures applicable only to particular tests are given in the appropriate sections.

9.2.1.2. Undisturbed specimens. Cylindrical test specimens can be taken from undisturbed samples of clay of a suitable consistency by using a specimen cutting tube of specific dimensions with a sharpened cutting edge. The tube should not be pushed into clay that is rigidly contained; instead the clay should be jacked out of the container so that it enters the specimen tube as it emerges from the container. A set of three tube specimens of 38 mm diameter for compression tests can be prepared simultaneously, at the same horizon, from a 100 mm diameter sampling tube by using a suitable jig of the type shown in Figure 9.1. The specimens are then jacked out of their tubes for final trimming.

For clays that are too stiff for the above procedure, or fissured clays that could disintegrate, careful hand

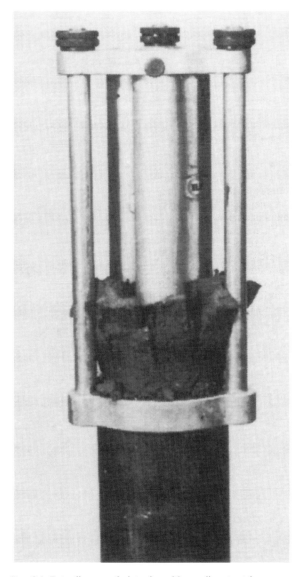

FIG. 9.1. Extruding sample into three 38 mm diameter tubes.

trimming is necessary, with the aid of the appropriate cutting tube or ring, as shown in Figure 9.2. Hand trimming is also used when specimens are to be prepared with their axes orientated at a specified angle to the vertical.

A soil lathe (Fig. 9.3) provides an alternative method of trimming cylindrical specimens of clay not containing hard particles, but care must be taken to avoid subjecting the clay to torsional stresses.

The largest size of particle contained in an undisturbed clay specimen, in relation to its dimensions, should not exceed the maximum size given in Table 9.1. If, after a test, any particles larger than this are found to be in the specimen, the fact should be reported.

During preparation, small portions of clay may be taken from adjacent to the specimen for determination of an initial moisture content value. The completed specimen is measured and weighed so that its volume and density can be calculated.

9.2.1.3. Recompacted specimens. To prepare samples of recompacted clay, the undried material is first shredded, or chopped into pieces to pass a 20 mm sieve. The appropriate amount of water is added and well mixed in, and the clay is sealed and stored for at least 24 hours before proceeding further. If a moisture content less than the 'as received' value is needed, this should be achieved by partial air drying, not by fully drying and re-wetting.

The prepared and matured sample is compacted with the required effort into the appropriate mould, which is sealed. The sample is allowed to stand for at least 24 hours to enable excess pore pressures to dissipate. Test specimens can then be prepared in the same way as from an undisturbed sample.

When small recompacted specimens, such as are used for oedometer or small shearbox tests, are required, a sample should first be prepared and matured in this way in a tube or compaction mould, from which the test specimens can be trimmed as if from an undisturbed sample. It is preferable not to form small specimens by compacting the clay directly into the specimen ring.

9.2.1.4. Remoulded specimens. To prepare a remoulded specimen, the clay is enclosed in waterproof film (such as a polythene bag) and is kneaded and squeezed with the fingers for a few minutes. It is then worked into the appropriate mould, without entrapping air pockets, to form the desired test specimen. There should be no loss of water from the clay during this process.

9.2.2. Laboratory environment

The laboratory environment in which sample preparation and tests are carried out should be clean and orderly, properly ventilated, well lit, and maintained at an appropriate working temperature. Unobstructed access to all working and storage areas is essential. National Health and Safety Regulations should be strictly adhered to, and appropriate Laboratory Rules should be enforced. Some activities and test procedures require more specific environmental control, as outlined below.

During the preparation of test specimens, loss of moisture from the clay should be kept to a minimum. A humidified preparation room, or a locally humidified zone, is desirable to achieve this. Some samples held in log-term storage might require a humidified storage area.

Temperature control to within $\pm 4°C$ of normal ambient temperature is required for consolidation tests and direct shear tests. Control to within $\pm 2°C$ is required for effective stress triaxial tests and other long-term tests in which pore pressure is measured.

Adequate ventilation, and in some cases a fume cupboard, should be provided in areas where chemical tests are performed, and where mercury is handled. Use of mercury also needs other special arrangements and safety precautions.

9.2.3. Calibration of equipment

The purpose of calibration is to enable the uncertainties in any measurement to be quantified. To achieve this,

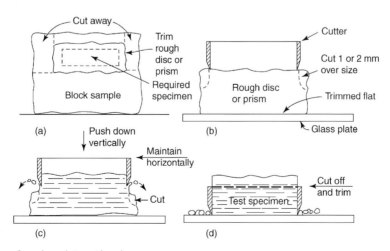

FIG. 9.2. Hand trimming of specimen into cutting ring.

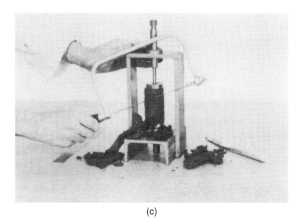

Fig. 9.3. Soil lathe.

Table 9.1. *Allowable size of particles in test specimens*

Type of test	Maximum size of particle
Oedometer consolidation	$H/5$
Direct shear (shearbox)	$H/10$
Compressive strength (cylinder, H/D about 2)	$D/5$
Permeability	$D/12$

H, height of specimen; D, diameter of specimen.

all measurements made in a testing laboratory need to be traceable to national or international standards of measurement. This requires that all instruments used in making test measurements should be calibrated, and records of all calibrations maintained. In addition, significant measurements of certain items of test equipment, and their performance, need to be verified to ensure that they lie within the specified working tolerances.

All measuring instruments must be calibrated initially by an accredited calibration organisation, and subsequently re-calibrated at specified regular intervals. The maximum intervals between calibrations and routine checks are specified in BS 1377: Part 1: 4.4. All calibrations should be carried out in accordance with written procedures, and details should be recorded. Detailed guidance is given in UKAS Publication ref: LAB 21.

Calibration of working instruments may be carried out either by a suitably accredited external calibration laboratory, or in-house by the laboratory's own suitably qualified and experienced staff. If an external laboratory is used, each calibration certificate must show that the calibration is traceable to relevant current standards. Calibration may be undertaken in-house if the laboratory maintains its own appropriate reference instruments which have valid traceable calibration. These instruments should be re-calibrated by an accredited external organisation at intervals specified by the relevant Standards. Reference instruments should be used for calibration purposes only, and should be properly maintained and kept separate from working instruments.

Notwithstanding the specified intervals between calibrations, re-calibration is necessary whenever a change in accuracy is suspected, or when an instrument has been repaired, dismantled, adjusted or overhauled.

In addition to the measuring instruments referred to above, numerous items of testing equipment need to be calibrated or checked at regular intervals, or at intervals depending on frequency of use. Maximum intervals between routine checks and calibrations are set out in BS 1377: Part 1: 4.2. Further requirements for some of these items, and the calibration and checking of items relevant to particular tests, are referred to in the appropriate Sections.

9.2.4. Precision of data

The calibration requirements outlined above, if correctly observed, should ensure that uncertainties in the data derived from measurements lie within known acceptable limits. The final results as reported need to be sufficiently precise for geotechnical design applications, but their precision should be realistic and compatible with the known uncertainties of measurement. Results from many tests on clays are reported to an accuracy of three significant figures, or in some cases the nearest whole number.

In making calculations from measured data, the values as measured should be used without rounding off until the final result has been derived. Only then should rounding off to the recommended precision be done.

TABLE 9.2. *Recommended accuracy for reporting laboratory test results*

Item	Symbol	Accuracy	Unit
Specimen dimensions		0.1	mm
Density	ρ, ρ_D	0.01	Mg/m³
Moisture content	w	<10 : 0.1	%
		>10 : 1	%
Liquid limit	w_L	1	%
Plastic limit	w_P	1	%
Plasticity index	I_p (or PI)	1	—
Particle density	ρ_s	0.01	Mg/m³
Clay, silt, sand or		generally: 1	%
gravel fraction		less than 5% : 0.1	%
Voids ratio	e	0.001	—
Porosity	n	1	%
Permeability	k	2 significant figures × 10^{-n} where n is an integer	m/s
CBR value	CBR	2 significant figures	%
Angle of shear resistance	ϕ', ϕ_r'	½	degree
Apparent cohesion	c', c_r', c_u	2 significant figures	kN/m²
Vane shear strength	τ_v	2 significant figures	kN/m²
Unconfined compressive strength	q_u	2 significant figures	kN/m²
ditto (autographic test)	q_u	<50 : 2	kN/m²
		50 to 100 : 5	kN/m²
		>100 : 10	kN/m²
Strain at failure	ε_f	0.2	%
Rate of strain		2 significant figures	% per min
Sensitivity to remoulding	S_t	1 decimal place	—
Degree of saturation	S_r	1	%
Swelling pressure	p_s	1	kN/m²
Applied pressures	p	1	kN/m²
Coefficient of volume compressibility	m_v	3 significant figures	m²/MN
Coefficient of consolidation	c_v	2 significant figures	m²/year
Coefficient of secondary compression	C_{sec}	2 significant figures	—
Compression index	C_c	2 significant figures	—
Swell index	C_s	2 significant figures	—

Recommended accuracies for reporting test results are summarised in Table 9.2, which also includes the customary units in which test data are reported, and the symbols used in BS 1377.

9.3. Classification tests

9.3.1. Scope

Soil classification tests are required in order to classify soils in general, and clays in particular, in accordance with recognized standards. In the UK the relevant standard is usually BS 5930. These tests provide an indication of the general type of soil and the engineering category to which it belongs. Data acquired from these tests can be applied to the identification of soil strata as part of a site investigation. Appropriate tests are usually carried out on any soil sample that is subjected to other tests for determination of engineering properties, and the data are reported with the test results.

The tests described in Sections 9.3.2 to 9.3.8 are those that are most commonly carried out for the engineering classification of clays, and comprise the following.

- moisture content (water content);
- Atterberg limits (liquid limit and plastic limit);
- shrinkage limit and linear shrinkage;
- density and dry density;
- particle density;
- particle size distribution (for determination of clay fraction);
- clay suction (filter paper method).

All these tests, except the last, are covered in BS 1377: Part 2: 1990.

9.3.2. Moisture content

Determination of moisture content (or water content) is perhaps the most frequently performed geotechnical test, and applies to all types of soil as well as clays. The test is performed not only in its own right but also as part of the standard procedure for many other tests.

The moisture content of a soil is defined as the amount of water removed by oven drying (usually overnight) at 105–110°C, expressed as a percentage of the dry mass of clay. In this context the clay is defined as being dry if continued drying in the oven for a further period of four hours results in a further loss of moisture not exceeding 0.1% of its original mass. Exposure to temperatures higher than 110°C can result in changes in the clay mineral structure which would affect the properties of the clay. For soils containing gypsum the temperature should not exceed 80°C, and for those containing organic matter the maximum is 60°C. Clays containing hydrated halloysite should be air-dried at room temperature. A longer drying period is needed in these cases, and repeated weighings should be performed to confirm that constant mass has been achieved.

A representative test specimen of at least 30 g is required and duplicate or triplicate determinations should be made on separate specimens taken from the same sample of soil. Undisturbed samples, or disturbed samples that have not been subject to a change in moisture content, are suitable. If the clay contains particles larger than 2 mm, larger test specimens should be used—at least 300 g for 'medium-grained' soils, or 3 kg for 'coarse-grained' soils, as defined by BS 1377: Part 1: Clause 3.1.

The moisture content is determined by weighing the specimen in a suitable weighed container, drying in a thermostatically controlled thermally heated oven, cooling in a desiccator and weighing again. The difference in mass is the mass of water lost by drying and is expressed as a percentage of the dry mass of soil. Where the pore water is saline a correction for concentration of dissolved solids is required.

Calibrated balances readable to 0.01 g are required for weighing and the oven temperatures should be verified periodically.

9.3.3. Liquid and plastic limits

Tests for the determination of liquid limit and plastic limit (known as the Atterberg limits), are the most frequently performed index tests for clays. These limits are defined by moisture contents, and provide essential data for the identification and engineering classification of clays. The natural moisture content of a clay, when compared to the Atterberg limits, indicates the consistency of the clay and provides a useful indication of certain engineering characteristics.

As mentioned in Chapter 4, clay mixed with sufficient water forms a slurry which behaves as a liquid, but as the moisture content is reduced the fluid becomes more viscous and at some stage the material begins to behave in a plastic manner. The moisture content at this point is known as the *liquid limit* (LL) of the clay. With further removal of water the clay remains in the plastic state, but with increasing stiffness, until it loses its plasticity and becomes brittle or begins to crumble. The moisture content at this point is known as the *plastic limit* (PL) of the clay, and with further drying the clay behaves as a solid. The difference between these moisture content values is known as the *plasticity index* (PI).

The moisture contents corresponding to the liquid limit and plastic limit are defined by standardized test procedures. At the liquid limit most clays have an undrained shear strength of about 1.5 to 2 kPa, and at the plastic limit the shear strength could be about 100 times this value.

The samples used for the determination of liquid and plastic limits should not be dried before testing, but used in their natural state wherever possible. Drying (even air drying at room temperature) can cause changes in properties that are not reversible on re-wetting, leading to incorrect classification data. (Fookes 1997). If the clay contains sand-size or coarser particles that cannot be picked out by hand, it should be soaked in distilled water until it disaggregates into slurry, which is washed through a 425 μm sieve. All the washings are retained, decanted and partially air-dried to bring the mixture to the desired consistency for starting the test, as described in BS 1377: Part 2: Clause 4.2.

At least 300 g of clay passing the 425 μm sieve is required for performing a set of liquid limit and plastic limit tests. The procedures for determining the moisture contents are as described in Section 9.3.2.

9.3.3.1. Liquid limit. The clay is first worked by mixing with a spatula on a glass plate to a uniform consistency before starting the test, and after the addition of each increment of water during the test. Distilled water or de-ionized water must be used.

The standard procedure now used in the UK for determination of the liquid limit is the four-point cone penetrometer method, as described in BS 1377: Part 2: Clause 4.3. The criterion for the liquid limit of the clay is a cone penetration of 20 mm in a duration of 5 seconds. Penetration of the cone is measured over a range of (usually four) moisture contents, and by plotting each penetration against the corresponding moisture content the moisture content (w_L) at 20 mm penetration is obtained by interpolation. The apparatus is shown in Figure 9.4.

The one-point method, given in BS 1377: Part 2: 4.4 enables a value of the liquid limit to be derived from a single determination, but this method should be used only if the amount of clay available is insufficient for the standard procedure.

The original procedure for determining liquid limit, known as the Casagrande method, depends on the flow of the clay paste when subjected to repeated blows in a metal cup. This procedure is still used in American countries, but it is now almost obsolete in UK and much of Europe. The cone penetrometer method is easier to perform and gives results that are more reproducible and less liable to operator error. There are variations in the cone angle, cone mass, and criterion for liquid limit between different countries.

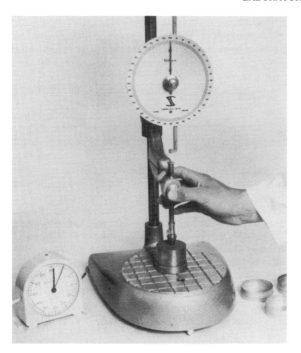

FIG. 9.4. Liquid limit apparatus (cone penetrometer).

9.3.3.2. Plastic limit. The plastic limit is determined in the traditional way by rolling a thread of remoulded clay between the fingers and a glass plate. When the thread crumbles or develops longitudinal or transverse cracks at a diameter of 3 mm, the clay is at the plastic limit and its moisture content (w_p) is then measured. This test is very subjective to the operator's technique, and it is essential to ensure that the correct state of crumbling has been achieved. Numerous alternative methods for determining the plastic limit have been suggested from time to time, but none has yet found general acceptance.

To classify the soil the Plasticity Index ($I_P = w_L - w_P$), is plotted against w_L on the plasticity, or A-line chart (Fig. 9.5). The plasticity range in which a clay falls provides a good initial indication of its likely engineering characteristics (see Chapter 4).

The relationship of the natural moisture content of a clay to its LL and PL values provides another parameter for indicating likely behaviour, known as the *liquidity index* (I_L). This is calculated from the equation:

$$I_L = \frac{w - w_P}{w_L - w_P}. \qquad (9.1)$$

When calculating this parameter the moisture content value w should relate to the same fraction of the clay as is used for determining the Atterberg limits, i.e. the fraction passing the 425 µm sieve. If the original soil included particles coarser than 425 µm, a correction needs to be applied to the moisture content w (%) measured on the whole sample. The corrected moisture content, w_a (%) is given in equation (9.2), where p_a is the percentage of coarse material:

$$w_a = \frac{100w}{p_a}. \qquad (9.2)$$

Some authorities use *relative consistency* (C_r) in place of liquidity index, calculated from the equation

$$C_r = \frac{w_L - w}{w_L - w_P}. \qquad (9.3)$$

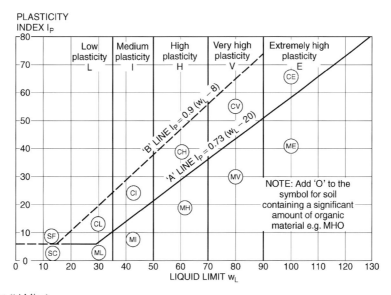

FIG. 9.5. Plasticity chart ('A' line).

The plasticity index of a clay soil depends not only on the properties of the clay minerals but also on the clay fraction, i.e. the percentage of particles finer than 2 μm present in the soil (see Section 9.3.7). For a given clay the ratio (plasticity index/clay fraction) is constant, and is called the *activity* of the clay. The activity value depends upon the type of clay mineral present (see Chapter 4), and provides a means of identifying the clay. To be consistent with the Atterberg limits, the clay fraction should be expressed as a percentage of that portion of the original soil which passes a 425 μm sieve. On the basis of activity, clays can be placed into five categories (Table 9.3).

9.3.4. Shrinkage tests

All clays decrease in volume as they lose moisture, but some clay soils can shrink more than others. Shrinkage tests are used indicate whether drying of a clay stratum is likely to cause a settlement problem and also whether there is a possibility of swelling when a clay becomes wetter.

In general, clays of high plasticity and high activity are likely to have a greater potential for shrinkage than clays of low plasticity or low activity. The standard laboratory tests to investigate shrinkage behaviour provide data, which indicate both the likely amount of shrinkage and the moisture content below which shrinkage does not occur.

TABLE 9.3. *Categories of clays*

Category	Activity value
Inactive clays	<0.75
Normal clays	0.75–1.25
Active clays	1.25–2
Highly active clays	>2
Bentonite	⩾6

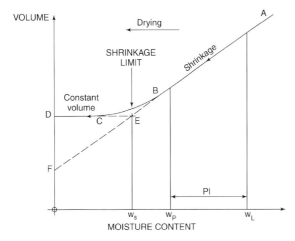

FIG. 9.6. Shrinkage curve for clay.

The shrinkage process of a volume of clay initially at a moisture content higher than the plastic limit is illustrated graphically in Figure 9.6, in which change in volume is plotted against decreasing moisture content. Initially (line AB), the volume decreases approximately linearly with loss of water. In the final stage (CD), below the plastic limit, there is no further decrease in volume as the clay dries. The portion BC represents the transition zone between these two types of behaviour. The point E, at the intersection of the two straight lines, defines the *shrinkage limit*, denoted by w_S, which is the conceptual moisture content below which it is assumed that there is no further decrease in volume due to loss of water.

The initial line AB can be extended back to meet the volume axis at the point F, which represents a hypothetical state in which there are no air voids. The value at F therefore indicates the volume of the dry solids in the soil mass. Expressed in ml, it is equal to the mass of dry soil divided by the particle density of the soil grains.

The ratio of the change in volume to the change in moisture content above the shrinkage limit defines the *shrinkage ratio*, denoted by R_S. If volume change is expressed as a percentage of the final dry volume, the shrinkage ratio is equal to the slope of the line AB in Figure 9.6. In SI units the shrinkage ratio is numerically equal to the mass of dry soil divided by the volume of dry soil.

Clay samples used for these tests should be representative, and should not be subjected to drying before testing. The definitive shrinkage limit test can be carried out on either undisturbed or disturbed samples. Remoulded samples are used for the other tests.

Two test methods for determination of the shrinkage limit are given in BS 1377:1990. The definitive method is often referred to as the TRL method (since it was there that it was developed). The subsidiary method is based on that given in the ASTM Standards (USA). The former also provides data for quantifying the amount of volumetric shrinkage from a graphical plot similar to that shown in Figure 9.6. Both the BS and ASTM methods require the use of mercury, which is a hazardous substance. These tests should be carried out only under properly controlled conditions, using a fume cupboard or similar approved facilities for extraction of mercury vapour.

The BS also gives a relatively simple test for measuring the percentage linear shrinkage of a clay due to drying.

9.3.4.1. Shrinkage limit—definitive method. For this test a cylindrical specimen of undisturbed or remoulded clay, typically 38 mm diameter and 76 mm high, is required. The clay specimen is weighed and measured, and immersed in a special tank containing mercury to determine its initial volume, see Figure 9.7. It is then allowed to lose moisture slowly by exposure to the atmosphere, and at suitable intervals it is weighed and its corresponding volume is determined by further immersions in the mercury tank. When no further decrease in volume is detected the specimen is dried in the oven and final measurements of mass and volume are made. The

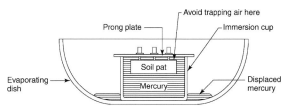

FIG. 9.8. Shrinkage limit apparatus (ASTM).

9.3.4.3. Linear shrinkage. About 150 g of clay is required, remoulded with distilled water to about the liquid limit. The clay is placed into a metal mould of semicircular cross section, 25 mm diameter and 140 mm long. The clay is allowed to dry, slowly by exposure to the atmosphere at first, and finally in the drying oven. The difference between the initial and final lengths, expressed as a percentage of the initial length, is calculated and reported as the *linear shrinkage* (L_s).

9.3.5. Density tests

Measurement of density, like moisture content, forms part of the routine procedure when undisturbed samples of clay are tested. The density of clay is an important property for many applications in geotechnical engineering such as the analysis of the stability of natural slopes, cuttings and foundations. Density of re-worked compacted clay is relevant to stability analysis control of embankments and earth dams. Measurement of density, and determination of dry density, provides an important means of control of compaction of clays in all aspects of earthworks construction, including sub-bases for roads, railways and airfield runways (see Chapter 10).

The *density* of a soil is the mass of bulk soil per unit volume, often called the *bulk density* and denoted by the symbol ρ. In SI units mass density is expressed in Mg/m^3, which is a convenient unit because it is numerically equal to grams per cubic centimetre and therefore the density of water is unity. If the force exerted by a mass of soil is relevant, the unit weight (γ), expressed in kN/m^3, is obtained by multiplying the mass density by g, the gravitational acceleration in m/s^2.

$$\gamma \text{ (kN/m}^3\text{)} = \rho \times g = 9.81 \times \rho \text{ or approximately } (10) \times \rho.$$

When clay is used as a construction material and is compacted, the mass of dry soil particles per unit volume is significant. This is known as the *dry density*, denoted by ρ_D (sometimes by ρ_d). Dry density can be calculated, if the moisture content $w\%$ is known, from equation (9.4)

$$\rho_D = \frac{100\rho}{100 + w}. \quad (9.4)$$

The density of water is denoted by ρ_w, and for most practical purposes can be assumed to be 1 Mg/m^3. Another parameter sometimes used is the *submerged*

FIG. 9.7. Shrinkage limit apparatus (TRL).

moisture contents corresponding to each measurement are then calculated, and a graphical plot of specimen volume against moisture content enables the shrinkage limit and shrinkage ratio to be evaluated.

9.3.4.2. Shrinkage limit—ASTM method. This test requires the clay to be remoulded at a moisture content higher than its liquid limit, using distilled water. The clay is placed into a small weighed metal dish of known volume. The soil pat is allowed to dry slowly, first by exposure to the atmosphere, and finally by oven drying. It is then weighed and its final volume is determined by immersion in mercury using a glass 'prong plate', see Figure 9.8. From the initial and final measurements the shrinkage limit and shrinkage ratio can be calculated. If the shrinkage curve is also required, the soil pat can be removed from the dish at suitable intervals for weighing and measurement of its corresponding volume by immersion in the mercury.

density, i.e. the apparent density of clay when submerged in water, denoted by ρ'.

Undisturbed samples of good quality are essential. Three methods for determination of the density of clays are given in BS 1377—linear measurement; immersion in water; and water displacement. For determination of specimen volume by linear measurement, test specimens are either hand trimmed to a regular shape or extruded from a sampling tube. Trimming needs to be carried out using sharp knives and cutting tools, a flat surface and suitable appliances such as a mitre-box and engineer's square. A steel rule, vernier calipers and micrometer dial gauge reading to 0.1 mm are required for measurements. Weighings to 0.01 g are needed. Specimens prepared for use in other tests are frequently used, and normally the moisture content is also determined. Water immersion or water displacement procedures are used for irregular lumps of soil.

9.3.5.1. Linear measurement. This procedure is given in BS 1377: Part 2: 1990, Clause 7.2. The main application is for specimens for shear strength and compressibility tests, which are usually in the form of right cylinders or rectangular prisms. The volume of such specimens is calculated from linear measurements made with vernier calipers. The specimen is weighed, usually to the nearest 0.01 g, and the bulk density (Mg/m^3) is calculated by dividing this mass (grams) by the volume (cubic centimetres). If the moisture is also determined the dry density can be calculated.

The density of an undisturbed sample in a tube can be measured before it is extruded from the tube. Each end of the sample is trimmed flat so that its length can be calculated from end measurements. The internal diameter of the tube is measured and the sample volume is calculated. The sample mass is obtained by difference after emptying the tube, and hence the bulk density can be calculated.

9.3.5.2. Immersion in water. This method, given in Clause 7.3 of BS 1377: Part 2: 1990, can be used for lumps of clay or specimens of irregular shape, as well as those of regular geometric shape. The largest practicable size of specimen should be used. It is preferable that no one dimension is significantly larger than the other two, and re-entrant angles should be avoided.

The specimen is weighed, coated with several layers of just-molten paraffin wax and weighed again so that the mass of wax, and hence its volume, can be determined. The coated specimen is placed in a cradle or cage in which it is lowered into a container of water while suspended from a suitably adapted balance as shown in Figure 9.9. This enables the apparent mass when immersed in water to be measured. The bulk density is then calculated using the principle of Archimedes.

9.3.5.3. Water displacement. This method is given in Clause 7.4 of BS 1377: Part 2: 1990, and makes use of a water displacement apparatus known as a siphon-can as illustrated in Figure 9.10. The clay specimen is prepared, weighed and coated with wax as for the weighing in water

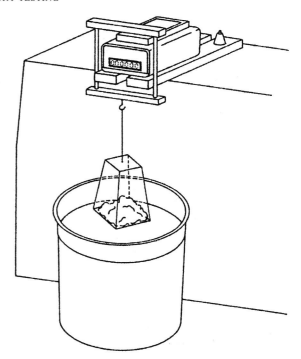

FIG. 9.9. Apparatus for weighing in water.

FIG. 9.10. Water displacement apparatus.

method. The volume of water displaced from the can when the specimen is lowered into the water is measured, from which the volume of the clay, and hence the density, is calculated.

9.3.6. Particle density

The various aspects of density covered in Section 9.3.4 relate to the density of a mass of clay as a whole, comprising solid particles and voids (water and air). The density of the solid particles themselves is known as the *particle density*, denoted by the symbol ρ_s. This is the current terminology, which replaces *specific gravity*. Whereas specific gravity is a ratio and therefore dimensionless, particle density is expressed in density units, usually Mg/m³. Because the density of water (denoted by ρ_w) is approximately 1 Mg/m³, the particle density is numerically the same as specific gravity.

The particle density of many clays typically lies in the range 2.68 to 2.72 Mg/m³. Higher values might indicate the presence of heavy metal oxides, or other compounds, in the clay. Unusually low values would suggest the presence of organic matter such as peat, or particles containing small cavities such as pumice. Tropical residual soils might show particle densities that are unexpectedly high or low. Particle density is essential for the analysis of compaction and consolidation behaviour, and in computations for particle size analysis by sedimentation. It is rarely used for classification of clays, but its value is needed for calculating voids ratio, specific volume and porosity, as follows:

$$\text{Voids ratio: } e = \frac{\rho_s}{\rho_D} - 1 \quad (9.5)$$

$$\text{Specific volume: } v = 1 + e. \quad (9.6)$$

Particle density depends upon the type of clay minerals present. Typical particle densities for some of the more common clay minerals are given in Table 9.4. However, clay soils can include more than one mineral type, and the value of particle density reflects the average value for all the materials present in the sample.

For clays, the test is carried out on at least two representative specimens, each of about 10 g, obtained by *riffling* from a sample of about 100 g. Since the test specimens are small it is essential that the sample is representative of the whole material. For clays containing particles up to gravel size, larger specimens (up to 400 g) are required. The proportion of included coarse particles should be properly representative of the material.

In BS 1377: Part 2: 1990, three methods for determination of particle density of soils are given, two of which are suitable for clay soils. For clays containing particles no larger than 2 mm the small (50 ml) pyknometer (density bottle) method (Clause 8.3) is relevant. The gas jar method (Clause 8.4) applies to soils (including clays) containing gravel-size particles.

9.3.6.1. Small pyknometer method. For the 50 ml pyknometer test, calibrated and numbered density bottles (pyknometers) with ground glass stoppers are required. An analytical balance reading to 0.001 g or better, and accurate weighing, are essential. A high vacuum is required for removal of air from the inundated specimen and the water bath temperature has to be maintained constant to within $+/-0.2°C$ of the stated value. Each dried specimen is placed in a 50 ml density bottle, weighed, covered with air-free distilled water, and placed in a vacuum desiccator to extract entrapped air. The bottles are filled, the ground glass stoppers fitted, and they are immersed in a constant-temperature water bath so that they can be weighed to an accuracy of 0.001 g or better when exactly full at a known temperature. The procedure is repeated using air-free distilled water only in each bottle. Care needs to be taken not to cause excessive bubbling of the liquid in the bottles when the vacuum is applied. It is usually necessary to fill the bottle in two stages. From the measured masses, the particle density can be calculated for each specimen. The average of the two (or more) values is reported as the particle density of the clay sample.

9.3.6.2. Methods for clays containing coarse particles. For clays containing sand- and gravel-size particles up to 37.5 mm the gas jar method (Clause 8.2) is used. The principle is the same as for the density bottle method, but about 200 to 400 g of dry soil is required, and the container used is a 1 litre gas jar with a ground glass plate for closing it. Air is removed by end-over-end shaking in a rotary mechanical device for about 30 minutes. The ground glass plate ensures that the level to which the jar is filled is repeatable. Weighing is carried out to an accuracy of 0.1 g. This method provides the mean particle density for all sizes of particles in the soil, not just that of the clay itself.

9.3.7. Particle size

In BS 1377, clay is defined as material with particles smaller than 0.002 mm (2 μm). In routine geotechnical testing the distribution of particle sizes within the clay range is not investigated. Most clays contain silt-size particles, and often some sand. Silt size particles fall in the range 2 μm to 63 μm, and sand from 63 μm to 2 mm. The proportion of clay-size material present, expressed as a percentage of the total material by dry mass, is called the *clay fraction*.

TABLE 9.4. *Absolute particle densities of some common minerals*

Mineral	Composition	Absolute particle density Mg/m³
Anhydrite	$CaSO_4$	2.9
Barytes	$BaSO_4$	4.5
Calcite, chalk	$CaCO_3$	2.71
Feldspar	$KAlSi_3O_8$, etc	2.6–2.7
Gypsum	$CaSO_4·2H_2O$	2.3
Haematite	Fe_2O_3	5.2
Kaolinite	$Al_4Si_4O_{10}(OH)_8$	2.6
Magnetite	Fe_3O_4	5.2
Quartz (silica)	SiO_2	2.65
Peat	Organic	1.0 or less
Diatomaceous earth	'Skeletal' remains of microscopic plants	2.00

Two processes are involved in determination of the clay fraction. The first is sieving by washing on a standard 63 μm sieve, which removes the silt and clay. The retained sand and gravel sizes, if present, can be analysed by further sieving. The combined total of silt and clay is obtained by difference. The second requires a separate identical specimen from which any particles coarser than 63 μm are removed. A sedimentation process is carried out on the fine material, from which the distribution of particle sizes through the silt range, and the proportion of clay in that portion, are obtained. The percentage clay in the original sample can then be calculated.

There are two methods of sedimentation for this purpose—the *hydrometer method*, and the *pipette method*. Before either procedure is started, the test specimen has to be made ready by a *pre-treatment* process, which is the same for both procedures. This prevents flocculation of particles, and removes organic matter and sesquioxides which would otherwise introduce errors. Both procedures require the use of a constant temperature bath in which the sedimentation cylinder is immersed.

9.3.7.1. Hydrometer test. In the hydrometer test the density of a suspension of the soil in water is measured at specified time intervals by means of a special soil hydrometer (Fig. 9.11). In the pipette test a small sample of the suspension from a fixed depth is taken at specified intervals using a special pipette (Fig. 9.12), and the mass of solid material is determined to the nearest 0.001 g. Each set of readings is used in calculations based on equations derived from Stokes' law, which states that the terminal velocity of a spherical particle falling freely in a known fluid is proportional to its density and the square of its diameter. These equations enable the relevant 'equivalent particle diameter', and the corresponding percentage smaller than that size, to be calculated.

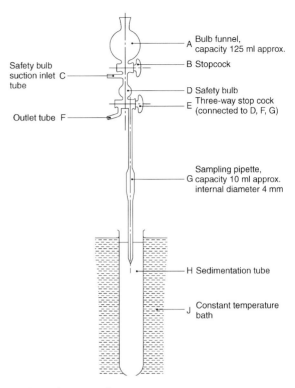

FIG. 9.12. Pipette sampling apparatus.

Hence the percentages of coarse, medium and fine silt can be derived, or the particle size distribution curve in the silt size range from 63 μm down to 2 μm can be plotted. Typical particle size distribution curves for four different types of soil containing clay are given in Figure 9.13.

9.3.8. Clay suction

The filter paper method of clay suction measurement is an empirical test for the determination of the state of desiccation of soils, and of clays in particular, especially if expansive behaviour is suspected. It is based on a relationship which has been established between the equilibrium water content of a certain type of filter paper, after a period in close contact with a clay specimen, and the pore water suction in that clay.

The normal test procedure applies to undisturbed samples 100 mm diameter, but samples of smaller diameter, and disturbed recompacted samples, can also be used. The undisturbed sample is sliced into about four discs of at least 10 mm diameter, in a controlled laboratory environment. The discs are fitted together again interspersed with discs of Whatman's No. 42. filter paper in close contact. The assembly is wrapped and sealed, and stored at constant temperature for 5 to 10 days. The wet and dry masses of each disc of filter paper are then determined using a high-precision analytical balance reading to 0.0001 g. The water content, by dry mass, of

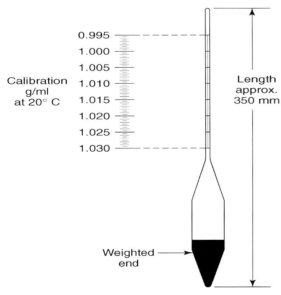

FIG. 9.11. Soil hydrometer.

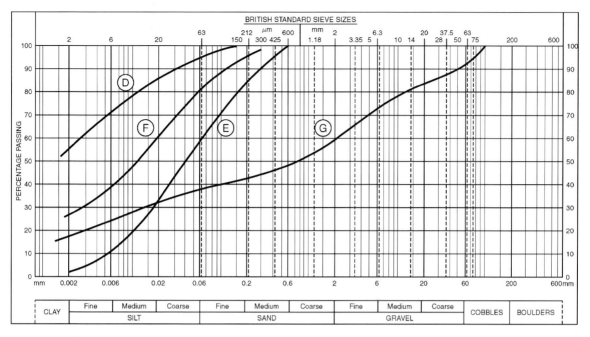

FIG. 9.13. Particle size distribution curves for soils containing clay: D, described as 'clay', although containing 44% silt; E, well-graded silt with sand and less than 2% (a trace of) clay; F, silty clay with sand (engineering properties dominated by the clay); G, well-graded soil with all sizes from cobbles to clay (glacial till).

each disc is calculated. Drying is carried out in a standard 105–110°C oven.

From the filter paper water contents, values of the soil suction are calculated, and these are averaged to give a value for the whole sample. There are two empirical equations relating suction to water content, depending on whether the latter is greater than or less than 47%.

9.4. Chemical tests

The detailed chemical composition of clay is of little interest for civil engineering purposes, but the presence of certain constituents in the clay can be very significant. The pH reaction (acidity or alkalinity) of the groundwater can also be of importance. The purposes of chemical tests are to classify the clay, and to assess any detrimental effect the clay or groundwater could have on concrete, steel or other materials buried in or in contact with the clay. The detailed chemical composition of clays is of great importance in some applications, for examples where they are being utilized in the manufacture of ceramics and cement (see Chapter 14 and 15). Further details of the chemical investigation of clays is contained in Chapter 8.

Chemical testing in a soil laboratory is usually limited to routine tests for determination of the following:

- acidity or alkalinity (pH value);
- sulphate content;
- organic content;
- carbonate content;
- chloride content;
- total dissolved solids (in water).

The chemical characterisation of contaminated soils is a specialist area of chemical testing. The reader is referred to Crompton (1999, 2001) for further information.

Chemical tests should be performed by a chemist or by a technician who has been specially trained in chemical testing procedures. Chemical tests for the presence of substances other than those listed above would normally be carried out by a specialist chemical-testing laboratory. Great care must be exercised over the specification of the soil chemical test to be carried out, and also with the interpretation of test results. It should be appreciated that exposing soil to changed environmental conditions may cause chemical change. One particular process that can affect clay soils is the oxidation of pyrite, which results in the production in acidity and increase in the sulphate present. Further details of the appropriate chemical testing procedures for assessing chemical aggressivity of the ground are given in BRE Special Digest 1 (2005), Reid et al. (2001), and (for highway structures) TRL Report 3/174.

Assessment of the possibility of the detrimental effects of sulphates and sulphides in the ground is a complex subject and involves knowledge of wider aspects such as the geology, groundwater regime and expected changes in ground conditions. The effects of sulphates on lime- and cement-stabilized clays, and guidance on testing, is discussed by Longworth (2004), who provides numerous references.

TABLE 9.5. *Chemical tests for clays and groundwaters*

Section in MSLT Vol 1	Type of test	Procedure	Reference ('BS' implies BS1377:1990)	Uses	Limitations and comments
5.5	pH value	Indicator papers	Supplier's instructions	Simple and quick. Useful for determining approximate pH range for a more sensitive test.	Gives approximate values only.
		Colorimetric (Kuhn's method) Lovibond Comparator	Manufacturer's instructions	Quick field test for soils. Apparatus available as a kit. Colour comparison with standard coloured discs gives pH to nearest 0.2. Range of indicators available.	Requires colour comparison with chart. Values given to nearest 0.5.
		Electrometric	BS Part 3:9	BS 'standard' method. Accurate to 0.1 pH unit or better.	Requires a special electrical apparatus, although low-priced portable battery models are available. Electrodes age slowly, and should be checked periodically with buffer solutions.
5.6	Sulphate content	Total sulphates in soils	BS Part 3:5.2, 5.5 BRE Special Digest 1 (2005)	Accurate if performed with care and with proper chemical testing facilities. Gives the total amount of sulphates present, including calcium sulphate, which is insoluble in water.	If the measured sulphate content is greater than 0.5%, the water-soluble sulphates should also be measured.
		Water-soluble sulphates in soils (gravimetric)	BS Part 3:5.3, 5.5	Accuracy as above. Gives the amount of water-soluble sulphates only, which are those most likely to attack concrete.	
		Water-soluble sulphates in soils (ion exchange)	BS Part 3:5.3, 5.6	Quick, easy	Cannot be used if chloride, nitrate or phosphate ions are present. Requires a special ion-exchange resin which needs reactivating frequently.
		Sulphates in groundwater (ion exchange)	BS Part 3:5.4, 5.6	As above.	As above.
		Sulphates in groundwater (gravimetric)	BS Part 3:5.4, 5.5	As for water-soluble sulphates in soils.	
5.7	Organic content	Peroxide oxidation	BS Part 3:3	Eliminates organic matter before sedimentation particle size tests.	Has limited action on undecomposed plant remains (e.g. roots and fibres).
		Dichromate oxidation		Accurate, if proper chemical testing facilities used. Suitable for all soils. Presence of carbon and carbonates does not affect results. Fairly rapid, suitable for small batches.	Presence of chlorides affects results but a correction can be applied if chlorides are measured separately. Their effect can be overcome by adding mercuric sulphate.
5.8	Carbonate content	Rapid titration Gravimetric	BS Part 3:6.3 BS Part 3:6.4	For carbonate content exceeding 10% Requires precision weighing and chemical testing facilities.	Accuracy no better than 1% carbonates Method as used for hardened concrete.
		Calcimeter (Collin's modification of Schleiber's apparatus)	BS1881: Part 124	Compact, simple, fairly quick. Measures the volume of carbon dioxide evolved.	An approximate method, but accurate enough for most engineering purposes. Atmospheric pressure must be known.

TABLE 9.5. *Continued*

Section in MSLT Vol 1	Type of test	Procedure	Reference ('BS' implies BS1377:1990)	Uses	Limitations and comments
5.9	Chloride content	Reaction with silver nitrate (Volhard's method):		Titration process requiring proper chemical testing facilities. Designed for concrete aggregates.	Several standardised reagents are required
		Water soluble	BS Part 3:7.2		
		Acid soluble	BS Part 3:7.3		
		Mohr's titration method	Bowley (1995)	Simpler than Volhard's method. Designed for concrete aggregates.	Both methods require an analytical balance.
5.10.2	Total dissolved solids	Evaporation	BS Part 3:8	Simple procedure	Requires very accurate weighing.
5.10.3	Loss on ignition	Ignition	BS Part 3:4	Destroys all organic matter. Suitable for sandy soils containing little or no clay or chalk.	High temperature breaks down certain minerals in clay, and carbonates, and removes water of crystallisation.
5.10.4	Concentration of certain salts	Indicator papers	Manufacturer's instructions	Very simple, quick, inexpensive.	Gives approximate indication only; not for accurate work. Presence of salts other than those being tested might affect readings.

The relevant tests are listed in Table 9.5, which also includes notes on their use and their limitations. These tests are usually carried out on very small duplicate or triplicate specimens of clay. Test specimens should be obtained from a larger sample of clay, which has been thoroughly mixed and riffled down to provide the required mass of material. Advice on the correct sampling and storage procedures for soils intended for chemical tests is given in Chapter 7. Reliance should not be placed on a single test, but several tests should be carried out on samples taken from different locations and horizons within the clay deposit, and care needs to be taken to ensure that the samples selected are representative of the deposit.

Results of chemical tests on soils should be regarded as indications of the orders of magnitude of constituents for classification purposes, rather than precise percentages. The procedures referred to in Table 9.5 provide accurate enough results for most soils, but with some soils (especially tropical and contaminated soils or made ground) there is the possibility that the presence of other constituents could have an undetermined effect on the chemical process for the measurement of a particular substance.

9.5. Compaction related tests

The tests described here cover the compaction of clay, and the determination of soil parameters that are related to the degree of compaction. The tests are as follows.

- compaction tests ('light' compaction and 'heavy' compaction) for deriving dry density/moisture content relationships;
- Moisture Condition Value (MCV);
- California Bearing Ratio (CBR).

All these tests are covered in BS 1377: Part 4: 1990. Similar tests are given in ASTM standards D 698, D1557 and D 1883.

Compaction is the mechanical process by which the solid particles are packed more closely together, thereby increasing the dry density of the soil. Laboratory compaction tests provide data which relate the achievable dry density to moisture content for a given compactive effort (the *moisture content—density relationship*). The MCV test is an extension of the standard compaction tests, and enables a rapid assessment to be made of the suitability of clay for placement and compaction *in situ*.

The CBR test is an empirical measurement of soil strength developed for estimating the bearing value of sub-bases and subgrades for roads and airfield runways. It is usually carried out on clay that has been compacted by one of the standard laboratory compaction procedures, but using a larger sample.

9.5.1. Compaction tests (moisture content—density relationship)

When clay soil is placed as an engineering fill, it is usually necessary to compact it to a dense state (the criterion being dry density) in order to obtain satisfactory engineering properties. The dry density that can be achieved by a given degree of compaction depends on the amount of water present in the soil, i.e. its moisture content. There is an *optimum moisture content* at which the dry density achievable by a given compactive effort reaches a maximum value, referred to as the *maximum dry density* for that effort. Laboratory compaction tests provide these values, which can be used as criteria for control of placement of fill material. Two standardized degrees of compactive effort are recognized, sometimes referred to as the 'light' and 'heavy' compaction procedures.

In these tests the clay is compacted into a rigid mould of known volume, using a standard number of blows from a hand or mechanical rammer of known mass with a controlled height of drop. The dry density achieved is calculated from the measured density and moisture content. From a series of at least five determinations with the clay at different moisture contents a graph of dry density against moisture content is plotted, from which the maximum dry density, and the corresponding optimum moisture content for that degree of compaction, can be derived. A typical graphical relationship is of the form shown in Figure 9.14. Lines indicating the percentage of air voids in the clay (*air voids lines*) are usually added for certain values (e.g. 0%, 5% and 10%) of air voids. These are sometimes referred to in specifications as criteria for *in situ* compaction control. Each voids line is derived and plotted from the following equation:

$$\rho_D = \frac{1 - \dfrac{V_a}{100}}{\dfrac{1}{\rho_s} + \dfrac{w}{100}} \, \text{Mg/m}^3 \qquad (9.7)$$

where ρ_D (Mg/m^3) is the dry density corresponding to moisture content w (%), ρ_s (Mg/m^3) is the particle density of the clay, and V_a is the volume of air voids, expressed as a percentage of the total volume.

Undisturbed or disturbed samples of clay can be used. A separate portion of at least 5 kg of soil at each of five or more moisture contents is prepared and matured as outlined in Section 9.1. If particles larger than 20 mm are present in significant quantity, either they should be removed first, or the larger CBR mould should be used, for which about 10 kg of soil is required for each point.

The main items of equipment for laboratory compaction tests comprise a one-litre compaction mould, or CBR mould, with appropriate detachable baseplate and extension collar; and a hand rammer, 2.5 kg ('light') or 4.5 kg ('heavy'). The height of drop (300 or 450 mm) is controlled by the steel tube in which each rammer is free to slide. Alternatively, an automatic compaction machine can be used if its compatibility has been verified. A hydraulic extruder for removing the compacted clay from the mould after weighing is also required.

The items for tests according to BS 1377 are shown in Figure 9.15. The equipment for ASTM tests is similar

LABORATORY TESTING

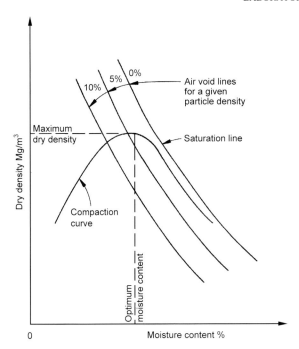

FIG. 9.14. Typical compaction curve.

in principle, but there are differences in detail. Dimensions of the two types of mould, both BS and ASTM specifications, are summarized in Table 9.6. Details of BS and ASTM compaction procedures are summarized in Table 9.7.

TABLE 9.6. *Details of compaction moulds*

Type of mould	Internal dimensions			Internal volume	
	Diameter (mm)	(in)	Height (mm)	(cm³)	(cu ft)
BS 1 litre	105		115.5	1000	
CBR	152		127	2305	
ASTM 4 in	101.6	(4)	116.4	944	(1/30)
6 in	152.4	(6)	116.4	2124	(0.075)

9.5.1.1. B.S. 'light compaction' test. The first prepared and matured clay specimen is placed and compacted in the mould in three equal layers, using the 2.5 kg rammer and applying 27 blows to each layer. The resulting density is determined and the moisture content is measured so that the dry density can be calculated.

The process is repeated with the other prepared batches of the sample, and each dry density is plotted against moisture content. The highest point of a smooth curve drawn through the plotted points gives the maximum dry density and the associated optimum moisture content for the clay. See Figure 9.14, and Figure 9.16 curve **A**.

9.5.1.2. B.S. 'heavy compaction' test. The procedure is similar to that described above, but a 4.5 kg rammer with a drop of 450 mm is used. The soil is placed in five layers instead of three, but the number of blows per layer (27) is the same. The dry density/moisture content curve is similar to that obtained from the 'light' compaction test, but is displaced upwards and to the left (see Fig. 9.16,

FIG. 9.15. Apparatus for compaction tests.

TABLE 9.7. *Compaction procedures*

Type of test	Mould	Rammer Mass (kg)	Rammer Drop (mm)	No of Layers	Blows per layer	Refer to section in MSLT Vol. 1
BS 'light'	One litre	2.5	300	3	27	6.5.3
	CBR	2.5	300	3	62	6.5.5
ASTM (5.5 lb)	4 in	2.49	305	3	25	6.5.7
	6 in	2.49	305	3	56	
BS 'heavy'	One litre	4.5	450	5	27	6.5.4
	CBR	4.5	450	5	62	6.5.5
ASTM (10 lb)	4 in	4.54	457	5	25	6.5.7
	6 in	4.54	457	5	56	

curve **C**). The maximum dry density is of course higher, but the corresponding optimum moisture content is lower.

9.5.1.3. Compaction in CBR mould. A CBR mould, with a volume of about 2.3 litres, is used for compaction tests on clay containing a significant quantity of coarse particles, up to a maximum size of 37.5 mm. Either the 'light' or the 'heavy' compaction test procedures can be carried out, in the same manner as described above except that the number of blows per layer is increased to 62 to compensate for the greater mass of soil used.

9.5.2. Moisture condition value

The moisture condition value (MCV) test provides a rapid means of assessing the suitability of a soil for use as compacted fill, in relation to the compaction criteria referred to above. The MCV value is a measure of the minimum compactive effort required to produce virtually full compaction of a clay. The test attempts to overcome some of the problems encountered in the control of quality of earthworks construction due to variability of the materials. Three aspects of the test, as given in BS 1377: Part 4: 1990, Clause 5, are as follows:

- determination of the MCV of a given clay at the moisture content 'as received';
- moisture condition calibration (MCC)—determination of the relationship between MCV and moisture content for a given clay;
- rapid assessment of the suitability of a clay for compaction.

In Figure 9.16, idealised moisture/density curves for three degrees of compaction are drawn. The lowest curve, A, could, for example, represent BS 'light' compaction; curve C, BS 'heavy' compaction; and curve B, an intermediate compactive effort. If the clay has a moisture content denoted by m_1, the lowest compactive effort, A, gives a dry density denoted by the point a. The higher effort, B, appreciably increases the dry density to point b, and the highest effort, C, results in another significant increase to the point c. So the dry density of a clay with a moisture content of m_1 can be increased appreciably by applying additional compactive effort.

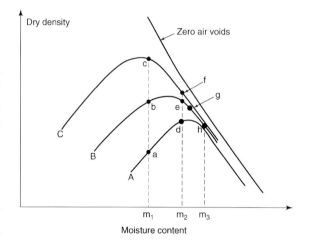

FIG. 9.16. Curves for three compactive efforts.

For clay at a higher moisture content denoted by m_2, it is still possible to increase the dry density by applying the additional compactive effort, B, (point d to point e), but increasing the effort to C makes little difference (e to f). At moisture content m_3 and above (point h) no significant increase in dry density can be achieved with additional compactive effort and full compaction is provided by effort A.

In the MCV test a sample of clay, similar to that required for a compaction test, in a rigid mould, 100 mm diameter by 200 mm high, with a permeable base, is subjected to blows which progressively increase the degree of compaction, until virtually full compaction is reached. The changes in volume resulting from increased compaction are continually measured. The test is carried out in a modified aggregate impact value apparatus (Geol. Soc., '*Aggregates*', Section 7.5.3.2), and is show in Figure 9.17. The rammer has a falling mass of 7 kg, and the height of drop can be maintained at 250 mm. A reading of penetration of the rammer is taken after each specified cumulative number of blows, as follows.

1, 2, 3, 4, 6, 8, 12, 16, 24, 32, 48, 64, 96, 128, 192, 256 blows.

LABORATORY TESTING

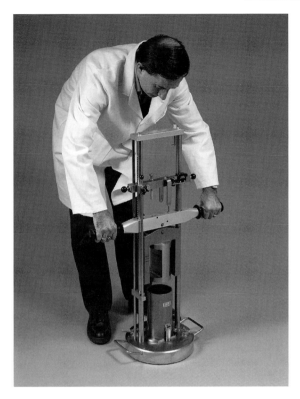

FIG. 9.17. Apparatus for moisture condition test.

The arbitrary criterion of 'no significant increase in penetration' is that the change of penetration between n blows and $4n$ blows is 5 mm or less. Every fourth number in the above sequence increases by a factor of 4.

The change in penetration between each pair of n blows and $4n$ blows is plotted on a graph against the number n, on a logarithmic scale, as shown in Figure 9.18. The graphical construction shown is used for reading off the abscissa representing 5 mm change in penetration from the horizontal scale (as a number N). The MCV value is equal to $10 \log N$. The numbers marked as an arithmetical scale on the zero abscissa in Figure 9.18 are the MCV values.

Different interpretations of the graph are sometimes necessary for deriving the MCV. For example the glacial tills found in Scotland give a slightly different shape of graph, and the intersection of the curve itself with the 5 mm penetration line may then be more appropriate.

For each type of clay a calibration relationship has to be established to enable an assessment to be made of the clay strength. The MCV test described above is carried out on at least four specimens of the same clay, each made up to a different moisture content, as in a compaction test, to cover the appropriate range of moisture contents.

The moisture content of each compacted specimen is plotted against its MCV, and the line of best fit (which might be curved) is drawn through the points, as shown in Figure 9.19. The 'effective' calibration line (MCC) is the portion of the relationship showing increasing MCV with decreasing moisture content.

This procedure can be carried out on site and provides a rapid assessment of the condition of the clay, i.e. whether or not it is suitable for use as compacted fill, but it does not indicate the extent to which it lies outside the specification requirements. It is based on MCV and MCC tests that have previously been carried out on similar clay.

In practice the appropriate upper limit of moisture content, w_u, for compaction is selected by the engineer

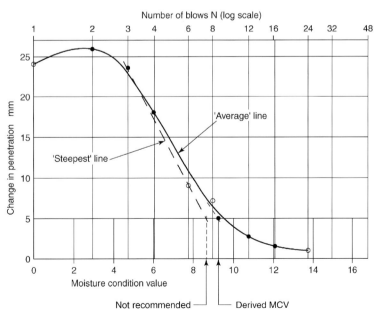

FIG. 9.18. Typical graph for derivation of MCV.

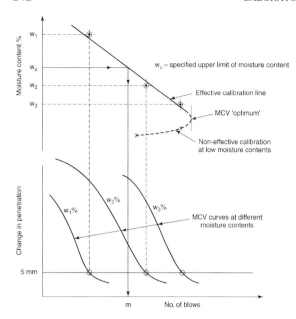

FIG. 9.19. Moisture condition calibration.

responsible. From the calibration graph already obtained the number of blows, m, corresponding to w_u, is read off, as shown in Figure 9.19.

The clay is placed in the mould, and m blows are applied as in the MCV test. The penetration of the rammer is measured and a further $3m$ blows are applied, making a total of $4m$, without adjusting the height of drop of the rammer. If the difference in penetration between m blows and $4m$ blows exceeds 5 mm, the clay is stronger than the pre-calibrated standard, and is in a suitable condition for compaction. If the difference is less than 5 mm the clay is weaker and is not suitable.

9.5.3. California bearing ratio

The California bearing ratio (CBR) test is an empirical test, which provides a rational method of design for flexible or rigid pavements. The CBR value is applicable only to the estimation of the thickness of sub-base construction to withstand the loading imposed by anticipated traffic conditions.

A standard plunger is pushed into the clay, which has usually been compacted into a mould, at a constant rate of penetration, while the force required to maintain that rate is measured. From the resulting load-penetration relationship the CBR value of the clay, at the condition at which it is tested, can be derived. The British Standard and ASTM test procedures are very similar in principle, but there are differences in detail in the apparatus and in the methods of sample preparation.

The clay for testing is prepared and matured as outlined in Section 9.1. If the clay contains particles larger than 20 mm, these are removed and their proportion is reported. The test specimen may be prepared in the CBR mould (details in Table 9.6), by either the light or the heavy compaction procedures. Alternatively, the clay can be compressed into the mould under static loading by a compression machine. After preparation the sample is sealed in the mould and allowed to stand for 24 hours, to enable any excess pore pressures to dissipate. The top surface is levelled off flat and smooth before testing.

When allowance has to be made for the effects of flooding or a rising water table, the prepared specimen is immersed in water under surcharge weights and the resulting swelling of the clay is measured before testing—this is the *soaked* procedure (see Fig. 9.20).

A typical arrangement for the CBR penetration test is shown in Figure 9.21. The plunger penetrates the clay at a constant rate of 1 mm per minute, up to a maximum penetration of 7.5 mm. At the end of the test the sample is removed for determination of moisture content, or it can be turned over for a second penetration test on the other end.

Readings of applied force (kN) are plotted against the corresponding penetration readings (mm) and a smooth curve is drawn through the points. The forces corresponding to penetrations of 2.5 mm and 5 mm are read off from the graph, as shown in Figure 9.22(a). If the initial portion is curved concave upwards, a modified 'zero' is substituted, as indicated in Figure 9.22(b). Each force is expressed as a percentage of the relevant 'standard' force (13.2 kN at 2.5 mm penetration, and 20 kN at 5 mm penetration). These percentages are the CBR values, and the greater of the two values is reported.

9.6. Dispersibility and durability tests

The three tests in this Section, called Dispersibility Tests, are relatively simple empirical procedures that were developed by Sherard *et al.* (1976), and subsequently Atkinson *et al.* (1990), for the identification of *dispersive*

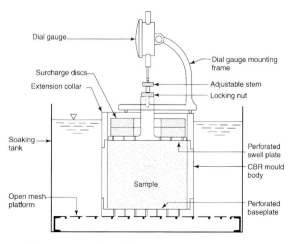

FIG. 9.20. Apparatus for soaking and swelling measurement.

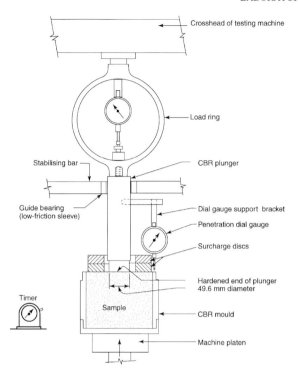

FIG. 9.21. Arrangement for CBR penetration test.

clays, which are clays with a high sodium content that are susceptible to rapid internal erosion by water. Other tests involving the chemical analysis of extracted pore water are outlined in Section 9.4. Another test (not covered here) provides a method for determining the extent to which a clay is affected by frost, and the amount of heave likely to result from freezing temperatures (BS812: Part 124).

These procedures were developed because dispersive clays cannot be identified from traditional soil classification tests. Where there is a flow of water through clay, such as occurs in the clay core of an earth dam, internal erosion can lead to disastrous failure.

9.6.1. Pinhole method

A specimen of recompacted clay is formed in a cylindrical container and is pierced with a 1 mm diameter hole. Distilled water from a constant-head supply is passed through the hole with inlet heads ranging from 50 mm to 1020 mm. The water emerging is examined for the presence of colloidal particles in suspension, which indicate that the clay is dispersive. The final diameter of the hole in the specimen is measured.

From the appearance and rate of flow of emerging water, and the final hole diameter, the clay is classified into 'dispersive soil' (category D1 or D2), or 'non-dispersive soil' (categories ND1 to ND4), in accordance with Table 9.8. The arrangement of the apparatus is illustrated in Figure 9.23.

9.6.2. Crumb and cylinder tests

For the Crumb test, a few 'crumbs' of clay, about 6 to 10 mm diameter, are dropped into a beaker containing sodium hydroxide solution (c (NaOH) = 0.001 mol/litre, or 0.04 g/litre). The reaction is observed after allowing the contents of the beaker to stand for 5 to 10 minutes.

The reaction is reported as one of four categories, as follows.

Grade 1: no reaction—perhaps some slaking but no cloudiness in the water.
Grade 2: slight reaction—some cloudiness in the water due to colloids in suspension.
Grade 3: moderate reaction—cloud of suspended colloids easily recognisable.
Grade 4: strong reaction—colloid cloud covers bottom of the beaker, or the whole solution is cloudy.

The Cylinder test was developed at City University as an extension to the Crumb test. Clay is remoulded to a slurry with de-aired distilled water and consolidated in stages under static pressure to form a cylinder about 38 mm diameter. The specimen is placed in a beaker containing the appropriate type of water, and its behaviour is observed during the ensuing few days.

For a non-dispersive clay some plastic deformation of the cylinder might occur but the water remains clear. Cloudy and opaque water indicates a dispersive clay.

9.6.3. Dispersion method

This method (also called the 'Double Hydrometer Test') requires two identical portions of the clay, designated A and B, on each of which a hydrometer sedimentation test is carried out using the procedure outlined in Section 9.3.7. For specimen A, the mechanical stirring is omitted and distilled water is used instead of the dispersant solution. The normal test is carried out on Specimen B.

The two particle size curves are plotted on the same sheet and the percentages of clay (A% and B%) are read off from the intercept with the 0.002 mm ordinate. The value (A/B × 100%) is reported as the 'Percentage Dispersion'.

9.7. Consolidation tests

Consolidation is the process whereby soil particles are brought more closely together over a period of time under the application of sustained pressure. It is accompanied by drainage of water from the pore spaces between solid particles.

When an increment of load is applied to a clay, the additional pressure is immediately carried by the resulting increase of the pore water pressure. Drainage of water from the clay enables the additional stress to be transferred to the solid particles, but this process

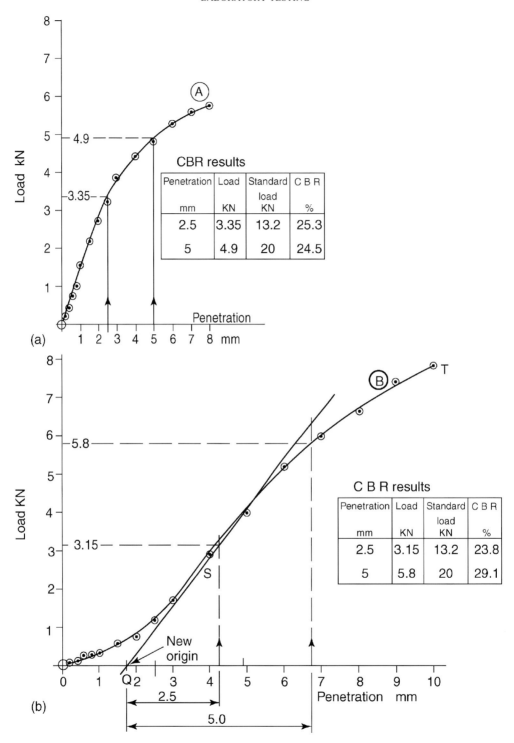

FIG. 9.22. Typical CBR curves (two types).

takes time because of the relatively low permeability of clay. Therefore two parameters are required to define the consolidation characteristics of a clay:

(1) the *compressibility* of the clay, i.e. the amount by which the clay will eventually compress when loaded and allowed to consolidate;

LABORATORY TESTING

TABLE 9.8. *Classification of clays from pinhole test data*

Dispersive classification	Head (mm)	Test time for given head (min)	Final flow rate through specimen (ml/s)	Cloudiness of flow at end of test		Hole size after test (mm)
				from side	from top	
D1	50	5	1.0 to 1.4	dark	very dark	2.0
D2	50	10	1.0 to 1.4	moderately dark	dark	>1.5
ND4	50	10	0.8 to 1.0	slightly dark	moderately dark	⩽1.5
ND3	180	5	1.4 to 2.7	barely visible	slightly dark	⩾1.5
	380	5	1.8 to 3.2			
ND2	1020	5	>3.0	clear	barely visible	<1.5
ND1	1020	5	⩽3.0	perfectly clear	perfectly clear	1.0

From Table 2 of BS 1377:Part 5:1990.

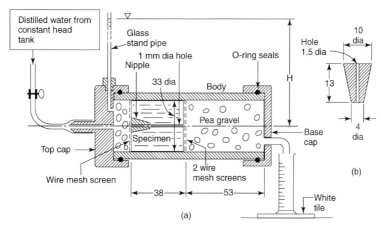

FIG. 9.23. Pinhole test apparatus.

(2) the rate at which that compression will take place—the *time-dependant* factor.

Four types of test for the determination of these characteristics are described in this Section, as follows:

- one-dimensional consolidation in a mechanical load frame (the *oedometer*), which requires relatively simple apparatus and procedures;
- continuous rate of strain one-dimensional consolidation, using a motorised loading device;
- one-dimensional consolidation in a hydraulic consolidation cell, which needs complex apparatus but provides for a variety of loading and drainage conditions;
- Three-dimensional consolidation in a triaxial cell, in which the loading and deformation are isotropic (i.e. the three principal stresses are equal).

The first is covered in BS 1377:1990, Part 5, and the third and fourth in Part 6. ASTM Standards cover the first (D 2435) and second (D 4186).

9.7.1. One-dimensional consolidation in an oedometer

The 'oedometer test' is the most common type of consolidation test procedure for saturated or near-saturated clay specimens, usually taken from undisturbed samples. The test specimen is in the form of a disc, confined laterally in a metal ring under the condition of zero lateral strain. Water is allowed to drain freely from the clay at each end (*double drainage*), and the flow is laminar. The specimen is subjected to typically six increments of vertical axial pressure, and the decrease in height resulting from drainage of water out of the voids in the clay (*consolidation settlement*) is measured. Each load increment is maintained until the *primary consolidation* phase is completed, leading into the *secondary compression* phase. (The primary and secondary consolidation phases are identified from graphical plots of settlement readings against time, as described below). This typically takes 24 hours, due to the slow rate at which water drains out of the clay, but some clays may need considerably longer for primary consolidation to be complete.

Data from consolidation tests enable estimates to be made of the likely amount of settlement under loading; the extent of differential settlements; and some indication of the rate at which settlement will take place.

9.7.1.1. Procedure. Undisturbed samples of good quality are necessary for consolidation tests. Disturbance of the soil structure, or disturbance due to change in moisture content, can have a significant effect on the test results. A typical specimen size is 75 mm diameter and 20 mm high.

When a consolidation test is to be carried on compacted clay, it should not be compacted directly into the ring, but the specimen should be cut from a larger prepared and matured sample.

The test specimen in its confining ring is placed between two saturated discs of rigid porous material, which provide drainage from each end, under a rigid loading cap, contained within a rigid cell which holds water. The cell is mounted on a loading frame beneath a lever, which applies a vertical force to the loading cap generated by hanger weights. The main features of the apparatus are shown in Figure 9.24.

The cell is filled with water, and any tendency to swell is resisted and the swelling pressure is determined. Settlement due to consolidation of the clay under the first applied stress is monitored at suitable time intervals from the instant of loading, normally for a period of up to 24 hours. Cumulative settlement can be plotted against elapsed time in two ways: settlement against log time (minutes) and settlement against square-root time (minutes). Inspection of the graphical plot or plots indicates whether the primary consolidation phase has been completed: if it has not, the load is maintained for a further 24 hours, or longer if necessary. The process is repeated for each of the subsequent applied loads. The specimen is then unloaded in a similar manner, with a lesser number of decrements. After reaching equilibrium under the final pressure the specimen is removed, weighed, and weighed again after drying.

A graph of settlement against log time from one load stage is typically of the form shown in Figure 9.25, and against square-root time as shown in Figure 9.26. The graphs from one stage must be plotted before continuing to the next stage of loading so that the completion of primary consolidation can be confirmed. This is shown in the log-time plot by the flattening curve leading out of the steepest part of the graph (at the point of inflection) into the final linear portion. This linear relationship represents the secondary compression phase.

From each of these graphs the range of the theoretical primary consolidation phase can be identified, using the empirical geometrical constructions indicated in Figure 9.25 and Figure 9.26. The point of theoretical 50% primary consolidation, denoted by d_{50}, can then be found, and the corresponding time t_{50} (minutes) is read off from the graph and is used to calculate the *coefficient of consolidation*, c_v, for the chosen pressure.

Measurements of density, moisture content and particle density enable the initial voids ratio of the clay, e_0, to be calculated, as described in Section 9.3.5, and also the voids ratio at the end of each load stage. A graph is then plotted of voids ratio against pressure (on a log scale), commonly referred to as the e-log p curve. A typical curve is shown in Figure 9.27.

For each increment of loading, the *coefficient of volume compressibility*, m_v (m²/MN) and the *coefficient*

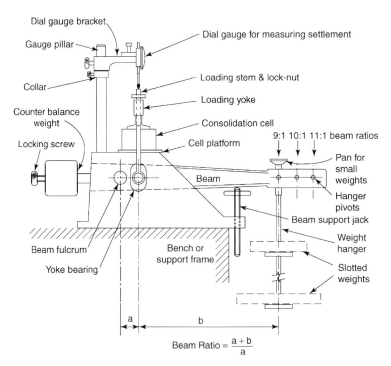

FIG. 9.24. Oedometer consolidation apparatus.

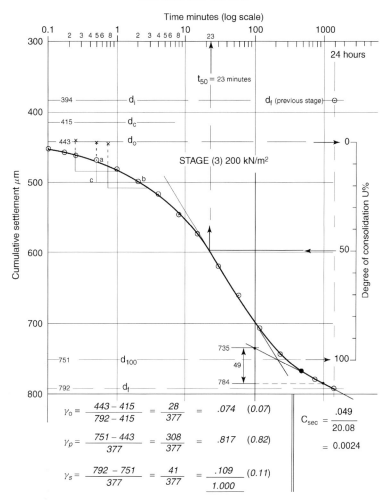

FIG. 9.25. Typical plot of settlement against log time.

of consolidation, c_v (m²/year), are calculated. These parameters relate to the compressibility of the clay specimen, and the rate at which it consolidates, respectively. The values are summarised and reported with the e-log p curve, as shown in Figure 9.27. The equations are:

$$m_v = \frac{\delta e \cdot 1000}{\delta p(1+e_1)} (\text{m}^2/\text{MN}) \quad (9.8)$$

$$c_v = \frac{0.026(H^*)^2}{t_{50}} (\text{m}^2/\text{year}) \quad (9.9)$$

in which δe is the change in voids ratio during the loading stage, δp is the increment of applied pressure (kPa), e_1 is the voids ratio at the start of the load stage, H^* is the mean height of the specimen during the stage (mm), t_{50} is the time for theoretical 50% consolidation (minutes), obtained from the settlement/time relationship as described above.

9.7.2. Constant, or controlled, rate of strain oedometer consolidation

A constant, also so known as a controlled, rate of strain (CRS) consolidation test is a one-dimensional test, similar in principle to the oedometer test described above except that the load on the specimen is increased continuously by applying a controlled rate of strain, instead of discrete loading increments. It is the best known of a number of continuous-loading consolidation tests, which are discussed by Davison & Atkinson (1990). The top loading cap is pushed down at a uniform slow speed, by means of a motorised load frame, applying a constant rate of vertical strain to the specimen. This does not necessarily imply a constant rate of increase of vertical stress. A major advantage of this type of test is that it can be completed in a much shorter time than is required for a conventional incrementally-loaded test.

Drainage of water, to dissipate the excess pore water pressure, is allowed from the top surface of the specimen

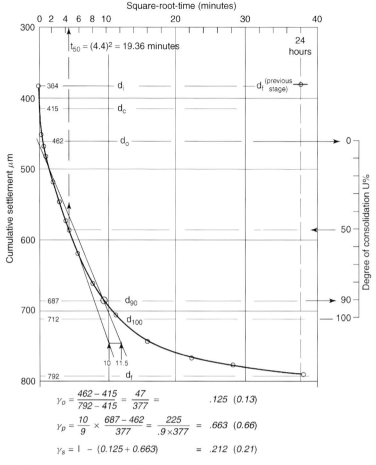

FIG. 9.26. Typical plot of settlement against square-root time.

only (single drainage), and pore water pressure is measured at the base. Continuous monitoring of the applied force and pore pressure enables a continuous record to be obtained of the effective stress.

A test procedure is given in ASTM D 4186. Further details are described by Smith & Wahls (1969); Wissa et al. (1971); Gorman et al. (1978). Application of this test to very soft clays was reported by Umehara & Zen (1980).

In the simplest form of the test, the top drainage surface is open to atmosphere, so the pore pressure near that surface is zero. A preferable arrangement (obligatory in the ASTM procedure) is to use a cell that can be completely sealed so that a back pressure can be applied to the water at the drainage surface (Fig. 9.28a).

A suitable rate of strain has to be selected before starting the test, to give excess pore pressures generally in the range 3% to 20% of the total applied vertical stress. Suggested rates of strain, based on the liquid limit of the clay, are summarized in Table 9.9. The initial estimate of strain rate might have to be revised in the light of experience.

Values of c_v and m_v can be calculated for each interval between successive sets of readings, enabling a continuous record of these parameters to be obtained, from the following equations. Further details are given in the references cited above.

$$c_v = 0.263 \frac{\delta p'}{\delta t} \cdot \frac{H^2}{\delta u} \left(m^2 / year \right) \qquad (9.10)$$

$$m_v = \frac{4.34 \, r \, \delta t}{\log_{10}(p_2/p_1) \times p'} \left(m^2 / MN \right) \qquad (9.11)$$

In these equations, p_1, p_2 (kPa) are the total vertical stresses at start and end of time interval δt minutes, p (kPa) is the average total vertical stress, $p' = p - \tfrac{2}{3} \delta u$ (kPa) (the average effective vertical stress), u (kPa) is the pore pressure at time t minutes, r is the rate of strain (% per min) and H (mm) is the mean specimen height during time interval δt minutes. Other ways of performing continuous-loading consolidation tests are summarized in Section 9.7.5.

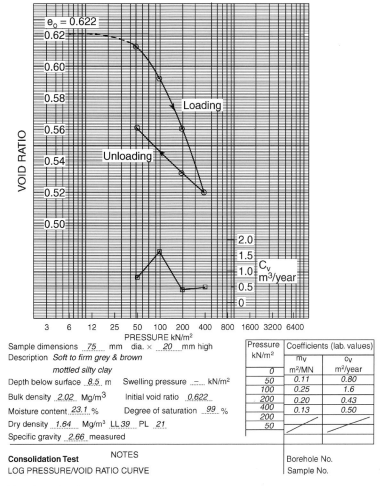

Fig. 9.27. Typical plot of voids ratio against log pressure.

9.7.3. Consolidation in a hydraulic consolidation cell

The principles of consolidation of clay referred to in Section 9.7.1 also apply to the tests described in this Section. The main differences lie in the mechanics of the tests, and in the greater degree of control and the much wider scope of the procedures that are available.

The hydraulic consolidation cell (also called the Rowe cell) was developed in order to overcome most of the disadvantages and limitations of the conventional oedometer apparatus described in Section 9.7.1. The main advantages of the hydraulic cell are summarized as follows:

1. Several sizes of hydraulic cell are available, enabling samples up to 250 mm diameter to be tested.
2. Pressures up to 1000 kPa can be applied, even to large samples.
3. Hydraulic loading is less susceptible to effects of vibration than mechanical lever-arm loading, and can be more closely regulated.
4. Drainage can be controlled, and several different vertical or horizontal drainage conditions can be imposed. See Figure 9.29.
5. The volume of water draining from the sample can be measured, as well as surface settlement.
6. Pore water pressure can be measured, enabling known *effective* pressures to be applied.
7. A back pressure can be applied to the sample, which enables the clay to be fully saturated under controlled conditions and tested in the saturated condition.
8. Loading can be applied either as a uniform pressure across the surface of the sample (*free strain*), or through a rigid plate which maintains the loaded surface plane (*equal strain*).
9. Use of large samples enables the effects of the soil fabric, such as laminations, or inclusions of organic matter, to be taken into account.

Specimens may consist of undisturbed, compacted or consolidated clay. An undisturbed specimen is prepared by cutting and trimming the clay from an undisturbed

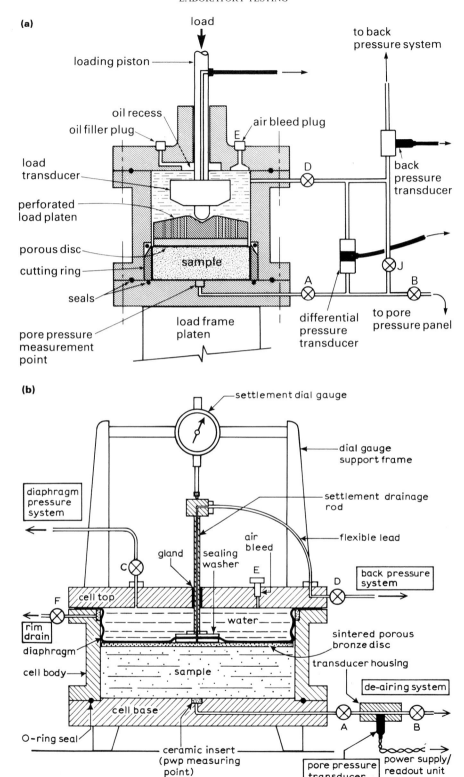

FIG. 9.28. (a) Apparatus for CRS test with back pressure. (b) Typical hydraulic consolidation cell.

TABLE 9.9. *Suggested rates of strain for CRS consolidation tests*

Liquid limit range %	Rate of strain % per minute
up to 40	0.04
40–60	0.01
60–80	0.004
80–100	0.001
100–120	0.0004
120–140	0.0001

(Based on ASTM D 4186).

sample so that it fits into the cell. On the other hand a compacted specimen is prepared in the cell using either dynamic compaction or static compression. A re-consolidated specimen is prepared from a slurry, and then drained and consolidated to replicate natural deposition (mostly used as a research procedure).

The main features of a typical 254 mm diameter hydraulic consolidation cell are shown in Figure 9.28b. Vertical stress is applied to the top of the clay specimen by means of pressurized water acting on a flexible bellows (the *diaphragm*), which can expand downwards. A relatively flexible porous disc, having a diameter a little less than that of the specimen, is placed on the top surface of the specimen to collect water draining from it. Drainage is taken via a tube seated on the disc to a back pressure system. The top end of the tube provides a seating for the stem of a displacement dial gauge or transducer.

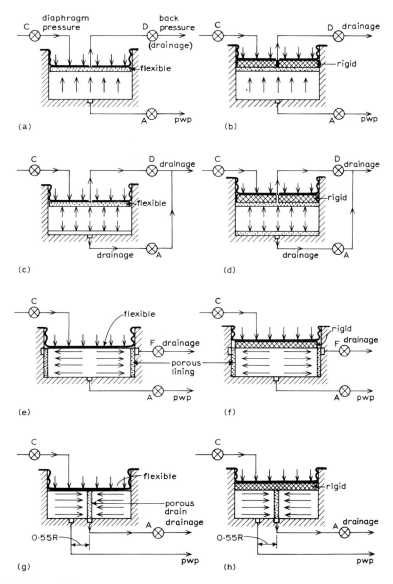

FIG. 9.29. Drainage and loading conditions.

At the centre of the cell base is a small porous ceramic disc, connected to a *pore pressure transducer* which measures the pore pressure at the base of the clay. Provision is made for draining water from the periphery of the specimen when horizontal drainage is required.

A constant pressure system is connected to valve *C* to provide the *diaphragm pressure*. A second constant pressure system, connected to valve *D*, provides the *back pressure*, and incorporates a volume-change measuring device to measure the volume of water draining out of the clay.

9.7.3.1. Test procedures. To carry out the test the clay is first brought to the state of full saturation by increasing the diaphragm pressure and the back pressure alternately in suitable increments, observing the pore pressure during each step and allowing sufficient time to enable the pore pressure to stabilize. An increment of vertical stress ($\Delta\sigma$) generates an increase in pore pressure (Δu) in the clay. As the pressures increase, the ratio ($\Delta u/\Delta\sigma$) increases towards a value of unity, which represents a theoretical saturation of 100%. In practice a value of about 0.96 is generally accepted as indicating that the clay is sufficiently saturated.

The specimen is subjected to a series of vertical pressure increments, each being equal to the existing pressure, but here it is the *effective* pressures (difference between diaphragm pressure and pore water pressure) that are significant. The clay is allowed to consolidate under each effective pressure until at least 90% of the excess pore water pressure has been dissipated by drainage, so that primary consolidation is virtually completed. After consolidating under the maximum desired effective pressure, the specimen is unloaded in a series of pressure decrements.

The pore pressure reading at any time from the start can be expressed as *percentage pore pressure dissipation*, denoted by *U*, which is the percentage of primary consolidation referred to in Section 9.7.1. This is calculated from the equation

$$U = \frac{u_i - u}{u_i - u_b} \times 100\% \quad (9.12)$$

where u_i is the initial pore pressure (i.e. at the start of drainage); u_b is the back pressure, which is the pore pressure that would eventually be reached at theoretical 100% primary consolidation; *u* is the pore pressure measured at a given time.

If *U* is plotted against a function of time, the time for theoretical 50% primary consolidation, t_{50} (minutes), can be read directly from this graph without the need for a geometric construction, as shown in Figure 9.30. The coefficient of consolidation for a stage is then calculated from the equation

$$c_v = \frac{0.197.(H^*)^2}{t_{50}} \left(m^2/year\right) \quad (9.13)$$

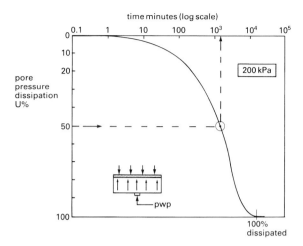

FIG. 9.30. Percentage pore pressure dissipation versus log time.

where H^* is the mean height of the specimen (mm) during the stage.

The above equation relates to the 'free strain' condition with one way vertical drainage. The same equations, but with a different numerical coefficient, are used for deriving the coefficient of consolidation for other loading and drainage conditions. The coefficients are summarized in Table 9.10 under theoretical time factor.

Volume change, and the measured vertical movement at the centre of the top surface, can be plotted against log time, and/or square-root time. Values of voids ratio at the end of each loading stage are calculated from changes in volume, using the equation given below. The relationship between the calculated voids ratios and log pressure is plotted graphically as in the oedometer test.

The voids ratio at the end of each consolidation stage (*e*) is calculated from the equation

$$e = e_0 - (1+e_0).\frac{\Delta V}{V_0}. \quad (9.14)$$

where e_0 is the initial voids ratio (calculated from dry density and particle density), V_0 is the initial volume of the specimen (cm³), ΔV is the cumulative change in volume from the start of the first stage to the end of the stage being considered (cm³).

The coefficient of volume compressibility (m_v) for a stage is calculated from the equation

$$m_v = \frac{\delta e.1000}{\delta p'(1+e_1)} \left(m^2/MN\right) \quad (9.15)$$

where δe is the change in voids ratio due to the pressure increment ($= e_1 - e_2$), $\delta p'$ is the increment of applied effective pressure (kPa), e_1 is the voids ratio at the start of the increment, e_2 is the voids ratio at the end of the increment.

TABLE 9.10. *Rowe cell consolidation tests—data for deriving coefficient of consolidation*

Test ref.	Drainage direction	Boundary strain	Consolidation location	T_{50}	Theoretical time factor	T_{90}
(a) and (b)	Vertical, one way	Free and equal	Average	0.197	(T_v)	0.848
			Centre of base	0.379		1.031
(c) and (d)	Vertical, two way	Free and equal	Average	0.197	(T_v)	0.848
(e)	Radial, outward	Free	Average	0.0632	(T_{ro})	0.335
			Central	0.200		0.479
(f)		Equal	Average	0.0866	(T_{ro})	0.288
			Central	0.173		0.374
(g)	Radial, inward†	Free	Average	0.771	(T_{ri})	2.631
			$r = 0.55R$	0.765		2.625
(h)		Equal	Average	0.781	(T_{ri})	2.595
			$r = 0.55R$	0.778		2.592

†Drain ratio 1/20.
T_v, T_{ro}, T_{ri} = theoretical time factors.
t = time (minutes).
H = specimen height (mm).
D = specimen diameter (mm).

9.7.4. Isotropic consolidation in a triaxial cell

In this type of test a cylindrical clay specimen is subjected to *isotropic stress* conditions in a triaxial cell, i.e. the stress is applied equally in all three dimensions, and the specimen deforms and consolidates three-dimensionally. The volumetric coefficients of compressibility and consolidation for a given clay are not the same as the equivalent coefficients for one-dimensional consolidation.

Isotropic stress is applied to the test specimen by means of a cell confining pressure. Drainage of water to relieve the excess pore pressure is normally vertical, to the top surface of the specimen. Otherwise, the basic principles and sequence of operations, and the measurements that are made, are similar to those for the tests outlined in Section 9.7.3 above.

The test is usually carried out in the same kind of cell, and using the same pressure systems and other apparatus, as is used in triaxial compression tests for measurement of effective shear strength (Section 9.10).

The test is normally carried out on cylindrical specimens of clay, nominally about 100 mm diameter and 100 mm high. Smaller specimens can be used, but the larger size is preferable because it can incorporate features of the soil fabric. Test specimens are normally prepared from undisturbed clay samples of good quality, as described for triaxial compression test specimens in Section 9.10.

9.7.4.1. Equipment and test procedure. The cell in which the specimen is confined is of the same type as those used for triaxial compression tests (see Section 9.10), but the loading piston is restrained from movement.

The specimen is contained in a tubular rubber membrane, which is sealed to the top cap and pedestal by rubber O-rings. Between the clay and the top cap is a disc of porous rigid material, which collects drainage water from the top surface and enables it to be connected to the back pressure system which incorporates a volume change indicator. A porous ceramic insert at the centre of the base is connected to a pore pressure transducer. The cell chamber is connected to its own constant pressure system. The arrangement is shown diagrammatically in Figure 9.31.

A test with horizontal drainage is possible in the triaxial cell by preventing drainage from the top surface and wrapping a layer of flexible porous plastic material around the curved surface of the specimen, between it and the enclosing rubber membrane.

The clay is first brought to the state of full saturation by increasing the cell pressure and the back pressure alternately in suitable increments, observing the pore pressure during each step and allowing sufficient time to enable the pore pressure to stabilize. An increment of cell pressure ($\Delta\sigma_3$) generates an increase in pore pressure (Δu) in the clay. The ratio ($\Delta u/\Delta\sigma_3$) is known as the B value, and indicates the response of the pore water pressure to change in confining pressure. As the pressures increase, the B value increases towards a value of unity, which represents a theoretical saturation of 100%. In practice, a B value of about 0.96 is generally accepted as indicating that the clay is sufficiently saturated.

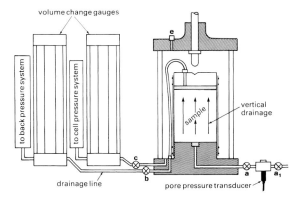

FIG. 9.31. Apparatus for triaxial consolidation test.

The saturated specimen is then subjected to a series of isotropic pressure increments of *effective* confining pressure (cell pressure minus pore pressure). Each increment is usually equal to the existing effective pressure. The clay is allowed to consolidate under each pressure, by means of drainage of water, until the excess pore pressure is virtually dissipated. In practice this means that at least 90% pore pressure dissipation should be achieved. After consolidation under the maximum desired effective pressure has been completed, the specimen is unloaded in a series of pressure decrements back to the initial pressure.

The voids ratio at the end of each consolidation stage (e) is calculated from the equation

$$e = e_0 - (1 + e_0) \cdot \frac{\Delta V}{V_0} \qquad (9.16)$$

where e_0 is the initial voids ratio (calculated from dry density and particle density), V_0 is the initial volume of the specimen (cm³) and ΔV is the cumulative change in volume from the start of the first stage (cm³).

The coefficient of volume compressibility (m_{vi}) for isotropic consolidation is calculated from the equation

$$m_{vi} = \frac{\delta e \cdot 1000}{\delta p'(1 + e_1)} \; (\mathrm{m^2/MN}) \qquad (9.17)$$

where δe is the change in voids ratio due to the pressure increment ($= e_1 - e_2$), $\delta p'$ is the increment of applied effective stress (kPa), e_1 is the voids ratio at the start of the increment and e_2 is the voids ratio at the end of the increment.

Pore pressure readings are expressed as percentage pore pressure dissipation (U), which is plotted against log time or square-root time from start of drainage. The time taken for 50% dissipation of pore-water pressure, t_{50} (minutes), may be read from the graph. Volume change is also plotted on similar time scales. Voids ratios calculated as above are plotted against log of cell pressure.

Changes in specimen height are not measured, but the height at the end of each loading stage is needed in order to calculate the coefficient of consolidation. The height, H, at the end of a consolidation stage is estimated from the equation

$$H = H_0 \left\{ 1 - \frac{\Delta V}{3 V_0} \right\} \qquad (9.18)$$

where H_0 is the initial height of the specimen (mm).

The value of the coefficient of consolidation, c_v, for the stage is calculated from the equation

$$c_v = \frac{0.199(H^*)^2}{t_{50}} \; (\mathrm{m^2/year}) \qquad (9.19)$$

where H^* is the mean specimen height during the stage.

For a test in which drainage takes place horizontally to the radial boundary, calculations are similar to the above except for the equation to derive the coefficient of consolidation, which becomes

$$c_v = \frac{0.026 D^2}{t_{50}} \; (\mathrm{m^2/year}) \qquad (9.20)$$

in which D is the specimen diameter (mm) during that stage.

9.7.5. Continuous-loading consolidation tests

In addition to the controlled rate of strain consolidation test referred to in Section 9.7.3, there are several other recognised procedures for deriving semi-continuous plots of voids ratio and consolidation parameters (m_v and c_v) against vertical effective stress. A significant advantage of these plots over those derived from conventional incremental tests is that they enable a better identification of the yield point or preconsolidation pressure to be obtained, together with continuous values for the constrained modulus for loading and unloading. They are mainly used for obtaining consolidation parameters appropriate for particular circumstances.

These procedures are not yet covered by Standards, and are regarded as 'special' or 'advanced' tests. They all require high-quality undisturbed samples, means of varying loading and pressures continuously, and automatic logging of test data. Some require automated computer-controlled systems using feedback data. Oedometers or hydraulic consolidation cells, with suitable modifications, are generally used.

9.7.5.1. Constant rate of loading. The vertical stress applied to the specimen is steadily increased at a constant rate, until the required maximum stress is reached. A period of a few hours to a few days might be appropriate, depending of the time required to establish pore pressure equalization.

9.7.5.2. Constant pore pressure gradient. The applied vertical stress is controlled so as to maintain the pore pressure at the non-drained face at a constant value. Since the pore pressure at the drained face is equal to the back pressure and is therefore constant, the gradient between the two faces remains constant.

9.7.5.3. Constant pressure ratio. The ratio between pore pressure and total vertical stress is maintained at a constant value. A back pressure is not used. As well as the usual consolidation parameters a value of the tangent modulus can be derived from the test data.

9.7.5.4. Consolidation with restricted flow. A flow restrictor is connected to the drainage line between the specimen and the back pressure system. Vertical load is applied in a single increment and the resulting pore pressure changes and settlement are recorded continuously. This type of test was devised for remoulded specimens of estuarine mud. A new design of cell was subsequently developed for these tests.

9.7.5.5. Back pressure control. The back pressure is controlled so that it decreases at a pre-determined rate. The rate could be constant, or non-linear with time if a computer controlled back pressure is used.

9.8. Permeability tests

Clays, like sands, are pervious to water; the difference between their characteristics is only one of degree. The capacity to allow flow of water through a soil is its *permeability*, which is expressed as a velocity of flow. The permeability of a relatively 'impervious' clay is very low, and may be ten million times less than that of a sand; nevertheless the clay has a permeability value which can be measured.

The *coefficient of permeability*, denoted by k and expressed in m/s, is the mean velocity of laminar flow of water through the soil under the action of a *unit hydraulic gradient*. Three methods for measuring the coefficient of permeability of clays are given in this Section, namely:

- falling head test in a falling head permeameter;
- constant head test in a triaxial cell;
- constant head test in a hydraulic consolidation cell.

There are two important assumptions relating to these tests. The first is that the flow of water through the pore spaces in the clay is laminar, or streamline, and not turbulent. The second is the validity of Darcy's Law: that the velocity of flow of water through any soil is directly proportional to the hydraulic gradient. The measured permeability depends on the temperature of the test, because the viscosity of water varies with temperature. Reported values are usually corrected to a standard temperature of 20°C.

9.8.1. Falling head test

In this test a specimen of clay is compacted at a suitable moisture content to the desired dry density, or with a specified compactive effort, into a rigid cylindrical container (the cell). This is fitted with a perforated base and a top plate with a central hole to which a standpipe is connected (see Fig. 9.32). The base is submerged, but water can drain through the perforations. Water in the

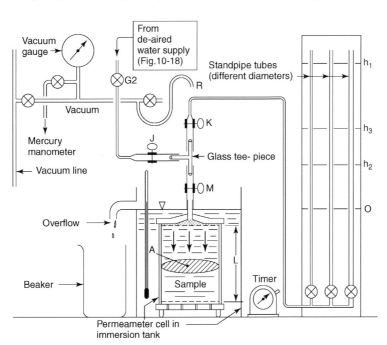

FIG. 9.32. Arrangement of falling head apparatus.

standpipe provides an excess pressure, and observations of the water level in the standpipe at measured intervals of time as the level falls provide data for deriving the coefficient of permeability. The equation used is based on Darcy's law.

This test is often used for determining the permeability of recompacted clays after saturation for assessment of their water-retaining properties. It is less reliable for tests on undisturbed clay because of the possibility of drainage paths between the specimen and the cell wall. In addition, the stresses acting on the specimen are not known, so for undisturbed clay it is preferable to perform permeability tests in hydraulic consolidation or triaxial cells (Sections 9.8.2 and 9.8.3).

The coefficient of permeability (k_T) of the clay at the temperature of the test (T°C) is calculated from the equation

$$k_T = 3.84 \frac{aL}{At} \log(h_1/h_2) \times 10^{-5} \, (\text{m/s}) \qquad (9.21)$$

in which a is the area of cross-section of the manometer tube (mm^2), A is the area of cross-section of the specimen (mm^2), L is the specimen height (mm), h_1 and h_2 are the heights of two reference levels above the outlet overflow level (mm) and t is the time between manometer readings h_1 and h_2 (minutes). The calculation can be repeated if the time interval to reach an intermediate level h_3 is also recorded.

If necessary the measured value is corrected to 20°C by applying the appropriate temperature correction factor for the viscosity of water.

9.8.2. Permeability in a hydraulic consolidation cell

9.8.2.1. Scope. This type of cell enables the axial or radial permeability of a laterally confined specimen of undisturbed of recompacted clay to be measured under a known vertical stress, and under the application of a back pressure. The pore pressure, and therefore the vertical effective stress, is known. The constant head procedure is used, in which the volume of water passing through the specimen under a known hydraulic gradient in a given time is measured.

The direction of flow can be either vertical (parallel to the specimen axis), usually downwards, or horizontal (radially inwards or radially outwards). Tests with horizontal flow are relevant when the effect of soil fabric (e.g. discontinuities or laminations), which can have an enormous effect on the permeability properties, needs investigation. For this purpose 250 mm diameter cells are preferable. Permeability can also be measured at any stage during a consolidation test (Section 9.7.3) in this type of cell.

Undisturbed or recompacted specimens are prepared and placed in the hydraulic cell as described in Section 9.7.3. If an undisturbed clay includes laminations or other discontinuities, the specimen may be orientated so that these features lie horizontally in the cell, regardless of their inclination *in situ*.

The cells are the same as described in Section 9.7.3 for hydraulic cell consolidation tests, but with some modifications and additions to the ancillary equipment. The 'equal strain' loading condition is usually applied.

For tests with vertical flow (downwards or upwards), the arrangement of the cell is similar to that used for a two-way vertical drainage consolidation test, i.e. with a porous drainage disc at the top and bottom faces of the specimen. Three constant pressure systems are required: one to provide the vertical stress, and one connected to each end of the specimen to provide known pressures in the inlet and outlet drainage lines. Both of the latter include a volume change indicator. A typical arrangement is shown in Figure 9.33. If a high back pressure at the outlet end is not necessary, the outlet can be connected to an elevated constant-level water reservoir instead.

For tests with horizontal (radial) flow (inwards or outwards), no drainage is allowed from the top and bottom surfaces. The cell is fitted with both a peripheral drainage layer, and a central drainage well (as described in Section 9.7.3), each of which is connected to a constant pressure system with volume change indicator.

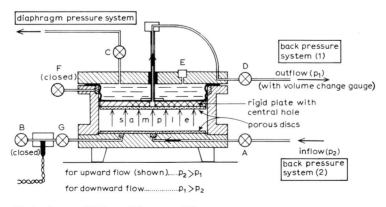

FIG. 9.33. Arrangement of hydraulic consolidation cell for permeability tests.

9.8.2.2. Procedure and calculations. The specimen is saturated, and consolidated to the required effective stress, as described in Section 9.7.3. Flow of water through the specimen under the required hydraulic gradient is timed until a steady rate of flow is reached. The pressure loss due to pipe friction is obtained from calibration data. The coefficient of permeability (k_V or k_H m/s) is determined from the appropriate equation.

For vertical flow:

$$k_V = \frac{1.63 qL}{A\left[(p_1 - p_2) - p_c\right]} \times R_t \times 10^{-4} \, (\text{m/s}). \quad (9.22)$$

For horizontal (radial) flow:

$$k_H = \frac{0.26 q \log_e\left(\frac{D}{d}\right)}{L\left[(p_1 - p_2) - p_c\right]} \times R_t \times 10^{-4} \, (\text{m/s}) \quad (9.23)$$

where L is the length of the specimen (mm), A is the area of cross-section of the specimen (mm²), D is the diameter of the specimen (mm), p_1 and p_2 are the inlet and outlet pressures (kPa), q is the mean steady rate of flow of water through the specimen (ml/minute), p_c is the pressure loss in the system for the mean rate of flow q (kPa) (usually negligible for the low rates of flow in clays) and R_t is the temperature correction factor for viscosity of water. The vertical effective stress during the test is equal to [diaphragm pressure $-(p_1 + p_2)/2$].

9.8.3. Permeability in a triaxial cell

The axial permeability of a cylindrical specimen can be measured under known isotropic stress conditions in a triaxial cell. Tests are carried out under a back pressure, so that pore pressure and the effective stress are also known. The constant head procedure is used, in which the volume of water passing through the specimen under a known hydraulic gradient in a given time is measured. Because the specimen is contained in a flexible membrane, there are no cell wall effects that could cause non-uniform flow conditions. The triaxial cell and pressurising equipment is the same as that described in Section 9.7.4 for triaxial consolidation tests, with some modifications. A porous disc of the same diameter as the specimen is fitted to the base pedestal, instead of the small porous insert used for pore pressure measurements. Permeability measurements can be made at any stage during a triaxial consolidation test (Section 9.7.4).

Undisturbed samples of 100 mm diameter and about 100 mm high are normally used for this test, but specimens from 38 mm diameter upwards can be used. The larger size is preferable because it can incorporate features of the soil fabric. When preparing the specimen for test, its axis should be orientated to take account of any fabric features.

Three constant pressure systems are required to be connected to the cell (Fig. 9.34)—one to provide the cell pressure, and one connected to each end of the specimen to provide known pressures in the inlet and outlet drainage lines. Both of the latter include a volume change indicator.

9.8.3.1. Procedure and calculations. The specimen is saturated, and consolidated to the required effective stress, as described in Section 9.7.4. Flow of water through the specimen under the required hydraulic gradient is timed until a constant rate of flow is reached. The pressure loss due to pipe friction can be obtained from calibration data.

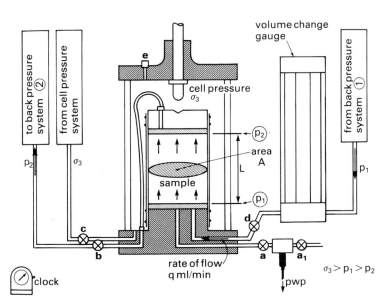

FIG. 9.34. Apparatus for permeability test in a triaxial cell.

The coefficient of permeability in the direction of the vertical axis of the specimen is calculated from the equation:

$$k_V = \frac{1.63qL}{A\left[(p_1 - p_2) - p_c\right]} \times R_t \times 10^{-4} \, \text{(m/s)} \quad (9.24)$$

where L is the length of the specimen (mm), A is the area of cross section of the specimen (mm^2), p_1 and p_2 are the inlet and outlet pressures (kPa), q is the mean steady rate of flow of water through the specimen (ml/minute), p_c is the pressure loss in the system for the mean rate of flow q (kPa) (usually negligible for the low rates of flow in clays) and R_t is the temperature correction factor for viscosity of water.

The isotropic effective stress during the test is theoretically equal to $[\sigma_3 - (2p_1 + p_2)/3]$, but for many practical purposes it can be assumed to be $[\sigma_3 - (p_1 + p_2)/2]$, where σ_3 is the cell confining pressure.

9.9. Shear strength tests: total stress

The tests described in this Section are for the determination of the shear strength of clays, from measurements of total stresses only. Pore pressures are not measured so effective stresses are not determined. However, in the direct shear tests (shearbox and ring shear tests) the procedures ensure that effective stresses are equal to total stresses because excess pore pressures are dissipated or not allowed to develop (see Chapter 4).

The test procedures covered are summarized in Table 9.11.

9.9.1. Vane shear test

This test provides a simple and quick method of determining the undrained (vane) shear strength of relatively soft clays. It is a rotational shear test, in which a cylindrical volume of clay is made to rotate by a cruciform vane pushed into the clay. The opposing shear resistance between the cylinder and the surrounding material is determined from the torque applied to the vane, which is measured by means of a calibrated torsion spring, and hence the shear strength of the clay can be calculated.

The test is the same in principle as the field vane test used on site in boreholes, but it is on a much smaller scale, the blades of the vane usually being 12.7 mm wide and 12.7 mm long. A typical laboratory vane apparatus is shown in Figure 9.35.

The vane shear strength is a measure of the undrained shear strength of soft to firm clay, from which no drainage of water takes place during the test. The test is useful for measuring the shear strength of clays that are too soft or too sensitive for the preparation of specimens for other types of strength tests, and can be carried out on clay within the sampling container so that it suffers no disturbance. Samples of this kind should also be protected from disturbance during transport, handling and storage.

The remoulded strength of the clay can be measured by repeating the test after rotating the vane rapidly through two revolutions. The ratio of undisturbed strength to the remoulded strength is a measure of the *sensitivity* of the clay.

The torque, M (N mm) which caused failure on the sheared surface of the clay is equal to the maximum observed angular deflection (degrees) of the torsion spring multiplied by the known spring constant (N mm/degree). The vane shear strength (τ_v) of the clay is calculated from the following equation:

TABLE 9.11. *Test procedures and measured parameters*

Test	Parameters measured
Vane shear	τ_v
Unconfined compression (field autographic apparatus)	c_u
Undrained triaxial compression	$c_u = (\sigma_1 - \sigma_3)_f/2$
Direct shear in shearbox	c', φ'
Direct shear (shearbox)—residual strength	c'_r, φ'_r
Ring shear—residual strength of remoulded clay	c'_r, φ'_r

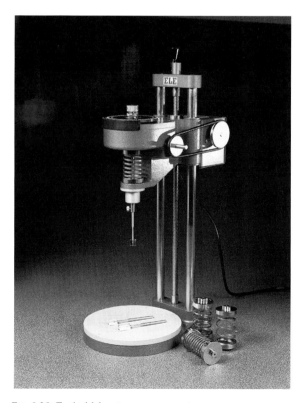

FIG. 9.35. Typical laboratory vane apparatus.

$$\tau_v = \frac{1000M}{\pi D^2 (H/2 + D/6)} \text{ (kPa)} \qquad (9.25)$$

where H is the height of the vane blades, (mm) and D is their width (i.e. the diameter of the sheared cylinder of clay) in mm.

For the usual size of laboratory vane in which $H = D = 12.7$ mm, this becomes

$$\tau_v = \frac{M}{4.29} \text{ (kPa)} \qquad (9.26)$$

9.9.2. Unconfined compression test

In this test a cylindrical specimen of clay is subjected to a steadily increasing axial compressive force until failure occurs. No other force is applied. The test is carried out quickly enough to ensure that no drainage of water takes place into or out of the clay. The test is suitable only for saturated clays that are not fissured.

The criterion of failure is either the condition in which the specimen can sustain no further increase in axial stress, i.e. the point of maximum axial stress, or an axial strain of 20%, whichever occurs first. The test can be carried out on site using a portable apparatus, as well as in the laboratory in a conventional loading frame. It is usually performed on undisturbed specimens, but remoulded specimens are also tested when the sensitivity of the clay is needed. For saturated clays the undrained shear strength, c_u, is equal to half the unconfined compressive strength. This parameter is applicable to the bearing capacity of intact clays immediately after construction, usually the most critical period.

Cylindrical specimens of clay having a height of about twice the diameter, taken from samples of good quality, are required. The portable apparatus used on site accepts specimens of 38 mm or 50 mm diameter. Larger specimens, up to 100 mm diameter, can be accommodated in the laboratory load frame.

9.9.2.1. Portable autographic apparatus. The principles of this apparatus, which was designed for use on site, are shown diagrammatically in Figure 9.36. The apparatus enables a stress/strain curve to be drawn automatically as the test proceeds.

The specimen is compressed by turning the handle at a rate appropriate to cause a rate of vertical strain of about 2% per minute (about one turn every two seconds) until either failure occurs, with a definite peak in the plotted curve, or a strain of 20% is reached.

A transparent mask placed over the graph enables the unconfined compressive strength of the clay (q_u) (kN/m^2), and the strain (%) at failure, to be determined.

9.9.2.2. Load frame. In the laboratory, unconfined compression tests are performed in a motorised load frame, as used for triaxial compression tests (Section 9.9.3). A calibrated load ring, or a load cell, is used to

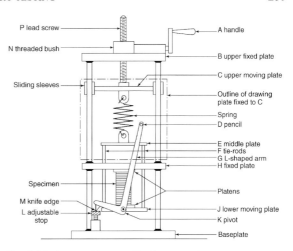

FIG. 9.36. Autographic unconfined compression apparatus.

measure the force applied to the specimen, and a dial gauge or transducer measures its the axial deformation. The specimen is loaded at a constant rate of strain and readings of both gauges are recorded at regular intervals, until either failure occurs, indicated by decreasing or constant load, or a strain of 20% is reached.

The axial compressive stress (σ_1) in the specimen for each set of readings is calculated from the equation

$$\sigma_1 = \frac{10 C_R (100 - \varepsilon) . R}{A_0} \text{ (kPa)} \qquad (9.27)$$

in which C_R is the calibration constant (N/division) of the load ring, R is the load ring reading (divisions), ε is the strain (%), equal to 100 ($\Delta L/L_0$), where ΔL is the change in length of the specimen, from the dial gauge reading (mm), L_0 is the original specimen length (mm) and A_0 is the original area of cross-section of the specimen (mm^2), calculated from the diameter. This equation takes into account the increasing area of the specimen as it undergoes axial compression, assuming that the geometry of the specimen remains a right-cylinder.

From a graphical plot of axial stress against percentage strain, the maximum compressive stress, and the corresponding strain, are determined. The maximum stress is reported as the unconfined compressive strength, q_u (kPa), of the clay, together with the strain at failure.

9.9.3. Undrained triaxial compression test

In this test a cylindrical specimen of clay is first subjected to an isotropic stress, which is maintained constant as a *confining pressure*, and then to strain-controlled axial loading to determine its undrained compressive strength. No drainage from the specimen is allowed, so that there is no change in moisture content of the clay. It is commonly referred to as a 'quick-undrained' triaxial test, because the confining and axial stresses are applied within a short time relative to the drainage characteristics of clay.

For saturated clays the undrained shear strength, c_u, is equal to half the compressive strength. This parameter is applicable to short-term stability analyses in which loads are imposed so rapidly that pore water pressures are not able to dissipate, as occurs during or immediately after construction. For clays containing large particles, fissures or other fabric features, triaxial tests should be performed on specimens that are large enough to incorporate such features.

Cylindrical specimens of clay having a height of about twice the diameter, taken from undisturbed samples of good quality, are required, as outlined in Section 9.2.1. A set of three specimens of 38 mm diameter can be taken at one horizon from a 100 mm diameter sampling tube.

The main items required for an undrained triaxial test are as follows:

- triaxial cell;
- constant pressure system for the cell, with pressure gauge;
- motorised load frame with platen speed control;
- axial force measuring device (load ring or load transducer);
- axial displacement dial gauge or transducer.

A typical load frame and cell, set up for an undrained triaxial test on a 38 mm diameter specimen, are shown in Figure 9.37. Details of the cell are shown in Figure 9.38.

The test specimen is sealed within a rubber membrane in the cell, which is filled with water and pressurized to the desired value of confining pressure (σ_3) for the test.

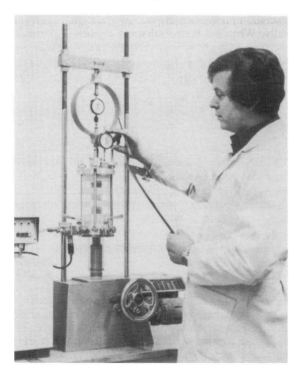

FIG. 9.37. Load frame and cell for triaxial test.

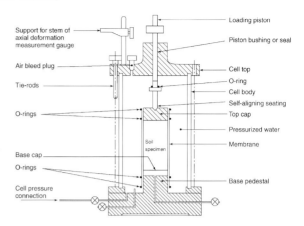

FIG. 9.38. Typical details of a triaxial cell.

Axial compression is applied to the specimen at a constant rate until either failure occurs, indicated by decreasing axial stress following a 'peak' value, or a strain of 20% is reached.

Calculation of the axial strain and the uncorrected compressive stress from each set of readings is the same as for the unconfined compression test in a load frame (Section 9.9.2). A similar graphical plot is made, and the maximum measured axial stress is calculated. However, this has to be corrected for the restraining effect of the membrane by making a deduction approximately equal to:

$$\frac{19 t \varepsilon_f}{D}$$

where ε_f is the strain (%) at failure, t (mm) is the membrane thickness and D (mm) is the specimen diameter. For specimens of 100 mm diameter or larger, the membrane correction is neglected.

The corrected maximum axial stress for each specimen is reported (in kPa) as the *deviator stress* ($\sigma_1 - \sigma_3$) at failure, i.e. the compressive strength, under its cell confining pressure σ_3 (kPa). The corresponding strains at failure are also reported.

Several specimens from one sample may be tested at different confining pressures. Alternatively, a single specimen can be tested under more than one confining pressure in a *multistage test*. In each stage, when failure is evident the cell pressure is increased, and loading is continued as the next stage. Although this procedure is given in BS 1377, it is not recommended practice but is sometimes used merely as an expedient when only one test specimen is available.

9.9.4. Direct shear test

In this test, usually known as the *shearbox test*, a square or circular prism of clay is confined under a normally applied pressure in a split container and is sheared horizontally so that one half of the clay is caused to slide on the other. The shearing resistance offered by the clay is measured at regular intervals of displacement until the maximum resistance which the clay can sustain is reached, which is the condition of failure. Usually, three similar specimens from the same sample of clay are tested under different normal stresses.

However, unlike free-draining soils such as sands, the time effect must be taken into account when testing clays. A rapidly run test would induce unknown pore pressures in clay, so only the total stresses would be known, which would have no meaning. Since there is no control of drainage in the shearbox apparatus, and pore pressures cannot be determined, the test has to be run so that effective stresses are equal to the known total stresses. This is achieved by first consolidating the specimen under the applied normal pressure until the excess pore pressure has dissipated, as indicated by readings of settlement or volume change against time. Then, the shear displacement is applied slowly enough to prevent further changes in pore pressure induced by the increasing applied shear stress. The maximum rate of displacement that may be applied to ensure this condition is derived from the consolidation stage data.

Results from a set of three similar specimens tested under different normal stresses are combined to derive the effective shear strength parameters c' and ϕ' (cohesion and angle of shearing resistance) for the fully drained condition.

The effective stress parameters derived from this test provide a measure of the shear strength of intact clay when subjected to first-time shearing. The parameters are applicable to long-term stability, i.e. after consolidation has taken place, of foundations and embankments in intact clay.

Undisturbed samples of good quality are necessary for shearbox tests. The usual sizes of test specimen used in Britain are 60 mm square and 100 mm square; both being about 20 mm high. Apparatus of higher capacity is used for larger specimens, e.g. 150 mm square and 300 mm square. Specimens of these sizes are usually hand trimmed from block samples. The test may also be carried out on remoulded specimens.

The general arrangement of a typical small shearbox apparatus is shown diagrammatically in Figure 9.39. The test specimen is fitted between porous plates in the shearbox, to allow drainage. The box is divided horizontally into two halves which can be locked together while the specimen is being placed and consolidated. A loading yoke, lever arm and suspended weights enable a vertical load (the *normal stress*) to be applied to the specimen.

The test is carried out in two stages: consolidation, followed by shearing. During shearing, the lower half is driven forward at a constant speed while the upper half is restrained by a force-measuring device. The driving force and the restraining force both act in the same horizontal plane as the surface of shear. The horizontal displacement of one half of the box relative to the other and the vertical movement of the top surface of the specimen are measured during shearing.

9.9.4.1. Consolidation stage. The specimen is allowed to consolidate under the normal (vertical) load as in the consolidation test (Section 9.7.1). Settlement is allowed to continue until it is evident that the end of primary consolidation has been reached. Settlement measurements are plotted against square-root time.

The theoretical time value designated by $\sqrt{t_{100}}$ is obtained from the graph, as indicated in Figure 9.40. The value of t_{100} is multiplied by 12.7 to give the minimum time to reach failure (t_f). By assuming a suitable displacement at which failure is likely to occur, the maximum rate of displacement (machine speed) can be estimated.

9.9.4.2. Shearing stage. The two halves of the shearbox are unclamped and shear displacement is applied at the appropriate speed. Readings of the horizontal shear force (converted to stress) and the vertical displacement (change in height of the specimen) are recorded at suitable intervals of horizontal displacement, and plotted graphically. When the horizontal stress reaches a maximum ('*peak*') value and begins to decrease, or the limit of travel of the apparatus is reached, the test is terminated.

Further specimens from the same sample are tested in the same way but under different vertical stresses, selected to cover the relevant stress range.

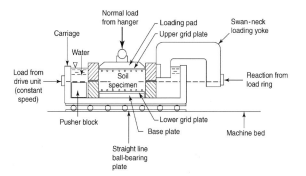

FIG. 9.39. Details of shearbox for 60 × 60 mm specimens.

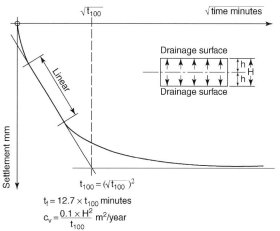

FIG. 9.40. Graph of settlement against square-root time.

The peak shear stress, or the stress at maximum displacement, from each test is plotted against the corresponding normal stress (both to the same scale in kPa), as shown by the upper set of points on the strength envelopes graph in Figure 9.41. If it is evident that the relationship is linear, a line of best fit is drawn through the points. The slope of the line is the angle of shearing resistance (ϕ'), and the intercept on the vertical axis gives the apparent cohesion (c'), both in terms of effective stress.

9.9.5. Direct shear test for residual strength

When a clay is subjected to a large shearing displacement under fully drained conditions, the shear resistance decreases after passing the peak strength until it reaches a virtually constant lower value, referred to as the *residual strength*. If a set of similar specimens are tested under different normal stresses the residual strength parameters, denoted by c'_r and ϕ'_r can be derived.

In fissured clays the shear strength along fissures is likely to be close to the residual strength because shearing beyond the peak has already taken place in nature.

Realistic residual strength parameters can be derived from shearbox tests similar to those described in Section 9.9.4 above. The effect of very large displacements is simulated by applying several stages of shearing, reversing and re-shearing, all under the same normal stress, until a constant shear resistance indicates that the residual condition has been reached.

The residual shear strength applies to the long-term stability of slopes in overconsolidated clays, and in fissured clays and clays which include surfaces of discontinuity due to previous movement of one mass of clay relative to another.

Test specimens of intact clay are prepared from undisturbed samples of good quality as described in Section 9.9.4 above. For clays containing discontinuities such as fissures or natural shear surfaces, test specimens may be orientated and carefully prepared so that the surface of a discontinuity lies on the plane of shear in the shearbox *after* consolidation.

The test equipment is as described in Section 9.9.4. The shearbox and drive unit are fitted with a coupling so that the displaced half of the split box can be pulled back to its starting position when the drive motor is run in reverse.

Prior to shearing, the specimen is consolidated, and the maximum rate of strain determined, as described in Section 9.9.4. Shearing is then applied and is continued beyond the 'peak' shear stress up to the limit of travel of the shearbox. The shearbox is then reversed back to the starting position over a period of time about equal to the time taken to reach the peak strength. The specimen is re-sheared, reversed again and re-sheared for as many times as necessary to obtain a steady and repeatable value of the shear resistance, from which the *residual strength* is derived.

The whole procedure is repeated on other specimens from the same sample, each under a different normal vertical stress. Typical graphical plots of shear stress against displacement, and consolidation settlement, for a set of three specimens are shown in Figure 9.41.

The peak shear stress (if applicable), and the residual stress, from each test are plotted against the corresponding normal stress (both to the same scale in kPa), as shown by the two strength envelopes in Figure 9.41. If it is evident that the relationships are linear, a line of best fit is drawn through each set of points. The line through the upper set of points gives the peak strength parameters, ϕ' and c', as described above. The lower set of points gives the residual strength parameters, ϕ'_r and c'_r. It is often reasonable to assume that the cohesion intercept, c'_r, is zero, and that the shear strength envelope at low stresses is curved, as shown by the dashed residual envelope in Figure 9.41.

9.9.6. Ring shear test for residual shear strength

In the ring shear apparatus, rotational shear is applied to an annular specimen of remoulded clay while subjected to a vertical stress. The rotational motion enables unlimited shear displacement to be applied continuously without the need for reversal of travel direction. Remoulded clay is used, so a peak strength is not measured.

The significance of the residual strength is outlined in Section 9.9.5. The ring shear test avoids the disturbance imposed on the test specimen by the reversal process of a shearbox test. This procedure is considered to provide an ultimately low value for the residual strength of remoulded clay.

A specimen of clay with any particles retained on a 1.18 mm sieve removed, is thoroughly remoulded and kneaded into the annular cavity of the cell. The specimen is confined vertically between annular porous plates with rough surfaces, in a cell comprising two concentric rings within a water bath. Vertical pressure is applied to the specimen through the upper plates by means of a hanger and lever-arm, similar in principle to that used for the shearbox (Section 9.9.4). The lower part of the cell can be rotated at a controlled speed, and the upper part is restrained by a torsion beam bearing against a matched pair of load-measuring devices which enable the resisting torque to be determined. The arrangement is shown in Figure 9.42.

The specimen is consolidated under the required normal pressure, the value of t_{100} is derived, and the appropriate rate of displacement, v mm/minute, is calculated, as in a shearbox test (Section 9.9.4). The maximum rate of angular displacement to apply for the test is equal to 57.3 v/r degrees per minute, where r is the mean radius of the specimen (mm).

The shear plane is pre-formed in the clay by rapidly rotating the lower half of the cell by several revolutions. Any excess pore pressures generated are dissipated by leaving the specimen to rest for a period not less than t_{100}.

The specimen is then sheared at the pre-determined rate of rotation until the plot of shear stress against angular displacement indicates that a steady state (the residual strength) has been reached. Readings of vertical movement are also recorded during shear. The vertical stress is increased to the next value and the above consolidation and shearing procedure are repeated.

LABORATORY TESTING

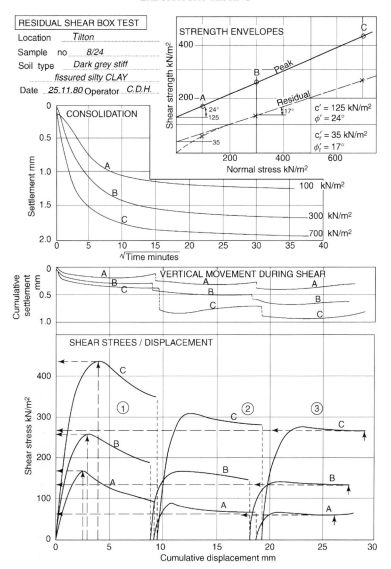

FIG. 9.41. Graphical data from a set of residual shear strength tests in the shearbox.

The area of the specimen is calculated from the inner and outer radii of the cell. The applied torque is calculated from the length of the torque arm and the sum of the two load measuring device readings. The average shear stress, τ, on the plane of shear is calculated from the equation

$$\tau = \frac{3(A+B)LR}{4\pi(r_2^3 - r_1^3)} \times 1000 \, (\text{kPa}) \qquad (9.28)$$

where A and B are the load-measuring device readings (div), L is the length of the torque arm, R is the mean load device calibration factor (N/div), r_2 and r_1 are the outer and inner cell radii (mm).

The load device readings are also taken into account in the calculation of angular displacement.

The residual shear strength from each of a set of three determinations is plotted against the respective vertical stress, and the slope of a mean envelope through the points gives the residual angle of shear resistance, ϕ'_r, similar to that shown in Figure 9.41.

9.10. Effective stress shear strength tests

This Section outlines the two most common routine tests for the determination of the shear strength of clays in terms of effective stress, which require the measurement of pore water pressures. They are designated fully as the 'Consolidated Undrained triaxial compression test with

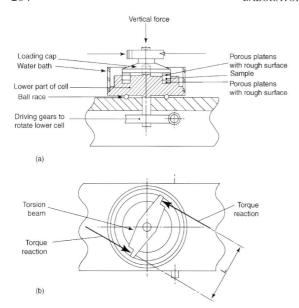

Fig. 9.42. Ring shear apparatus.

measurement of pore water pressure' (the CU test), and the 'Consolidated Drained triaxial compression test with measurement of volume change' (the CD test). The procedures which precede the compression test stage are the same for both types of test, so they are described only once.

A set of three tests on three similar specimens of a clay enable its effective stress shear strength parameters, ϕ' (the angle of shearing resistance, expressed in degrees), and c' (the apparent cohesion, kPa) to be derived. These parameters are necessary for many geotechnical applications, such as the stability of earth dams, embankments, cut slopes, and the long-term stability of foundations.

The procedures given in British Standards and in ASTM Standards are similar in principle but there are differences in some details.

9.10.1. Principles

As stated in Chapter 4, 'All measurable effects of a change of stress in soils, such as compression, distortion and change of shearing resistance, are exclusively due to changes in the effective stress' (Terzaghi 1936). The *effective stress* (σ') in a soil is the excess of the total stress (σ) over the pressure of the water in the pore spaces, known as the *pore pressure* (u). Thus the most fundamental equation in soil mechanics is

$$\sigma' = \sigma - u.$$

In order to determine the effective stress in a clay, it is necessary to measure the pore pressure as well as the total stress. Measurement of pore pressure is a vital part of all stages of the two types of test outlined below.

Each of these tests is carried out in three stages, namely saturation, consolidation, and shearing (compression). The first two stages are common to both types of test. The saturation process brings the specimen to a state of full saturation, with no free air in the void spaces. Consolidation brings the specimen to the desired state of effective stress for the appropriate compression test. Only isotropic consolidation, in which the same pressure acts in all directions, is considered here.

In the undrained test, no further water is allowed to drain out of (or into) the specimen after consolidation. Therefore the volume remains constant, but the pore pressure can change due to compression and shearing. In the drained test, water is allowed to drain out of (or into) the specimen to maintain the pore pressure at a constant level, and the volume of the specimen therefore changes during compression. Both tests need to be run slowly enough to ensure that uniformity of conditions within the specimen is maintained.

9.10.2. Samples and equipment

Undisturbed samples of Quality Class 1 are required for these tests. Cylindrical specimens with a height of about twice the diameter are required. Usually a set of three specimens 38 mm diameter are prepared from the initial sample, but larger specimens (typically 100 mm diameter) are desirable if the clay contains fabric features or large particles. The diameter of the largest particle should not exceed 1/6 of the specimen diameter.

For clays of low permeability, side drains consisting of strips of filter paper may be fitted between the specimen and the enclosing rubber membrane to accelerate drainage, but they should be used only if the testing time would otherwise be of unacceptably long duration.

The main items required for effective stress triaxial tests are as follows:

- triaxial cell, fitted with: valve and connection to the cell pressures system; base pedestal with porous insert connected via a valve to a pressure transducer for measuring pore pressure; top loading cap with drainage line connected via a valve to the back pressure system;
- two independent constant pressure systems, one for the pressurising the cell, the other for providing a back pressure (the latter also including a volume change indicator), and a pressure gauge;
- control panel with pressure gauge and rotary hand pump;
- motorised load frame with platen speed control;
- axial force measuring device (load ring or load transducer);
- axial displacement dial gauge or transducer.

Essential details of the cell are shown in Figure 9.43, and a typical arrangement of the necessary ancillary equipment is shown in Figure 9.44. (The latter diagram shows traditional pressure gauges and burette volume-change indicators, which are now being superseded by transducers).

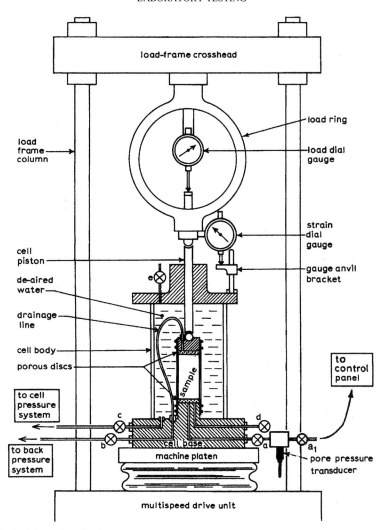

FIG. 9.43. Triaxial cell in load frame for effective stress tests.

9.10.3. Saturation and consolidation stages

Saturation is necessary to ensure that no free air remains in the voids of the clay during the subsequent stages. Saturation is achieved by alternately increasing the cell pressure and back pressure in suitable increments, allowing sufficient time for steady conditions to be achieved. Pore pressure, and sometimes the volume of water entering the specimen during the back pressure increment stages, are measured.

An increment of cell pressure ($\delta\sigma_3$) generates an increase in pore pressure (δu) in the clay. The ratio ($\Delta u / \Delta\sigma_3$) is known as the B value, and indicates the response of the pore water pressure to change in confining pressure. As the pressures increase, the B value increases towards a value of unity, which represents a theoretical saturation of 100%. In practice, a B value of about 0.96 is generally accepted as indicating that the clay is sufficiently saturated.

Once saturation is complete, the cell pressure and the back pressure are adjusted so that their difference is equal to the effective stress required for the test. When conditions are stabilised, consolidation is started by opening valve b (Fig. 9.43) to allow water to drain from the specimen, and consequently the excess pore pressure to be dissipated. Readings of pore pressure and of the volume change indicator (which indicates the change in volume of the specimen) are recorded at appropriate intervals of time from the start of drainage. At least 95% pore pressure dissipation should be achieved.

Volume change readings are plotted against square-root time (minutes from start of consolidation). The linear portion of the graph is extended, and the value designated $\sqrt{t_{100}}$ is read off, as shown in Figure 9.45. The square

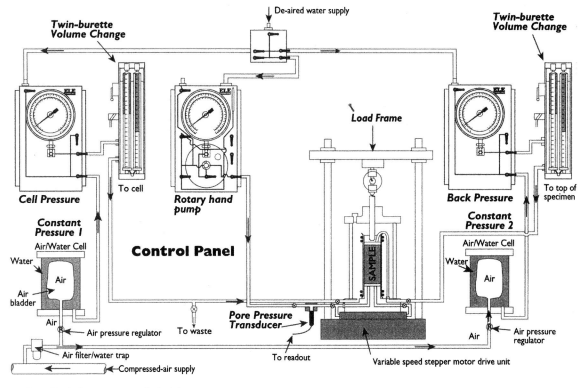

FIG. 9.44. Typical arrangement of triaxial test apparatus.

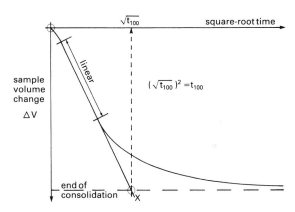

FIG. 9.45. Graph of volume change against square-root time.

of this value gives the theoretical value t_{100}, which is used for estimating the rate of strain to be applied in the compression test.

The specimen dimensions after consolidation are calculated from the following equations:

$$\text{Volume: } V_c = V_0 - \Delta V_c \text{ (cm}^3\text{)} \quad (9.29)$$

$$\text{Area: } A_c = A_0\left(1 - \frac{2\Delta V_c}{3V_0}\right) \text{(mm}^2\text{)} \quad (9.30)$$

$$\text{Length: } L_c = L_0\left(1 - \frac{\Delta V_c}{3V_0}\right) \text{mm} \quad (9.31)$$

in which ΔV_c is the change in volume due to consolidation.

9.10.4. Consolidated undrained compression (CU) test

The drainage valve is closed, so that no drainage can take place during the compression test.

The minimum time to reach failure to ensure at least 95% equalisation of pore pressure, t_f, is obtained by multiplying t_{100} by the relevant factor given in Table 9.12. By assuming an appropriate strain at which failure is likely to occur the rate of axial compression can be estimated.

TABLE 9.12. *Factors for calculating time to failure*

Type of test (any diameter) (2:1 ratio)	NO side drains	WITH side drains
Undrained (CU)*	$0.53 \times t_{100}$	$1.8 \times t_{100}$
Drained (CD)	$8.5 \times t_{100}$	$14 \times t_{100}$

* For plastic deformation of non-sensitive soils only. (Based on Table 1 of BS 1377: Part 8: 1990).

During the compression test, readings of the axial force and pore pressure are recorded at prescribed intervals of axial displacement. The test is terminated when it is evident that the calculated deviator stress, or the principal effective stress ratio, has passed a maximum ('peak') value, or when an axial displacement equal to 20% strain has been reached.

The deviator stress $(\sigma_1 - \sigma_3)$ is the vertical stress applied to the specimen by the axial force, additional to that due to the cell pressure (σ_3) It is calculated for each set of readings from the equation

$$(\sigma_1 - \sigma_3) = \frac{10 C_R (100 - \varepsilon).R}{100 A_c} \text{(kPa)} \quad (9.32)$$

in which C_R is the calibration constant (N/division) of the load ring, R is the load ring reading (divisions), ε is the axial strain (%), equal to 100 $(\Delta L / L_c)$, where ΔL is the change in length of the specimen, from the dial gauge reading (mm) and L_c is the specimen length at start of compression (mm) (i. e. as at the end of consolidation), A_c is the area of cross section of the specimen at start of compression (mm²). This equation allows for the increase in area of the specimen as it is compressed axially.

Deviator stress, and pore pressure change due to shearing, are plotted against percentage strain, to provide relationships of the form shown in Figure 9.46. The value of the deviator stress at the peak point of the curve, or at 20% strain, is recorded as the measured value at failure, $(\sigma_1 - \sigma_3)_m$. The failure strain, ε_f, and the pore pressure at that strain, u, are also recorded. A correction for the restraining effect of the rubber membrane (obtained from Fig. 9.51), and if appropriate the filter-paper side drains (Table 9.13), is deducted from $(\sigma_1 - \sigma_3)_m$ to give the corrected deviator stress at failure, $(\sigma_1 - \sigma_3)_f$.

TABLE 9.13. *Corrections for vertical side drains*

Specimen diameter	Drain correction, σ_{dr}
mm	kPa
38	10
50	7
70	5
100	3.5
150	2.5

NOTE. Corrections for specimens of intermediate diameters may be obtained by interpolation.

The principal stresses at failure, σ_{1f}' and σ_{3f}', are calculated from the equations:

$$\sigma_{3f}' = \sigma_3 - u \quad (9.33)$$
$$\sigma_{1f}' = (\sigma_1 - \sigma_3)_f + \sigma_{3f}' - u. \quad (9.34)$$

Instead of, or in addition to, plotting deviator stress against strain, the principal stress ratio σ_1'/σ_3' may be plotted against strain. This curve usually peaks at a lower strain than that of the peak deviator stress. Alternatively, the data may be plotted as a stress path of t against s', or q against p', where

$$s' = \frac{\sigma_1' + \sigma_3'}{2} \quad t = \frac{\sigma_1 - \sigma_3}{2} \quad (9.35)$$

$$p' = \frac{\sigma_1' + 2\sigma_3'}{3} \quad q = \sigma_1 - \sigma_3. \quad (9.36)$$

The form of the plot of t against s' is shown by the curve marked ESP (the effective stress path) in Figure 9.47.

If a set of three similar specimens are tested at three different confining pressures, three Mohr circles of effective stress at failure are drawn, from which the effective shear strength parameters c' and ϕ' can be derived, as shown in Figure 9.48. Alternatively, the parameters can be derived from the corresponding set of effective stress paths, as shown in Figure 9.49. A line of best fit is drawn through the points representing failure and its slope, θ, is measured and its intercept t_0 determined. The values of ϕ' and c' are obtained from the equations:

$$\sin\phi' = \tan\theta \quad (9.37)$$

$$c' = \frac{t_0}{\cos\phi'} \text{(kPa)} \quad (9.38)$$

9.10.5. Consolidated drained compression (CD) test

The drainage valve **b** remains open, so that water can drain out of (or into) the specimen and thus maintain the pore pressure substantially constant.

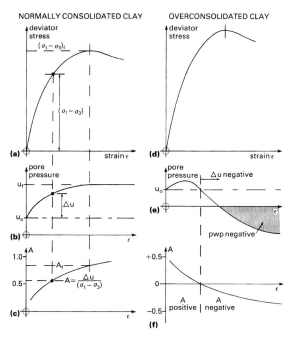

FIG. 9.46. Typical plots of deviator stress and pore pressure against strain (CU test).

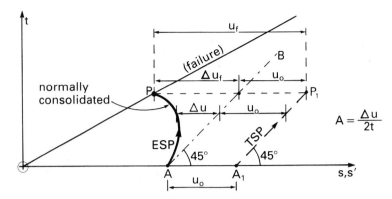

FIG. 9.47. Typical stress paths from a CU triaxial test: ESP, effective stress path; TSP, total stress path.

The minimum time to reach failure to ensure at least 95% dissipation of any excess pore pressure, t_f, is obtained by multiplying t_{100} by the relevant factor given in Table 9.12. By assuming an appropriate strain at which failure is likely to occur, the rate of axial compression can be estimated. The rate of strain is appreciably lower than for an undrained test on similar material because physical drainage of water takes longer than equalisation of pressure.

During the compression test, readings of the axial force and the volume change indicator in the back pressure line are recorded at prescribed intervals of axial displacement. The pore pressure is kept under observation, and if a significant change (in excess of 4% of the effective confining pressure) is observed the test is being run too fast and the rate of strain needs to be reduced. The test is terminated when it is evident that the calculated deviator stress has passed a maximum ('peak') value, or when an axial displacement equal to 20% strain has been reached.

Readings of the volume change indicator are used to calculate values of the volumetric strain (ε_v), from the equation:

$$\varepsilon_v = \frac{\Delta V}{V_c} \times 100\% \qquad (9.39)$$

in which ΔV is the change in volume of the specimen from the start of compression and V_c is the specimen volume at the start of compression (i.e. after consolidation). The sign convention is that volume decrease is positive, because this results from compressive stresses which are taken to be positive.

The deviator stress ($\sigma_1 - \sigma_3$) is calculated for each set of readings from the equation:

$$(\sigma_1 - \sigma_3) = \frac{10 C_R (100 - \varepsilon) \cdot R}{A_c (100 - \varepsilon_v)} \text{(kPa)} \qquad (9.40)$$

in which C_R is the calibration constant (N/division) of the load ring, R is the load ring reading (divisions), ε is the axial strain (%), equal to 100 ($\Delta L / L_c$) where ΔL is the change in length of the specimen, from the dial gauge reading (mm), L_c is the specimen length at start of compression (mm), ε_v is the volumetric strain (%), derived as above and A_c is the area of cross-section of the specimen at start of compression (mm^2). This equation allows for the increase in area of the specimen as it is compressed axially, and the change in area due to its change in volume.

Deviator stress and volume change during shearing are plotted against percentage strain, to provide relationships of the form shown in Figure 9.50. The value of the deviator stress at the peak point of the curve, or at 20% strain, is recorded as the measured value at failure, $(\sigma_1 - \sigma_3)_m$. The relevant strain, ε_f, and the pore pressure at that strain, u, are also recorded. A correction for the restraining effect of the rubber membrane (Fig. 9.51), and if appropriate the filter-paper side drains (Table 9.13), is deducted from $(\sigma_1 - \sigma_3)_m$ to give the corrected deviator stress at failure, $(\sigma_1 - \sigma_3)_f$.

The principal stresses at failure, σ_{1f}' and σ_{3f}', are calculated from the equations

$$\sigma_{3f}' = \sigma_3 - u \qquad (9.41)$$

$$\sigma_{1f}' = (\sigma_1 - \sigma_3)_f + \sigma_{3f}' - u. \qquad (9.42)$$

If a set of three similar specimens are tested at three different confining pressures, three Mohr circles of effective stress at failure are drawn, from which the effective shear strength parameters c', ϕ' can be derived, similar to those in Figure 9.48. The parameters derived from drained tests are more correctly referred to as c_D, ϕ_D, but for most practical purposes they can be assumed to be the same as c', ϕ' derived from undrained tests. There is little point in plotting stress paths for drained tests because with no significant change in pore pressure the effective stress path is a straight line parallel to the total stress path.

9.11. 'Advanced' tests

The tests referred to in Sections 9.3 to 9.10, which are described in British Standards and other literature, are carried out routinely by numerous commercial laboratories and have now become 'routine' procedures. But there

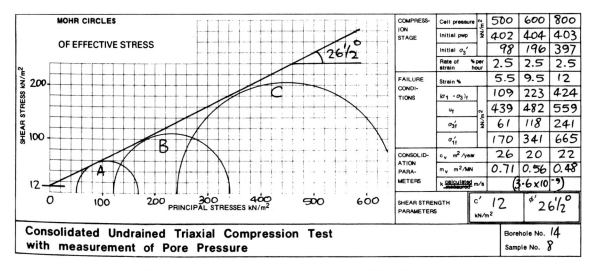

FIG. 9.48. Derivation of c' and ϕ' from a set of three Mohr circles of effective stress (CU or CD test).

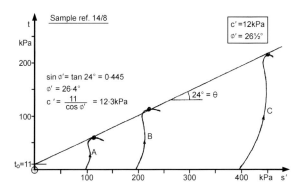

FIG. 9.49. Set of three stress path plots and derivation of c' and ϕ'.

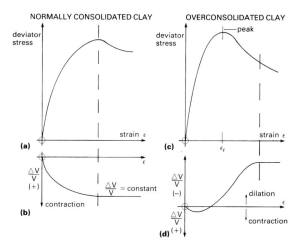

FIG. 9.50. Typical plots of deviator stress and volume change against strain (CD test).

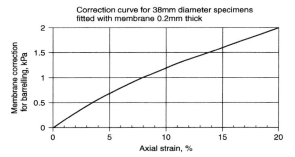

FIG. 9.51. Correction for rubber membrane.

are numerous other test procedures for soils, especially clays, which have been developed for determining soil properties under specific conditions. These are more complex, and more expensive, than routine tests, and many make use of advances in measurement technology. Only a few testing laboratories have the facilities and capabilities to perform these tests. The objective has been not only to extend the scope of laboratory tests, including the determination of fundamental soil properties, but also to relate to *in situ* soil behaviour and properties more realistically. Samples of the highest quality, and first-class testing facilities, are essential for these tests.

Procedural details of these 'special' or 'advanced' tests are usually formulated to meet the requirements of a particular design problem, and require close liaison between the client and the laboratory. The test objectives, and how the test data are to be applied, are among the issues that should be discussed and agreed before a testing programme is started. Some non-standard procedures using consolidation test equipment are referred to in Section 9.7.5. The procedures outlined below are performed in triaxial test apparatus.

9.11.1. Measuring devices

Many of the 'advanced' tests now available for soils have been made practicable only by relatively recent developments in instrumentation techniques. The most significant of these measuring devices are as follows:

- Pore pressure measurement using a probe, typically at mid-height. This enables pore pressure to be measured within the specimen, instead of being limited to the ends.
- Local strain measurement using strain gauge devices attached directly to the specimen. This provides for more accurate measurement of strains, especially small strains for assessment of soil stiffness, than can be derived from traditional end cap measurements.
- Use of bender elements, which measure the velocity of propagation of a shear wave through the specimen. Using these devices the soil shear modulus at very low strains can be inferred.
- Lateral strain indicators, which can be used to monitor lateral strain or to maintain the condition of zero lateral strain of the specimen.

9.11.2. Stress path testing

One of the most significant developments in triaxial testing is the use of stress paths for defining loading sequences. This procedure offers opportunities for reproducing a realistic stress history in the test specimen. Use of pore pressure probes and local strain measurements are essential to take full advantage of this approach. One example of such a procedure is given in Section 9.11.3.

A *stress path* is a continuous graphical representation of the relationship between the components of stress in a soil specimen in a laboratory test, or a soil element in the ground, when it is subjected to changing external forces. The relevant stress components are the principal stresses, which are the normal stresses on the three mutually perpendicular planes on which shear stresses are zero. In laboratory triaxial tests the major principal stress (σ_1) is often the vertical (axial) stress, and the intermediate and minor principal stresses (σ_2 and σ_3) are horizontal and equal, and are induced by the cell pressure.

Laboratory stress paths depict the relationship between the total or effective principal stresses, using a graphical plot on a *stress field*. The two types of stress field most often used are the 'Cambridge' stress field and the 'MIT' stress field (indicative of their places of origin). In the 'Cambridge' field the deviator stress ($\sigma_1-\sigma_3$), denoted by q, is plotted against the mean of all three principal stresses ($\sigma_1+2\sigma_3$)/3 or ($\sigma_1'+2\sigma_3'$)/3, denoted by p or p'. In the 'MIT' stress field, half the deviator stress, ($\sigma_1-\sigma_3$)/2, denoted by t, is plotted against the mean of the major and minor principal stresses, ($\sigma_1+\sigma_3$)/2 or ($\sigma_1'+\sigma_3'$)/2, denoted by s or s'. The former is more fundamentally representative, but the latter is more convenient for many laboratory tests and is commonly used in practice. Unfortunately some authors use the symbols p and q with the MIT plot, which can cause confusion, so the actual equations should always be presented.

Data from drained and undrained triaxial compression tests outlined in Section 9.10 can be plotted in the form of stress paths, in addition to the conventional plots of deviator stress against strain. Plots on the MIT field are generally of the form shown in Figure 9.52, in which the total stress paths are denoted by TSP, and the effective stress paths by ESP. However, the rates of strain used in conventional tests allow for pore pressure equalisation only at failure, not throughout the test.

Stress path plots are used because they can provide much more than data relating to failure conditions, provided that a test is run slowly enough to ensure that realistic pore pressure readings can be observed. It is possible to model the present *in situ* stress conditions, as well as past stress changes and those which will occur under future loadings due to construction. Stress paths are applied to the study of soils at the *critical state*, which is covered by Schofield & Wroth (1968), and Atkinson & Bransby (1978).

Stress path tests can have many objectives and can take many forms. One example is the Stress History and Normalised Soil Engineering Properties (SHANSEP)

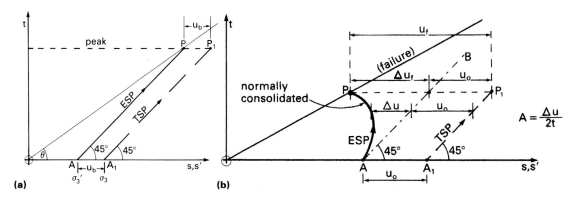

FIG. 9.52. Stress paths of total and effective stresses for a drained (a) and undrained (b) triaxial compression test.

approach in which a series of soil specimens from a single geologically similar layer are 'normally consolidated' far beyond any stresses previously experienced and then unloaded to various overconsolidation ratios (OCRs) before shearing to failure. By comparing measured shear strength with applied OCR it is possible to formulate a layer-specific relationship between OCR and shear strength. Shear strength profiles can then be estimated from the estimated geological history. However, this approach is not normally favoured by UK practitioners as all time-dependent, or 'ageing', effects are removed during the initial consolidation process. Further details of the uses of stress path tests for the design of slopes, foundations and retaining walls are provided by Atkinson (1993).

9.11.3. B-bar test

This is a special type of stress path test determining the pore pressure coefficient $\bar{B}$ for a partially saturated clay, under conditions in which the vertical and horizontal principal stresses change simultaneously, maintaining a constant ratio. These conditions apply for instance in an earth embankment during construction. The test provides parameters that can be used for the control of pore pressures during construction.

The pore pressure coefficient $\bar{B}$ is similar to the pore pressure coefficient B (defined in Section 9.10.2) but relates the pore pressure change to the change in the major principal stress as follows:

$$\delta u = \bar{B}.\delta\sigma_1. \qquad (9.43)$$

The aim of the test is to determine this relationship as the applied stresses are increased, while maintaining a pre-determined margin between the current state of stress and the failure condition. To do this the ratio of the vertical effective stress to the horizontal effective stress is kept constant, at a value that is derived from the selected factor of safety against failure. This requires that the pore pressure, as well as axial load, volume change and cell pressure, are all monitored continuously.

The test is usually carried out on a 100 mm diameter specimen of clay that has been re-compacted to the condition in which it is to be placed in the structure, for example an embankment. Triaxial test equipment as described in Section 9.10.2 is used, with the following modifications (see Figs 9.43 and 9.44):

- a back pressure system is not required, and no drainage is allowed from the specimen;
- a porous disc of high air-entry value is fitted on the base pedestal, where pore pressure is measured;
- a volume-change indicator is included in the cell pressure line, and the triaxial cell should be calibrated for volume change with changing pressure.

Before carrying out the test it is first necessary to determine the shear strength parameters c', ϕ', from a set of conventional consolidated-drained triaxial compression tests (Section 9.10.5) on specimens similar to the clay being tested. The factor of safety, F, at which the test is to be run is selected by the Engineer, and the corresponding limiting angle of shear resistance, ϕ'_m, is calculated from the equation

$$\tan\phi'_m = \frac{\tan\phi'}{F}. \qquad (9.44)$$

The traditional test procedure, as described by Bishop & Henkel (1964) uses a stress field in which deviator stress $(\sigma_1 - \sigma_3)$ is plotted against effective horizontal stress, i.e. effective cell pressure (σ_3'). In this stress field, the envelope representing the failure condition rises from the origin at an angle denoted by α, and the envelope representing the working stress (factor of safety F) is at an angle denoted by β, these angles being derived from the equations

$$\tan\alpha = \frac{2\sin\phi'}{1-\sin\phi'} \quad \tan\beta = \frac{2\sin\phi'_m}{1-\sin\phi'_m}. \qquad (9.45)$$

These lines are shown in Figure 9.53.

If at all times during the test the plotted stresses lie on or below the lower line, the factor of safety with respect to failure will never be less than the specified value F. Ideally, the stresses should be increased continuously so that this line is followed, but in practice, with manual control, the horizontal and vertical stresses are applied alternately in discrete steps.

The first increment of cell pressure (e.g. 50 kPa) is applied, and when the pore pressure becomes steady, axial loading is commenced. The rate of strain is as derived from the consolidation stage of the pre-test specimens, except that the strain interval used in the equation is the anticipated change in strain during each load increment.

When the plotted points reach the working stress envelope, loading is discontinued and the next cell pressure increment is applied, followed by further axial loading. The cycle is repeated until the applied stresses reach the maximum desired values. Further axial loading can then be imposed until failure is reached, if required. A typical test plot is shown in Figure 9.54.

For each point on the envelope, pore pressure is plotted against total vertical stress, σ_1, which is equal to the applied axial stress plus cell pressure. The slope of this line is equal to the coefficient $\bar{B}$. More than one value might be appropriate, over different stress ranges, if the relationship is not linear. The graph derived from the above test plot is shown in Figure 9.55.

Other customary graphical plots are of volume change and axial strain against effective vertical stress, σ_1', and the Mohr circles of effective stress at each point on the working stress envelope and at failure.

9.11.4. Other 'advanced' test procedures

Some other test procedures that are offered by specialist laboratories are listed below, with brief notes on their

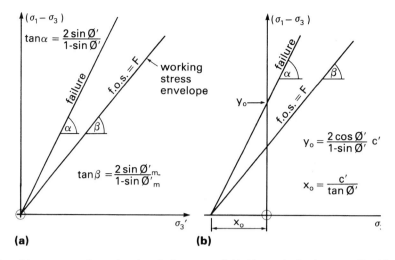

FIG. 9.53. Failure and working stress envelopes plotted on deviator stress field: (a) no cohesion intercept, (b) with cohesion intercept.

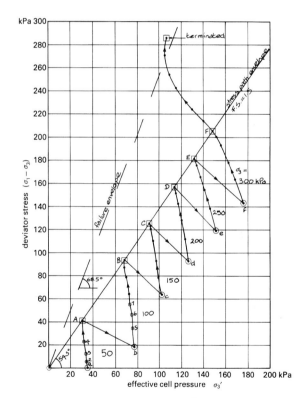

FIG. 9.54. Typical plot of stress path for a $\bar{B}$ test on compacted clay, related to the working stress envelope.

methods and applications. Further details can be found in the Appendix D.

9.11.4.1. Initial effective stress. The initial effective stress in a triaxial test specimen is derived from the measured initial pore pressure. The pore pressure at mid height, measured by a probe, is required to provide a representative value.

9.11.4.2. Triaxial extension. To simulate the condition in which the vertical stress decreases, such as at the base of an excavation, the vertical stress on a triaxial specimen is progressively decreased while the confining pressure remains constant. A specially designed top loading cap is needed to achieve this. However, the specimen is never subjected to actual tension, but fails because of the shear stress induced by the difference in the radial and axial stresses.

9.11.4.3. Stress ratio K_o. The *in situ* ratio of the horizontal to the vertical effective stresses, if there is no lateral yield, is defined as the *coefficient of earth pressure at rest* (K_o). This ratio can be estimated in the triaxial cell by increasing the vertical stresses and controlling the horizontal stress so that no lateral deformation of the specimen occurs.

9.11.4.4. Anisotropic consolidation. In routine triaxial compression tests the consolidation stage is carried out under isotropic loading in which the vertical and horizontal stresses are assumed equal. However, in many cases the *in situ* radial and vertical effective stress are not equal. These conditions may be modelled by applying independent, anisotropic, radial and vertical effective consolidation pressures, and used to estimate anisotropic consolidation movements or soil strength.

9.11.4.5. Resonant column. A soil specimen is saturated and consolidated in a similar manner to a triaxial test specimen. After consolidation an electromagnetic oscillator is used to apply a torque to the top of the soil specimen. The specimen is subjected to shear strain amplitudes

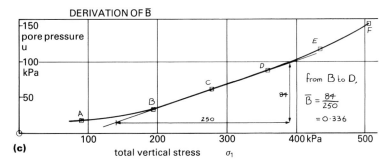

FIG. 9.55. Graphical plots from a $\bar{B}$ tent on compacted clay.

varying from 0.0001% to 0.1%. From the peak acceleration and frequency of the top cap at resonance, the shear strain amplitude and shear modulus of the specimen can be obtained. The material damping of the soil may also be investigated by suddenly stopping the oscillator and observing the rate at which the test specimen returns to its steady state. By noting the percentage reduction in amplitude between successive peaks it is possible to calculate the logarithmic decrement and hence the viscous damping of the soil.

9.11.4.6. Simple shear. A circular soil specimen is surrounded by a wire-wound reinforced membrane. Typically, the specimen is consolidated in a manner intended to simulate the previous stress history of the soil, i.e. it is initially consolidated to the estimated maximum past vertical effective stress, then unloaded to the estimated current effective overburden pressure. After consolidation, the specimen is sheared in the horizontal plane, but is constrained, using computer control of the vertical stress to remain a constant-height parallelogram. This process is intended to produce a constant shear strain throughout the specimen (unlike a conventional shearbox test), which approximates to a state of pure shear. Test data are applicable to potential failure zones beneath a structure where simple shear conditions would be expected to dominate. An example of such a situation would be beneath a structure with a very large base area, but with a relatively shallow critical yield surface. Shearing can be performed statically to model static conditions or cyclically to model environmental or seismic loading.

9.11.4.7. Cyclic triaxial. Standard triaxial testing is performed to assess soil behaviour under static loading. However, there are numerous engineering situations where the loading fluctuates with time, either in a predefined or random manner. Examples include soils beneath structures exposed to significant environmental loading, structures designed to resist seismic loading, or road and railway foundations. To model these situations, cyclic triaxial tests can be performed. A triaxial specimen is first consolidated to the expected *in situ* effective radial and vertical stresses. Cyclic loading is then applied, vertically, horizontally or in combination, to model the type of loading that the soil is expected to experience. The changes in displacement and pore pressure that occur during the cyclic loading stage are recorded and used to assess the rate at which the soil stiffness is likely to degrade *in situ*. Hysteretic soil damping can also be investigated by applying the cyclic loading in a strain-controlled manner and observing the stress–strain behaviour over individual cycles of load.

References

ATKINSON, J. H. 1993. *An Introduction to the Mechanics of Soils and Foundations*. McGraw-Hill, London.
ATIKINSON, J. H. & BRANSBY, P. L. 1978. *The Mechanics of Soils. An Introduction to Critical State Soil Mechanics*. McGraw Hill Book Co., Maidenhead, Berks.
ATIKINSON, J. H., CHARLES, J. A. & MHACH, H. K. 1990. Examination of erosion resistance of clays in embankment dams. *Quarterly Journal of Engineering Geology*, **23**, 103–108.
BISHOP, A. W. & HENKEL, D. J. 1964. *The Measurement of Soil Properties in the Triaxial Test*, 2nd edn. Edward Arnold, London.
BOWLEY, M. J. 1995. *Sulphate and Acid Attack on Concrete in the Ground: Recommended Procedures for Soil Analysis*. Building Research Establishment Report. BRE, Garston, Watford, Herts.
BRE Special Digest 1. (revised 2005) *Concrete in Aggressive Ground, Part 1: Assessing the Aggressive Chemical Environment*. Construction Research Communications, London.
BS812: Part 124: 1989. *Method for the Determination of Frost Heave*. British Standards Institution, London.
BS 1377: 1990. British Standard Methods of Test for Soils for Civil Engineering Purposes. British Standards Institution, London.
CROMPTON, T. R. 1999. *Determination of Organic Compounds in Soils, Sediments and Sludges*. Spon, London.
CROMPTON, T. R. 2001. *Determination of Metals and Anions in Soils, Sediments and Sludges*. Spon, London.
FOOKES, P. G. (ed.) 1997. *Tropical Residual Soils*, Chapter 5. Geological Society, London.
GORMAN C. T., HOPKINS T. C., DEEN R. C. & DRNEVICH, V. P. 1978. Constant rate of strain and controlled gradient consolidation testing. *ASTM Geotech. Testing J.*, **1**, 1, 3–15.
HEAD, K. H. 1992. Manual of Soil Laboratory Testing, Volume 1, second edition. John Wiley and Sons, Chichester.
HEAD, K. H. 1994. Manual of Soil Laboratory Testing, Volume 2, second edition. John Wiley and Sons, Chichester.
HEAD, K. H. 1998. Manual of Soil Laboratory Testing, Volume 3, second edition. John Wiley and Sons, Chichester.

Longworth, I. 2004. Assessment of sulfate-bearing ground for soil stabilisation for built development. Technical Note. *Ground Engineering*, May 2004, **37**, No. 5.

Reid, J. M., Czerewko, M. A. & Cripps, J. C. 2001. *Sulfate Specification for Structural Backfills*. TRL report 447, Transport Research Laboratory, Crowthorne.

Schofield, A. N. & Wroth, C. P. 1968. *Critical State Soil Mechanics*. McGraw Hill Book Co., London.

Sherard, J. L., Dunnigan, L. P. & Decker, R. S. 1976. Identification and nature of dispersive soils. *Journal of the Geotechnical Engineering Division*, ASCE, Paper 12052.

Smith, R. E. & Wahls, H. E. 1969. Consolidation under constant rates of strain. ASCE *J. Soil Mechanics and Foundation Div.*, **95**, SM2, 519–539.

Terzaghi, K. 1936. Presidential address to the first International Conference on Soil Mechanics and Foundation Engineering. Cambridge, Massachussets, USA.

Umehara, Y. & Zen, K. 1980. Constant rate of strain consolidation for very soft clayey soils. *Soils and Foundations*, **20**, No 2. Japanese Society of Soil Mechanics and Foundation Engineering.

Wissa, A., Christian, J., Davis, E. & Heiberg, S. 1971. Consolidation at constant rate of strain. ASCE *J. Soil Mechanics & Foundation Div.* (97), **10**, 1393.

Further Reading

BS1881: Part 124: 1988. *Testing Concrete—Methods for Analysis of Hardened Concrete*. British Standards Institution, London.

BS3978: *Specification for Water for Laboratory Use*. British Standards Institution, London.

BS812: Part 117: 1990. *Testing Aggregates—Methods for Determination of Water-soluble Chloride Salts*. British Standards Institution, London.

Collins, S. H. 1906. Scheibler's apparatus for the determination of carbonic acid in carbonates; an improved construction and use for accurate analysis. *J. Soc. Chem. Ind.*, 25.

Cumming, A. C. & Kay, S. A. 1948. *Quantitative Chemical Analysis*. Gurney and Jackson, London.

Davison, L. R. & Atkinson, J. H. 1990. Continuous loading oedometer testing of soils. *Quarterly Journal of Engineering Geology*, **23**, 347.

Gutt, W. H. & Harrison, W. H. 1977. *Chemical Resistance of Concrete*. BRE Current Paper CP 23/77, May 1977. Building Research Establishment, Garston, Watford. (Reprint from *Concrete* 1977, **11**, No. 5).

Hoare, S. D. 1980. *Consolidation with Flow Restriction; an Improved Laboratory Test*. University of Oxford Department of Engineering Science, Report No. SM014/ELE/80. (Paper submitted to the BGS for the Cooling Prize Competition, October 1980).

Kuhn, S. 1930. Eine neue kolorimetrische Schnellmethode zur Bestimmung des pH von Boden. *Z. Pflernahr. Dung.*, Vol. 18A.

Seah, T. H. & Juirnarongrit. Constant rate of strain consolidation with radial drainage, *ASTM Geotech. Testing J.*, **26**, 432–443.

Sheahan, T. C. & Watters, P. J. Using an automated Rowe cell for constant rate of strain consolidation testing, *ASTM Geotech. Testing J.*, **19**, 354–363.

Sills, G. C., Hoare, S. D. & Baker, N. 1984. An experimental assessment of the restricted flow consolidation test. *Proc. ASTM Symp. on Consolidation Behaviour of Soils, Fort Lauderdale, Florida, USA, October 1985*.

United Kingdom Accreditation Service. 2000. *Calibration and Measurement Traceability for Construction Materials Testing Equipment*. UKAS Publication ref: LAB 21, 2nd edn, October 2000.

Vogel, A. I. 1961, *A Textbook of Qualitative Inorganic Analysis*, 3rd edn. Longmans, London.

10. Earthworks

Ever since man began establishing settlements in prehistoric times, he has been reshaping his built environment. The placement of clay has been used to form embankments, dams, quarry tips, mounds or levelled areas. This process is referred to as 'filling' and the final structure is known as a 'fill'. By excavating clays, engineering structures such as cuttings, ditches and moats have been constructed. Clay has also been excavated as a raw material in manufacturing, for example for the pottery and building industries. The resulting fill structures and excavations are collectively referred to as earthworks. Horner (1988) and Trenter (2001) provide a thorough summary of British practice and the somewhat dated *'Earth Manual'* (US Bureau of Reclamation 1974) describes American practice. A particular useful reference on the use of soil and rock in construction with an emphasis on Australian experience is given by McNally (1998) and Fookes (1997) covers tropical residual clay soils in terms of characteristics, description and engineering properties.

It is likely that the earliest earthworks were for the purpose of living and defence but as populations increased water supply and irrigation became important. Although many of the earliest earthworks have become obscured with time, canals have been dated from about 4000 BC in Iran and earth dams in Jordan from about 3200 BC (McFarlan 1989). Clays have been excavated for the production of pottery for thousands of years and the occurrence of kaolin-rich clays probably led to the development of porcelain in China. The use of clay to make earthen walls and to produce bricks and other products was first developed in hot countries where clays could be baked in the sun and later clays were fired to produce more stable clay ware products for building (see Chapter 15). Latterly the demand for and the range of items manufactured using clay has rapidly increased, as has the use of particular clays, such as bentonite in civil engineering, because of their mineralogy and properties (see Chapters 3 and 4).

In this Chapter only the use of clay in fills and the excavating process required to win clay is considered. The resulting clay excavation itself, either to form an engineered structure (i.e. a cutting) or resulting from the winning of clay for manufacturing, is not considered part of the use of clay. This chapter concentrates on the design, planning, performance and use of clay in construction of fills. Chapter 11 covers the construction aspects including the excavation and transportation of clay for use in fills and as a quarried material. Chapter 12 considers the specialist applications of clay.

Clay fills are engineered using the knowledge available at the time of construction. In the context of this publication, engineered clay fills are structures that are composed of clay that has been excavated, transported and placed, where an assessment of the clay's properties and behaviour has been undertaken and a design prepared which is subsequently implemented in the construction of the structure.

The use of clay fill can vary but is generally intended to:

- raise the level of the ground where it is too low for an engineering function; e.g. embankments for infrastructure;
- raise the level of or infill depressions in the ground as the consequence of another process e.g. forming tips in quarries;
- impound water; such as earth dams;
- raise the level of the ground where it is too low for the environmental function; e.g. environmental bunds;
- establish a level for large non-linear structures (substructure fills); e.g. fill beneath a concrete floor for an industrial development on uneven ground;
- be treated for uses that require higher performance; e.g. the replacement of granular materials with treated clay in the upper layers of weak road foundations.

In civil engineering construction works due to the high cost of transportation, clay tends to be obtained from locations in close proximity to where it is to be utilised. It can be brought from adjacent borrow pits should there be a deficit of material for filling on the site and the distances are usually small. On larger projects, particularly infrastructure ones, clay may be imported on to the site, where this is more economical than transporting material longer distance within the site. Chapters 5 and 6 describe the wide geographical extent of clay in the United Kingdom (UK) and the rest of the world. It is not a 'manufactured' material for civil engineering purposes, like processed aggregate or bricks, and for use as a fill best use is made of what is available. The following are useful for further reading: British Standards Institution (1981), Horner (1988), Trenter (2001), US Bureau of Reclamation (1974), McNally (1998), Fookes (1997), Institution of Civil Engineers (1979) and Clarke *et al*. (1993).

Besides civil engineering applications, clay may occur as the mineral being won (e.g. brick or china clay) or as a waste material that has to be excavated to remove the mineral deposit (e.g. clay overlying sand and gravel deposits). In the latter case the clay waste is often deposited in tips, either as a temporary stockpile before being deposited back into the worked out quarry or as a permanent means of disposing of the material. In either case the processes may result in the formation of clay fills.

The knowledge of clay properties and behaviour, the understanding of compaction, the development of representative testing and the available construction machines has improved the performance of clay fills during construction and in use, although this science is still young compared to other physical sciences.

Design has developed from paper to computer analysis allowing more rapid assessment. In the hands of the experienced, these modern techniques allow greater assessment of a far greater number of design options. Observational information with an understanding of the theory are now an increasing part of the design procedure.

Like all engineering structures, fills require maintenance and the need to undertake it has become increasingly apparent as the materials within these structures age and weaken. This can lead to instability with both economic and safety implications. It is important to know the engineering history of fills so that the properties and behaviour can be understood and interpreted. Canal fills, for example, are 200 years of age or older and methods of construction have changed considerably over this period.

The understanding of clay properties has benefited from research and development. The sciences of soil mechanics and engineering geology have provided the necessary information to ensure greater stability of clay fills. The understanding of compaction, clay strength and pore water pressures (discussed in Chapter 4), have influenced the design of fills and allowed their performance to be improved. In practice, new information gained from experience influences research and a cycle of improvement can be established.

10.1. Types of engineered fill

10.1.1. Infrastructure embankments

The greatest quantities of clay are, or have been, used in constructing embankments for railways, highways and canals. For example, the amount of clay used for engineered fills on a 15 km two-lane dual carriageway highway traversing ground of moderate relief can be of the order of three million cubic metres.

Embankments are made from materials placed on the natural ground (examples are given in Figure 10.1 from Perry *et al*. 2003). The predominant use of infrastructure embankments is to carry rail, highway and canal traffic across relatively low lying natural ground or over other structures and to maintain the vertical alignment required (Fig. 10.2). Cuttings are constructed through high ground in order to maintain vertical alignment. In areas of undulating countryside, a balance has to be attempted between the amount of material needed to form an embankment and that produced by excavations to form a cutting. The intention of the balance is to avoid the need to dispose spoil off-site or to import material for embankments.

Where the transport infrastructure follows the contours of the land, 'side-long ground', the earth structure may be a combination of cutting and embankment (Fig. 10.3). Minimum amounts of excavation, haulage and filling are required as the material on the upper slope is excavated, and placed on the lower slope to bring the ground to the required level.

Figure 10.4 (Perry *et al*. 2003) shows the development of embankment construction with time. The early canals tended to follow ground contours with the earthworks being constructed largely on 'side-long ground'. By the 1790s, the success of this means of transport led to greater demand of canals more independent of terrain. As a result newer canals were constructed with embankments, cuttings and tunnels. Existing contour canals were shortened, sometimes by as much as 35%, by using these structures to carry the existing canal through hills and across valleys. Labourers, known as 'Navigators' or 'Navvies', using the same techniques they were later to employ on the railway lines, constructed substantial embankments with the added complication of carrying large volumes of water. The water was retained within the canal trough on the embankment by a layer of impermeable clay known as 'puddle clay' (see Chapter 12 for details). The integrity of this layer is paramount for the canal operation. The rate of canal construction slowed considerably after 1830, although a number incorporating embankments of considerable size were not completed until as late as 1835. After this time the canal systems of the world were largely complete (Gascoigne 1994).

Following the advent of the railways, from the mid-19th century, the use of the canal system for freight transport gradually declined. During this period a reduction in the network took place, largely by infilling or redevelopment, and the use of canals is now primarily for leisure purposes although many strategic commercial canals still exists worldwide.

The railway networks of the world were mainly constructed in the mid-19th century and is on average about 150 years old. Clay cuttings were excavated by shovel and horse drawn wagons transported the material to the fill area where the clay was end or side tipped to form the embankment (Wiseman 1888; Skempton 1996). The amount of clay worked is impressive. For example, between 1834 and 1841 nine main line railways were built in England totalling 1060 km in length with some 54 million cubic metres of excavation. This was a remarkable quantity of excavation that was not achieved again on works of a comparable nature until the introduction of modern earthmoving plant on the first motorway contracts, over a century later.

Some of the later railways were constructed with 'steam navvies'; a steam driven excavator that replaced much of the hand-dug work of the Navvies. Materials were transported using steam locomotives and side or end tipped to form the embankment. The material was tipped and placed as excavated although occasionally, where weak materials were encountered, mixing with coal and burning baked the clay.

It is important to note that nearly all historic railway and canal embankments are made of relatively uncompacted material. Prior to the 1930s, little or no compaction

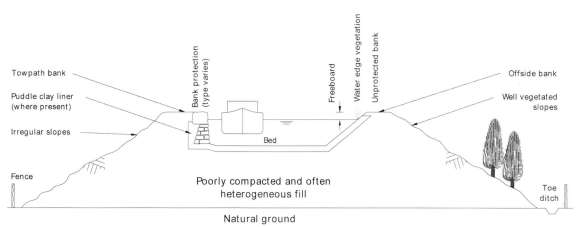

FIG. 10.1. (a) Canal embankment. Water is typically retained by a lining. Traditionally, puddle clay was used for this purpose, but replaced often utilised modern materials (Perry *et al.* 2003).

was routinely undertaken for any type of embankment as the construction plant was not then available and the process not well understood. In addition, the angles of the embankment slopes were based on short-term angles of repose attained during the labour intensive construction. These would be considered oversteep in modern practice. As a result large settlements were common after construction and have continued up to the present time. Also failures of the slope occurred after construction and are still a major hazard today.

Highway embankments are generally newer structures although some highways are located on embankments constructed for horse-and-carts in the late 19th century. Like modern high-speed railways, the flat gradients of

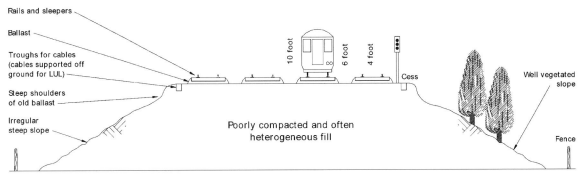

FIG. 10.1. (b) Railway embankment. Poor compaction and steep slopes are characteristic of this type of embankment (Perry *et al.* 2003).

these new high-speed transport modes required a major development in the use of the embankment. With the introduction of modern road and rail design standards, gradients reduced and average journey times decreased. Some major trunk roads were up-graded and new roads, motorways and railways were constructed with more embankments and larger quantities of materials. The construction was undertaken more quickly and with less embankment instability due to the development of new construction plant and the development of the relatively new discipline of geotechnical engineering. In the most recent past, highways have required greater capacity and extra lanes have been added. This has led to embankment widening commonly within the same highway land boundary. As a result steeper slopes have been used and the use of reinforcement within the clay is needed for the structure to remain stable.

Railway, highway or canal operations will depend on the integrity of the embankment. The change in condition of materials with time and rate of deformation of embankments are critical influences on their safe and efficient use. Large slope movements (Fig. 10.5) or settlements can occur and lead to traffic speed restrictions or route closure, and in some critical circumstances may affect the

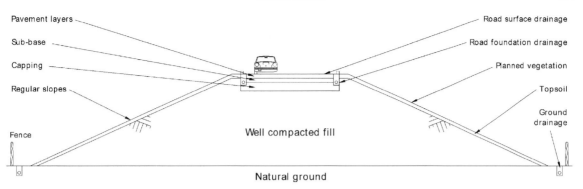

FIG. 10.1. (c) Highway embankment. Construction and design is to modern standards with adequate compaction and flatter slopes (Perry *et al.* 2003).

safety of users. Embankment instability affects the very foundation of the infrastructure and can undermine any efforts to maintain other assets that depend on it.

A particularly useful concise source on earthworks concepts and practice is given in Transportation Research Board (1990).

10.1.2. Earth dams

Clay has been recognized for many centuries as the 'ideal' material for the construction of earth dams principally for the retention of water. The inherent properties of clay, characterised by low permeability and resistance to

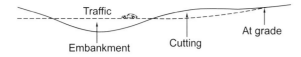

FIG. 10.2. Vertical alignment of a transport infrastructure requires construction of embankment and cuttings.

erosion, meet the basic requirements of a material for the construction of a water retaining structure. However, not all clays have suitable properties, in particular some tropical clays by virtue of their genesis are naturally dispersive. Such clays are susceptible to internal erosion and in embankment dams special techniques and design approaches are required to ensure their erosive tendencies

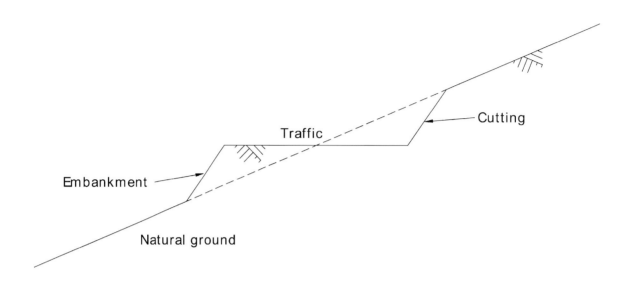

FIG. 10.3. An example of construction on side-long ground; in this case a railway with tracks supported on a repaired slope failure.

EARTHWORKS 281

FIG. 10.4. Timeline of embankment construction in the UK (Perry et al. 2003).

do not result in piping or other material stability problems, which could lead to failure.

In general natural clays provide suitable embankment construction materials and earth dams can be structures of a few metres to hundreds of metres in height. In

FIG. 10.5. Failure of a clay fill.

the ICOLD (International Commission on Large Dams) Register of Dams (ICOLD 1998), the league table of the world's highest dams is headed by an earth and rock dam at a height of 335 m whilst in second place is an earth-only dam at 300 m height. The ability to build safe earth dam structures to these heights has only become possible in the last 60 years or so owing to advances in engineering analysis and an in-depth understanding of the mechanical, physical and chemical properties of clay. These advances, particularly the recognition of effective stresses in laboratory testing techniques and slope stability analyses which thus model pore pressure changes, can be readily seen in the development of multiple zoned embankment dam cross sections, of which the central and water-retaining element is a clay or low permeability clay core (Fig. 10.6).

As for infrastructure embankments, the sources of material, for cost-effective earth dam construction, have to be found relatively close to the site for a project to be viable. However, not all dam sites have suitable material located in the immediate vicinity. The availability of a suitable clay construction material is inextricably linked to the regional geology and where the source is a residual deposit to the weathering processes. Alternatively, the source of clay may be sedimentary or be derived from solifluucted or transported materials in which cases the transportation process over distances of between a few hundred metres and many kilometres results in breakdown by both physical or chemical processes and re-deposition (see Chapter 3). Each potential dam site therefore has to be investigated in detail to determine the geological origin, geotechnical characteristics and the range of materials available for construction. In many

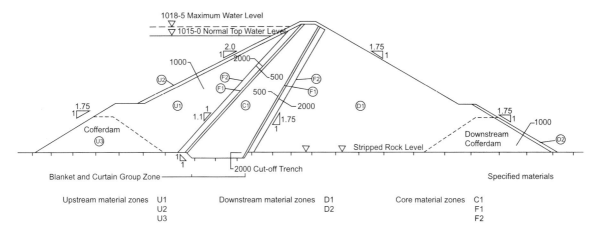

FIG. 10.6. Typical cross section through a multi-zonaled dam showing the variety of specified materials.

cases a broad range of material types and sizes occur locally providing the designer with a range of options for embankment construction. Provided that there is sufficient clay to form an impermeable barrier a range of different sized materials from boulders and gravels to sands and silts can be used to produce a water retaining to appropriate design standards.

The use of clay in the construction of earth dams requires particular consideration of the geotechnical properties of the material and its engineering behaviour in order to be able to design and build a safe structure. Failure to address design issues associated with these characteristics in dam design, in construction methods and programming with respect to the behaviour of clay, can dramatically affect both short and long-term security of an embankment. A lack of understanding of the characteristics of the material both within the dam and in the foundations can and has led to catastrophic failure and to loss of life.

The history and development of earth dams in the UK can be traced to surviving structures of the 12th century and earlier (Bridle & Sims 1999). However, worldwide it is likely that the construction of earth dams for the capture and retention of water has a history stretching back over a number of millennia with very early examples from both the Egyptian and Babylonian civilisations being recorded. The purposes of earth dams (reservoirs) have developed from the low earth embankments, homogeneous style dam (Fig. 10.7) of the earliest monastic fishponds and ornamental dams for pleasure parks etc, to later and larger dams for industrial uses, such as mills, canals, domestic and industrial water supply, and more recently for hydroelectric power, flood control, quarry and mine waste storage and irrigation dams. In recent years in the UK there have been increasing numbers of earth dams associated with leisure activities, particularly for fishing. A historical review is presented by Skempton (1989).

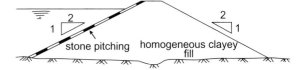

FIG. 10.7. Typical cross section through homogeneous-style dam.

The various purposes for which earth dams have been built has often required the construction of embankments of only modest height. However, height alone is not necessarily a measure of the significance of a reservoir. In the UK, the Reservoirs Act (1975) regulates dam safety. This Act applies to 'large raised reservoirs', which are defined as reservoirs that hold, or are capable of holding, more than 25 000 metres cubed of water above the natural level of any part of the land adjoining the reservoir. Reservoirs that fall into this category must be registered and are subject to the rules on dam safety, inspection and monitoring that are embodied within the Act. The associated Register currently lists about 2650 such reservoirs, of which 83% are retained by earth dams. If the ICOLD definition of 'large dams' (height greater than 15 m) were applied to the Register then about 530 of these reservoirs would be classified as large dams. Of these 530 dams, 368 are earth dams and 19 earth and rock fill. The predominant purpose for which these larger dams have been constructed is for the supply of domestic water, with the remainder generally being built for the generation of power.

The frequency of construction of dams in the UK has been surprisingly consistent over the last 200 years. However, the distribution of dam construction has not been evenly spread across the country. This variation in distribution can be attributed in particular to industrial and urban development but is also influenced by regional

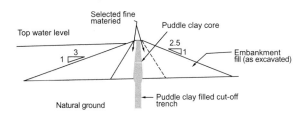

FIG. 10.8. Typical cross section through 'Pennines'-style dam.

differences such as the availability of water from alternative sources such as the chalk aquifers of southern England and hydro-power potential associated with the Scottish Highlands. These influences have resulted in the greatest incidence of dams being within the geographic mid-third of the country, and hence associated primarily with the industrial revolution.

In a presentation to the British Dam Society, Moffat (1999) provided a resume of the form of construction of many of the earth embankments in the UK. A key feature of these dams is the manner in which clay has been used to construct the impervious core. During the period 1850 to 1960 most dam construction adopted a generic form of design known as the 'Pennines' profile (Fig. 10.8). The paper describes the development of dam construction within a historic framework centred on the Pennines Era. Earth dams constructed before about 1850, were typically modest in size and were either of homogeneous clay section, or included a rammed or punned clay core of relatively generous width or a puddle clay upstream blanket.

The terms 'punned' (or 'rammed') and 'puddled' refer to different methods of working the clay, initially using bullocks and subsequently steam driven plant, in order to produce a homogeneous zone of material of high water content and low permeability. Whilst the terms punned and puddled refer to the intensive reworking of the clay, the process of puddling differed in that it specifically included the addition of measured amounts of water to the clay. Clays suitable for puddling included those from the Weald, Oxford Clay and London Clay, and also some types of glacial till. In the Pennines Era, identified as between 1850 and about 1950 to 1960, larger dams were constructed which incorporated a slender puddle clay core with puddled cut-off trench or later with a concrete filled cut-off trench. The embankment section relied upon the clay core being supported by competent shoulders and protected by zones of fine material upstream and downstream, the stability of the section being reliant upon local geology and the competence of the bulk of the fill material derived from excavations from the immediate surroundings of the site. Puddled or 'puddle' clay is dealt with more thoroughly in Section 12.3.5.

Moffat (1999 and 2002) presents the main technical weaknesses of the 'Pennines' dam, which relate primarily to the quality of the clay included in the core and its geotechnical characteristics and include:

- slenderness of the core;
- susceptibility to hydraulic fracture;
- potential for internal core erosion, dependent on downstream shoulder material;
- vulnerability of the puddle trench to erosion;
- progressive consolidation and settlement of the core;
- susceptibility to drying out and cracking on draw down.

In the Post-Pennines Era, identified as from 1950 to 1960 onwards, earth dam cross sections have placed less reliance on narrow zones and the impervious zones now comprise wider rolled clay cores. Principally economic pressures brought about this change in design, as the construction methods to produce and place the punned and puddle cores were labour intensive and hence less cost effective for modern contractors. Additionally the wider embankment section was seen as more robust against the perceived technical weaknesses of the more slender 'Pennine' construction.

Most but not all of these weaknesses have been addressed by advances in geotechnical knowledge (in particular, developments in the understanding of clay properties and earth dam design) and in the selection of material for the impervious zones of modern earth dams. Further, the development of effective stress testing (as described in Chapter 4) and limit state stability analyses have enabled the development of pore pressures during construction and under loading and unloading to be understood and accommodated in design leading to more secure embankment performance. As experience has been gained from the performance of embankments both nationally and internationally, in a range of geological environments and with a range of geotechnical materials, so a greater understanding of the behaviour of clays has been developed and the importance of clay mineralogy recognized.

An example of the development of earth dam design related to a greater understanding of clay clay behaviour is with regard to the use of naturally dispersive materials. In the mid-1970s, a greater awareness of the properties of some clays exhibiting dispersive behaviour, often associated with tropical clay profiles, was being developed through work carried out in the discipline of clay science in agriculture. In the engineering profession the matter was raised and researched by Sherard et al. (1972). Naturally dispersive clays, exhibit finer grading and are prone to erosion and piping as individual clay platelets do not naturally aggregate or floc to form larger composite particles and can thus pass through fine fissures and cracks and conventionally designed filter systems under seepage pressure flow.

Such behaviour is clearly undesirable in a material that is to form the impervious zone of an earth dam. However, having identified that a material is naturally dispersive, geotechnical developments have made it possible to introduce safeguards into a dam design, such as additives and appropriately designed downstream filters, to enable such material to be safely used. In addition, laboratory tests have been developed to aid identification and degree of dispersivity such as the Emerson Class Number, Soil Conservation Service test and Pinhole Dispersion Classification, (Fell *et al.* 1992). Chemical tests comparing relative ion concentrations of clay and of the clay pore water have also been developed. Safe use has required the 'protection' of these clays by appropriately designed and tested filters. Techniques such as these are described by Forbes *et al.* (1980).

The development and understanding of the importance of filters to control seepages and their design was an early contribution to geotechnical knowledge (through work by Terzaghi and Peck (1948) and Casagrande (1937)) that predated the development of an understanding of dispersive clays and was an important scientific contribution to the advancement of earth dam design. Through experience, engineers realised that seepage control is a primary design requirement in all embankment dams and that erosion can occur in zoned dams irrespective of the dispersive tendency of the material. This experience led to the development of filtration criteria and to the publication of guidelines and rules on filter design predominately based on research in the USA (US Corps of Engineers 1953) and in the UK (Spalding 1970). The initial guidelines were based on specially designed laboratory tests which continue in use today to verify the embankment design, particularly where the standard criteria may not be met as a result of clay mineralogy or in the case of critical filter zoning.

Other advances in dam design have resulted from improved laboratory testing methods and computer based analysis techniques. These provide a further insight into dam behaviour, enabling embankment dams to be both constructed from a wide range of materials and to resist extreme loadings. These advances include the use of mid-plane pore pressure monitoring and small strain measurement devices.

The lessons learnt from failures of earth embankments has generally, but not always, led to a better understanding of the special class of engineering that dam design represents. There has thus been a gradual increase in the knowledge of both short- and long-term behaviour leading to more cost effective design. In addition, observations made of embankment performance under extreme events, particularly where dams have withstood overtopping as a result of untoward flood flows, has provided an insight into the benefits of features and details that have lead to an improvement in safety. A simple example is the protection provided to the downstream face of an earth dam by a well-established grass sward, now known to provide a significant degree of resistance to erosion during overtopping events (Hewlett *et al.* 1987). Such performance data, which have been enhanced by improvements in legislation requiring inspection and monitoring, together with advances in fundamental research into the properties of clay as a construction material have added to the now extensive knowledge base that has developed over the years, demonstrating that earth dams can be a safe and cost effective means of construction for the retention of water.

10.1.3. Environmental mitigation structures

Increasingly in recent years it has been recognized that engineered fills can be used to form bunds to hide quarries, engineering works and developments or to use as landscape areas to improve visual appearance. With increasing noise levels and interaction between people and engineering there is now a routine need to consider these mitigation measures.

An environmental bund provides a visual and noise barrier. The bund is constructed between source and target of intrusion, and consists of a mount of clay material of sufficient height to hide the source (Fig. 10.9). The source may, for example, be an industrial estate, a highway or a railway. The target may be a housing estate, a sensitive fauna or an attractive area of countryside. On an embankment where a bund is constructed immediately adjacent to the source, as is commonly the case, the slope of the earthwork will rise above the source, a situation usually associated with cuttings rather than embankments.

Unlike embankments, bunds do not form a foundation for any other works such as a highway or development. They can therefore be composed of more voided, wetter and weaker clays provided there is sufficient room to accommodate the flatter side-slopes. They are usually well vegetated to provide a further barrier against noise and to provide an aesthetically pleasing appearance, and in some cases, where they are of insufficient height, may have an acoustic fence founded at the crest.

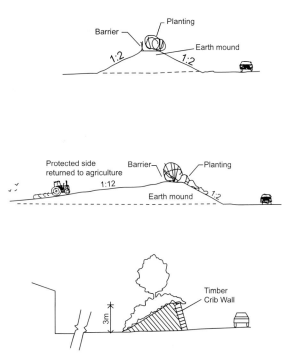

FIG. 10.9. Environmental bunds in use to provide a visual and noise barrier. (Design manual for Roads and Bridges Volume 10, Section 5).

Although the materials used in environmental bunds are usually of poorer quality than in embankments, the restraints of site area may require steeper slopes and hence a higher shear strength. Recently, reinforced clay bunds have become popular to provide this strength improvement (see Section 10.6.1). The bunds are often vegetated with grasses or trees that also reinforce the slope and add aesthetic value.

Landscaping to improve the aesthetic appearance of an embankment involves contouring the slopes of the embankment to mimic the surrounding landform (Fig. 10.10). This requires additional fill to the embankment and flatter slopes. Planting is an integral part of the landscaping design. Topsoil thicknesses can vary and are influenced by indigenous flora. Landscaping areas have little structural function and hence the requirements of the clay fill are much less. Landscape architects should be employed to assist in the landscape design.

10.1.4. Substructure fills

Substructure fill refers to filling over non-linear areas for use primarily as a foundation for a development, and usually provides the material upon which a granular layer is placed prior to casting a large concrete floor slab. The fill is therefore rarely of as great a thickness as an embankment. Similar design properties are considered but the requirements of the fill are different due to the type of applied loadings and the life expectancy of the development. The source of the fill is invariably within the site as the level of the floor is usually set to allow excavation on the higher side of rising ground level. Clay is rarely imported in these situations as the deficit of material volume required is usually small and commonly is made up of material that is needed elsewhere on the site. This keeps the number of types of imported material small. So, for example, if aggregate is required for a layer beneath the floor slab, the aggregate may be used for fill instead of clay where there is a deficit of clay volume from the cutting area.

The development of design and construction for substructure fills follows that of embankments but in general the development is more greatly influenced by movements due to consolidation of the clay, compaction of the clay and seasonal shrink and swell cycles within the clay. In the last two decades there has developed a greater awareness of clay shrinkage and swelling under buildings. This is particularly prevalent in areas where the moisture regime within the undisturbed ground is changed. Vegetation clearance, a common stage in any construction works, allows a greater presence of moisture that causes the clay to swell and heave of the building can follow.

10.1.5. Quarry tips

Waste materials arise during mineral exploitation from removal of the overburden above the deposit, unusable or

FIG. 10.10. Contouring of slopes to mimic the surrounding landform.

poor quality materials from and around the deposit and as solid waste from processing. These would normally be placed in temporary or permanent tips and occasionally placed back in the exploited workings. Water from dewatering of workings, from run-off etc. and aqueous wastes from processing that often contain suspended solids may also need to be stored in reservoirs known as silt ponds or tailings lagoons until the suspended solids have settled out and the water can be treated, re-used or disposed of in water courses or by percolation into the ground. These lagoons can be retained by earth dams (Leeming *et al.* 2001).

The growing need to reuse brownfield sites, and in particular disused quarry workings, has stimulated research into techniques that can be used to stabilize uncompacted fills and has lead to many modern backfills in quarries being compacted to minimise settlements (Charles & Watts 1996; Charles *et al.* 1998).

As a consequence of the tragic failure of a colliery spoil tip at Aberfan in 1966, the Mines and Quarries (Tips) Act 1969 and the associated Regulations 1971 were introduced and have more recently been replaced and reinforced by the Quarries Regulations 1999. In effect these require that new tips are investigated, designed and properly constructed to the design, that all tips associated with quarries are regularly inspected and, where potential hazards are identified, they must be investigated and dealt with appropriately. This also applies to mine and quarry lagoons that were excluded from the Reservoir Act 1975. Thus modern tips should be engineered fills. The requirement is to provide a stable embankment, which usually determines its basic design and methods of construction, including the need to compact the fill. The end use of a quarry may also influence the design of the tips. For example, if the worked out pit is to be used for waste disposal, clay waste may be placed and systematically compacted into the base and perimeter of the void to provide a seal and stockpiled to provide an eventual capping layer. Where the land is to be reused for development, a high degree of compaction may be given to reduce long-term settlements of the surface of the restored site (Burford & Charles 1992; Charles *et al.* 1993).

Silt ponds and tailings lagoons are similar to reservoirs except that the former retains particles in suspension. Retention is achieved with tailings or by earth fill embankments. Failures are not uncommon, especially in areas of low population where the consequences of failure may not be deemed significant and where low factors of safety have been adopted. However on failure, the tailings often liquefy and can travel considerable distances. Recent European failures have resulted in flows over large distances and have severely affected the local environment. Such lagoons can also pose a long-term safety hazard unless they are adequately restored, particularly following the sudden closure of the quarry or mine (Leeming *et al.* 2001).

10.2. Planning requirements

10.2.1. Planning considerations

In most parts of the world, it is necessary to obtain permits or licences from the local, regional, or national government to construct buildings and engineering structures, to extract minerals and to dispose of waste in worked out quarries. Generally these permits are required to ensure orderly and planned developments that are carried out within acceptable local or national design standards or guidelines to meet a local or national need.

The information required in a planning application will depend on whether it is an 'outline' or 'full' application. It is generally advisable to liase with the planning authority at an early stage to help identify and overcome any potential problems. At the 'full' application stage, plans, designs and details of the development will need to be submitted together with a justification of their need. The need may already be demonstrated in local structure or development plans or national resource studies but this may be more difficult to justify where the land is not zoned for that type of use, there is no local precedent or the development is in a greenfield site. In the case of minerals and some construction projects, geological

---

UK legislation—I

In England and Wales, the main legislation covering planning is the Town and Country Planning Act 1990. Further legislation is included in the Planning and Compensation Act 1991 and various Regulations, including the Town and Country Planning (Minerals) Regulations 1995 that covers the development of mineral deposits. Applications are first made to the local planning authority, which, for construction, is usually the district and metropolitan district councils, London Boroughs and unitary authorities. For planning applications relating to minerals development (the Minerals Planning Authority (MPA)) and waste disposal, the local authority would normally be the County Council instead of the district councils. Similar legislation and arrangements apply in Scotland. The Environmental Protection Act 1990 and the Environment Act 1995 particularly apply to waste disposal. In addition to planning permission, waste disposal sites require a licence from the Environment Agency.

Planning applications on larger schemes or schemes with significant public interest may be 'called in' by the Secretary of State and a public planning inquiry held. Such inquiries may also be held as part of the appeals procedure when applications have been refused or when permissions have unacceptable conditions attached.

information, calculations of reserves and working methods may be required and the Department of the Environment has issued various guidance notes. During the planning process, the application will normally be put out to public consultation. It may be useful to discuss the scheme in advance with local interest groups and residents to identify means of ameliorating any potential difficulties or objections.

Larger construction schemes and mineral workings will often have an impact on the environment, particularly in the case of greenfield sites. In developed areas, such as within the European Union (EU), USA etc., it is necessary to carry out an environmental impact assessment of the scheme. This is included in the 1988 Town and County Planning Regulations and given statutory recognition in the 1991 Act. Guidance is given in DoE Circular 97/11.

It is not necessary to own the land before a planning application is submitted. However before construction works can start, ownership or a lease on the land should be obtained. In the UK and many other countries, the ownership of the land and the mineral right are frequently separable. In many countries, the mineral rights are vested with the government, although in the UK this is not the case except for coal, oil, gas, gold and seabed minerals and other local exceptions. Both the land ownership and mineral rights need to be obtained either by purchase or lease. It is often necessary to pay royalties, usually based on the value, volume or weight of the mineral deposit extracted.

10.2.2. Environmental and safety considerations

Construction and mineral extraction projects may have far reaching temporary and permanent effects on the environment, which may affect the local residents and those working on site to different extents. These may include the effects of or on the following:

- visual intrusion;
- stability of slopes;
- waste disposal;
- vehicular traffic;
- noise;
- dust;
- ground and surface waters;
- flora and fauna;
- amenity;
- archaeology;
- restoration.

10.2.2.1. Visual intrusion. The landscape of the area should be taken into consideration in the design of the works and the restoration measures, particularly in areas of high landscape value, such as in National Parks. In urban areas, visual intrusion may not be as important, particularly when brownfield sites are being redeveloped. Excavations, embankments, tips, lagoons, structures and moving vehicular traffic may cause visual intrusion. Ideally the works should not disfigure the landscape any more than is unavoidable. Whilst overall visual intrusion should generally be minimized, it may be more important to minimize the impact for particular viewpoints such as windows of dwellings, public roads and footpaths. The effects of the development can be minimised by careful selection of the earthworks profile, the use of landscaping or screening embankments or bunds, reducing the level or height of structures, tips, embankment or cuttings, topsoiling and grass planting to reduce the impact of the earthworks, tree planting to provide wooded environments or as long term screens and the use of timber fences and screens. It may be appropriate to adopt flatter slopes (1 vertical to 6 horizontal or less) to enable land to be returned to agriculture or to regrade pit sides to provide safer gradients. In some cases, ponds or lakes that increase the amenity value or provide new habitats within the area may be provided in restoration schemes. The advice of a landscape architect at an early stage is recommended. These measures may increase the land required for the project or minimize the amount of usable space and may temporarily or permanently sterilize mineral reserves in quarries.

10.2.2.2. Stability of slopes. In both construction and minerals extraction projects, the side slopes of cuttings and embankments may be required depending on the situation to be stable for periods of between hours and years. Slope failure can provide a potentially life-threatening hazard and may result in the damage or loss of property and equipment within or beyond the boundaries of the site. The slope and geometry of all significant permanent excavations, embankments, tips, lagoons etc. should be considered when planning and designing the works to ensure that they are stable. Temporary slopes are formed in trenches, foundation excavations and working faces. These should be excavated such that there is either no significant risk of failure or the consequences of failure do not provide a hazard. This may require temporary support to the sidewalls or slopes of excavations or the design of the slopes so that they are stable for the required period. (Department of the Environment 1991; HSE 1997; HSC 1999).

10.2.2.3. Waste disposal. Whilst it is desirable to try to balance the earthworks quantities of excavation and filling on construction contracts, surplus materials or

---

UK legislation—II

The design of the works must take into account relevant legislation covering the scheme. This would include the Health and Safety at Work Act 1974, the Construction (Design and Management) Regulations 1994 (CDM), the Mines and Quarries Act 1954 and the Regulations 1999.

materials that are unacceptable for use are often produced. The latter may include weak or organic rich clays that cannot be used in structural embankments and contaminated clays that can only be disposed of in licensed tips. In recent years there has been a tendency to use waste materials in landscaping embankments and areas. Where earthworks materials are imported from borrow pits close to the site, it may be possible to dispose of unacceptable materials in these pits, thus helping to restore the pits. Materials in quarries that are difficult to sell generally consist of overburden, unusable materials within the deposit and waste from processing the minerals. If the processing waste is contaminated with chemicals, special measures will generally be required for its disposal. Other materials would be stockpiled or disposed of in bunds, tips or in the case of wet materials in tailings lagoons. In older quarries the tips would often remain a permanent feature of the landscape, but in modern quarries, they are often used to minimize visual intrusion and eventually to restore the site.

10.2.2.4. Vehicular traffic. Whilst in the long term, a construction project may increase or decrease traffic in an area, it usual effect is to increase traffic in the short term. Vehicles are required to take staff and plant to and from the site, deliver construction materials and remove waste and surplus materials. Vehicles are also used to transport staff and plant to quarries but, more importantly, the products are often transported from the quarries by road, except in the less common instance of a direct connection to a railway, canal or seaport. Increased vehicular traffic, can potentially increase flows on narrow local roads and increase noise and vibrations, all of which will cause a nuisance to the residents of the area and visitors and, in addition, is likely to increase the risk of road accidents. In order that these can be reduced, restrictions may be imposed in the construction contract or by planning conditions which may, for example, limit working hours, production output or vehicle sizes and impose obligatory routes for heavy good vehicles. It may be necessary to strengthen roads or structures on these routes and to provide safe junctions, particularly near site entrances.

10.2.2.5. Noise. Noise can be defined as unwanted sound and can be a health hazard and a nuisance. Noise is covered by the Environmental Protection Act. Noise is measured on a logarithmic scale in decibels (dB). Typically rural background noise would be about 40 dB and speech 60 dB. The threshold of pain in the ears is about 120 dB. Exposure to loud noise for long periods can induce hearing loss and EU regulations limit exposure to an equivalent level of 85 dB over 8 hours and a maximum of 130 dB (Gregory 1987). Engine and motor manufactures are endeavouring to reduce noise with better silencing and it can be reduced for drivers and other machinery operatives with soundproofed cabs or cabins. Others who work in the vicinity of noisy plant and equipment, such as in vehicles and processing plant should be equipped with and be instructed to use ear defenders. The temporary or permanent effects of construction and quarrying projects will often be to increase general noise levels. This can be minimized by ensuring that equipment is soundproofed, noisy plant and haul routes are located as far away as possible from habitation, plant and equipment is regularly maintained, by limiting working hours and with the use of noise bunds or screens. In extreme cases it may be desirable or necessary to install double-glazing in the windows of nearby buildings.

10.2.2.6. Dust. Dust, like noise, is a nuisance that can become a health hazard. Dust is covered by the Control of Substances Hazardous to Health (COSHH) Regulations 1994. Dust may occur from the operation of traffic, in the transference of materials (e.g. to stockpiles) and in processing plants. Of particular concern are silica-rich dusts that are particularly hazardous. Dust from haul and access roads is a particular problem in hot climates and during the summer in temperate climates. It can be minimized by ensuring that haul roads are properly maintained and access roads are cleaned and unpaved roads are watered in dry weather. Wheel washers will minimize both dust and mud on roads and dusty loads should be sheeted when leaving the site. It may be possible to use atomizing water sprays in the vicinity of transfer points (e.g. stockpiles), although this may be deleterious to some clays, and conveyer transfer points can be fully enclosed. Where dust is unavoidable, operatives should be supplied with facemasks or filtered helmets in extreme cases.

10.2.2.7. Ground and surface waters. The effect of natural or induced drainage on the groundwater table from excavations in clays is likely only to be local to the edges of the excavation. However more extensive reductions in groundwater levels may occur when associated granular or rock deposits are drained, which can lead to disruption of ground and surface water supplies. Excavations can also cut across land drains and watercourses. Conversely embankments may cause disruption to natural drainage or pore pressure dissipation either by filling or obstructing watercourses and springs or by acting as a barrier to surface flow. The former may lead to instability of embankments and tips especially those founded on slopes and both can lead to ponding. Construction of embankments and tips on clays can lead to the development of high pore pressures and failure within these soils, particularly on slopes, in areas of solifluction or past slope failure and where natural drainage has been impeded.

If uncontrolled, quarrying and construction works can increase and concentrate the amount of surface run-off and contaminate surface waters with suspended solids collected from the works, by increasing surface erosion, and from processing or drilling wastes which may be acidic and rich in heavy metals and other contaminants (see Section 10.4.4). If untreated, this can cause pollution if it is allowed to discharge directly into watercourses. The use of silt traps and settling lagoons will often be required to reduce the load of suspended solids to acceptable levels, whilst treatment plants may be required to remove unacceptable chemicals. The former may be less of a problem in some tropical areas where the suspended solid content of the rivers is inherently high (e.g. the Yellow River in China). Leaking chemicals or hydrocarbons and chemical stores can contaminate groundwater and fuelling points should be designed to minimize such risks.

10.2.2.8. Flora and fauna. There is increasing concern in the developed countries about the loss of wildlife habitats. In the UK, sites of particular geological importance or containing particular habitats may be classified as Sites of Special Scientific Interest and the development of such sites is likely to be strongly opposed. Unless the national interest is at stake or acceptable proposals are made to preserve the site habitat, the application will probably fail. However, less important examples of conservation are commonly raised. These may include ponds containing rare fauna such as great crested newts, nesting sites of rare or migratory birds, the presence of badger sets or paths, amphibian trails, areas of rare flora and bat colonies. Consideration may be given or required to accommodate or relocate the particular features. In the case of quarries this may be incorporated into the restoration plan.

10.2.2.9. Amenity. Construction and quarrying will usually affect the environment but may contribute to or reduce the amenities in an area. The redevelopment of run down urban areas will usually increase the overall amenity value and some mineral workings may provide areas of water that can be used for fishing, water sports or other amenities. Other projects may result in the loss of amenities such as playing field and footpaths, although it may be possible to relocate the latter.

10.2.2.10. Archaeology. Modern archaeological studies cover the period from pre-historic to modern times. Whilst some features may still be evident, older features may not be visible from the surface or included in the historical record. In certain areas, local knowledge may indicate the chances of encountering archaeological remains within the site. Planning consent may be conditional on having an archaeological desk study and/or walk over survey carried out and where remains are known to be present, objections to the development may be made. Agreement to carrying out exploratory or more detailed excavations may reduce these objections, particularly if they include part or total funding by the developer. In many cases, there are inadequate funds or time to properly excavate archaeological sites and means will be considered to preserve the site for the future. However, in some cases archaeological remains may be so important that they need to be preserved or removed before development work proceeds.

10.2.2.11. Restoration. Since 1948 planning permissions for mineral workings have inevitably been given with conditions relating to restoration following the completion of extraction. Since 1982, Mineral Planning Authorities (MPAs) have been able to impose 'aftercare conditions' for a stated period to bring the land up to the required standard for agricultural, forestry or amenity use. Thus it is important that topsoil and overburden should be stockpiled separately for use in restoration. This may be carried out in phases as parts of the quarry become worked out or on completion of all of the extraction works. At present, there is little pressure to restore land to agricultural use, but there are financial incentives to develop sites as landfills. Restoration to amenity use such as reservoirs, water sports areas, wildlife reserves and recreational areas are not unusual and occasionally it is possible to utilize the area for domestic housing, industrial or retail parks. In most cases, regrading of the excavated slopes, amenity embankments, stockpiles and waste tips will be required, together with the removal of equipment and demolition of structures. Professional assistance will often be required to restore the site to the state where its final use can be achieved.

General safety legislation—UK

The two regulations that are pertinent to and have an impact on construction using clay as fill are given below.

The Management of Health and Safety at Work Regulations (1992) requires all employers to make a *suitable and sufficient* assessment of the risks to employees and others from operations or undertakings. The Regulations, however, do not stipulate a particular methodology for such assessments nor the measures required as a result. Consequently embankment condition appraisal and that of other slopes has developed in different forms. Employers with five or more employers must record their assessment.

The Construction, Design and Management Regulations (1994) imposes a duty on designers to have adequate regard to the possible effects of a design on the health and safety of those who will carry out the construction, maintenance or demolition of the works. It also covers those who may use the infrastructure or be within the vicinity, such as the general public. The Regulations recommend that designers, if possible, alter their design to avoid envisaged health and safety hazards, or where this is not reasonably practicable follow a hierarchy of risk control. A *Planning Supervisor* is appointed by the client, usually from within the design organisation, to co-ordinate health and safety throughout the design stage.

10.2.3. Parties to the development

The different nature of construction projects and the extraction of minerals lead to many completely different contractual relationships, methods of payment for carrying out work and methods of working for those involved in earthworks. The common factors between the two are that excavation of clay is required and that similar types of plant may be used.

The developer in construction may be a public body or company, and in the case of house builders, may also be a contractor. The developer will generally appoint a team consisting of architects, engineers, quantity surveyors and related professionals to design and supervise the construction of the project, although in 'design and build' contracts the professionals will generally be appointed or provided by the contractor. At some time the main contractor will be appointed to carry out the work, and may carry out the earthmoving himself or sub-contract part or all to other, usually specialist, contractors. The main and sub-contractor would normally be paid by the developer either a lump sum for the work or an amount determined by the quantities and tendered or agreed bill of rates.

The developer of mineral deposits would normally be a company and it (or a subsidiary) would normally be the operator of the site. The developer may or may not own the site and the mineral rights. In cases in which the mineral rights are not owned then the developer may need to lease these and would then generally pay royalties on the minerals won. The operator will often carry out all or part of the earthmoving work with the remainder contracted out, usually on a rate basis. Once excavated, the clay would be transported to a central area for stockpiling, sale, processing or manufacture or both into a product on or off site. Eventually the processed material or product would be sold to generate income for the work.

A disused quarry may become a waste disposal site. The developer may be the original developer of the quarry, a subsidiary, an unconnected company or a public body. If developed by a public body, relationships would probably be more akin to the construction model, although the operator may be a company. Income would be generated by charges for the use of the facility. If developed by a company either the construction or minerals model could apply with income being generated by user fees.

10.3. Material characteristics of new fills

10.3.1. Earthmoving in clay

Earthmoving practice is covered in Chapter 11, but it needs to be considered in the design and planning aspects of earthworks, hence a short explanation is given here. Clays may occur as superficial deposits such as alluvial, glacial and lacustrine clays, head, mudflow debris, and weathered deposits such as laterites. Older clay deposits, which may often extend to greater depths, include sedimentary deposits such as Carboniferous mudstones, Gault, London Clay and china clay from altered granite (see Chapter 6).

Earthmoving may be carried out in construction works to form a large variety of structures of widely differing sizes. These may include the construction of harbours, coastal defences, dams and other water-retaining banks, embankments or cuttings for buildings, roads, railways, canals and watercourses and excavations for foundations and pipelines. Because of the widespread occurrence of clays, many earthworks are constructed partially or wholly of clays. Some earthworks are designed to retain water or other fluids and therefore require that the permeability of at least part of the structure is low and clays are therefore selectively used to form the cores of earth dams, the banks of lagoons, the linings of canals and sometimes impermeable barriers to the movement of groundwater or leachate, particularly in or adjacent to areas of contaminated ground. Clays may also be excavated in mines and tunnels, which are beyond the scope of this book, although the disposal of the spoil will often be carried out in a similar manner to construction or in mineral working. Weak clays may be unsuitable or unacceptable for use in structural embankments and may need to be disposed of in tips or landscaping areas, together with any surplus materials.

In quarries, earthmoving may be carried out in clays that form the overburden, the unproductive ground within or the waste from processing the mineral deposit. In addition, clays may occur within the earthworks associated with the development of the quarry. In order to reach the mineral deposit, it is generally necessary to first remove topsoil and overburden that are often stockpiled separately in tips outside the area to be worked or placed in disused parts of the quarry workings. Unproductive ground from within the mineral deposits may be similarly stockpiled or placed in disused workings. In some cases, clays may be use to form low-permeability barriers for reservoirs or lagoons associated with the processing plant or disposal of tailings.

Examples of clays won for their mineral properties include Fuller's Earth and China Clay and a wide variety of clays including Carboniferous mudstones and Oxford Clay which are quarried for the manufacture of ceramics including pottery and bricks. The main by-product of the china clay workings is granitic sand and gravel which can be disposed of or used as a construction material (for example as aggregate for concrete or as bulk fill). Fine-grained by-products of processing include scalping from screening aggregates, some of which may not be saleable.

In many countries, disused quarries will need to be restored to an acceptable profile. This may involve demolishing structures and regrading the tips and excavated faces to an agreed profile. Some quarries or parts of quarries may be returned to agriculture, used for recreation or developed into particular habitats. Other quarry workings, especially those within clay deposits may be used for the disposal of solid or liquid wastes and clays may be used to construct basal, lateral and top seals to minimise the migration of leachates from and the ingress of surface or ground water to the tip. Such areas may eventually be restored to agricultural or amenity use.

Consideration is given below to fill characteristics, design and chemical effects.

10.3.2. Fill characteristics

A clay soil has three phases and the extent of each determines its engineering characteristics (Chapter 4). Soils of various types have different capacities for absorbing water. Thus a high plasticity overconsolidated clay may have a moisture content in excess of 30% and have a high undrained strength, whereas a similar undrained strength may be attained at moisture contents of only 15 to 20% with a low plasticity clay. The surface area of the coarser particles and the plasticity of the fines determine the moisture capacity and this, in turn, for a well-graded soil, determines the potential dry density that can be achieved with a given compactive effort. Figure 10.11 shows typical laboratory results using the British Standard 2.5 kg rammer compaction test (British Standards Institution 1990) for various types of soil. As the soil becomes less plastic and more granular the relation between dry density and moisture content generally moves upwards and to the left, with increased values of maximum dry density.

Shear stresses exerted by the compaction plant reduce with increasing depth in the layer being compacted. Thus the shear strengths mobilized to equalise the shear stresses will also decrease with increasing depth and a gradient of dry density is produced. The general form of the relation between dry density and depth in a compacted layer is given in Figure 10.12. The effect of increasing the thickness of compacted layer, and hence of increasing the effective volume to be compacted, will, therefore, generally be to reduce the average state of compaction in the layer (unless additional compactive effort is given).

Increase in the compacted volume, due to either thicker layers in the field or to larger moulds in the laboratory, results in movements of the relation between dry density and moisture content downwards and to the right, giving rise to reduced values of maximum dry density and increased values of optimum moisture content. The interrelation between depth in the compacted layer, dry density and moisture content is illustrated in Figure 10.13, where dry densities were determined at four increments of depth in the layer, each at three values of moisture content (Lewis 1954). As only three points were available for each relation between dry density and moisture content in this figure, the most likely curves are drawn.

With clays, the aggregations of particles to form lumps can influence the level of compaction attained. For example, clay soils excavated from trenches may prove difficult to recompact if large lumps are replaced for compaction by typical small trench compactors. Figure 10.14 compares results obtained when replacing material for

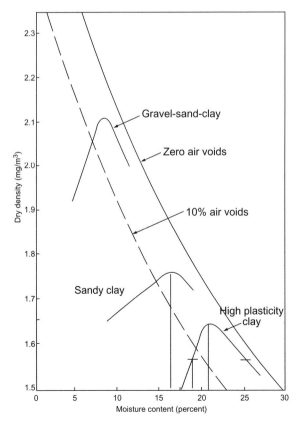

FIG. 10.11. Relations between dry density and moisture content for various soils and aggregates when compacted in the BS 2.5 kg rammer test (Part 4, BS 1377: 1990).

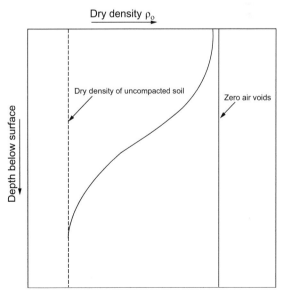

FIG. 10.12. General form of relation between dry density and depth in a compacted layer of soil.

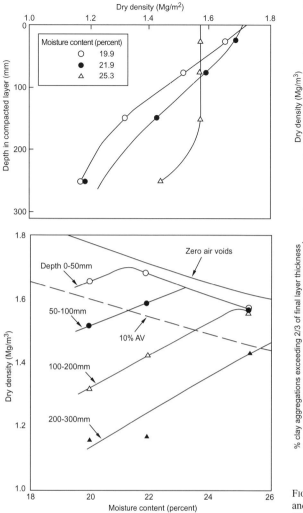

FIG. 10.13. Relations between dry density, depth in the compacted layer and moisture content. These results were obtained with 32 passes of an 8-tonne smooth-wheeled roller on high plasticity clay (Lewis 1954).

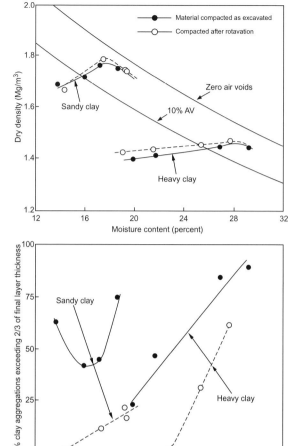

FIG. 10.14. Effect of 'lumps' on the relations between dry density and moisture content of sandy clay and heavy clay. Results for 100 mm thick layers compacted with 6 passes of a 55 kg vibrotamper (Parsons 1992).

compaction in a trench in an 'as excavated' condition with those obtained after treatment of the clay with a rotovator to produce a finer tilth. With sandy clay a reduction of about 1% in dry density occurred when the lumps were not broken down, whereas for high-plasticity clay the loss in dry density was a maximum of about 2.5%. The lower part of Figure 10.14 shows the relation between the proportion of the clay with a lump size in excess of about two-thirds of the thickness of the compacted layer and the moisture content for the tests. This indicates that the main trend was for the proportion of larger lumps to increase with increase in moisture content; this also applies, but to a lesser extent, for material which had been treated with the rotavator cultivator. However, at the higher moisture contents such lumps will be weaker and more easily deformed by the compactor. The increase in the proportion of large lumps at low values of moisture content, as shown for the untreated sandy clay, might also be expected as this material would be stronger.

On a larger scale, over-consolidated clay from a deep cutting or borrow pit, excavated by large earthmoving equipment, can also be deposited in fill areas in very large, 'boulder-size' lumps which are difficult to negotiate and compact by even the largest compactors. It is generally considered that the achievement of a relatively fine tilth with no particle aggregation exceeding about two-thirds of the final compacted thickness of the layer is the best means of ensuring uniformly well-compacted clay soils. These large clay lumps also retain any *in situ* suction present before excavation.

Settlement of fill occurs due to compression. The main process is the reduction in pore volume and the expulsion of air and water from voids. Loosely placed (low density) unsaturated fills may undergo large settlements due to infiltration or saturation with water.

Shear strength of the clay fill is important for slope stability and trafficability. This is usually characterised using interparticle cohesion, friction and pore water pressure (effective stress) (Chapter 4). The measurement of stiffness of the clay near trafficked surfaces may also be measured as it governs the ride quality of traffic such as rail and road.

10.3.3. Design

Design is a mixture of computation, observation, experience and judgement. The desk study, ground investigation and interpretation (Chapter 7) and testing (Chapter 8) provide the design input parameters for design. Feedback cycles of variation of parameters judge sensitivity until a confidence is reached between experience and research. The following are the common parameters that are considered in design. (For further explanation see Chapter 4.)

10.3.3.1. Shear strength. In the classic limit equilibrium model, the strength of the clay, and hence its resistance to failure, is given in terms of effective stress by:

$$\tau = c' + (\sigma_n - u)\tan\varnothing'$$

where τ is the resistance to shear along an actual or potential rupture surface; c' is the cohesion of the clay in terms of effective stress; σ_n is the total normal stress on the rupture surface under consideration; u is the pore water pressure and $\varnothing'$ is the angle of shearing resistance with respect to effective stress.

The value of τ thus depends on the value of u. If u is positive, the $(\sigma_n - u)\tan\varnothing'$ term will decrease, which may lead to failure of the slope in clays with low values of c' and $\varnothing'$; in clays with a high clay content, $\varnothing'$ may be lower than the angle of the slope and the influence of c' is large. For low clay content soils, c' will be low and the strength will depend principally on $\varnothing'$.

10.3.3.2. Compaction. Parsons (1992) provides a detailed assessment and reference document for compaction, particularly for highways. Trenter & Charles (1996) cover compaction for low-rise structures.

The design for compaction is based on:

- moisture content;
- compactive effort (i.e. energy applied and depth of soil).

Shear planes can develop at the interfaces between compacted layers due to the orientation of clay particles in materials, particularly those with a liquid limit greater than 50%. To prevent this occurring it is usual to use a tamping roller or grid roller (see Chapter 11 for a description of these machines) to provide a better key between layers.

The compaction requirements are based on either an 'end-product' or a 'method' specification. The end-product type does not specify how an end result is achieved but includes testing to measure the final result. The method type requires the specification of the means to achieve a required result.

10.3.3.3. End-product compaction specifications. The basic measurement of the *in situ* state of compaction, i.e. its end-product, is in terms of the dry density of the soil. However, because of the effects of soil characteristics, dry density alone is not indicative of the state of compaction and interpretations of the dry density have to be made for purposes of determining the levels of compaction achieved in relation to the levels that are potentially attainable or that are necessary to attain the required engineering properties. Two basic forms of end-product specification are commonly used for clays:

- relative compaction;
- air voids.

Relative compaction is defined as the dry density achieved in the compaction process in the field expressed as a percentage of the maximum dry density obtained by using a specified compactive effort in the laboratory. The most suitable laboratory compaction test for clays, including tropical residual soils, is the 2.5 kg rammer test (see Chapter 9). The actual percentage specified can vary depending on the application of the earthworks being constructed and the soil type, but for the 2.5 kg rammer test it is usually in the range of 90 to 100%. Where soil type varies from place to place in a compacted layer it is often necessary to establish the maximum dry density at each location of a measurement of the *in situ* dry density. To reduce the effort required in the laboratory compaction tests under these circumstances, rapid methods have been evolved for controlling compaction to this form of specification (Hilf 1957).

The measurement of the air remaining in the compacted clay is a logical expression of the state of compaction, considering that the reduction of air voids is implicit in the compaction process. Air void content is determined from the equation:

$$\text{air voids} = 100\left\{1 - \rho_d\left(\frac{1+w}{\rho_s + 100\rho_w}\right)\right\} \text{(per cent) (10.1)}$$

where ρ_d is the dry density of the compacted clay; ρ_w is the density of water; ρ_s is the particle density; w is the moisture content (per cent).

As indicated by equation (10.1), an increase in moisture content at constant dry density leads to a reduction in air voids and it is necessary to ensure that a low air content is not achieved at the expense of using clay of excessively high moisture content. Commonly specified values for clays are a maximum of 10% air voids for bulk earthworks and 5% air voids for high quality applications, such as the final upper layers of an embankment. Normally these values have to be achieved at moisture contents within specified limits or below a specified upper limit.

For most of the 1950s and 1960s an end-product specification was in use for the compaction of earthworks in road construction in the UK. The specification for the bulk earthworks required that a state of compaction of 10% air voids or better had to be achieved in at least nine results out of every ten consecutive measurements made in the compacted material. The required air content was decreased to 5% air voids in the upper 600 mm of the embankment. However, the testing required provided too onerous and method specifications were developed.

10.3.3.4. Method compaction specifications. In a method specification for compaction the precise procedure to be used is laid down. Thus the type of compactor, mass, relevant dimensions and any other factors influencing performance, together with the thickness of layer to be compacted and the number of passes of the machine, are all specified. The principle of method specification for earthwork compaction is that the contractor should have as free a choice of compaction plant as possible. Having selected the type of plant to be used, then the maximum depth of layer and the minimum number of passes to be used are given for that item of plant in combination with the clay type encountered. For the highways specification in the UK (Manual of Contract Documents for Highway Works (MCHW) Volume 1), the compactive effort that was used to formulate the specification was that which would achieve an average state of compaction of 10% air voids at a moisture content at the bottom of the estimated range of natural moisture contents in the UK. This moisture content was considered to represent a condition in which the clay would be at its most difficult to compact and is estimated for the near-surface condition of *in situ* materials. However, clay soils can have moisture contents dry of this condition when excavated from near the bottom of deep cuttings.

10.3.3.5. Slope stability. Slope stability design of clay fills is a well documented and researched subject and it is not the purpose of this chapter to discuss stability analysis in any great depth. The reader is referred to the following: Bromhead (1986); Anderson & Richards (1987); Vaughan (1994). Slope stability design involves an assessment of the disturbing forces and moments and the restraining forces and moments, the ratio of which yields a factor known as 'Factor of Safety'. Embankment slopes are usually designed to have a factor of safety in excess of 1.3 (British Standards Institution 1981) in the long term. As explained earlier, pore water pressures have a great influence on shear strength and their values need to be carefully considered based on any monitoring of similar slopes, local water regimes, weather conditions and drainage conditions. A pore water pressure coefficient is commonly used in fills to give a general figure. However, Crabb & Atkinson (1991) have shown that in the longer term, pore water pressures on slopes can change seasonally owing to water ingress.

Computers are regularly used, except for very simple situations where relatively little iteration is required. They allow a sensitivity analysis to be undertaken using the rapid calculation of differing scenarios. The type of analysis falls into two types: limit equilibrium and finite element. Limit equilibrium considers the equilibrium of disturbing and restraining forces and moments. Finite element segregates the clay slope into a large number of blocks. Properties are ascribed to the blocks and the program then determines the stability of the blocks.

In all design, however, emphasis must be placed on field observation and consideration of similar slopes. Based on empirical information, guidelines for the angle of slopes at various heights and for a comprehensive list of differing materials are given in Perry (1989).

10.3.3.6. Setting moisture content limits for the construction of fills. Moisture content has an influence on the strength of clays which become weaker as their moisture content increases. This improves their ability to compact adequately. In design it is convenient to relate these properties on graphs (see Fig. 10.15) to give the limits of moisture content for construction. This is termed 'relationship testing'.

The relation between dry density and moisture content which passes through the intersection of the design

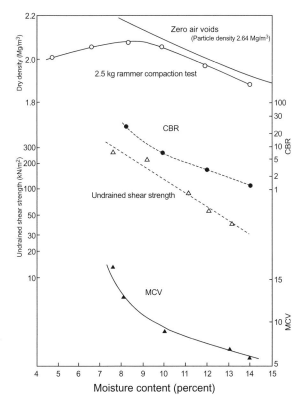

FIG. 10.15. Results of laboratory tests on a bulk sample of boulder clay from which limits of acceptability may be set.

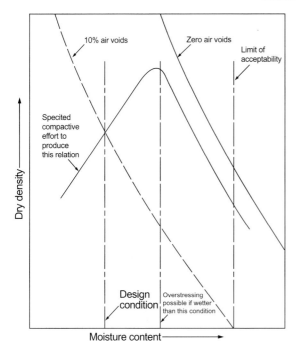

FIG. 10.16. Principles for compaction.

moisture content and the 10% air voids line is shown in Figure 10.16. For soils dry of a design condition an air content higher than 10% will be produced using the specified compactive effort, whilst as the moisture content increases beyond the design condition the air content will decrease, with increasing dry density, until the minimum air content is reached. With increasing moisture content beyond the optimum moisture content/maximum dry density condition for the specified compactive effort, dry density will decrease at an approximately constant air content, as shown.

The lower moisture content (dry end) is based on compaction requirements, either the intersection of the 10% air voids line with the compaction curve, or the intersection of the 95%, or higher, of laboratory maximum dry density with the compaction curve. Setting the lower moisture content by 10% air voids is the preferred method in the UK for clays in bulk fills. In shallow foundations, 5% air voids is preferred (Charles et al. 1998).

The upper limit of moisture content is based on the minimum strength requirements (California Bearing Ratio (CBR) or shear strength) in the fill material. The tests for these properties are described in Chapter 9. The traffickability of earthmoving plant and of compaction plant is an important consideration in setting the limit of acceptability (Symons 1970; Parsons & Toombs 1988) and is usually set by the undrained shear strength or CBR, commonly taken as 45 kPa to 50 kPa or 2% respectively, but it does depend on the type of construction plant and its capability. For high embankments, the upper limit is also governed by the shear strength requirements for stability of the slope. However, compaction in the field needs to be considered as water in the soil can take up some of the stress exerted by the compactor and excess pore water pressures can be generated and show symptoms known as 'overstressing'. With increasing moisture content the excess pore water pressures will be generated at an earlier stage in the compaction process. The symptoms of this condition are the remoulding of the surface of the compacted layer and permanent deformation in the form of rutting under the wheels or rolls of the compactor. Elastic deformation of the surface of the layer is not indicative of excess pore water pressures in the surface layer, but possibly in one or more of the underlying layers. Overstressing in this way is not serious with most soil types encountered, although a temporary reduction in compactive effort, by reducing the number of passes of the compactor or increasing the thickness of layer, may be considered.

In the past, two other forms of upper moisture content have been used: (a) a factor times Plastic Limit, e.g. 1.2 x PL; (b) a factor times optimum moisture content. The reliance on the use of the Plastic Limit test is endemic in the first method. However, this test has been shown to be one of the most inaccurate (Sherwood 1971). Also its use with stoney clays with a high coarse fraction can be unrepresentative. Hence this practice is now less popular except for differentiating between wet and dry clays as a classification in specifications. The use of a factor times optimum moisture content has been used mostly for Glacial Tills. However, since this factor to be applied is based on shear strength results and moisture content, it is the same as a direct comparison of moisture content. Also the use of moisture content only requires considerably less testing and is quite properly related to shear strength and not compaction at the wet limit.

Both limits on moisture content (dry and wet) can be controlled on site by the third relationship test for Moisture Condition value (MCV) (see Chapter 9) and moisture content. For clays there should be a good correlation between the two and hence MCV, which is quick to measure on site, can be used for site control of moisture. The preferred method of specifying moisture limits on clays in the UK is to use MCV. For stoney clays of large particle size (greater 20 mm), the MCV can be inappropriate if there is insufficient matrix (less than 55 to 50%) for the test. In these cases reliance on moisture content is necessary (Oliphant & Winter, 1997).

10.3.3.7. Sub grade properties for infrastructure embankments and substructure fills. The surface, and immediately below it, of the fill on which loadings are to be applied is required to have sufficient strength, durability and stiffness to carry the static and live stresses imposed by the embankment. For most transport infrastructure, the surface is required to have a design long-term laboratory CBR value of more than 15% or more. Increasingly in modern design the stiffness of the material rather than its CBR value is being used and is an area of continuing research. A useful discussion is given in Brown et al. (1987). These criteria cannot always be achieved with clay fills and so a layer of more competent material is placed beneath the pavement layers with a thickness dependent on the properties of the clay material. In most cases the upper part of the fill receives more compaction, usually twice that of the bulk fill, reflecting the importance of the properties required in the upper part of the fill. A similar approach is adopted for substructure fills.

10.3.3.8. Making maximum use of clay in construction. With increasing demands to protect the environment, both in terms of reducing export of materials and scarce natural resources, and the recognition of financial benefits, there is an increasing demand for the maximum use of clay materials. Treatment of clay with lime and cement is a common method that because of its specialised nature is covered in a section of its own (Section 10.6.2). However, there are other ways of employing clay in construction depending on the site circumstances.

On-site disposal is the most obvious means of using clay which is surplus to requirements or unsuitable for structural use. On-site disposal encompasses the placing of clay in environmental bunds or enlarged landscaping areas in which large volumes of material can be used. It can, however, require additional land and so this must be recognized early in the construction process so that land ownership issues can be resolved.

Topsoil is usually in short supply on a construction site. One method of increasing the volume of topsoil is to mix clay with peat excavated from within the site. The mix will vary depending on the local soils, which should be used as a guide, but a one to one mix is not uncommon.

Careful earthworks management on the site, with an understanding of materials available and their possible application, can lead to best use of materials. In areas proposed for embankment construction, good quality materials such as gravels that would otherwise be built on by the embankment can be excavated and refilled with properly engineered clay fills. This is useful where clay is a surplus and good quality gravels are not. The embankment or bund can then be constructed on the filled excavation.

Low-angle slopes should be designed and constructed where areas of fill can be used for agricultural purposes. This allows clay to be used on the outer zone of embankments with little or no maintenance in the future. The opposite is also possible. Poor quality clay, i.e. high moisture content material or high plasticity, can be used in the central core of fills. The outer and upper zones should then be constructed with better quality materials to ensure a stable and trafficable fill. However, in densely populated and intensely farmed areas, such as Java, this approach is rarely possible. The type of agriculture is obviously critical as methods such as rice growing in paddy fields can lead to greater stability. Making maximum use of clay reduces the import of higher performance materials and reduces the export of surplus materials.

10.3.4. Deleterious chemical effects

The chemistry of clays (Chapter 2) can have a detrimental effect on buried engineering structures. The three most common areas of concern are:

- the corrosion of buried steel structures due to the development of acidic conditions;
- thaumasite attack of concrete;
- the development of expansive minerals when certain clays are mixed with lime or cement (see Chapter 12).

Recent experience with the rapid corrosion of buried steel culvert structures in highway schemes (Reid *et al.* 2001) and concrete bridge structures (DETR 1999) have highlighted the need for careful consideration to be given to the chemistry of construction materials and the ground in the vicinity of buried structures. Naturally occurring sulphur and chloride compounds, in particular, are among a group of soil chemical constituents that may be aggressive to construction materials. Volume change due to hydration, solution and precipitation processes may also be responsible for damage to structures (Hawkins & Pinches 1987; Nixon 1978). BRE (2005) provides guidance as to the assessment of the ground and allowable amounts of potentially reactive constituents in construction materials and the ground adjacent to structures.

As explained in BRE (2005), sulphur compounds are usually present as either sulphides, sulphates or as a component of organic matter. Sulphates such as gypsum and epsomite are readily soluble in water, and so give rise to reactive groundwater conditions. On the other hand, in the presence of oxygen and water, that occur in weathering environments, sulphides may to be oxidized to sulphate. As this process also produces acid, extremely aggressive conditions can be produced. Czerewko *et al.* (2002) indicate that sulphates occur naturally in a number of UK mudrock formations, including Mercia Mudstone. Furthermore, sulphides such as pyrite commonly occur in unweathered, dark coloured, organic rich overconsolidated (stiff) clays, mudrocks and coal. Typically such deposits contain dark coloured pyrite of very small grain size, which makes it difficult to detect. Such pyrite is in fact much more reactive than the more familiar brass yellow cubes and octahedra, otherwise known as fools' gold, that occur in geologically older slates and mineral deposits. Marcasite is a potentially troublesome reactive monosulpfide mineral, that commonly occurs in chalk. Sulphides of very small grain size may also be present in organic rich deposits formed within stagnant water bodies. BRE (2005) recommends that the total potential sulphate be determined, which assumes that any sulphide mineral present will be oxidized to sulphate. This is then used to determine the Aggressive Chemical Environment for Concrete (ACEC) Classification for the ground. BRE (2005) contains recommended actions, depending upon this classification.

In the lime and cement stabilization process (discussed in more detail in Chapter 12), the presence of sulphur and organic material within the clay can lead to a detrimental effect on strength. Sulphates, although not affecting the reduction in plasticity associated with the process, react with lime and cause swelling in the longer term that can weaken the stabilized soil and cause heave of structures. The reactions are fully described by Sherwood (1993). The sulphates may be present either within the soil already, be produced by the oxidation of sulphides, including pyrite within the clay, or be introduced by groundwater.

Organic materials may also prevent the stabilisation process occurring unless sufficient lime is added. Lime stabilization relies on an increase in the pH of the soil that may not be possible with organic (acid) soils unless large and, possibly uneconomic, amounts of lime are added. The change in pH depends on the type and amount of organic materials present (Sherwood 1992).

These effects can have particular relevance for engineering works which rely on a significant increase in strength for successful construction and good performance of earthworks, for example for use as stabilized capping layer under a road pavement.

10.4. Long-term material characteristics and maintenance

10.4.1. Clay degradation

The principal causes of loss of performance of clay fill are:

- a change in the nature of the embankment materials;
- the presence of water;
- shrink and swell cycles induced by seasonal moisture changes and vegetation;
- the slope geometry, angle and height;
- poor method of construction;
- the age of the embankment;
- foundation inadequacy;
- changes in traffic loading or frequency;
- extraneous factors, such as vandalism;
- internal erosion.

As transport use increases, more frequent, faster and heavier traffic will impose high dynamic loads on embankments. Thus embankments need to be designed to meet the new demands and performance loss has more serious consequences.

The most common causes of loss of performance are described in the following sections and further details can be found in Institution of Civil Engineers (1985). In each case, an assessment is made of the potential effect on the serviceability, the level of performance that can be sustained by routine maintenance and, for catastrophic loss, the ultimate condition when capital maintenance is required.

Many embankments suffer from serviceability failure in which the loss of performance is slow and insidious, and is associated with excessive movement rather than overall instability. Catastrophic failure is less frequent but the consequences are dramatic, usually resulting in traffic being halted or severely restricted. Catastrophic failure may develop if serviceability failure is not addressed. Catastrophic failure is the only loss in performance that has a consequence for environmental bunds.

There are several types of slope failure that can affect infrastructure embankments, ranging from small-scale shallow translational slides to major deep rotational slips that run from the crest through the embankment and the underlying foundation material to emerge beyond the toe. The types of failure that occur tend to be different in canal, railway and road embankments, as a result of the different materials and their distribution, the methods of construction and the functions which the embankments perform. Each type of infrastructure embankment is therefore discussed separately below.

All the types of slope failure can be analysed in terms of soil mechanics principles. Pore water pressures and their variation with time are major factors in the stability of infrastructure embankments.

Shallow failures are those that are contained entirely within the embankment side slopes and where the maximum depth of the rupture surface does not usually exceed 2.0 metres (Coppin & Richards 1990). These failures are commonly translational although some shallow circular slips do occur.

Deep failures are those where the rupture surface exceeds 2.0 metres in depth and may extend into the natural ground beneath the embankment. They can be a variety of shapes depending on the ground conditions. The rupture surface may daylight on the embankment crest, beyond the toe or be entirely contained within the slope.

Failures on sidelong ground are a third major failure type. Failure occurs near the interface between the natural ground and the embankment fill. This can be exacerbated by seepage in the natural ground. For railways, canals and older roads, the embankments were not generally benched into the existing ground, and in many cases the original topsoil would have been left in place, forming a potential rupture surface.

In addition, some embankments have been constructed on ground that contains pre-existing rupture surfaces that can promote deep failures and failures on sidelong ground. A further problem for all embankments is burrowing by animals. Many embankments are home to colonies of rabbits, badgers, foxes, moles and other creatures. Whilst this is an ecological habitat and consideration should be given to this, this activity may have implications for the stability of the embankment and relations with neighbours.

Slope failures that begin as shallow failures may progress to deep failures. Shallow failures are generally serviceability failures, whereas deep failures are almost always catastrophic.

Most overconsolidated clays (Chapter 4) when freshly excavated, have negative pore water pressures, i.e. suctions, or low positive values. Embankments constructed with these materials will be initially stable at slopes of up

TABLE 10.1. *UK Clays with a high percentage of failure (Perry 1989)*

Geology	Percentage of failure	Predominant slope angle ($v:h$)
Embankments		
Gault Clay	8.2	1:2.5
Reading Beds	7.6	1:2
Kimmeridge Clay	6.1	1:2
Oxford Clay	5.7	1:2
London Clay	4.4	1:2
Cuttings		
Gault Clay	9.6	1:2.5
Oxford Clay	3.2	1:2
Reading Beds	2.9	1:3

to 1:2 (vertical: horizontal) (26°) if the material is well compacted, and 1:2.5 (22°) to 1:3 (18.4°) if loosely tipped. Hence at the construction stage the embankment appears stable, although this stability is sometimes marginal and failures may occur. With time, however, the negative pore pressures decrease or the positive pore water pressures increase as water percolates through the fill. Perry (1989) describes a survey of slope condition. Table 10.1 shows the typical slope angles for those clays most susceptible to failure and the extent of failure. Failures are largely restricted to the outer 1.0 to 1.5 metres of the slope in well-compacted highway embankments and occur deeper within loosely tipped railway and canal embankments. Softening of the clay occurs because of the uptake of water, shrink-swell movements in response to seasonal changes in moisture content and vegetation. Detrimental changes in the values of c' and $\varnothing'$, as the effects of overconsolidation on the clay are removed, lead to movement on a rupture surface. Once movement is initiated, the shear strength along the rupture surface will rapidly decrease towards a lower residual value (see Chapter 4).

Many old railway and canal embankments contain rupture surfaces and zones of failed material. The associated movements may have been exacerbated by the addition of ballast or other material to the top of the embankment. They may be in a metastable condition such that further movements can be initiated by changes in pore water pressure or a small increase in loading. A particularly critical time is when a period of intense rain follows a long dry spell. Monitoring of railway embankments has indicated that there may be a very rapid increase in pore water pressures some time after the wet spell commences. Catastrophic deep slope failures may be initiated by this rapid change. It is apparent therefore that pore water pressure exerts a critical control on slope stability. It can be subject to rapid change, and the effect on stability may occur almost immediately.

For further information on the loss in performance of infrastructure embankments see Vaughan (1994), Perry (1991) and Crabb & Atkinson (1991).

ICOLD (1995) provides a summary of large dam failure statistics that concluded that:

- The percentage of failures of large dams has been falling over the last four decades.
- Most failures have involved newly built dams. The greatest proportion (70%) of failures occurred chiefly in the first ten years and more especially in the first year after commissioning.
- The highest rate of failure was found in dams built in the ten years between 1910 and 1920 i.e. prior to the important developments in soil mechanics.
- The most common causes of failure in earth and rock fill dams in descending order are overtopping, internal erosion in the body of the dam and internal erosion in the foundation.

The most common cause of failure, which can be controlled by design, is internal erosion. In a number of cases, catastrophic failures of earth dams have been caused due to the formation of transverse cracks within the core zone. Vassiades (1984) presents examples of earth fill dams that have been affected by such cracking but noted that none of the examples of zoned earth fill dams presented experienced catastrophic failure. However, they all experienced marked increases in seepage flow representing a failure in design, extensive remedial works were required in order to control the elevated seepage flow to return the dams to serviceability. The examples sited were Balderhead, Hyttejuvet, Matahina, Viddalsvatn and Yards Creek dams.

Factors that were attributed to the formation of transverse cracks through cores included pronounced load transfer from core to shell leading to hydraulic fracture, rapid reservoir filling, variable foundation conditions and seismic activity. Whilst there has always been a requirement to limit the quantity of seepage through a dam, consistent with limiting water losses, it is of paramount importance to prevent associated erosion and migration of material. If seepage through a core is uncontrolled and leads to internal erosion, failure is a likely consequence unless early remedial measures are taken. Measures to control seepage and prevent such an occurrence are primarily related to the design of the downstream filter or transition zone, and upstream zones for embankment dams that operate with a significant drawdown regime. In the case of an embankment with earth fill shoulders; this criterion should extend to the provision of an erosion resistant filter outlet zone. Each zone has to be designed such that an appropriate filter relationship exists to prevent migration of core material particularly in the case where the core develops a crack. In addition, sufficient flow capacity is required to discharge the elevated seepage flow safely and maintain the phreatic surface within the seepage control zones. The development of what are now considered to be standard design practices has only come about by the study and understanding of past failures.

Legislation to regulate the design and construction of dams is being developed worldwide. For example, the

Reservoirs Act 1975 in the UK which imposes both design and construction auditing of embankment dams and probably more importantly, imposes a regular inspection and monitoring regime on the owner. Experienced dam inspection engineers report that one of the most important indicators of the continuing integrity of such structures is the absence of seepage.

It should also be noted that embankment dam failures have occurred for many reasons not associated with dam cores or due to erosion and seepage. Some spectacular failures have been attributed to clay layers within foundation zones, which have highlighted modes of failure that have required detailed investigations and advanced analysis to fully explain. The lessons learnt from such failures are not only technical in nature but can influence the process of design whereby for large projects it is these days considered mandatory for expert panel and peer reviews to be undertaken.

Since substructure fills may carry building developments on shallow foundations, sensitivity to differential settlement can be greater. Also, since these are usually small-scale earthmoving operations, a site quality control system may not be in place and the correct plant and suitable quality materials may not be available. Charles (1993) and Charles et al. (1998) are useful references on settlement causes.

For substructure fills, where vegetation clearance has occurred where clay is present beneath the fill this can swell as the clay becomes wetter and cause heave of thin layers which can distort floor slabs. The opposite is also possible where newly planted trees grow adjacent to structures, reducing the moisture content beneath the fill and causing settlement. In hot climates or in periods of sustained hot weather, the shrinkage of expansive clays (fills and *in situ*) can be a major problem that must be considered in the design process.

Other causes of settlement are loading from buildings, self-weight of fill and dissipation of excessive pore water pressures. Collapse compression is an important cause of settlement and is caused by water inundation of poorly compacted fill. The inundation may be due to rising water level or from the infiltration of surface water (Charles et al. 1993; Burford & Charles 1992). Engineered fills with proper compaction should not suffer from this type of volume loss.

10.4.2. Types and effect of maintenance

The purpose of maintenance is to ensure the design life of the fill is achieved or exceeded. In certain circumstances, such as quarry tips, no maintenance may be required to achieve the life expectancy of the tip; once this life is exceeded, failure may not have any significant consequences. The design may include for maintenance to be cost effective, or, perhaps for safety and access reasons, the design must require very little maintenance.

The maintenance of clay fills can be divided into two main types:

- routine, or current, maintenance.
- capital, or major, maintenance.

Routine maintenance is more frequent than capital and less costly. Works to maintain the integrity of the fill may be to provide additional areas of slope drainage, address faulty drainage, either on or within the fill or at the top of the fill, and small-scale stabilization works such as low height retaining walls at the embankment toe. These works tend to be of a preventative nature to maintain or improve the condition of the clay fill.

Capital maintenance involves large-scale repair works, such as embankment repairs where slope failures have occurred, and usually major drainage works. The works usually concentrate on one embankment, or part thereof, and the costs are usually significant, requiring separate tendering procedures of a shorter period than the more general routine maintenance contract.

If maintenance is not carried out then the clay will lose strength and major problems can develop leading to large costs and consequential damage, which far exceed the total of smaller more frequent payments.

10.4.3. Long-term requirements of clay fills

Efficient and effective management of an infrastructure asset will help to ensure that performance requirements are met. Typical performance requirements of owners and operators of infrastructure embankments include:

- safety and reliability;
- operational efficiency;
- value for money and business improvement;
- minimizing environmental impact;
- satisfying customer and employee expectations and perceptions;
- satisfying statutory and regulatory obligations.

In addition to achieving performance requirements an effective asset management system also has the benefits of:

- converting owner policy and objectives into appropriate actions;
- assisting with the prioritization of expenditure on the basis of need at regional and national level;
- providing comparative analysis between regional and nationwide assets;
- supporting submissions to financial sponsors;
- allowing progress against strategic and financial targets to be monitored and reported;
- providing information regarding the serviceability and rate of improvement or deterioration of the asset;
- assisting with the quantification and mitigation of risks;
- identifying immediate and future investment requirements.

Asset management follows a cycle of inspection, assessment and improvement or repair leading to a continuous awareness of, and improvement in, asset quality (Perry *et al.* 2003). An integral part of this process is the management of asset data and the provision of information links between inspections and planning. Effective asset management will ensure that all aspects of clay fills are allocated the maintenance resources, within budgetary constraints, required for efficient operation. Historically, infrastructure embankments and other earth structures have not been acknowledged as an asset and have consequently received a lower priority for maintenance and renewal funding than other elements of the infrastructure they support. The loss of performance suffered as a result of their lack of recognition as assets has manifested itself in many ways and some of the actual or potential consequences of this are discussed below.

10.4.3.1. Operational safety. Factors such as age, increased traffic loading and inadequate or poor maintenance can ultimately reduce the performance of an embankment and may compromise operational safety. When such a situation arises it is often necessary to impose temporary speed restrictions on railway embankments (Perry *et al.* 1999), close traffic lanes on road embankments and possibly drain sections of canals. In extreme cases where public safety is jeopardised, complete closure of the embankment may be necessary. In such circumstances the cost of delays and disruption can be significant and where safety regulations have been breached the consequences of a prosecution in the courts may be severe.

10.4.3.2. Synergy with other assets. Infrastructure embankments are often used to support other assets (track bed and rolling stock for railways and road pavement, lighting and drains for roads) that require the embankment to act as a firm and stable foundation. Third parties may also utilise embankments for their services (e.g. telecommunications and cable companies). A loss in embankment condition may lead to damage of these assets and increased expenditure on maintenance will result. The long-term costs associated with additional repair and renewal of other assets may exceed by far the cost of carrying out a single stabilisation of the embankment that is causing the problem.

10.4.3.3. Costs of repair. When the condition of an embankment has degraded to such a degree that repair becomes necessary, significant costs will usually be incurred. These costs must be considered in terms of their benefit. The cost of the repair may extend beyond the direct costs of employing designers and contractors and include the provision of access tracks, temporary speed restrictions, line and lane closures and reduced revenue. However, the cost of unplanned repairs can also be extremely high. McGinnity *et al.* (1998) state that typical direct costs for London Underground Limited (LUL) embankment remedial works fell from between £3000 to £5000 per metre to between £1000 to £2000 per metre when carried out as a part of a proactive maintenance and renewal strategy. Furthermore, the need for expensive emergency stabilisation works has now been significantly reduced. Patterson & Perry (1998) report that a proactive maintenance strategy on the M23 motorway in Surrey has resulted in significant cost savings compared with carrying out the works in an unplanned reactionary manner. Planned repair and renewal clearly benefits owners as it permits more efficient use of resources and control of expenditure.

10.4.3.4. Disruption and customer satisfaction. Where loss of condition leads to traffic disruption, speed restriction, delay, closure or deterioration of ride quality, customers will inevitably express dissatisfaction. This is particularly the case where unplanned closures occur and disruption is severe. Whilst it is impossible to eliminate completely the occurrence of unplanned closures, the use of planned maintenance and renewal strategies for embankments significantly contributes to avoiding them. Finally, with the increasing demand to improve public transportation the influence and power of government regulators will place greater pressures, and possibly penalties for failure to comply, on infrastructure owners to ensure the smooth operation of services.

10.4.3.5. Costs of failure. Many of the implications of a loss of condition resulting in embankment failure have already been described above. However, it should be emphasised that consequential costs may be a high multiple of the direct cost of the failure. For example, the cost of repairing a 25 m section of failed embankment may be small compared with the cost of the closure of several kilometres of embankment to traffic, the repair of signals equipment, flooding and third party claims or regulatory fines.

Infrastructure embankments must perform to specific requirements that depend upon their end use. However, irrespective of end use there are some basic generic performance requirements that can be summarized as an ability to:

- maintain their design dimensions, line and level within specified tolerances and those of the assets supported by them;
- support their own weight and current and reasonably foreseeable loads, forces and pressures;
- achieve the foregoing for their specified design lives, subject to routine specified maintenance;
- ensure the safety of those using, maintaining and eventually decommissioning the embankment;
- whenever possible, enhance the local environment and to avoid negative impact on it;
- satisfy the statutory and regulatory obligations of the owner and to ensure the owner's operational requirements are met.

Specific end performance requirements of the major infrastructure embankment owners are summarized below for each infrastructure type.

For railway embankments it is necessary to ensure that ride safety and track quality are maintained for a specified line speed. This is achieved by maintaining 'cant' on curves (the lateral difference in level between two rails); minimizing 'twist' (the rate of change of cant on the two rails) and maintaining 'top and line' (vertical and lateral position). There are limits set that relate to speed and loading (Cope 1993; Railtrack Group Standards (Railtrack 1999*a*, 1999*b*); and LUL Standard E8404 (London Underground Limited 1998)).

Roads also have ride quality requirements (Standard HD 29/94 in the Design Manual for Roads and Bridges (DMRB) Part 7.3.2). These are based on the variance of profile level from a datum derived from a moving average. A high-speed (95 km/hour) road monitor mounted to the rear of a vehicle measures the profile.

Canal embankments. The need to retain water, navigation depth and freeboard are the main end performance criteria for canal embankments. Typically a freeboard of between 300 and 600 mm is required, depending mainly on the navigation size, depth, craft dimensions and the degree of control over water level. Freeboard is defined as the distance between water level and the top of the canal bank, or the bank protection, whichever is the lesser.

The Quarries Regulations (1999) require that the quarry operator ensures that tips are designed, constructed, operated and maintained to ensure that instability or movements do not give rise to a safety hazard. This involves a regime of regular inspections and appraisals and geotechnical assessments must be carried out if there is cause for concern and remedial measures carried out where required.

10.4.4. Remedial and preventative measures

To combat the diverse range of problems leading to the loss of embankment performance, asset and maintenance managers, engineers and contractors require a variety of remedial and preventative techniques that address the causes.

A number of the techniques are available. Historically notable repair methods include: granular replacement of embankment slopes for highways; the use of lightweight ash fills to compensate for track settlement; regrading, and toe berms and grouting of railway embankments; excavation and reconstruction of waterway embankments; and the use of trenched and deep internal wellscreen drains for all embankment types. Hutchinson (1977) gives an overview of these methods in the general context of slope stabilization. However, the improvement in geotechnical knowledge and the introduction of new materials, such as geosynthetics, coupled with the desire to carry out repair and maintenance work in a more proactive, systematic and cost-effective manner has increased the range of techniques available for each transport type.

Remedial techniques are defined as those that are used to repair an embankment where there has already been a loss of performance. Preventative measures are defined as those that are used either to maintain or improve the current level of serviceability of an embankment.

Remedial and preventative techniques can also be described as either hard or soft. The term *hard* is often used to describe structural techniques that are visually obvious. The term *soft* is usually used to describe techniques that are visually less intrusive and improve the strength or other properties of the ground, such as its drainage capability. Table 10.2 summarizes the most common forms of remedial and preventative methods and designates them as either hard or soft. It also indicates their suitability for use in emergency or permanent works, or both. Table 10.3 sets out the principal advantages and disadvantages of the various techniques and suitable conditions for their use.

A combination, rather than a single technique, is normally necessary to stabilize or remediate an embankment and the main considerations in proposing and choosing either an individual or combination of techniques include:

- safety;
- maintaining traffic operation;
- applicability to soil and groundwater conditions;
- access, temporary works and ease of construction;
- cost and long-term maintainability;
- design life and performance requirements;
- environmental, aesthetic and sustainability issues.

It is first necessary to emphasize the importance of safety considerations and planning for infrastructure embankments in particular:

- the hazardous nature of construction and quarrying works in general;
- confined works areas;
- sloping site conditions;
- the proximity of high-speed traffic in the case of road and rail;
- the likelihood of high-voltage cables and other services;
- water in the case of canals.

All works should be planned and carried out in accordance with the Construction (Design and Management) Regulations (1994) or The Quarries Regulations (1999) as appropriate to ensure, where practicable, hazards and risks are dealt with at an early stage. In addition to mandatory compliance with the Health and Safety at Work Act (1974) and its associated regulations, compliance with specific client's requirements and those laid down by Her Majesty's Railway Inspectorate will also be necessary. For guidance on site safety the reader is referred to the CIRIA publications by Bielby (1997) and Sir William Halcrow and Partners (1997) that review general site safety and specific water industry safety hazards and precautions.

10.5. Improved clay

In some new fill applications, the strength of the clay alone is insufficient for use. To increase the strength properties of the clay either mechanical improvement, where the mass strength is increased by the addition of

TABLE 10.2. *Summary of remedial and preventative techniques (Perry et al. 2003)*

Technique	Remedial or Preventative		Hard or Soft		Emergency or Permanent		Comments
	R	P	H	S	E	P	
Granular replacement	•		•			•	Historically popular for repair of shallow slips on highways
Lime and cement treated fill	•		•			•	Relatively new for embankment stabilization
Reinforced soil	•		•			•	Reinforced soil methods are now widely accepted
Regrading	•			•	•	•	
Mass concrete and gabion retaining walls	•		•		•	•	Gabions less costly than cast *in situ* mass walls, suitable for emergency repairs
Bored and minipile retaining walls	•		•			•	Often used near the shoulder of an embankment
Reticulated pile walls	•		•			•	Often used near the shoulder of an embankment or on sidelong ground
Sheetpiles	•		•		•	•	Suitable for either scour or retaining wall purposes
Ground anchors and piles	•		•			•	Used in combination with retaining wall
Anchor ties	•		•		•	•	Requires two parallel walls
Surface drainage		•		•		•	Relevant to highways only
Slope drainage		•		•		•	Important feature of embankments
Internal drainage		•		•		•	Installation can be difficult
Shear dowels	•	•	•			•	Stabilization may take several years
Shear trenches	•	•		•		•	Simple but effective
Soil nails	•	•	•		•	•	Can provide green finish
Lime nails and piles	•	•	•			•	Handling of lime requires caution
Grouting	•		•			•	Formerly popular for railway works
Mix in place methods	•	•	•			•	Works well in correct conditions
Vegetation		•		•		•	Use of green solution becoming increasingly more important
Membranes		•	•	•		•	

reinforcement or improved drainage, or chemical treatment, where the strength is improved by treating the clay with an additive, is undertaken. A particularly useful reference is Bell (1993).

10.5.1. Mechanical improvement

As explained earlier, compaction improves the mass strength of clay fill by densification. It is a standard requirement for embankment and substructure fills, although bunds may require only minimal compaction as they do not carry loads other than their own self-weight. Compaction is a routine necessity in modern fills and has been described above and is mentioned here for completeness only. Similarly, drainage is routine good practice.

The most common form of artificial mechanical stabilisation is the incorporation of geosynthetics to reinforce the clay and increase its mass strength. Geosynthetics are included in the fill between each compacted layer and provide tensile strength (Fig. 10.17). Their most common usage is to provide sufficient support for steepened slopes and for basal reinforcement of the fill when placed on soft ground. British Standard 8002 (British Standards Institution 1994) and British Standard 8006 (British Standards Institution 1995) include information on design.

The most common type of geosynthetic used is a 'geogrid' that has strength in one direction, which is known as 'uniaxial'. The geogrid has therefore to be orientated so that the direction of greatest strength is toward the slope surface. A layer of clay is placed on the geogrid and compacted. Each layer is typically one metre thick before the next geogrid is laid, or the lower geogrid wrapped around the face of the slope and returned on the top of the layer. The length and strength of the geogrid will depend on the height and angle of the fill, the moisture present and the clay shear strength.

TABLE 10.3. *Principal advantages and disadvantages of remedial and preventative techniques (Perry et al. 2003)*

Technique	Advantages	Disadvantages	Suitable conditions
Granular replacement	Well proven technique. Simple to design and supervise.	Cost of importing suitable fill may be high. Disposal of excavated material required. Increased lorry movements. Requires considerable excavation beyond shallow rupture surface	All soil conditions although typically used in high plasticity clays
Lime and cement stabilized fill	Allows reuse of *in situ* materials	Specialist knowledge required. Requires considerable excavation beyond shallow rupture surface	Generally recommended in clays with organic content <2% (some exceptions)
Reinforced soil	Steep slopes can be achieved. No importation required if slope angle maintained	Not always possible to reuse in situ soils for steep slopes. Requires considerable excavation beyond shallow rupture surface	Can be used for the repair of shallow or deep slip failures
Regrading and toe berms	Simple and effective measure and immediate effect	Sometimes difficult to achieve with site boundaries	All soil conditions
Mass concrete and gabion retaining walls	Highly effective stabilization method	Cost of mass walls can be high	All soil conditions
Bored and minipile retaining walls	Generally a cost effective repair method	Access platforms normally required for installation	All soil conditions
Reticulated pile walls	Result in formation of *in situ* mass wall	Access platforms normally required for installation	Requires strong subsoil if used to stabilize sidelong ground
Sheetpiles	Well proven technology	Cannot penetrate stiff or dense ground	Where piles can be driven
Ground anchors and piles	Can provide high lateral support forces	Require specialist installation methods	Soils where high friction forces possible
Tied walls	Anchor method does not depend on ground conditions	Estimation of in service tie force difficult	Use not dependent on ground conditions
Surface drainage	Removes potential erosive action of stormwater	Requires maintenance	Suitable for all soil conditions and road pavements
Slope drainage	Simple to install	Difficult to inspect	All embankments
Internal drainage	Controls piping or ponding	Requires maintenance	Where ponding or risk of piping exists
Shear dowels	Can solve major slope instability problems	High installation cost. Requires some soil deformation to be effective	Applied to deep seated major slope failures
Shear trenches	Simple skills necessary	Concrete costs high	Shallow and deep failures
Soil nails	Relatively cheap to install	Access platforms required for installation of drill and grouted nails	Most effective in granular soils and low plasticity clays
Lime nails and piles	Provide shear increase and improve ground conditions	Long-term performance not proven	Best in overconsolidated clays e.g. London Clay/Gault
Grouting	Wide range of techniques can be used	Long-term effectiveness is not certain	Technique should be appropriate to soil conditions
Mix in place methods	Provides overall increase in shear resistance	Specialist plant and skills necessary	Cement methods for granular soils and lime for clay soils
Vegetation	Aesthetically pleasing and enhancement of soil stability near surface of slope	Not suitable in all environments and can mask underlying problems. Can present a safety and maintenance liability	Success dependent on plant selection, soil type, orientation and maintenance
Membranes	May be used to protect surface from erosion or as a barrier to prevent water flow (e.g. canal lining)	Can mask underlying problems and may have unacceptable visual appearance	All soil conditions

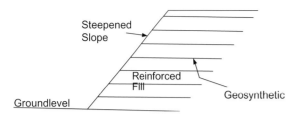

Fig. 10.17. Use of geosynthetic (in this case a geogrid) to reinforce soils for higher performance.

The face of the slope requires consideration for local stability. The wrapped face (where the geogrid is returned back into the fill) is inherently stable but where the geogrid is laid in single unreturned layers, secondary reinforcement of less strength is used between each primary layer of geogrid. Grass is sown on both types of design of slope to establish an erosion resistant surface. The advantage of a non-wrapped face type of reinforced slope is ease of construction, as there is no sheet of geogrid interfering with the deposition and compaction of the clay layer, and the face can be smoothed to a planar finish.

Vegetation plays an important part in stabilizing clay fills (Coppin & Richards 1990). The roots of the vegetation provide a reinforcing action due to their tensile strength and fibrous fabric, and the plant's need for water reduces the moisture content of the clay. Vegetation principally works at the surface of the fill, although larger trees develop roots that can extend to a considerable depth. Grasses are the most useful form of erosion control and their sowing and seed mix are commonly included in earthworks specifications as a geotechnical requirement. The establishment of grass should follow immediately the spreading of topsoil, commonly 100 to 150 mm thick, on the clay fill. However, the environmental requirements can be accommodated within the engineering need. Different types of seed mix, reflecting local species, can be used to ensure an environmentally sympathetic appearance.

However, there are disadvantages to vegetation. The action of changing moisture content of the seasons is exacerbated by vegetation's varying need for water. During the dry months of the year there is less precipitation and the vegetation's demand is highest; during the wetter months the vegetation's demand is easily satisfied. This cycle has a pronounced effect on clays that, as a consequence, shrink as they dry and swell as they become wetter. This movement leads to distortion of the trafficked surface of embankments and the margins of buildings. For this reason planting within two metres of a trafficked or load bearing surface, or further for larger shrubs and trees, is usually prohibited and vegetation management procedures need to be provided to ensure this state is maintained. The positioning of large trees at the top of clay fills is discouraged further because of the loading the tree can exert on the top of any failing clay mass. Trees at the base of a slope, however, improve stability by loading the toe of any failure and improving shear strength.

There are other influences on vegetation such as maintaining a line of sight for traffic and adjacent neighbours to the embankment that also need to be considered.

10.5.2. Chemical treatment

The use of lime for treating cohesive materials is now well established. However, a two-stage process of mixing lime then cement has become more prevalent on recent projects. These processes have the advantage of increasing the range of uses of clay by improving its material properties.

The applications include:

- widened fills, for example for an additional highway lane. Additional land is normally not available and hence the angle of the fill slope may be required to be steep; this slope angle may be too severe for the shear strength of the clay;
- where the applied load exceeds the bearing capacity of the clay. This can occur where clays are used near or on the upper surface of a fill and the surface is exposed to loadings such as from rail or road traffic;
- bulk fills, where lime is used to reduce moisture content and plasticity primarily as a site expedient and is known as improvement;
- capping fill, where lime or lime and Portland cement are used to improve long-term performance and which requires design. This is known as stabilization.

Lime has two functions:

- to improve the clay by moisture content and plasticity reduction;
- to increase bearing ratio by the generation of cementitious products.

The first process is known as 'improvement' and both together are referred to as 'stabilization'. Improvement precedes stabilization once lime is mixed. Not all clays can be stabilized but all can be improved.

Improvement is a reaction used to render unacceptable bulk fills acceptable. Stabilization is used for higher performance uses such as capping and slope repairs. Testing in the laboratory is necessary to see how the clay will react with lime.

Some cohesive materials do not achieve the required long-term bearing ratio when lime is added so Portland cement is added as a second process to achieve the required strength. In this two-stage process the initial addition of lime is primarily to make the cohesive material friable so that the cement can be added to increase bearing ratio.

10.5.2.1. Lime. For engineering purposes, there are two types of lime: quicklime (CaO) and hydrated lime ($Ca(OH)_2$). Compared to hydrated lime, quicklime's advantages are that it has a higher available lime content and a very fast drying action on wet clays. The disadvantage is that it requires considerable amounts of water to hydrate, which although an advantage for wet clays, can be a problem in dry clays. Quicklime is also available in granular form that has the advantage of having fewer dust particles but the disadvantage of a less well-distributed mix. Granulated quicklime is particularly useful on works next to live carriageways or where drifting dust is a sensitive issue. It does not, however, distribute as well as powder, per volume of lime, and capping stabilisation in new works uses a powder. Hydrated lime, or slaked lime, comes in the form of a fine, dry powder. The main advantage is that it requires less water to react and so may be desirable for drier sites. However, this may be counteracted by the cost of transporting water, albeit chemically bonded to the quicklime, with the lime. It may be cheaper to transport quicklime, with its higher available lime content, and add water on site.

10.5.2.2. Lime improvement. The improvement of a cohesive material using lime changes the clay properties in two ways if quicklime is used, or in one way if hydrated lime is used and is generally only applied in rendering unacceptable materials acceptable. Mixing quicklime with wet clay immediately causes the lime to start to hydrate via an exothermic reaction as water is absorbed from the clay into the lime. The heat produced is sufficient to drive off some of the moisture within the clay as vapour and hence further reduce the moisture content. The second effect is a reduction in plasticity as the clay particles flocculate. Figure 10.18 illustrates the effect of lime on plasticity.

A clay at a moisture content of 35% with no lime added will be above its plastic limit of 25%. The addition of 2% of lime will increase the plastic limit to 40% and the clay will be 5% below the plastic limit, dimension 'y' in the figure, rather than 10% above, dimension 'x' in the figure even if there is no change in the moisture content of the clay. The effect of lime on liquid limit is much less marked. This improvement in plasticity also occurs with hydrated lime (slaked lime), but since the lime is already hydrated, no exothermic reaction occurs. These effects are immediate, little affected by temperature providing the material is above freezing and are complete within 24 to 72 hours. They have been used to gain access to and work materials on a number of waterlogged sites (Sherwood 1992).

10.5.2.3. Lime stabilization. Lime improvement of clay soils has an almost immediate effect with significant improvement on mixing and with some remaining improvement occurring up to 72 hours later. In the longer term, the lime reacts with the silica and alumina in clay particles to produce cementitious products that then bind the clay together if sufficient lime above that required for full improvement (termed the fixation level) is present in the mix. This is termed stabilization. The key reference for UK highway work is HA 74/00 (DMRB 4.1.6).

Stabilization using lime depends on temperature and should only be carried out at certain times of the year. In the UK this is usually between March and September and when the shade temperature is above 7°C.

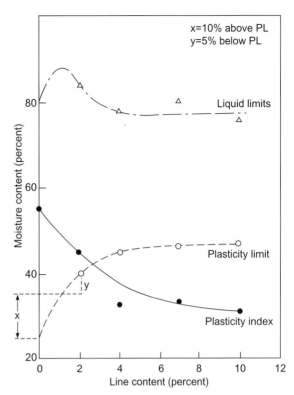

FIG. 10.18. Effect of the addition of lime on the plasticity properties of London Clay (Sherwood 1967).

Sulphates react with lime and, although not affecting the reduction in moisture content and plasticity, cause swelling in the longer term which can be detrimental to the pavement's strength and cause deformations of the road surface. The reactions are fully described in Sherwood (1993). The sulphates may be present either within the clay already, be produced by the oxidation of sulphides, or be introduced by groundwater.

Organic materials may also inhibit the stabilization process occurring unless sufficient lime is added. Lime stabilization relies on an increase in the pH of the clay that may not be possible in organic (acid) clays unless large and, possibly uneconomic, amounts of lime are added. The change in pH depends on the type and amount of organic materials.

Lime stabilization requires the clay and stabilizer to be thoroughly pulverised and mixed on site using mobile stabilizing machines.

A period of between 24 and 72 hours is allowed for the material to mellow although this practice is not common in Europe where it is considered that inter particle bonds are broken when the material is remixed. However, this period allows for considerable slaking, if quicklime is used, provides time for the lime to migrate through the material clods and, as a result of the plasticity changes, makes mixing of the material before final compaction easier. The clay is principally undergoing improvement during this time although some cement bonds are forming. The layer is sealed by one pass of a smooth wheeled roller prior to mellowing. The purpose of the sealing is to reduce carbonation of the lime. Carbonation occurs when the lime reacts with carbon dioxide in the air and reverts to calcium carbonate on long-term exposure. Clearly any carbonation that occurs before the lime has reacted with the clay reduces the amount of lime available. Sealing the layer reduces the air content and prevents significant carbonation. In practice, no problems have been reported with carbonation in the UK although it can be a problem in the tropics. The material is then remixed, adding water to achieve the correct moisture content for adequate final compaction.

10.5.2.4. Lime and cement stabilization. This two stage process is principally used in capping where either lime alone mixed with clay does not achieve the required strength or mixtures of cohesive and granular materials occur and it is not feasible to use lime only or cement-only stabilization. Lime is mixed first with the material to improve it and then Portland cement is added and mixed to provide the strength required.

10.5.2.5. Stabilizing materials for use in highways and substructure capping. Material used in capping must be of sufficient strength and stiffness to provide a working platform for construction of the pavement layers and act as a structural layer in the longer term. The formation should not be operated on unless adequate protection from trafficking is provided in addition to that required for weather. For highways, the minimum strength and stiffness of a capping is a laboratory CBR value in excess of 15%. Sub-grade CBR values of 15% or less will require a design with either a thicker sub-base than for sub-grades with higher CBR values, or the use of capping of differing thickness depending on the value of CBR obtained and the type of pavements being considered. Details are given in UK highways document HD 25 (DMRB 7.2.2). Also from this same source, HA 44 (DMRB 4.1.1) provides guidance on the assessment of CBR values for construction and in the longer term. The requirement to have CBR greater than 15% can be achieved by using clay stabilized with either lime or cement or both. The varying circumstances on different sites will make the use of either granular capping or stabilized capping more appropriate. Availability of granular material and the impact of importing this material is one consideration. On-site materials may not have properties suitable for stabilization, in which cases granular material may be the only option. However, stabilization may produce substantial benefits. These include the maximum use of on-site materials with less haulage off-site and reduced need for spoil tips, the saving of scarce resources and reduced impact of the construction on the surrounding environment.

Stabilized materials will require final compaction according to a method specification. Considerable amounts of water may need to be added during mixing in order to achieve adequate compaction. The water should be introduced in the hood of the rotary mixer. It is essential that the moisture content of lime-stabilized materials is wet of optimum moisture content, although below the high moisture content limit, otherwise inadequate compaction will occur.

The amount of lime added for lime-only stabilization of capping should be subject to a minimum of 2.5% by weight of available lime expressed as a percentage of the dry weight of the material. During the design stage, it should be ascertained if higher percentages of lime are required to achieve an adequate bearing ratio. For lime and cement stabilization a minimum addition of 1% lime and 2% cement is required subject to testing during the design stage.

The addition of lime can render unacceptable materials, on the basis of a Plasticity Index greater than 65% or wet of acceptable limits, to a state of being acceptable as structural fill. Research shows (Perry *et al.* 1996) that cohesive materials are unlikely to alter to a granular state with lime improvement and that the MCV limits for the untreated clay can still be used as they will fall within the limits for modified materials (i.e. be conservative).

A change in compactive effort may be required if the clay becomes very granular in properties. This can be observed and controlled on-site using compliance testing.

10.5.2.6. Treating materials for use in bulk fill. The use of lime for general fill improvement does not lend itself to a prescriptive approach, as it is a method of 'achieving' material requirements. The addition of lime to cohesive materials causes some reduction in moisture content owing to the exothermic hydration reaction. But of more significance is the improvement in engineering characteristics as indicated by changes in compaction curves, MCV calibrations, plastic limit and bearing capacity.

The ground investigation and interpretation will give an indication of the quantities of unacceptable materials, their moisture content and plasticity. The contractor may intend to use the improved material when bidding or use improvement as a site expedient. Past experience, particularly in France, Germany, Italy and America, has highlighted the benefits of the contractor adopting a flexible approach to the use of lime. The contractor will require ready access to lime stores, either on site or by tanker. The unacceptable material is likely to be exposed in a cutting where material handling and trafficking is a problem. The use of lime to remove small amounts of standing water, and reduce moisture content and plasticity, will improve handling and render the material acceptable ready for transportation and placement. The spreading of the lime is by tractor-based spreaders or, for small projects, by hand from bags. The vehicles obviously have to be able to cope with adverse ground conditions. It is possible to spread lime by shovel from bags but for most projects this is inefficient and raises health and safety issues. It is important that the quicklime is fresh, i.e. not partially hydrated or carbonated, and delivered to site in sealed tankers or bags to remove almost any chance of slaking. Mixing the lime does not have to be as thorough as for capping materials and the techniques used for mixing capping materials would be inappropriate for the quantities of material used for general fill. In most cases disc harrows or ploughs or a combination of both are used to mix the lime. Ploughs are effective for turning the clay, especially the larger ploughs mounted on large crawler tractors. The harrow is more appropriate for breaking up clay clods.

The amount of quicklime required to be added is difficult to specify owing to the expedient nature of the process but normally only 1% to 2% by dry weight of available lime is required for all general fill material types requiring rendering. The amount of lime required for general fill is therefore much less than for capping.

Once the material is mixed at the cutting site, it can be excavated, transported and deposited at the fill site. During this period the material will mellow and become more friable making deposition easier. The material will still need to meet the compaction and moisture content requirements of the general fill classes and the testing for this should be undertaken at the place of deposition. It is important to beware of changing an excessively wet fill problem to one where the material is too dry for compaction.

10.5.2.7. Materials characteristics. It is recommended that the MCV is used for both capping and bulk fills to control moisture content, bearing ratio and shear strength of clay fills. This is because the MCV remains constant with time after lime mixing whereas the optimum moisture content changes. As the time after mixing, and before compacting, bulk fills are likely to be greater than for capping, the MCV is even more appropriate. Plastic Limit varies with time and lime addition and so testing for it is inappropriate.

Rigorous sulphate and sulphide testing of the clay to be stabilized is required for capping to assess swelling potential but are considered unnecessary to the same extent for general fill materials. A site trial is advisable in both cases. In HA 74/00 (DMRB 4.1.6) considerable guidance on testing to ensure materials are suitable for lime treatment and limits on swelling criteria are given (ie promoting the soaked laboratory CBR tests, with the maximum mean swell limit set at 5 mm with no individual value greater than 10 mm).

Materials such as pyritic clays and sulphate bearing strata will be particularly susceptible to expansion and a similar approach to that given in the laboratory testing for suitability of capping will be necessary especially within two metres of pavements and structures. Pyritic argillaceous materials, such as colliery shales, will not be suitable for lime improvement. Other forms of rendering such as stockpiling or spreading are better ways of improving these and any other wet materials unsuitable for lime improvement.

The change in compaction curve for clays is to increase the optimum moisture content as Plastic Limit increases and to produce a flatter compaction curve. Hence, a maximum dry density can be achieved at wetter moisture contents.

10.5.2.8. Lime piles and lime-stabilized clay columns. Potentially unstable clay embankment slopes may have their stability improved by installing in them either lime piles or lime-stabilized clay columns. Lime piles are simply boreholes filled with quicklime, and lime-stabilized clay columns are columns of lime-stabilized clay produced by deep mixing methods. Both provide simple and cheap solutions for the problem of maintaining embankment slopes by providing increased shear strength in the clay in which they are installed. A review of lime piles and lime-stabilised clay columns, including design methods, is provided by West & Carder (1997).

10.5.2.9. Safety and the environment. Both lime and cement powders and granulated lime are used in the *in situ* improvement of cohesive and granular soils using purpose built spreaders and rotovators. As strong alkalis, lime and Portland cement require that operators working with them have adequate personal protection (dust masks, gloves, goggles, etc).

The possibility of lime or Portland cement leachate or run off is unlikely as far as is currently known (Reid & Clark 1999), because they are effectively consumed within the improved clay layer to form a cemented material with cementitious gels. If leaching were to occur from the stabilized clay, then the lime or cement leachate would be quickly consumed within the surrounding ground due to its reaction with the clay minerals. Recent studies (Rogers & Glendinning 1993, 1997) of the potential for quicklime to migrate through clay soils have found that migration is less than 50 mm.

Measures are adopted in the process of *in situ* improvement to reduce airborne dust to a level, which will not present a risk either to health or the environment outside the confines of the construction area.

Principally there are two effects, which must be guarded against in order to protect the health and safety of personnel:

- When quicklime comes into contact with water, considerable heat is produced as the quicklime hydrates; this can result in burns.
- In the presence of water, hydrated lime produces a caustic alkali, which can cause chemical burns. Contact of hydrated lime with the skin can be sufficient to result in such burns.

The same effects can happen with Portland cement, although to a lesser degree.

In view of the requirements of the Health and Safety Commission Chemicals (Hazard Information and Packaging) Regulations and the Health and Safety at Work Act, specific advice on handling lime and Portland cement must be given by suppliers in the form of Chemical Safety Data Sheets. These safety data sheets cover the handling of quicklime and hydrated lime as well as cement.

10.6. Case histories

10.6.1. A34 Newbury Bypass

The A34 Newbury Bypass is a useful case history to illustrate the innovative use of clay fill within an environmentally constrained project (Perry et al. 2000). The scheme is described and two of the major aspects relating to clay fills are presented.

The A34 Newbury Bypass formed the last principal connection between the Midlands and North of England and the south coast docks at Southampton, Portsmouth and Poole. Internationally, the section of the A34 at Newbury is part of Euroroute E05 that extends from Greenock on the Clyde to Algeciras in Southern Spain. The need for a bypass was considered a high priority, as the existing route through Newbury was a notorious bottleneck with some 40 000 vehicles passing per day of which 20% were heavy goods vehicles. The route of the bypass is approximately 13.5 km long. The construction works consisted of a dual two-lane carriageway trunk road, together with associated side roads, four grade separated junctions and 29 major structures. Construction began with site clearance early in 1996 and the highway was opened on 17 November 1998.

The project required extensive consultation with all interested parties both during the initial route selection stage and during construction. Comprehensive environmental surveys and assessment work were undertaken in consultation with specialists in a wide range of fields. This approach was the key to the satisfactory construction of the scheme, which was one of the most environmentally sensitive construction projects in the UK.

A comprehensive desk study and preliminary ground investigation for the route were undertaken in 1985. Indications of ground conditions likely to be encountered during construction were established by reference to a variety of sources of published geotechnical information. Additional data were also sought concerning the physical conditions and the historical development of the site. This information was compiled as a geotechnical desk study and was used to determine the scope of the ground investigation required to supplement the data already available. A ground investigation was carried out along the whole route during the spring and summer of 1991. This largely confirmed the predictions made in the geotechnical desk study. The subsequent interpretation of this additional data formed the basis for the detailed design of the geotechnical aspects of the scheme. A supplementary ground investigation was carried out in the autumn of 1993 to cover modifications made to the scheme during the Public Inquiry.

The principal clay bearing strata were London Clay, basal units of the Bagshot Beds and Lambeth Group deposits (formerly Woolwich and Reading Beds). The London Clay outcropped at ground level; the upper two to three metres comprised a weathered firm to stiff orange brown silty clay. At depth, fresh London Clay was typically stiff to very stiff highly fissured occasionally finely laminated dark grey clay. Basal units of the Bagshot Beds formed soft to firm clay horizons up to five metres thick. The Lambeth Group comprised an interbedded sequence of medium dense to dense light brown and grey silty fine and medium sands and stiff to very stiff mottled grey, orange, brown and red silty clays. The recent deposits consisted predominantly of soft dark brown to black organic clay and fibrous clayey peat with occasional sand partings.

Modern vertical highway alignments are kept low relative to ground level in order to minimise the visual and noise impact of the route. However, in hilly areas, this generates considerable amounts of material as the highway is inevitably in cutting over a great length. The original works for the A34 Newbury Bypass showed a large surplus of material, principally clay, which was environmentally and financially unacceptable. Therefore measures were taken to mitigate this position that allowed this surplus to remain on site and to reduce the environmental impact of the road. The provision of additional environmental bunding was conceived to enhance the environmental measures throughout the route, to provide more aesthetically pleasing landscaping, to reduce noise levels and to allow greater use of surplus on-site materials. This was achieved within the highway boundary by the redesign of earthwork slopes and availability of additional land. This single action produced a significant saving in disposal of material. This is an extremely simple method of disposing of clay with minimal effort provided additional land can be made to accommodate the flat slopes of the clay filled bunds. It may in some cases prove advantageous to use stabilized clay to allow construction within the site and to use clay surpluses. Care must also be taken to ensure that the impermeable clay fill placed as landscaping over natural ground does not act as a seal to any water issuing from the ground. (This may only become apparent in winter months with no sign in summer months when earthwork construction is at its peak.) The clay will allow water pressures to develop behind it until the slope of the landscape area fails probably taking existing natural ground with it. Drainage may therefore need to be installed before the clay is placed.

A more technically demanding use of clay fill was in an area known as the Kennet Valley. Parts of the Kennet Valley adjacent to watercourses are designated as Sites of Special Scientific Interest and this was a major influence on the construction methods adopted.

The original embankment design in the Kennet Valley comprised the removal of the peat and alluvial deposits below the embankment footprint down to the top of an extensive and thick gravel layer (Fig. 10.19). The void was then to be backfilled with very coarse granular material, effectively rock fill (ten per cent maximum passing the 125 mm British Standard Series sieve) placed below groundwater and general well graded granular fill placed above groundwater level prior to embankment construction. This is the traditional approach to embankment construction across soft ground areas where the thickness of soft material is typically less than three or four metres thick. However, only a small volume of rock fill was available on site and some of this, the harder chalk,

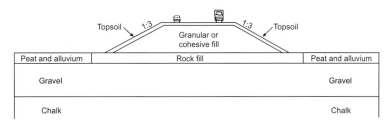

Fig. 10.19. Conventional rock fill replacement below a water table (Perry *et al.* 2000).

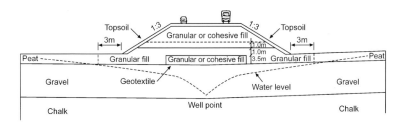

Fig. 10.20. Dewatering method for replacement with surplus materials in areas of no landscaping (Perry *et al.* 2000).

was too alkaline to be placed below water in the acidic meadows. It is important to remember that placing extreme pH materials below water level can lead to leaching and dissolution which can affect surrounding flora and fauna and lead to the damage of important natural environments. Due to the lack of rock fill the bulk of the fill required for construction would require importing material from off site.

To reduce the environmental impact and the cost of importing this material, and to save on off-site disposal costs of other site-won materials, the engineer and the contractor consulted with Environment Agency and English Nature, undertook measurements of water levels and moisture contents of the peat, and devised an alternative embankment that is illustrated in Figure 10.20.

The embankment footprint was divided into fifteen zones, each being around 50 m wide by 50 m long (Fig. 10.21). Each zone was dewatered by means of a 100 mm diameter pump in a 5 m deep well located in the centre of the zone so that the drawdown of the water was less towards the site boundary. It was recognized that the pumping rate required to lower the water table may have caused problems with the wildlife in the immediate vicinity of the site. As a result, piezometers were installed at 50 m centres along both east and west boundaries of the Kennet Valley to record variations in groundwater levels before and during earthworks. The extracted water was discharged through filters and settling basins to the River Kennet for which a discharge licence was obtained. River gauges were installed upstream and downstream of bridge locations to monitor discharge volumes and quality, and peat moisture contents were recorded weekly before and during earthworks.

Where the depth of the peat and alluvium layer was relatively shallow it was possible to reduce the surface of the groundwater to levels below the top of the underlying gravel such that the excavation proceeded in a dry environment. The excavation was then filled and compacted with general granular material. Where the peat and alluvium layer increased in thickness greater than 4 m it was not possible to achieve full drawdown across the zone so undertaking the earthworks in completely dry conditions was not possible. In these locations, the water table was lowered to as low a level as possible and the peat and alluvium were then removed. Initial excavation was in dry conditions but the lower levels were removed under water. The void was backfilled with rock fill until a level above the suppressed water table was achieved from which point general granular fill materials were used. On higher embankments, a central longitudinal zone of general cohesive fill with general granular material on the outer zones, directly under the slopes, was used to replace the peat and alluvium.

Following the replacement of peat and alluvium, after dewatering, a minimum height of one metre of general granular fill material was placed above existing ground level, to act as a capillary break. A further one metre of general granular or general cohesive fill material was then placed before the pumps dewatering the gravel beneath were switched off. The two metres of fill provided a surcharge to load the sands, and clay central zone where placed, as the water level recovered once the pumps were switched off. This was a minimum height and the whole of the fill to formation level could have been placed in one continual operation if the contractor wished to do so.

The environmental planning of, and site clearance for, the A34 Newbury Bypass has received considerable publicity both in the technical and general press. However, geotechnical engineering had a major influence on reducing the environmental impact of the construction and in the provision of earthworks designed to reduce the highway's intrusion into the surrounding countryside.

Fig. 10.21. A34 Newbury Bypass. Construction across soft ground, dewatered to allow use of clay fill. The works involved a series of bays with due consideration of surrounding wetlands.

The employment of a wide range of varying geotechnical engineering solutions was successful in reducing the amount of surplus from the site and the amount of material imported for earthworks. The final volume of surplus material was reduced from about 854 000 m^3 to about 4500 m^3, all of which was contaminated material and had to be removed from site. The final volume of imported materials was also considerably reduced by about two-thirds. Converting these volumes into estimated lorry movements, the effective use of materials has resulted in the elimination of about 250 000 lorry journeys through the local area. The approach adopted for the A34 Newbury Bypass resulted in a figure of only 0.01 m^3 of surplus material per m^2 of road pavement surface that is considered very low given the terrain along the route.

The main methods to reduce the surplus and import included:

- on-site disposal and larger environmental bunds;
- maximizing the use of on-site materials by material handling to ensure proper deployment of low and high performance materials;
- removal of requirement for capping on chalk by using an end-product specification;
- peat replacement alternative which ensured, by dewatering, that surplus materials could be used rather than large imports of rock fill.

Other geotechnical methods to accommodate environmental influences were steep reinforced slopes for environmental bunds and reinforced slopes to reduce land take in areas of rare fauna.

10.6.2. Orville Dam

The United States of America (USA) has a rich history of earth dam construction. The US Army Corps of Engineer's 1981 inventory recorded 61 321 earth fill dams of over 7.6 m in height (Kollgaard *et al.* 1988). One of the listed 'landmark' earth fill dams was the Orville Dam built in 1967 with a height of 234 m. The dam is located in California on the Feather River and was built for the purposes of water conservation, power generation, flood control, recreation and fishing. The embankment is of a zoned earth fill configuration that in total required 61 × 10^6 m^3 of material for construction. A feature of the design was a concrete core block that supported the relatively narrow core in the bottom of the river channel. The core material was derived from alluvial deposits adjacent to areas where the more pervious shell material was 'borrowed'. The material is described as a well-graded mixture of clays, silts, sands and gravels, to 75 mm maximum size. Compaction was carried out in 250 mm layers by four passes of a 100 tonne pneumatic roller. The core was protected by transition zones both upstream and downstream consisting of a well-graded mixture of sands, gravels, cobbles and boulders to 375 mm. Compaction was carried out in 375 mm lifts using two passes of a smooth-drum vibratory roller. Extensive trial embankment works were carried out to determine the practicalities of achieving the required densities and over 150 different combinations of compaction conditions and equipment were tried. Since completion of construction in 1967 there had been three major incidents:

- Cracking of the unreinforced core block parapet occurred during construction in 1965, which was corrected by grouting.
- In 1975, the dam was subjected to a 5.7 magnitude earthquake (Richter Scale) located about 7.5 miles from the dam site. Accelerations of $0.1g$ were measured at the base of the dam. No significant damage occurred and the dam was re-evaluated and found to be capable of withstanding a magnitude 6.5 with peak acceleration of $0.6g$.
- The third major incident was the failure of thirty-nine twin tube hydraulic piezometers installed in the core. These malfunctioned over a period of several days, in December 1978, recording sudden and large increases in pore pressure on the downstream side of the core. Extensive investigations found that the piezometer tubes were broken where they were bundled into vertical risers through the core. No further action was taken except to drain the piezometer lines and maintain a record of flow and silt carried through the broken tubes.

The 1988 reference records that apart from the above incidents the dam had performed 'very well'. Further monitoring data, and time will tell, if this situation is to continue.

References

ANDERSON, M. G. & RICHARDS, K. S. 1987. Slope stability. Wiley, New York.
BELL, F. G. 1993. *Engineering Treatment of Soils.* Spon, London.
BIELBY, S. C. 1997. *Site Safety Handbook,* 2nd edition. CIRIA Special Publication 130. Construction Industry Research and Information Association, London.
BRE 2005. *Concrete in Aggressive Ground.* Special Digest 1. BRE, London.
BRIDLE, R. & SIMS, G. 1999. *The Benefits of Dams to British Society.* Workshop on Benefits and Concerns about Dams, ICOLD, Antalya, Turkey.
BRITISH STANDARDS INSTITUTION 1981. *Code of Practice for Earthworks.* BS 6031: BSI, London.
BRITISH STANDARDS INSTITUTION 1990. *Methods of Test for Soils for Civil Engineering Purposes.* BS 1377: BSI, London.
BRITISH STANDARDS INSTITUTION 1994. *Code of Practice for Earth Retaining Structures.* BS 8002: BSI, London.
BRITISH STANDARDS INSTITUTION 1995. *Code of Practice for Strenghtened/Reinforced Soils and Other Fills.* BS 8006: BSI, London.
BROMHEAD, E. N. 1986. *The Stability of Slopes.* Surrey University Press, London.
BROWN, S. F., LOACH, S. C. & O'REILLY, M. P. O. 1987. *Repeated Loading of Fine Grained Soils.* TRL Contractor Report 72. Transport Research Laboratory, Crowthorne.
BURFORD, D. & CHARLES, J. A. 1992. Long term performance of houses built on opencast ironstone mining backfill at Corby 1975–1990. In: Proceedings of the 4th International Conference on Ground Movements and Structures, Cardiff, July 1991. Pentech, London, 54–67.
CASAGRANDE, A. 1937. *Seepage Through Earth Dams.* New England Water Works Association, 51 No. 2.
CHARLES, J. A. 1993. *Building on Fill: Geotechnical Aspects.* BRE Report BR 230. Building Research Establishment, Garston, Watford.
CHARLES, J. A. & WATTS, K. S. 1996. The assessment of the collapse potential of fills and its significance for building on fill. *Proceedings of the Institution of Civil Engineers, Geotechnical Engineering.* **119.1**.
CHARLES, J. A., BURFORD, D. & HUGHES, D. B. 1993. Settlement of opencast coal mining backfill at Horsley 1973–1992. In: CLARKE, B. G., JONES, C. J. F. P. & MOFFAT, A. I. B. (eds) Engineered fills. *Proceedings of the Conference Engineered fills '93, 15–17 September 1993, Newcastle-upon-Tyne.* Thomas Telford, London, 429–440.
CHARLES, J. A., SKINNER, H. D. & WATTS, K. S. 1998. The specification of fills to support buildings on shallow foundations: the '95% fixation'. *Ground Engineering,* January 1998. Emap, London.
CLARKE, B. G., JONES, C. J. F. P. & MOFFAT, A. I. B. 1993. Engineered fills. *Proceedings of the Conference Engineered Fills '93, 15–17 September 1993, Newcastle-upon-Tyne.* Thomas Telford, London.
COPE, G. H. (ed.) 1993. *British Railway Track: Design, Construction and Maintenance.* The Permanent Way Institution, London.
COPPIN, N. J. & RICHARDS, I. G. 1990. *Use of Vegetation in Civil Engineering.* Construction Industry and Research Information Association. Butterworths, London.
CRABB, G. & ATKINSON. J. 1991. Determination of soil strength parameters for the analysis of highway slope failures. In: CHANDLER, R. J. (ed.) *Proceedings of the International Conference on Slope stability, 'Slope stability engineering: developments and applications'.* Isle of Wight, 15–18 April 1991. Thomas Telford, London, 13–18.
CZEREWKO, M. A., CRIPPS, J. C., DUFFELL, C. G. & REID, J. M. 2002. The distribution and evaluation of sulfur species in geological materials and manmade fills. *First International Conference on Thaumasite in Cementitious Material.* Garston, June, 2002.
DEPARTMENT OF THE ENVIRONMENT/GEOFFREY WALTON PRACTICE 1991. Handbook on the Design of Tips and Related Structures. The Stationery Office, London.
DETR 1999. The Thuamasite Form of Sulfate Attack: Risks, Diagnosis, Remedial Works and Guidance on New Construction. Report of the thaumasite expert group. Department of the Environment Transport and the Regions, London.
DESIGN MANUAL FOR ROADS AND BRIDGES (DMRB). The Stationery Office, London. Vol. 4, Section 1, Part 6; Vol. 7, Section 2, Part 2; Vol. 7, Section 3, Part 2; Vol. 10, Section 5, Part 1.
FELL. R., MACGREGOR. P. & STAPLEDON, D. 1992. *Geotechnical Engineering of Embankment Dams.* Balkema, Rotterdam.
FOOKES, P. G. (ed.) 1997. *Tropical Residual Soils.* Geological Society Professional Handbooks. The Geological Society, London.
FORBES, P. J., SHEERMAN-CHASE, A. & BIRRELL, J. 1980. Control of dispersion in the Mnjoli Dam. *Water Power & Dam Construction,* Dec 1980.
GASCOIGNE, B. 1994. *Encyclopedia of Britain.* MacMillan, London.
GREGORY. J. 1987. The measurement and control of industrial noise. *Quarry Management,* 25–33, January.
HAWKINS, A. B. & PINCHES, G. M. 1987. Causes and significance of heave at Llandough Hospital, Cardiff—a case study of ground floor heave due to gypsum growth. *Quarterly Journal of Engineering Geology,* **20**, 41–58.
HEALTH AND SAFETY COMMISSION (HSC) 1999. Health and Safety at Quarries. Quarries Regulations 1999. Approved Code of Practice. L118. HSE Books.
HEALTH AND SAFETY EXECUTIVE (HSE) 1997. *Health and Safety in Construction.* HSE Books.
HEWLETT, H. W. M., BOORMAN, L. A. & BRAMLEY, M. E. 1987. *Design of Reinforced Grass Waterways.* CIRIA Report 116. Construction Industry Research and Information Association, London.

HILF 1957. *A Rapid Method of Construction Control for Embankments of Cohesive Soils. Conference on Soils for Engineering Purposes*. ASTM Special Technical Publication STP No 232. American Society for Testing and Materials, Philadelphia.

HORNER, P. C. 1988. Earthworks. I.C.E. Works Construction Guide. Second edition. Thomas Telford, London.

HUTCHISON, J. N. 1977. Assessment of the effectiveness of corrective measures in relation to geological conditions and types of slope movement: General Report. *Bulletin of the International Association of Engineering Geology*, **16**, 131–155.

ICOLD 1995. *Dam Failures—Statistical Analysis*. Bulletin 99, Commission Internationale des Grands Barrages, Paris.

ICOLD 1998. *World Register of Dams*. International Commission on Large Dams, Paris.

INSTITUTION OF CIVIL ENGINEERS 1979. *Clay Fills. Proceedings of the Conference Held at the Institution of Civil Engineers, 14–15 November 1978*. Institution of Civil Engineers, London.

INSTITUTION OF CIVIL ENGINEERS 1985. *Failures in Earthworks. Proceedings of the Conference Held at the Institution of Civil Engineers, 6–7 March 1985*. Institution of Civil Engineers, London.

KOLLGAARD, E. B. & CHADWICK, W. L. (ed.) 1988. *Development of Dam Engineering in the United States*. Pergamon, New York.

LEEMING, J. R., LAMONT, D. R. & BRUMBY 2001. Essential engineering criteria for the abandonment of tailings lagoons as wetland features. *Transactions of the Institution of Mining and Metallurgy, Section A*, **109.1**.

LEWIS, W. A. 1954. *Further Studies in the Compaction of Soil and the Performance of Compaction Plant*. Road Research Technical Paper No 33. HMSO, London.

LONDON UNDERGROUND LIMITED 1998. *Track Geometry and Condition Standards*. Engineering Standard E8404 A2 (March 1998). London Underground Limited, London.

MANUAL OF CONTRACT DOCUMENTS FOR HIGHWAY WORKS (MCHW). Volume 1 Specification for Highway Works Series 600: *Earthworks*. Stationery Office, London.

MCFARLAN 1989. *The Guiness Book of Records*. Guiness Publishing.

MCGINNITY, B. T., FITCH, T. R. & RANKIN, W. J. 1998. A systematic and cost-effective approach to inspecting, prioritising and upgrading London Underground's earth structures. *In*: *Proceedings of the Seminar 'Value of Geotechnics in Construction'. The Institution of Civil Engineers, 4 November 1998*, 309–332.

MCNALLY, G. H. 1998. *Soil and Rock Construction Materials*. Spon, London.

MOFFAT, A. I. B. 1999. Puddle Clay and the Pennines Embankment Dam. *Journal of the British Dam Society*, **9**, 1.

MOFFAT, A. I. B. 2002. The characteristics of UK puddle clay cores – a review. *Proc. 12th Conf. British Dam Society*, Dublin Thomas Telford, London, 581–601.

NIXON, P. 1978. Floor heave in buildings due to the use of pyritic shales as fill materials. *Chemistry and Industry*, **4**, 160–164.

OLIPHANT, J. & WINTER, M. G. 1997. Limits of use of the moisture condition apparatus. *Proceedings of the Institution of Civil Engineers*, **123**, 17–29.

PARSONS, A. W. 1992. Compaction of soils and granular materials: a review of research performed at the Transport Research Laboratory. HMSO, London.

PARSONS, A. W. & TOOMBES, A. F. 1988. Pilot-scale studies of the trafficability of soil by earthmoving vehicles. Department of Transport, TRRL Research Report 130. Transport Research Laboratory, Crowthorne.

PATTERSON, D. & PERRY, J. 1998. Geotechnical data and asset management systems for highways. *In*: *Proceedings of the Maintenance Engineers Conference*. Nottingham, 9–11 September 1998. Surveyor and Municipal Journal, London.

PERRY, J. 1989. *A Survey of Slope Condition on Motorway Earthworks in England and Wales*. TRRL Research Report 199. Transport Research Laboratory, Crowthorne.

PERRY, J. 1991. *The Extent and an Analysis of Shallow Failures on the Slopes of Highway Earthworks*. PhD Thesis, University of Durham.

PERRY, J., MACNEIL, D. J. & WILSON, P. E. 1996. *The Uses of Lime in Ground Engineering: A Review of Recent Work Undertaken at the Transport Research Laboratory. In*: 'Lime stabilisation' Proceedings of the seminar held at Loughborough University, 25 September 1996. Thomas Telford, London, 27–45.

PERRY, J., MCGINNITY, B. & RUSSELL, D. 1999. Railway earth structures: their condition, management and influence on track quality. *In*: FORDE, M. C. (ed.) *Proceedings of the 2nd International Conference on Maintenance and Renewal of Permanent Way and Structures*, 'Railway Engineering '99', London, 25–26 May 1999. Keynote Paper. Engineering Technics Press, Edinburgh.

PERRY, J., FIELD, M., DAVIDSON, W. & THOMPSON, D. 2000. The benefits from geotechnics in construction of the A34 Newbury Bypass. *In: Proceedings of the Institution of Civil Engineers Geotechnical Engineering*, **143**, 83–92.

PERRY, J., PEDLEY, M. & REID, J. M. 2003. *Infrastructure Embankments: Condition Appraisal and Remedial Treatment*. CIRIA Report C592. Construction Industry Research and Information Association, London.

RAILTRACK 1999a. *Track Construction Standards*. Railtrack Line Specification RT/CE/S/102. Railtrack plc, London.

RAILTRACK 1999b. *Track Maintenance Requirements*. Railtrack Line Specification RT/CE/S/104. Railtrack plc, London.

REID, J. M. & CLARK, G. T. 1999. Remedial treatment of contaminated materials with inorganic cementitious agents: case studies. *In: Geoenvironmental Engineering*. Thomas Telford, London.

REID, J. M., CZEREWKO, M. A. & CRIPPS, J. C. 2001. *Sulfate Specification for Structural Backfills*. Transport Research Laboratory Report TRL 447. Transport Research Laboratory, Crowthorne.

ROGERS, C. D. F. & GLENDINNING, S. 1993. Stabilisation of embankment clay fills using lime piles. *In: Proceedings of the International Conference on Engineering Fills*. Thomas Telford, London.

ROGERS, C. D. F. & GLENDINNING, S. 1997. Slope stabilisation using lime piles. *In: Proceedings of the Conference on Ground Improvement Systems: Densification and Reinforcement*. Thomas Telford, London.

SHERARD, J. L., DECKER, R. S. & RYKER, N. L. 1972. *Piping in Earth Dams of Dispersive Clay*. Proceedings, Speciality Conference on Performance of Earth and Earth Supported Structures. American Society of Civil Engineers, **1**, 1.

SHERWOOD, P. T. 1967. Views of the Road Research Laboratory on Soil Stabilisation in the United Kingdom. *Cement, Lime and Gravel*, **42** (a), 277–280.

SHERWOOD, P. T. 1971. *The Reproducibility of the Results of Soil Classification and Compaction Tests*. Ministry of Transport RRL Report LR 339. Transport Research Laboratory, Crowthorne.

SHERWOOD, P. T. 1992. Stabilized capping layers using either lime, or cement, or lime and cement. TRL Contractor Report 151. Transport Research Laboratory. Crowthorne.

SHERWOOD, P. T. 1993. Soil stabilisation with cement and lime. Transport Research Laboratory. A State of the Art Review. HMSO, London.

SIR WILLIAM HALCROW & PARTNERS 1997. *Site Safety for the Water Industry*. CIRIA Special Publication 137. Construction Industry Research and Information Association, London.

SKEMPTON, A. W. 1996. Embankments and cuttings on the early railways. *Construction History*, **11**, 33–39.

SKEMPTON, A. W. 1989. Historical development of British embankment dams to 1960. Keynote Address Proc. Conf. On 'Clay barriers for embankment dams', London. Thomas Telford, London, 15–52.

SPALDING, R. 1970. *Selection of Materials for Sub-Surface Drains*. TRRL Laboratory Report LR346. Transport Road Research Laboratory, Crowthorne.

SYMONS, I. F. 1970. *The Magnitude and Cost of Minor Instability in the Side Slopes of Earthworks on Major Roads*. RRL Laboratory Report LR331. Road Research Laboratory, Crowthorne.

TERZAGHI, K. & PECK, R. B. 1948. *Soil Mechanics in Engineering Practice*. Wiley, London.

TRANSPORTATION RESEARCH BOARD 1990. Guide to earthwork construction. State of the Art Report 8. Transportation Research Board, 1990.

TRENTER, N. A. 2001. *Earthworks: A Guide*. Thomas Telford, London.

TRENTER, N. A. & CHARLES, J. A. 1996. *A Model Specification for Engineered Fills for Building Purposes*. Proc. Institution of Civil Engineers. *Geotechnical Engineering*, 119 Oct, 219–230.

US BUREAU OF RECLAMATION (USBR) 1974. *Earth Manual*. US Bureau of Reclamation, Department of the Interior, Washington D C.

US CORPS OF ENGINEERS 1953. *Filter Experiments and Design Criteria*. Technical Memorandum 3–360. Waterways Experiment Station, Vicksburg Mass.

VASSIADES, C. S. 1984. *Leakage Through the Damaged Cores of Zoned Embankment Dams*. Paper presented to BNCOLD, London, Nov 1984.

VAUGHAN, P. 1994. Assumption, prediction and reality in geotechnical engineering. *Geotechnique*, **44**, 573–609.

WEST, G. & CARDER, D. R. 1997. *Review of Lime Piles and Lime-Stabilised Soil Columns*. TRL Report 305. Transport Research Laboratory, Crowthorne.

WISEMAN, W. 1888. Railway embankments. *The Railway Engineer*, **9**, 343–350.

11. Earthmoving

This chapter follows on from the design, planning, performance and use of clay in the construction of fills and in the quarrying of clay as a mineral for use in construction presented in Chapter 10 by considering the practical aspects involved in the excavation, loading, transportation, modification, placement and compaction of clays in construction and mineral extraction projects.

The various operations involved in the excavation, loading, transportation, modification, placement and compaction of materials in or on the ground are termed 'earthmoving'. The structures that are formed as the result of these operations are known as 'earthworks'. Some earthworks are constructed to carry out a particular function, such as cuttings and embankments for roads or railways, whereas others, such as quarries and borrow pits, are formed as a consequence of the removal of materials for other uses. Some quarries may be re-used as landfill sites for the disposal of waste materials. In recent times, landfill sites have become engineered structures in their own right and quarries in clay are especially valued owing to the low permeability of the underlying clays, which serve to limit the movement of leachate or liquid wastes away from the site.

In earthmoving practice, clays may be considered as those materials containing predominantly clay- and silt-size materials (i.e. fine-grained soils); however they also include materials that contain sufficient fine material that cause them to behave as fine-grained soils and weak rocks, including clayey sands, glacial till and some highly to completely weathered rocks. They may vary considerably in strength from slurries to weak rocks, such as mudstone and weathered slate. In this chapter, all these materials are referred to as 'clays'. In addition, many projects involving earthmoving in clays will also involve earthmoving in other materials that lie close to or that must be excavated in order to gain access to clays. Few items of earthmoving plant have been specifically developed for use in clays and most may be used in a variety of materials and circumstances.

Humans have been earthmoving to form structures and to extract minerals for well over 6000 years. Over most of this time, excavation was carried out manually using simple tools, such as picks and shovels. Excavated materials were transported manually in baskets, barrows and similar containers or using pack animals, carts, rafts or wooden ships. Apart from effects of the self-weight of the placed material, compaction was given by the feet of humans or animals or by dropping weights on to the surface of the material. These simple techniques are still used for local projects in developing countries. Indeed one of the greatest feats of manual excavation in the modern era is claimed to be the construction of the North Kiangsu Canal in China, in which 69.9×10^6 m^3 was excavated in 80 days without the use of mechanical plant (Stephens 1976). The development of steam power in the 18th century enabled the processes of earthmoving to be mechanized. Steam powered railways were developed in the coalfields of northern England in the 1820s and the first steam shovels were produced in the 1830s. Both were being extensively used in construction and mineral extraction projects by the 1880s. Mechanization increased further with the invention and development of the internal combustion engine, which resulted in the growth of a wide variety of different types of earthmoving plant for different conditions and applications and in the increase in the size, power and output of the plant. In turn, this has enabled larger and more complex earthworks to be formed both in construction and mineral extraction projects. Examples of different types of earthworks have been given in Chapter 10.

In this chapter, different operations that may be carried out during earthmoving in quarries and construction sites are described, together with the main types of the plant and equipment that are used. Many of the types of plant used in the minerals extraction and construction industries are similar, although they may vary in size. Underground earthmoving operations that are carried out in mining and tunnelling are only briefly referred to, as these constitute subjects in their own right. This is followed by consideration of the factors controlling the use of plant and equipment in earthmoving projects and examples of earthmoving schemes that have involved clays on construction sites and in mineral extraction. The chapter concludes with comments on some of the more common safety issues on earthmoving sites.

11.1. Earthmoving practice

Earthmoving in construction and mineral exploitation frequently involves the excavation, loading, transportation, and deposition of materials, including clays and deposits that are associated with them. When forming embankments in construction projects (e.g. for roads, dams, railways etc.), tips in quarries, and backfilling or restoring quarry workings, it is usually necessary to spread and compact the clay (and associated materials) to provide adequate stability of the earthworks or to optimise the amount of material which has to be disposed.

In developed countries, most earthmoving is carried out using mechanical plant, except where small quantities or small areas are involved. However, manual labour is still extensively used in developing countries where labour is readily available and inexpensive and mechanical plant or fuel may be in short supply and/or expensive. Clays often occur with other soils and/or rock and constitute either the materials that are to be used or those to be discarded. Earthmoving plant that is used for the clay deposits may or may not be used for the other deposits, depending on the suitability of the plant, its availability and the costs of earthmoving with that plant. In the following sections, the various means of earthmoving in clays and their associated deposits are described and this is followed by consideration of the main factors that may influence the operation and cost of earthmoving plant. Examples are given of the main types of plant and their roles in earthmoving on construction sites and in mineral workings and how different types of plant are typically used together. Photographs of some of the main types of earthmoving and compaction plant are given in Figures 11.1 and 11.2. A comprehensive treatise on the main operations and the types of plant involved in earthmoving may be found in Nichols & Day (1998). Other useful information is produced by plant manufacturers, such as in the annual Caterpillar Performance Handbook.

11.1.1. Preparation works for earthworks in construction and for mineral deposits

These may include:

- obtaining access to the site and constructing haulage and access roads/routes within the site;

(a)

(b)

(c)

(d)

(e)

(f)

FIG. 11.1. Common earthmoving plant (continued on p. 317).

(g)  (h)

(i)  (j)

FIG. 11.1. Common earthmoving plant. (a) Bulldozer with single shank ripper; (b) bulldozer; (c) motor grader; (d) face shovel loading a rigid dump truck; (e) tracked back-acter loading an articulated dump truck; (f) backhoe-loader; (g) long reach back-acter; (h) dragline loading to a dump truck; (i) rigid dump truck being loaded by a wheeled (forward) loader; (j) twin engine wheeled motor scraper. Photographs reproduced with permission: Photographs a, b, c, d and j from Caterpillar Inc.; photographs e, g and i from the Volvo Construction Europe Ltd.; photograph f from the Case Corporation; and photograph h from Philip C. Horner.

- diverting existing roads, footpaths, watercourses and public utility services;
- constructing permanent and/or temporary fences or barriers around the site or sections of the site;
- removing trees and other vegetation and their disposal;
- demolishing existing structures;
- processing demolition rubble and disposing of unusable materials;
- stripping and stockpiling topsoil;
- constructing buildings such as offices, stores, workshops, processing plant and weighbridges required to support the earthworks and other operations.

11.1.2. Bulk earthmoving for construction

This covers a wide variety of operations including the excavation of materials to reduce ground levels, to form cuttings and to obtain earthworks materials, the excavation of borrow pits to obtain earthworks and other construction materials, the excavation of trenches and pits for public utility services, foundations etc. and the re-excavation of materials from stockpiles. The excavated materials that can be reused are transported to and deposited on embankments and other areas of filling, or in stockpiles for future use, and surplus or otherwise unusable materials are transported to and deposited in disposal areas (dumps or tips). The fill in structural embankments is spread and systematically compacted in layers in order to obtain optimum strength and stability and to minimize long-term settlements, although it may not be necessary or possible systematically to compact unusable, poor quality or surplus fills to the same extent in disposal or landscaping areas.

Once the bulk earthworks have been completed, they are trimmed, topsoil is placed, vegetation planted where required and temporary structures, fences, diversions, access roads etc. are removed.

11.1.3. Bulk earthmoving in quarries and other mineral workings

This commences with the excavation of the overburden and other unusable materials to expose the mineral deposit and the transportation of these materials to and their deposition on amenity embankments, stockpiles or tips. In former times, tips were usually built outwards in one or more thick layers by tipping over the end face (end tipping) and little compaction was given except from the

Fig. 11.2. Common compaction plant. (a) Towed smooth-wheeled roller; (b) pneumatic-tyred roller; (c) towed grid roller; (d) self-propelled tamping roller; (e) single drum pedestrian vibrating roller; (f) single drum self-propelled vibrating roller; (g) double drum self-propelled vibrating roller; (h) small double drum vibrating tamping roller; (i) single drum vibrating tamping roller; (j) vibrating plate compactor; (k) vibro-tamper. Photographs reproduced with permission: photographs a–c and i from Philip C. Horner; photograph d from Caterpillar Inc.; and photographs e–h, j & k from Bomag (Great Britain) Ltd).

plant running on the tip and the self-weight of the tipped materials. However, following the disastrous failure of a colliery spoil tip at Aberfan, South Wales in 1966 and other large failures elsewhere (Bishop et al. 1969; Bishop 1973), it has become increasingly realized that waste tips should be treated as engineered earthworks. In the UK, legislation to regulate tips was introduced in the Mines and Quarries (Tips) Act 1969 and Regulations 1971, and more recently in the Quarry Regulations 1999. It has now become more common to spread materials in thin layers and to provide at least nominal systematic compaction to obtain optimum strength and stability and sometimes to increase the density of the fill to allow more material to be accommodated within the available volume of tips. The construction of amenity embankments to reduce the impact of the workings on the local environment may also be considered as preparation works and may be followed by placing topsoil and the planting of vegetation on these embankments.

Once the overburden has been removed, excavation of the mineral deposit can be carried out, together with its transportation to and deposition at the processing plant or on to stockpiles close to the plant. Solid wastes from processing are deposited on tips unless they can be sold, in which case they would be stockpiled or loaded and transported off site. Liquid wastes should be discharged

FIG. 11.2. Continued.

into lagoons to allow the solids to settle out and to enable further treatment to be carried out if required; the purified water will often be reused within the works.

Most mineral workings must be restored in a manner agreed with the local planning authority. The land may be restored to agricultural or sometimes commercial uses or as wet and/or dry amenity areas. This will commonly involve regrading part or all of the excavations, amenity embankments and tips (which is often equivalent to bulk earthworks in construction), demolishing or removing temporary structures, fences, road and stream diversions, access roads etc., placing topsoil and planting vegetation. Some mineral workings, particularly those in clay, may be converted for use as landfill sites, in which case the operations required to restore the site may be relaxed, removed or delayed. The works required to restore the mineral workings will often be co-ordinated with the preparation of the site for waste disposal. These will depend to a large extent on the planning requirements, the nature of the waste and the design of the landfill and are not considered further.

These stages are often carried out sequentially in construction works. By contrast, mineral workings are often developed in stages over a longer time scale and different stages may need to be carried out at the same time in different parts of a quarry. For example, new sections of mineral workings may be started at the same time as disused workings are being restored.

11.2. Methods of earthmoving

The methods of earthmoving can be categorised as using the following:

- *manual work* in excavation, loading, transportation, deposition and compaction;
- *mechanical plant that can only excavate materials*, such as rippers, scarifiers, rooters, graders, and monitors;
- *plant that can excavate and load materials*, such as draglines, grabs, face shovels, back-acters, forward loaders, and bucket wheel excavators;
- *plant that hauls and deposits materials*, such as dumpers, dump trucks, lorries and conveyors;
- *plant that excavates, loads, hauls and deposits materials*, such as motor and tractor-drawn scrapers, dozers and dredgers;
- *compaction plant* such as deadweight and impact rollers, vibrators and rammers;
- *special plant*, to excavate tunnels and mines, construct piles etc. and stabilize soils.

11.2.1. Manual work

In developing countries, manual work may be used for small to large-scale local projects where labour is inexpensive and readily available. This may involve the use of simple tools such as picks and spades for excavation and levelling, baskets, barrows or animal-drawn carts for transportation, and the feet of humans or animals or simple dropping weights for compaction. These operations may be assisted by mechanical plant where it is available. Manual work is also carried out in small areas in developed countries, particularly where machines cannot be used, although some small mechanical tools and equipment are often used, including clay spades, power breakers and manually steered compactors.

11.2.2. Plant that can only excavate

Rippers are attached to the rear of bulldozers, and occasionally to other plant, to loosen hard or dense soils and weak rock so that they can be more easily excavated. They would not normally be used in the excavation of clays, except possibly in advance of scrapers in hard clays and weak mudstones, but may be used on associated deposits of dense sands and gravels and weak rocks such as weathered granite, cemented sand and sandstone, which also has to be removed. Larger rippers consist of one to three steel shanks with a tooth at one end that are set at the rear of a heavy tractor unit on a mounting that can be lifted and lowered by hydraulic rams (Fig. 11.1a). The ripper teeth are lowered into and dragged through the ground by the tractor, breaking up the material and therefore the rear of the tractor needs to be robust to take the large induced stresses. Straight or curved shanks are available and the teeth can be replaced when worn.

Scarifiers and rooters are similar in concept to rippers, although they are less robust and may be pulled by or fitted to lighter plant than dozers. They are used mainly for ripping paved surfaces and removing tree roots respectively.

Graders consist of a wide controllable blade that is mounted below and in the centre of a motorized long wheelbase rubber-tyred chassis. The blade can be set at an angle to the direction of movement and to the vertical, elevated and lowered and extended beyond the sides of the machine. It is fitted with a replaceable wearing edge. Graders vary from 26 kW to over 450 kW (35 hp to over 600 hp) and operate at speeds up to 50 km/hr. The front wheels have the ability to lean to provide resistance against sliding sideways on slopes (Fig. 11.1c). Graders are primarily used to trim and spread soil in levelling, finishing and trimming operations and two of the more common specific uses are in trimming and regrading haul roads and shaping side slopes.

Monitors are special water cannon used in hydraulic mining of clays and placer deposits and are extensively used for excavating kaolinized granite to obtain china clay in southwest England. Excavation is primarily carried out by the erosive action of water under hydrostatic or pumped pressure of up to about 2100 kN/m^2 (300 psi) being directed at a bank or face of material through a swivel mounted nozzle or 'monitor'. The monitors can be directed to cover the face manually or automatically from distance of between about 6 and 60 m, depending on the size of the monitor and the face. Operators can control a number of monitors from a central cabin. Nozzle sizes between 40 and 150 mm are used and water flows of between about 500 and 15 000 litres/min can be achieved. Drilling and blasting and/or ripping may be required to loosen harder material such as cemented, less weathered or less altered strata and veins and dozers may be used to push the material towards the monitors. In the china clay workings in southwest England, the jet of water from the monitor washes out clay, silt and sand size materials, which flow in suspension under gravity to a low point. The suspension is then pumped to the processing area, where the clay minerals are separated out and treated. The coarse sand and gravel debris remaining after the granite has been 'washed' is excavated by back-acters or face shovels, loaded in to dump trucks and taken to tip.

Clean waste water from the processing area is usually recycled back to the monitors and the sand waste is taken to tips by conveyors or dump trucks (Thurlow 1997; Anon 2000).

11.2.3. Plant that can excavate and load materials

Face, front or *loading shovel.* The classic face shovel consists of an inclined boom on which is pivoted the 'dipper' or 'bucket' arm which is attached to a bucket. The boom and dipper arm are usually set on a revolving superstructure unit which houses the operator, controls, motors, cable fixings, winches and counterweight. The whole unit is set on a crawler tractor unit and can rotate through 360°. By control of the boom, the dipper arm and bucket with cables, the bucket is pushed in a vertical arc upwards into the material to be excavated. When the bucket is full the boom is retracted and rotated to over the unloading point adjacent to the machine, the bucket lowered and emptied by opening a rear 'door' or by moving the front part of the bucket away from the rear (Fig. 11.1d).

Medium to very large cable operated shovels are still used in quarrying and are still cable-controlled with the same basic design, but whilst they were originally steam powered they now mainly use electric motors or diesel engines. Cable controlled shovels with buckets up to about 107 m^3 have been manufactured for use in open cast coal mining. Small to medium size shovels are used in bulk earthmoving in construction and quarrying and have largely been replaced by their hydraulic equivalents.

The smaller hydraulic machines consist of an articulated boom at the end of which is a bucket. With most modern machines, the boom, bucket arm or stick and bucket are operated by hydraulic rams and the unit is set on a rubber-tyred or a crawler tractor unit, which can rotate through 360°. Machines of up to about 280 kW (375 hp) are in common use and most are diesel powered. Hydraulic shovels operate in a similar way to the cable operated machines except that, on most machines, the bucket is rotated downwards to empty it. Buckets up to about 4 m^3 are in general use and up to 35 m^3 on the largest hydraulic machines. Buckets for all types of face shovels often have replaceable teeth on the leading edge. The main uses of face shovels are to excavate near-vertical or steep faces and stockpiles in a wide variety of materials including clays. They excavate upwards from near the toe of the face/stockpile and load to trucks, railcars, conveyors or stockpile alongside. Consequently, the strata at the level of the base of the working face needs to be above water and of sufficient bearing capacity for the shovel and the haulage plant.

Back-acters or *back-hoes.* Originally the back-acter was an alternative digging unit to the classic steam-powered face shovel and was used to excavate below the level of the tractor unit. Now they are far more common than face shovels. The basic machine typically consists of an articulated boom and bucket arm at the end of which is a bucket that is operated by hydraulic rams. The unit is set upon a rubber-tyred or crawler tractor unit. Smaller back-acters may be set on rigid units that can only operate over an arc of about 180°, whereas others, particularly the larger machines, are able to rotate fully on the tractor unit. Machines of up to about 270 kW (360 hp) are in common use and are usually diesel powered (Fig. 11.1e)

Back-acters operate by digging in a vertical arc from above the surface on which the tractor unit stands to a position beneath the outer edge of the tractor unit. The boom is then raised and rotated to the side or front of the machine and the bucket emptied by rotating it over the unloading point. The maximum depth and reach depends on the size of the machine and its use (which is reflected in the configuration of boom, stick and bucket) and is commonly between about 2.5 m and 8 m, although 'long-reach' machines are available with reaches up to about 18 m. The buckets are related to the size of the machine and may be obtained in different styles from 0.3 m to 2.4 m wide with heaped capacities up to about 4 m^3, although very large excavators have capacities up to 33 m^3. Most buckets for excavating soils and weak rocks have replaceable teeth on the leading edge but smooth buckets may be used for grading and ditching. Some of the larger back-acters can be converted to face shovels. Smaller back-acter units can be attached to other plant and the combination with a forward loader has produced the versatile 'back-hoe loader' (Fig. 11.1f), which is much used in the construction industry.

Back-acters are used for a wide variety of excavation work in construction including general excavation work, excavating faces and trenching. They are also used widely used for excavating overburden or mineral deposits in quarries, particularly from above. They can be used for a wide variety of materials including clays and can load to trucks, railcars, conveyors or stockpile. When excavating faces, trucks etc. may be loaded at the base of the face below, or loaded at the same level as the back-acter. As the machine operates best from the top of the working face, the strata at this level need to be above water and of sufficient bearing capacity for the back-acter. The haulage plant may be loaded below or at the same level as the back-acter, depending on the particular layout and the bearing capacity of the ground forming the access. Long-reach back-acters can excavate materials some distance from the machine and are useful in excavating in wet or soft soils from a firm platform (Fig. 11.1g).

Draglines are operated from a crane or similar plant with a long boom. A bucket is suspended near its base by a (hoist) cable and chains from the boom. A second (drag) cable runs from the top of the bucket via a pulley on the first cable to the base of the machine. A set of chains stretch from near the end of the drag cable to the bucket. If there is no tension on the drag cable the bucket will hang in the emptying position but, on tensioning the drag cable, the bucket will first rotate and then move towards the machine. The crane is set on crawler tracks or a 'walking' mechanism and the superstructure is able to rotate as with a face shovel. Draglines are especially suited to the excavation of material at a distance from the machine below to slightly above the level of the tracks. Excavation is carried out by dropping the bucket with the boom and hoist cable onto the material to be excavated and then pulling the bucket towards the machine with the drag cable. Once full, the bucket is lifted up and the boom rotated to the unloading point. When the tension on the drag cable is released, the bucket turns downwards and the material falls out. The excavated material can be discharged directly into vehicles, into conveyor hoppers or onto the ground (Fig. 11.1h). Excavation can be carried out underwater and in areas where other plant may not be able to reach. Buckets can be perforated to let water out or enable soft soils to be released more easily and bottom-dump buckets can also be used.

A wide variety of materials from soft clay to loose or blasted rock can be excavated, although different types of bucket would be used. On construction sites, draglines are often used for excavating 'unacceptable' materials, such as soft clays, or below the water table, and for removing overburden or minerals in quarries. In construction and in many quarries, draglines are operated from cranes with buckets up to a few cubic metres capacity. However, one of the world's largest dragline was used in excavating overburden in opencast coal workings and had a 94 m long boom and bucket capacity of 168 m^3.

A grab is a cable or hydraulically controlled bottom-opening bucket that is suspended from a crane or lifting arm. The bucket is opened and dropped onto the material to be excavated. It is then closed and the grab with the material caught inside is lifted up, rotated to over the unloading point, the bucket opened and the material discharged into a waiting vehicle or to stockpile. Grabs are typically used for the excavation of pits and trenches (including diaphragm walls) and loading to/from stockpiles. They can be fitted to many lifting devices and may be fitted to trucks. They can be used to excavate a variety of materials from soft clays to weak rocks.

A bucket wheel excavator consists of a series of buckets set on a wheel or closed loop, which is itself set on a boom that can move laterally and vertically. Material is continuously excavated during the upward movement of each bucket and is discharged onto a conveyor once the bucket has passed beyond the top of the wheel or loop. The material is then taken on a second conveyor to a discharge point, which may be a truck, rail car, conveyor system or stockpile. The boom is usually able to swing in an arc and the whole machine is set on crawler tracks for lateral movement, although in some machines the boom lies parallel to the slope and the buckets 'skim' the surface of the slope. Bucket wheels vary between less than 1.5 m to about 21 m diameter and the largest machines may excavate up to 16 000 m^3/hour. Bucket wheel excavators are not extensively used in construction operations, except in trenching work. However, they are particularly suited to work in thick consistent deposits above the water table and are sometimes used in quarries for the excavation of mineral deposits, particularly clays, or the removal of overburden soils.

11.2.4. Plant that hauls and deposits

Road lorries/trucks. Road lorries are used to transport materials on public roads. These are normally available in the UK in various sizes up to 40 tonnes gross vehicle weight and have steel or aluminium sheeted bodies. Most lorries need to be loaded by other plant, although some are fitted with a grab to enable unassisted loading. Unloading is usually carried out by elevating the body and tipping to the rear or, less commonly, the side. In order to operate efficiently off road, they require haul roads or ground with a California Bearing Ratio (CBR) of about 5% or an undrained shear strength of about 100 kN/m^2 and without large ruts (further information on the CBR and the undrained shear strength is given in Chapters 4 and 9).

Untaxed road lorries are road lorries that cannot legally be used on public roads. They may be older vehicles that have been purchased or hired in for the scheme and may be in a relatively poor state of repair. If this is the case, they may represent a safety hazard to the driver and others using the site roads and, if so, ideally they should not be used, however this may not be a realistic option in some developing countries.

Dump trucks and dumpers are purpose-made off-road haulage vehicles of which most vary from about 1 tonne to 80 tonnes capacity, although trucks of over 300 tonnes are used in large mines and quarries. They may be powered by diesel, petrol or, in large vehicles, by electric drive powered by diesel or turbine engines. Economic hauls typically vary between 150 m and 10 km each way for rigid trucks (Figs. 11.1d and i). Articulated dump trucks with capacities up to about 40 tonnes have become popular in construction and some quarries (Fig. 11.1e). They operate with lower tyre pressures than rigid trucks or motor scrapers and they are more suitable for hauling on softer sub-grades. They have proved to be more versatile in contractor's plant fleets than rigid dump trucks of equivalent size or outputs although economic hauls are typically limited to up to about 3.5 km. The bodies of dump trucks are made of steel and may be heated by exhaust gases. The sides are often splayed out to increase the target area during loading and have an upward sloping rear which eliminates the need for a tailgate, thus increasing the speed of unloading. Dumpers are smaller trucks that carry the load in front of the driver and are available up to about 7 tonnes capacity. They are commonly used for hauling relatively small loads on smaller or confined sites. The size and type of dump truck will determine the strength of the sub-grade required for it to operate efficiently. Rigid trucks typically require sub-grades with a CBR of at least 5% but articulated trucks can be operated on ground with a CBR as low as about 2%. Crawler mounted dump trucks are also available with load capacities of up to about 23 tonnes for use in areas of low bearing capacity.

Railways. Light railways were used in former times in quarries and construction but they are now generally only used in specialist operations. Wagons were first moved manually or using horses, which significantly limited the gradients on which they could be used. Steam locomotives began to be used in the early part of the 19th century and subsequently diesel and electric powered locomotives took their place. In recent years, the greater versatility of dump trucks has largely displaced the use of light railways, although they are still widely used in the construction of tunnels and in underground mines.

Conveyor, conveyor belt. Conveyors are used in relatively short sections as part of a fixed or mobile plant (e.g. for loading a processing plant), as part of other earthmoving equipment (e.g. within a bucket wheel excavator (Fig. 11.1i)) or in longer sections to transport materials. They consist of a continuous flat belt, which is supported by idlers set on a frame and powered at one end. Materials are carried on the upper surface of the belt, which may be formed from rubber covered cotton or rayon. Belts can be obtained from about 0.2 m to 2.4 m wide but more commonly 0.3 m to 1.5 m wide and up to about 1.5 km or more in length, although up to 400 m is more usual. Greater distance can be achieved by linking conveyors in series and dumping the material from one conveyor on to the next through a hopper or baffle plate.

Belts may be open, partially or totally enclosed, depending on the sensitivity of the material to the weather conditions, dust generation or other environmental considerations. Loading is carried out from a hopper, which will often have a sloping coarse screen on top to remove oversize material that could damage the belt. If clay is particularly cohesive and produces large lumps on excavation, it may be necessary to break it down so that the lumps will fit on the conveyor, in which case the clay can be first loaded into mobile 'clay crushers' which discharge fragmented clay on to the conveyor. If the face is moving away from the end of the conveyor, shorter movable extensions or extensible conveyors may be added between the working face and the fixed conveyor until a fixed extension can be installed.

Soil can be transported over gradients up to about 28° on standard belts and the use of metal cleats can increase this up to about 45°. Vertical slopes can be traversed using buckets attached to the belt or to chains. Soil may be deposited from the conveyor into lorries, railcars and hoppers and onto stockpiles from either the fixed end of the conveyor or from a movable end length of conveyor called a 'stacker'. Fixed conveyors and associated installations are usually electric powered, whilst short movable sections are typically diesel or petrol powered. Longer conveyors need to be custom made and have a relatively high installation cost but low operating costs, of which the repair and replacement of the belts are the largest maintenance costs.

Conveyors are especially useful where large volumes of excavated soil or blasted rock must be transported along a single route, especially where the load must be lifted steeply or carried across difficult terrain, through narrow openings or across obstructions where road construction would be difficult. They may also be used where the source is some distance from the processing plant. Conveyors are not commonly used in construction, except for removing the spoil in tunnelling and in material processing areas.

11.2.5. Plant that can excavate, load, haul and deposit materials

Bulldozer or 'dozer'. A bulldozer is a tractor unit equipped with a front pusher blade, which can be raised or lowered by hydraulic rams or, less commonly, cables and is used for digging and pushing. An angle dozer has a blade that is capable of being set at an angle to push material sideways whilst moving forward. The tractor unit is usually set on crawler tracks, which allow the unit to travel over and push a wide variety of materials, although rubber-tyred tractors are also available. Traction can be increased with grousers or cleats on the tracks and wide-tracked dozers are available for use on soft ground. Blades are manufactured in a number of different styles but all need to be of heavy-duty construction and to have a hardened steel basal leading edge, which is driven into the ground to excavate and push material (Figs. 11.1a and b). Bulldozers have a variety of roles including excavating soils and weak rocks, ripping, moving excavated materials over short distances (typically up to about 150 m), spreading materials, trimming earthworks, acting as pusher units to boost the effective power of motor scrapers and towing rollers. Crawler dozers are available in a wide variety of sizes ranging from about 30 kW to over 750 kW (40 hp to over 1000 hp). Depending on their size, bulldozers can typically operate on ground with undrained shear strengths as low as about 30 kN/m^2 and lower if wide tracks are used (Farrar & Darley 1975; Caterpillar 2004). Pusher blades can be attached to other plant (such as self-propelled compactors) so that they can operate without the need of a bulldozer.

Forward loader, front end loader, loader, dozer shovel. This consists of a rubber-tyred or crawler tractor with a pair of lift arms (or boom) on the end of which is a bucket. The bucket is approximately the width of the tractor and can be rotated in the vertical plane. Excavation is carried out driving the tractor and bucket forward into the material and then rotating and lifting the bucket upwards. The tractor is then reversed away from the excavation and driven to discharge point. The bucket is lifted over discharge point and the material is deposited by rotating the bucket downwards (Fig. 11.1i). Loaders are best suited for excavating material above and for a short distance below ground level. They are most commonly used for excavating loose or worked materials such as in stockpiles and blasted rock with a reasonably firm base on which they and the hauling vehicles can operate, although they can be used for excavating faces of *in situ* soils. They can also be used to push or carry soil over short distances (typically up to about 200 m) and some can be equipped with a ripper or scarifier. They can load to a hopper, truck, railcar or stockpile. Loaders with buckets up to about 18 m^3 heaped (14.5 m^3 struck) capacity are available and smaller units may be equipped with a back-hoe (back-hoe loaders) (Fig. 11.1f).

Tractor scraper, motor scraper, scraper. A motor scraper is a highly mobile excavator consisting of a wheel-mounted tractor unit connected to a centrally-mounted excavating and transporting bowl by a vertical swivel connector or 'gooseneck'. The bowl is suspended between the gooseneck and rear wheels and has a hardened steel cutting edge or 'knife' at the front of the floor. The bowl is lowered into the ground during loading and the basal part of the front, the 'apron', is raised. As the machine moves forward, the top layer of the soil is scraped into the bowl. Additional force for loading can be provided by a second engine at the rear of the scraper and one or two dozers pushing on a steel pad at the rear. The bowl is raised and the apron lowered when transporting. To deposit the material the bowl is partially lowered, the apron raised and the rear wall, the 'ejector', moved forwards to push the material out. The thickness of the deposited layer is regulated by control of the ejector, apron and bowl (Fig. 11.1j).

Motor scrapers are designed to transport material quickly at speeds up to about 50 km/h and ideally haul roads should be well maintained with the minimum of steep gradients or sharp turns. They may be used for the excavation of soils and ripped rock, particularly clays and weak mudstones, provided that they do not contain boulders of hard material. They are very susceptible to rainfall, particularly on cohesive soils and are unsuited to the excavation of weak cohesive soils and loose granular soils as the machines tend to generate large ruts and the operating efficiency is low. For efficient operation, the undrained shear strength of cohesive sub-grades need to be at least about 60 kN/m^2 for small motor scrapers and over about 100 kN/m^2 for medium to large scrapers (Farrar & Darley 1975).

Due to a general reduction in lower limit of the shear strength of acceptable fill in the last 30 years, changes in market conditions, and their relatively low rate of utilization, the use of motor scrapers for bulk excavation in construction and mineral workings in the UK and Europe has decreased and the use of back-acters and dump trucks, particularly articulated dump trucks, has increased.

Single and double engine scrapers typically range in size from about 11 m^3 to 25 m^3 struck (15 m^3 to 34 m^3 heaped) capacity. They can be economically operated on fast hauls of up to about 2.5 km each way, with optimum hauls of about 800 m (Fig. 11.1k).

Elevating and auger scrapers are self-loading scrapers and are similar to conventional motor scrapers except that a rotating elevator or auger moves the excavated material from the front of the bowl, breaks it down and deposits it within the bowl. This decreases the resistance of loading and reduces the power required for or increases the speed of loading. Elevating scrapers typically have capacities from 7 m^3 to 26 m^3 heaped and auger scrapers from 17 m^3 to 34 m^3 heaped.

Towed scrapers typically range in size from about 5 m^3 to 17 m^3 struck capacity. The driving unit of towed scrapers consists of a crawler tractor and they can be economically operated on each way hauls of up to about 400 m. They are slower, have greater traction, tend to be smaller than motor scrapers and can operate on weaker soils with less marked reduction in efficiency. They are often used for stripping topsoil and overburden where hauls are short.

Dredgers. Dredgers are designed to excavate material from below water and most are purpose-made floating vessels. Several different types are available including cutter-suction, bucket wheel, grab and dipper (face shovel). Material can be pumped or transported by barge or the dredger itself to the area of deposition or off-loading point. The types and application of dredgers is a specialist subject and is not here considered further.

11.2.6. Plant that can compact materials

Although considerable compactive effort may be given by earthmoving plant, it is more efficient to provide systematic compaction with purpose-made compaction plant. These can be sub-divided into three main groups, consisting of deadweight rollers, vibrating rollers and plates and rammers.

Dead-weight rollers include smooth-wheeled, rubber-tyred and tamping rollers and construction traffic.

Smooth-wheeled rollers are widely used on large earthmoving schemes and consist of smooth-wheeled drum rollers that are manufactured as three-roll or tandem-roll self-propelled units or as single towed rollers that can be connected in series. Because self-propelled rollers with only steel drums have relatively poor traction characteristics, they are not generally used for large-scale compaction of earthworks materials, particularly clays, although they may be used for selected granular fills and stabilized soils. Therefore towed rollers are often used. They typically range in size from about 1.7 to 17 tonnes, which can often be increased by adding sand or water ballast to the rolls or by placing concrete weights on the frame (Fig. 11.2a) In the UK they are rated by the mass per unit width. Optimum speeds are from 2.5 to 5 km/h. Smooth-wheeled rollers are not suited to the compaction of soft or wet clays.

Pneumatic-tyred rollers are not often used on earthworks in the UK but are more commonly used in other European countries. They consist of one or two axles on which are set a number of rubber-tyred wheels. On two-axle machines, one axle usually has one less tyre and the tires are offset so that the ground is completely covered by the passage of the roller. The mass of the roller can be increased by applying ballast to a box or platform above the axles and the compactive effort can also be altered by adjusting the inflation pressure of the tyres. A certain amount of vertical movement is allowed for between individual and pairs of wheels to enable the roller to negotiate uneven ground whilst maintaining constant contact pressure (Fig. 11.2b). 'Wobbly-wheel' pneumatic-tyred rollers have wheels that move from side to side, thus providing additional kneading action to the soil. In the UK pneumatic-tyred rollers are rated by the mass per wheel. Optimum speeds are from 1.5 to 24 km/h.

Grid rollers are similar to smooth-wheel rollers except that the surface of the roll consists of a heavy steel mesh and they can also be ballasted with concrete weights. They are towed units and weigh between about 5.5 tonnes net to 15 tonnes ballasted (Fig. 11.2c). They are rated by the mass per unit width and have optimum speeds from 5 to 24 km/h. Grid rollers are not suited to the compaction of weak or wet clays.

Tamping rollers consist of steel drums fitted with projecting feet. They may be self-propelled or towed with single or double axles and the mass may be increased with ballast. There are wide variations in the shape and size of the projecting feet, which may be circular, square or rectangular in plan with different lengths and tapers. They are sometimes called 'sheepsfoot' rollers but this term is more correctly used for a particular variety with long slender feet. Optimum speeds are from 4 to 10 km/h (Fig. 11.2d). Tamping rollers leave 'footprints', which can collect rainfall, particularly with cohesive soils, which if later incorporated in the embankment can lead to reduction in the strength and bearing capacity of the fill. Therefore it is advisable to complete the compaction of the last layer of fill at the end of each shift with a smooth-wheeled roller if there is a possibility of rainfall. Tamping rollers are particularly suited to the compaction of clays.

Construction traffic. The general effect of heavy construction traffic is similar to that of rubber-tyred rollers and result in very effective compaction. However, in practice, it is difficult to ensure that construction traffic is directed over the area to cover completely and evenly compact the fill. Therefore the compaction achieved by construction traffic should be regarded as a bonus, although excessive or concentrated compaction by construction traffic can lead to over-compaction, rutting and degradation of the fill (see Section 11.3.3). This can be minimized by providing a haul road or protective layer on which the construction traffic can operate or by spreading the traffic over a wide area, although such traffic movements may create a safety hazard. The only compaction given to many quarry waste tips and backfill areas is from the vehicles delivering the materials and the bulldozers spreading the tipped materials.

Vibrating compactors apply both mass and vibrational energy to the ground and may be either rollers or plates. Over the last 30 years they have progressively become the dominant type of compactors in use at the expense of deadweight rollers.

Vibrating rollers are vibrating smooth-wheeled rollers that range in size from about 0.5 to 17 tonnes and may be towed, self-propelled or manually guided (Fig. 11.2e). Self-propelled rollers may have a single (Fig. 11.2f) or two (tandem) (Fig. 11.2g) vibrating drums although (as with smooth-wheeled deadweight rollers) traction on the latter is poor on clays. The larger machines operate best at between about 1.5 and 2.5 km/h and manually-guided machines between about 0.5 and 1 km/h. Repeated passes at fast speed are not as effective as fewer passes at low speed. The frequency of vibration is usually fixed by the manufactures and typically varies between 20 to 35 Hz for larger rollers and 45 to 75 Hz for smaller rollers, although the frequency of some rollers can be varied. When the vibration is not switched on, the roller should be regarded as a deadweight roller. Vibrating rollers are rated by the mass per unit width. They are suitable for a wide range of soils but have performed poorly on wet and soft clays.

Vibrating tamping rollers are similar to tamping rollers except that the drums can be vibrated and the feet tend to be more blocky and robust. Smaller machines are available for use in trenches and on smaller sites (Fig. 11.2h). As with tamping rollers, the large rollers (Fig. 11.2i) may need to be used with a smooth-wheeled roller at the end of each shift. They can be used over a wide range of clay soils, although they are less effective on soft and wet clays.

Vibrating plate compactors vary in mass between about 100 kg and 2 tonnes with plate areas between 0.15 and 1.6 m^2. Most are manually-guided machines appropriate for compacting small or awkwardly shaped areas. They operate best at about 0.7 km/h and are classified by the mass per unit base area (Fig. 11.2j).

Vibro-tampers (vibro-rammers) are manually-guided machines that usually weigh between about 50 and 100 kg. Compaction is induced by vibrations that are set up in a base plate through a spring activated by an engine-driven reciprocating mechanism. They are rated by the static mass of the machine and are appropriate for compacting very small or awkwardly shaped areas (Fig. 11.2k).

Compaction by impact including compaction by impact rollers, power rammers and falling weights.

Impact rollers have been developed recently but are relatively rare. Present varieties consist of 3- or 5-sided twin smooth drums with a mass of between 10 and 12 tonnes that are towed by a rubber-tracked tractor. As the drums are pulled forward they rotate and drop on to adjacent face of the drum through a distance of between 150 and 230 mm. The energy is typically between 15 and 25 kJ and it is claimed that ground improvement can be observed up to depths of 4 m.

Power rammers are manually driven machines suitable for compacting small or awkwardly shaped areas. Explosions in an internal combustion engine cause the machine to be driven upward and the subsequent impact of the base plate on the ground causes the compaction. Most weigh about 100 kg. They are rated by the static mass of the machine.

Dropping weight compactors consist of machines that drop a mass of 500 kg or more from a controlled height of about 3 m. This has been developed into *dynamic compaction* in which a mass of up to 40 tonnes is dropped from heights of up to about 20 m. The technique uses large cranes and weights or purpose-made machines and involves monitoring water levels, densities, strengths and bearing capacity of the ground during and after compaction. It is suitable for the compaction of larger areas of compressible materials to depth of about 30 m.

11.2.7. Specialist earthworks plant

Special plant may be used for particular specialist operations, such as the construction of piles, slurry walls and tunnels and in the modification of or stabilization of soils. Augers and shells are used for the excavation of piles, and augers, drills, grabs and other plant used in the excavation of slurry walls. These are not considered in detail further here but further information is given in Chapter 12.

Specialist plant is used for regulating the moisture content of soils by the addition of water or to carry out lime or cement modification or stabilization. As described in Chapter 10, quicklime and hydrated lime can be added to modify and improve the properties of clays, often to make weak clays useable. Quicklime, hydrated lime and cement, alone or together can be used to stabilize clays, for example to permanently improve the bearing capacity of sub-grade clays.

Lime/cement spreaders. In the UK, lime or cement is added to clay soils by first spreading it on the material to be treated in granular and powder form respectively. This is carried out by purpose-made towed or self-propelled spreaders that place a line of the additive as they move forward. A flexible skirt is fixed around or behind the spreader to minimize the generation of airborne dust. The mass of additive placed per square metre can be regulated. Agricultural spreaders have been used but they spread at a low rate, thus requiring multiple passes, and generating much airborne dust. Where less control of the spreading rate is tolerable, spreading can be carried out using the bucket of a forward loader and small areas can be spread manually by spotting bags on the ground and spreading the lime with shovels. In areas of high temperature and high evaporation, such as the southern states of the USA, it may be more efficient to apply the lime as a water-based slurry that is sprayed directly on the ground from tankers equipped with internal agitators. This also requires a high capacity mixing plant to make the slurry.

Rotovators. Lime or cement is most effectively mixed into the ground using a purpose-made self-propelled rotovator. This typically consists of a horizontal shaft containing numerous long teeth or tines that rotate in a vertical plane and which is mounted at the rear of a tractor unit. As the shaft is rotated and the tractor moves forward, the tines are lowered to and into the ground. The rotating tines dig into, excavate and break up the soil and mix it with the additive and/or water. The rotary mixer is confined within a hood to prevent the rotovated material being thrown beyond the limit of the machine and as a safety guard over the tines. A number of passes of the rotovator are required to ensure full mixing to the required depth, which is typically up to 0.35 m. As the reactions of the lime with clay are time-dependant, it is usual to carry out initial rotovation to mix the soil and lime. The lime is allowed to react with the clays for 2 to 3 days and the mixture is rotovated again, usually with significantly improved degree of mixing and pulverization. Water can be added to the treated soil to increase the moisture content for compaction during the mixing processes through an integral spray bar in the hood of the rotovator from a water tanker running in tandem with the rotovator. Alternatively, it can be sprayed directly on to the soil from a water bowser equipped with a spray bar. Whilst larger areas of modification and all stabilisation are best carried out using purpose-made rotovators, smaller areas of modification can be carried out to depths of about 200 mm using standard agricultural rotovators.

Water bowsers, water tankers. These consist of lorry or scraper mounted tankers or water tanks mounted on towed units. They are commonly used to spray water on to haul and access routes to suppress dust during dry periods. Occasionally they may be used to add water to dry or excessively strong cohesive fill to enable compaction to be carried out. These operations are best carried out via a spray bar at the rear of the bowser. They can be used to regulate the moisture content of stabilized soils either through a spray bar or by introducing water into the hood of the rotovator.

Ploughs and harrows. Summer sunshine can be used to reduce the moisture content of wet clay fills, although the unreliability of the weather limits it use in some areas. Given favourable conditions, up to 0.5% of moisture content can be lost per hour but this requires that the soils be frequently turned to optimize exposure. This can be carried out using agricultural ploughs and harrows. Prior to the introduction of the method specification for compaction in the UK, harrows were sometimes used to scarify the surface of plastic clays between the placement of layers to bond the layers, improve compaction and minimize the formation of weakness planes between layers.

11.2.8. Plant used in excavation and spoil removal in tunnelling

Tunnels in clay can vary in size from microtunnels that are too small for man entry (down to 250 mm diameter) to road, rail and water tunnels and structures in excess of 10 m wide. Some of the methods of excavation and removal of clay are similar to those used in above-ground earthworks, but may have to be modified to fit into the limited space available in smaller diameter tunnels. In addition, it is often necessary to provide temporary and/or permanent support to the walls of tunnels at the same time as or shortly after they are excavated to prevent their collapse or to limit the amount of settlement above the tunnel. Permanent linings are usually required for civil engineering tunnels in clays, whereas they may not be required in mines.

With an open face, excavation may be carried out using hand tools, by mechanical excavators similar to a backhoe or forward loader, or by specialist tunnelling machines incorporating a rotary cutter, or a full-diameter cutting head consisting of various tools mounted on spokes. Excavation is usually carried out at the front of a 'shield' consisting of a metal drum that houses the tunnelling equipment, gives protection to workers, and provides temporary support to the ground until a permanent lining is installed behind. In stiff clays, mudstones and shales, excavation of large diameter tunnels may proceed without a shield, with excavation taking place in stages (in two or three horizontal benches, and/or two or three vertical sections with temporary supports in between). Initial support to the ground may be provided by sprayed concrete reinforced with carbon or steel fibres or steel mesh and/or arch ribs. Steel or timber sheets may be installed at the crown to minimize the possibility effects of collapse. These tunnels can have a variety of different cross sections, whilst shield-driven tunnels are generally circular in section, although a number of ingenious tunnelling machines have been built in Japan for non-circular sections.

Where ground conditions are more uncertain or the soils weaker, tunnel boring machines (TBMs) will normally be used. These incorporate a shield with a largely enclosed cutting head, which can be driven against the face to provide mechanical support to the ground. Excavated soil enters the head through relatively small apertures, which may be closed by doors in the event of unstable ground or excessive groundwater flows being encountered. The problematic ground can then be treated, for example by grouting or installation of reinforcement through ports in the cutting head, before excavation continues.

If such conditions are likely to occur for much of the tunnel length, a system which provides continuous support to the ground will normally be chosen; the support may be by compressed air, slurry pressure, or by the pressure from excavated spoil in a head chamber. Further information on the latter two is given in Chapter 12. The excavated material must then be removed from the head of the TBM without dissipating the pressure in the head chamber. With slurry tunnelling, this is done by also using the slurry as a medium to transport the spoil by pumping through pipes. With the other methods, the primary transfer of soil out of the head chamber is most often via a screw conveyor, which either provides sufficient back-pressure due to soil resistance within the screw or has a lock system at the outer end through which the spoil passes.

In soft uniform clays, as are widespread for example in Scandinavia, Japan and parts of Malaysia, excavation may be by extrusion through controlled apertures in the face of a 'blind' shield; the cutting head does not rotate, but simply pushed forward into the ground.

Once excavated, the spoil must generally be transported out of the tunnel to a disposal point either through a main tunnel entrance or an intermediate shaft. In small diameter tunnels (such as for sewers) the distance between the face and the exit point may be tens of metres, whereas it may be up to several kilometres in tunnels beneath water or mountain ranges. The most common methods of transport are by railcars drawn by electric or diesel locomotives (or in small tunnels by winches), belt conveyors, or in larger tunnels by the use of dump trucks. Conveyor systems have become increasingly popular in the UK in recent years, partly as a result of second-hand conveyor systems becoming available due to the run-down of coal mining. They may be extended beyond the tunnel portal to the disposal area, or to convenient road or railhead for onward transport of the spoil.

The other common method of transport is by pumping, either with a slurry transport system or, less commonly, using sludge pumps to move soft cohesive spoil. Vacuum systems using air as the transport medium through pipelines have also occasionally been used. In microtunnels, soil transport is either by a scaled-down version of a slurry system, or using a screw conveyor consisting of an auger within a casing for the full length of the drive. The shaft of the auger is then normally also used to drive the cutting head at the tunnel face.

The choice of excavation and transport system will be strongly influenced by the ground conditions and the length of the tunnel drive. Hand excavation is only economic for short drives of modest diameter, typically up to about 100 m for a 1.5 m diameter sewer tunnel in good ground conditions of firm to stiff clay. It may also be necessary when local areas of poor ground, such as large clay-bearing faults, are found in otherwise intact material. Simple shield machines quickly become more economic and safer for longer drives, larger diameters and poorer ground conditions. A large tunnelling machine providing full ground support, along with all the back-up systems including drive units, power modules, segment-erector devices, grouting equipment etc, will involve an investment of several million pounds sterling and will usually be purchased for a specific scheme.

For smaller schemes, with drives of only a few hundred metres, secondhand machines are often used, having been refurbished and perhaps 'reskinned' to a slightly different (usually larger) diameter. Typically smaller diameter machines will be used on several different projects during their lifetime, with shield diameter and cutter head arrangement altered to suit the project and ground conditions. Similarly, specialist underground dump trucks (e.g. with low headroom), rail locomotives and skips may be used on different projects.

For further details of tunnelling equipment, reference should be made to specialist publications, such as Maidl *et al.* (1996) and, for smaller diameters, Thomson (1993).

11.2.9. Plant teams

Earthmoving may involve the excavation, loading, transport, deposition and compaction of different materials in different parts of the site and, in addition, materials may be exported from or imported to the site. This will usually involve different types of earthmoving plant working on different parts of the work at the same time, sometimes in teams and sometimes independently. Examples of plant and plant teams that may be used in the various earthworks operations on construction and mineral extraction projects are given in Tables 11.1 and 11.2. A summary of the plant appropriate for compacting different earthworks materials is given in Table 11.3.

TABLE 11.1. *Examples of typical plant combinations used on construction sites*

Operation	Excavator	Transport	Transport to	Spreading/Compaction
Topsoil strip	Motor scraper/tractor and box		Stockpile	Dozer assistance/None
	Dozer pushing to heaps for back-acter/forward loader to load	Dump trucks		Dozer to spread/None
Place topsoil	Motor scraper/tractor and box		Cutting/embankment/ landscape area	Dozer assistance. Dozer or grader to spread/None
	Dozer pushing to heaps for back-acter/forward loader to load	Dump trucks		Dozer and/or grader to spread/None
Acceptable fill	Motor scraper/tractor and box with/without dozer/ripper assistance		Embankment or landscape area	Dozer assistance and to spread/towed or self propelled rollers
	Back-acter or face shovel	Dump trucks		Dozer to spread/towed or self propelled rollers
	Blasting and/or dozer with ripper for breaking up rock for loading by back-acter/forward loader			
Unacceptable fill	Dozer		Local stockpile	None
	Face shovel/Back-acter/ forward loader	Dump trucks	Internal tip	Dozer to spread/towed or self propelled rollers if possible, otherwise construction plant
		Road lorries	External tip	Not applicable
	Dragline	Dump trucks	Internal tip	Dozer to spread/towed or self propelled rollers if possible, otherwise construction plant
		Road lorries	External tip	Not applicable
Imported fill	Not applicable	Road lorries	Embankment or other filling area	Dozer to spread/towed or self propelled rollers

TABLE 11.2. *Examples of typical plant combinations used in mineral extraction*

Operation	Excavator	Transport	Spreading/compaction
Topsoil strip (for reuse)	Motor scraper/tractor and box		Dozer assistance/None
	Dozer pushing to heaps for back-acter/forward loader to load	Dump trucks	Dozer to spread/None
Overburden (to tip)	Motor scraper/tractor and box with/without dozer/ripper assistance		Dozer assistance and to spread/towed or self propelled rollers
	Back-acter or face shovel	Dump trucks	Dozer to spread/towed or self propelled rollers
	Dragline		
	Blasting and/or dozer with ripper for breaking up rock for loading by back-acter/forward loader		
Clay mineral deposit	Face shovel/Back-acter	Dump trucks	Load direct to stockpile or processing plant/None
	Bucket wheel excavator	Conveyor, rail or dump trucks	
	Dragline		
	(Hydraulic) Monitor usually with dozer assistance	Hydraulic under gravity	Collection in settlement lagoons at start of processing/None
Processing waste	Face shovel from stockpile (solid)	Dump trucks to tip	Dozer to spread/towed or self propelled rollers
	Pump to settling pond. Recycle water		–
Restoration	As equivalent construction operations		

11.3. Selection of earthmoving and compaction plant

The planning for an earthmoving project will often start at the tender stage of a construction contract or in the viability assessment for mineral exploitation in order that the anticipated methods and programme of working and the likely costs can be assessed. As the project develops, more detailed planning will be carried out, as the result of which specific plant will be purchased and/or hired and/or sub-contractors employed. These will need to be

TABLE 11.3. *Suitability of compaction plant on different soil and rock fill types (after Horner 1988)*

Compactor	Principal soil types						Rock	
	Cohesive		Granular					
	Wet	Other	Well graded		Uniform		Soft	Hard
			Coarse	Fine	Coarse	Fine		
Smooth wheeled roller	N				N	N		N
Grid roller	N							
Pneumatic-tyred roller		N						
Tamping roller						N		N
Vibrating roller								
Vibrating plate	N							
Vibro-tamper	N							
Power rammer					N	N		
Dropping weight	N				N	N		
Dynamic consolidation								

N = Not generally suitable   Can be used but less efficiently   Most suited

constantly reviewed and amended when required during the various stages of earthmoving in response to current information, experience and circumstances.

The main purposes of the planning stage are to identify and analyse the items, order, methods and duration of works that are to be carried out, the plant, labour, materials and other resources required to carry the works in the most efficient and cost-effective manner and to determine the likely operating costs. As no two construction contracts or mineral workings are the same, the nature and extent of the work, the plant required and the production rates of the plant can vary considerably and thus they will need to be estimated for each scheme. This is usually carried out from first principles, using experience from past schemes where appropriate, and will involve the consideration of a number of factors which will influence how the various works are best carried out and what plant is best suited for the works. These factors are often inter-related and include:

- *material factors*, such as the types, properties and variability of the different deposits that are to be excavated, transported, deposited and compacted;
- the *quantities* of the various materials to be excavated and/or the production rates to be achieved for profitable operation;
- *spatial factors*, such as the locations and elevations of the various materials to be excavated and those at which they must be stockpiled, placed and compacted;
- *the geography of the area*, such as the features and relief of the area, the timing and means of access to and within the site, the anticipated weather conditions, the proximity to major highway and rail links;
- *hydrological and hydrogeological factors* including the nature and distribution of the surface water drainage and the groundwater conditions;
- *plant factors*, such as the types of plant that are suitable and available to carry out the works, their operating characteristics, output rates and operating costs;
- *other factors* such as economical, legal, contractual, political or planning restraints that may influence the selection, use or access of plant, the use and availability of labour and materials and the operation of the works.

11.3.1. Material factors

During an early stage in the project, ground investigations should have been carried out to determine the nature, distribution and properties of the ground within the scheme to enable the scheme to be evaluated, designed and costed. In particular, the distribution, properties and variability of a mineral deposit should have been determined, together with (at least) the general properties of the overburden. In addition to information on the properties needed to assess the quality of the mineral deposit, information is required on the geotechnical properties of the materials (e.g. undrained and drained strengths) in order that preliminary designs of the works can be prepared including the design of the slope angles of working faces, edges of quarries, accesses etc. In some cases, information on the overburden may be required to assess its potential economic value (e.g. granular deposits overlying economically viable clay). It will be necessary to dispose of waste materials and sufficient information should be obtained to design and construct the tips and lagoons that may be required.

In construction projects, enough information should be available to enable the design of the works to be carried out and for the contractor to identify from where the different materials required in the earthworks design are to be obtained, whether materials are unusable or surplus and must be disposed of, and whether materials need to be imported.

In both cases, the general geological information should be sufficient to allow earthmoving specialists to assess the types of plant and equipment that are likely to be required (e.g. rock and hard strata may need blasting or ripping, plant for operating on soft or weak soils with low bearing capacity etc.). Additional geotechnical information such as systematic descriptions and laboratory data on shear strengths, CBRs, densities etc. will provide more detailed information for planning. In some areas, particularly brownfield sites, detailed investigations may be required to determine the presence, nature and extent of any gas emissions or chemical and biological contaminants in the soils or groundwater. These may require particular safe working practices, special tips may be required for the disposal of contaminated ground, techniques may need to be adopted to remediate or immobilize the contaminants or contaminated groundwater may need to be controlled and/or treated.

Care should be taken when locating temporary stockpiles or waste tips, especially when they are underlain by cohesive soils or lie close to the edge of active workings. Near-surface clays on slopes may be close to their limit of equilibrium and may have been weakened by previously landslips, solifluction or periglacial processes as occur, for example, in many parts of southern England (see Chapter 6). Similarly, temporary excavations at the toe of clay slopes can lead to instability of the slopes. Examples of the destabilisation of slopes or reactivation of old landslides by the construction of stockpiles of topsoil, overburden or surplus excavated materials placed on or at the top of slopes and by uncontrolled excavations for minerals, trenches or foundations at the toe of slopes are not unusual and the consequences can be fatal and expensive.

The stability of tips, stockpiles, lagoons, faces etc. in quarries in the UK is covered by the Quarry Regulations 1999. They are required to be designed, constructed, operated and maintained to avoid any instability or movement that can pose a risk to the health and safety of any person. Site investigations are required before starting new excavations or tips. The operator must carry out an appraisal to identify those new and existing tips and excavations that pose a significant hazard if they failed. If the hazard is significant, the stability and safety of the excavation or tip must be assessed by a geotechnical specialist. The Regulations also require work to be carried out in accordance with excavation and tipping rules prepared for the quarry.

In construction works, the Construction (Design and Management) (CDM) Regulations 1994 require designers to foresee health and safety risks in their designs and where possible avoid them. The principal contractor is responsible for identifying the hazards and allowing for these in a health and safety plan. In earthworks in clay, these should include the stability of temporary excavations, the sides of which should be battered back to a safe angle or adequately supported.

Topsoil is a resource that can be reused and should be treated separately from the other materials. It is often billed separately in construction contracts and reused for topsoiling cuttings, embankments and landscaping areas. Therefore it is usually excavated separately (stripped) by scrapers, dozers, loaders or back-acters, and stockpiled in heaps or windrows, usually close to the areas where it will eventually be reused. Surplus topsoil may be sold, given away or disposed of in tips, but in practice it is often 'lost' in overfilling and in landscaping areas.

The materials to be excavated will be treated differently according to their use and therefore they will need to be classified. In the UK, earthworks materials in construction are classified into those that are 'acceptable' (or 'suitable') and 'unacceptable' (or 'unsuitable') for reuse. Acceptable fills may be subdivided further according to their properties and potential use within the works. For example, the *Specification for Highway and Bridge Works* (Highways Agency 2005) has nine classes and 50 sub-classes of acceptable materials (see Table 11.4). Materials within mineral deposits will be primarily classified into those for which the deposit is being worked (e.g. clay for brickmaking), subsidiary deposits of value for sale (e.g. the overlying gravel for aggregates) and the unsaleable materials (waste). The materials within the prime mineral deposits may then be further classified according to their composition or quality and excavated separately and transported to the processing plant. The saleable secondary deposits will be excavated and transported to a processing works or off site, sometimes via a weighbridge. The unusable materials will consist of the overburden and other unusable materials associated with the mineral deposits (e.g. unaltered granite within china clay deposits) and the waste from the processing plant that also must be disposed of. Some of these deposits may be required for specific uses, such as construction of amenity bunds or haul roads, and the remainder will be placed in tips, some of which may be special tips (e.g. for weak or contaminated materials) or lagoons (for slurries and waste water).

The excavatability of soils and rock is determined by the properties of the soil and the type, size and power of the excavating plant. Most plant excavates by pushing, pulling or driving a blade or cutter into the material to loosen, move and/or pick it up. In order to achieve this, the resistance to penetration provided by the shear strength of the material (measured as the cohesion, frictional strength or unconfined compressive strength) must be overcome. Further resistance to excavation is provided between the cutting device and the material and the mass of the overlying material. The presence of cobbles, boulders, hard masses and cemented zones tend to increase the resistance, whilst discontinuities such as bedding planes, joints, fissures and faults tend to reduce resistance.

There are various means of defining the ease of excavation. A simple classification can be used which categorises as follows:

- *easy digging*, such as very soft to firm clays, very loose to dense sands and gravels;
- *hard digging*, such as stiff to hard clays, dense to very dense sands and gravels, blasted rock, weak rocks such as shale, slate and weathered harder rocks;
- *rock*, such as unweathered limestone, sandstone, granite and dolerite.

TABLE 11.4. *Classification of earthworks materials in the UK by the Highways Agency*
Extract from Table 6/1, Specification for Highway Works, Highways Agency 2005. © Crown copyright material is produced with the permission of the Controller of HMSO and the Queen's Printer for Scotland. Clays and cohesive soils are shown in shaded boxes

Type	Class	Description	Typical Use
General Granular Fill	1A	Well graded granular material	General fill
	1B	Uniformly graded granular material	
	1C	Coarse granular material	
General Cohesive Fill	2A	Wet cohesive material	General fill
	2B	Dry cohesive material	
	2C	Stony cohesive material	
	2D	Silty cohesive material	
	2E	Reclaimed pulverised fuel ash (pfa) cohesive material	
General Chalk Fill	3	Chalk	General fill
Landscape Fill	4	Various	Fill for landscape areas
Topsoil	5A	Topsoil or turf existing on site	Topsoiling
	5B	Imported topsoil	
Selected Granular Fill	6A	Selected well graded granular material	Below water
	6B	Selected coarse granular material	Starter layer
	6C	Selected uniformly graded granular material	Starter layer
	6D	Selected uniformly graded granular material	Starter layer below pfa
	6E	Selected granular material	For cement stabilisation to form capping—Class 9A
	6F1	Selected granular material (fine grading)	Capping
	6F2	Selected granular material (coarse grading)	Capping
	6F3	Selected granular material (recycled bituminous/asphaltic materials etc.)	Capping
	6F4	Selected/imported (unbound) granular material complying with BS EN 13285 (fine grading)	Capping
	6F5	Selected/imported (unbound) granular material complying with BS EN 13285 (coarse grading)	Capping
	6G	Selected granular material	Gabion filling
	6H	Selected granular material	Drainage layer — For reinforced soil and anchored earth structures
	6I	Selected well graded granular material	Fill
	6J	Selected uniformly graded granular material	Fill
	6K	Selected granular material	Lower bedding for — Corrugated steel buried structures
	6L	Selected uniformly graded granular material	Upper bedding for
	6M	Selected granular material	Surround to
	6N	Selected well graded granular material	Fill to structures
	6P	Selected granular material	Fill to structures
	6Q	Well graded, uniformly graded or coarse granular material	Overlying fill for corrugated steel buried structures
	6R	Selected granular material	For stabilisation with lime and cement to form capping—Class 9F
	6S	Selected well graded granular material	Filter layer below subbase
Selected Cohesive Fill	7A	Selected cohesive material	Fill to structures
	7B	Selected conditioned pfa cohesive material	Fill to structures and reinforced soil
	7C	Selected wet cohesive material	Fill to reinforce soil
	7D	Selected stony cohesive material	Fill to reinforce soil
	7E	Selected cohesive material	For stabilization to form capping with lime—Class 9D, cement—Class 9B, cement—Class 9C
	7F	Selected silty cohesive material	
	7G	Selected conditioned pfa cohesive material	
	7H	Wet, dry, stony or silty cohesive material and chalk	Overlying fill for corrugated steel buried structures
	7I	Selected cohesive material	For stabilisation with lime and cement to form capping—Class 9E
Miscellaneous Fill	8	Class 1, Class 2 or Class 3 material	Lower trench fill
Stabilised Materials	9A	Cement stabilized well graded granular material	Capping
	9B	Cement stabilised silty cohesive material	Capping
	9C	Cement stabilised conditioned pfa cohesive material	Capping
	9D	Lime stabilised cohesive material	Capping
	9E	Lime and cement stabilised cohesive material	Capping
	9F	Lime and cement stabilised well graded granular material	Capping

Easily dug materials can be excavated by most excavation plant including dozers, scrapers, draglines and excavators. As the deposits become harder to excavate, larger or differently designed machines are required to maintain output rates and/or additional plant may be required. For example, teeth would be required on buckets, motor scrapers will require dozers for push loading and the harder materials may need to be ripped. As the material becomes more rock-like, production output will tend to fall. An example of the effects are shown in the guidance given by Caterpillar (2004) on easy to hard digging by back-acters which is reproduced in Table 11.5. The excavation of 'rock' will require heavy ripping and/or blasting. The assessment of the rippability of rock can be carried out by experienced engineers and geologists from the rock properties by using a geomechanical classification (e.g. MacGregor *et al.* 1994; Pettifer & Fookes 1994) or by using seismic velocities with published rippability charts such as that reproduced in Table 11.6 (Caterpillar 2000, 2004).

The density of material affects the power required to lift or push an equivalent volume of material and may limit the amount of material that can be hauled by a vehicle. During excavation, soils and rock are loosened and sometimes broken down resulting in an increase in volume and a decrease in density (bulk-up or swell). This increase in volume must be allowed for in determining the amount of plant required for hauling, spreading and compacting the excavated materials and may be defined as:

$$\text{bulk up} = \frac{\text{loose volume} - \textit{in situ} \text{ volume}}{\textit{in situ} \text{ volume}} \times 100\% \quad (11.1)$$

$$\text{load factor} = \frac{\text{compacted volume}}{\text{loose volume}}$$

$$\text{load factor} = \frac{100}{100 + \text{bulk-up}} \times \text{compacted vol}/\textit{in situ} \text{ vol.} \quad (11.2)$$

Typical *in situ* densities and bulking of common earthworks materials are included in Table 11.7.

When material is placed and compacted the process is reversed and the material reduces in volume. Depending on the nature of the soils and the amount of compaction given, there may be a net bulk-up or shrinkage on completion of compaction. In construction with full systematic compaction, the net bulk-up is typically between 0 and 10% for most soils and weak rocks (and is usually small enough to be ignored) and between 5% and 20% for stronger rocks. A net shrinkage may occur where water drains from the soil (e.g. 0% to 10% for sand) or where the structure is broken down (e.g. 0% to 15% for chalk). This is not to be confused with wastage of earthworks materials, which may occur, for example, in creating haul roads, overfilling embankments and rendering material unusable; this can typically be about 5% of the volume of fill.

The nature of the *in situ* ground and the fill can affect the trafficking of the site by plant. The presence of weak or soft soils may lead to excessive sinking of plant and consequent rutting and reduced operational efficiency. This may warrant the use of tracked plant such as excavators and dozers with wide tracks or plant that can stand off for excavation, such as draglines and long reach back-acters, and articulated dump trucks for haulage. Larger motor scrapers and dump trucks require a sub-grade with an undrained shear strength of at least 100 kN/m^2 or a CBR of 5% for efficient operation. In addition, the quality of the haul road affects the energy required to power the vehicles. This may be measured as the rolling resistance, which is the force required to roll or pull a wheel over the ground and which is approximately 2% of the gross machine mass plus 0.6% per centimetre of tyre penetration (or equivalent road flexure). Thus the rolling resistance on a hard well maintained rockfill haul road will be significantly less than on soft very rutted clay. Typical values of rolling resistance for different surfaces are given in Table 11.8.

Tyre life can significantly affect the cost of earthmoving and depends on a number of factors including the nature of the sub-grade and the type of tyre. Tyre life is

TABLE 11.5. *Classification of excavatability for estimation of excavator cycle times (reproduced with permission of Caterpillar Inc.)*

Excavatability	Typical Materials	Typical operating conditions
Easy digging	Unpacked earth, sand and gravel, ditch cleaning etc.	Digging less than 40% of machine's maximum depth capacity. Swing angle less than 30°. Dump onto spoil pile or truck in excavation. No obstructions. Good operator.
Medium digging	Packed earth, tough dry clay, soil with less than 25% rock content	Digging up to 50% of machine's maximum depth capacity. Swing angle to 60°. Large dump target. Few obstructions.
Medium to hard digging	Hard packed soil with up to 50% rock content.	Digging up to 70% of machine's maximum depth capacity. Swing angle to 90°. Loading trucks spotted close to excavator.
Hard digging	Shot rock or tough soil with up to 75% rock content	Digging up to 90% of machine's maximum depth capacity. Swing angle to 120°. Shored trench. Small dump target. Working over pipe crew.
Toughest digging	Sandstone, caliche, shale, certain limestones, hard frost	Digging over 90% of machine's maximum depth capacity. Swing angle over 120°. Loading bucket in man box (drag) box. Dump into small target requiring maximum excavator reach. People and obstructions in the work area.

TABLE 11.6. *Ripper performance chart for D11R multi or single shank ripper (reproduced with the permission of Caterpillar Inc.)*

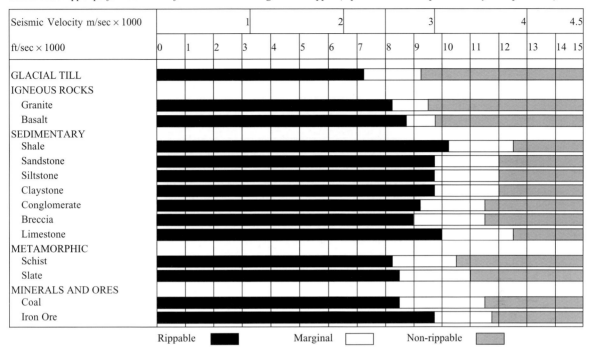

greater on soft clay sub-grades and well-maintained haul roads but decreases with increasing rock content, particularly inherently sharp rock such as flints and blasted rock. Similarly track life can be significantly affected by ground conditions. Sand, particularly sharp silica sand in water, is relatively abrasive whereas most clays only provide low to medium wear and the cumulative effects of impacts are least on penetrable surfaces such as clay and greatest on bumpy surfaces such as rockfill.

The nature of the materials to be compacted is often the most important factor determining the choice of the compaction and associated plant, of which their type, moisture content, strength, grading and variability tend to have the most influence. These determine the most appropriate type and size of the compactor, together with the layer thickness and the number of passes of the compactor. They may also determine the ancillary plant that may be required.

In the UK, most compaction is carried out in accordance with the *Specification for Highway Works* (Highways Agency 2005) in which most materials are compacted according to a 'method' specification (see Chapter 10.5.2). 'Performance' or 'end-product' specifications are included for eight sub-classes of materials in the *Specification* but are more widely used in specifications derived from American practice (Monahan 1994). With a performance specification, the contractor will provisionally determine his compaction methods from past experience on similar materials. At an early stage of earthmoving he will generally carry out full-scale site trials, using different combinations of compaction plant, layer thicknesses and number of passes of the plant, to determine the most practical and economical methods of working to achieve the required degree of compaction for each of the main types of material that are to be compacted. Performance is usually measured by a minimum relative compaction or maximum percentage air voids that must be achieved, but other criteria may be specified, such as a minimum dry density, CBR or undrained shear strength. The types of compaction plant appropriate for different materials are summarized in Table 11.3.

Thus for any material there will be a range of compaction plant that could be used, of which some will be more or less appropriate or available. In some cases, the choice of plant may be limited by the specification as, for example, in the *Specification for Highway Works* (2005) for wet cohesive material with liquid limits over 50, for which only deadweight tamping, vibratory tamping or grid rollers are permitted. If the quantities are known, the likely costs of compaction using the different plant can be assessed. It may be found that a range of different compactors are most suited and economical for the different materials present but, in practice, these would be rationalized so that a limited range of plant would be used with varying degrees of efficiency on the different materials, thus minimizing the amount of unproductive standing or down time. Even so, on larger projects a number of different types of compaction plant may be used, whereas on small sites only a limited amount of plant would be

TABLE 11.7. *Typical in situ bulk densities, bulking factors and load factors of common soils and rocks (after Horner 1988)*

Material		Bulk density (Mg/m^3)	Bulk-up on excavation %	Load factor
Soils				
Sand: uniform	loose	1.60–1.90	10–15	0.91–0.87
	compact	1.65–2.10	10–15	0.91–0.87
Sand: well graded	loose	1.75–2.20	10–15	0.91–0.87
	compact	1.90–2.25	10–15	0.91–0.87
Gravel		1.70–2.25	10–15	0.91–0.87
Sandy gravel		1.90–2.25	10–15	0.91–0.87
Clay: soft		1.65–1.95	20–40	0.83–0.71
firm		1.75–2.10	20–40	0.83–0.71
stiff		1.80–2.25	20–40	0.83–0.71
gravelly		1.65–2.30	20–40	0.83–0.71
organic		1.40–1.60	20–40	0.83–0.71
glacial		1.75–2.10	20–40	0.83–0.71
Loam		1.50–1.60	25–35	0.80–0.74
Peat		1.05–1.40	25–45	0.80–0.70
Topsoil		1.35–1.40	25–45	0.80–0.70
Rocks				
Granite		2.60–2.70	50–80	0.67–0.56
Basalt/Dolerite		2.70–2.90	50–80	0.67–0.56
Gabbro		2.80–3.00	50–80	0.67–0.56
Gneiss		2.70–2.90	30–65	0.77–0.61
Schist and Slate		2.70–2.90	30–65	0.77–0.61
Quartzite		2.60–2.75	40–70	0.71–0.59
Sandstone		2.45–2.65	40–70	0.71–0.59
Limestone		2.40–2.70	45–75	0.69–0.57
Marble		2.60–2.80	45–75	0.69–0.57
Chalk: Upper and Middle		1.65–2.05	30–40	0.77–0.71
Lower		2.00–2.40	30–40	0.77–0.71
Chert and Flint		2.50–2.60	40–70	0.71–0.59
Marl		1.90–2.35	25–40	0.80–0.71
Shale		2.15–2.60	30–65	0.77–0.61
Coal		1.25–2.60	35	0.741

TABLE 11.8. *Typical rolling resistance values for wheeled and tracked plant on hauling surfaces (reproduced with the permission of Caterpillar Inc.)*

Haul road condition	Rolling resistance factor % Tyred	Tracked
Hard stabilised surfaced well maintained roadway without penetration under load.	2.0	0
Firm smooth rolling roadway with dust or light surfacing, flexing slightly under load or undulating, maintained fairly regularly, watered.	3.0	0
Dirt roadway, rutted or flexing under load, little maintenance, not watered,		
25 mm tyre penetration/flexing	4.0	0
50 mm tyre penetration/flexing	5.0	0
Rutted dirt roadway, soft under travel, no maintenance, no stabilisation,		
100 mm tyre penetration/flexing	8.0	0
200 mm tyre penetration/flexing	14	5
Very soft, muddy, rutted roadway, 300 mm tyre penetration, no flexing	20	8
Loose sand and gravel	10	2

employed. Developments in the design of vibrating rollers have led to their adoption over a wide range of materials, largely at the expense of deadweight rollers. They may also be used in conjunction with other plant, such as with tamping rollers on cohesive soils to seal the surface at the end of shifts.

11.3.2. Quantities

It will often necessary to carry out an assessment of the quantities of the various materials that must be excavated, hauled to the different locations, placed and compacted or stockpiled in order that the numbers and types of plant required to carry out the various operations in the time allowed can be determined, or conversely to determine the time required to carry out the operations, and to establish the costs of the various operations. These quantities will usually be determined from survey data and from interpretation of the available information on the ground conditions and properties. The accuracy of these will depend to a large degree on the quality and extent of the available survey data and the ground investigations.

11.3.3. Spatial factors

The size of the site will strongly affect the size and amount of plant that can be used. In general, smaller plant will be used on smaller sites, as access within the site is more restricted. For example, large towed rollers may be used for compacting material on a major highway construction site, whereas smaller self-propelled or even pedestrian rollers will often be used on a small factory development. Space and duration may also dictate the type of plant that may best be adopted. For example, if excavated clay must be transported from a quarry across a major road to a brickworks over a long period, it may be more cost effective and safer to use a conveyor system as this could be installed in a narrow tunnel or overbridge that would not be affected by the road traffic. However in the short term, lorries or dump trucks could be used, provided that they could cross the road safely by a temporary bridge or through a traffic light controlled crossing.

Other major factors include the distances and gradients between the excavations and areas of deposition of the various materials in tips, lagoons, embankments, stockpiles and processing plants. The distance and gradient will affect the choice, cycle time, amount and operating cost of the haulage plant. If different hauls are required for different materials or at different times, it may be cost effective to used different haulage plant. For example, a tractor and box may be needed to excavate topsoil from above mineral workings and stockpile it nearby for use later in restoration, whereas the overburden may be best excavated by a back-acter and hauled in articulated or rigid dump trucks at different times to amenity embankments, tips and other areas being restored. In addition, if the different tipping areas are located at different distances, the trucks will take different times to complete the round trip and therefore different numbers and types of trucks may be required at different times as the haul distances change, to ensure that the excavator is operating efficiently. Similarly when excavating the underlying mineral deposit, most of the material will be taken to the processing plant. This may be carried out with dump trucks or conveyors but account should be taken of the increasing depth of the quarry and the changing haulage distance, possibly involving a circuitous route, on the operation of the plant. In construction projects, fill from a large cutting may be required initially in the adjacent embankments and later in embankments some distance away. The economics of the excavation and haulage may dictate that the plant for excavation and haulage is changed for the longer haul.

The geography of the area may provide a variety of controls that could affect earthmoving. The relief may affect the amount of excavation and filling required, and the length and gradient of haulage routes, which will affect the choice, cycle time, amount and operating cost of the haulage plant. In mineral working, and less commonly in construction projects, access to the site is often made in phases and this may affect the means of accessing the site at different times. The weather conditions may significantly affect the choice of plant to be used or the time that certain operations are carried out. For example, the efficiency of motor scrapers operating on clay is very dependent on the rainfall and it may be inappropriate to use them at all during the wet season, whereas articulated dump trucks loaded with a back-acter may be able to keep working for longer. Good quality haul roads are especially important, especially in and immediately after periods of wet weather if efficient earthmoving is to be carried out. The distance from existing major road or rail links may significantly affect the operating costs of mineral operations, particularly in the initial development phases and in transporting the products from the site. In addition, they may affect the costs of servicing the site. In some cases, set access routes between the main roads and the site may be imposed as a planning or contractual requirement or may be desirable to ensure optimum safety, minimum disruption to residents and to minimize wear on minor roads.

11.3.4. Surface water and groundwater

Earthworks may have temporary or permanent effects on surface and groundwater conditions. Excavations can cause significant reductions in groundwater levels, which may extend considerable distances from the works in permeable deposits. Streams and rivers may need to be diverted. Sidelong embankments and tips may inhibit surface water run-off and ponding can occur on their upslope sides. This could cause the basal soils or foundations to soften and lead to instability of the embankment. Embankments and tips founded on uncontrolled groundwater seepages or springs can become unstable, as can those in which high pore pressures are allowed to develop in the base or in underlying cohesive soils (Bishop 1973) These factors should be considered in the design of embankments and tips (HSC 1999).

Control of surface and groundwater is often important during in earthworks. During earthmoving in wet climates, the earthworks should be continually shaped to shed and collect the run-off and groundwater and to take it, via temporary ditches, drains, pipelines etc. to disposal locations. Run-off from bare soil surfaces can become polluted with suspended solids. If this water is to be discharged into water courses, intermediate settling ponds or lagoons may be required to reduce the amount of suspended solids to an acceptable level. If present, chemically contaminated water or leachate may need to be stored or treated. Temporary diversions of water courses should be carried out with care to ensure that expected flows are maintained and that flooding does not occur. In some cases it may necessary to ensure that fish can still migrate up or down the watercourse. Uncontrolled groundwater in excavations may cause flooding and/or instability and it may be necessary to de-water the ground by pumping from sumps, wellpoints or wells. The effluent may be discharged in ponds or lagoons elsewhere in the works, into watercourses or reinjected into the aquifer upstream or downstream. Water that has become contaminated during earthmoving or processing, should be collected and treated or safely disposed of and should not be allowed to escape into surface waters or groundwater. For example, oxidation of pyrite in Carboniferous shales and mudstones may generate sulphuric acid and there are examples of surface water on or discharging from earthworks with pH levels as low as 2.4 (e.g. Aitkenhead 1984; Pye & Miller 1990). In addition, earthworks may encounter groundwater from old mineworkings which may already be contaminated (e.g. Wood et al. 1999). The problem may be reduced when the earthworks are completed, graded, surfaced and vegetated and the amount of surface water coming into contact with the clays is reduced.

11.3.5. Plant and other factors

Once the nature and extent of the work has been determined and the controls and limitations determined, the output rates and operating costs of appropriate plant or plant teams for the different operations can be determined. Output rates can be calculated from a critical appraisal of the operations using manufacturer's output rates, preferably aided by data obtained from the use of the plant in similar situations. From this appraisal, and an assessment of the potential variables, the earthmoving and associated plant required to carry out each part of the works can be determined and costed. Some rationalisation may be required to optimize the use of plant by using the plant that is available on site at a lesser efficiency rather than mobilizing the more appropriate plant for small amounts of work. Some of the work may be subcontracted, in which case those costs would be included. Once this has been carried out, the programme and plant requirements can be finalized and costed. The programme and detailed plant requirements will need to be regularly reviewed during the works and modified where required. In order that earthmoving can be carried out efficiently and safely, the plant should be designed and constructed for the purpose for which it is to be used, well maintained and operated skillfully and correctly. Therefore allowance should be made for adequate maintenance, operator training and supervision.

Other factors may need to be considered such as new or local legislation, regulations and working practices. In some cases restrictions may be placed on the types of plant that may be used or the hours during which work may be carried out.

11.4. The control of earthworks

Wherever possible, earthworks are designed to perform their function using materials that are readily available on, or close to, a site, although special fills may need to be imported from further afield. The design may utilize the properties of the majority of the available materials (as with bulk fills) or require the materials to have particular properties (as with special or selected fills). Costs tend to increase when on site or local materials cannot be used in the construction of the earthworks.

Thus in construction, most embankments are designed to utilize the materials excavated from within the site or from nearby borrow pits or quarries. These materials need to have a minimum shear strength appropriate to the design, adequate bearing capacity to enable the embankment to be trafficked by construction plant, and sufficiently compacted to achieve the required strength and to minimise the amount of settlement of the fill. In certain circumstances the design may require that the fill has other properties, such as a minimum value of shear strength or CBR or minimum permeability. In order that materials that satisfy the design can be selected, it is necessary to define acceptability or suitability criteria for the materials to be used in the construction of the earthworks and/or performance criteria for the compaction of the embankments. This is discussed further in Chapter 10.

Structural embankments may be required in quarries for haul roads and water or slurry retaining structures. These may be designed in a similar manner to dams or other construction earthworks and the same considerations should apply. In addition, particular materials may be used to construct ancillary earthworks within quarries, such as amenity embankments and areas of landscaping which will require the selection of earthworks materials that will achieve the design (e.g. stronger earthworks materials and topsoil). However many of the embankments will consist of tips in which materials must be disposed of or stored. All tips should be sufficiently stable as not to pose any risk to life or property and modern tips should be designed accordingly. This may involve different tips to take different categories of waste from the quarry, including solid waste such as overburden, other unusable materials and dry processing waste, for topsoil and for contaminated waste, and lagoons for liquid waste including slurries and recycled water (HSC 1999).

11.4.1. Acceptability criteria for earthworks materials

In the UK, many earthworks contracts are carried out to the *Specification for Highway Works* (Highways Agency 2005) or similar, in which earthworks materials are classified into groups and subgroups based on material use and properties. Clays for general or bulk fill are included as Class 2 and as special fills in Classes 4, 7, 8 and 9 (see

Table 11.4). Clays are first subdivided by particle size and then by properties defined by the specifier. These may include plasticity and moisture content, however both of these tests take some hours to carry out. By contrast, the Moisture Condition Value (MCV) can be related to the undrained shear strength or moisture content and can be determined in minutes and the undrained shear strength of clays with relatively low gravel content can be measured in cuttings and compacted embankments with a hand vane. These tests provide the means for rapidly assessing the acceptability of many types of clay. Typically, for clays with liquid limits less than 50%, a MCV of between 7 and 8 would provide a lower limit of acceptability equivalent to an undrained shear strength of about 40 kN/m^2 and between 7 and 9 for a liquid limit over 50% (see Chapter 9). In order that the material is not too stiff to be compacted, MCVs of between 14 and 18 have been specified as upper limits typically equivalent to an undrained shear strength of about 150 kN/m^2. For dams, Kennard *et al.* (1978) reported that the specified strength range of clay cores in the UK was commonly between 45 and 110 kN/m^2. In many countries the lower limit of acceptability of materials in highway embankments is defined by the soaked or unsoaked CBR value.

By contrast, some specifications define the acceptability of earthworks materials by their ability to be compacted to a particular dry density or percentage air voids. Relative (dry) densities of between 95% and 100% of the Maximum Dry Density derived from the Standard Proctor Compaction Test which is equivalent to the British Standard Test using the 2.5 kg rammer (BS 1377:1990, Part 4, Methods 3.3 and 3.4) or between 90% and 95% of the Modified Proctor Test and the B.S. Test using the 4.5 kg rammer (BS 1377:1990, Part 4, Methods 3.5 and 3.6) have been used for the compaction of clay (Equation 11.4). In effect this also defines upper and lower limits of moisture content (see Chapter 9). As an alternative to the dry density, a minimum percentage air voids (often between 5% and 10%) may be adopted (Equation 11.5). This is calculated from the compacted dry density and moisture content and the particle density (specific gravity). However it should be noted that the measurement or assumption of a value of particle density may be prone to error and the air voids criteria may be satisfied simply by adding water to the soil during or after compaction. The upper and lower limits may therefore be further limited by minimum and maximum shear strength or minimum bearing capacity values. In the UK the use of relative compaction or air voids specifications require a large amount of relatively slow testing. This can be accelerated by the use of nuclear density gauges, but usually this is only justified on large earthworks such as earth dams.

11.4.2. Control during earthmoving

During construction the contractor is responsible for carrying out the works in accordance with the design and specification and a supervisor is often present to ensure that the design is viable and that work is carried out satisfactorily. Therefore the contractor and the supervisor have different but often overlapping roles in the construction process.

The contract should define who is responsible for classifying the earthworks materials during the works and both parties will need access to this information. The contractor needs to be able to classify the material so that he can carry out the works, as does the supervisor so that he can ensure that the design is being complied with and to check payment applications from the contractor.

By the start of the earthworks the contractor will usually have been provided with site plans and the records of the ground investigations with which he can assess the distribution of the various earthworks materials. The contractor will often supplement and update this information by carrying out his own detailed ground investigations, for example by using systematically distributed trial pits, backed up with laboratory testing, to classify and determine the distribution of the various materials in greater detail. The classification would define the nature of the materials and the likely compaction requirements. This enables the contractor to plan and resource the works in more detail and to sort out for himself and with the supervisor any problems that may be evident at this stage. The supervisor may also require further investigations at this stage to check or finalize the design assumptions. Once this initial stage of classification has been carried out, the contractor and/or supervisor would usually try to classify the majority of the materials in the excavations visually or by using simple rapid tests. Laboratory tests are then only carried out to confirm the visual classification and to classify marginal materials that could not be satisfactorily classified visually.

With a method specification for compaction, the contractor needs to define the nature of the different materials being delivered to embankments, spread them in layers of such thickness that comply with the specification for a specified number of passes of the plant chosen. He would use an engineer or foreman to direct the work, who would regularly check that the plant operators are spreading to the correct thickness and the compactor operators are uniformly providing the required number of passes. Problems often occur when either there is not enough spreading or compaction plant for the volume of fill being delivered, in which case there can be a temptation to increase the layer thicknesses or reduce the number of passes to complete the works. By contrast, many compactor operators believe incorrectly that additional compaction is always beneficial, so that when there is not enough fill to keep the compactors fully occupied, the operators increase the number of passes, but this can lead to deterioration of some materials, particularly weaker clay fills.

The extent to which the compaction plant is over or under resourced can often be determined by watching the operations and from a rapid assessment of the productivity of the haulage plant and the compaction plant.

Definition of terms used in the control of compaction

Compaction of earthworks is controlled by measuring the bulk density but calculating and comparing using the dry density

$$\text{Dry density} = \frac{\text{Bulk Density}}{1 + w/100} = \frac{\gamma_b}{1 + w/100}. \qquad (11.3)$$

The **relative compaction** is the dry density achieved in the field compared to the maximum dry density measured in one of the three possible compaction tests. It is imperative the particular test is defined (e.g. BS 1377:1990, Part 4, Method 3.3/Proctor Test) as each test is carried out with different amounts of compactive effort.

$$\text{Relative compaction} = \frac{\gamma_d}{\gamma_{md}} \times 100 \ (\%) \qquad (11.4)$$

With increasing compaction the **Percentage air voids** reduce. This is defined as:

$$\text{Percentage air voids} = 1 - \frac{\gamma_d}{\gamma_w}\left(\frac{1}{G} + \frac{w}{100}\right) \times 100 \ (\%) \qquad (11.5)$$

where:
w is the Moisture content (%)
γ_d is the Dry density (Mg/m³)
γ_b is the Bulk density (Mg/m³)
γ_{md} is the Maximum dry density (Mg/m³)
γ_w is the Density of water (Mg/m³)
G_s is the Particle density of soil
The main compaction tests are included in BS 1377:1990, Part 4, but are often referred to as:
Method 3.3: Standard, BS with 2.5 kg rammer or Proctor
Method 3.4: BS with 4.5 kg rammer or Modified Proctor.
Method 3.5: BS Vibrating Hammer

(see Chapters 9 and 10)

The potential output of a compactor can be determined from the following:

$$P = \frac{0.85WVD}{N} \qquad (11.6)$$

where P is the output of the compactor of compacted fill (m³/h); W is the effective width of the compactor (m); V is the speed of travel (km/h); D is the maximum depth of the compacted layer permitted by the specification (mm); and N is the minimum number of passes required in the specification.

The factor of 0.85 is allowed for overlap between adjacent compacted strips, turns and minor stoppages. If productivity is being monitored on a more extensive basis, further factors may be required to allow for inevitable delays such as meal breaks, maintenance, breakdowns and inclement weather.

The amount of fill being delivered to the embankment can be estimated from the number of loads being delivered by the haulage plant, which must be corrected for the reduction in volume caused by the compaction (after the load factor is taken into account) which usually will be slightly less than the bulk up (see Table 11.7).

In addition to ensuring that the overall design was being complied with, the supervisor should check that the layer thicknesses of the material spread by the contractor and the number of passes of the compactors are correct. Initially the contractor may carry out density tests to check the methods adopted were achieving the desired level of compaction assumed in the design but thereafter would only carry out or instruct tests if he believed that compaction was not adequate either due to the nature of the materials or the failure of the contractor to carry out the works correctly. Engineers and inspectors would normally be used to supervise the works.

Prior to compaction with a performance specification, the contractor will need to carry out a sufficient number of compaction tests to define the optimum moisture content and maximum dry density for the different materials to be compacted (see Chapters 4 and 9). He will usually carry out full-scale compaction trials to determine the optimum plant combinations, layer thicknesses and number of passes. It may be necessary to carry out further compaction trials in their full or abridged form during the works if the materials change or other plant becomes available. Both the contractor and supervisor would control the works as described above but, in addition, regular monitoring of the *in situ* density will be carried out, often jointly, by the contractor and supervisor to ensure that the target densities are being achieved.

In mineral workings, the earthworks are a consequence of the removal of the minerals and need only to remain safe and stable and satisfy the planning requirements. The materials are often classified on the basis of their potential end use to identify, for example, the topsoil, overburden, usable minerals of different types, materials suitable for the construction of amenity embankments and excessively wet materials. Once the extent of the mineral deposit or deposits have been defined by exploratory holes and laboratory testing, most of the classification is usually carried out visually at the time of excavation and the materials taken for processing or to the appropriate tip or embankment. Samples may be taken during excavation to check on the composition or quality of the mineral deposits. Poor control during excavation of the deposits can potentially lead to wastage of the minerals or increase the amount of extraneous materials taken for processing, which would increase processing costs. Modern tips should be constructed to a verified design. Supervision of the works should ensure that the tip is constructed to the design and that only materials allowed for in the design are incorporated in the tip. Systematic compaction is given to many modern tips although this is often not as stringent as for embankments in construction. Site control should ensure that the assumed degree of compaction is being provided uniformly across the tip.

Over-compaction can occur when the moisture content of the clay is in excess of the optimum moisture content for the compactive effort being applied (see Chapters 4 and 10). Typical symptoms are the remoulding of the surface of the compacted soil or severe permanent deformation upon passage of the compactor. It can occur with wet clays, especially in areas of heavy trafficking or when compaction is poorly controlled.

As modern specifications encourage the use of relatively wet clays, over-compaction should not be assumed before adequately investigating the problem. It can be useful to excavate a few trial holes to the base of the layer and/or carry out density tests at different levels. If the base of the layer is not saturated then over-compaction has not taken place and the compactive effort should be maintained at its original level. If this is not possible and it is necessary to reduce the compactive effort to minimize the symptoms of over-compaction, this should be considered as a temporary expedient and the compactive effort should be increased to the original level as soon as possible, even periodically on a trial basis.

Under-compaction occurs when the soil is drier than its design moisture content so that the material is left with a high percentage air voids content after compaction. As the material is dry, it will appear to be relatively strong, the surface will often appear well compacted and it will not deform under traffic loading. Therefore the condition is often difficult to identify. However trial holes will often reveal open voids between the stiff dry clay particles, particularly near the base of the layers. Under-compaction is serious if the soils are subject to the ingress of water from the surface or from below. The increase in moisture content will result in weakening and settlement of the soil. This can lead to instability of the side slopes and/or differential settlement of the surface some time after the completion of construction. In general, clay with a shear strength of over about 150 kN/m^2 or an MCV over about 14 may be susceptible to under-compaction. Under-compaction can be avoided by increasing the compactive effort by increasing the rating of the roller, reducing the layer thickness or, less efficiently, by increasing the number of passes of the roller. Alternatively, the material could be made easier to compact by increasing the moisture content by spraying the fill with water and mixing or turning prior to compaction. This may be appropriate for local areas involving relatively small volumes in moderate ambient temperatures. However the addition of water to 'heavy' clays of high plasticity and low permeability is not straightforward. The operation requires the breaking down of larger lumps and time for water to penetrate the material, together with suitable equipment and storage facilities for these processes.

11.4.3. Behaviour of clays during earthmoving

With good construction practices, firm and stiff clays can be compacted to produce stable fills. Most problems with clays during earthmoving occur due to inappropriate designs, with weak fills or due to excessive moisture within the fill or foundations.

Although the failure of the slopes of cuttings and excavations is beyond the scope of this review, it is important that the slopes of cuttings are properly investigated and designed to take into account all of the conditions that can realistically occur within their lifetime. In the case of temporary works in construction and slopes within quarries, this may involve assessment and reassessment of their stability prior to and during the works. Whilst analysis may show that 'safe' slope angles for temporary slopes in clay (based on the undrained strength) are steeper than for long term slopes (based on the drained strength), particular care should be taken during wet weather in areas of adverse groundwater conditions or in areas of past instability (see Chapter 4). Even instability of modest size can lead to injury or loss of life.

The foundations of embankments should be investigated prior to construction to ensure that they are capable of supporting the embankment. This is particularly important on areas of soft or potentially unstable ground. Soft weak foundations can lead to bearing capacity failure or excessive settlements beneath embankments and compressive materials such as peat can lead to very large total and differential settlements. It is sometimes economical to excavate soft and compressible soils to depths of about 3 m and replace them with acceptable fill. In excess of this depth or if particularly extensive, it may be more economical to incorporate measures within the design of the embankment to overcome the problems (e.g. by using geotextiles, piles, granular mattresses etc.) and/or by allowing the deformation to occur (e.g. with drainage below and surcharge on the embankment). Topsoil is usually removed for reuse from beneath embankments or tips and stockpiles. Local weak and compressible deposits should also be removed. Springs, drains and other water courses should be intercepted and discharged to a suitable outfall to ensure that unnecessary pore water pressures are not generated beneath the embankment or tip.

If embankments are constructed too quickly with wet fill, large positive pore pressures can develop in the lower layers and in highly stressed areas. As the pore pressure rises, the effective shearing resistance falls, which can lead to failure. Wet fills can be used for relatively high embankments if they are constructed more slowly or if drainage layers are incorporated to accelerate the dissipation of pore pressures. Alternatively the clays can be 'modified' using quick or hydrated lime to reduce their moisture content and increase their strength (see Chapter 10).

Deterioration of the working surfaces of clay embankments can occur for a variety of causes. Clay is particularly susceptible to damage due to wetting, desiccation and the passage of heavily loaded plant, such as motor scrapers and rigid dump trucks. Heavy rainfall can rapidly soften the surface of the clay and make it very slippery, thus reducing the grip of the tyres. It is desirable to cease work in heavy rain, except on dry soils, as, even if the plant can operate, it is likely to cause the material to degrade during excavation, transportation, placing and/or compaction. To prevent problems, earthworks should be constructed with regular falls to shed the surface run-off to the edge of the earthworks or drains. It may be necessary to place additional or sacrificial fill or replace areas of fill in areas where run-off is concentrated over long periods during construction. Similarly if an embankment is required to stand uncompleted, it may be necessary to place a protection layer over the top, which is often best constructed from clay.

Movement of traffic on wet fill can readily cause rutting, which can damage the long-term integrity of an embankment and reduce the efficiency of earthmoving. Wherever possible, purpose-made haul roads should be used. If this is not practical, heavy traffic should be distributed over the whole width of the embankment so that large ruts are not formed, although this needs careful control to ensure that the varying routes of the vehicles do not increase the safety hazards to pedestrians and other vehicles. Where deep rutting occurs, it should not be buried by additional fill but the soft material should be removed or allowed to dry out then and regraded and recompacted. Otherwise, where earthmoving must be carried out over soft ground, articulated dump trucks will often be able to access the area with less damage than other haulage plant.

11.5. Examples of earthmoving schemes

11.5.1. Clay and sand quarry

Miocene ball clays are quarried in conjunction with silica sand in the Donbas region of eastern Ukraine. The clays are quarried mainly for the production of white bodied fully vitrified (gres porcellanato) floor tiles, whilst the silica sand is presently stockpiled in anticipation of it being used mainly for glass making. The strata typically consist of 0.5 m of black friable topsoil, up to 25 m of stiff loam, up to 10 m of unconsolidated silica sand, about 5 m of interbedded damp unconsolidated sands, siliceous clays and thin silcrete overlying 2 m of stiff plastic ball clay. The main deposit of ball clay occurs within this 2 m sequence at a depth of between 25 m to 38 m, although there is also selective clay extraction from within the siliceous clay horizon located at the base of the overburden.

The quarry is 400 m to 500 m long, 300 m to 350 m wide and up to 40 m deep. In 2002, it was proposed to excavate 2.9×10^6 m^3 of overburden to release 370 000 tonnes of clay and advance the faces by approximately 200 m. As the operations are to a certain extent weather dependent, productivity diminishes during the winter months (especially in very wet periods). Overburden is stripped continuously to expose firstly the silica sand and then the ball clay. The waste material is then transported up to about 0.7 km to the worked out end of the quarry where it is placed in benches. When backfilling is complete, the backfilled workings are progressively restored to agricultural use (17 Ha were restored in 2001).

Within the overburden volumes, approximately 60 000 –90 000 m^3 of silica sand is being extracted per month and taken to stockpiles between 0.75 km and 1.25 km from the quarry face. The silica sand is selected on the basis of its Fe_2O_3 content and is worked in one or, more commonly, two benches. Up to 40 000 tonnes of ball clays are extracted per month. They are selected on the basis of their Al_2O_3 and Fe_2O_3 contents, and up to five seams are being extracted at any given time from the one main bench, and sometimes from within the thin siliceous clay seams at the base of the overburden. The excavated ball clays are transported 15 km to a covered clay storage area and shredding plant with railway sidings. Here the clay is blended and mixed to provide the required sales blend and is shredded if necessary before loading into railway wagons for onward transport.

Rainfall run-off from within the quarry and water derived from the overburden is directed via ditches to a sump within the workings. From here, a pump transfers the water to a settlement lagoon, from where clean water is released into a local water course. The earthmoving plant involved in these various operations is shown in Table 11.9.

11.5.2. Leighton–Linslade Bypass

The Leighton–Linslade Bypass consists of approximately 10.5 km of wide single carriageway road, typical of a modern rural by-pass. It runs to the south of Leighton Buzzard providing an east–west link and connecting a series of rural roads to the A5. At the western end it lies on an escarpment underlain by Glacial Sands, Gravels and Clays and falls eastward onto a wide vale that is underlain by Gault Clay and Gault Clay-derived Head deposits. The road profile is shown on Figure 11.3a.

The Gault Clay is typically a highly plastic to extremely highly plastic clay with a liquid limit typically between 40 and 90 but locally rises to about 120. The fact that much of the road was underlain by clay and the effects of glacial activity controlled much of the design and construction of the earthworks, particularly over the westernmost 3 km. The main features of the earthworks were as follows:

- Glacial sands and gravels with subsidiary clays were present above the escarpment in the first 0.5 km. The ground on the underlying escarpment contained a

TABLE 11.9.

Material	Excavation	Transportation	To
Overburden-topsoil and loam	Formerly as Silica Sand (see below) but scraper units have been recently been introduced consisting of an American 306 kW (410 hp) rubber-tracked tractor towing two scraper boxes each with a capacity of 13.3 m^3.		Worked out quarry. The material is spread by Russian bulldozers (being replaced by 2 American 130 kW (175 hp) dozers) and compacted by the earthmoving plant
Overburden – Silica Sand	Small Russian built draglines, being progressively replaced by American 215 kW (290 hp) back-acters with 2.5 m^3 buckets	Each excavator loads to up to three Ukrainian or Czech lorries with load capacity of 12.5 to 13.2 tonnes	Stockpile
Ball Clay (in situ)	Using a 96 kW (128 hp) and a 165 kW (222 hp) back-acter with a 1.0 m^3 and 1.5 m^3 bucket respectively	Up to ten Ukranian, Czech and Belorussian lorries per excavator, each transporting between 12 and 13.5 tonnes of clay. These are being replaced by new Swedish 25 tonnes capacity lorries	To stockpile area at railhead 15 km distant
Ball clay storage sheds	Four front end loaders used to transfer the clay		Shredders and/or railway wagons

spring line associated with peat deposits, shallow landslips and solifluction shear planes. Below the escarpment, the Head and upper part of the Gault Clay generally contained polished surfaces thought to be formed from solifluction or freeze–thaw activity, thus weakening the soil mass. This shearing became less common in the flatter area beyond the first 3 km.

- Plant movement along the scheme was inhibited by a main railway line, road, river and canal between 1.15 km and 1.45 km. Although there was a narrow bridge under the railway, these prevent the free flow of earthworks materials along the scheme. In order that fill materials could be brought from the east across to the area between the railway, the contractor constructed a temporary bridge across the river and canal.
- Two embankments up to 12 m high lay between 0.5 km and 1.4 km, which were underlain by sheared clay. As the factor of safety of these embankments was likely to be lowest during and immediately after construction, and in order to maintain a factor of safety of at least 1.3, the design of these required a basal drainage layer incorporating a high strength geotextile layer, a rate of construction limited to 1 m height per week to minimize pore pressure rises in the foundations and a rest period of three months after completion to capping layer to enable settlements to take place. In addition more extensive drainage and stabilization works were required in the embankment foundations on and immediately below the escarpment. Extensive instrumentation was installed at an early stage and during embankment construction to monitor the stability of these embankments.
- The scheme required about 550 000 m^3 of bulk fill plus a further 100 000 m^3 of special granular fill for the basal drainage layer and road capping layer. The main three cuttings at 0–500 m, 1665–2820 m and 3490–5670 m would provide about 400 000 m^3 of the bulk fill. The slopes of the first cutting in predominantly sands and gravels were flattened from the initial design slope of 1:2.5 to 1:8 to provide 50 000 m^3 of additional fill, which enabled the land forming the cutting slopes to be returned to agricultural use (in this case forestry). The fill obtained from the other two cuttings was predominantly cohesive Head and Gault Clay. The shortfall of bulk fill had to be imported. The contractor found much of the additional bulk fill on a local site that had a surplus of Gault Clay fill and imported it to the road/river corridor. The remainder of the fill was obtained from a chalk quarry a short distance north of the site and was taken on a haul across agricultural land and along the site to the eastern end of the site, thus eliminating the need to use public roads. The volume balance of the earthworks materials is shown diagrammatically in Figure 11.3b.
- In addition to the usable clay, the works produced about 160 000 m^3 of unacceptable material that needed to be disposed of, most of which occurred in the three large cuttings. The contractor was able to tip the materials in the disused workings of active sand quarries which lay immediately to the north of the route at 1.0 km and south of the route at 2.2 km from the western end of the site. Planning permissions were readily obtained for this as it assisted in the restoration of the sand quarries, which are relatively common in this part of Bedfordshire.
- Some of the materials that would normally be classified as 'unacceptable' (for use in the construction of embankments) were extremely plastic clays with liquid limits above 90. Approximately 10 000 m^3 of clay with liquid limits between 90 and 120 were used in the high embankment between 1.15 km and 1.4 km but were placed in the core at least 6 m below the surface of the embankment.

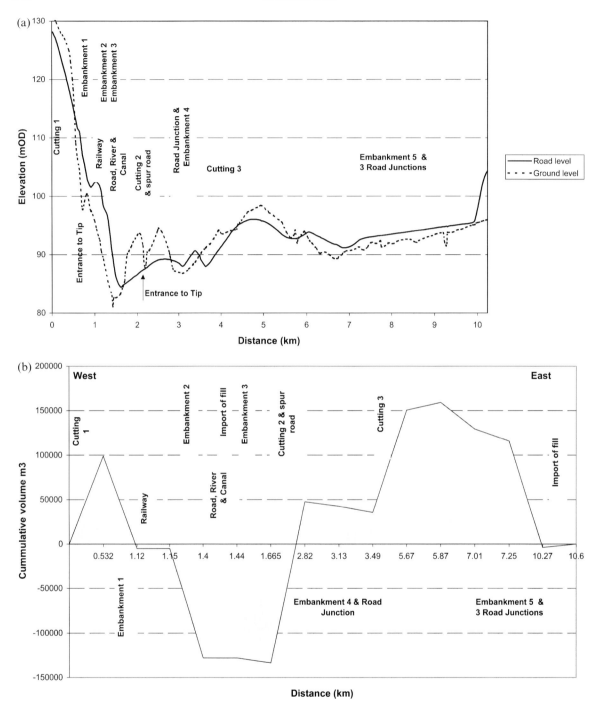

FIG. 11.3. (a) By-pass — longitudinal profile, (b) By-pass — mass haul diagram.

- In view of the limited access across the railway and canal the large amounts of unacceptable materials, the restricted rates of working on the high embankments and the lengths of the hauls, the contractor opted to carry out most of the bulk earthworks using teams of articulated dump trucks loaded by large back-acters. Motor scrapers were used to strip most of the topsoil and were used for some of the bulk earthworks during peak periods, particularly at the western end of the site. Compaction of the mainly granular fill to the west of the railway and the chalk at the eastern end of the scheme was carried out by towed and self-propelled vibrating rollers, whilst the clays were compacted using self-propelled tamping rollers.

11.6. Safety on earthmoving sites

Construction sites and quarries are inherently dangerous working environments. Accidents are still numerous and cause death, injuries, physical suffering and mental anguish, in addition to losses in productivity, earnings and morale. The Health and Safety at Work Act 1974 provides the framework of legislation in the UK and a number of derivative Regulations have been since introduced to improve safety in different areas of work. These place obligations on everyone in the work place from the company to individual employees and, in general, the greater a person's knowledge or authority the more is expected of him. In many cases, individual sites and quarries will have one or more people specifically responsible for health and safety aspects, including risk assessment, safety monitoring and training. If in doubt, advice should be sought from specialist safety officers or more senior personnel. Everyone working in these industries needs to make themselves aware of the potential hazards that exist so that they can plan their work in such a way that prevents them causing or receiving harm. When dealing with the great forces of nature or large earthmoving plant, one may not get a second chance. Comments on some of the main safety hazards associated with earthworks and earthmoving are given in the following paragraphs and further guidance is given in publications such as Health and Safety in Construction (HSE 1996) and Health and safety at quarries (HSC 1999).

11.6.1. Safety of earthworks

The failure of temporary slopes and trenches is a common cause of accidents on construction sites. An appraisal of the safety risk should always be made and if necessary specialist geotechnical advice obtained. Trenches over 1.2 m deep that must be entered should always be properly supported, which in very poor ground may involve close sheeting with timber and/or steel or the use of a drag box. Special care should be taken when groundwater is encountered as this may reduce the stability of slopes and trenches.

Unless specifically allowed for in the design, heavy plant should not be permitted close to the edge of cuttings, trenches, embankments, tips and the like. Where these edges are more permanent, such as on haul roads or when tipping from a ramp into a hopper, crash barriers or steep edged bunds should be provided to prevent vehicles from driving over the edge.

Care should be taken when excavating near to existing structures or earthworks. They may need special support to ensure their stability. Excavation at the toe of existing tips and embankments has caused significant numbers of slope failures and has led to injury.

Care should be taken when excavating at the base of steep or overhanging faces. The slope should not be higher than the machine can reach and the machine should not be in a position that failure could cause a hazard to the driver, the excavator or the haulage plant.

Where bench working is being carried out, care should be taken to ensure that higher benches are not later undermined or that there is a risk of material falling or rolling from the upper benches.

Barriers should be placed round construction sites and quarries to discourage trespass on to the site. In low risk areas this could consist of mounds or simple fences, stock-proof fencing adjacent to grazing animals and high metal paling fencing in or near urban areas where there is persistent trespass by children and the hazards are significant. Particularly dangerous areas that people could access should be identified with signs and barriers to prevent inadvertent entry. These may include areas at the top or base of steep faces, unstable areas, areas of water, particularly noisy areas or in areas with large amounts of dust. In some circumstances, patrolling security guards may be needed to ensure that unauthorized people do not enter potentially hazardous areas.

The use of explosives should be carried out in accordance with the appropriate regulations and the site rules. In particular, only authorized personnel, such as the shotfirer and those under his direction, should handle explosives. A safe system of work is required to ensure that people do not remain within the danger zone when blasting is being carried out. If excessive fly rock is created, the blasting techniques should be revised and the use of mats and other precautions considered.

Where contaminated ground, ground containing noxious or potentially explosive gases, unexploded bombs or similar hazards are suspected or unexpectedly encountered, the immediate area should be cleared and expert advice should be obtained before continuing with any work to ensure that it can be carried out safely. When working in areas of contaminated ground, operatives and those working in the area should be provided with appropriate clothing and personal protection equipment. Adequate changing, showering and washing facilities should be provided.

11.6.2. Safety related to earthmoving plant

Many injuries and deaths still occur from the use of mobile plant in quarries and on construction sites. Plant and equipment should only be used in places, operations and under conditions for which they are suitable. Particular care should be taken when operating plant during periods of inclement weather and defective plant should not be used. Many injuries have been caused from vehicles overturning and, where appropriate, sturdy cabs should be provided to protect the operator from the risks of rolling over and from the impact of falling objects (e.g. face collapse). Seat belts should be worn by drivers and by passengers when practical.

Visibility from large vehicles is often poor, particularly when reversing, and pedestrians should keep away from them wherever possible. If this is difficult, for example when carrying out *in situ* tests or monitoring on the earthworks, drivers of heavy plant should be briefed to expect the hazard and the potentially hazardous working areas clearly signed in advance and at the location. Sitting in a light vehicle may not ensure safety as it can be easily crushed by large plant such as a bulldozer or dump truck. If light vehicles need to be near heavy plant, they should be clearly identified by distinctive colours, warning lights and other means.

Heavy plant should be directed along clearly defined and marked haul routes. Wherever possible pedestrians should be kept away, preferably along clearly identified pedestrian routes. The use of one-way traffic systems should be used if this can minimize the amount of reversing required.

Haul roads and benches should be sufficiently wide for the size and volume of traffic expected. They should be regularly maintained in good condition as this will minimize braking distances, optimize the operator's control of the vehicle and minimize long-term health problems due to whole body vibrations. Bunds or barriers should be provided in those areas where there would be a significant risk to the driver or others if the vehicle left the road, such as above steep faces or lagoons.

Tipping on embankments should be carried out short of the edge and the fill then dozed to the edge. If tipping is required at the edge, stop chocks should be provided.

Plant is more stable running up and down a slope than across it and the latter should be avoided whenever possible.

The existence of public utility apparatus and services should be checked before the start of earthmoving and, where found, marked, protected and diverted as required. Areas of restricted access, such as beneath overhead cables and soffits on narrow bridges or between temporary works, should be identified and clearly indicated by signs, tapes, restraints or frameworks.

Where mud could be or is spread on public road by the wheels of vehicles leaving construction sites or quarries, there may be an increased risk of accidents. The use of wheel washers and/or road cleaning equipment may be required to minimize this possibility, particularly in winter.

In conclusion, everyone working on construction sites, in mines and in quarries should be aware of the inherent general risks of the working environment and the potential dangers of that particular site.

References

AITKENHEAD, N., RILEY, N., BALL, T. K., NICHOLSON, R. A., PEACHEY, D., BLOODWORTH, A. J., ROUSE, J. E., MILLER, M. F. & THRIFT, L. 1984. *Carsington Dam—Geological, Geomorphological and Geochemical Aspects of the Site*. Final Report. British Geological Survey, Keyworth.

ANON 2000. UK's Biggest Mineral Exporter is English China Clays. *International Mining and Minerals*, 3, May.

BISHOP, A. W. 1973. The stability of tips and spoil heaps. *Quarterly Journal of Engineering Geology*, **6**, 335–376.

BISHOP, A. W., HUTCHISON, J. N., PENMAN, A. D. M. & EVANS, H. E. 1969. Geotechnical investigations into the causes and circumstances of the disaster of 21st October 1966. A selection of technical reports submitted to the Aberfan Tribunal, Welsh Office. London; HMSO.

CATERPILLAR INC. 2000. *Handbook of Ripping*. Twelfth edition.

CATERPILLAR INC. 2004. *Caterpillar Performance Handbook*. Edition 35.

DEPARTMENT OF THE ENVIRONMENT/GEOFFREY WALTON PRACTICE. 1991. *Handbook on the Design of Tips and Related Structures*. HMSO, London.

FARRAR, D. M. & DARLEY, P. 1975. *The Operation of Earthmoving Plant on Wet Fill*. Department of the Environment. Transport and Road Research Laboratory Report 688, Crowthorne.

HEALTH AND SAFETY EXECUTIVE (HSE). 1997. *Health and Safety in Construction*. HSE, London.

HEALTH AND SAFETY COMMISSION (HSC). 1999. *Health and Safety at Quarries*. Quarries Regulations 1999. Approved Code of Practice. L118. HSE, London.

HIGHWAYS AGENCY. 2005. *Manual of Contract Documents for Highway Works*. Volume 1. Specification for Highway Works. The Stationery Office, London.

HORNER, P. C. *Earthworks*. 1988. I.C.E. Works Construction Guide. Second edition. Thomas Telford, London.

KENNARD, M. F., LOVENBURY, H. T., CHARTRES, F. R. D. & HOSKINS, C. G. 1978. Shear strength specification for clay fills. Clay Fills. Institution of Civil Engineers, London.

MACGREGOR, F., FELL, R., MOSTYN, G. R., HOCKING, G. & MCNALLY, G. 1994. The estimation of rock rippability. *Quarterly Journal of Engineering Geology*, **27**, 123–144.

MAIDL, B., HERRENKNECHT, M. & ANHEUSER, L. 1996. *Mechanised Shield Tunnelling*. Ernst and Sohn, Berlin.

MONAHAN, E. J. 1994. *Construction of Fills*. Wiley, New York.

NICHOLS, H. L. & DAY, D. A. 1998. *Moving the Earth: the Workbook of Excavation*. Fourth edition, McGraw Hill, New York.

PETTIFER, G. S. & FOOKES, P. G. 1994. A revision of the graphical method for assessing the excavatability of rock. *Quarterly Journal of Engineering Geology*, **27**, 145–164.

PYE, K. & MILLER, J. A. Chemical and biochemical weathering of pyritic mudrocks in a shale embankment. *Quarterly Journal of Engineering Geology*, **23**, 365–381.

STEPHENS, J. H. 1976. *The Guiness Book of Structures*. Guiness Superlatives.

THOMPSON, J. 1993. *Pipejacking and Microtunnelling*. Blackie/Chapman and Hall, Glasgow.

THURLOW, C. 1997. *China Clay from Cornwall and Devon*. An illustrated account of the modern China Clay Industry. Cornish Hillside Publications. Second edition.

WOOD, S. C., YOUNGER, P. L. & ROBINS, N. S. 1999. Long-term changes in the quality of polluted minewater discharges from abandoned underground coal workings in Scotland. *Quarterly Journal of Engineering Geology*, **32**, 69–79.

Additional Reading

BRITISH LIME ASSOCIATION. 1990. Lime Stabilisation Manual.

BRITISH STANDARDS INSTITUTION. 1990. *British Standard Methods of Test for Soils for Civil Engineering Purposes*. BS 1377.

BRITISH STANDARDS INSTITUTION. 1990. *Stabilised Materials for Civil Engineering Purposes*. BS 1924.

BRITISH STANDARDS INSTITUTION. 1999. *Code of Practice for Site Investigations*. BS 5930.
BRITISH STANDARDS INSTITUTION. 1981. *Code of Practice for Earthworks*. BS 6031.
BRITISH STANDARDS INSTITUTION. 2003. Unbound Mixtures. Specifications. BS EN 13285.
BRYAN, H. 1996. *Planning Applications and Appeals*. Butterworth Heinmann, London.
BUDLEIGH, J. K. 1989. *Trench Excavation and Support*. Thomas Telford, London.
CIRIA. 1994. *Environmental Assessment*. Special Publication 96. CIRIA, London.
CIRIA. 1998. *Site Safety* (CD Rom). Special Publication 130-CD. CIRIA, London.
BIELBY, S. C. rev READ, J. A. 2001. *Site Safety Handbook*. 3rd Edition. Special Publication 151. CIRIA, London.
DENNEHY, J. O. 1978. The remoulded undrained shear strength of cohesive soils and its influence on the suitability of embankment fill. *Clay Fills*. Institution of Civil Engineers, London.
DUXBURY, R. M. C. 1996. Telling and Duxbury's Planning Law and Procedure, Tenth edition. Butterworths, London.
EKINS, J. D. K., CATER, R. & HOUNSHAM, A. D. 1993. Highway embankments—their design, construction and performance. *Engineered Fills*. Thomas Telford, London.
ERVIN, M. C. 1993. Specification and control of earthworks. *Engineered Fills*. Thomas Telford, London.
HODGETTS, S. J., HOLDEN, J. M. W., MORGAN, C. S. & ADAMS, J. N. 1993. Specifications for and performance of compacted opencast backfills. *Engineered Fills*. Thomas Telford, London.
JONES, R. H. & GREENWOOD, J. R. 1993. Relationship testing for acceptability assessment of cohesive fills. *Engineered Fills*. Thomas Telford, London.
KWAN, J. C. T., SCEAL, J. S., BRYSON, F. E., STANBURY, J., BICKERDIKE, J. & JARDINE, F. M. 1977. Ground engineering spoil: good management practice. Report 179. CIRIA, London.
MAIDL, B., HERRENKNECHT, M. & ANHEUSER, L. 1996. *Mechanised Shield Tunnelling*. Ernst and Sohn, Berlin.
PARSONS, A. W. & DARLEY, P. 1982. The effect of soil condition on the operation of earthmoving plant. Department of the Environment. Transport and Road Research Laboratory Report 1034, Crowthorne.
PARSONS, A. W. 1992. Compaction of Soils and Granular Materials: A review of research performed at the Transport Research Laboratory. Department of Transport. Transport Research Laboratory. HMSO, London.
RYDIN, Y. 1998. Urban and Environmental Planning in the UK. McMillan Press, London.
SOMMERVILLE, S. H. 1986. Control of groundwater for temporary works. CIRIA Report No. R113, CIRIA, London. 88pp.
STEEDS, J. E., SHEPHERD, E. & BARRY, D. L. 1996. A guide for safe working on contaminated sites. Report R132. CIRIA, London.
THOMPSON, J. 1993. *Pipejacking and Microtunnelling*. Blackie/Chapman and Hall, Glasgow.
TRENTER, N. A. 1999. Engineering in glacial tills. Report C504. CIRIA, London.
TRENTER, N. A. 2001. Earthworks: a guide. Thomas Telford, London.
WHYTE, I. L. & VAKILIS, I. G. 1985. Smooth slip planes in clay fills resulting from soil machine interaction. *Failures in Earthworks*. Thomas Telford, London.

Relevant Legislation

Relevant legislation includes:
Construction (Design and Management) Regulations 1994 (CDM)
Construction (Health, Safety and Welfare) Regulations 1996
Control of Substances Hazardous to Health (COSHH) Regulations 1999
Environment Protection Act 1990
Control of Explosives Regulations 1991
Health and Safety at Work etc. Act 1974
Management of Health and Safety at Work Regulations 1992
Mines and Quarries Act 1954
Mines and Quarries (Tips) Act 1969
Mines and Quarries (Tips) Regulations 1971
Noise at Work Regulations 1989
Personal Protective Equipment at Work Regulations 1992
Provision and Use of Work Equipment Regulations 1998
Quarries Regulations 1999
Workplace (Health, Safety and Welfare) Regulations 1992

12. Specialized applications

12.1. Principles

12.1.1. Scope of chapter

This chapter essentially covers the specialized uses of clay in construction not included in the more routine applications covered in other chapters. There is a very wide range of such specialized applications, so none can be treated in great depth in this book. In addition, some of the applications, especially in the areas of environmental engineering, are undergoing rapid development. Readers interested in particular applications should therefore consult the appropriate references given and be prepared to search for more recent publications.

The applications discussed in this chapter may be divided into two main categories, though there is some overlap between the two. The first category includes the use of clay slurries in drilling, piling, diaphragm wall construction and tunnelling. In most cases the slurry is used as a construction expedient to provide fluid pressure, support soil particles in suspension to prevent sedimentation, and to act as a medium of transport for excavated material. In some cases the slurry may be left in place to form an impermeable barrier; in such cases the clay slurry may be mixed with natural soil and/or cement to achieve a semi-solid final state. Clay or clay/cement slurries may also be used as grouts to seal permeable natural ground for either short- or long-term purposes.

The other principal category includes uses where plastic solid clay is employed to form impermeable barriers or waterproof layers, most commonly in the construction of engineered landfill facilities or for the containment of hazardous solids or liquids. In these cases the applications are making use of the low hydraulic conductivity of clays, which is maintained even when the material is deformed, due to the clay's ability to strain plastically without cracking. A traditional form of such material is 'puddle clay', widely used in the past for lining canals and forming the core of earth embankment dams.

In the great majority of cases the clay used in these applications is bentonite, although attapulgite (syn. palygorskite) has occasionally been used for slurries mixed with salt water. The nature, origin and properties of bentonites are covered in Chapters 2 to 4, while important aspects of their behaviour in the context of this chapter are covered in the next three sections. In some cases, natural locally occurring clays may be employed instead of processed bentonite; such clay will usually be of high plasticity and have a significant content of bentonite-type minerals.

12.1.2. Bentonite

The name bentonite is popularly used for a range of natural clay minerals of the smectite group, principally potassium, calcium and sodium montmorillonites derived from the weathering of feldspars. The name derives from the discovery of large deposits near Fort Benton in Wyoming, USA. Because of the chemistry and microstructure of the clay particles, they have a strong ability to absorb water and are able to hold up to ten times their dry volume by absorbtion of water. Montmorillonite (after Montmorillon, southwest of Paris) consists of very thin flat crystalline sheets of clay minerals which are negatively charged and are held together in 'stacks' by positively charged sodium or calcium ions in a layer of adsorbed water. In particular the soil particles comprising a stack of sheets of sodium montmorillonite form extremely small and thin platelets, being typically of the order of 1.0 μm or less in length and 0.001 μm thick. The ability to absorb water comes from the relatively low bonding energy of the sheets, which allows water molecules to be adsorbed onto the internal and external sheet surfaces. Calcium ions provide a stronger bond than sodium, so that calcium montmorillonite swells less readily than sodium montmorillonite. Potassium ions provide much stronger bonding between clay sheets as the potassium ion is of exactly the right diameter to fit between atoms in the sheet structure with negligible gap between the clay sheets. A similar material to montmorillonite but with potassium bonding is the non-swelling clay mineral known as illite. The substitution of sodium by calcium or potassium ions in montmorillonite greatly reduces the ability of the clay structure to hold water.

The very small particle size of bentonites results in an extremely low hydraulic conductivity for intact clay, with a coefficient of permeability of typically less than 10^{-10} m/s. This allows the clay to be used to form 'impermeable' or 'waterproof' layers and sustain high hydraulic gradients across thin layers with negligible water flow. The swelling property is also important in such applications, since should water permeate a layer of dry bentonite it will swell even against high pressures and tend to seal any crack or fault which might otherwise develop into a leakage path. The volumetric swelling of particles can be up to 13%, but that of an agglomeration of particles is somewhat less depending on their packing.

Many applications of bentonite involve the use of slurry. Mineral particles in a slurry generally carry electrical charges, the nature and intensity of which vary

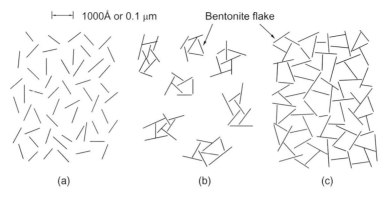

FIG. 12.1. Structures of bentonite slurry: (a) dispersed, (b) flocculated, (c) gel.

with the particle surface characteristics and the chemistry of the liquid phase. Polar water molecules may then be adsorbed on to the particle surface, forming a layer of 'bound' water surrounding each particle. The result of the two effects is to produce repulsive forces between particles, which are greater than attractive Van der Waal's forces except when the particles are very close together. The particles in a slurry therefore tend to keep apart from each other in a 'dispersed' condition (Fig. 12.1a). The effects are most noticeable with small particles (clay/silt rather than sand/gravel, and in practical terms only with finer clay particles) since the relative surface areas are much larger, and gravitational forces are much smaller. Under some conditions the plate-like particles of clay minerals may have different charges on the edges and faces of the particles, and are able to clump together in a 'flocculated' structure (Fig. 12.1b). The large flocs settle out of the slurry much more readily than the small individual particles.

Some slurries demonstrate the effect known as thixotropy, whereby they 'set' into a gel if left undisturbed, but revert to a viscous fluid (sol) when sheared. The alternation between sol and gel may take place any number of times. The phenomenon is well known in 'non-drip' paints. A gelled 'house-of-cards' type of structure with edge to face connections is illustrated in Figure 12.1c; gels of thin clay particles may contain only a few per cent of solid material. The gelled structure is also able to support larger soil particles and prevent them from settling out. Bentonite slurries are thixotropic and typically form a gel at concentrations of a few per cent by mass in water; this is an important property of bentonite slurries in many applications. For a more detailed discussion of the nature and properties of bentonite slurries see Jefferis (1992).

Bentonite clays occur, and are mined and processed commercially, in many parts of the world. Some natural deposits, notably those from Wyoming, have a high proportion of sodium. These tend to produce slurries with high viscosity but relatively low gel strength. The deposits mined in the UK, near Woburn, are mainly of the calcium form, and these are converted by ion exchange to the sodium form by ball-milling with sodium carbonate. These materials tend to be less dispersive and give lower viscosities for the same slurry density, but higher gel strengths. As natural products, bentonites vary widely around the world in quality and content of other minerals, even after commercial processing, and these variations must be taken account of in their specification and use.

Bentonite is available commercially in a variety of forms, but nearly always in a dry state, as powder (in bulk or bags, like cement), pellets or blocks. For applications in construction it will usually be hydrated, although in some waterproofing materials the hydration is allowed to occur *in situ*. For use as a slurry, the bentonite is mixed with water at a rate of a few per cent of solids by mass. The aim is normally to produce a slurry in which the bentonite particles are well dispersed and fully hydrated. For good mixing and rapid hydration, a high-shear colloidal mixer (shear rate $>900/s$) should be used, and the slurry then left to stand for some time while the clay particles hydrate. The quality of the slurry obtained depends on the hydrogen ion concentration (pH) of the water used in mixing; saline or acidic water or water containing impurities may cause the clay particles in the slurry to flocculate. This may initially cause the slurry to 'thicken', but there will then be a tendency for the flocculated particles to settle out of suspension and form a sludge. However there is not normally a practical problem with seawater coming into contact with a slurry, provided the slurry cannot mix freely with the seawater and has previously been fully hydrated with fresh water. Deliberate flocculation with flocculating agents may be used to help remove bentonite from suspension when the slurry is no longer required or has become too contaminated with cement, clay or silt. A combination of low hydraulic flow into the slurry (so long as hydraulic heads are low), and long diffusion times for salt compared with exposure times, usually causes few problems in the presence of seawater.

Bentonite is also used in combination with other materials, in particular other soil materials and Portland cement. At one extreme a small quantity of bentonite may be added to a concrete mix to produce highly plastic concrete able to undergo quite large deformations without cracking; while a small quantity of cement in a bentonite slurry can produce a hardening slurry with a small shear strength. Natural clay, silt and sand may be used as 'fillers' to produce cheaper material while keeping most of the benefits of the sealing ability and low permeability of the bentonite. Gleason et al. (1997) found that about 5% of sodium bentonite and 10–15% of calcium bentonite had to be added to fine sands to achieve a sand–bentonite mix with a permeability of less than 10^{-9} m/s. Hardened bentonite-cement slurry mixes containing 180 kg/m³ of cement and 60 kg/m³ of bentonite had permeabilities of about 10^{-7} m/s with calcium bentonite and 10^{-8} m/s with sodium bentonite. These mixtures are discussed further below in relation to various different applications. Small quantities of polymers and other chemical additives may also be used to enhance or modify the properties of bentonite slurries for particular applications. These are also discussed further below.

12.1.3. Interaction of slurries and natural ground

The interaction of slurries with the ground is considered in this introductory section because it is important in many different applications. It may be required for a slurry to permeate the ground when it is being used as a grout to reduce the permeability or increase the strength of the ground. On the other hand, the slurry may be required to seal the ground at the interface so that the fluid pressure of the slurry may be transmitted to the soil particles within the natural ground and provide support. Penetration of slurry into the ground, or lack of it, is controlled by two principal effects known as pore blocking and rheological blocking. Pore blocking is the mechanical effect whereby, as the slurry tries to infiltrate the ground, agglomerations of particles from the slurry become wedged in the pore channels of the ground, thereby blocking the channels and preventing further inflow of slurry. Rheological blocking occurs when slurry has penetrated more deeply into the ground, until the pressure gradient becomes too small to maintain flow of the slurry through the pore channels, the slurry gels, and no further flow can occur. In practice the two effects act together and their relative importance depends on the nature of the slurry, the nature of the ground, and the pressure difference applied.

Penetration of slurries has been studied in relation to grouting of soils by Raffle & Greenwood (1961). They quote the expression:

$$s = \frac{\Delta p \cdot \alpha}{2.0 \tau_s}$$

where s is the penetration distance, α is the average minimum pore size (1/10 of the average minimum particle size), τ_s is the shear resistance of the slurry, and Δp is the difference between the slurry pressure p and the ground water pressure u. The German Standard DIN 4126 1986 substitutes d_{10} for α for tunnelling, where d_{10} is the particle size such that 10% by mass of soil is of smaller size, but this is not consistent with experience from grouting applications. Jancsecz & Steiner (1994) produced a similar formula but with the number 2.0 replaced by 3.5, giving almost half the penetration distance, more consistent with grouting experience. The penetration seems likely to be influenced by soil density, grading and particle shape as well as by d_{10} particle size. Jefferis (1992) provides an expression which includes the porosity of the soil n:

$$s = \frac{\Delta p \cdot d_{10}}{\tau_s} \cdot \frac{n}{(1-n)} \cdot f$$

where f is a factor to take account of the geometry and tortuosity of the flow paths within the soil, and may be about 0.3. With clean bentonite slurries, typical values of Δp, and typical shear strengths of the slurry of 20 to 50 Pa, penetrations of several metres result in soils coarser than medium sand. However, if the slurry contains larger particles of cement, silt or fine sand, which help to block the pores, penetration is greatly reduced.

The influence of cement on the shear resistance of hydrated bentonite slurries can be judged from Figure 12.2. The curve for 0% bentonite represents neat cement grouts. Even small additions of bentonite to grouts increase their shear resistance dramatically, thus reducing their ability to permeate fine soil pores. The suspension/cement rates approximating to the shear resistance minima are typical of diaphragm walling and piling hole support slurries. The minima arise relative to clean bentonite (shown as higher suspension/cement ratios) because of the degrading of the bentonite by free calcium from the cement. Thus the supporting capacity of bentonite for soil particles is reduced causing increased sedimentation and 'bleed' of clear water.

Where the slurry is required to support the ground by fluid pressure, it is best if the interface is effectively sealed so that the fluid pressure is transmitted with a large pressure gradient within the natural ground. This is achieved by the formation of a 'filter cake', a thin layer of highly impermeable bentonite 'caked' or 'plastered' on to the interface (Fig. 12.3). This occurs when (as usually) there is a range of particle sizes in both soil and bentonite; the coarser particles in the slurry filter out in the finer pore sizes of the soil. As long as some flow continues into the coarser pores of the soil the filtration also continues, gradually thickening to cake the soil surface with a mixture of silt sized particles in a matrix of bentonite clay from the slurry. This occurs with clean bentonite slurries in sand and silt; in coarser soils the slurry will dissipate into the ground without forming a filter cake, while in clays there will be insufficient inflow for a thick cake to be formed (and support pressure may be provided simply by water without need for a slurry). Filter cake formation in coarser soils may be encouraged by inclusion of silt and clay particles in the slurry to act as pore-blockers.

High molecular weight long-chain polymers may also be incorporated in the slurry to perform a blocking function; the long molecules also act as reinforcing fibres in the filter cake and help to form a net to hold the clay particles. A slurry of high density (around 1.2 Mg/m³) containing some fine sand and polymer can effectively seal sand-free fine gravels or sandy cobbles.

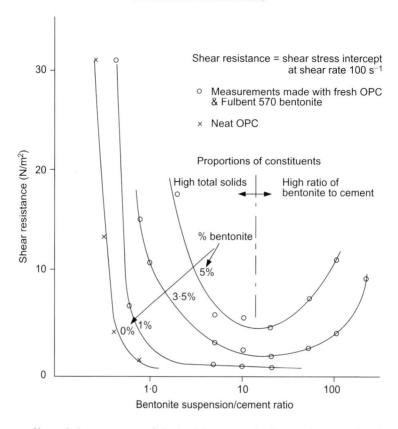

FIG. 12.2. Shear resistance of bentonite/cement grouts and slurries. Measurement by Cementation Research Ltd.

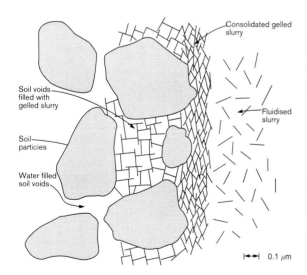

FIG. 12.3. Formation of filter cake.

Under favourable conditions, a filter cake will form very rapidly, for instance during the rotation of the cutter head in a tunnelling machine or the cut of a diaphragm walling grab. A typical thickness might be less than 1 mm with high-quality bentonite and up to 5 mm with bentonite of lower quality. If a differential pressure is maintained across the filter cake it will gradually increase in thickness, at a reducing rate, due to filtration of the slurry through the cake (see Text Box on p. 351). If the filter cake is insufficiently impermeable, the fluid loss through it may contribute significantly to increasing the pore pressures in the ground (or reducing negative pore pressures induced by excavation) and allowing localised swelling and softening at the contact with soils with significant clay content. The best filter cake in most cases is therefore one that is thin but highly impermeable; however under less onerous conditions the thicker, more permeable filter cake formed by lower quality (calcium) bentonite may be perfectly satisfactory.

12.1.4. Test procedures

A number of test procedures are used to categorize the quality and performance of bentonite slurries. These relate to the viscosity, gel strength, density, sand content, pH, and filtration characteristics of the slurry. Many of the tests used have been developed by the oil industry in relation to drilling of deep wells with bentonite (mud) flush. They are covered by American Petroleum Institute

SPECIALIZED APPLICATIONS

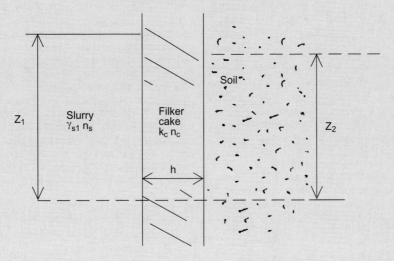

Formation of filter cake

Consider a filter cake layer of porosity n_c, and thickness h varying with time t, building by filtration against the wall of a trench or pile. By equating the rate of build up of the filter cake to the rate of loss of solids from the slurry by water permeating through the filter layer due to the difference in pressure head between the slurry in the trench and the water in the ground, the following equation is obtained:

$$h = \left(\frac{2k_c (1-n_s)(\gamma_s z_1 - \gamma_w z_2)}{(n_s - n_c)\gamma_w} \right)^{1/2} \sqrt{t}$$

where n_s and γ_s are the porosity and unit weight of the slurry, k_c is the permeability of the filter cake, γ_w is the unit weight of water, z_1 is the depth below the surface of the slurry and z_2 the depth below the groundwater surface. Since the pressure differential across the filter cake is constant, the flow rate will reduce in inverse proportion to the thickness of the filter cake, and will thus decrease with the square root of time.

(API) Publication RP 13B, and are also described in detail by the Federation of Piling Specialists (FPS 2000). Details of the tests are not repeated here, but the purpose and limitations of the tests are discussed.

It is not normal to explore the full rheological behaviour of a slurry over a range of shearing rates. It has been found generally acceptable to treat a slurry as a Bingham fluid with thixotropic properties. A Bingham fluid is one in which the viscosity rises linearly with shear rate from an initial non-zero value at zero shear rate (Fig. 12.4). The initial value is known as the yield strength and the rate of increase with shear rate as the plastic viscosity. Both can be derived provided measurements are made of viscosity at two different rates of shear. This is most conveniently done with a Fann viscometer. The relative performance of slurries is often checked on site using a Marsh funnel, in which an apparent viscosity may be determined by the rate of flow of slurry from a standard funnel. Its main use is as a quality control test, to check that slurry being used on site is consistent.

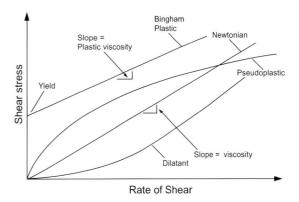

FIG. 12.4. Models of viscous behaviour.

The density of a slurry is measured simply using a 'mud balance'; the density is important both in its own right, by creating hydrostatic pressures somewhat in excess of groundwater pressures, and as a controlling factor in many of the other aspects of slurry performance. While it is common to allow some silt and sand to remain in the slurry, to improve pore blocking and filter cake formation, excessive amounts of sand lead to problems with material settling out and in excessive wear of pumps and other equipment. The sand content is measured by screening out the sand from a small sample of the slurry using a 75 micron sieve.

Bentonite slurries disperse best in neutral or alkaline conditions, and it is sometimes necessary to increase the alkalinity artificially. Values of pH in the range 9 to 12 are generally considered satisfactory, and may be measured using a glass-electrode meter or, less accurately, with pH papers. Higher pH values are usually the result of Portland cement contamination, which can have a severely deleterious effect on slurry behaviour.

The ability of a slurry to form a filter cake and the rate of fluid loss through the cake is measured in a test developed by API using a fluid loss apparatus. In this a sample of slurry contained in a cell is pressurized and forced through a filter paper. The volume of filtrate collected after 30 minutes and the thickness of the resulting filter cake are measured. The test was devised for drilling fluids and the applied pressure of 100 lbf/in^2 (690 kPa) is high for civil engineering applications, with the result that specifications may be unnecessarily onerous.

Grouts are sometimes tested for bleeding, the tendency for water to separate from a grout due to settlement of the solids. The tendency for bentonite–cement grouts to bleed will be partially or wholly offset by the setting process, depending on the depth of the sample tested, and results of bleed tests need to be interpreted with caution. Grouts also need to be tested for set time and increase of viscosity with time, since these control the period for which it is possible to inject the grout. Set time may be difficult to define for grouts which have a gel strength immediately after mixing which then increases with time until a fully hardened state is reached.

Hardened strength of bentonite–cement may be measured in unconfined compression tests, or in triaxial tests with a confining pressure. The former is a useful quality control procedure, but gives no indication of the stress–strain behaviour of the material in use, when there will nearly always be confining stresses. Drained tests carried out in a triaxial cell allow the long-term stress–strain behaviour to be investigated; typically the behaviour will be brittle at low confining pressure and ductile at higher pressures. Hydraulic conductivity may also be measured in a triaxial cell; it often reduces significantly with time and it may be necessary for tests to be continued for several days when the results are critical.

Specifications for testing are required to cover three different aspects of the materials to be used: the quality of the raw materials as supplied; the behaviour of the newly mixed materials in a fluid or plastic state; and for permanent installation the long-term properties affecting performance. The second of these aspects may cover either the suitability of the material for immediate construction purposes, or as an indicator of the long-term behaviour (or in some cases both). Results of tests for long-term performance will not normally be available until it is too late for them to be used to control construction; their purpose will be to confirm that adequate properties have been achieved. An anology is the use of slump tests as a site control procedure for concrete quality followed by compression tests on cubes or cores to confirm that adequate concrete strengths have been achieved. It must be commented, however, that, depending on the mixture proportions, this can be a weakly cemented material with residual swelling capacity, so there are many issues (such as curing time, sampling disturbance, and effective stress changes) that need careful consideration in order to obtain high-quality results. In addition, the necessary test regime and the use made of the test results will vary with the application, and are considered further in relation to different uses in the following sections.

12.2. Clay slurries

12.2.1. Introduction

Clay slurries are most commonly used as construction aids in the formation of diaphragm walls, bored piles, vertical and horizontal boreholes, tunnels, pipe jacks and caissons. The slurry may be required to seal permeable ground to prevent water inflow and/or allow fluid pressures to be applied to the ground; it may be needed to 'hold' excavated spoil, prevent it settling out and allow it to be transported away from the point of excavation; or it may be used to provide lubrication. In many cases the slurry may be required to do several or all of these things, and the ideal specification of the slurry may be different for each. The slurry design will nearly always be a compromise between different technical requirements, with the usual addition of the need for the process in which it is used to be as economical as possible. For instance, a decision may be required between cleaning the slurry of accumulated sand and silt for re-use or dumping it and supplying fresh slurry. In the following section the main applications are considered in turn.

12.2.2. Applications

12.2.2.1. Diaphragm walls and piles. In these applications the slurry is used to support the sides of the excavation for a diaphragm wall or uncased bored pile (Fig. 12.5). The slurry pressure must exceed the groundwater pressure and the excess pressures must be transmitted via a filter cake to the soil to provide sufficient additional effective stress to maintain stability of the trench or pile sides. The slurry must also be able to keep particles of soil in suspension so that they do not settle to the base of the trench, yet must be fluid enough to be easily displaced by concrete placed by tremie pipe and not to adhere to the reinforcement to an extent that would impair the bond between reinforcement and concrete. With some types of excavator using reverse circulation

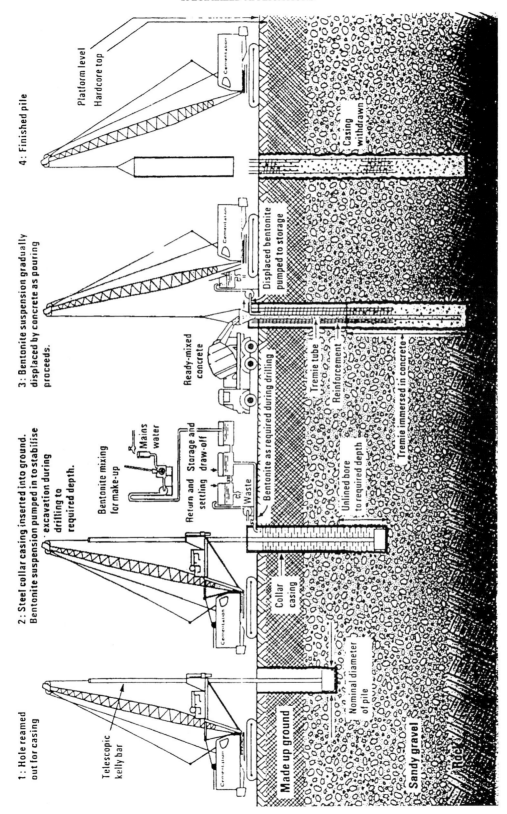

Fig. 12.5. Construction of bored pile using bentonite slurry. Figure courtesy of Cementation Foundations Skanska Ltd.

Effect of finite trench length on stability

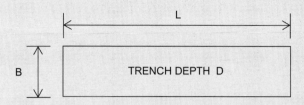

For cohesionless soils, after Huder (1972), the pressure required to maintain stability of the trench wall is the sum of the ground water pressure and the effective active earth pressure, with the latter multiplied by a factor A given by:

$$A = \frac{1 - e^{-2nK_a \tan\phi}}{2nK_a \tan\phi}$$

where $n = D/L$, D is the depth of panel, L is the length of panel. For cohesive soils, after Meyerhof (1972), the factor of safety F against collapse of the trench is given by

$$F = \frac{N_c \cdot c_u}{H(\gamma - \gamma_f)}$$

where N_c is a bearing capacity factor given in the plot below, γ is the total unit weight of the soil and γ_f is the unit weight of the slurry.

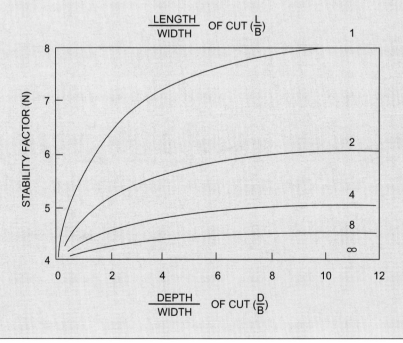

cutters the slurry is also used to transport the cut material to the surface.

The technique has been in use since the 1950s, and a major milestone in its development in the UK was the conference on diaphragm walls and anchorages held at the Institution of Civil Engineers in 1974 (ICE 1975). At this conference, Sliwinski & Fleming (1975) provided an overview of practical considerations affecting the construction of diaphragm walls, while Hutchison et al. (1975) discussed the properties required of bentonite slurries and their control. Many of their comments are still relevant, though improved equipment has allowed the use of the technique to be extended; for instance more powerful excavators can now cope with very stiff soils and soft rocks. Diaphragm walls may be routinely formed to a depth of 60 m and thickness 1.2 m, and in special cases to a depth of 120 m and thickness of 2.0 m.

Practical limitations on the use of the method are:

- in soil with permeability of greater than about 10^{-3} m/s loss of slurry may be excessive;
- cavities in the ground may lead to sudden loss of slurry;
- in very weak strata, with shear strength less than about 10 kPa, the ground may not be able to withstand the pressure of fresh concrete and a casing may be needed;
- water pressures under artesian head cannot be controlled.

Apart from these limitations, the conclusions of CIRIA Report PG3 are that the use of bentonite slurry has no significant adverse influence on eventual pile performance, provided that the technique is properly understood and good materials and workmanship are employed.

In the UK, use of the technique is usually required to be in accordance with the Specification for Piling and Embedded Retaining Walls (ICE 1996). This document requires that the slurry level is kept at all times a minimum of 2.0 m above groundwater level in pile bores and 1.5 m in diaphragm wall trenches (the reason for this difference is not clear). The excess fluid pressure will not normally be sufficient to provide the full active pressure required to support the sides of a long trench, and stability can only be achieved by taking advantage of the reduction in required support pressure due to horizontal arching around panels of relatively short length in plan (see Nash 1974a,b) (see Text Box on p. 354). Depending on ground conditions, typical panel lengths are in the range 2 to 7 m. Curved panels or panels with T, L or X shapes in plan are also possible (Fig. 12.6).

In the Guidance Notes to the ICE specification, it is suggested that a slurry with the required properties is likely to contain between 3% bentonite for natural sodium montmorillonite and 7% or more for some manufactured bentonites, with a figure of 5% appropriate for activated (ion exchanged) calcium montmorillonite of UK manufacture. In this application, the thickness of the filter cake is not usually of great importance and the thicker layer resulting from the use of activated rather than natural sodium bentonite is acceptable. The bentonite is required to be of a quality in accordance with Publication 163: *Drilling Fluid Materials* of the Engineering Equipment and Materials Users Association, last reprinted in 1988. An alternative is the API Specification 13A, 15th edition, May 1993, Section 6 (OCMA grade bentonite). There are some differences between these specifications, which are discussed in detail by the Federation of Piling Specialists (2000).

A table is provided in the ICE Guidance Notes (reproduced in Table 12.1a) giving suggested test procedures and compliance values for the slurry as supplied to the pile or trench, and as sampled from the pile or trench immediately prior to concreting. Similar but generally less restrictive criteria are included in the draft European Standards for bored piles and diaphragm walling, prEN 1536 and 1538 respectively—see Table 12.1b. The restriction on the density and viscosity of the slurry in the trench or pile bore, especially at the base, immediately prior to placing concrete is to ensure that the fresh concrete is able to displace the slurry without trapping any of it. A simple device is used to take samples of the slurry from the required depth. More detailed requirements used by one of the main companies specializing in diaphragm walling in the UK are given in Appendix 12.1, along with some comments on the tests used. The values quoted are based on the use of UK manufactured bentonite and might have to be modified for other types. Further guidance on the use and limitations of bentonite slurry in these applications is given in CIRIA Report PG3 and by the Federation of Piling Specialists (2000).

Successful formation of piles or diaphragm walls under slurry requires the use of a suitable concrete mix, one that is sufficiently fluid to displace the slurry at the base of the trench when placed through a tremie pipe, but is also cohesive and not prone to segregation. Requirements for concrete mixes are given in the ICE Specification for Piling and Embedded Retaining Walls. Typically the minimum cement content will be 400 kg/m³, the water/cement ratio below 0.6, and the aggregate will be naturally rounded sand and gravel, well graded with maximum aggregate size 20 mm and with about 40% sand content. Slump will be in excess of 175 mm, and workability therefore best measured using a flow table, with a target flow of 500–600 mm.

12.2.2.2. Slurry tunnelling. In this application, slurry is pumped to a chamber in the head of a tunnelling shield, where it has two main functions: to apply pressure to the excavated soil face and thereby help to maintain its stability and reduce ground movements into the tunnel; and to act as a transport medium back to the ground surface

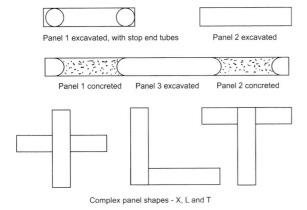

FIG. 12.6. Construction of diaphragm walls (in plan).

TABLE 12.1(a). *Tests and compliance values for bentonite support fluids (from ICE 1996)*

Property to be measured	API RP13* Test method and	Section No.	Compliance values at 20°C	
			As supplied to pile	Sample prior to placing concrete
Density	Mud balance	1	<1.10 g/ml	<1.15 g/ml
Fluid loss (30 min test)	Fluid loss test (low temperature)	3	<40 ml	<60 ml
Viscosity	Marsh cone	2	30–70 s	<90 s
Shear strength (10 min gel strength)	Fann viscometer	2	4 to 40 N/m^2	4 to 40 N/m^2
Sand content	Sand screen test	4	<2%	<2%
pH	Electrical pH meter		9.5 to 10.8	9.5 to 11.7

* American Petroleum Institute: *Recommended practice standard procedure for field testing water-based drilling fluids.*

TABLE 12.1(b). *Characteristics for bentonite suspensions (from prEN 1538:1996)*

Property	Stages		
	Fresh	Ready for re-use	Before concreting
Density in g/ml	<1.10	<1.25	<1.15
Marsh value in s	32 to 50	32 to 60	32 to 50
Fluid loss in ml	<30	<50	n.a.
pH	7 to 11	7 to 12	n.a.
Sand content in %	n.a.	n.a.	<4*
Filter cake in mm	<3	<6	n.a.

n.a., not applicable. Requirements for prEN1536 are similar but omit requirements for filter cake.
* Sand content may be increased to 6% before concreting in special cases. Sufficient gel strength is required, and may be checked with rotational viscometers or other suitable equipment.

for the soil material cut from the face (Fig. 12.7a). For the latter purpose the slurry is continuously circulated and carries the excavated material in suspension. Since relatively large volumes of excavation are involved, except in short tunnels of small diameter, it is usually economic to separate the excavated material from the slurry on the surface, and recirculate the cleaned slurry.

In naturally cohesive ground, the 'slurry' may initially be pure water. When returned to the surface, this is only partially cleaned, leaving some clayey material in suspension. Where there are fissures or more permeable zones of ground, some larger sand-sized particles will also be left in suspension to act as pore-blockers in the formation of a filter cake. However excessive quantities of sand in the slurry cause excessive wear to the pumps used to circulate the slurry. In cohesionless soils, or ground containing inadequate amounts of clay to form a natural slurry, a slurry based on bentonite is required. Again this is usually allowed to retain some of the excavated material; the quantity of suspended material may gradually build up until it becomes necessary to dispose of some of the slurry and replace it with fresh bentonite. The original concentration of bentonite may range between zero (pure water) and about 6 or 7%, giving a slurry density of 1000 to about 1040 kg/m^3, while the maximum slurry density still capable of being pumped is 1400–1500 kg/m^3. Small quantities of polymer, typically about 0.5% by mass of

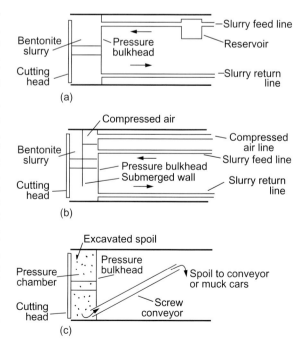

FIG. 12.7. Tunnelling shields (schematic): (a) slurry shield; (b) hydroshield; (c) earth pressure balance shield.

slurry, may also be included to help with filter cake formation; suitable long-chain polymers are sodium carboxy methyl cellulose (CMC), polyanionic cellulose (PAC) or polyacrylamides and their derivatives.

Support of the tunnel face requires formation of a filter cake as an impermeable layer, then sufficient fluid pressure to oppose any groundwater pressure and provide a positive effective pressure on the soil skeleton sufficient to maintain stability of the ground. The extent to which a filter cake has time to form during active tunnelling is debatable, since the face is cut back two, three or more times (depending on the cutter head design) for each revolution of the cutter head. However, excessive penetration of the slurry must be avoided, otherwise large quantities of slurry may be lost into the ground. A slurry that very rapidly (within a few seconds) forms a thin, highly impermeable cake is therefore preferred. Face support is particularly critical when the tunnelling machine is stationary. For long tunnels it is usually necessary to get access to the cutter head on one or more occasions during the drive to change cutter tools for varying ground conditions or replace worn tools. One way of achieving this while maintaining face support is to replace the slurry pressure by compressed air, in which case the filter cake must be effective enough to transmit the air pressure to the ground without excessive losses. Workmen are then able to access the face through an air lock in the shield or tunnel. In some slurry tunnelling machines a 'bubble' of air is trapped in the shield to provide an elastic cushion which helps to minimize changes in pressure caused by variations in tunnelling and slurry pumping rates (Fig. 12.7b).

Penetration distances for slurry into the ground are discussed above in Section 12.1.3. In practice slurry shields are best adapted for use in reasonably well-graded sands and gravels (Fig. 12.8); more open ground will allow excessive penetration of the slurry, though appropriate use of polymer additives may allow this range to be extended. In finer-grained soils, problems arise in the sufficiently rapid removal of the excavated material from the slurry in the separation plant, and in the tendency for the more plastic clays to clog the openings in the cutter head. The maximum size of particle that can be transported is limited by the diameter of the slurry pipes, and the cutter head openings are often limited to exclude particles too large to be handled by the machine. Alternatively a crusher unit can be fitted just ahead of the spoil intake to reduce cobbles and boulders to sizes with which the machine can cope.

The slurry may also have a lubricating action on the cutter tools and cutter head face, reducing wear and the power required to drive the head. Small quantities of natural oils such as palm or jute oil may be added to the slurry to increase its lubricity.

When acting as the medium for transporting the spoil back to the surface, the slurry must be sufficiently viscous to stop material falling to the bottom of the machine head chamber or settling out in the pipes, and have sufficient gel strength to hold material in suspension if slurry circulation is halted for any reason. High concentrations of bentonite, of up to 12 or even 15%, may be used. However excessive gel strength may make it difficult to restart circulation after a stoppage, while excessive viscosity increases power requirements for pumping. A natural Wyoming bentonite may be preferred, its superior properties for this application justifying its greater cost over an activated bentonite.

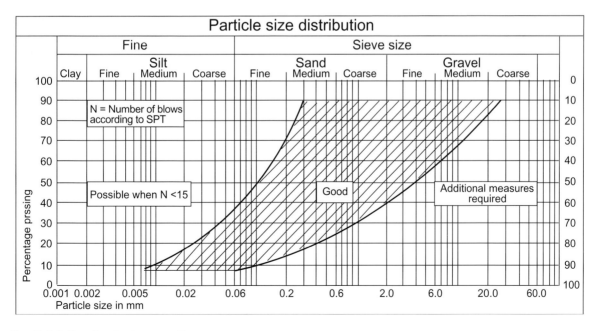

FIG. 12.8. Soil grading for slurry tunnelling.

12.2.2.3. Pipe jacking and microtunnelling; shaft sinking. Pipe jacking is the technique of forming tunnel linings by pushing a 'string' of pipes in behind the tunnelling shield, additional pipes being added at the launch pit as excavation proceeds at the shield (Fig. 12.9). The term 'microtunnelling' is variously used, in the USA to describe all types of pipe jacking in which the tunnelling process is remotely controlled from the surface, but in the UK the term is normally restricted to tunnel diameters too small for man entry (and therefore perforce remotely controlled), or less than about 1.0 m internal diameter.

Pipe jacking and microtunnelling often use slurry tunnelling machines, for which the same comments apply as for large scale machines. However slurry is often also used as a 'lubricant' to reduce the frictional resistance to forward movement of the pipe string and hence limit the total jacking force which has to be supplied by the hydraulic rams, resisted by the back wall of the jacking pit, and transmitted through the pipes.

The tunnel bore is usually excavated to a slightly larger diameter than the outside diameter of the pipes, the 'overcut' being typically 10 to 20 mm on diameter. In stable ground, including all except very soft clays and fine sands above the water table supported by capillary suction, the pipes are then able to slide along the bottom of the bore. In unstable ground, particularly sands and gravels below the water table, the ground will collapse on to the pipes and generate large frictional resistance to jacking. The use of lubricant slurry within the overcut can have three effects (Milligan & Marshall 1998). The first, and most important, is to support unstable ground. This it does by the same process as at the face of a slurry machine, by creating a filter cake and then transmitting radial stresses to the ground. The pressure required to maintain stability is not greatly in excess of the groundwater pressure (see Text Boxes on p. 360 and p. 361). The second effect is that the pipe becomes more or less buoyant within the fluid; small, relatively thick-walled microtunnelling pipes may not become fully buoyant, while larger diameter reinforced concrete jacking pipes may become positively buoyant, contacting the ground at the crown rather than invert of the tunnel. In either case the contact forces between pipe and ground are greatly reduced, and the frictional resistance correspondingly lessened. Finally, the slurry may reduce the coefficient of friction between pipe and ground, though this effect is generally the least important of the three. Overall, the jacking resistance may be reduced by over 90% in unstable sand and gravel, and by typically about 50% in stable ground.

In swelling clays, the use of simple bentonite slurry for lubrication may be counter-productive. The ground takes in water from the slurry and swells, 'squeezing' on to the pipes and increasing the jacking resistance substantially, even though the interface friction coefficient may be very low. In these cases a slurry designed to inhibit swelling may be used; this is achieved either by the incorporation of potassium salts such as potassium chloride, causing ion-exchange with the swelling clay minerals and rendering the near-surface soil non-swelling, or by including a polymer such as partially hydrolysed polyacrylamide (PHPA) which is highly anionic, binds to the clay particles, and prevents water penetration of the clay mineral. Polymers are however considered environmentally harmful in the oil industry, especially when used in aquatic environments (see Sections 12.3.1 to 12.3.3).

Slurries may similarly be used as stabilising and lubricating agents around the outside of caissons being sunk to form vertical shafts. In this case the excavation takes place at the base of the shaft. In soft soil the weight of the caisson may be sufficient to penetrate the bottom edge into the ground ahead of excavation, but in stronger soils the excavation is made slightly larger than the outside diameter of the shaft, leaving a small annulus which is filled with slurry. The action of the slurry is very similar to that in pipe jacking lubrication.

12.2.2.4. Soil conditioning in earth-pressure-balance tunnelling. An earth-pressure-balance machine (EPBM) is an alternative to a slurry tunnelling machine. It has a working (pressure) chamber immediately behind the cutter head in which the excavated soil is remoulded into a plastic mass which provides the support to the tunnel face provided by the slurry in a slurry machine (Fig. 12.7c). The pressure is maintained by balancing the shield advance rate, the excavation rate and the rate at which spoil is removed from the pressure chamber, usually by a screw conveyor. The EPBM has the advantage over the slurry machine of not requiring a separation plant and of producing spoil in a condition suitable for disposal as normal landfill. However EPB shields only work effectively in reasonably fine-grained soils which remould to a soft plastic consistency and have sufficiently low permeability to control inflow of water through the working chamber and screw conveyor. Typically this requires a fines content (<63 μm) of more than 30%, and less than 30% greater than sand-size (2.0 mm). The natural water content of the ground needs to be such as to give a liquidity index in the range 0.4 to 0.75. EPB shields were mainly developed during the 1970s in Japan, where natural ground conditions were often close to ideal.

For coarser or dryer soils the excavated soil must be 'conditioned' by the addition of water, clay or other material such as polymers or foam to provide the required consistency. Bentonite slurry is suitable for this, sometimes improved by the addition of polymers; a relatively small quantity gives a substantial increase in plasticity and reduction in permeability of the spoil. It is best injected through ports in the cutter head, so as to have the maximum chance to mix thoroughly with the excavated material, but as an emergency measure may also be injected into the screw conveyor. The former requires the presence of hydraulic slip rings to deliver slurry to the cutter face.

12.2.2.5. Vertical and horizontal (directional) drilling. Bentonite slurries (muds) have for long been used as a stabilizing, lubricating, cooling and spoil transport medium in drilling both vertical boreholes and near-horizontal bores for pipes and ducts. The action of the

SPECIALIZED APPLICATIONS

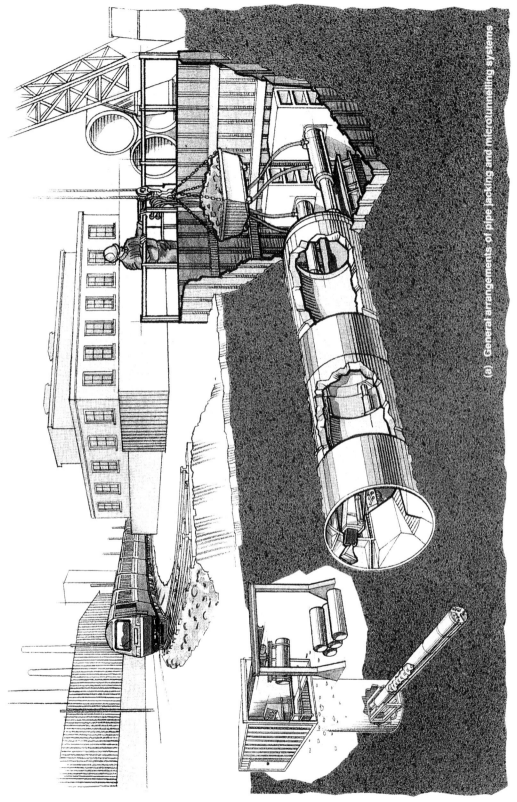

FIG. 12.9. Pipe jacking and microtunnelling. Figure courtesy of The Pipe Jacking Association.

(a) General arrangements of pipe jacking and microtunnelling systems

Slurry pressures required for tunnel stability

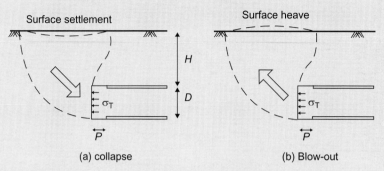

(a) collapse (b) Blow-out

In cohesive soils the slurry pressure σ_T must lie between the limits given by

$$\gamma(H+D/2) - T_c c_u \leq \sigma_T \leq \gamma(H+D/2) + T_c c_u$$

where H is the soil depth above the tunnel, D the tunnel diameter, c_u the undrained strength of the soil, γ the unit weight of the soil, and T_c a stability number given in the plot below from Atkinson & Mair (1981). The value of T_c depends on the value of P, the unsupported length of tunnel ahead of the tunnel lining. The lower and upper limits of the slurry presure given in this expression are for tunnel collapse and blow-out respectively. To assess a safe range for σ_T, with acceptably small ground movements, a factor of safety of 2 is usually applied to the soil strength. For a long unsupported bore, as may occur during a pipe jack, the appropriate stability number is that for an infinite value of P.

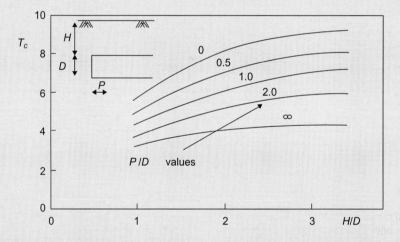

In cohesionless soil, the assessment of face stability is more complex, though analytical and numerical solutions have been presented by Anagnostou & Kovari (1996) and Leca & Dormieux (1990). Solutions for a long tunnel applicable to the pipe jack condition have been obtained by Atkinson & Potts (1977). For a deep tunnel, the pressure required to prevent collapse is

$$\sigma_T \geq \gamma D T_\gamma$$

and for a shallow tunnel with a significant surcharge on the ground surface

$$\sigma_T \geq \sigma_s T_s$$

where σ_s is the surcharge pressure and T_γ and T_s are stability numbers given by the plots below. Note that T_γ is independent of depth, and that both these solutions apply to dry soil. Below the water table, water pressure must be added and the buoyant weight of soil used in these equations.

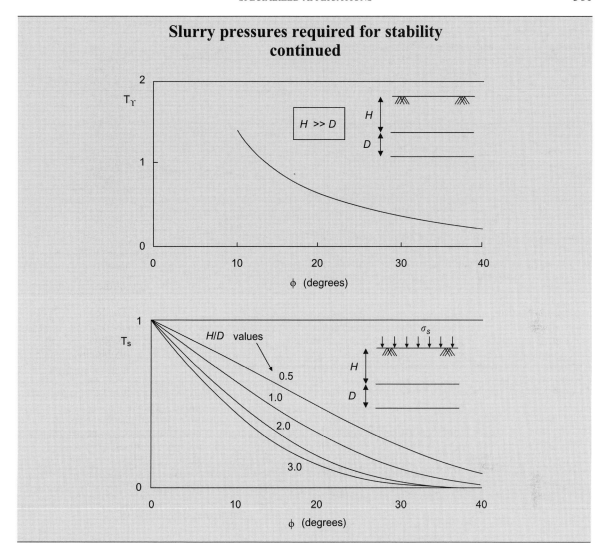

Slurry pressures required for stability continued

slurry is essentially the same as in other applications considered above, but the performance requirements may be considerably more onerous where boreholes penetrate to considerable depths through variable strata. The oil exploration industry has developed a wide range of additives to and replacements for simple bentonite slurries, which have had a major impact on drilling rates. Some of these are gradually finding their way into the construction industry, but are beyond the scope of this report. More conventional bentonite-based slurries used in horizontal drilling differ somewhat from those used in vertical drilling. In vertical drilling the cuttings are returned to the surface using a combination of high viscosity and high flow rates in the slurry, and because of the large depths to which drilling sometimes occurs the density of the slurry and the pressure head generated are important. In horizontal drilling the flow rates are much lower, excessive viscosity is unacceptable as the resulting pressures may become excessive at shallow depth leading to danger of blow-out, but gel strength is necessary to prevent settlement of excavated material in the bore. In cohesionless soils the drilling fluid is required to form a filter cake and support the ground in the same way as described for pipe jacking and microtunnelling.

12.2.3. Re-use and disposal of slurries

Where relatively small quantities of slurry are required, it will normally be used until it no longer fits its purpose, and then removed from site for disposal. However slurries are becoming increasingly difficult to dispose of, particularly when affected by cement or contaminated soil. It is therefore better practice whenever possible to clean the slurry on site for re-use. With piling and diaphragm walling the slurry is used in batches, and provided sufficient space is available for storage tanks the cleaning of the slurry is an off-line process and the rate at which it can be carried out is not critical. However in a large-volume continuous usage such as slurry tunnelling it is necessary to process most of the slurry on a continuous basis, the rate of which may well be a limiting factor in the rate of tunnelling progress. In fact the decision whether or not to use a slurry tunnelling machine on a particular job is often controlled by the performance of the separation plant. Large-scale separation plants are therefore a feature of slurry tunnelling, and require a major investment and considerable space. Smaller modular units are available for pipe jacking, microtunnelling and directional drilling projects.

Separation plants typically include a combination of screens, settling tanks and filters (Fig. 12.10). Control of the amount of sand and silt to be left in the slurry is usually by passing through hydrocyclones; centrifuges may be required to reduce excessive amounts of silt-sized particles. Simple vibrating screens will typically be used to remove particles larger than about 3–5 mm. Finer material may then be allowed to settle out in settlement tanks, probably with the assistance of flocculating agents; however this is a slow process which is usually only suitable for the final stage of treatment of waste water before disposal to drains. Cyclones may be used for accelerated removal of particles down to about 0.1 mm in a single stage or 0.02 mm in two stages. Centrifuges may be used to remove particles down to 5 µm (fine silt) or smaller, but can only handle relatively small throughputs. They may be used to clean part of the carrying fluid (slurry) for re-use in the machine, while the remainder is re-circulated without treatment. The sludge from settlement tanks or cyclones may be further dewatered using belt presses to produce a material more suitable for tipping. Figure 12.10 shows a flow diagram for a typical separation plant, but considerable variations are possible depending on the size and nature of the project.

The coarser fractions from separation plants cause no problems in disposal, but the final products containing the finer fractions from the excavated ground (and possibly significant quantities of bentonite) may be marginal for disposal as land fill. The residual water content of the filter cake material from belt presses may still exceed 100%, although the material may seem drier if flocculants have been used. Gradual degradation of the flocculant may allow free water to be released and the material to become more fluid again. Criteria for the acceptability of spoil may not be well defined, and lead to conflict with owners of landfill facilities or environmental agencies. Simple criteria combining minimum shear strength and minimum solids content have been suggested (Fig. 12.11), in which material with shear strength in excess of 10 kPa and solids content greater than 35% would be considered acceptable.

Highly contaminated slurry may be treated with Portland cement or lime to produce a mix stiff enough to be transported by lorry. However there is an increasing tendency for such mixtures to be treated as special wastes, with the attendant high costs of disposal. For further discussion of the cleaning, re-use and disposal of bentonite slurries, reference should be made to the Federation of Piling Specialists (2000).

12.3. Plastic clay

12.3.1. Introduction

In this section the use of clays (or clay mixed with other materials) in a plastic state is considered. In most cases the purpose is to exclude or retain water or other fluids, and the prime property of the material is its hydraulic conductivity. Clays have probably been used in these ways since man first tried to build rain-proof shelters or collect or control water for domestic or agricultural purposes. However in terms of relatively recent large scale use in construction, the first really major development was in the canal systems developed in the UK and elsewhere from the 18th century. Where these did not run through naturally impermeable terrain, they were lined with 'puddle clay', natural clay of high plasticity reworked and compacted into place to remove all natural fabric or structure (sand layers, fissures etc.) and so create homogeneous material of low hydraulic conductivity.

In recent years the most important applications have been the development of engineered waste facilities, for domestic and industrial wastes (Section 12.3.2), and for radioactive wastes (Section 12.3.3). A parallel development has been in the treatment of previously contaminated land for re-use, by containment of the contaminants, *in situ* treatment of the contaminated ground, or a combination of the two (Section 12.3.4). The formation of bentonite-cement cut-off walls and piles, and the use of clay in grouts, are included in this section rather than Section 12.2. Although the materials may initially be used in slurry form they are designed to set and form structural materials on their own or in combination with the ground. Past and present usage of puddle clay for dam cores and canal work is covered in Section 12.3.5, and other sealing and waterproofing systems in Section 12.3.6.

12.3.2. Waste disposal facilities

12.3.2.1. Landfill liners and covers. The use of clay in landfill liners and covers is a vast subject, which can only be covered briefly here. The principle is straightforward and now generally accepted in developed countries; any waste material to be disposed of that is not naturally inert and non-hazardous is to be encapsulated in engineered landfills to control aqueous and gaseous emissions and prevent them from polluting the environment. It is not necessary or possible to reduce emissions to zero, but they should be reduced to such levels that with the help of natural dilution and attenuation they pose no threat to plant, fish or animal life (including of course humans). The complexity of the subject arises from the wide range of materials to be disposed of and the potential pollutants produced by them, the various interactions between pollutants and the environment and the materials used to contain the waste, and the variations in regulations and waste management strategies in different countries. Jessberger (1994) provides a useful suite of papers on geotechnical aspects of landfill design and construction, based mainly on German practice, while Street (1994)

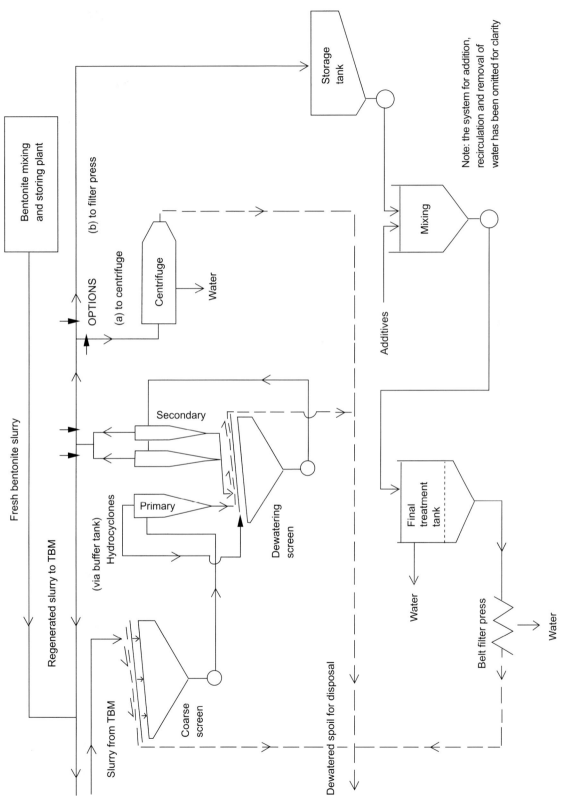

FIG. 12.10. Simplified flow diagrams for typical separation plants: (a) using vibrating screens, hydrocyclones and centrifuge; (b) using a belt filter press in place of a centrifuge.

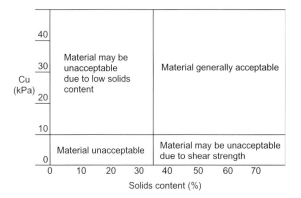

FIG. 12.11. Criteria for acceptability of tunnel spoil for disposal.

compares UK practice with that of a number of other European states. Much of the early development work on landfill design was done in the USA and Canada, and is covered in detail by Rowe *et al.* (1995). Chapuis (2002) provides practical guidelines for the use of soil-bentonite and compacted clay liners, including quality control and performance test procedures, and case histories of successful and unsuccessful projects.

A typical engineered landfill has a basal lining system below the waste, and a capping system above the waste. The purpose of the former is to prevent the migration of waste material or leachates derived from it into the subsoil or groundwater. It will do so by a combination of one or more sealing layers, made from mineral (soils) or polymer membranes, with drainage layers to collect leachate for disposal or re-circulation through the waste heap. The purpose of the capping layer or cover is to prevent excessive infiltration of water from rainfall, which would increase the volume of leachate to be handled, and to control emissions of gas produced as the waste decomposes (Fig. 12.12). Barrier systems of increasing complexity are shown in Figure 12.13. Unless the subsoil is of very low hydraulic conductivity, and can be included as part of the containment system, the basal lining will usually have primary and secondary defences; both may be mineral layers of compacted natural clay, compacted waste materials such as burnt oil shale, or a sand/bentonite mix, or one may be a flexible geomembrane made of high density polyethylene (HDPE). Layers of free-draining material are placed above and sometimes also between the low-permeability layers, to collect the leachate above the upper layer and any leakage through it and allow it to be pumped away for treatment and disposal. When geomembranes are used, a protective layer of soil or a thick geotextile (or both) is usually placed above the membrane; this layer may double as the drainage layer. Similar combinations within the capping system provide a single low-permeability layer to reduce infiltration and a high-permeability layer to allow the collection and venting of gases (Fig. 12.14). In the design of waste disposal facilities consideration should be given to the interaction between the natural hydrogeology and the engineered systems. For instance, depending on the relative levels of the leachate in the waste facility, the ground water nearby, and the piezometric head in an underlying aquifer, there may be a tendency for flow either into or out of the facility. The former condition is known as a hydraulic trap, since it tends to prevent escape of pollutants from the facility by advection (movement of contaminants due to water movements), though pollutant transfer out may

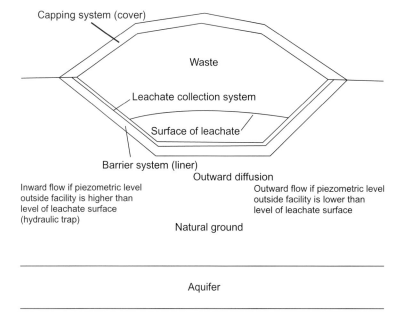

FIG. 12.12. Typical waste containment.

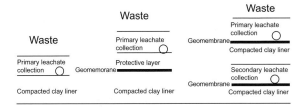

FIG. 12.13. Typical base lining system for waste facilities.

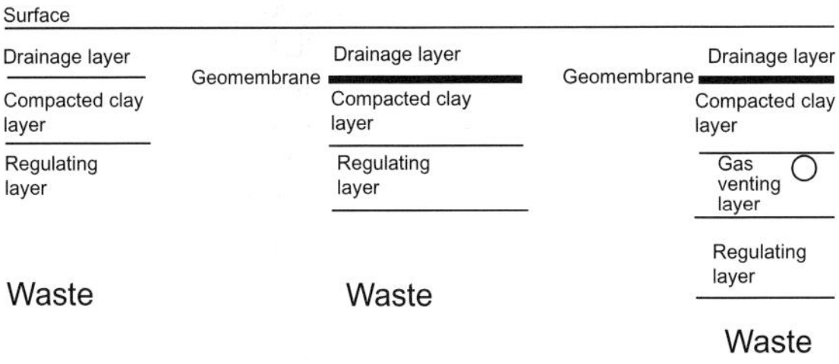

FIG. 12.14. Typical cover systems for waste facilities.

still occur by diffusion (movement from areas of high concentration to areas of low concentration).

Where natural clay strata form part of the engineered containment system, the effects of natural structure and fabric within the clay are often of great importance. In particular stiff clays, while highly impermeable in small samples, are often much more permeable in the mass due to the presence of fissures. It may be necessary to rework the surface layers to achieve sufficiently low hydraulic conductivity. Compacted clay or bentonite-sand barriers are typically a minimum of 1.0 m thick, but may be thinner (down to 300 mm) when used in conjunction with a geomembrane. With industrial or toxic wastes the thickness may be 3 to 4 m or more. A typical specification requires a final coefficient of permeability of 10^{-9} m/s or less; this should not increase significantly after permeation by the leachate. The soil should have a minimum of 10% of its particles smaller than 2 μm and a plasticity index greater than 7%. Alternatively a minimum cation exchange capacity of 10 meq per 100 g may be specified.

Hydraulic conductivity may be measured in the laboratory using rigid-walled permeameter cells or in triaxial cells. It is important that it is measured at appropriate stress levels, since results from tests at low confining pressures may be misleading, and realistic hydraulic gradient. In the field, the hydraulic conductivity of natural clays or compacted layers may be determined from conventional variable- or constant-head tests in boreholes or by pumping tests.

In compacted clays the hydraulic conductivity depends mainly on the 'secondary porosity' associated with poor interlocking of clods and layers, rather than with the porosity of the intact clay. It has been found that the lowest hydraulic conductivity is obtained by compacting the material at a water content 2–4 percentage points higher than the Standard Procter optimum value (Fig. 12.15),

with as much kneading energy as possible to break down soil lumps and get rid of drainage paths between the lumps. The moulding water content should also be above the plastic limit for the soil. The undrained strength of the clay should typically be in the range of about 30 to 140 kPa, the lower limit to ensure trafficability and the upper to allow adequate deformation by a sheeps-foot roller.

Where clay barriers have been unsuccessful in practice, the cause of failure has been either under-compaction or cracking of the completed layer due to desiccation in hot weather or formation of ice lenses in very cold weather. Compacted layers should be covered immediately after placing to prevent the detrimental effects of drying or freezing, which can increase the hydraulic conductivity by two orders of magnitude.

Generally clays with predominantly chlorite and illite minerals have been found to be best for liners. Kaolinitic clays cannot easily achieve the low permeability required, while bentonites are difficult to control adequately. In addition, because they are highly impermeable, there is a tendency to use them in thin layers which are easily damaged and which may allow excessive pollutant transfer by diffusion.

There have been desiccation problems caused by covering mineral liners with black HDPE liners (Hewitt & Philip 1997), which then absorb solar energy, heat-up, and dry the mineral layer. If the warm moist air created by this heating is unable to escape, further problems are cause by condensation of the moisture on the underside of the liner at night because the water can dribble down the underside of the membrane on the side slopes and accumulate at the bottom of the landfill).

Regulatory systems are commonly based on a prescriptive approach, whereby particular barrier designs are required in terms of materials, layer thicknesses, compaction procedures, etc. This approach allows for simple control, but takes no account of the particular nature of the facility or of its interaction with its surroundings. The prescription may be unsafe at one location, but over-conservative in another. It also does not allow the use of hydraulic containment systems as a possible alternative to barrier systems. The alternative approach is to define a required end-product in terms of acceptable levels of pollutants in an aquifer or at the boundary of the site, either for a specified period or indefinitely. Rowe *et al.* (1995) and Rowe (1997) argue for the last of these, on the basis that a waste facility designed to meet regulatory requirements for 30 or even 100 years may still pose a threat of pollution in the longer term, and that it is not right to bequeath such potential problems to future generations.

The concept of indefinite protection against pollution is possible because the facility will contain a finite quantity of actual or potential pollutants, though the larger the facility the greater this amount will be. The pollutants will gradually dissipate by movement out of the facility and physical and chemical actions within the waste heap. Where leachates are extracted from the base of the heap for treatment, an interesting question arises as to whether it is better to have a very permeable cover to the heap,

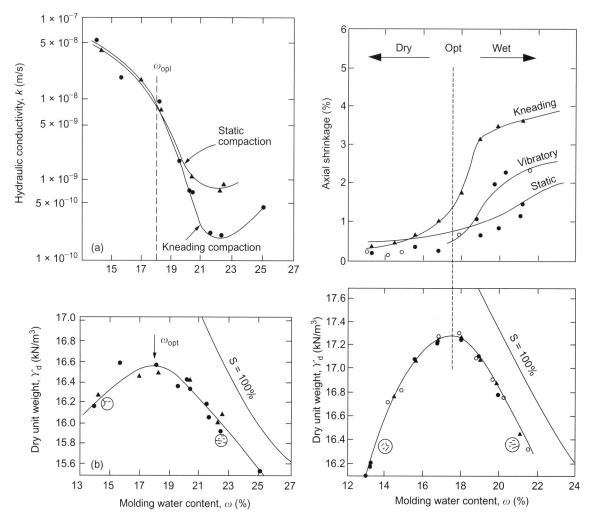

FIG. 12.15. Relation between permeability and compaction for clay barriers.

to leach out pollutants as rapidly as possible, or a highly impermeable one to keep leachate production to a minimum. The former may involve high costs in the short term to cope with treating large volumes of leachate. The latter postpones dealing with the problem, but runs the risk that pollution will occur after the period over which proper monitoring and control of the facility is exercised, and that leachate removal systems will become ineffective in time owing to clogging or other failure. Capping layers will themselves deteriorate in time due to differential settlement of the waste, environmental effects such as freezing, drying or erosion, degradation of synthetic materials, vegetation, etc., and may need to be renewed. A good compromise might be to allow rapid infiltration and leaching initially, then place a highly impermeable cover to keep leachate production to a low level which the system can handle in the long term.

The quantity and nature of leachate will of course depend on the nature of the waste material. Household waste, usually referred to as municipal solid waste or MSW, tends initially to produce acidic leachates due to oxidation of organic materials, sometimes leading to quite high temperatures (the acetogenic stage). These tend to mobilize heavy metals. On depletion of available oxygen, decomposition continues under anaerobic conditions, favourable to the production of methane gas (the methogenic stage); iron tends under these conditions to form iron sulphide, a black slime which may clog sand filter layers. Removal before deposition of some of the materials from waste for recycling, such as metals, paper, textiles and organic material (for composting) can be very beneficial in reducing the quantity and toxicity of leachate. The nature of the leachate is important because of possible interaction between it and the clay in a clayey barrier or the natural ground. Tests are done in constant-flow consolidation permeameters whereby leachate is forced through a sample of the soil until the effluent is of the same composition as the initial liquid. It is found that most inactive clay minerals are relatively insensitive to leachate from MSW; the hydraulic conductivity may actually reduce due to adsorption of sodium ions and expansion of the double layer on the clay particles, or perhaps by bacterial clogging.

On the other hand the interaction of clays and liquid hydrocarbon pollutants is very complex. When dry clay is mixed with a hydrocarbon, the permeability of the clay for that hydrocarbon is very much higher than for water in hydrated clay. With normally hydrated clay, the permeability under some conditions for water-soluble hydrocarbons may be high, though it is reduced by confining stress. Water-compacted clay barriers are highly resistant to penetration by insoluble organic liquids due to surface tension effects, but this resistance may disappear in the presence of surfactants or mutually soluble organic liquids which destroy the surface tension. Again, high confining stresses are beneficial, and it is a good idea to combine a clay barrier with a geomembrane which will provide initial protection to the clay until the loading from the waste heap has built up. Clay and HDPE barriers are also complementary in terms of resistance to diffusion, which is likely to be the primary process for pollutant transfer in a well-designed facility. Clay is highly resistant to organics, but poor for ionic chemicals, while HDPE membranes are poor with regard to organics but resistant to ionic materials. Clay should always be hydrated with clean water, as it then performs much better than when it has been hydrated with contaminated water. Rowe et al. (1995) give a detailed discussion of clay-leachate compatibility. They note that smectite clays (bentonite) appear to be the most affected by leachates, and that in the current state of knowledge the susceptibility of sand-bentonite mixes to damage by saline and organic leachates has not been fully clarified.

Sand–bentonite mixes, usually referred to as bentonite-enhanced (or enriched) sands or BES, are nevertheless increasing in popularity for use in liners; they can be engineered to have the necessary low hydraulic conductivity and are less susceptible to frost damage or drying shrinkage than compacted clays. Originally mixes were only made with clean sand and gravel, with less than 10% of non-plastic fines, but these soils had relatively high natural porosity and required high bentonite contents. There is also a danger of the bentonite being washed out from the matrix of coarser particles (Chapuis 2002). In the 1990s, mixes started to be used with soils with more than 20% fines (Alston et al. 1997), which had lower natural porosity and required smaller bentonite contents.

Sand–bentonite mixes have relatively good mechanical properties to provide structural integrity during construction and operation. Dineen et al. (1999) have studied the compaction characteristics of a BES comprising well-graded sand with 10% of bentonite (by mass at 14% moisture content); the combined soil had a liquid limit of 37, and plastic limit of 15. Material that was poorly compacted, or compacted at moisture contents much less than optimum, were found to have high initial suctions and produced large reductions in volume on wetting (Fig. 12.16). Stewart et al. (1999) report on the shrinkage behaviour of BES mixtures with 10 and 20% bentonite. While some shrinkage was observed for all specimens, with shrinkage increasing with moulding water content, volumetric shrinkage was always less than 4% (Fig. 12.17). Shrinkage in excess of 4% is required to cause significant cracking of mineral liners. Cracking therefore seems unlikely for BES compacted at moisture contents typically specified for liners (between optimum and optimum $+2\%$). Hydraulic conductivity was in the range 10^{-12} to 10^{-9} m/s for distilled water, and about an order of magnitude higher with salt solution. Peak friction angles were higher for BES with 5 to 10% bentonite than for sand alone; this appears to be due to improved compactability with the bentonite, allowing a denser packing of the sand grains. BES with 20% bentonite had lower peak friction angles, but still in excess of 25°. The swelling behaviour of dry BES mixes under a wide range of vertical stresses was also studied. A model for the soil which combined the load-deformation behaviour of the sand, the swelling behaviour of the bentonite, the hydraulic conductivity of the bentonite, and the porosity and tortuosity of the sand matrix, was able to match most of the experimental results quite closely.

12.3.2.2. Geosynthetic clay liners. Geosynthetic clay liners (GCL) are thin sheets of bentonite (usually sodium bentonite but sometimes modified calcium bentonite) attached to or encased by geosynthetic materials. They are made under factory conditions and typically consist of a layer of bentonite granules or powder either sandwiched between two layers of geotextile, the strength and other properties of which may be varied for particular applications, or bonded to a geomembrane. The geotextile layers may be mechanically connected by stitching or needle punching fibres from one material through to the other, or the clay may be mixed with an adhesive to provide some structural strength during placing (Fig. 12.18). The material is laid like a carpet, with simple overlaps at the joints sealed by loose bentonite. After laying, the bentonite is activated by hydration, which causes the bentonite to swell and seal the joints and any accidental damage inflicted during installation. These liners are typically only a few millimetres thick, but the clay has a permeability below 10^{-10} m/s (10^{-11} m/s under typical confining pressures). The hydraulic resistance is therefore theoretically the same as a clay layer about a metre thick, but the diffusion resistance of the thin layer is not as great. As with clay barrier layers, it is important that the bentonite is hydrated with clean water, but under a confining pressure, before coming into contact with leachate or hydrocarbon fluids (Petrov & Rowe 1997; Ruhl & Daniel 1997).

In simple single-layer applications the liner should be covered with a layer of soil to protect the material from damage or desiccation and provide some confining pressure. The layer thickness will depend on the nature of the application, but should be at least 150 mm. The geotextile in a needle-punched GCL may provide some confining

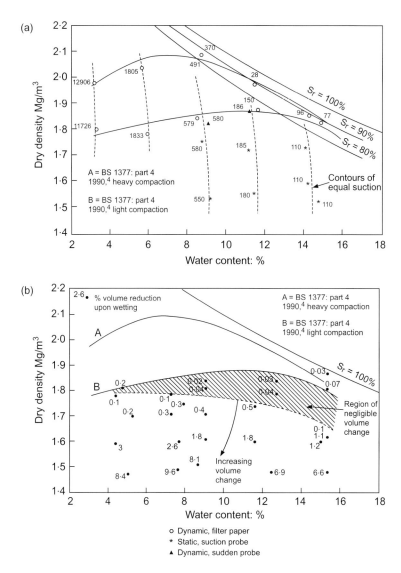

FIG. 12.16. Compaction characteristics of bentonitite-enriched sand (from Dineen *et al.* 1999). (a) Suction characteristics for the bentonite-enriched sand; (b) volume change characteristics on wetting for the bentonite-enriched sand.

pressure. For engineered landfills the material will be used as one element of a multilayer system as discussed above. It may be possible to substitute one of the geotextile layers by a drainage geosynthetic with large in-plane hydraulic conductivity to act as a leachate-collection layer. Similarly for landfill covers such a system can provide a gas-collecting layer in combination with an impermeable layer to control infiltration. Although they are very flexible and can to some extent 'self-heal' minor punctures, GCLs are easy to damage during placement and subsequent operations. They may also have quite low resistance to in-plane shear, and care has to be taken in their use on the side slopes of waste facilities, although problems do not seem to arise until side slopes are steeper than 1 in 3 (Daniel *et al.* 1998).

Needle-punched GCLs may be used on slopes of 1 in 2.5 or even 1 in 2, depending on the slope length. For further general information and discussion of their use see Browning (1998).

12.3.3. Radioactive waste storage and disposal

12.3.3.1. Introduction and background. The use of clay-based materials, predominantly bentonite, is common in many countries' radioactive waste disposal concepts, strategies or programmes. Natural or artificially produced processed bentonite is the most commonly considered material, either used as a grout with or without added cement, or as cast blocks of pure bentonite,

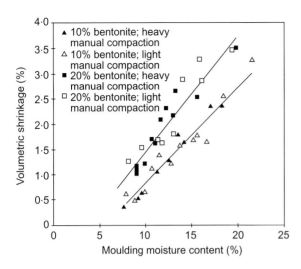

FIG. 12.17. Volumetric shrinkage of bentonite-enriched sand (from Stewart *et al.* 1999).

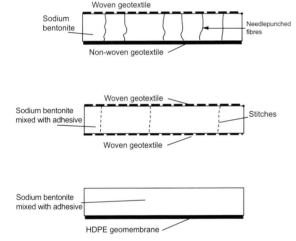

FIG. 12.18. Typical geosynthetic clay liners (GCLs).

surrounding canisters of reprocessed wastes or unprocessed waste fuel bundles in engineered containers. Cryoptinolite and other zeolite rich clays are also components of other strategies, usually for the shallow disposal of low-level radioactive wastes (LLW).

Recent developments have tended towards the concept of monitored retrievable waste disposal, either shallow or deeply sited, often in natural clay horizons or crystalline rock (ENRESA 2000; NAGRA 1996; AECL 1994; Allinson 1989; Allinson *et al.* 1991, 1994). The concepts discussed below are largely based on these countries' programmes (especially those of AECL 1994 and NIREX 1995), but reference is made to other international programmes, where appropriate.

Many countries follow the pattern developed by British Nuclear Fuels Ltd (BNFL) for LLW disposal, of supercompaction of the wastes (in 50 gallon mild steel drums) into 'pucks', which are then stacked in half-height ISO containers. The pucks are then grouted in the container with a lean bentonite/cement mix. Much use is also made of bentonite and other adsorbing clays in the containment, backfilling, grouting and isolation of wastes from the environment. The special case of cryoptinolite, selected by AECL from a US supplier for its radionuclide absorbing properties is worthy of specific mention (AECL 1994).

In the original concept for the disposal of UK-generated intermediate-level wastes (ILW) in a deep geological repository in volcanic or crystalline rock (the RCF laboratory/disposal concept in the Borrowdale Volcanic Group in W Cumbria), the buffer/backfilling around wastes emplaced in disposal galleries was to be a bentonite-cement mixed grout (NIREX 1995). Heater experiments carried out in Spain, Switzerland, Canada and the UK, using simulated wastes (heaters) in cast blocks of bentonite (ENRESA 2000; AECL 1994; NAGRA 1996), demonstrated that the self-healing properties of bentonite, together with the adsorbing potential for both toxic and radionuclide-rich leachates, made bentonite an excellent choice for sealing galleries and vaults.

High-level wastes (HLW) are different from LLW and ILW in that they are heat-generating. Therefore most radioactive waste disposal concepts have these materials stored either in air- or water-cooled storage systems, with preference given to passive maintenance storage in air-cooled stores. At the end of the fifty-year cooling period, it is then proposed to dispose of such materials in specially designed and engineered containers, in a deep geological disposal vault, with similar buffer/backfill emplacement as for ILW, possibly in another area of the same repository.

12.3.3.2. Functions and materials for disposal vaults.
To help achieve the overall objective of limiting the release of contaminants from a disposal vault, a number of different types of vault seals would be needed, as summarized in Table 12.2 and Figure 12.19:

- buffer surrounding the disposal container;
- backfill filling the remainder of the excavated openings of rooms, tunnels, and shafts;
- bulkheads and plugs at the entrance to rooms or at intervals along tunnels and shaft;
- grout injected into the rock mass at selected locations;
- seals for the exploration boreholes that would be used to investigate the site and define the geomechanical, geochemical and hydrogeological conditions.

If the disposal concept were implemented, vault seals suitable for the particular site characteristics would be specifically designed and detailed.

TABLE 12.2. *Applications of clay materials in radioactive waste disposal vaults*

Location	Function
Buffer	To limit the container corrosion rate by inhibiting the movement and modifying the chemistry of groundwater near each container
	To limit the waste form dissolution rate by inhibiting the movement and modifying the chemistry of groundwater near each container
	To retard the movement of contaminants by limiting the movement to the slow process of diffusion, by enhancing sorption of contaminants, and by modifying the chemistry of the groundwater.
Backfill	To fill the space in disposal rooms in order to keep the buffer and containers securely in place.
	To fill the space in tunnels and shafts in order to keep people away from the waste.
	To retard the movement of contaminants by slowing any movement of groundwater, by enhancing sorption of contaminants, and by chemically conditioning the groundwater.
Bulkheads	To inhibit groundwater movement at the entrances to disposal rooms.
Plugs	To inhibit groundwater movement at hydraulically critical points in the vault, such as locations where tunnels and shafts intersect fracture zones.
Grout	To inhibit groundwater movement at hydraulically critical points in the vault, for example at bulkheads, around plugs, and in the zone of rock damaged by excavation.
Shaft seals	To prevent the shafts from being preferential pathways for the movement of groundwater and contaminants.
Exploration borehole completion	To prevent the exploration boreholes from being preferential pathways for the movement of groundwater and contaminants.

To be considered suitable for vault sealing, the materials chosen for this purpose need to exhibit the following characteristics:

- be available in adequate quantities, because several million cubic metres of excavation would need to be filled;
- be workable, capable of being placed to the design specification, using available technology;
- have predictable long-term performance.

A number of screening studies (Coons 1987; Mott *et al.* 1984) identified clay-based, cement-based, and bitumen-based materials as having the highest likelihood of being suitable for vault sealing. Although the possible advantages of using bituminous materials to seal vaults in crystalline rock have been recognized (Allinson 1989), bituminous materials would introduce into the vault large quantities of organic matter, whose effects on total system performance cannot at present be reliably estimated with available information. Studies in Canada and abroad have focused on clay-based and cement-based materials, which are used extensively in civil engineering practice. A wealth of information on their properties, performance, and use is available.

Bentonite is widely used to retard the movement of water in underground structures. Advantage is taken of its low hydraulic conductivity, plasticity and ability to absorb water and swell. If confined, bentonite in contact with water will develop a swelling pressure that enhances its ability to seal open fractures and produce tight contact with interfaces. Fine- to medium-grained aggregates (silt, sand and gravel) can be added to clays to improve some characteristics, such as resistance to erosion by flowing water, density, thermal properties, and sorption. Because the primary function of clay seals is to limit groundwater movement, it is important that the hydraulic properties of the clays are not adversely affected by the inclusion of either granular or other additives.

Provided the aggregate content does not exceed about 80% by total dry mass, at sufficiently high densities the hydraulic conductivities of clay–aggregate mixtures are virtually the same as those of the clay alone (Yong *et al.* 1986; Dixon *et al.* 1985, 1987). Such densities can be achieved by *in situ* compaction or by pre-compacting blocks of material. For less surface-active clays, such as kaolinite or illite, the hydraulic conductivity of clay–sand mixtures compacted *in situ* would be of the order of 10^{-10} m/s; for bentonite, the hydraulic conductivity of clay–sand mixtures compacted *in situ* would be of the order of 10^{-12} m/s. These values are in the same range as those observed in sparsely fractured granitic rock.

12.3.3.3. Buffer materials. Because of its low hydraulic conductivity and its ability to self-seal deformation and shrinkage cracks upon rewetting, compacted bentonite is favoured by researchers in Canada, Finland, Sweden, and Switzerland for use as a buffer. In addition, the near-neutral (pH 7 to 9) pore water of bentonite provides a more benign environment for the disposal container than the alkaline (pH 11 to 13) pore water of cements.

Swelling pressure developed by confined bentonite buffer material would act as an imposed load on the container; thus the designs for the buffer and container must be developed together. Designs have been developed in this way for the pre-closure and post-closure conditions (AECL 1994). Swelling pressures in excess of 10 MPa can be generated by high-density pre-compacted blocks of bentonite, but such high swelling pressures could be avoided by using buffer material with somewhat lower clay densities. For example, studies have shown that reducing the clay density can reduce the swelling pressure

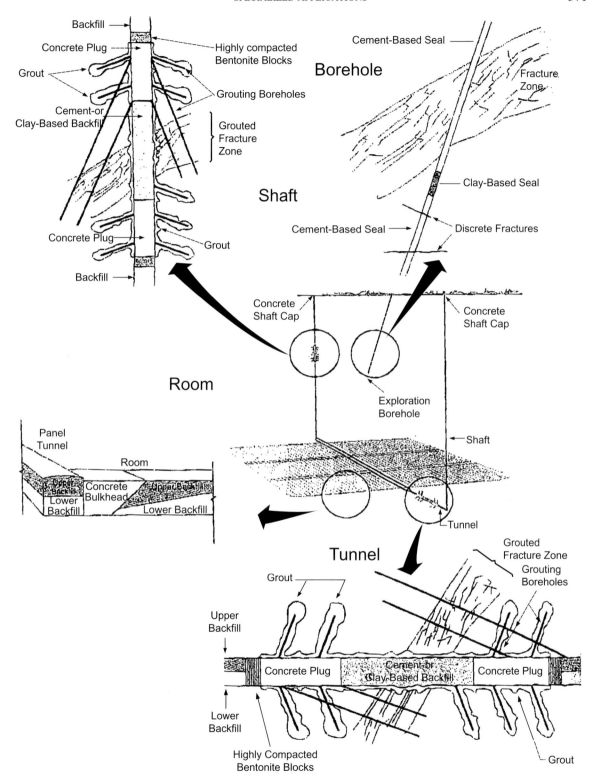

FIG. 12.19. Uses of bentonite in nuclear disposal facilities (after AECLI 1994).

to 1 to 2 MPa while maintaining a hydraulic conductivity of less than 10^{-12} m/s. Thus reducing the swelling pressure would not significantly increase contaminant diffusion through the buffer.

To limit the hydraulic conductivity and contaminant diffusion rates, a pure bentonite buffer should be compacted to a minimum density of approximately 1.2 Mg/m^3 (Cheung *et al.* 1987; Cheung & Gray 1989). Densities in excess of this value can be achieved when the buffer is placed as pre-compacted blocks (SKBF/KBS 1983) and when the clay is compacted *in situ* (Dixon *et al.* 1985). Similarly, clay densities of 1.2 Mg/m^3 or higher can be achieved in sand-bentonite mixtures provided the clay content exceeds approximately 40% by dry mass of the mixture (Dixon *et al.* 1985). In a clay–sand mixture, the clay density is defined as the ratio of the dry mass of clay to the sum of the volumes of the dry clay and the voids; that is, the mass and volume of the sand are not included when calculating the clay density. A sand–bentonite mixture with a clay density of 1.2 Mg/m^3 would have mass transport properties similar to those of a bentonite clay compacted to the same density. Totally confined, both materials would exert similar pressures. Allowed to swell freely, the clay–sand mixture would expand less than the clay. The clay–sand mixture would suffer less drying shrinkage than the clay. Moreover, the clay–sand mixture would deform less under load and, when water-saturated, would have higher thermal conductivity and heat capacity than the bentonite.

The effect of elevated temperatures on the integrity of bentonite-based buffers has been studied extensively. At the temperatures expected in a disposal vault (<1000°C), the alteration of bentonite would be negligible over a period of tens of thousands of years. Also, bentonite samples containing natural open fractures have been obtained from bentonite deposits and exposed to water. The ability of these fractures to self-seal as the bentonite swells on exposure to water has been demonstrated.

12.3.3.4. Room and tunnel backfills. Room and tunnel backfills could be composed of either clay- or cement-based materials. Clay-based backfills would likely be preferred for use in disposal rooms because of the desire to ensure a near-neutral pH near the containers. Various mixtures have been considered, including 90/10 sand/bentonite in Sweden (SKBF/KBS 1983) and 75/25 crushed granite/lake clay in Canada. Such backfills have considerably lower swelling pressures than more highly compacted bentonites but still have a relatively low hydraulic conductivity of about 10^{-10} m/s. The machinery for *in situ* compaction of backfill would require about 2 m of headroom to operate. This space could be filled with clay-sand mixtures using remotely operated machines, as demonstrated by experiments for the International Stripa Project in Sweden (SKBF/KBS 1983). The swelling properties of these upper backfill materials would result in complete closure of the room cavity. The resulting hydraulic conductivity would be about 10^{-10} m/s.

The access tunnels could be backfilled using the clay-based backfills and associated technologies described for the disposal rooms. For operational safety, rock bolts would likely have been installed in tunnels and ramps. Where required by rock conditions, wire mesh would have been installed to prevent the adverse effects of rock spalling. Where possible, the bolts and mesh would be removed and the rock surfaces would be scaled before the seals were constructed. These measures would increase the probability of effectively closing the rock–backfill interface and limiting the hydraulic conductivity of the excavation-damaged zone.

12.3.3.5. Bulkheads, plugs and grout. After successful backfilling of the excavations, the principal flow of water would be confined to the open fractures in the rock and to the interfaces between the vault seals and the rock. These flows would be restricted as much as possible by the construction of concrete bulkheads at the ends of disposal rooms. The bulkheads would also confine the clay-based sealing materials in the room and allow for their full swelling pressure to develop. The clay-based sealing materials would thus be firmly pressed against the rock, and water flow along the interface would be restricted.

Tests at AECL's Underground Research Laboratory have shown that careful preparation of the rock surfaces, such as scaling, can substantially reduce enhanced water flow along the bulkhead–rock interface and through the excavation-damaged zone. Tests undertaken for the International Stripa Project in Sweden have shown that gaskets of pre-compacted blocks of bentonite placed next to concrete bulkheads can virtually eliminate water flow along the bulkhead–rock interface (Pusch *et al.* 1987). Under the high swelling pressures that would be developed, the bentonite would tend to penetrate and seal open fractures in the rock that were exposed or generated by the excavations.

Experiments for the International Stripa Project in Sweden and at AECL's Underground Research Laboratory have shown that rock with hydraulic conductivity as low as 10^{-8} m/s can be sealed by grouting (Gray & Keil 1989). Grouting of rock with lower hydraulic conductivity has not been demonstrated. Should excavation-damaged zones with lower hydraulic conductivity be formed, a cut-off wall that extended from the bulkhead through the excavation-damaged zone could be considered. Cutoff walls are used extensively in civil engineering practice, and experience has shown them to be effective at preventing groundwater flow (Casagrande 1961). The decision to use a cutoff wall and/or grouting to seal the excavation-damaged zone around a bulkhead could only be made when the location of the bulkhead and the rock conditions were known through field observations and measurements.

Massive, high-performance concrete plugs, with associated grouting and bentonite gaskets, could be used to seal tunnels where they passed through hydraulically active fracture zones. The ability to grout fracture zones would depend on their hydraulic properties. Observations at AECL's Underground Research Laboratory have shown that where tunnels intersect fracture zones, the hydraulic conductivity of the zone may be from 10^{-1} to 10^{-7} m/s. Experiments there have demonstrated that high-performance cement-based grouts can seal fracture zones with hydraulic conductivities as low as 10^{-8} m/s (Gray & Keil 1989) and that cement-based grouts can be injected into and seal rock fissures as narrow as 20 μm. It is likely that grouting would improve the system performance; thus grouting where fracture zones are intersected is recommended. However, the long-term benefits of grout injection into the fracture zones have not been quantified, and therefore such benefits are not assumed when assessing post-closure safety.

12.3.3.6. Shaft seals. The materials previously described for sealing rooms and tunnels could also be used for shafts, but the methods for emplacing them would have to be different. Some of the shafts might be fitted with a concrete liner during the construction stage. Golder Associates (Golder/Dynatec Mining Ltd. 1993) recommend that during decommissioning, such liners be removed and the shafts be reamed to enhance the effectiveness of shaft seals. In appropriate geological conditions where there were no long-term adverse influences on the conditions near the emplaced waste, the reamed shafts could be completely backfilled with high-performance concrete. Studies have shown that these materials could maintain hydraulic conductivities less than those of sparsely fractured rock for millions of years (Coons & Alcorn 1989). Alternatively, the technologies used in the rooms and tunnels could be adopted for the shafts. The major fraction of the shafts could be filled with clay-based backfilling materials, and special plugs could be built where the shafts intersected major water-bearing fracture zones in the rock mass. Headroom would not be a problem when backfilling shafts (as it would be when backfilling rooms and tunnels), and a non-swelling backfill would be appropriate. Composite seals consisting of clay-based backfills and concrete plugs with highly compacted bentonite gaskets would likely be more effective than backfill alone.

12.3.3.7. Exploration borehole completions. Exploration boreholes would likely have diameters of less than 200 mm. Any that could compromise the safety of the disposal vault if left open would be sealed. An example of such a borehole might be one that penetrated a water-bearing fracture zone and also penetrated the rock between two disposal rooms in the vault. Others could be left open for monitoring if desired. The requirement for sealing would be site-specific.

Methods for plugging both horizontal and vertical exploration boreholes up to 100 m long with highly compacted bentonite have been demonstrated at the International Stripa Project in Sweden (Pusch *et al.* 1987). Also, the effectiveness of sealing short vertical boreholes by packing them with highly compacted bentonite pellets has been measured (Daeman *et al.* 1983). This latter method is not appropriate for horizontal or inclined boreholes. Moreover, to function effectively, bentonite needs to be confined. Cement-filled sections of the borehole would confine the bentonite and allow for full development of its swelling properties and sealing potential.

High-performance, cement-based plugs and grouts could be used to seal boreholes at any angle. Standard grouting methods could be used to fill a borehole. These include retreat pressure grout injection through specially designed, moveable, mechanical or hydraulic borehole packers. In combination with clay seals, long sections of cement seals could be placed in vertical boreholes (Seymour *et al.* 1987) using the 'balance method' used to place cement slurries when sealing oil and gas wells (Smith 1976). This method allows for controlled placement and for subsequent pressurization to limit the presence of voids in the plug and to grout fissures in the rock.

12.3.4. Bentonite-cement and grouts

12.3.4.1. Bentonite-cement soft piles and cut-off walls. Bentonite and Portland cement may be combined in fluid mixes that harden to form plastic material of low permeability. This may be used to form the primary (female) piles in hard/soft secant pile walls, in which the secondary piles are of normal concrete construction (Fig. 12.20). The soft piles are typically designed to have a strength of 0.2 to 0.3 N/mm² (200–300 kPa).

Continuous diaphragm walls may be constructed with similar material, usually for the purpose of forming an underground cut-off to contain pollutants, control seepage under or through embankment dams, or reduce inflow of groundwater to a dewatered excavation. A typical thickness for such walls is in the range 450 to 1000 mm, and they may include a vertical geomembrane, usually 2 mm thick HDPE. Where they are required to seal into an

Stage 1 - cement-bentonite 'soft' piles

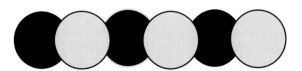

Stage 2 - then concrete 'hard' piles
(for a cut-off wall, both sets of piles may be soft)

FIG. 12.20. Plan view of construction of secant pile wall.

underlying aquitard stratum, typical toe-in depths are 1.0 to 1.5 m. Excavation of trenches up to and sometimes over 30 m is possible with backacter machines; for greater depths diaphragm walling excavation clamshell buckets are normally employed.

A wide range of material composition is possible, from relatively strong and stiff plastic concrete containing a small proportion of bentonite to essentially bentonite slurries with Portland cement added to give a set consistency similar to that of a soft to firm clay. The latter have much lower strength than the plastic concrete, but are able to deform substantially without cracking and maintain their low permeability. The appropriate material will obviously depend on the design circumstances, but is also influenced by local experience and availability of materials. The slurries may be used from the start of excavation, initially as the stabilizing fluid, and left in place at the end to set and harden. For greater control of the final composition, or with the stiffer mixes, the supporting fluid may be replaced by the designed mix on completion of excavation, in a similar manner to that for normal concrete diaphragm walls. There must be a sufficient difference in density between the initial slurry and the final mix to allow the latter to displace the former. Some examples of cut-off walls are given in the Text Box below.

Typical compositions for plastic concrete are given in the European Standard for diaphragm walls (BSEN 1538:1999) as:

	water	400–500 litres
	cement or binder	50–200 kg
	aggregates	1200–1500 kg
	sodium bentonite	12–30 kg
or	calcium bentonite	30–90 kg
or	clay	30–250 kg.

Examples of cut-off walls

- Sheth & Coker (1998), describe a wall up to 44 m deep, 0.76 m thick, constructed of plastic concrete containing a minimum of 20.4 kg of bentonite per cubic metre. The coefficient of permeability achieved was 0.62×10^{-9} m/s for samples of the mix, increasing to 1.8×10^{-8} m/s for the whole wall due to additional leakage through joints between panels.
- Little (1975) gives four examples of slurry trench diaphragms used as cut-offs in embankment dams. The first three consisted of plastic concrete, the fourth a hardening slurry. Some details of the projects are given in Table 12.3. At Withens Clough the concrete mix was specially developed with fly ash replacement of the cement to increase resistance to very acidic water. Hardening slurry was used instead of plastic concrete at Upper Peirce dam because substantial settlements of the dam were expected. Measurements of strength and strain to failure after different periods showed that higher water/cement ratios produced more plastic and weaker concrete. Plasticity fell and strength rose with the age of the concrete. Strength and plasticity did not correlate well with flyash content or bentonite content.
- Puller (1996) refers to three examples of bentonite-cement cut-offs: one from Kielder Water dam in the UK, which was 19 m deep and 600 mm wide; and two from nuclear power station complexes in Italy, one near Milan being 450 mm thick and 23.5 m deep and the other near Rome being 34 m deep and 800 mm thick.

TABLE 12.3. *Details of cut-off diaphragm walls (from Little 1975)*

Site	Balderhead	Lluest Wen	Withens Clough	Upper Peirce
Cut-off depth (m)	46.4	36	15	
Thickness (m)	0.6	0.6		
Mix Unit weight (%) (kg/m³)	2039	1956	1845	1252
Lignosulphite	–	–	–	0.1
Bentonite	2.2	1.2	1.3	2.6
Water	19.6	20.7	22.2	67.1
Cement	9.6	11.6	3.3	20.1
Fly ash	–	–	16.3	10.1
Aggregate	68.6	66.5	56.9	–
W/C ratio	2.05	1.78	6.7	3.35
W/solids ratio	0.24	0.26	0.28	2.06
Permeability (m/s)	0.6 to 2×10^{-9}	10^{-5} to 10^{-6}	–	1.26 to 2.07×10^{-10}

Typical compositions for bentonite-cement walls include about 50 kg/m^3 (45–55) of bentonite, cementitious material about 110 kg/m^3 (70–150), and water 950 kg/m^3, with a typical cement replacement by GGBS or PFA of 60–80%. Mixes with high GGBS content are very susceptible to sulphate attack; bentonite-cement or bentonite-cement-PFA mixes are less susceptible.

It is common in the UK to specify cut-off walls by performance rather than prescriptive specifications. The properties normally specified for set bentonite-cement are unconfined compression strength, permeability and strain at failure. The attainment of sufficient strength is not usually a problem, values of 100 to 500 kPa at 28 days being typical. Target permeability used typically to be 1×10^{-8} m/s, but is now usually 1×10^{-9} m/s, measured in a triaxial cell at 90 days. Because of the inevitable variability of results, some latitude is allowed. A typical specification requiring 80% of results to be less than 10^{-9}, 95% less than 10^{-8}, and 100% less than 5×10^{-8} m/s (ICE/CIRIA/BRE Specification for the construction of slurry trench cut-off walls, 1999) can be cited. The requirement for testing at 90 days results from the slow development of the set slurry properties; if 28-day values are used the specification may be unduly conservative. However it may cause problems on small projects in which the work is complete and the contractor off site before the 90 days have elapsed. In this case 28-day values may be used as a guide to the values likely to be achieved at 90 days.

The specification of a minimum strain to failure in consolidated drained triaxial tests is intended to ensure that the resulting wall is ductile rather than brittle in behaviour, to accommodate long-term movements of the surrounding ground without cracking. A value of 5% is often required. However in the Guidance Notes to the ICE/CIRIA/BRE specification it is argued that the combination of 5% strain at failure and a permeability of less than 10^{-9} m/s is barely achievable, unless unrealistically high confining stresses are used in the tests. No such requirement is therefore included in this specification. If cracking due to excessive strain is likely to be a serious problem, the cut-off should include a geomembrane. It is also noted that BSEN 1538:1999 requires that the stiffness of the set material should satisfy the functional requirements of the wall, but test conditions are not specified.

Construction requirements for bentonite–cement walls are similar to those for normal diaphragm walls; they are discussed in the ICE/CIRIA/BRE specification along with details of testing for control purposes. Fluid loss by filtration from bentonite-cement slurries is usually significantly greater than from pure bentonite slurries, leading to thicker filter cake layers. The filter cake and gelled slurry which has permeated the ground provide additional low-permeability components to the cut-off.

Slurry trench cut-offs were originally largely developed in the USA by the Corps of Engineers during the 1940s. These trenches were up to 3 m wide, excavated by dragline and backfilled with excavated soil to form soil-bentonite cut-offs. This form of construction has remained popular in the USA, where it is more commonly used than bentonite-cement. For less strict permeability requirements, bentonite-cement slurries may also be partially backfilled with soil to achieve economies in construction.

12.3.4.2. Grouting. Clays are frequently used as grout constituents. They have value in one or more of the following ways:

- The weak gel structure or highly porous mechanical structure of some clay minerals can support much larger particles of the grout constituents (cement, PFA or sand), to inhibit their sedimentation from dilute suspensions at or near rest conditions until such time as chemical reactions produce a self-supporting structure. Thus they resist sedimentation and segregation of disparate particle sizes and the consequent appearance of 'bleed' liquid at the upper surface of the grout. The static suspension retains the more-or-less uniform distribution of particle sizes which it had on mixing.
- The corollary of anti-bleed action is that dilute suspensions can be maintained. Thus where low-strength grouts are acceptable, economy of the relatively expensive reactive materials, in particular cement, results from addition of small quantities of active clays.
- The clays can be used to modify the rheological properties of a fluid grout. They can encourage or discourage flow of grouts through small pore spaces or fissures in the target material according to the clay mineral and its concentration.
- Clays can enhance the static Bingham yield strength of grouts and so help to resist wash-out of grouts from coarse-grained soil structures.

Examples of applications of each of these categories are given below.

Any clay material with a high proportion of montmorillonite may be used. Clays occurring naturally, however, are usually contaminated with silts and sands and occasional cemented zones or gravel particles. For works requiring only small quantities of clay, the cost of extracting and processing the raw material outweighs transport costs and it is normal to use only processed bagged clays which have been subject to quality controls for uniformity of characteristics. Only for large works such as dams in remote places is it worth opening a pit and processing near site, for example Mangla Dam in Pakistan (Skempton & Cattin 1963).

For grouting small voids in soils and fissures in rocks, the presence of coarse silt and sand particles in a grout can be highly detrimental to penetration by permeation as the coarser particles rapidly promote filtering blockages. The limitations on such blockages are given by CIRIA Project Report 62, 1997. The processed clays, which are usually bentonites, contain little silt and virtually no sand. They are commonly available and usually produced to conform with national and international standards; clays for grouts are almost invariably of this type. As noted previously, however, quality does vary according to source, and for critical applications selection of an appropriate clay can be important.

There are sometimes conflicting physical requirements of the grout. Maximum fluidity is sought, to allow penetration of the narrowest void openings and complete filling of pores, but a concentrated suspension may be needed to avoid sedimentation and maximize set strength. Weak suspensions are subject to sedimentation when at rest after injection into void spaces, and water is expelled and bleeds along still narrower spaces too small for particles to penetrate. The sediment then does not fill the upper parts of near horizontal fissures, leaving water passages that reduce the effectiveness of water cut-off properties. These residual voids are much more inaccessible to secondary injections and cannot always be filled. Sedimentation in vertical fissures does not present the same problem, as closure is aided by particle settlement. This is well illustrated by Houlsby (1982).

Strength is rarely important in a grout. It is of much more value to fill all the target space by ensuring the requisite fluidity of the grout than to aim for a concentrated stronger but thicker mix. Raffle & Greenwood (1961) showed how little strength in the grout can resist large pressure gradients across narrow openings.

Accordingly, the incorporation of a few per cent by weight of hydrated bentonite or other similar clay can stabilise an otherwise unstable suspension such as ordinary Portland cement with a water/cement ratio by weight greater than 0.4 to 0.45. This results in a grout classified by Greenwood & Raffle (1963) as a cement-bentonite, typically with a proportion of clay to cement of 1:20 and water/cement ratio of 1.0. The shear resistance shows a marked increase on that of neat cement. Set strength depends essentially on water/cement ratio and is therefore reduced. However the mix is stable with respect to sedimentation. In the void sizes amenable to passage by cement particles, the increased shear resistance to injection is of little consequence as filtration is the dominant inhibitor to penetration. The clay thus has an overall beneficial effect for rock fissure injection or similar uses.

If immediate set strength is required to resist adverse water gradients until cement hardening has progressed sufficiently, the addition of a little sodium silicate can improve the yield strength without detriment to other characteristics. The mixture can also be used to restrict the travel of grout in large voids and to lend cohesion to inhibit intermixing and dilution when the voids are filled with water. In this situation the clay content might be increased somewhat.

Those grouts that use clays to extend the volume and cheapen the grout are termed clay-cements; in these the clay content is typically 10 to 50% of the weight of cement. The water content may be up to ten times that of cement, resulting in significant loss of set strength, but also a lower shear resistance during injection than the cement-bentonite. The volume per unit of cement is extended by about 4 to 8 times relative to neat cement or cement-bentonites whilst retaining a stable grout. Such grouts are suitable for injecting into open sand-free gravel beds or for fracturing injections into gap-graded sand-filled cobbles.

For grouts that must permeate under non-fracturing pressures, the penetration of grout through tortuous soil voids requires that both the rheological properties and particle size during injection be strictly controlled. Clay is rarely used for such grouting, but one example relates to its use to facilitate the construction of the second crossing at the Blackwall tunnel in 1961. Here the coefficient of permeability of target river gravels was about 10^{-3} to 10^{-4} m/s with corresponding pore entrances of about 0.03 mm. Consequently the silt and sand particles contained in the clay had to be removed during site processing in multiple banks of hydrocyclones (Greenwood 1963). The concentration of bentonite was about 5% by weight, and sodium silicate and sodium hexametaphosphate were used to modify its rheological properties. Both shear strength and plastic viscosity were reduced and also the Bingham yield strength. In effect the thixotropy of the clay suspension was deliberately destroyed so that any temporary cessation of pumping would not result in enhanced yield strength due to standing which would inhibit displacement and further penetration of grout on resumption of pumping. The gelation of the grout was thus controlled only by the chemical reactions. These are very complex and the role of the clay minerals is not entirely passive.

The closure of cavernous voids—natural or in mine workings—by grouts, affords an example of the use of clays to enhance the yield value of a grout. For complete void filling and economy it is usual to employ a rather fluid grout as the bulk filler of large volumes. To prevent this being wasted to open galleries beyond the zone to be filled it is often necessary to create a perimeter barrier in the void to contain the fluid infill grout. Since conventional grout pumping equipment is designed for quasi-liquid suspensions, these will stand at an angle of repose of only 10 to 12°. Coarse gravel by contrast stands at 40° or more when tipped in a heap, but if combined with fines, especially if introduced in a water-filled void, adopts a flatter slope. To confine the volume of a barrier and at the steepest possible slope the technique was devised of pouring 5 to 10 mm gravel via a 100 mm diameter tremie on to the floor of the void whilst simultaneously pumping a bentonite–cement mixture, sometimes with sodium silicate added, down the same hole. This was first tried in disused mine workings below Wishaw church, Scotland, in 1956. As the gravel pile accumulated, the grout was trapped within its pore spaces, where the high yield strength of the grout held it in place to make the gravel effectively impervious.

Finally mention should be made of the incorporation of clays into soil-cement pastes sometimes used for filling mine workings, notably in the West Midlands. The clay is used to modify the particle size distribution of the pastes to make them more amenable to pumping by reciprocating mortar or 'sludge' pumps. Too high an internal friction makes pumping difficult and inclusion of clays facilitates sliding in pipelines by plug-type flow. Clays for this use need not be refined and are incorporated in the mix in a concrete type mixer.

12.3.4.3. Treatment of contaminated land. Clay may be used to form capping layers for contaminated land, usually in combination with cut-off walls around the site, to encapsulate zones of contaminated land, in much the same way as with engineered landfill waste facilities. Techniques are also being developed in which cementitious and modified bentonite materials are mixed with the contaminated soil to absorb pollutants, including heavy metals and inorganic compounds as well as stable organic compounds. The bentonite in this application is modified to provide sites within the clay particle structure which are attractive to the molecules of the pollutants. Metal ions are adsorbed by ionic exchange, while adsorption of organic pollutants depend on hydrophobic bonding to other non-polar substances such as organic carbon. The modified bentonite is mixed in place with the contaminated soil by auger. Addition of Portland cement also allows the strength of the ground to be increased and hence improved as a foundation material at the same time as being remediated.

12.3.5. Puddle Clay

12.3.5.1. Introduction. Puddle clay, briefly identified in Section 10.2.2, was used extensively in UK canal and dam construction in the period *c.* 1780–1960. First used to form an 'impermeable' liner for canals constructed in permeable ground, it was subsequently more widely employed for the water-retaining element or core within earthfill embankment dams until finally superseded in the latter application by the rolled clay core from *c.*1960. In the context of dam construction satisfactory long-term performance of the puddle core is critical to embankment integrity and hence to safety.

The earthfill embankment dam with a central puddle clay core, commonly referred to in the UK as the 'Pennines-type' dam, has in general performed satisfactorily in operational service. Instances of excessive seepage, settlement and/or other regressive changes have, however, been identified at a significant proportion of the large UK stock of dams of this type. These occurrences have frequently necessitated extensive, costly and sometimes technically difficult remedial action. In the case of canals, cases of serious deterioration or occasional failure of the relatively thin (*c.*300 mm) puddle clay liner have similarly required expensive remedial work. Such events must be viewed in context with a heightened public and political awareness of all issues which may have implications with regard to public safety. They must also be considered against a background perspective of the rundown state of much of the UK canal system and, in the case of the 'Pennines' embankments, both the size and high median age (*c.* 100 years) of the stock of dams of this type. The critical importance of many of these dams as major elements within the national water infrastructure is self-evident. It is also noteworthy that there is considerable current interest in proposals for the selective regeneration of sections of the UK canal system for recreational or industrial purposes and, in some instances, to provide an inter-regional water transfer capability.

Modern remedial technologies such as the *in situ* construction of bentonite-cement diaphragms or the use of geosynthetics, grouting etc. are not necessarily the most appropriate or cost-effective solution to the problems which can arise with puddle clay cores or liners, e.g. when it is necessary to restore embankment or canal bank freeboard lost through long-term settlement. In these and in other instances it may be preferable for reasons of cost and/or material compatibility to effect repairs using fresh puddle clay or an appropriate substitute fill. It will also be apparent that the effective routine surveillance and assessment of older embankment dams and canal banks requires an appreciation of the nature and characteristics of puddle clay.

Puddle clay (or more accurately 'puddled' clay) is neither a unique clay type nor does it derive from such. It can be defined as a plastic soil, i.e. a soil with significant clay content, which has had its natural fabric and structure destroyed by intensive working and remoulding at a water content, w, such that the resulting clay 'puddle' is soft, plastic, dense, homogeneous and of very low permeability. It is thus in principle a fill material well-suited to use in water retaining barriers such as dam cores or canal liners.

12.3.5.2. Puddle clay and the 'Pennines' embankment dam. Puddle clay is particularly associated with the era of the 'Pennines' embankment dam, i.e. from *c.* 1840– *c.* 1960. The 'Pennines' dam was characterized by a central slender vertical clay core of puddle clay supported on either side by an earthfill shoulder, the latter formed using such relatively random and heterogeneous soil fills as were available. Control of underseepage through the natural foundation below the 'Pennines' embankment was generally effected by construction of a puddle clay-filled (or, later, concrete-filled) trench cutoff directly below the central core. It can be inferred that up to 60% of the 2200 UK embankment dams subject to safety legislation (The Reservoirs Act 1975 – see Section 10.2.2), i.e. some 1300 dams, may be of 'Pennines' type (Moffat 2002).

The evolutionary development of the UK 'Pennines' embankment can be traced through the four Phases identified in Table 12.4 before the type was finally displaced by the multi-zoned embankment with rolled clay core from *c.* 1960. Figures 12.21 and 12.22 present illustrative profiles of 'Pennines' dams from Phases 1 and 3 as defined in Table 12.4.

The primary functions of puddle clay core and trench cutoff in the 'Pennines' dam profile can be summarized as:

- control of seepage and head loss through and under the dam;
- limitation of water loss.

The puddle core also serves to provide a more plastic zone capable of absorbing a degree of deformation without serious cracking and loss of integrity. Examined in the specific context of the core and cutoff the 'Pennines' dam profile can be criticised on several points:

- the nominal hydraulic gradients across the slender core and cutoff are relatively high, and both elements may also prove susceptible to hydraulic fracture;
- the effects of long-term consolidation and settlement of the crest;
- possible local drying out and cracking of the core at or near crest level during a prolonged drawdown.

The evolution of the 'Pennines' embankment is described in greater detail in Skempton (1989) and in Moffat (1999), (2002).

12.3.5.3. Puddle clay: nature and characteristics. A satisfactory puddle clay has been produced from cohesive source soils of widely different clay content (from c. 20–70% clay fraction), and thus from across a correspondingly great range as regards soil plasticity, e.g. from a low plasticity (CL) clay through to an extremely high (CE) plasticity clay. Representative generic puddle core clays are characterized in Table 12.5 in terms of their liquid and plastic consistency limits, w_L and w_p. Liquidity indices, defined as

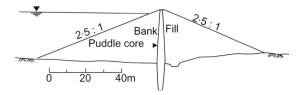

FIG. 12.21. Pennines embankment: c.1860 (Phase 1) (after Skempton 1989).

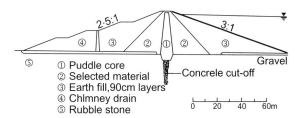

FIG. 12.22. Pennines embankment: c.1900 (Phase 3) (after Skempton 1989).

$$I_L = \frac{w - w_p}{I_p}$$

where plasticity index $I_p = w_L - w_p$, are also tabulated, and express the water content, w, of the puddle clay at placing relative to its consistency limits. It will be noted that I_L frequently lies around the 40–50% mark, i.e. those examples of puddle clay are at a consistency corresponding to a water content approximately midway between that of their liquid and plastic limits. Clays with $w_L < 80$

TABLE 12.4. *Evolution of the 'Pennines' embankment dam in the United Kingdom*

Period[2]	Representative dimensions		Impervious element	Cutoff provision	Supporting shoulders
	max. height (m)	core H/b ratio[1]			
Phase 1 1840–1865	30	3.5 to 6	slender central clay core; puddled *in situ*	puddle clay : deep trench as necessary	random fill; lifts up to 1.2 m; no compaction
Phase 2 1865–1880	35	3 to 5	slender central puddle clay core	puddle clay : deep trench as necessary	select fill against core (Moody zoning); lifts to 1.2 m; toe drainage; compaction incidental
Phase 3 1880–1945	35	3 to 4	slender central core; 'pugged' puddle clay	puddle clay: deep trench as necessary	select fill against core; lifts to 1.2 m; compaction limited
Phase 4 1945–1960	45	3 to 5.5	slender central core; 'pugged' puddle clay	concrete in deeper trenches and grout curtain	select fill against core; drainage provision etc; controlled compaction

[1] H = max. core height above base; b = max. core width at base.
[2] Phase dates indicative only; changes generally progressive and not universal.

TABLE 12.5. *Representative generic puddle core clays: consistency limits and shrinkage*

Generic Core Clay/Dams (BS Classn.)		Liquid Limit w_L%	Plasticity Index I_p%	Water Content w%	Liquidity Index I_L%	Shrinkage Limit w_S
Boulder clays (CL/CI/CH)						
Coulter	(CH)	63	39	35	28	—
Hallington West	(CI)	48	24	31	29	—
Muirhead	(CI)	43	26	28	42	13
Selset	(CL)	30	15	—	—	—
Silent Valley	(CI)	38	21	27	47	13
London clay (CV)						
King George V		80	50	50	40	14
Queen Mary		75	50	44	38	14
William Girling		73	52	43	42	15
Lias clay (CH/CV)						
Barrow No. 3	(CH)	66	41	36	27	—
Monkswood	(CV)	85	60	50	58	14
Alluvial clays (CE)						
Banbury		132	92	75	40	12
King George V		125	80	70	32	11

are considered unlikely to display excessive shrinkage under UK climatic conditions.

Clay soils at their plastic limit, i.e. $w = w_p$, generally display comparable values of undrained shear strength, c_u, in the range 100–200 kPa. On intensive remoulding strength will decrease across the plastic consistency range to some 1–2 kPa at $w = w_L$. Plasticity index I_p is thus a measure of the change in water content required to reduce strength by some two orders of magnitude. For soils of relatively low plasticity (e.g. CL, CI) a comparatively small change in water content effects a significant change in strength, while for clays of higher plasticity (CH, CV etc.) a considerable change in water content is necessary to effect a comparable change in strength. Plasticity index is sensitive to clay content and clay mineralogy, parameters which also control soil permeability, k, and hence consolidation and rate of settlement for clay fills.

12.3.5.4. Puddle clay cores: construction practice and specification. In the earlier 'Pennines' dams raw clay excavated from the borrow area was placed in location in the core in thin layers and water added as necessary to permit working and puddling *in situ* by the feet and spades of the puddling gang. From c.1860 steam pugmills were increasingly employed, allowing effective preparation of stiffer clays, this now being carried out close by the borrow area. Raw clay was generally stockpiled and allowed to weather ('sour') for a period at the clay field before being conditioned with water ('toned') and pugged. If necessary the clay was also 'tempered' by the addition of a small proportion of sand and/or gravel to moderate shrinkage. The processed clay was then transferred to the core for final working *in situ* in layers of 150–200 mm by the puddling gang. This change in practice led to a significantly more uniform core clay.

The final manual working and puddling of the core clay *in situ* was never successfully mechanised, and puddle clay fill was therefore always relatively costly. This encouraged the adoption of slender core geometries, i.e. with height/base thickness ratios of six and more, most notably in some of the earlier 'Pennines' dams (see Table 12.4). There is compelling evidence that such slim cores are more susceptible to hydraulic fracturing.

Traditional puddle clay core specifications, i.e. those which predated any understanding of soil mechanics, developed progressively from the rather vague to the highly prescriptive. Examples are quoted in Moffat (1999) and (2002), and in Johnston *et al.* (1999). Simple and essentially qualitative index tests to establish the suitability of a puddle clay were widely recognised, particularly from c.1900, and embraced the primary characteristics of 'strength', ductility, and susceptibility to cracking and dispersion. These tests, which have a residual relevance, are detailed in the Text Box on p. 380 Head (1980).

British Waterways Board have developed a Standard Specification for puddle clay which effectively writes traditional good practice into a soil mechanics context. The Specification, which specifies 'toning' at $w > 1.3\, w_p$ and a final coefficient of permeability, k, not exceeding 1×10^{-9} m/s, is reproduced as Appendix 12.2.

12.3.5.5. Puddle clay in cores: characteristics. Illustrative values for the primary characteristics of core puddle clays are presented below in the context of assisting with the selection of puddle-compatible remedial methods and materials.

12.3.5.6. Consistency. Freshly prepared puddle clay has the consistency of a very soft clay and can be exuded

> **Site tests for puddle clay**
>
> (a) Tenacity Test
> A 300 mm long, 25 mm diameter cylinder of clay is held vertically for 15 seconds so that at least 200 mm is unsupported and in tension under its own weight. If the cylinder breaks the clay will be rejected as unsuitable.
>
> (b) Pinch Test
> A 75 mm diameter ball of remoulded clay is squeezed into a 25 cm flat disc. If any cracks appear the clay will be rejected as unsuitable.
>
> (c) Slaking Test
> A 50 mm diameter ball of remoulded clay is placed in a 600 ml beaker and covered with water. If the ball disintegrates within 24 hrs the clay will be rejected as unsuitable.
>
> (d) Permeability Test
> A sample of remoulded clay shall be formed into a tray to hold 20 litres of water, and the loss measured after 24 hrs. This shall be compared with the water loss from a metal tray of the same surface area holding the same quantity of water. If the difference is greater than 1% the clay will be rejected.

between the fingers on squeezing. Such a consistency will generally require a water content significantly above the plastic limit (e.g. $w > 1.3\, w_p$).

12.3.5.7. Shear strength. Undrained shear strength, c_u, of a fresh puddle clay at placing lies in the range $c_u = 8.0$–10.0 kPa irrespective of the source clay. This value equated historically to the minimum strength found necessary to support the traditional puddling gang. (The modern stiff rolled clay core, by comparison, will typically have an undrained shear strength in the range $c_u = 60$–120 kPa.) The undrained shear strength of fresh puddle clay increases significantly with time. A modest thixotropic gain in strength over the first 12–18 months, more apparent for source clays with a high liquid limit, is subsequently much enhanced as a consequence of long-term consolidation of the puddle core. A mature puddle clay core will frequently attain a long-term undrained shear strength in excess of 20 kPa at or near its crest, and will characteristically demonstrate a near-linear increase with depth below that level. Some illustrative examples of measured shear strengths are detailed in Table 12.6, and further data is presented in Moffat (2002).

12.3.5.8. In situ permeability. Recorded values of coefficient of permeability, k, for mature puddle clay in cores differ by many orders of magnitude. Moffat (2002) tabulates several examples lying broadly in the range $k = 1 \times 10^{-7}$ m/s to 1×10^{-10} m/s alongside other cases where significantly higher and less satisfactory values, e.g. $k = 1 \times 10^{-6}$ have been reported. The intrinsic permeability of a processed puddle clay is a function of clay content and plasticity, diminishing significantly as the latter increases. The *in situ* mass permeability of puddle

TABLE 12.6. *Representative puddle clay cores: undrained shear strength*

			Mature Undrained Shear Strength cu (kPa)		
Dam	Completion	Core Clay BS Classn.	c_u	triaxial (T) vane (V)	variation c_u with depth z
Blackmoorfoot	1876	CH	26	T	–
Coulter	1907	CH	4.8–74 (mean 28)	V	–
			3.5–86 (mean 33)	T	–
Cwmwernderi	1901	CI	10–65 (SBP (T+10))	T	(20+2 z)
Hallington West	1889	CI	19–50	T	–
March Ghyll	1906	CH	10–76	–	(30+0.7 z)
Ramsden	1883	MI/MH	5–38 (mean 30)	T	'no pattern'

clay within a core (or liner) will be further influenced by factors such as puddling or working techniques, service conditions etc. A value in the range $k = (1–100) \times 10^{-10}$ m/s may be considered desirable and is not atypical for a well-constructed and satisfactory puddle clay.

12.3.5.9. Compressibility. Relatively little data has been published on the compressibility and consolidation characteristics of puddle clay cores, but illustrative values for the principal parameters are presented in Moffat (2002). Example values of coefficient of vertical consolidation, c_v, lying in the range 0.5–10.0 m/year are also reported, together with corresponding estimates of compression index, C_c, of the order of 0.15–0.60. (It may be noted here that the median age of the UK stock of 'Pennines' dams is such that many if not most of these dams have required some modest raising of the core crest to restore freeboard lost by long-term consolidation, and some have required attention to this on several occasions. Recorded settlements well in excess of 0.5 m are not exceptional.)

12.3.5.10. Erodibility and dispersivity. Erodibility and dispersivity of clay soils are functions of the electrochemistry of the clay mineral/soil water complex. For a puddle clay they are thus characteristics which are clearly site-specific. Little quantitative evidence has been published with regard to the erodibility of puddle clays under the seepage conditions prevailing within a dam, i.e. relatively modest seepage velocities and pressures, but the erodibility of most UK clays under the less arduous conditions seen in nature is characteristically very low. Clay dispersivity is a useful but not infallible pointer to erodibility, but while the latter is of prime concern in a dam core no fully satisfactory quantitative measure of this characteristic is yet available. With respect to dispersivity, most of the limited number of UK puddle clays tested to date have been categorised as non-dispersive, but there is evidence of certain inconsistencies which cast doubt over some of the reported findings. A simple qualitative test for puddle clay dispersivity is described in Atkinson *et al.* (1980). Tests for the estimation of erodibility are detailed in Head (1980) and in BSI (1990). The results of a limited study of puddle clay dispersivity and erodibility are summarized in Moffat (2002).

12.3.5.11. Puddle clay service performance. The service of performance puddle clay cores in UK 'Pennines' dams is discussed in some detail in Charles *et al.* (1996) and in Johnston *et al.* (1999). An analysis of the long-term behaviour of a sample 200 UK dams, of which 84% were embankments, is published in Moffat (1982). A review of the behaviour and deterioration of puddle cores in specific dams is presented in Charles (1989). Haider (1989) provides an account of the deterioration and failure of a very old canal bank constructed on sidelong ground which includes discussion of the puddle clay liner.

12.3.6. Other applications

12.3.6.1. Sealing of lagoons, ponds and reservoirs. Lagoons for the storage or treatment of liquid wastes, irrigation lakes, reservoirs etc. may all be sealed by similar methods to those for waste facilities. As an example, Alston *et al.* (1997) describe an effluent treatment lagoon at a pulp mill in Ontario, Canada. A 300 mm thick base layer was required to have a coefficient of permeability of less than 10^{-9} m/s and was constructed from compacted sand–bentonite mix, in the absence of suitable local clay deposits. Extensive testing was conducted to arrive at an optimum mix, which was subjected to permeation for over a year with pulp mill effluent. No loss in performance of the material was recorded.

Existing facilities that are found to be leaking may sometimes be sealed by simply adding bentonite to the water, and allowing it to be drawn into the leaks.

12.3.6.2. Tanking systems. Thin-layer systems for tanking building structures are also available. Flexible geotextile–bentonite sandwich materials as discussed above can be used for horizontal applications, or nailed to the external walls of basements—the swelling clay seals the punctures formed by the nails. Alternatively bentonite can be combined with corrugated cardboard panels which gradually biodegrade, leaving a layer of bentonite, typically a few millimetres thick.

12.3.6.3. Borehole sealing. Pure bentonite is used to seal well points or piezometer tips into boreholes by forming a low-permeability plug below and above the response zone of the instrument. The remainder of the bore will usually be backfilled with arisings or a cement–bentonite grout. A further sealing zone may be installed higher up the bore to prevent cross-contamination of groundwater by surface water. The bentonite is added as a fluid grout of high solids content, or as dry pellets, tablets or chips. The grout is typically of about 20% solids content, but if there is the likelihood of grout loss into very open ground this can be increased to about 30% without the grout becoming too viscous to pump. It should be introduced using a tremie pipe to the base of the zone to be grouted, to ensure that air is not trapped.

Dry pellets or tablets, up to about 12 mm in diameter, or dry chips, up to about 20 mm, are simply poured down the hole if the depth is small, or placed using a tremie pipe and, if necessary, forced down by water or air. In permeable ground below the water table, the groundwater hydrates the dry bentonite; in other cases sufficient water must usually also be added to allow full hydration. The dry bentonite can swell by 10 to 15 times during hydration, ensuring good contact between grout and soil.

Instruments such as inclinometers and extensometers which measure below-ground movements are usually grouted into boreholes using cement–bentonite mixtures; the proportions should be adjusted to give a set stiffness of the grout similar to that of the ground in which the instruments are installed.

Appendix 12.1 Bentonite slurry for diaphragm walling

Test	Range Fresh bentonite	Range Desanded and blended	Comments
1. Mud density (g/ml)	1.025–1.03	1.025–1.06	Mud balance is reliable but not very sensitive. Check calibration weekly. For fresh bentonite, would only indicate gross errors
2. Marshfunnel(seconds)	30–35	30–50	Empirical test, influenced by both density and gel strength. Reproducible, simple to carry out in field. In sands/gravels, limit should be reduced to approx. 40 secs.
3. pH	9.5–10.0	9.5–11.5	pH indicator papers are not very sensitive pH unlikely to vary for fresh bentonite, reduced test frequency possible. In sands/gravel limit pH to 11.5 approx.
4. 10-minute gel strength (N/m^2)	1.4–5.0	1.4–10.0	Simple test but not very reproducible. Insensitive at low gel strength (shearometer)
5. 10-minute gel strength (N/m^2)	2–10	2–20	Sensitive at low gel strength but gives higher values than shearometer readings. Reproducible (Baroid rheometer)
6. Plastic viscosity (cp)	4–10	4–15	Reliable, reproducible. Gives indication of actual viscosity, pump rates etc.
7. Fluid loss (ml)	18–20	Below 25	In granular strata this test gives indication of the ability of bentonite to seal the pores
8. Filter thickness (mm)	1–2	<4	Thickness of filter should be proportional to the fluid loss (measured in same test as fluid loss)
9. Sand content (200 BSS)	<0.5%	<0.5%	Should be less than 0.5% for fresh bentonite. Can be correlated with mud density but of less importance
10. Bingham yield strength (N/m^2)	1.5–5	1.5–25	Reliable measurement but the result is difficult to interpret in relation to other physical properties
11. Apparent viscosity (cp)	8–20	8–40	As above

Appendix 12.2 British Waterways: Technical Services—Standard Specification for Puddle Clay

1. **General**
1.1 Material to be used as puddle clay shall be naturally occurring homogeneous plastic material. It shall be free from deleterious matter such as sand, stones and organic material. The use of lime stabilised clays shall not be allowed.

2. **Properties**
2.1 More than 65% of the natural material shall be finer than 0.06 mm, and more than 40% shall be finer than 0.002 mm.
2.2 The natural material shall be defined as firm clay in accordance with BS5930:1981, Table 8 (C_u 40–75 kPa).
2.3 The natural material shall be defined as clay of intermediate to extremely high plasticity in accordance with BS5930:1981, Figure 31, and the liquid limit shall be not less than 35%.
2.4 The coefficient of permeability (k) of the remoulded material shall not be greater than 10^{-9} ms^{-1}.
2.5 The remoulded material shall be defined as Non-dispersive (ND1) in accordance with BS1377: Part5: 1990, Table 2.

3. **Approval**
3.1 A representative sample of the proposed clay material, not less than 25 kg in weight, together with appropriate test results shall be supplied to the Engineer for his approval not more than two weeks after acceptance of the Tender, and at least 4 weeks in advance of any proposed change in source or quality of the material.
3.2 Test results to BS1377 are required as follows:-
   i) Grading
   ii) Liquid and Plastic Limits
   iii) Natural Moisture Content
   iv) Coefficient of permeability of remoulded clay
   v) Compaction (2.5 kg rammer)

4. **Emplacement**
4.1 The clay should be reworked in a stockpile on site and water added as necessary to destroy the original structure of the clay and produce a smooth plastic homogeneous puddle clay with a moisture content of a minimum of 1.3 times the plastic limit. Reworking of the clay should be carried out in such a manner as to prevent contamination.

4.2 The method of working the clay shall be agreed by the Engineer before work commences. Whatever means are adopted they shall produce a continuous plastic mass of puddle clay effectively free from voids, laminations or imperfections which could affect its water retaining properties.

4.3 The clay shall be placed in horizontal layers not exceeding 150 mm consolidated thickness and compacted by an approved method to an air void content not exceeding 5%.

4.4 Unless agreed otherwise with the Engineer, the type of compaction plant and number of passes shall conform with the requirements of Clause 608 and Tables 6/1 and 6/4 for material class 7C (selected wet cohesive material) of the DoT Specification for Highway Works Part 2.

4.5 Before placing a further layer of puddle, the surface of the previous layer shall be cleansed of all slurry and surplus water and the surface prepared to ensure that the clay to be placed shall be integrated with that already placed. Preparation of surfaces between successive layers shall be formed by frequent non-continuous spade cuts into the upper surface of the clay to a depth of 75 mm.

4.6 Where puddle clay is to be joined with existing clay puddle, the existing clay shall be cut back and stepped to form a good key between the existing and new clay puddle over a distance to be agreed by the Engineer, but not less than 1000 mm. All trace of junction marks shall be wholly eliminated.

4.7 Precautions shall be taken to ensure any puddle clay awaiting placing, puddle which has been placed and any puddle clay in dry areas shall be kept continuously wet to prevent it drying out, and covered by waterproof sheet to protect it from rain damage. Precautions shall be taken to prevent the material freezing.

References

AECL (Atomic Energy of Canada Limited) 1994. Summary of the environmental impact statement on the concept for disposal of Canada's nuclear fuel waste. Atomic Energy of Canada Limited Report, AECL-10721, COG-93-l 1. Available in French and English.

ALLINSON, J. A. 1989. Development of an Engineered Backfilling and Sealing Concept Incorporating Swelling Clay and Bitumen Proc NEA/CEC Workshop, Paris 1989.

ALLINSON, J. A. et al. 1991. Research on Swelling Clays and Bitumen as Sealing Materials for Radioactive Waste Repositories. CEC Report EUR 13522 EN 1991.

ALLINSON, J. A. et al. 1994. Quality Assurance Aspects of Waste Emplacement and Backfilling in ILW and LLW Radioactive Waste Respositories. Report to the Commission of the European Communities REF JAA/87065/001/C/JJ 1994.

ALSTON, C., DANIEL, D. E. & DEVROY, D. J. 1997. Design and construction of sand bentonite liner for effluent treatment lagoon, Marathon, Ontario. *Canadian Geotechnical Journal*, **34**, 841–852.

ANAGNOSTOU, G. & KOVARI, K. 1996. Face stability in slurry and EPB shield tunnelling. *In*: Mair, R. J. & Taylor, R. N. (eds) *Geotechnical Aspects of Underground Construction in Soft Ground* (Proceedings of the International Symposium, London). Balkema, Rotterdam, 453–458. [Also in *Tunnels and Tunnelling*, Dec. 1996, **28**, 27–29.]

ATKINSON, J. H. & MAIR, R. J. 1981. Soil mechanics aspects of soft ground tunnelling. *Ground Engineering*, **14** (5), 20–26.

ATKINSON, J. H. & POTTS, D. M. 1977. Stability of a shallow circular tunnel in cohesionless soil. *Geotechnique*, **27** (2), 203–216.

ATKINSON, J. H., CHARLES, J. A. & MHACH, H. K. 1990. Examination of erosion resistance of clays in embankment dams. *Quarterley Journal of Engineering Geology*, **23**, 103–108.

BRITISH STANDARDS INSTITUTION. British Standards methods of test for soils for civil engineering purposes: BS1377 (in Parts). British Standards Institution, London.

BROWNING, G. R. J. 1998. Geosynthetic clay liners: a review and evaluation. *Transactions of the Institution of Mining and Metallurgy (Section B Applied Earth Science)*, **107**, Sept.-Dec. 1998, B120–129.

CASAGRANDE, A. 1961. Control of seepage through foundations and abutments of dams. The 1961 Rankine Lecture. Milestones in Soil Mechanics, The First Ten Rankine Lectures. Published by Thomas Telford Ltd for the Institution of Civil Engineers, in 1975.

CHAPUIS, R. P. 2002. The 2000 R. M. Hardy lecture: Full-scale hydraulic performance of soil-bentonite and compacted clay liners. *Canadian Geotechnical Journal*, **39**, April, 417–439.

CHARLES, J. A. 1989. Deterioration of clay barriers: case histories. *Proceedings of the Conference on Clay Barriers for Embankment Dams, London*, 109–129. Thomas Telford, London.

CHARLES, J. A. TEDD, P. HUGHES, A. K. & LOVENBURG, H. T. 1996. Investigating embankment dams: a guide to the identification and repair of defects. Building Research Establishment Report. Construction Research Communications, Watford, 88pp.

CHEUNG, S. C. H. & GRAY, M. N. 1989. Mechanism of ionic diffusion in dense bentonite. *Materials Research Society Symposium Proceedings 127 (Scientific Basis for Nuclear Waste Management XII)*, 677–681. Also Atomic Energy of Canada Limited Reprint, AECL-9809.

CHEUNG, S. C. H., GRAY, M. N. & DIXON, D. A. 1987. Hydraulic and ionic diffusion properties of bentonite-sand buffer materials. *In*: TSANG, C. F. (ed) *Coupled Processes Associated with Nuclear Waste Repositories*. Academic Press Inc., Orlando.

COONS, W. E. (ed) 1987. State-of-the-art report on potentially useful materials for sealing nuclear waste repositories. Stripa Project Report, TR 87-12, Swedish Nuclear Fuel and Waste Management Company, Stockholm.

COONS, W. E. & ALCORN, S. R. 1989. Estimated longevity of performance of Portland cement grout seal materials. Sealing of Radioactive Waste Repositories, Proceedings of an NEACEC Workshop, Braunschweig, Germany, 280–296.

DAEMEN, J. J. K., STORMONT, J. C., COLBURN, N. I., SOUTH, D. L. & DISCHLER, S. A. 1983. Rock-mass sealing—Experimental assessment of borehole plug performance. Nuclear Regulatory Commission Report, NUREGCR-3473, Office of Nuclear Regulatory Research, Washington, DC.

DANIEL, D. E., KOERNER, R. M., BONAPARTE, R., LANDRETH, R. E. CARSON, D. A. & SCRANTON, H. B. 1998. Slope stability of geosynthetic clay liner test plots. *Journal of Geotechnical and Geoenvironmental Engineering*, **124**, 628–637.

DINEEN, K., COLMENARES, J. E., RIDLEY, A. M. & BURLAND, J. B. 1999. Suction and volume changes of a bentonite-enriched sand. *Proceedings of the Institution of Civil Engineers, Geotechnical Engineering*, **137**, Oct. 1999, 197–201.

DIXON, D. A., GRAY, M. N. & THOMAS, A. W. 1985. A study of the compaction properties of potential clay-sand buffer mixtures for use in nuclear fuel waste disposal. *Engineering Geology*, 247–255. Also Atomic Energy of Canada Limited Reprint, AECL-8684.

DIXON, D. A., GRAY, M. N., CHEUNG, S. C. H. & DAVIDSON, B. 1987. The hydraulic conductivity of dense clay soils. *Proceedings of the 40th Canadian Geotechnical Society Conference*, Regina, Saskatchewan, 1987, 389–396.

ENRESA 2000. 'FEBEX Project- full scale engineered barriers experiment for a deep geological repository for high level radioactive waste in crystalline host rock' Final Report, Publicaciones tecnicas 1/2000 Enresa, Madrid.

GLEASON, M. H., DANIEL, D. E. & EYKHOLT, G. R. 1997. Calcium and sodium bentonite for hydraulic containment applications. *Journal of Geotechnical and Geoenvironmental Engineering*, **123**, 438–445.

GOLDER ASSOCIATES (in association with Dynatec Mining Ltd) 1993. Conceptual design of shaft seals for a nuclear waste disposal vault. Atomic Energy of Canada Limited Report, AECL- 10047, COG-93- 162.

GRAY, M. N. & KEIL, L. D. 1989. Field trials of superplasticized grout at AECL's Underground Research Laboratory. *In*: *Superplasticizers and Other Chemical Admixtures in Concrete*, Proceedings, Third International Conference, Ottawa, 1989, 605624.

GREENWOOD, D. A. 1963. Discussion, *Proceedings of the Symposium on Grouts and Drilling Muds in Engineering Practice*, pp. 166–167. ICE, Butterworths.

GREENWOOD, D. A. & RAFFLE, J. F. 1963. Formulation of grouts containing clay. *Proceedings of the Symposium on Grouts and Drilling Muds in Engineering Practice*, ICE, Butterworths.

HAIDER, G. 1989. Reconstruction of Llangollen canal. *Proceedings of the Institution of Civil Engineers, Part 1*, **86**, Dec., 1047–1066.

HEAD, K. H. 1980. *Manual of Soil Laboratory Testing* (3 vols), 1238pp. Pentech Press, London.

HEWITT, P. & PHILIP, L. 1997. Problems of clay desiccation in composite lining systems. Geoenvironmental Engineering. Contaminated ground: fate of pollutants and remediation. *Proceedings of the BGA Geoenvironmental Engineering Conference. 1997.*

HOULSBY, A. C. 1982. Cement grouting for dams. *Proceedings of the ASCE Conference on Grouting in Geotechnical Engineering, New Orleans.*

HUDER, J. 1972. Stability of bentonite slurry trenches with some experience in Swiss practice. *Proceedings of 5th European Conference on Soil Mechanics and Foundation Engineering. (Madrid)*, **1**, 517–522.

HUTCHISON, M. T., DAW, G. P., SHOTTON, P. G. & JAMES, A. N. 1975. The properties of bentonite slurries used in diaphragm walling and their control. *Diaphragm Walls and Anchorages*, ICE, London, 33–39.

JANCZESZ, S. & STEINER, W. 1994. Face support for a large mix-shield in heterogeneous ground conditions. *Tunnelling 94, of Mining and Metallurgy and British Tunnelling Society*, Chapman & Hall, London, 531–549.

JEFFERIS, S. A. 1992. Slurries and Grouts. *In*: DORAN, K. (ed) *Construction Materials Reference Book*. Butterworth-Heinemann, Oxford, Ch. 48.

JESSBERGER, H. L. 1994. Geotechnical aspects of landfill design and construction. Part 1: principles and requirements; Part 2: material parameters and test methods; Part 3: selected calculation methods for geotechnical landfill design. *Proceedings of the Institution of Civil Engineers Geotechnical Engineering*, **107**, 99–104, 105–113, 115–122.

JOHNSTON, T. A., MILLMORE, J. P., CHARLES, J. A. & TEDD, P. (1999). *An Engineering Guide to the Safety of Embankment Dams in the United Kingdom*, 2nd edn. Construction Research Communications, Watford, 102pp.

LECA, E. & DORMIEUX, K. 1990. Upper and lower bound solutions for the face stability of shallow circular tunnels in frictional soil. *Geotechnique*, **40**, 581–606.

LITTLE, A. L. 1975. In situ diaphragm walls for embankment dams. *Diaphragm Walls and Anchorages*, ICE, London, 23–26.

MEYERHOF, G. G. 1972. Stability of slurry trenches in saturated clay. *Proceedings of the Conference on Performance of Earth and Earth-supported Structures*, **1**, Part 2, ASCE-Purdue University, Lafayette, Ind., USA, 1451–1466.

MILLIGAN, G. W. E. & MARSHALL, M. A. 1998. The functions and effects of lubrication in pipe jacking. *In*: Arsenio Negro Jr & Argimiro Alvarez Ferreira (eds), *Tunnels and Metropolises*, vol. 2. Balkema, Rotterdam, 739–744.

MOFFAT, A. I. B. 1982. Dam deterioration – a British perspective. *Proceedings of the 2nd Conference of the British National Committee on Large Dams, Keele*, 103–115. British Dam Society, London.

MOFFAT, A. I. B. 1999. Presentation on Puddle clay and the Pennines embankment dam. *Dams and Reservoirs*, **9**, 9–15.

MOFFAT, A. I. B. 2002. The characteristics of UK puddle clay cores —a review. *Proc. 12th Conf. British Dam Society, Dublin*, 581–601. Thomas Telford, London.

MOTT, HAY & ANDERSON (Consulting Engineers) 1984. The backfilling and sealing of radioactive waste repositories. Commission of the European Communities Report, EUR 9115 EN (2 Volumes), Luxembourg.

NAGRA 1996. Grimsel Test Site (GTS)—1996. *NAGRA Bulletin* **27**.

NASH, K. L. 1974*a*. Stability of trenches filled with fluids. *Journal of the Construction Division, Proceedings of the ASCE*, **100** (CO4), 533–542.

NASH, K. L. 1974*b*. Diaphragm wall construction techniques. *Journal of the Construction Division, Proceedings of the ASCE*, 100 (CO4), 605–622.

NIREX 1995. *Nirex Safety Assessment Research Programme*: Nirex Near-Field Research: Report on Current Status in 1994. Report No. S/95/011: NIREX, UK.

PETROV, R. J. & ROWE, R. K. 1997. Geosynthetic clay liner (GCL)—chemical compatability by hydraulic conductivity testing and factors impacting its performance. *Canadian Geotechnical Journal*, **34**, 863–885.

PULLER, M. 1996. Control of groundwater, section on groundwater control by cut-offs. *Deep Excavations: a practice manual*. Thomas Telford, London.

PUSCH, R., BORGESSON, L. & RAMQVIST, G. 1987. Final report of the borehole, shaft and tunnel sealing test (3 Volumes). Swedish Nuclear Fuel and Waste Management Company Stockholm, Stripa Project Reports, SKB-SP-TR-87-01, -02 and -03.

RAFFLE, J. F. & GREENWOOD, D. A. 1961. The relation between the rheological characteristics of grouts and their capacity to permeate soil. *Proceedings of the 5th International Conference of Soil Mechanics and Foundation Engineering, Paris*, **2**, 789.

ROWE, R. K., QUIGLEY, R. M. & BOOKER, J. R. 1995. *Clayey Barrier Systems for Waste Disposal Facilities*. E & FN Spon, London.

ROWE, K. R. 1997. The design of landfill barrier systems: should there be a choice. *Ground Engineering*, **30**, 36–39

RUHL, J. L. & DANIEL, D. E. 1997. Geosynthetic clay liners permeated with chemical solutions and leachates. *Journal of Geotechnical and Geoenvironmental Engineering*, **123**, 369–381.

SEYMOUR, P. H., GRAY, M. N. & CHEUNG, S. C. H. 1987. Design considerations for borehole and shaft seals for a nuclear fuel waste disposal vault. Geotechnical and Geohydrological Aspects of Waste Management, Proceedings of a Symposium, Ft. Collins, CO, 1986. A. A. Balkema Publications, Boston, MA, 317–328.

SHETH, J. N., & COKER, R. L. 1998. Innovative use of slurry walls at dam number 2 hydropower project. *Journal of Geotechnical and Geoenvironmental Engineering*, **124**, 518.

SKBF/KBS (Swedish Nuclear Fuel Supply Company, Division KBS). 1983. Final storage of spent nuclear fuel-KBS-3. Swedish Nuclear Fuel Supply Company Report, Volumes I–IV.

SKEMPTON, A. W. 1989. Historical development of British embankment dams to 1960. Keynote address, *Proceedings of the Conference on 'Clay Barriers for Embankment Dams'*, London, 15–52. Thomas Telford, London.

SKEMPTON, A. W. & CATTIN, P. 1963. A full scale alluvial grouting test at the site of Mangla Dam. *Proceedings of the Symposium on Grouts and Drilling Muds in Engineering Practice*, ICE, Butterworths.

SLIWINSKI, Z. & FLEMING, W. G. K. 1975. Practical considerations affecting the construction of diaphragm walls. *Diaphragm Walls and Anchorages*, ICE, London, 1–10.

SMITH, D. K. 1976. Cementing. Society of Petroleum Engineers of AIME, Dallas, TX.

STEWART, D. I., COUSENS, T. W., STUDDS, P. G. & TAY, Y. Y. 1999. Design parameters for bentonite-enhanced sand as a landfill liner. *Proceedings of the Institution of Civil Engineers, Geotechnical Engineering*, **137**, 189–195.

STREET, A. 1994. Landfilling: the difference between Continental European and British practice. *Proceedings of the Institution of Civil Engineers, Geotechnical Engineering*, **107**, 41–46.

YONG, R. N., BOONSINSUK, P. & WONG, G. 1986. Formulation of backfill material for nuclear waste disposal. *Canadian Geotechnical Journal*, 216–228.

Additional Reading

AL-MANASEER, A. & KEIL, L. D. 1990. Physical properties of cement grout containing silica fume and superplasticizer. *Ground Water and Engineering Materials, Proceedings of the Annual Conference and 1st Biennial Environmental Specialty Conference*, Hamilton, Ontario, 1990, Supplementary Volume, 1–17.

CIRIA 1977. *The use and influence of bentonite in bored pile construction*. W. K. Fleming & Z. J. Sliwinski, Construction Industry Research and Information Association Report PG3/Property Services Agency Civil Engineering Technical Guide No.8, September 1977.

CIRIA 1997. *Fundamental basis of grout injection for ground treatment*. Construction Industry Research and Information Association, Project Report 62.

CIRIA 2001. *Infrastructure embankments—condition appraisal and remedial treatment*. J. Perry, M. Pedley & M. Reid. Construction Industry Research and Information Association Report C550.

DIXON, D. A., HNATIW, D. S. J. & WALKER, B. T. 1992. *The bentonite industry in North America: Suppliers, reserves, processing capacity and products*. Atomic Energy of Canada Limited Report, AECL-10587, COG-92-80.

FPS 2000. *Bentonite support fluids in civil engineering*. Federation of Piling Specialists, Kent.

ICE 1963. *Grouts and drilling muds in engineering practice*, Symposium Proceedings. Butterworths, London.

ICE 1975. *Diaphragm walls and anchorages*. Proceedings of a Conference organized by the Institution of Civil Engineers, London, September 1974. ICE, London.

ICE 1996. *Specification for Piling and Embedded Retaining Walls*. The Institution of Civil Engineers, Thomas Telford, London.

ICE/CIRIA/BRE 1999. *Specification for the construction of slurry trench cut-off walls*. The Institution of Civil Engineers, Construction Industry Research and Information Association, and Building Research Establishment, Thomas Telford, London.

JUHLIN, C. & SANDSTEDT, H. 1989. Storage of nuclear waste in very deep boreholes: Feasibility study and assessment of economic potential. Swedish Nuclear Fuel and Waste Management Company Report, SKB-TR-89-39.

MERRETT, G. J. & GILLESPIE, P. A. 1983. Nuclear fuel waste disposal: Long-term stability analysis. Atomic Energy of Canada Limited Report, AECL-6820.

MERRITT, W. F. 1967. Permanent disposal by burial of highly radioactive wastes incorporated into glass. *Disposal of Radioactive Wastes into the Ground, Proceedings of a Joint IAEA/ENEA Symposium*, Vienna, 1967, CONF-6705 12,403-408. Also Atomic Energy of Canada Limited Reprint, AECL-3014.

ONOFREI, M., MATHEW, P. M., MCKAY, P., HOSALUK, L. J. & OSCARSON, D. W. 1986. Advanced containment research for the Canadian Nuclear Fuel Waste Management Program. Atomic Energy of Canada Limited Report, AECL-8402.

ONOFREI, M., GRAY, M. N., KEIL, L. D. & PUSCH, R. 1988. Studies of cement grouts and grouting techniques for sealing a nuclear fuel waste disposal vault. *Materials Research Society Symposium Proceedings 137 (Pore Structure and Permeability of Cementitious Materials)*, 349–358. Also Atomic Energy of Canada Limited Reprint, AECL-9854.

ROWE, R. K. (ed.) 2001. *Geotechnical and geoenvironmental engineering handbook*. Kluwer Academic, Boston.

WARDROP, W. L. & Associates (in association with Canadian Mine Services Ltd and Hardy Associates, Ltd) 1985. Buffer and backfilling systems for a nuclear fuel waste disposal vault. Atomic Energy of Canada Limited Technical Record, TR-341.

13. Earthen architecture

Earthen architecture can be defined as building where the main constructional material is unfired earth. Load-bearing walls, infilling of walls, roof structures, roof finishes and furniture can be constructed from earth.

Earth suitable for building is generally a well-graded subsoil with a good distribution of clay, silt, sand and aggregate. It should be noted that the clay component is essential for providing cohesion and plasticity during construction, and strength during service. The inevitable variation in subsoils including the moisture content has resulted in a number of manufacturing and construction techniques. This versatility of earth makes it possible to build with earth in cold wet climates such as Britain and hot dry climates such as Morocco.

Variations in soil types, diverse climatic conditions and a wide range of building techniques and numerous architectural details have resulted in culturally and geographically distinctive architecture. These differences are precious for the survival of local distinctiveness. To match this diversity earthen architecture has generated variation within its nomenclature. Clay, loam, soil, mud or earthen architecture are interchangeable terms used to describe the same type of building, where earth is the major constructional material. The name of the same construction techniques also differs from country to country and sometime across regions. For instance the earth building technique known as *cob* in Devon is called *clom* in South Wales, *clob* in Cornwall and *witchert* in Buckinghamshire, *bauge* in France, *tapis* in Spain and *zabour* (or *zabur*) in North Yemen.

This chapter presents an overview of earthen architecture and includes the use of earth as a global building material; applications of earth in building construction; characterization of earth for building; the performance of earth in building; engineered soils; specification and practice, the tradition of earth building in the UK; new earth building in the UK and advantages of earth as a contemporary building material.

The use of building materials available from the site or nearby creates a strong visual bond between buildings within a settlement and between settlements and their landscapes.

13.1. The use of earth as a global building material

In its raw state earth has been a material from which to construct buildings for as long as buildings have been erected to sustain the lives of people and animals. Estimates of at least one-third the world's population occupying earth buildings today suggest the popularity of this very basic building material. Studies of the enormous legacy of international earth building which survives shows that numerous building typologies exist from mosques and palaces to humble dwellings and agricultural shelters in at least forty countries of five continents (Dethier 1982, p7). (Fig. 13.1) Some countries have long forgotten the skills of earth buildings whilst others have maintained or reintroduced traditional earth building techniques, to ensure the conservation of their existing buildings and the creation of sustainable contemporary architecture. However lack of skilled builders and of technical knowledge has meant new buildings are rarely of good construction and often deteriorate very soon after their completion (Olivier & Mesbah 1994).

The significance of the wide geographical distribution in developed and developing countries and the enormous quantity of earth building is that it still represents a major construction material at a global level. Entire settlements can be made from earth, possibly some of the most astonishing are the 'mud manhattans' of the Yemen, where cities such as Shibam and Tarim have earth built towers over six storeys in height (Plate 1). Each tower accommodates a family with strict hierarchy of uses on different floors and is unlikely to have changed during the four hundred years of occupation. Unlike the slick surfaces of contemporary skyscrapers, Yemeni examples are decorated and painted with lime wash to provide individuality to each family.

Equally extraordinary are the Dogon villages of Mali to the South of Timbuktu. These settlements continuously occupied for up to one thousand years are symbolically arrange with square dwellings alongside circular granaries with their distinctive conical thatched roofs (Knevitt 1994). The Great Mosque, Djenne, reconstructed in 1905, is the world's largest earth construction (Plate 2). The well heeled global travellers could likely add many more equally exquisite earthen architectures from many more countries. The emerging international intelligence concerning earth building technology generously disseminated by experts can be used to inspire and support new earth building and conserve the earthen heritage in every country. Although global diversity inevitably means transfer of knowledge requires modification to suit local circumstances to guarantee successful results. Improved communications, publications and websites; earth research programmes; information and training centres;

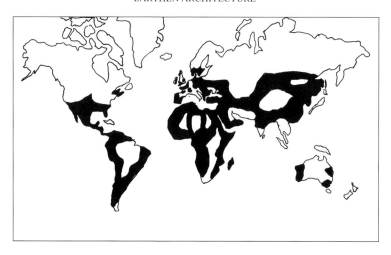

FIG. 13.1. Global distribution of earthen building as a common technique in use.

courses and conferences; and regional, national and international networks provide accessible mechanisms by which earth building will eventually flourish. Already excellent publications with a comprehensive range of illustrations show the global inheritance and explain the manufacturing and construction, maintenance and repair techniques. Also shown are the varying stages of development of earth building in different countries and the associated perceptions of earth as a building material.

Australia and France are good examples of the many countries which have introduced earth as a contemporary building material. In Australia it has taken almost two decades to change scepticism to the point of earth being a fashionable status material, the building 'to be seen in' today. In Australia, earth is used for a wide range of typologies from speculative mass housing developments to hotels and warehouses (Fig. 13.2). What the buildings all share is a contemporary architectural language showing that there is no need to rely on traditional aesthetics. To satisfy the cynics in Australia, recent earth walls have normally been stabilized with the addition of cement in the mix. This has caused purists to consider Australian earth in a derogatory way; such buildings have been christened 'brown concrete'. The skills of earth building have improved over two decades in Australia, when twenty year old earth buildings are compared to those completed recently. This is due to the investment, which has been made by contractors in plant, the adoption of good aspects from the past, together with new methods to develop more commercial procedures (Dobson 2000). The removal of cement stabilization is being considered by those intending to achieve more sustainable building.

Many French regions have inherited large numbers of historic earth buildings constructed from different techniques. For instance bauge has been used in Brittany, Normandy and the Loire; adobe in the Southwest; and rammed earth (pisé de terre) in the Rhone Alps region. In the new town of l'Isle d'Abeau to the east of Lyon, Domaine de la Terre (1982–85), a social housing estate

FIG. 13.2. Speculative housing at Margaret River, Western Australia (1997).

of 61 dwellings is an important milestone in the recent history of earth building. Architects were invited to design houses using whatever earth building techniques they felt appropriate and not restricted to the rammed earth technique, which has traditionally fashioned much of the region's architecture. This was to test the viability of using this natural material in a modern manner (Philippe, Mayor of Ville Fontaine 1990). As a result rammed earth created some of the dwellings (Plate 3) but others used compressed block (Plate 4) or clay straw blocks with external timber cladding. Constructing sixty one new buildings provided an excellent learning experience for their designers and contractors and they have become seminal to those studying earth building techniques. The success of this development is proven, with residents speaking very highly of their houses. To the north of Lyon, rammed earth was used to erect a collegiate church at Anse (Plate 5).

Some countries including Britain are more concerned to ensure the survival of the existing heritage with very few new earth buildings being erected to continue the use of traditional techniques. The general lack of interest in Britain in earth as a contemporary building material could also threaten the earthen heritage due to lack of need to develop skills in the current labour force. This shortage has been identified generally through Heritage Lottery Fund supported research.

13.2. Applications of earth in building construction

Earth has been used in many situations within buildings, including for a major structural walling material, infilling material for walls such as daub on wattle, as a mortar for both stone and earth brick walls, render for both stone and brick walls, flooring and roofing material, an insulator and to construct furniture and fittings. This section considers examples where earth has been used to produce successful wall structures.

Numerous techniques exist to convert subsoil to a walling material. These techniques can be categorized in different ways, for instance those that require a dry earth (8 to 15% moisture content) and those which need a wetter material (18% to 30% moisture content) The former is compactable and the latter is mouldable. Those that produce monolithic components constructed on site can be distinguished from those which require the manufacture of prefabricated small units to construct the building to achieve a further classification system. The main techniques currently applicable to the British construction industry are generally categorized using this latter monolithic/block system.

Cob is the technique whereby subsoil, straw and water are mixed by foot, animals or a tractor and placed to produce an unshuttered, monolithic wall capable of supporting floor and roof loads (Fig. 13.3). This technique has produced thousands of buildings in the West Country, but can also be found in other areas of Britain, throughout Europe and beyond.

FIG. 13.3. University of Plymouth students demonstrating mixing for the cob building technique.

A masonry plinth was generally built to elevate the cob above ground level. From the plinth, which can vary between 30 cm to an entire storey in height, the cob wall is built in lifts. Cob walls are at least 60 cm thick. The height of lifts also vary and represent the work achieved in one day by the cob masons. The wet mix is placed on the wall with a fork and compacted by the foot of the cob mason standing on the wall: one reason for the wall to be thick. In addition, the mix should not be allowed to become too wet, as it would not be able to take the bodyweight of the mason. The next lift cannot commence until that laid previously is sufficiently dry. The time it takes to dry depends upon both the weather conditions and the wetness of the construction mix. It is when the initial drying process is complete that cob walls develop their strength, durability and other properties of performance.

To achieve a good surface finish the wall in a semi-wet state is pared back with a spade. Cob walls are generally protected with an external lime render and either a lime or earth internal plaster. Nevertheless, the handmade nature of cob walling remains visually apparent despite the surface finish. Occasionally cob was constructed with shuttering to achieve the flat surfaces and sharp arrises of Georgian style architecture. This should not be confused with rammed earth.

Rammed earth is the technique which produces a monolithic wall by ramming earth without fibre reinforcement in shallow layers between shutters (Plate 6). Again the wall is achieved in lifts, whose height relates to the dimensions of shuttering. The mix is relatively dry, so when the shuttering is removed there is no need to wait for the lift to dry out.

Considerable force needs to be exerted to compact the earth to approximately half its volume. Hand, mechanical and pneumatic rammers are available to achieve compaction. Rigid strong shuttering is essential to withstand the force of the ramming. When the shuttering is removed the surface has a good quality finish which can be left exposed, avoiding the arduous paring back of cob, and the application of surface finishes.

This traditional technique can be found in France and Morocco where it is known as *pise*. It has been developed by the contemporary construction industry especially in Australia where large investment in plant and equipment has made rammed earth a semi-industrial technique, competitive with other forms of construction.

Adobe are unfired earth bricks (Fig. 13.4). They are manufactured from a wet mix (20–30% moisture content) similar to cob with chopped straw to manufacture bricks from which walls, vaults and domes can be constructed. Adobes can be hand formed, or made in timber or metal moulds into a variety of shapes and sizes. The wet mix means the adobes must be air dried before they can be used in construction. Again, like cob, the time this takes can vary with climate, moisture content of the mix and size of the adobe. The manufacturing process can occur in advance of the building process on an off site location unlike cob. It is easily sheltered so production can take place independent of weather. Centralized manufacture is possible as in the county of Devon, where adobes are known as *cob blocks*, Here these are used in the partial reconstruction of cob walls in certain situations.

Once dried, the adobes achieve their performance properties and construction can commence following the same principles as for fired earth bricks such as observing good bonding. The mortar is similar to the adobe mix although, in preparing the earth, aggregates above 5 mm are removed by sieving.

Alternative construction techniques have been developed using wet adobes without mortar. In New Zealand abodes have been manufactured in their position on the wall.

Adobe is a popular technique from the Ancient Egyptian and Mesopotamian cultures to the contemporary builders of the Southern States of the USA where dwellings are also referred to as adobes. Adobes are frequently disguised by renders and plasters.

Compressed blocks are manufactured from dry earth without fibre compacted in a machine, using the same principle as rammed earth (Fig. 13.5). The machines can

FIG. 13.4. The manufacture of adobe blocks by University of Plymouth Students.

FIG. 13.5. Compressed blocks being manufactured at University of Plymouth.

PLATE 1. Shibam, Yemen. Photograph by Howard Meadowcroft.

PLATE 2. Great Mosque, Djenne, Mali. Photograph by Mike Smallcombe.

PLATE 3. Rammed Earth Housing, Domaine de la Terre at l'Isle d'Abeau, Lyon, France (1982–85).

PLATE 4. Compressed block housing, Domaine de la Terre, l'Isle d'Abeau, Lyon, France (1982–85).

PLATE 5. New church at Anse, near Lyon, France.

PLATE 7. Claystraw house under construction in Sweden by Helsinki University of Technology.

PLATE 6. Rammed earth being constructed in Morocco.

PLATE 8. Typical Devon earth hedgebank defining narrow lanes across the county.

PLATE 9. Sandford Primary School, Devon.

PLATE 10. Boundary walls of Haddenham, Buckinghamshire.

PLATE 11. Lower Tricombe, Near Honiton, East Devon.

PLATE 12. Keppel Gate cob house under construction, Ottery St. Mary, East Devon.

PLATE 13. Public Toilets, Eden Project, St. Austell, Cornwall.

PLATE 14. Building erected at the annual cob building course at the CEA, University of Plymouth.

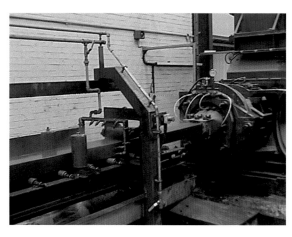

PLATE 17. Extruding of clay body for facing bricks. Hanson Brick, Kings Dyke, Peterborough, UK.

PLATE 15. Extraction of red clay for use in the brick industry. Hanson Brick, Torun, Poland.

PLATE 18. Firing bricks in tunnel kiln. Hanson Brick, Pine Hall kiln, USA.

PLATE 16. Extraction of brick clays by shale planer and dragline. Hanson Brick, Kings Dyke West Quarry, Peterborough, UK.

PLATE 19. Ball clay extraction, WBB Minerals Westbeare Quarry, Petrockstowe Basin, Devon. The dark seams are interbedded lignites.

PLATE 20. Ball clay extraction in Donbas Clays JSC South October (southern) quarry, Donbas Region, eastern Ukraine. Quarry is advancing to left, and being backfilled on the right.

PLATE 21. Clay shredding to create a homogenised, ball clay blend. Donbas Clays JSC, Mertsalovo, Donbas region, eastern Ukraine.

PLATE 23. Hydraulic mining of kaolinized granite (growan) using a high-pressure water monitor. The figure also shows the clearing of waste rock material (stent). WBB Minerals Headon Quarry, south-west Dartmoor, Devon, UK.

PLATE 22. Loading railway wagons with blended ball clay. WBB Fuchs Gmbh, Westerwald region, western Germany.

PLATE 24. Hydrocyclones are used to separate out the fine (kaolin) fraction from the growan (kaolinized granite) slurry. WBB Minerals Cornwood processing plant, Devon, UK.

PLATE 25. Filter presses remove the water from the thickened kaolin slurry and create a kaolin filter-cake. WBB Minerals Cornwood processing plant, south-west Dartmoor, Devon, UK.

PLATE 26. De-moulding of sanitaryware.

PLATE 27. Fired sanitaryware emerging from kiln.

PLATE 28. Automated tile production.

vary from simple hand-operated presses producing three hundred blocks per day to motorized presses manufacturing up to fifty thousand per day. This technique was first developed in the 1950s. The blocks are frequently stabilized with cement.

Blocks can be manufactured in a wide variety of shapes and dimensions to suit differing architectural requirements. Perforation can be incorporated in order that the block uses less material and can be lighter to handle. Indentations are also easy to incorporate to integrate beams, reinforcement and services. Blocks able to withstand earthquakes have also been developed. Compressed blocks are used in a similar way to adobe, although their appearance reflects their machine manufacture with smooth surfaces and sharp arises.

Earth as an infilling material can also be found in use in a timber framework. Wattle and daub was a common technique where daub, a wet earth, is pressed between woven wattles in turn supported by a timber frame. A similar contemporary techniques is where earth is mixed with wood chippings, straw or other additive for use as a infilling material constructed either as rammed earth or dried as non-load-bearing blockwork within a load-bearing timber frame (Plate 7). Light clay and clay straw are terms used to describe this technique.

Table 13.1 shows the relationship between soil characteristics and construction techniques. Traditionally, the technique used in a particular region was likely to be the technique which was always used by the builders. There was little incentive to reject a local technique which had been successful for centuries as the technique suited the local condition in terms of soil type, climate and local building skills. Thus architecture was created which belonged to its location. For instance, cob requires a wet mix, easy to achieve in the rainy climate of England, whilst the desert condition of North Africa, with little water best suits the indigenous rammed earth technique.

Today increased awareness of the numerous techniques together with varying architectural ambitions means a particular technique need no longer be local. Although difficulties could be experienced in achieving a suitable mix and finding appropriate skills in the local labour force when using non local techniques. Local distinctiveness may be eroded, although a balance has to be achieved. For earth to be a viable contemporary material it must be capable of development, but this must be achieved cautiously considering the impact of every decision.

13.3. Characterization of earth for building

Traditionally construction was empirically based with simple hands on field tests, today laboratories can characterize the subsoil proposed for construction. The results of these tests can predict the suitability of subsoil for the building, the most appropriate construction technique and whether modification will be necessary to achieve the necessary performance requirement.

However, the best starting point is through the observation of buildings which surround the site if new build is proposed. Where the surroundings comprise earth building, not only will the onsite or nearby materials be appropriate, existing buildings will suggest the most suitable construction technique. This is certainly true when extending an existing earth building.

In the case of repairing an old building, ideally the material which has fallen from the walls should be reconstituted to provide the repair material. Failing that, materials with comparable constituents should be used to guarantee the existing walling and repair material perform sympathetically.

New buildings in Britain together with those in many countries represent an enormous investment for their owners, so where earth is being considered for construction the suitability of the subsoil must be established. Initially, the subsoil should be examined *in situ* followed by a series of laboratory tests. Grading or particle size distribution and clay content, the former determined through wet sieving, the latter by sedimentation or laser analysis are important tests to establish suitability. A recommended zone of grading or particle size distribution has been presented by recent experts to allow interpretation of results together with optimum clay contents for various techniques (Middleton 1992; Houben & Guillaud 1994; Norton 1997; Harries *et al.* 2000). Cohesiveness directly determines the load bearing capacity of the soil and it will vary according to clay binder present and how it is distributed among the aggregates in the mix.

Where the initial material falls outside the guidelines for the intended construction technique appropriate amounts of gravel, sand or clay can be added or removed to modify the material. In addition, water may be added or the subsoil allowed to dry to achieve optimum moisture content.

TABLE 13.1. *Relationship between soil characteristics and construction techniques (Harries et al. 2000)*

	Cob	Rammed earth	Adobe	Compressed block
Clay content	10–25%	7–15%	15–30%	10–25%
Silt content	10–20%	15–30%	10–30%	15–30%
Sand/gravel content	55–70%	45–75%	40–75%	45–75%
Moisture content	18–25%	10–16%	15–30%	8–16%
Maximum gram size	50 mm	50 mm	10 mm	15 mm

13.4. The performance of earth in building

Although earth has been used empirically in traditional in British buildings, today inspectors, insurers, engineers and designers expect to have quantitative measurements of earth walls' performance. Average values are available, but the variation in raw material, manufacturing and construction techniques means these figures can only be guidelines. *Cob and the Building Regulations*, published by DEBA, is particularly useful for those considering cob in a new building (Devon Earth Building Association 1996).

There is much to be learnt in Britain numerically about the performance of earth in building, particularly where the application is in a non-standard design situation, a non-traditional construction technique or derives from an unusual subsoil. In cases such as these, our existing earthen heritage can provide only limited assurance in knowing whether to expect a successful solution. Where traditional techniques are to continue in conventional situations the outcome is more reliable. The large range of subsoils, natural and artificial additives, manufacture and construction techniques and the degree of quality control mean that several tests are necessary to numerically quantify each soil's performance. Fortunately some laboratory investigation has already been undertaken, so in some instances it is a matter of transference and modification of the existing data. Although the precise data from rigorous laboratory tests can only be reliably transferred to on-site production with good quality control mechanisms.

13.4.1. Structural performance

Available literature and laboratory testing supports the idea that earth walling, if designed and constructed correctly, will achieve the required structural performance required of a two storey dwelling with more than adequate factors of safety. In fact multi-storey buildings, as are found traditionally in countries such as the Yemen and today in Australia, could mean there is no need to be restricted to low-rise developments. The unconfined compressive strength of cob is between 400–1000 kN/m^2 and that of rammed earth 800–2000 kN/m^2 (Harries *et al.* 2000). Research into both test samples and collapsed buildings show that increase in moisture content can drastically effect the compressive strength of earth with disastrous results (Greer 1996; Keefe 1998). Simple principles such as protecting earth walls in wet climates built from plinths with over-hanging eaves need to be observed together with continued use of lime based finishes for protection. Also the combination of load bearing earth walls and concrete columns should be avoided as this will lead to dangerous cracks due to the low deformation modulus (200 to 2000 Mpa) compared to concrete (Olivier & Mesbah 1994).

13.4.2. Durability

The longevity of surviving buildings demonstrates the durability of the material if appropriately detailed and well maintained. Standard tests have been developed to quantify this durability in countries such as Australia. Coatings and sealants have been produced, which may improve that durability of exposed earth walls, although their effectiveness has been questioned in some situations.

13.4.3. Thermal performance

The thermal performance of earth in buildings is frequently questioned and although there is some information on earth's capacity and its improvement through additives, no definitive documentation is currently available for use in Britain. However, research at the University of Plymouth has established testing procedures, (Goodhew 2001). Much information on thermal characteristics of earth for warm countries exists, together with studies in cold climates from France, northern Europe and Canada.

13.4.4. Fire resistance

Fire resistance has also to be quantified for British. Again work overseas is already underway. For instance Australian practice is that any earth construction will achieve at least a two-hour resistance to fire (Dobson 2000).

13.5. Engineered earth

Many campaigners for the use of earth believe it should be used in its natural state as found on a site and the technique chosen a result of soil analysis, local economy and existing skills within the work force. However it can be modified to provide an appropriate material for the intended construction technique or its performance in the proposed building. In some locations the subsoil may not be suited to any technique, so modification is essential or a suitable soil may be sourced from elsewhere. Here the environmental and financial costs cost of transporting the optimum material needs to be balanced against the environmental and cultural gains of using earth.

Engineered earth for building refers to the optimum material which has been modified by either removal or addition of materials for instance sand may be added to clay rich soil prone to excessive cracking. In Germany the production of engineered earth in industrial plants has become a significant characteristic of their earth industry, with their products available outside Germany. To improve strength and durability, stabilizers in the form of cement, lime or bitumen may be added. The clay type will determine the type of stabiliser used. An incorrect additive can be disastrous (Mesbah). Ensuring a consistent mix in engineered soils is vital, although not easily achieved with on-site mixing, particularly when adding clay although the addition of sand can be difficult when using a dry technique such as rammed earth or compressed block because it is important that the mix does not exceed + or −1% to ensure compressive strength (Olivier & Mesbah 1994).

13.6. Specification and practice

Currently no British Standard or Building Code covers the use of earth as a building material in Britain. Although on-site materials are characterized using British Standard (BS) 1377:1995. Building codes exist in many countries where earth is a significant current building material. For instance in New Mexico, New Zealand and Germany their existence shows an acceptance of the material in the main stream industry. New Zealand's N25 4297:1998 Standard for Engineering Design of Earth Buildings contains formula and data required for engineers to design earth-earth walled buildings from first principles for specific design (North 2000). In Germany, Dachverband Lehmev (DVL) has produced Regulations for building with Earth Materials which was accepted and published in 1999 by ARGEAU, a special co-ordinating council of the sixteen German federal regions (Schroeder 2000).

13.7. The tradition of earth building in the UK

The lineage of earth construction in the British Isles can be traced back to the enormous earthworks of the Prehistoric period, such as was used in the banks at Stonehenge; the Neolithic burial mounds, where earth was piled in a configuration of different shapes to conceal tombs, as at West Kennet long barrow and Silbury Hill (the largest Prehistoric man-made hill in Europe), and Iron Age fortifications such as Maiden Castle (Dyer 1981). Earth continued as an important material to construct banks to contain and fortify sites throughout the country, as can be seen at the Norman Castle Rising in West Norfolk and Portsmouth's Tudor fortifications (c.1500) (McCann 1995). Earth hedge banks continue to delineate the patchwork of fields and roadways in the rural South West of England, contributing to the area's distinctive character (Plate 8). At a much greater scale today engineers can move a vast tonnage of earth to reform gradients and create enormous embankments. Artists, such as Lorna Green are taking this as an opportunity to use the surplus material to create works of art in the landscape, e.g. at Cribbs Causeway, Bristol.

The English have always occupied buildings made from earth and continue to do so in their thousands. Whilst it is hard to date the earliest earth building surviving in Britain, remnants of ancient Roman (AD65–300) earth walls survive as at Verulamium, St Albans, Hertfordshire (McCann 1995). Scant remains of any Pre Norman Conquest architecture (1066) leaves the occurrence of earth in building in the realms of the archaeologist carefully reconstructing the remnants of that millennium.

Earth has certainly occupied a significant position in Britain as a building material of the last thousand years. Whilst few earth buildings of the early part of this period survive to modern times, many earth buildings dating from Medieval times (c.1500) to the 19th century are currently in occupation. Further research, in particular the identification and dating of earth buildings (Ford 2002a, b) is required to reveal the true status of earth in construction. However, it is known to have had many applications in walls, both as a structural and infilling material and to make internal floors. In addition, earth was used as a mortar in stone construction and as an internal plaster on both earth and stone walls.

Where timber was available in large quantities, it readily made an excellent building material, but it was used as a frame in combination with other materials infilling between and later over these structural members. It is in the daubs that sealed the wattle, woven between these frames, that earth was used as the major ingredient well mixed with dung and chopped straw (Harris 1979). Some of the earliest daub had been discovered in the excavations of the Viking remains of York (c.800). (McCann 1995). The daub was finished with a smooth coat of white lime plaster if this was available or a sieved earth render (Clifton-Taylor 1987). This infilling technique was replaced in some regions in the 16th century but survived in other regions until the late 18th century (Harris 1979). Little remains of this original infilling material as these timber framed buildings were continuously occupied and when infilling techniques changed and aesthetic ambitions developed, the fabric of the building was updated. Fired brick noggings became a popular infilling material, readily and cheaply available in many regions (Clifton-Taylor 1987) whilst other owners chose to cover their frame in plaster, weatherboarding or tile hanging, possibly leaving remnants of the wattle and daub within the walls. Total encasement of timber frames with brick or stone also relieved the necessity of any infilling material although these earlier materials may still survive buried within the construction.

The geographical occurrence of timber-framed building, and hence the likely existence of wattle and daub, was widespread in England and has yet to be thoroughly studied with reference to earth used in wattle. Even less is known of earth renders and mortars, although successive identification of examples show it is likely that there was considerable use of the material for these purposes, especially if lime was not available or too expensive for the manufacture of renders and plasters.

The use of earth as a structural walling material is better understood, although not comprehensively inventoried in terms of historical and architectural development, quantity or geographical distribution. This is no easy task, because earth used as a walling material is almost always concealed by protective coverings of renders, panels or outer leaves of brick or stone. It appears that the use of this material may have been widespread in both geographical and typological situations. The majority of identified earth building survives in the southwest, central and to the east of the country. This is likely because good building stone was not readily available for use by indigenous builders.

It appears that the most popular traditional construction technique was cob when the number of buildings, which have survived, is taken into consideration. It is believed that the densest survival is in the county of Devon, where it is estimated that at least 20 000 cob dwellings remain in addition to 20 000 agricultural buildings (Devon Historic Building Trust 1992*a,b*). This method of construction exists to the west in Cornwall, to the east across Dorset into Hampshire and north into Somerset and Wiltshire. In Devon remains of cob can be dated back to 1200 and it appears to have been used continuously until the end of the 19th century with a few examples in the 20th century (Child 1995). Analysis of the Devonian building stock shows cob to be used in many different building types from simple agricultural buildings and humble dwellings, to significant houses of landowners and public buildings such as Sandford Primary School (Plate 9). Its use is not confined to rural areas, but has helped form many of the Devon villages, e.g. Broadhembury (Fig. 13.6) and Sandford, and towns, e.g. Dawlish and Crediton.

Earth mortars and renders are generally identified during repair or demolition, often covered by later lime or cement plasters and renders.

Whilst it is likely that many indigenous builders had little choice of materials from which to construct their architecture or knowledge of alternatives, the widespread use of cob in a variety of buildings for different social classes dispels the myth of earth being restricted to the poor. Buildings with clear aesthetic ambitions and whose owners were likely to have had the funds to employ other materials are nevertheless built from cob. Good examples are the early 19th century picturesque villas at Dawlish built from shuttered cob (Fig. 13.7). This demonstrates the possibility of earth being deliberately chosen to achieve internal comfort conditions a characteristic that is beginning to be recognized again today. The typological variation in Devon is similar to the occurrence of earth in other regions, for instance the Rhone Alps of central France.

Fig. 13.7. Early 19th century cob villas in Dawlish, Devon.

The cob technique in England is not restricted to the southwest region, but has been identified elsewhere, for instance in East Midlands and East Anglia. In Buckinghamshire for instance, a group of villages utilized this technique, although the term witchert, meaning white earth, describes the cob walls made from the locally occurring lime rich subsoil. Again the walls are built in lifts known as berries from a stone plinth called a griumnpling (Andrew 1986). Again, the material was consistently used on many building types until the 20th century, creating visual cohesion and giving a strong identity to these settlements. Of the group, Haddenham displays the finest use of curvilinear boundary walls defining intriguing paths throughout the village together with many fine buildings (Plate 10). Bone House, 1807, displays tools which might have been used to make witchert in its external render decoration (Fig. 13.8).

East Anglia can boast a number of earth walling techniques, but the most notable is the use of adobe-type earth blocks locally referred to as clay lump (Bouwens 1994).

Fig. 13.6. Broadhembury village, Devon.

Fig. 13.8. Bone House, Haddenham, Buckinghamshire.

FIG. 13.9. Local authority housing built using claylump at East Harling, Norfolk, 1930.

FIG. 13.10. Nant Wallter, rebuilt at the Museum of Welsh Life, St. Fagans, Cardiff.

This technique was introduced into the region in the 1800s and can be found extensively on the boundaries between the two counties of Norfolk and Suffolk and around Cambridge. It also occurs in other parts of East Anglia. Clay lump is represented in many building typologies and was used to construct entire towns and villages e.g. Watton and Attleborough. However, identification of clay lump is difficult as it was frequently used as an inner leaf with fired brick or flint forming the outer wall surface or it was entirely rendered (Fig. 13.9).

The earth building found to the north of East Anglia in Lincolnshire is characterized by the use of relatively slender walls. This is made possible through the use of a technique called mud and stud, where a timber armature is contained within the earth wall. Mud and stud is an ancient construction technique, although nothing representative survives before the 17th century (Hurd 1994). Of the thousands of Lincolnshire earth buildings which survived until the 20th century, only 200 remain today, many of which have been faced with brick.

Clay dabbs or dabbin is the regional term given to the earth buildings of the Solway Plain. Originally these buildings could be found on both sides of the Scottish English border which runs through the Plain, although today the majority of the survivals are to the southern part in the English county of Cumberland. Domestic and agricultural structures were constructed from a technique similar to cob, although in lifts of between 50 and 150 mm high. A thin bed of straw separates each layer of these 600 mm wide walls. As the wall dries, inconsistencies can cause wave-like layers further characterizing the earth buildings of this region. Like many regions the traditional roofing material was thatch supported by cruck roof trusses. Some survivals dating back to the 14th century (Messenger 1994).

Scotland is rich in building stone, but where it was unavailable or difficult to work, earth provided the material to build walls. Known as mudwalls, they date from prehistoric times to the nineteenth century. A wide range of different building techniques have been identified, the most common is the cob technique. Built by the community, it was the introduction of the construction industry in the 19th century and its preference for the use of stone which was the demise of earth construction, although earth and traditional mortars continued in use with stone until 1920s (Walker 1994).

Prehistoric earth structures were also built in Wales 4000 years ago including Neolithic cromlechi (burial chambers) and defensive ramparts in Bronze and Iron Age settlements. Clom (cob) dwellings for both farm labourers and estate workers were built through Wales with the highest concentrations in SW and NW of the country (Fig. 13.10). Earth too was a mortar and a render as well as the universal flooring material in nearly all cottages and small houses (Nash 1994).

A similar pattern existed in Ireland with earth structures dating back to 4000 BC. Although many different earth building techniques were represented in Irish dwellings including rammed earth, cob, irregularly formed clay lumps and un-burnt bricks to give regions their own style and diversity.

Knowledge regarding traditional British earth building is incomplete. It is possible that 100 years ago, earth buildings were present throughout Britain, but many of these have been lost in recent times, or that the material was not widely used in all the British counties. However, previously unrecognized earth buildings are continuously being identified and further historic references to the use of earth are being discovered. The national network represented by the ICOMOS (International Council on Monuments and Sites) (UK) Earth Structures Committee and its outreach activities is facilitating this exchange of knowledge.

What is beyond dispute is that earth was an important British building material contributing to the distinctive characteristics of the regions, particularly so in walls, both to infill timber frames and in loadbearing situations.

Today there is a preoccupation is Britain with past built heritage. However, knowledge of earth buildings is poor and the methods to ensure their survival sometimes misguided. It is therefore essential that greater public awareness of earth in building is developed and appropriate methods of maintenance and repair designed and adopted to ensure the survival of this heritage. Already great efforts have been made through primary research at universities (Watson & Harding 1994; Watson & Harries 1995) monitored conservation in practice (Harrison 1999) and collaboration with experts from other countries.

Commonly occurring examples of misguided practice include the removal of wide protective eaves replaced by those of slight dimensions, the use of impervious cement finishes and plastic based paints instead of lime render and washes and the repair of cracks and holes with unsuitable alternative materials. These practices may accelerate the deterioration of earth building possibly causing eventual collapse (Keefe 1998).

Unfortunately much of this maintenance and repair is not controlled by legislation, supervised by informed professionals or undertaken by skilled craftspeople so encouraging greater public awareness and basic knowledge is vital in the current building industry.

Fortunately earth is a tolerant building material and appears able to withstand some abuse although current inappropriate repair and maintenance may be storing up enormous problems for the future. Certainly analysing the variation in the nature of the subsoils from which the earth walls are constructed has shown some are less able to withstand high levels of moisture. This together with the use of cement renders and poor eaves details has already caused serious structural failures (Keefe 1998).

A series of interconnected parameters, must be understood, to ensure good conservation. Whilst principles, methods of testing materials and repair techniques can be shared throughout Britain and Europe, it is the peculiarities of individual locations which must also be understood. This site-specific architecture arises because buildings were constructed from local materials whose performance is sensitive to the microclimate and other site conditions. Such specialist knowledge may have been passed down through generations of families, builders and close communities.

Agenda 21 is the international programme of actions to achieve sustainable development in the 21st century agreed at the United Nations Summit in Rio de Janeiro, Brazil, 1996. It makes reference to the need to maintain traditional skills such as those that have created our earthen inheritance. In Britain these skills have been almost lost, but their manifestations in built form remain and it is only when those skills become common to the 'everyday' builder again, that we can feel certain that earth building from the past will survive. The knowledge of those builders who can remember the techniques must be cherished and encouragement given to passing on these skill to a younger generation. Whilst this seems an obvious goal, the current construction industry does little to assist in this campaign, appearing to believe only in contemporary building technology. It is a current conundrum that a country so keen on its past is investing so little in understanding its traditional architecture that the industry may be better informed and controlled. There is little government support in terms of grants for research into earth buildings or in developing teaching material. Organisations, such as COTAC (Conference on the Training in Architectural Conservation) struggle at a national level to ensure that at least some members of the industry are familiar with craft skills including earth building. CITB (Construction Industry Training Board) with COTAC are currently responsible for the development of (NVQs) National Vocational Qualifications levels II & III on traditional walling, which includes a section on earth building.

Nevertheless, knowledge about earth construction is accumulating. Regional groups have formed including EARTHA (East Anglian Earth Building Association), DEBA (Devon Earth Building Association), EMESS (East Midlands Earth Structures Society), HADES (Harborough and Daventry Earth Society) and the Scottish Earth Building Forum to share knowledge. These groups collectively create the ICOMOS(UK) Earth Structures Committee.

Recently this committee, with English Heritage and the Centre for Earthen Architecture at the University of Plymouth, hosted TERRA 2000, the Eighth International Conference on the Study and Conservation of Earthen Architecture. This brought over 400 delegates, representing over fifty countries together in Torquay to share experiences of conserving earth building and the creation of contemporary architecture across the world. The accompanying conference papers provide a lasting record of this significant event. Concurrently with the conference, a new publication on British earth building, titled *Terra Britannica*, was launched, together with a travelling exhibition (Sterry 2000).

13.8. New earth building in the UK

Whilst the past use of earth presents conservation issues to today's guardians of the heritage, a more current and broader conservation issue considers the viability of earth from which to build today's architecture.

A few enlightened British designers and building clients are beginning to recognize that earth has great potential and that the modern building construction industry, which led to this ancestral technique becoming obsolete, may well experience a shift in the light of the need to produce sustainable development and relearn the skills of building from earth. Fortunately those skills are applicable to both new building and the care and repair of existing earth building and hence the modern construction industry is well placed to support conservation.

Britain is not unique in this reawakening but lags considerably behind many countries. Elsewhere considerable advances have been made in the use of earth, with some countries having over two decades of experience in achieving contemporary forms. Other countries have always used earth in the creation of timeless appropriate buildings in the most hostile of climates. This list of countries already involved in earth building is long, and the international networks created can use communication technology to share experiences, disseminate research findings and give advice. This support is very welcome in Britain's current rediscovery of the material. Elsewhere, successful contemporary buildings show what can be achieved.

Some recent achievements from various parts of the UK in earth building include:

- bus shelter at Trusham Devon by Alfred Howard;
- bus shelter at Down St. Mary, Mid Devon by Alfred Howard;
- shelter at Starcross, Torbay by Larry Keefe;
- major extensions to dwelling at Down St. Mary, Mid Devon by Alfred Howard;
- house, studio and garage at Lower Tricombe, East Devon by Kevin McCabe;
- house at Trowleidgh, Dartmoor;
- walls of West Lake Brake house, at Heybrook Bay, Plymouth designed by David Shepard and built by Cob Construction Company;
- Visitors Centre, Eden Project, St. Austell designed by Grimshaw and Partners and built by Insitu Rammed Earth;
- Visitors Centre, Centre for Alternative Technology, Machynlleth, Mid Wales;
- Park and Ride Building, Norden, near Corfe Castle, Dorset, designed by Robert Nother;
- house at Basingstoke, Berkshire, by Ken and Carole Neal;
- Maharishi School Sports and Arts Centre, Skelmersdale, Cheshire, designed by John Renwick and built by Insitu Rammed Earth;
- Visitors Centre, Pitlochry, Scotland, Designed by Tom Morton and built by Becky Little;
- Keppel Gate, Ottery St. Mary, Devon by Kevin McCabe;
- Public Toilets, Eden Project, St. Austell designed by Jill Smallcombe and Jackie Abey and built by Chris Brookman.

This shows that there are signs of the recovery of earth building is beginning in Britain. Yet it was to make any significant impact on our contemporary building stock. The contribution of these pioneers has provided the basis for earth building to take hold in the new century, providing important examples for those who are to follow in achieving successful applications.

The design of a building is a complex process with a wide range of parameters determining the final solution. A significant factor is the choice of the walling material, which make a considerable contribution to the quality of the architecture. In building with earth, the point in the design process at which the decision to use earth is taken is critical, especially if the designer is unfamiliar with the characteristics of the materials. Ideally, the decision should be taken at the outset and the designer should develop the scheme as information accumulates through the architect's own endeavours or through collaboration with consultants.

The first, fundamental question the source. Enquiries received by the Centre of Earthen Architecture (CEA) at the University of Plymouth are generated mainly through interest in the utilization of on-site materials. Where the proposed site is surrounded by existing earth buildings there are likely to be suitable materials. This is not always the case as sometimes much of the surrounding earthen heritage has been lost. On other occasions it may never have existed, suggesting difficulties with on-site materials may arise.

CEA is a multidisciplinary group of academics working in collaboration with the DEBA. Test procedures have been established by CEA, based on established and internationally agreed techniques, to take samples from sites and undertake laboratory tests. CEA research programmes and continuing collaboration with organisations such as ENTPE (École National Travaux Public État), Lyon, France, support CEA in the interpretation of test results and advise on suitability and the need for modification. These test procedures were initially developed by CEA's consultancy at the Earth Centre, Doncaster. Both the name and the ethos of the buildings and the site, suggested earth would be an appropriate walling material. Once characterized it became apparent that the on-site material at the Earth Centre was unsuitable. Bringing other material from which the walls could be constructed from further afield was considered and a suitable nearby source located. However, questioning the sustainable nature of earth when transportation is taken into consideration, together with a number of other problems meant that earth was dismissed as unsuitable (Harries and Clark 1996). Even so, this consultancy did demonstrate the important contribution of laboratory testing on design decisions and material specification.

It can be noted that recent research at CEA has shown that, traditionally, suitable building earth may have been taken from a pit, in a similar way to the quarrying of building stone, to supply the builders of the village (Ford 2002a,b). Transport did not appear to be an issue, as the earth would be transported only a short distance from the periphery of the settlement. Whilst there is no reason why this practice cannot be continued, the need for excessive transport, with its associated environmental pollution and cost, would deny the inherent sustainable characteristic of earth. Generally, where on-site materials are appropriate, landscaping and engineering or building excavation on the site provides sufficient surplus material for building purposes. This is demonstrated by Kevin McCabe at his new earth building complex, Lower Tricombe, near Honiton East Devon (1993–94) (Plate 11).

Once the suitability of materials for construction is established through characterization, the techniques with which the earth can be utilized in the construction of walls must be determined. The nature of the site material may well indicate the appropriate technique, likewise the architectural style of the final solution. The latter may be generated by the need to sympathize with the existing building of the area to maintain regional identity. A continuation of the traditional technique may achieve a traditional form or a contemporary form (Plate 12). It may be that the continued use of the material, but a different technique, would achieve a sympathetic result. Alternatively, the architectural style may not be derived from contextual principles, and other constructional techniques could be considered appropriate; for instance, to achieve a more 'high-tech' solution utilizing the smooth, machine-like visual characteristics of rammed earth with steel and glass. This form is popular in Western Australia. Rammed earth is the technique used to achieve a major wall at the Visitors' Centre of the Eden Project, St Austell, Cornwall built by In Situ Rammed Earth Company. The same technique created the walls of West Lake Brake, a house at Heybrook Bay near Plymouth. The diversity of architectural styles possible with the use of earth is an important characteristic of the material.

The range of techniques possible using earth is large, particularly if the combination of earth with other materials is taken into consideration (Houben & Guillaud 1994). Each technique has its own manufacturing and design considerations, but it must be recognized that constructional issues require that appropriate skills need to be present within the local construction industry for the required result to be achieved. The client may be prepared to pay extra to use a national or even an international labour force if there are deficiencies in the local labour force. This can be beneficial if the 'imported' labour teaches the local builders. The decisions concerning techniques are very complex and may be generated by a strong belief in using local labour and traditional practice. Specialist contractors have been established in this country to practice the rammed earth technique, including In Situ Rammed Earth, London, Earth Structures (Europe) Ltd, Market Harborough and Earth Materials Consultancy, Plymouth. Whilst few traditional buildings have been built using this technique, it can appeal to current designers wishing for a contemporary style.

Low technology, together with the availability of free on-site materials, makes cob an ideal material for those wishing to construct their own dwellings. In Berkshire, Ken and Carole Neal used the cob techniques to build their own home, helped by a mortgage from the Ecology Building Society. Maharishi School Sports and Arts centres, another self-building community project, in Skelmersdale. Here, rammed earth was used under the guidance of Roland Keable of In situ Rammed Earth, an earth construction company. Becky Little has built a cob outbuilding at Pitlochry in Scotland, and planning permission has been obtained for a mud-and stud building in Lincolnshire and for a rammed earth structure in Norfolk.

13.9. Advantages of earth as a contemporary building material

Cynics may question the need to reintroduce what they believe to be an inferior material best left in the past. Earth when used in building is recognized as having a number of remarkable characteristics important for the realization of sustainable development. Its considered inferiority in relation to other materials is irrational when the numerous cases of surviving buildings are considered. The Great Wall of China, the Alhambra in Granada, Spain the city walls of Rabat, Morocco are testimony to the longevity of earth in building, showing the material's ability to survive aggressive attacks from bombarding armies and centuries of weather. Continuous occupation of a living heritage of cob buildings in Devon, some over six centuries is additional evidence that earth is a durable building material given appropriate maintenance.

Earth is one of the few building materials which is free, providing the on-site subsoil is suitable for building. Therefore earth walls avoid costly and time-consuming transport of materials which increase environmental pollution. Generally landscaping, engineering or building excavation on site provides sufficient surplus material for building purposes. In addition the use of on-site earth avoids the large-scale, centralized extraction of alternative materials that contributes to landscape degradation and ecological imbalance (Houben 1995). Whilst it is likely that many British soils can be used in building, as demonstrated at the Earth Centre, Doncaster, not every site can claim such opportunities.

The preparation of the material and the manufacture of the building components require 'low technology' and little energy. Architect Gernot Minke, researching at University of Kassel, Germany, through his tabulation of the amount of energy required in the manufacture of building products, shows earth to require only one percent of that of burnt brick or concrete elements (Table 13.2; Minke 2000). ITDG (Intermediate Technology Development Group) have supported Minke's conclusions comparing earth with a number of building materials.

The recyclable characteristic of the material, means there is no industrial waste and machinery is easily cleaned (Wimmel 1995). This is very important today with our problems of waste disposable and the relatively high percentage of overall waste, which can be attributed to the building industry. Neither does the process produce any toxic chemicals or gases (Houben 1995).

Financial investment in manufacturing equipment can be low and this, together with the simplicity of the production makes the material very accessible. There is no mystery involved, as delightfully described by Alfred Howard, who equates building with cob in Devon with swallows building their nests. Both use the same materials—earth and straw. Mr Howard also advises that the optimum time to build cob is when swallows are in Britain, as the days are long and warm to help dry the walls.

TABLE 13.2. *Comparative energy compution during manufacture and construction (Minke 2000)*

Material	Energy consumption (kWh/m^3)
Earth	5–10
Fired brick	1140
Concrete	550
Non-imported timber	600 (including growing, felling, transportation and conversion)

Time has to be invested in the relearning of old skills, and new ones also if the contemporary techniques utilized overseas are to have an impact on British architecture. These skills are appropriate to mass building industry through to the do-it-yourself enthusiasts as can be demonstrated by current earth building in Australia and southern states of North America. Many questions have to be answered as to how new earthen architecture should be expressed. Should the local traditional techniques be continued? It may represent the most suitable building method in material terms and ensure historic continuity. But is it appropriate for the current construction industry and if facsimiles of the past are created as a consequence, are these appropriate to our current culture? For example, cob is a technique, indigenous to southwest England, but should traditional cob and thatch cottages continue to be constricted in this region?

Collections of photographs of international earth buildings show there to be endless possibilities in formal and textural terms from the organic cottage of cob to the classical linear houses of rammed earth; from the textured and profiled decoration incised into forms to the smooth, flat surfaces of rammed earth. All appears possible with this versatile material provided a few basic principles are understood about suitable subsoils, detailed design and construction techniques. The designer is not restricted to recreating historic architecture (Plate 13).

Many inhabitants of earth buildings claim the 'feelgood factor' to be one of its greatest assets. Gernot Minke believes this to be related to the constant relative humidity of 50% within earth buildings achieved because earth is a porous breathable material. This is the optimum level for healthy respiratory system. Over eight years, measurements of relative humidity in Minke's own house showed a variation of only five percent throughout the year (Minke 2000). The porosity of earth also allows it to absorb other toxins yet insulate from electromagnetism in the external environment. Others claim there is great satisfaction gained from the occupation of a building erected from natural material taken from beneath the ground upon which their home rests. Further more, when there is no longer any need for the building the earth walls can easily return to the ground from which they came. The reactions which cause subsoil to change into a suitable building material are reversible, creating no redundant material to be disposed. Thus the material is ideal for sustainable building construction (Fig. 13.11).

13.10. Conclusions

The legacy left by past generations of British earth builders has not inspired recent architecture, although it provides a valuable, useable living heritage. This is particularly curious when the British interest in heritage is taken into consideration: why has this not influenced new earth buildings to sympathize with existing settlements? Nor has our growing concern for environmental issues and the pursuit of sustainable architecture generated much activity in earth building. This could be a consequence of

FIG. 13.11. Children from Dartington Primary School constructing an earth building with Rainer Warzecha, easily erected and disposed.

the lack of public and professional awareness in Britain. However, the situation is changing as professionals attend conferences, courses and seminars and earthen architecture is integrated into the educational programmes at British Universities (Plate 14). The development of contemporary earthen architecture is more advanced in other countries confirming the acceptance that earth is an accessible, durable and sustainable material.

References

ANDREW, M. 1986. Walls with hats on: Witchert buildings of Buckinghamshire. *Country Life*, 2nd October.
BOUWENS, D. 1994. Clay Lump Walls. In: *Out of Earth I: National Conference of Earth Building*. University of Plymouth, UK.
CHILD, P. 1995. Cob Buildings in Devon. In: *Out of Earth II: National Conference of Earth Building*. University of Plymouth, UK.
CLIFTON-TAYLOR, A. 1987. *The Pattern of English Building*. Faber & Faber, London.
DETHIER, J. 1982. *Down to Earth. Mud Architecture: An Old Idea, a New Future*. Thames & Hudson, London.

DEVON HISTORIC BUILDINGS TRUST 1992a. *The Cob Buildings of Devon. 1. History, Building Methods and Conservation*. DHBT, Exeter.

DEVON HISTORIC BUILDINGS TRUST 1992b. *The Cob Buildings of Devon. 2. Repair & Maintenance*. DHBT, Exeter.

DEVON EARTH BUILDING ASSOCIATION 1994. Appropriate Plaster, Renders and Finishes for Cob and Random Stone Walls in Devon. DEBA, Devon.

DEVON EARTH BUILDING ASSOCIATION 1996. Cob and the Building Regulations. DEBA, Devon.

DOBSON, S. 2000. Continuity of Tradition: New Earth Building. *In*: *Terra 2000: Eighth International Conference on the Study and Conservation of Earthen Architecture*. James & James, London.

DYER, J. 1981. *Prehistoric England and Wales*. Penguin, London.

FORD, M. 2002a. *The Development of a Methodology for Creating an Earthen Building Inventory*. PhD thesis, University of Plymouth, UK.

FORD, M. 2002b. *Sources of Literature Relevant to the Study of Cob Buildings*. Unpublished report, University of Plymouth. UK.

GOODHEW, S. 2001. *The Thermal Properties of Cob Buildings in Devon*. PhD thesis, University of Plymouth, UK.

GREER, M. J. A. 1996. *The Effect of Moisture Content and Composition on the Compressive Strength and Rigidity of Cob Made from Soil of the Breccia Measures Near Teignmouth, Devon*. PhD thesis, University of Plymouth, UK.

HARRIES, R. & CLARK, D. 1996. *An Investigation of an 'Aggregate, Claystone' Mix from Cadeby Quarry, Doncaster as a Material for 'Rammed Earth' Construction*. Unpublished report. Centre for Earthen Architecture, University of Plymouth.

HARRIES, R., CLARK, D. & WATSON, L. 2000. A rational return to earth as a contemporary building material. *In*: *TERRA 2000: Eighth International Conference on the Study and Conservation of Earthen Architecture*. James & James, London.

HARRIS, R. 1979. *Discovering Timber Framed Buildings*. Shire Publications Ltd, Buckinghamshire, UK.

HARRISON, R. 1999. *Earth—The Conservation and repair of Bowhill, Exeter: Working with Cob*, English Heritage Research Transactions Vol. 3. English Heritage, UK.

HOUBEN, H. 1995. Earthen Architecture and Modernity. *In*: *Out of Earth II: National Conference of Earth Building*. University of Plymouth, UK.

HOUBEN, H. & GUILLAUD, G. 1994. *Earth Construction: A Comprehensive Guide*. Intermediate Technology Publication/CRATerre-EAG.

HURD, J. 1994. Mud and stud in Lincolnshire. *In*: *Out of Earth I: National Conference of Earth Building*, University of Plymouth, UK.

KEEFE, L. N. 1998. *An Investigation into the Causes of Structural Damage in Traditional Cob Buildings*. M.Phil thesis, University of Plymouth, UK.

KNEVITT, C. 1994. *Shelter: Human Habits from Around the World*. Polymath Ltd, Berkshire.

MCCANN, J. 1983. *Clay and Cob Buildings*, Album No. 105. Shire Books, Buckinghamshire.

MESSENGER, P. 1994. The Clay Dabbins of Cumbria. *In*: *Out of Earth I: National Conference of Earth Building*, University of Plymouth, UK.

MIDDLETON, G. F. 1992. *Earth Wall Construction*. Commonwealth Scientific and Research Organisation, Sydney, Australia.

MINKE. G. 2000. *Earth Construction Handbook*. WIT Press, Southampton, UK.

NASH, G. 1994. Earth built structures in Wales. *In*: *Out of Earth I: National Conference of Earth Building*, University of Plymouth, UK.

NORTH, G. 2000. The earth architecture of Graeme North: The work of a leading New Zealand earth architect from 1971–2000. *In*: *Terra 2000, Eighth International Conference on the Study and Conservation of Earthen Architecture*. James & James, London.

NORTON, J. 2003. *Building with Earth: A Handbook*. Intermediate Technology Publications, London.

OLIVIER, M. & MESBAH, A. 1994. Earth as a building material. *In*: *Out of Earth I: National Conference of Earth Building*, University of Plymouth, UK.

PHILIPPE, J. P. 1990. Two decision-makers see strong potential for earth architecture and back major international project. *In*: The Renaissance of Architecture in France. France Informations No 137.

SCHROEDER, H. 2000. Building with earth in Germany: current tendencies. *In*: *Terra 2000, Eighth International Conference on the Study and Conservation of Earthen Architecture*. James & James, London.

STERRY, N. (ed.) 2000. *Terra 2000: Eighth International Conference on the Study and Conservation of Earthen Architecture*. James & James, London.

WALKER, B. 1994. Earth building of Scotland and Ireland. *In*: *Out of Earth I: National Conference of Earth Building*, University of Plymouth, UK.

WALKER, B. & MCGREGOR, C. 1996. *Earth Structures and Construction in Scotland: A Guide to the Recognition and Conservation of Earth Technology in Scottish Buildings*. Technical Advice Note No. 6. Historic Scotland, Edinburgh.

WATSON, L. & HARDING, S. (eds) 1994. *Out of Earth I: National Conference of Earth Building*, University of Plymouth, UK.

WATSON, L. & HARRIES, R. (eds) 1995. *Out of Earth II: National Conference of Earth Building*, University of Plymouth, UK.

WIMMEL, B. 1995 *Prima Materia: The Earth Building Material Rediscovered*. A video by Nord Film, ETH Zurich.

Further Reading

ASHURST, J. & ASHURST, N. 1988. *Practical Building Conservation. Vol. 2. Brick, Terracotta and Earth*. English Heritage Technical Handbook. Gower Technical Press, London.

EASTON, D. 1996. The Rammed Earth House, Chelsea Green Publishing Co., Vermont, USA.

ELIZABETH, L. & ADAMS, C. 2000. *Alternative Construction Contemporary Natural Building Methods*, John Wiley, New York.

HURD, J. & GOURLEY, B. (eds) 2000. *Terra Britannica*. James & James, London.

KEABLE, J, 1996. *Rammed Earth Structures*. Intermediate Technology Publication, London.

MACDONALD, F. & DOYLE, P. 1997. *Ireland's Earthen Houses*. A. & A. Farmar, Dublin.

OLIVER, P. 2003. *Dwellings: The Vernacular House World Wide*. Phaidon Press, London.

PEARSON, G. T. 1992. *Conservation of Clay and Chalk Buildings*. Donhead Publishing, London.

ROMERO, O. & LARKIN, D. 1994. *Adobe*. Houghton Mifflin Co., Boston, USA.

TIBBETS, J. M. 1988. *The Earth Builders' Encyclopaedia*. Southwest Solaradobe School, New Mexico, USA.

WARREN, J. 1999. *Conservation of Earth Structure*, Butterworth, London.

WILLIAMS-ELLIS, C. 1999. *Building in Cob, Pise and Stabilised Earth*. Donhead Publishing, Dorset.

WRIGHT, A. 1991. *Craft Techniques for Traditional Buildings*. Batsford, London.

14. Brick and other ceramic products

Clay has been used in a wide range of ceramic products for thousands of years and continues to be a major component in most ceramic bodies today. Fiebinger (1997) states that the annual worldwide production of clay is nearly 400 million tonnes. In the top 50 ranking of extracted minerals and materials, clay is placed 8th with respect to quantity, and 19th in terms of value. Over 90% of the annual tonnage is utilized in the heavy or structural clay sector, with the remainder (predominantly ball clays and plastic clays) being used for higher quality or fine ceramic products. The value of the worldwide ceramic production (of all varieties) was estimated at US$ 113 billion in 2000 (Reh 2000). From 1991 to 1999, growth was significant in the advanced ceramics, tile and sanitaryware sectors, but less so in structural ceramics and refractory products, as detailed below:

- **wall and floor tiles**; 70% growth with US$ 14 billion turnover in 1999;
- **advanced ceramics** (including the carbon and graphite sector); 63% growth with US$ 40 billion turnover in 1999;
- **sanitaryware**; 50% growth with US$ 12 billion turnover in 1999;
- **structural ceramic sector** (bricks, roofing tiles, pipes); 17.5% growth with US$ 23 billion turnover in 1999;
- **refractory products**; <10% growth with US$ 12 billion turnover in 1999.

The technical definition of a ceramic product is 'a product that is composed of polycrystalline, inorganic and non-metallic materials that have been subjected to a temperature of 540°C or more during manufacture or use' (O'Bannon 1984). In spite of the availability of modern, alternative products manufactured from plastic, steel, paper, glass-fibre/resin composites, borosilicate glass etc., the demand for ceramic products in all sectors remains relatively strong. This chapter describes how the chemical, physical and ceramic properties of the clay components have a predominant influence on the characteristics of both the unfired and fired product, and also considers the role and importance of the other body materials.

Clay is an integral part of many products and materials used in house construction, and also in many items utilised within our homes. Whereas bricks, roof tiles and pipes are the principal heavy or structural clay products, clay is used in numerous internal home items including sanitaryware, wall and floor tiles, acoustic ceiling tiles, storage heater blocks, gas fire radiants, tableware etc.

Understanding the requirements of both the ceramic manufacturer and the customer is fundamental to achieving and maintaining an effective clay raw material supply. There are both similarities and some differences when comparing the requirements or 'likes and dislikes' of a heavy or structural ceramic producer with the manufacturer of fine ceramic ware. All producers require high green (unfired) strength, plasticity and consistency in their plastic clay components. For some fine ceramic applications (slip casting and spray drying), the rheological or fluid properties are important. Other ceramic bodies require a strong fluxing or early vitrification properties in the clay. Structural ceramic producers can tolerate or even need iron minerals in the clay body to influence the fired colour or provide specking effects, and they also utilize clays with both a higher organic and sand content (proportion $>125\,\mu m$). In contrast, iron minerals (especially pyrite and siderite) are regarded as contaminants in most fine ceramic bodies and high residue and organic clays are also avoided, the latter due to the constraints imposed by fast firing technology. Soluble salts are generally avoided as they can lead to deflocculation problems and salt migrating to the surface as the product dries. Salts can also be a problem in the fired product when they are in use. The price of the clay is important in both sectors, but especially so when it is a component in a comparatively low value product. The customer is primarily interested in appearance (attractiveness, shape, style, colour and decoration, lustre and translucency), thickness, durability and few or no flaws. The ability to match and replace products is important, as is their cost and value.

This chapter covers the genesis and occurrence of the principal UK and European ceramic clay deposits, along with the associated production and refining technology. Through its durable and resilient nature, ceramic ware has proved to be a very important tool in the dating and categorizing of earlier cultures. The historical development of ceramic styles and technology are summarized, along with an appraisal of future trends.

14.1. Historical utilization of clay in structural and fine ceramics

The utilization of different types of clay has been an integral part of human society for many millennia. The initial use was in sculpted figurines and household ware (pots, storage jars, beakers, tableware, etc). The development of structural ceramics commenced between 6000 and 8000 years ago, when bricks, roofing tiles, and decorative wall and floor tiles were first used in building construction. In the intervening years, clay has become very widely employed in many other aspects of domestic life. Numerous familiar items and products now contain clay, including paint, rubber and PVC products, toothpaste, pet litter, acoustic ceiling tiles, pharmaceuticals, wallpaper, cosmetics, etc.

Tables 14.1a and 14.1b chart the significant historical developments in the production and utilisation of structural and fine ceramics.

14.2. Ceramic product range

Ceramic products can be broadly divided into four main varieties:

- **structural or heavy clay products** (bricks, roofing tiles, pipes and some floor tiles);
- **fine ceramics or whitewares** (wall tiles, floor tiles, sanitaryware and tableware);
- **refractories** (ceramics used in high temperature applications e.g. kiln linings, steel and glass plant refractories etc);
- **electrical and special ceramics** (electro-porcelain used for electrical insulators and modern ceramic materials).

This chapter focuses on the ceramic products that are more closely associated with construction, especially the structural or heavy clay products (bricks, pipes and roof tiles). Parts of the fine ceramic sector (wall tiles and sanitaryware) are also considered in detail.

14.2.1. Ceramic bodies

Every type of ceramic product must have a range of properties that meets the conditions presented by the environment in which it is to be utilized e.g. wall tiles normally have a porous body, whereas floor tiles are fully vitrified, resulting in a glassy non-porous body. A variety of properties can be created in a ceramic body by adjusting the proportions and types of the component materials. Clays are an essential and predominant component in most ceramic products. The clay materials which are most frequently utilized are ball (or plastic) clay, kaolin (or china clay), fireclay, common clay and shale, flint clay and more rarely bentonite. In fine whiteware ceramics, the basic body components traditionally consist of ball clay, kaolin (or china clay), a filler (quartz) and a flux (typically feldspar).

Common clay and shale, and to a lesser extent fireclay and ball clay, are the clays used in the structural or heavy clay sector. The major body components in structural ceramic products are normally common clay and shale and quartz sand, but other frequently used non-clay raw materials include finely ground calcium carbonate, a carbon containing material, pulverized fuel ash (PFA), sawdust and chopped straw (Šveda 2000). Chemical additives or colouring agents are also widely used to

TABLE 14.1a. *Historical developments in structural ceramics*

Date	Historical event or development
~6,000 BC	Sun-dried mud-bricks reinforced with straw first used in Mesopotamia.
~4,000 BC	Kiln-fired, sun-dried bricks composed of mud and straw first produced in Middle East. Brick technology and usage develops and spreads to Mediterranean countries.
250 BC	Romans develop widespread use of thin-fired 'biscuit' bricks and roof tiles.
43 AD	Romans introduce bricks and tiles to UK, and they become widely used in construction.
400–~14th C	In the UK, use of bricks and tiles in construction is largely discontinued, as stone, timber and wattle and daub predominate.
12th–13th C	Tile manufacture re-commences in UK (plain rectangular tiles).
14th–15th C	Brick-making and utilization is re-established in south-east England, due to immigrant Dutch and Flemish brick-makers. Bricks used in Eton College, Hampton Court Palace etc.
15th–16th C	English buildings are influenced by various elements of continental brick building styles (including curved Dutch gables).
17th C	Bricks are used widely in the construction of ordinary houses. London is largely rebuilt with bricks after the Great Fire of 1666.
17th C	Flemish bond brickwork (where headers and stretchers alternate within courses) becomes popular.
1630s	Introduction of pantiles in UK (S-shaped with horizontally overlapping lips) from Flanders. Roofs become cheaper and lighter.
1824	Portland cement patented by Joseph Aspdin.
1830s	Development of first major network of brick-built sewers in London.
16th C	English bond brickwork (consisting of alternating courses of headers and stretchers) is widely used.
1850s	Development of mechanical brick-making (pressing of the brick body into moulds) leads to the mass-production of relatively cheap standard bricks of good quality. Rapid expansion of brick utilization.
1854	Reinforced concrete first utilized in building.
1858	Invention of Hoffman kiln is significant factor in rapid expansion of brick industry.
19th and 20th C	Stretcher bond brickwork (where only stretchers are outward facing) becomes widely used in brick constructions.
Mid-20th C onwards	Increasing use of alternative materials in some structural sectors, such as plastic piping and concrete blocks.

TABLE 14.1b. *Historical developments in fine ceramics*

Date	Historical event or development
22,000 BC	First artistic figure made out of clay (discovered in Czech Republic).
10,000 BC	First tableware ceramics produced in Japan from a mixture of plastic clays and sand.
4,000 BC	Wall and floor tiles used in Egypt.
3,700 BC	Potters wheel developed in Mesopotamia.
3,000–2,000 BC	Neolithic Period: In the UK there is a gradual change from round-based, plain pottery bowls derived from Western Europe, to locally produced flat based highly decorated ware.
2,000 BC	Early Bronze Age: In the UK, there is an enlarged range of pottery, most of which is highly decorated and incorporates a distinct collar.
2,000 BC	China, Yanshao and Longshan cultures. White firing pottery contains some kaolin.
1,000 BC	China, Shang Dynasty. Pottery made entirely of kaolin.
300 BC	Kaolin first worked extracted on a high ridge (Kau-ling) near Jaucha Fu in Jiangxi Province, China. The name 'kaolin' is derived from this ridge.
900 AD	Decorative tiles widely used in Persia, Syria and Turkey (initially in mosques and palaces but use quickly spreads to other buildings).
1300	Manufacture and use of ceramic tiles spreads initially to Italy and Spain and latterly into the rest of Europe.
1500s	Manufacture of clay pipes (for smoking tobacco) from ball clay.
1710	Commencement of porcelain manufacture in Meissen (Johann Friedreich Böttger).
1720	Calcined crushed flint used in ceramic bodies to improve the whiteness and vitrification.
1727	Limoges, France. Commercial production of 'hard-paste' porcelain from clay derived from a deposit near Limoges.
1739	Savannah, Georgia, USA. Porcelain manufactured from local clay.
1768	Plymouth, Devon, UK. William Cookworthy patented method to manufacture porcelain from a mixture of kaolin and china stone derived from Cornwall.
1780s	Ceramic tiles imported into UK (mainly from Holland). Dutch potters establish factories in London, Bristol and Liverpool producing blue-patterned delft tiles.
1799	First use of steam power in ceramic manufacturing.
1800s	Pottery industry concentrates in Staffordshire, and many companies start to produce wall and floor tiles. By end of 19th century most buildings in the UK contained some ceramic wall or floor tiling.
1885	First electric motor used in the ceramics industry.
1921	Over 1 million square yards of ceramic tiles produced in the UK per annum, with over 30% exported.
~1945	Commencement of shredding and blending of ball clays.
1950	Semi-automation in tableware industry achieved through use of roller flatware machines and profile tools.
~1960	Commencement of inter-regional blending of ball clays.
From 1970	Quality, and cost effectiveness improved through the introduction of tightly controlled blends and body compositions and automated or semi-automated production lines.
1975	Development of refined ball clay for sanitaryware.
1980	Introduction of slurried sanitaryware ball clay blends.
1980s and 1990s	Efficiency maximized in some sectors through the introduction of pressure casting and capillary mould systems, microwave drying, fast (and once) firing, and robotic and automated systems.
1991	Sales of ceramic tiles in the UK exceed 38 million m^2 per annum.

ameliorate potential contaminant issues, aid the manufacturing process or provide a desired aesthetic characteristic. The components in structural ceramic bodies perform the following roles:

- **clay minerals** are used to provide plasticity, bind the other materials together and to vitrify on firing;
- **quartz** gives the product strength and durability;
- **CaCO$_3$** is added to provide a buff/yellow colour;
- **iron oxides** give a red colour in an oxidising kiln atmosphere, or dark blue under reducing conditions;
- **carbon** in a finely dispersed form acts as an internal body fuel and can improve the appearance of the brick;
- **secondary (by-product) materials**, including pulverised fuel ash (PFA), steelmaking slags, sawdust and straw, are added to provide both an internal body fuel and to improve the technical and aesthetic properties of the product. Sawdust and straw are also used to increase the porosity of a brick and enhance the thermal insulation properties.

Clay and quartz perform broadly similar roles in both structural and fine ceramic bodies. However the fine ceramic sector usually requires material with a low colouring oxide content and low impurity levels. A ceramic body must also have good rheological or fluid properties if the manufacturing process involves the utilisation of a slip or slurry (casting or spray-drying) and the particle size distribution may also be important. The functions of the major components in fine ceramic bodies are summarized as follows:

- **Ball clays** impart plasticity and strength to the body, and help to bind the other components together. Some ball clays have favourable rheological properties and the best qualities have a light or white fired colour.

- **Kaolin** is principally used to give high-fired whiteness to the product. A coarse particle size is important for use in casting slips, and a fine particle size or the presence of mixed–layer or smectitic minerals will result in higher strength and plasticity.
- **A filler** is the term used to describe the silica, in the form of ground flint, cristobalite or quartz sand, which is added to control the thermal expansion of the fired body, reduce the drying shrinkage, promote the drying operation and enhance the fired whiteness.
- **A flux** is necessary to lower the fusion temperature and to create a glassy or vitreous bond with the other body components. The higher the alkali content of the fluxing material, the better the fluxing action, and orthoclase feldspar and nepheline syenite (principally derived from Norway and Canada) are the main fluxes used today. Formerly 'Cornish Stone' (a naturally occurring mixture of kaolin and feldspar) was the predominant flux in the UK fine ceramic industry, but the high fluorine content of the emissions during firing has resulted in a marked decline in its use.

Although the non-plastic components i.e. the filler and flux are relatively coarse when compared with the kaolin and ball clay, they are sufficiently fine to ensure good dispersion and body packing in the ceramic product. A range of fine ceramic body compositions is displayed in Table 14.2 (in wt% of solids). Table 14.3 displays the typical chemical analyses for a variety of ceramic bodies.

14.3. Clay minerals in ceramic products

The clay mineralogy of the ball clay or kaolin is one of the key factors determining the behaviour of a ceramic body during production and processing, and the properties of the fired piece. The proportion, form (crystalline structure) and particle size of the clay minerals will directly influence the characteristics of the ceramic body. The interlayer site in the crystalline structure plays the most important role in determining the properties of the clay mineral. The clay minerals considered in this section are kaolinite, illite, halloysite, montmorillonite (or smectite) and chlorite. The following discussion provides details of the properties and behaviour of clay minerals in ceramic products.

14.3.1. Kaolinite

Kaolinite occurs in kaolins, ball clays, fire clays and many of the clays and shales utilized in the heavy ceramic sector. There is a major morphological difference

TABLE 14.2. *Fine ceramic body compositions*

Body	Ball Clay %	Kaolin %	Flux* %	Quartz %	$CaCO_3$ %	'Pitchers' %	Calcined Bone %
Wall tile	30–48	10–20	–	20–40	10–12	0–10	–
Floor tile	40	10	20	20	–	10	–
Gres porcellanato	40	15	25	20	–	–	–
Sanitaryware	23–25	27	18–22	21–26	–	0–5	–
Vitreous china tableware	25	30	30	15	–	–	–
Stoneware	60–100	–	5	0–35	–	0–10	–
Dolomitic earthenware	25	25	–	35	10–15	–	–
Feldspathic earthenware	25	30	12	33	–	–	–
Porcelain	0–5	50–55	15–25	25	–	–	–
Electro–porcelain	30	20	25	25	–	–	–
Bone china	–	35	15–25	–	–	–	50

* Predominantly feldspar, nepheline syenite or pegmatite e.g. 'Cornish Stone' or 'Potters Stone'.

TABLE 14.3. *Typical chemical analyses for a variety of fine ceramic and structural ceramic bodies*

Body	SiO_2	TiO_2	Al_2O_3	Fe_2O_3	CaO	MgO	K_2O	Na_2O	L.O.I.
Wall tile	65.4	0.7	14.1	0.5	8.7	0.2	1.3	0.4	9.4
Floor tile	67.0	0.5	19.0	0.5	0.4	0.3	3.1	1.0	5.4
Floor tile (gres porcellanato)	66.8	0.2	23.5	0.5	0.6	0.3	0.6	3.4	5.6
Sanitaryware	65.0	0.4	22.5	0.5	0.3	0.2	3.0	1.7	6.0
Vitreous china tableware	66.0	0.2	20.0	0.5	0.3	0.2	3.0	1.7	6.0
Feldspathic earthenware	67.0	0.2	20.0	0.5	0.9	0.7	1.6	0.7	7.1
Brick etruria marl*	60.6	1.2	20.6	7.4	0.3	0.7	1.6	0.1	7.6
Brick mercia mudstone†	48.7	0.7	13.3	5.6	5.6	8.4	5.1	0.1	11.8

* Ridgeway (1982). Average of 6 samples from top 92 m of Mercia Mudstone from Birmingham, Nottinghamshire, Warwickshire and Leicestershire areas, after Bonnell and Butterworth (1950).
† Ridgeway (1982). Average of 18 basal Etruria Marl samples from Goldendale Quarry North Staffs, after Holdridge (1959).

between the kaolinite occurring in kaolins and that occurring in the other raw materials (ball clays, clays, shales etc). The refined or processed kaolin derived from a primary kaolin deposit is composed predominantly of kaolinite with a well ordered crystallinity. Crystallinity indices derived from X-ray diffraction traces are normally expressed on a scale from 0 to 2.0 (Hinckley 1963). Most kaolins have a crystallinity index lying between 1.0 and 1.3, whilst ball clays range between 0.1 and 0.9. Kaolinite derived from kaolin has a coarse to medium particle size (typically ranging from 30 to 70% <2μ), and is often low in green strength and plasticity. Kaolinite occurring in sedimentary or secondary kaolin deposits also has a moderate to well ordered crystallinity, but generally has a finer particle size. The most important role of the kaolinite derived from kaolin in a fine ceramic body is to impart whiteness. The particle size and shape of the kaolinite also has a strong influence on the rheological behaviour of a casting slip.

The most abundant clay mineral in ball clays is usually kaolinite with a b-axis disordered crystallinity. However in some ball clays, the disordered kaolinite and illite occur in equal quantities. The kaolinite in ball clays typically has a much finer particle size (ranging between 70 and 95% <2μ). Increased green strength and plasticity are positive attributes of the kaolinite, but there is generally a poorer fired colour due to the inclusion of iron in the crystal lattice. The firing behaviour is characterized by broad reaction intervals, increased refractoriness, and the formation of mullite and cristobalite. However vitrification can be difficult.

14.3.2. Illite

Illite, illitic mica or sericite is the second most abundant clay mineral in most ball clays, and often the predominant clay mineral in many clays and shales used in the structural ceramic sector. Compared with kaolinite, illite has a higher alkali content, and an off-white fired colour. It imparts high green strength and plasticity to the unfired body, and its firing characteristics are early sintering (or vitrification) and a broad vitrification range. Illite performs an important role in assisting in the fluxing process and ensuring that a glassy bond develops with the other body components. Illite forms mullite, orthoclase feldpar and leucite during the firing cycle. The type and concentration of the cations occupying the interlayer site may have an important role in determining the properties of illite, especially the rheological behaviour.

14.3.3. Halloysite

Halloysite has a similar chemical composition to kaolinite, but differs in that it has a higher water content (2H$_2$O as interlayer water) and the crystalline structure is quite different. Whereas kaolinite is typified by 'books' or 'stacks' of platelets, the structure of halloysite is in the form of tightly rolled scroll-like tubes. The most important characteristic of halloysite is the high degree of whiteness, and it has developed a niche market in high quality porcelain and bone china. However, the unusual crystalline structure of halloysite often imparts poor rheological properties, and with the exception of some refractory products, it is not commonly used in other ceramic sectors.

14.3.4. Montmorillonite (smectite)

The three-layer structure of montmorillonite (and associated swelling characteristics) produces problematic ceramic properties. The extremely poor rheology, high drying shrinkage, very high cation exchange capacity and very fine particle size result in this mineral being largely avoided in most fine ceramic applications. However, small quantities are sometimes added to ceramic bodies to create a significant increase in strength and plasticity. The fired colour of montmorillonite is creamy-white. Montmorillonite can be a common constituent mineral in some brick clays, and upon firing transforms to mullite, cristobalite, anorthite and cordierite.

14.3.5. Chlorite

Chlorite does not occur in most ball clays and china clays and is therefore not a normal constituent clay mineral in fine ceramic bodies. However, chlorite is a common mineral in some shales and clays utilized in structural ceramics. Its influence on the structural ceramic body is largely subordinate to that of illite and kaolinite.

Table 14.4 summarizes some of the ceramic properties and behaviour of a range of clay minerals commonly found in ceramic products. The role and behaviour of non-clay minerals and impurities in a fired ceramic product are listed in Table 14.5.

14.4. Structural or heavy clay products

The structural or heavy clay sector is the major consumer of common clays and shales produced in the UK, with approximately 7.6 million tonnes per annum (67%) being used in the production of bricks, pipes and roof tiles (Hillier et al. 1997). Of the tonnage utilized in the structural or heavy clay product sector, over 94% is destined for bricks. The 2001 industry energy survey data shows similar figures, with nearly 8 million tonnes of clay being used in the structural ceramic sector, of which 95% is for bricks. As the volume and type of brick production is the predominant factor in the mining and utilization of common clays and shales in the UK, much of this section is devoted to clay materials suited to bricks and an explanation of the brick manufacturing process. Fireclays and to a lesser extent ball clays may also be used in structural clay products.

Many structural clay products have to comply with stringent specifications and tolerances governing dimensions, water absorption, compressive strength and durability. There is also the aesthetic appearance, which is an important factor in determining market share for many

TABLE 14.4. *Ceramic properties of the major clay minerals*

Clay mineral	Fired colour	Green stength and plasticity	Vitrification
Kaolinite	White	Often low in green strength and plasticity	Difficult to vitrify
Halloysite	Whitest firing		Difficult to vitrify
Illite	Off-white	Usually gives high plasticity and green strength	Easiest to vitrify
Montmorillonite	Off-white	Very strong and plastic	Easy to vitrify
Chlorite	Off-white	Moderate	Easy to vitrify

TABLE 14.5. *Influence of non-clay minerals and impurities in a fine ceramic body*

Mineral or impurity	Behaviour in fired product
Quartz	Acts as a filler and reduces fired shrinkage; can be a microcrack generator and reduces fired strength
Tourmaline	Slight discoloration. Some HF generated in flue gases
Haematite, Pyrite, Magnetite, Siderite	Black spots. Pyrite can contribute to SO_2 flue gases
Goethite	Important in defining colour hues in structural products—particularly strong red colour in roof tiles and bricks
Gibbsite, Gypsum	Strong flocculant
Lignite and Organic Matter	Can improve green strength and rheological properties if in a very fine or colloidal from; can leave holes on burn out; may result in 'black hearting' in a fast firing process
Rutile	Gives buff fired colour
Anatase	Gives slight yellow fired colour

products. The following discussion provides details of the product types, application and characteristics of the raw materials.

14.4.1. Bricks

A brick may have to comply with a range of parameters including durability, strength, desired aesthetic appearance, chemical and/or frost resistance etc. Bricks may also be expected to meet specific water absorption, fire resistance, sound-proofing and thermal insulation criteria. Bricks are utilized in numerous different applications in the structural sector, and this is reflected in the following terminology.

Internal bricks have low durability to exposure and frost and are only suitable for internal use. Ordinary or common bricks have a moderate to low strength and are suitable for low-rise constructions. This variety of brick has to comply with minimum compressive strength specifications. Concrete blocks have to a large extent replaced this type of brick in construction. Facing bricks are used to clad the exterior of buildings. The aesthetic appearance and in particular the fired colour and texture are important factors and these bricks are now the most important type with respect to the numbers produced. Engineering bricks have high strength and low porosity, and are used in load-bearing and high-rise applications. Reinforced concrete has long since replaced engineering bricks in most constructions. Special bricks fall into two categories: (i) bricks with a non-uniform shape (as per BS 4729; 1990) and (ii) bricks which are employed in extreme environments, i.e. areas which require acid-resistance or are susceptible to extreme freezing or wet conditions.

Calcium silicate bricks are uncommon in the UK but can form up to 50% of brick production in some European countries where suitable clay deposits are rare. The brick is produced by subjecting a mixture of quartz, sand and lime to high pressure and steam.

A variety of building blocks manufactured from concrete, and in some cases incorporating light-weight aggregates have become increasingly popular. Internal bricks now face strong competition from plasterboard partitioning and lightweight blocks.

14.4.2. Pipes

The ceramic pipes currently used in the UK are divided into the porous and vitreous categories. Porous pipes are largely used as in the land drain sector and must have a high tolerance to compression and low temperatures. Vitreous pipes are dense, unglazed impermeable pipes that have been widely utilized for sewage, chemical effluents, surface water drainage and as conduits for electrical cabling. The pipes must have a very low water permeability and be highly resistant to chemical corrosion.

Plastic pipes have made strong in-roads into this sector, but there can be longer-term problems with durability.

14.4.3. Tiles and other applications

The main structural ceramic application for tiles is in roofing and flooring. Fine ceramic tiles are covered in Section 14.8. The durability and aesthetic appearance of roof tiles are the two most important characteristics (Molders & Binder 1998). Roof tiles are generally finer grained and denser than most bricks. The use of roof tiles in the UK sector has declined due to the popularity of surface stained and textured concrete tiles. However, the demand for clay roof tiles in Europe is still strong.

Unglazed floor tiles can be used in kitchens, patios, etc. where both aesthetic appearance and ware resistance are important factors. They are normally dense, fully vitrified and stain resistant. However, their use has been severely curtailed by alternative products including fine ceramic tiles, wood blocks, carpet and plastics for interior use and stained concrete slabs for external use.

Other applications for clays in the structural and heavy clay sector include hollow clay blocks, conduits, cowls, flue liners and flower pots. However the total clay tonnage utilized in these sectors is very small.

14.5. Clays utilized in structural ceramic products

Clays are a major constituent of most structural ceramic products, and this section describes the most commonly used clay materials.

14.5.1. Common clay and shale

In the past, a wide range of common clays and shales, from all parts of the geological column, were utilized by hundreds of small structural ceramic companies. Small manufacturing sites and their associated quarries were widely scattered throughout the UK, and their products (bricks, pipes and tiles) catered mainly for a local market. The colour and style of these locally produced products often resulted in the buildings within a particular area or district having a distinctive character and appearance. Although the predominant modern brick colour is a dull red, the Cambridge area traditionally had yellow bricks and the London area grey.

The dominant trend in the 20th century has been for a rationalization of the production units, and an increasing concentration and reliance on a few specific clay deposits. The current major UK resources are largely restricted to the following units: the Weald Clay, the Wadhurst Clay, the Peterborough Member ('Lower Oxford Clay' or 'Fletton Clay'), the Mercia Mudstone and various Carboniferous mudstones and fireclays, including the economically important Etruria Formation, some Silurian shales, Brick Earths and Pleistocene glacial and aluvioglacial deposits. The ages and location of the most important formations are shown in Table 14.6. Further background reading on the evaluation and geology of brick clays can be found in Prentice (1988, 1990).

The suitability of a common clay or shale for utilization in the manufacture of structural clay products is determined by its mineralogy, chemistry and physical properties. These factors will dictate the behaviour of the clay during forming, drying and firing, and ultimately have a direct influence on the final properties of the fired product. These properties include the strength, porosity, durability and aesthetic appearance. The fired ware has to comply with stringent UK and European standards, which provide the specifications for dimensions, shape, strength, water absorption, soluble-salt content and durability (frost resistance). These standards can only be met through carefully controlling both the feedstock materials and the manufacturing process. The optimum raw material for a structural clay product should have a moderate plasticity (Atterberg plasticity index ranging between 10 and 20), good workability, high dry strength, a long vitrification range, a total shrinkage of less than 10% and an acceptable fired colour. Many brickworks still use indigenous clays, although to achieve the desirable physical properties and utilize lower grade clays, clay blending and the utilization of other materials is becoming more common.

Adequate strength and durability are relatively easy to obtain in facing bricks, and the key parameters are more usually linked to the aesthetic appearance of the brick, especially the shape, colour and texture. Each structural product also has an ideal particle size range, in terms of the proportions of sand and coarse silt, fine silt and clay fraction in the body. This is displayed in Figure 14.1.

The UK production of common clay and shale was 13.9 million tonnes in 1995, declining to 11 355 000 tonnes in 1999. Of the 1997 production, 67% was used in bricks, pipes and tiles, 20% in cement, 10% in construction and 3% for other uses (Hillier *et al.* 1999, 2001). Of the clays and shales produced for structural ceramics in the UK, approximately 94% are utilized in the brick sector.

14.5.2. Fireclay

Fireclay (sometimes referred to as refractory clay) is commonly associated with coal bearing sequences and

TABLE 14.6. *The major structural clay producing formations in the UK*

Geological unit	Age	Occurrence
Weald Clay	Lower Cretaceous	Surrey, Kent, Sussex
Wadhurst Clay	Lower Cretaceous	Kent, Sussex
Peterborough Member (Lower Oxford Clay)	Upper Jurassic	Buckinghamshire, Bedfordshire
Mercia Mudstone Group (Keuper Marl)	Triassic	Worcestershire, Warwickshire, Leicestershire, Cheshire, Nottinghamshire and S. Yorkshire
Middle & Lower Coal Measures	Carboniferous	Midlands and Northeast England, and Central Scottish Lowlands
Etruria Marl	Carboniferous	Staffordshire, Derbyshire

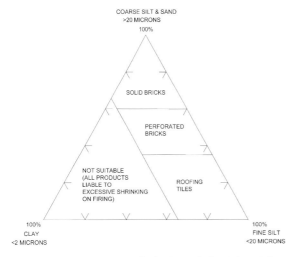

Fig. 14.1. Ideal particle size for bricks and tiles (adapted from McNally 1998).

may be produced as a by-product of coal mining. It is normally but not always associated with the 'seatclay' or 'underclay' occurring beneath a coal seam.

The term 'fireclay' is used to describe a seatearth or seatclay which can be used commercially in refractory or non-refractory ceramic applications (Highley 1982). Thus while most fireclays are also seatclays, not all seatclays can be used commercially as fireclays. The refractory fire clays have a higher content of kaolinite, while those suitable for non-refractory applications have approximately equal proportions of kaolinite, mica and quartz.

Fireclay has been traditionally utilized in the refractory industry (associated with steel production), vitrified clay pipes, facing bricks, and to a limited extent in the production of fireclay sanitaryware and stoneware. Fireclay bricks used in the refractory industry have a relatively low melting point (1600–1750°C) compared with high alumina clay (1802–1880°C) or chrome (1950–2200°C) bricks (Manning 1995). As this limits their use to less demanding applications, e.g. structural work supporting higher grade materials, the use of fireclays in refractories is declining in importance.

In structural ceramics, fireclays are predominantly used to produce buff or cream firing low water absorption bricks with high levels of durability suitable for construction in exposed locations (north and west of England and Scotland), where the normal buff or cream firing Mercia Mudstone bricks are not durable. Brick making fireclays are almost exclusively sourced from opencast coal operations, and the majority of these fireclays now occur in either the Northumbrian or Scottish coalfields.

The UK production of fireclay was 708 000 tonnes in 1995, declining to 338 000 tonnes in 1997, although this figure may be too low (Hillier et al. 1999). The utilization of fireclay in the UK ceramic sector is 50% for refractory purposes and 50% for bricks, pipes and tiles.

14.5.3. Ball clay

A relatively small tonnage of ball clay is used in the UK brick sector, although the tonnage was significantly greater when brickworks operated close to the ball clay deposits in south-west England and Dorset. The grades currently utilized in structural ceramics are usually not suitable for fine ceramic applications due to the higher content of iron minerals, carbon or quartz. Ball clay is covered in detail in the fine ceramic section. Ball clays, especially lower grade varieties, are now being used as alternatives to fireclays because of their colour and fired property characteristics.

14.6. Structural clay extraction, processing and manufacture

The following section describes the procedures most commonly adopted for the extraction of structural clays. The preparation and utilization of these clays within the heavy or structural ceramic sector is explained through the steps commonly undertaken in brick manufacturing. Table 14.7 summarizes the clay extraction and manufacturing process associated with heavy clay products.

14.6.1. Extraction

Large back-actor excavators are now commonly used to selectively extract the more plastic clay sequences (see Chapter 11 and Plate 15). Ripping and blasting of the harder and more indurated claystones and shales may be necessary at some sites. The different clay selections are then often placed on a layered stockpile. The creation of the stockpile is carefully controlled to ensure that a vertical slice (produced by an excavator or face shovel loading the feedstock bunkers) will produce a homogeneous material and consistently give the required clay properties.

Draglines and shale planers are used at some sites to achieve homogeneous extraction by scraping through the complete clay sequence (see Plate 16). Clays used in brick production may also occur as important by-products of open-cast coal production (especially fireclays).

14.6.2. Pre-treatment and preparation

Souring or weathering of the clays may also occur in the stockpiles, where over a period of several months the clays become more plastic and malleable.

The first stage of clay preparation is size reduction. The excavated clay undergoes primary crushing (through a kibbler or jaw crusher) which reduces the lump size to approximately 75 mm. Secondary crushing is achieved through a combination of hammer mills, wet or dry pan mills, or smooth crushing rolls. The latter process is widely used for clay material, which will be 'formed' by the extrusion process. Secondary crushing results in a lump size ranging from 10 mm to 25 mm. Screens

TABLE 14.7. *Production and manufacturing processes for heavy clay products*

	Process	Stages
1	Clay extraction	• Extraction of individual seams or homogenised multi-seam sequence • Creation of layered and/or blended stockpiles
2	Clay pre-treatment	• Clay stockpile weathering (or souring) • Primary crushing (kibblers or jaw crushers) • Secondary crushing (hammer mills, wet or dry pan mills, smooth crushing-mills) • Screening
3	Body preparation	• Addition and mixing of other raw materials and additives • Homogenization • Water addition (if required) • De-airing
4	Body forming	Production of unfired or green ware through one of the following processes: • Granulate pressing (tiles and bricks), moisture content 16–18% • Soft mud process and hand thrown items (bricks), moisture content up to 30% • Extrusion and cutting (bricks), moisture content 12–15% • Extrusion (pipes)
5	Drying	• Stacking and air drying • Heat conditioning • Tunnel or chamber driers (operating at 110°C) Drying cycle time between 18 and 60 hours
6	Firing	• Firing in one of the following types of kiln: • Intermittent kilns • Chamber or Hoffman Kilns • Tunnel Kilns Maximum temperature between 1050 and 1150°C, cycle time between 2 and 5 days
7	Finished product	Bricks Roof Tiles Pipes

scalp off over-size material, which is then returned for re-grinding.

The clay is thoroughly blended and mixed with the other body components, which can include quartz sand, sawdust, coal slurry, pulverized fly ash (PFA), anti-scumming agents and chemical additives (colorants, binders and plasticisers). The additives introduced at this stage can improve the plasticity of the clay mix, leading to a reduction in water addition and savings in the energy and time required for drying the ware. Water is added to achieve the desired moisture content for forming.

14.6.3. Forming

Forming is the process in which the structural clay product is shaped, textured (and sometimes coloured) prior to firing, and at the same time, as much air as possible is removed from the raw material mixture. The three methods used in the UK for forming are the soft mud process, extrusion (and wire-cutting) and to a lesser extent the semi-dry granulate pressing (Fletton process). The initial moisture content of the feedstock clay has a major influence in determining its suitability for a particular forming process. Extrusion requires the feedstock clay to have a moisture content of 12–15%, and the harder, more consolidated shales, mudstones and marls are more suited to this process. The clay slurry utilized in the 'soft mud' process can have a moisture content as high as 30%, and this forming process is best suited to the stratigraphically younger formations located in south-east England, e.g. the Weald Clay, Brick Earths etc, where the natural moisture content often exceeds 15%.

The mineralogy of the clay is also an important factor in determining which forming process is applicable. Extrusion requires a clay with an inherently low moisture content and a high plasticity. Clays that have a higher content of b-axis disordered kaolinite are more plastic. A higher amount of smectite in the clay is normally associated with an increase in the natural moisture content.

The soft mud process is employed in the production of handmade bricks and the body components form a wet, free-flowing slurry with a high moisture content of up to 30%. Clots of clay are thrown into mould boxes that have been pre-coated with sand or sawdust to ensure easy release. The surplus clay is skimmed from the mould box and the brick is then carefully de-moulded onto a drying rack. Although the high moisture content is effective at displacing air from the body, the resultant brick requires a much longer drying period. Sand is often added to the soft mud bodies to reduce the firing shrinkage and they can also accommodate more non-plastic additives e.g. pulverized-fuel ash (PFA). The only difference between the soft mud and handmade processes is that the former is totally mechanised and the latter is handmade as the name suggests. Handmade brick making is still a widespread technique employed for the production of premium-quality facing bricks or complicated special shapes.

Bricks that have been formed by this process have a more traditional appearance, with a slightly irregular form, surface creasing and a sandy finish. The aesthetic characteristics of these bricks have recently led to a significant rise in their popularity.

The extrusion process is used in the manufacture of wire-cut bricks and a significant amount of clay preparation is required prior to forming. This includes primary crushing, milling in a dry pan, and grinding through medium to high-speed rollers, in order to produce clay pieces of <1 mm.

Extrusion is now the most widely employed forming process, and it involves forcing the clay mix through a lubricated die. The clay is extruded in a partial vacuum to facilitate de-airing and emerges as a continuous, consolidated column. In order to obtain the optimum physical properties for extrusion, the moisture content of the clay mix is kept to 12–15%. The required aesthetic appearance of the brick is achieved through surface treatment of the clay column (texturing, sand blasting and pigment spraying). The column is cut by piano wire (hence the term 'wire-cut') to produce bricks of the required size, which are then transferred to the drying racks. Plate 17 shows an extruded clay column prior to being cut.

The clay column is normally penetrated by 'core-bars' to produce the characteristic perforations or pattern of holes. These help to reduce drying and firing times and also minimise the thermal gradients within the brick. The extrusion process produces bricks with superior physical properties and is also the most widespread method for forming other structural clay products, e.g. pipes and tiles. The popularity of this process has been ensured due to the formed product being ideally suited for firing in modern tunnel kilns.

Although the semi-dry granulate pressing or Fletton process was widespread in the past, the only semi-dry pressed bricks currently produced in the UK are the 'Flettons', derived from the Peterborough Member of the Oxford Clay. The Fletton process is a semi-dry pressing technique developed in the village of Fletton in Cambridgeshire. The pressed bricks are produced from relatively dry (16–18% moisture content) granulated clay. The 1–3 mm granules are poured into an oiled mould-box and mechanically pressed into shape. All air has to be removed from the material to ensure that there is full bonding between the clay granules.

The pressed brick is smooth, with a very regular shape and straight edges. A facing texture is applied with rollers and brushes and this is followed by the surface application of sands, stains and pigments. Through these means, a single brick body formulation or clay mix (with standard technical specifications) can be used to produce a wide range of coloured and textured bricks.

14.6.4. Drying

Structural ceramic ware has to be dried before firing and drying times can vary from 18 to 60 hours (depending on the nature of the ware and the moisture content of the body). Distortion and cracking of the ware can result if either the drying process is too rapid or the internal thermal gradient is too high. The drying process is carefully controlled within tunnel or chamber dryers, within which there is an initial warming-up period at high humidity, followed by exposure to hot dry air at a temperature of approximately 110°C. Excess or waste heat from the forming kilns is exploited for drying wherever possible.

14.6.5. Firing

The clay minerals break down and sinter during the firing process, forming a vitreous (or glassy bond) between the other minerals and materials in the body. The firing behaviour of the body, and the physical properties of the fired ware, are to a large extent determined by the clay mineralogy, the quantity and particle size of the quartz and the carbon and sulphur content. A clay mix with a relatively long vitrification range (of at least 100°C) is desirable, as it can tolerate variable kiln temperatures, and the fired product will still comply with physical property specifications. The firing of a clay body with a narrow vitrification range requires very careful kiln temperature control. Too high and the ware will deform, while too low a temperature will result in insufficient sintering and weak products.

A combination of the clay minerals b-axis disordered kaolinite and illite give a brick clay optimum firing characteristics. Whereas kaolinite vitrifies at a relatively high temperature and extends the vitrification range of the brick clay body, the higher potassium content in the illite has the opposite effect, promoting fluxing and reducing the temperature for glass formation.

An idealized brick body would consist of 20% illite, 25% kaolinite and 55% quartz sand. The quartz sand is used as a stabilizer and filler in the body, helping to prevent cracking due to excessive firing shrinkage and imparting gas permeability. The particle size of the quartz also influences the firing characteristics of the clay body. Care is required in the firing process if the quartz content is high or of a particle size that may cause internal structural stresses, due to the thermal expansion which occurs with the transformation from the α to β state at 573°C. Most care is required on the cooling cycle as the β to α quartz transformation, if too quick, can result in 'dunting' (cooling cracking). The influence that time and temperature has on the firing characteristics of ceramic clay minerals is explained in detail in Barlow & Manning (1997, 1999).

The carbon and sulphur contents are also important factors in the firing performance of the clay. A small amount of carbon in the clay body is beneficial as it will act as an internal fuel and reduce the overall energy requirement. However, it is preferable to have low carbon clay to which finely ground carbonaceous material is added up to a maximum of 1.5%. A carbon content in excess of 1.5% can result in the interior of the fired body turning blue due to the formation of ferrous iron, or when firing has been rapid, 'black cores' can develop in the product. In severe cases trapped gases will bloat and distort the ware. Low carbon clays are essential for clay bodies undergoing fast firing. Sulphur in the form of both sulphides and sulphates will lead to the generation of undesirable SO_2 and SO_3 rich flue gases and a maximum sulphur level of 0.2% is recommended in the raw materials.

As illite melts between 1050 and 1150°C, the peak firing temperature for bricks and pipes is normally kept to between 1000 and 1100°C. A peak temperature of 1200°C is used for firing roof tiles and paving bricks, to impart additional strength to the ware. The kiln firing cycle may range from 30 to 150 hours, but 48 to 72 hours (2 to 3 days) is now common in most modern tunnel kilns. The firing cycle normally consists of a gradual temperature increase, an extended period at peak (or firing) temperature, followed by gradual cooling and is described in Table 14.8. Further information on the mineralogical assemblages formed during brick making can be found in Dunham (1992) and Dunham *et al.* (2001).

A variety of kilns and firing techniques are used in the production of structural ware and chamber and tunnel kilns represent two radically different approaches to the mechanics of the firing process. Whereas in chamber kilns, the bricks remain static and the combusting gases move, the opposite occurs in tunnel kilns.

Clamp Firing. This is a traditional technique for producing bricks that does not involve the use of a kiln, and is employed only on a limited basis. Carbonaceous dried bricks are built into a rectangular block with interbedded layers of coke. The block is covered by fired bricks, and then ignited. The firing process is slow and the variable temperatures within the clamp produce a range of brick types.

Chamber (or Hoffman) Kilns. Such kilns consist of linked rooms or chambers that are filled with dried bricks. The drying, firing and cooling process moves successively from chamber to chamber, with the hot combustion gases being utilized for drying the ware in adjacent chambers.

Tunnel Kilns. These are now the most widely utilized method for firing structural ware. Dried products enter the refractory lined and insulated tunnel on pallets or kiln cars that run on rails. The ware is moved through the tunnel, undergoing pre-heating, firing to maximum temperature in the centre of the tunnel and then gradual cooling towards the exit end. Plate 18 shows fired bricks emerging from the Hanson Brick Pipe Hall tunnel kiln, USA.

Rapid Firing (Roller-Hearth) Kilns. This technology is widely employed in the floor and wall tile industries. A continuous conveyor transports the ware through the kiln, and the complete firing cycle can be as short as 30 minutes. There are marked energy cost savings and a reduction in overall emissions with this method but the rapid heating, firing and cooling cycle imposes significant constraints on the body composition.

Many clays (as dug) contain material or minerals that will generate defects in a fast-fired body, e.g. if the carbon content is too high, black coring and bloating will develop. The fast firing process dictates highly stringent selection of raw materials, especially the clay components, and the demand for the premium clays (Etruria Formation, Carboniferous mudstones and some fireclays) will increase. Fast firing has already been utilized for brick production in Australia, and for vitrified pipe production in the UK and is likely to become more popular in the structural clay sector.

14.7. Engineering properties of bricks

Porosity and strength are the most important properties of fired clay bricks. These are described below.

14.7.1. Porosity

The fired properties of a brick will depend on the unfired body composition and the firing cycle and conditions. The porosity can vary from 1 to 50%, and will have an important affect on the durability and application of the brick. Water is absorbed readily by high porosity bricks and frost damage can result from water freezing in the pores

TABLE 14.8. *Heavy ceramic ware firing cycle*

Firing cycle	Temperature (°C)	Event/process
Heating	Ambient–peak	Slow heating.
	Ambient–200	Drying, free water evaporation.
	150–650	Removal of mineral bound water, decomposition of gypsum.
	200–900	Burning of carbonaceous matter in air, main carbon burnout occurs 600–900. Important that all carbon is burnt out before onset of vitrification, as either bloating will occur due to trapped gas or 'black hearting' will develop due to the reduction of iron oxides by retained carbon.
	400–950	Decomposition of sulphides and carbonates.
	900+	Sulphate minerals decompose.
	900–1100	Sintering and vitrification. Excessive vitrification or melting should be avoided as piece will deform.
Peak	Between 1050 and 1150	Soak period is normal.
Cooling	Peak–ambient	Slow cooling.

and fissures (dense, low porosity bricks are less susceptible). Efflorescence creates an unsightly appearance and is the result of salt derived from the brick or groundwater migrating to the surface. This phenomenon can become excessive in moderate to higher porosity bricks.

A brick's porosity also influences the way in which it is treated during the construction process. Moderate to highly absorbent bricks need wetting or 'docking' to limit the transfer of moisture from the mortar into the brick. The rapid removal of moisture from mortar can adversely affect the overall strength of the brickwork. However, the natural porosity of brickwork is important as it permits both the absorption of rainwater and its gradual release through evaporation, thus preventing the ingress of moisture into a building.

New bricks expand or 'grow' due to the absorption of moisture by the new minerals formed during the firing process. The initial rapid growth of about 0.4 mm/m during the first week, is followed by a much slower expansion of a further 0.4 to 1.0 mm/m over the following 8 to 10 years. Over the last 10–15 years it has been shown that irreversible moisture expansion is linked more to the degree of firing (normally under fired) than the clay type used to make the brick.

14.7.2. Strength

Compressive strength is one of the key properties of a brick, and although it can range widely from 10 to 130 Mpa, most bricks have values between 20 and 60 Mpa. There is a close link between brick density, porosity and strength. Bricks with a higher porosity are generally weaker, whereas 'hard-fired' dense bricks with a low porosity and a high level of glass formation have the highest strength. The method of brick forming can also influence their strength, with extruded bricks being stronger than those formed from either the soft mud or fletton processes.

Water absorption is a useful indicator of brick strength, durability and structural integrity. Lower percentages of water absorption (in the region of 4.5%) are associated with the highest strength, with extruded bricks have approximately half the water absorption of pressed or fletton bricks. Higher levels indicate a porous and probably weak structure.

The strength of brickwork is dependant on the compressive strength of the bricks, the strength of the mortar and the bonding pattern. The bonding pattern refers to the way in which bricks are inserted into a layer or course of brickwork. In the UK, the English bond (consisting of alternating courses of headers and stretchers) was the earliest pattern introduced in late Medieval times. The 'stretcher' is a term applied to the long side of the brick and the 'header' is the short side. The Flemish bond (where headers and stretchers alternate within courses) became popular in the UK in the 17th Century. Most modern brickwork utilizes a stretcher bond pattern, in which only the long sides of the bricks are outward facing.

There is also a proven link between the water absorption of bricks and the flexural strength of the brickwork, with the optimum flexural strength being achieved when the bricks have a water absorption ranging between 7 and 12%.

14.8. Fine ceramic or whiteware products

This section describes the composition, properties and manufacturing process of some fine ceramic or whiteware products, namely wall tiles, floor tiles and sanitaryware. The other ceramic items containing clay and used frequently in homes (tableware, earthenware, stoneware, porcelain etc) are described briefly. Fine ceramic manufacturers generally use three varieties of mixed ceramic body, namely plastic, casting slip or a dry dust or granulate. Ball clays and kaolins are the two most widely used clays in fine ceramic bodies, and this section focuses on the function and performance of these clays, in particular in tile and sanitaryware products. Globally, the demand for ball and plastic clays for fine ceramic applications is increasing at the rate of approximately 2% per annum.

Each product-type must conform to a specific range of characteristics or properties, which are detailed below. Table 14.9 displays the relationship between raw material properties and manufacturing parameters for different types of fine ceramic or whiteware products.

14.8.1. Wall tiles

A modern wall tile has a relatively high porosity (in the region of 18% water absorption), must be craze resistant and have a low die to fired contraction of approximately 1%. Tiles produced from a typical body containing $CaCO_3$ have a low contraction on firing, and hence the tiles are of a constant size. The tile must also have a low moisture expansion and this is imparted by the calcium (magnesium) alumino-silicates formed during the firing process. The clays must also be low in alkalis in order to avoid moisture expansion problems, and be relatively fine grained (80% < 2 μm) so that, when combined with a coarse kaolin, a high packing density is achieved.

14.8.2. Floor tiles

Floor tiles typically have a much lower porosity than wall tiles, and some, e.g. gres porcellanato tiles, are fully vitrified with a water absorption of <0.05%. The vitrification is achieved through having up to 25% feldspar in the ceramic body and by utilising a plastic clay component containing illite. The higher alkali content in the illite contributes to the fluxing and early vitrification (glass formation) in the ceramic body and thus the avoidance of a porous body structure. The ball clay component must be strong and plastic, and for tile varieties such as gres porcellanato, have a low colouring oxide content to ensure a near white fired colour. The relationship between raw material properties and manufacturing parameters for dry pressed tiles is shown in Table 14.9a.

The global demand for ball and plastic clays for the floor and wall tile industry now exceeds 20 million tonnes per annum. In addition, some of the clay content in the tile would be derived from lower quality, indigenous materials not included in this tonnage.

14.8.3. Tableware and electrical porcelain

Tableware is a generic term that is applied to a wide range of products with very different properties. The worldwide tableware industry consumes about 400 000 tonnes of ball and plastic clays annually. Earthenware is one of the most commonly produced tableware products and is used extensively for standard household tableware items. The fired body is porous (water absorption is approximately 8%), opaque and easy to produce. Vitreous hotelware has a higher strength than earthenware and is generally a far more durable product, whereas stoneware is the term applied to a more rustic style of traditional pottery, with a body formulation consisting of a blend of clays and a small amount of quartz. The fired body is normally vitreous (<1% water absorption). Porcelain is a hard vitreous product, with a white translucent fired body, used in high quality table and ornamental ware. The relationship between raw material properties and manufacturing parameters for fine stoneware is shown in Table 14.9b.

Electrical porcelain is also a hard vitreous product predominantly used in electrical insulators. The worldwide demand for ball and plastic clays for electrical porcelain ware is about 100 000 tonnes per annum.

14.8.4. Sanitaryware

Modern sanitaryware has a vitreous body with very low permeability. The key criteria in the production of sanitaryware relate to the rheological properties of the casting slip. Slip density, fluidity and thixotropy have to be closely controlled to ensure appropriate casting rates, cast strength and mould life. Casting properties are largely dependant on the permeability and particle 'packing' in the casting slip. The mineralogy, surface chemistry and particle size of the clay fraction in the slip is one of the major factors influencing slip rheology. Glasson & Forbes (2001) describe the influence of clay minerals in a casting slip, through the effect that they have on the quantity of deflocculant needed to retain a constant slip thixotropy. A casting slip typically has a fluidity ranging between 290 and 340° (measured by a torsion viscometer with 11/16″ bob and 30 swg wire), a 5 minute thixotropy ranging from 50 to 90 (measured at 40°C) and a density of 1.8–1.83 g/ml. The relationship between raw material properties and manufacturing parameters for sanitaryware is shown in Table 14.9c. Bougher (1999) examines the utilization of us ball clays in sanitaryware bodies.

The annual global demand for ball and plastic clays for the sanitaryware sector is approximately 1 million tonnes.

14.9. Clay raw materials for fine ceramic products

The following sections describe the formation, properties, ceramic applications, major European localities and the recent production statistics for ball clay, kaolin and flint clay.

14.9.1. Ball clay

Ball clay is a term used to describe a plastic sedimentary clay associated with fresh-water, lacustrine, fluviatile or overbank deposits (also see Chapter 3 for further information regarding ball clays). Other terms used for ball clay are 'fine ceramic and refractory clay', and in some countries (particularly Germany), 'plastische ton' or 'plastic clay'. In North Devon (in the UK), ball clay was originally known as 'pipe clay', as it was used in the manufacture of pipes for smoking tobacco. The term 'ball clay' is derived from the old method of working. The clay was cut into cubes about 9 inches (23 cm) square, each one weighing between 13 and 15 kg, and these rapidly assumed a spherical shape during handling. Clay was sold in this form by the 'ball' and hence the name became widely used.

Most ball clay production is destined for use in the ceramic product sector, with relatively minor quantities being sold into applications such as landfill sealing, acoustic or insulating tiles, animal feed-stuffs, fertilizers and pet litter.

Mineralogy. The mineralogy of ball clays can be quite diverse, with most containing disordered kaolinite, illite or sericite, and fine quartz. Kaolinite is usually the dominant clay mineral, although some ball clays have equal quantities of kaolinite and illite. Some ball clays may also contain a small amount of mixed layer or smectitic clay minerals, feldspar and carbonaceous material. Impurities found in ball clays may include small quantities of iron minerals (siderite, goethite, hematite, pyrite) and/or titanium minerals (anatase, rutile) which can restrict their use or application in ceramics.

Although the most abundant mineral in both ball clay and kaolin (china clay) is kaolinite, there is normally a difference in its crystallinity. The kaolinite in kaolin has a generally well-ordered crystallinity, whereas most ball clays contain disordered kaolinite (see Chapter 8 for further information on kaolinite mineralogy). The crystal structure of both types is similar (stacks of pseudo-hexagonal platelets) but the kaolinite in ball clay is disordered, principally by random layer displacement parallel to the crystallographic *b*-axis. Ball clays have a Hinckley crystallinity index ranging between 0.1 and 0.9. The Hinckley index is explained in Section 14.2.1. Well ordered kaolinite normally gives a whiter fired colour and lower unfired (or green) strength and plasticity. Disordered kaolinite usually contains some iron in the crystal lattice, which results in a poorer fired colour but increased green strength and plasticity.

TABLE 14.9a. *Relationship between raw material properties and manufacturing parameters for dry pressed tiles*

Raw material properties	Solubility	Screening behaviour	Deflocculant added	Viscosity	Density	Thixotropy	Casting rate	Hardening Plasticity	Plasticity	De-moulding	Drying	Efflorescence	Refractoriness	Vitrification	Pinholes	Colour	Spots	Black cores	Piece Strength
Particle size	3	2	2	2					2					2				2	2
Residue	1	3																	2
BET / Particle size distribution	3	1	3	3	3	3			3		3			2				2	1
Dilation			2																
Loss on ignition	2			2	1	1			1		1								
Water absorption														3					
Drying shrinkage																			
Rheology	3	3																	
Dry bending strength																			
3 layer : 2 layer	3		3	3	3	3			2		3			1					2
Insoluble impurities									3		1				1	1			
Inorganic salts			2	2	2	3									3	1			
Organic compounds											1								
Carbon														3	3			3	2
Alkali content														3					2
Fe + Ti																3			
Importance or priority rating	**2**	**2**	**2**	**2**	**2**				**3**		**2**			**3**	**2**	**2**		**3**	**3**

3, major influence; 2, moderate influence; 1, some influence.
Adapted from Fiebinger (1997).

TABLE 14.9b. *Relationship between raw material properties and manufacturing parameters for fine stoneware*

Raw material properties	Manufacturing parameters																		
	Solubility	Screening behaviour	Deflocculant added	Viscosity	Density	Thixotropy	Casting rate	Hardening	Plasticity	De-moulding	Drying	Efflorescence	Refractoriness	Vitrification	Pinholes	Colour	Spots	Black cores	Piece Strength
Particle size									2		1			1				2	
Residue									3		3		3	3					3
BET / Particle size distribution																			
Dilation									3		3		3	2					3
Loss on ignition													2	3					2
Water absorption											3								
Drying shrinkage																			
Rheology									3										
Dry bending strength									2		2		2	2					3
3 layer : 2 layer																			
Insoluble impurities																	3		
Inorganic salts																	3		
Organic compounds									3										
Carbon													3	3					
Alkali content													1	1		3			
Fe + Ti																3			
Importance or priority rating									3		3		2	3		2	2	2	3

3, major influence; 2, moderate influence; 1, some influence.
Adapted from Fiebinger (1997).

TABLE 14.9c. *Relationship between raw material properties and manufacturing parameters for sanitaryware*

Raw material properties	Manufacturing parameters																		
	Solubility	Screening behaviour	Deflocculant added	Viscosity	Density	Thixotropy	Casting rate	Hardening	Plasticity	De-moulding	Drying	Efflorescence	Refractoriness	Vitrification	Pinholes	Colour	Spots	Black cores	Piece Strength
Particle size	2	1	3	3		3	3	3		3	3		1	2	1				
Residue	1	2													3	1	3		3
BET / Particle size distribution	2	1	3	3		3	3	3	3	3	3			2	1				
Dilation											1		1						
Loss on ignition															1				
Water absorption													1	3	1	2			
Drying shrinkage											3								
Rheology	3	1	3	3		3	3	1		2									
Dry bending strength																			
3 layer : 2 layer	2		2	2	1	2	2	2	2	3	2		1	1					
Insoluble impurities		1									3				3		3		
Inorganic salts	3		3	3	1	1				1		3			3		2		
Organic compounds			1	1	1	1									2		1		
Carbon	1														2			3	
Alkali content													2	3					2
Fe + Ti													1			3			
Importance or priority rating	**2**	**2**	**2**	**3**	**2**	**3**	**3**	**2**	**1**	**2**	**2**	**1**	**3**	**3**	**3**	**2**	**2**	**1**	**2**

3, major influence; 2, moderate influence; 1, some influence.
Adapted from Fiebinger (1997).

Deposit form and depositional environment. Ball clay deposits can occur in a variety of forms. In Devon (UK), the clays are associated with fault-bounded, tectonic basins which has resulted in the accumulation and preservation of several hundred metres of ball clay. While the clay extracted from the Wareham area in Dorset occurs as lenses within sandy deposits, the Ukrainian deposits form a laterally extensive but thin sheet only a few metres thick. The ball clay within the economically important Westerwald area in Germany occurs in an extensive network of faulted basins and ranges in thickness from 15 m to a maximum of 90 m. The English and German ball clays were deposited in the Eocene and Oligocene, while the Ukrainian ball clays are Miocene in age.

Ball clays are characteristically associated with lignites, sands and silty clays and deposits may often exhibit rapid lateral and vertical facies changes. The clays are largely derived from strata which had been subjected to intense chemical weathering in the Cretaceous and Tertiary periods. The mineralogy (and to a certain extent the ceramic application) of a ball clay is governed by the derivation of the weathered material. In Devon, the clays are largely formed from material derived from weathered Dartmoor granite and/or Carboniferous slates (Bristow 1968, 1988). Clays which contain a high proportion of material derived from the granite have a higher content of moderately well ordered kaolinite, are whiter firing and are more refractory. However, the ball clays which are derived from the weathered Carboniferous slates contain fine disordered kaolinite and higher levels of illite, giving a cream fired colour and more vitreous firing properties.

Ball clay properties and ceramic applications. Ball clays are usually fine grained with a particle size ranging from 50–90% <2 μm. The coarser varieties normally contain a higher quantity of fine quartz. The best quality ball clays contain low levels of iron and titanium oxides and have a white or near-white fired colour. Ball clay also exhibits some or all of the following properties: high plasticity and/or refractoriness, high green strength and thixotropy. The principal functions of ball clay in a ceramic body are to provide plasticity and green strength and to bind the other body components together. This allows a ceramic body to be shaped (or formed) and handled prior to firing. The ball clays with good rheological or fluid properties are particularly important when the manufacturing process involves casting a piece from a liquid body or 'slip'. Rheological properties are governed by the mineralogy, particle size and surface chemistry of the clay.

The higher quality, light or white firing ball clays are used in many types of whiteware ceramics. This includes tableware and ornamentalware, electro-porcelain, sanitaryware and white bodied floor and wall tiles. Ball clays with a higher content of the colouring oxides Fe_2O_3 and TiO_2 can be utilized in monogres floor tile bodies (which are glazed) and also in the heavy or structural clay sector, including pipes, bricks and roof tiles. The content of ball clay in whiteware ceramic bodies (with the exception of porcelain) ranges from 20 to 100%.

Production localities and statistics. The most important commercial deposits of ball clay in Europe are located in Devon and Dorset in the UK, the Westerwald area in Germany, eastern Ukraine, the Czech Republic and France. The UK production of ball clay in 2000 was 1 075 000 tonnes of which 82% was exported (Hillier *et al.* 2001). The companies operating in the Westerwald area in Germany, currently have a combined annual production of approximately 4 million tonnes, the majority of which is sold into the floor tile, wall tile, brick, pipe and roof tile sectors. The East Ukrainian annual ball clay production is between 2.5 and 3 million tonnes with a substantial proportion being exported to tile producers in the Mediterranean area (especially Italy).

14.9.2. Kaolin (china clay)

The term kaolin is derived from the Chinese word 'Kau-ling' (or Geo-ling) meaning the 'high ridge' situated near the town of Jaucha Fu in Jiangxi Province, China. It was from this area that kaolin was extracted for ceramic use in the 3rd Century BC. The Chinese were the first to utilize kaolin in white firing pottery over 4000 years ago.

Whereas the ceramic product market is by far the most important for ball clays, kaolin is utilized in a far wider spectrum of products as a white inert filler. The most important application for kaolin remains in paper manufacture where it is used for coating and filling, although sales have been declining in recent years due to the increasing popularity of PCC (precipitated calcium carbonate) and GCC (ground calcium carbonate). The demand for high-quality kaolin for the ceramic whiteware market remains strong and is likely to increase.

Mineralogy. Kaolin is a white soft friable clay composed predominantly of well-ordered kaolinite, $Al_4Si_4O_{10}(OH)_4$. In the unprocessed state (and depending on the source rock), kaolin can also contain quartz, mica, feldspar, mixed-layer minerals, montmorillonite and tourmaline. Most raw kaolins undergo refining or beneficiation, which concentrates the kaolinite content and results in a saleable product. Kaolin deposits are normally divided into primary and secondary (or sedimentary) types.

Kaolin formation. Primary kaolin deposits result from the kaolinization of rocks rich in potassium feldpar. These include leucocratic or felsic igneous rocks such as granite, granodiorite, syenite, rhyolite and quartz-porphyry, arkose (feldpathic sandstone) and gneiss. The processes which result in kaolinization can act singly or in combination and include hydrothermal alteration and deep weathering. Hydrothermal alteration is normally associated with the circulation of hot waters where the heat source can be due to a cooling igneous body, volcanic activity or a concentration of radiogenic elements (also see Chapter 3). Chemical weathering can occur to considerable depths in sub-tropical and tropical environments and results in the formation of halloysite and/or kaolinite. In both cases the K feldspar undergoes argillization, initially into sericitic mica and ultimately into kaolinite. Recent work has shown that much of the kaolinization in south-west England is due to the circulation of relatively low temperature meteoric water (Psyrillos et al. 1998). Further reading on kaolin formation and processing, particularly in the UK can be found in Highley (1984) and Bloodworth et al. (1993).

The quality and characteristics of a kaolin deposit are largely dependant on the mineralogy and physical properties of the original source rock. The kaolinization of a granodiorite is likely to result in a white firing, but comparatively weak kaolin (the Kemmlitz deposit in Saxony). This is in marked contrast to a kaolinized pitchstone, which will be rich in smectite and/or mixed layer minerals, and thus will produce a strong, plastic kaolin but with an off-white or cream fired colour. The rheological properties of both kaolins will also differ markedly.

The intensity and depth of the kaolinization is often enhanced in shear or fracture zones where circulating waters can penetrate more readily and also by the permeability (porosity and micro-fracturing) of the rock. Examples of primary kaolin deposits are those associated with the Cornubian granite batholith in Devon and Cornwall, and the Glookhovetskoe and Prosyanovskoe deposits in the Ukrainian Shield (deep weathering). Kaolin yields in the raw matrix (or growan) normally range from 10% to 55%.

Secondary or sedimentary kaolin deposits are derived from the erosion of primary kaolins and their subsequent transport and re-deposition in a freshwater environment. Such deposits are not as common as primary kaolins, and occur as beds or lenses, which are commonly intercalated with other sediments, e.g. sands or sandstones, lignites, clays etc. Typical examples of sedimentary kaolins are the extensive deposits occurring in Georgia and Carolina in the USA, and the Pology deposit occurring in central Ukraine. The kaolin yield of the raw material commonly exceeds 50%, and in the case of the Georgia kaolin can be as high as 95%.

Ceramic applications. The principal function of kaolin in a ceramic body is to impart whiteness. The percentage of Fe_2O_3 and TiO_2 in the kaolin has a major influence on the raw and fired colour, and the content must be kept at a relatively low level. Other deleterious elements, which affect the fired appearance of the ceramic product, are copper, chrome and manganese. If these are present within the kaolinite crystal lattice, there is a general deterioration in whiteness, but if they occur in particulate form, there will be damaging and unsightly specking. As the alkali content will affect both fired shrinkage and colour, it is also desirable to keep the K_2O level to $<1.5\%$ for high quality whiteware applications.

Another key parameter is particle size. While a fine particle size will impart good plasticity and higher strength to the kaolin, which are important factors in the porcelain and tableware sectors, there are disadvantages with slow casting and higher shrinkage rates upon firing. Kaolins with a coarser particle size cast faster and are more suitable for sanitaryware applications. The coarse particle size is related to stacks of individual kaolinite platelets. Over-blunging of the china clay slurry will lead to de-lamination of these stacks, a closer packing of the platelets and reduce the casting rate (Psyrillos et al. 1999). The proportion of china clay in a whiteware ceramic body can range from 5 to 55%.

Production localities and statistics. The principal European kaolin production areas are located in south-west England, the German States of Bavaria and Saxony, northwest Bohemia in the Czech Republic, Ukraine, France and Spain. UK kaolin production in 2000 was 2 420 000 dry tonnes of which 87% was exported (Hillier et al. 2001).

14.9.3. Flint clay

Flint clay is a smooth, compact microcrystalline clay rock which is predominantly composed of well crystallized kaolinite and minor quantities of illite and quartz. It has a relatively high density, a flint-like appearance and breaks with a pronounced conchoidal fracture. However, the name does not imply any relation to flint. Flint clay has a low fired shrinkage (about 12%), resists slaking in water and has almost no plasticity. A flint clay becomes plastic following prolonged grinding in water.

Flint clays can occur in basins underlain by carbonate strata or in depressions within mudstones or sandstones. The flint clay deposits can be basin or funnel shaped or they occur as beds or lenses. Flint clay is probably formed from an illitic soil or regolith which has been subjected to processes which have removed the potassium and silica. These processes include dissolution, hydrolysis and dialysis and the result is kaolinite enrichment. If continued, these processes will ultimately result in an enrichment in alumina and the formation of oolitic or nodular boehmite and/or diaspore.

14.10. Fine ceramic clay production and refining

There is a wide range of mobile plant and production techniques used in the extraction of ceramic clays. The production regime employed at a particular site is normally influenced by many of the following factors, including bed thickness and the degree of selectivity required, the physical properties of the raw material, i.e. plasticity, friability, hardness etc., local custom or best practice, the degree of production flexibility required, the purchase price and operating costs and in the more remote areas, the availability of service personnel and spares for the mobile plant. The appropriate production plant and methods are described in detail in Chapter 10. Many of the standard extraction methods are common to all types of ceramic clay. This section will describe them briefly, but focus on the areas of clay selection, quality control, storage and refining or processing.

14.10.1. Ball clay

The following sections describe the extraction, production and processing of ball clays utilized in the manufacture of fine ceramic ware.

Seam selection and quality control. Before extraction commences within any ball clay deposit, a comprehensive analysis of the sequence should be carried out. This will not only determine which markets can be supplied but also enable a clay-seam blending programme to be developed. It will also be an initial guide in deciding whether clay seams should be extracted individually or in multiple units. The comprehensive analysis will seek to evaluate the chemical, mineralogical, physical and ceramic properties of the individual clay seams.

When production commences, it is standard practice to use a limited number of definitive tests to achieve accurate clay selection and quality control. These may include a combination of the following tests: chemical analysis (usually by XRF), loss on ignition, residue content at a specified sieve spacing, fired colour and linear shrinkage (see also Chapters 4 and 8). For clays destined for use in slips, it may also be necessary to include a fluid property test. The essence of efficient quality control is to define a range of tests, which although relatively cheap, are rapid and simple to perform and provide accurate guidance for clay selection and blending. On many large extraction sites, the rapid delivery, testing and appraisal of high numbers of quality control samples is a vital part of the business.

Accurate categorization and selection is achieved by channel sampling across the complete thickness of the seam (or sub-sampling where appropriate). This is often augmented with data from short-term, production-planning boreholes, which are relatively shallow and drilled on closely spaced grids. The clay sequence data can then be tabulated or incorporated into photographs or bench sections and made available to the production staff.

Production techniques. The vast majority of ball clay deposits are now worked by opencast extraction methods. The high operating costs of underground mining compared with modern opencast techniques has resulted in the closure of all underground mines, except those producing clay for very high value applications. The last of the underground ball clay mines in Devon closed in 1999. However there are still several remaining in Germany, one of which supplies a rare engobe clay rich in celadonite.

The methods employed to remove overburden are dependent on a number of factors, which include thickness, annual volume and stripping rate, type of material and haulage distance to the backfilling or tip site. Many companies employ contractors who utilize large back actor excavators and high-capacity dumpers or lorries. Face shovels, bucket wheel excavators and draglines of varying sizes are also used, especially when stripping rates are high. Scrapers can be a viable option where the overburden material is friable and unconsolidated, and if the transport distance is not excessive. For further information, see Chapter 11.

The soft, plastic nature of ball clays makes them suitable for extraction by a machine equipped with a blade and cutting action. The most widely employed extraction method consists of a backactor excavator operating with a team of dumpers or lorries. A backactor can achieve accurate seam extraction while retaining good production rates and flexibility. The disadvantage with this method is that the clay is often extracted in large lumps (especially when higher capacity buckets are utilized) that require breaking down by a shredding or kibbling process. Figure 14.2 is a flowchart displaying the production and processing methods employed in the WBB Minerals ball operations in south Devon (UK). Plate 19 shows ball extraction in north Devon, UK, and Plate 20 shows the extraction process in eastern Ukraine.

Bucket wheel excavators are still used widely in many East European countries. Most are electrically powered, and while production rates can exceed those achieved by a backactor, the accuracy of seam selection is diminished. They are more suited to the extraction of thicker clay seams or multi-seam units. There is also the advantage that the excavation process naturally produces far smaller clay pieces. Some companies have conveyor systems to transport the ball clay from the quarry. When a backactor excavator is being used, the clay must be fed through an in-quarry mobile shredder or kibbler (which contains large toothed rollers) to reduce the size of the clay pieces before they are discharged onto the conveyor belt.

Production rates can vary widely from <10 000 tonnes per annum at some sites, to approximately 1 million tonnes per annum in the largest of the east Ukrainian ball clay quarries.

Storage. In most ball clay operations it is common to transfer the excavated clay either to external stockpiles, large open sheds or bulk (single component) storage bays or boxes. Covered storage is particularly important during the winter months to avoid the clay acquiring an excessive moisture content and to enable the shredding and blending operations to function effectively.

Blending and refining. It is now very rare for clay selections or seams to be sold individually. Most ball clay producers blend a variety of selections in specific proportions to achieve a consistent and homogeneous sales blend, which is suitable for a particular application or market. A variety of formulations or recipes can be used to achieve the same sales blend. This is an effective mechanism for ensuring that the full clay sequence is utilized as much as possible, and gives greater flexibility in short-term production planning. The blending or mixing is often carried out at the same time as the size reduction of the clay pieces through a cutting and/or crushing process (shredding and kibbling). The blended and shredded (or kibbled) clay pieces, which range in size from 2.5 cm to 6.5 cm, are then placed in covered storage. Approximately 70% of clay sales from the UK ball clay producer WBB Devon Clays are in a shredded form, with the major markets being in the UK and Western Europe. Plate 21 shows the shredding process in the Donbas Clays JSC storage facility at Mertsalovo, eastern Ukraine. Plate 22 shows the loading of railway wagons by an overhead conveyor system installed at a WBB Fuchs facility in the Westerwald region of western Germany.

The shredded clay still retains its original moisture content, which normally ranges between 16% and 19%. Some customers may wish to reduce freight and processing costs by purchasing the clay in a powder form with a moisture content of 1% to 2%. This can be achieved through a drying and grinding process (often incorporating an attrition mill). The powdered clay is either bagged or transferred directly into road tankers.

Other ball clay refining and beneficiation processes can include wet screening, de-watering, extrusion and drying. The sales products can be supplied as a slurry, dried noodles, granulate or spray-dried material. Slurry, dried noodles or granulate are the preferred forms for refined ball clays used in the sanitaryware industry. Spray-dried material is used widely in the floor and wall tile industries.

There is also a developing demand from ceramic producers for 'prepared body', which is a ready-to-use, homogenized material containing all the components required to create a ceramic product. The ceramic producer is not required to undertake any body formulation or blending and can use the prepared body directly for pressing, forming or casting. Many ball clay producers are now developing their own prepared body plants, where their ball clays can be added to other standard body components e.g. kaolin, ground quartz and fluxing materials (feldspar, nepheline syenite etc).

14.10.2. Kaolin

The selection, production and refining techniques for kaolin can differ markedly from those used for ball clays. They are described below.

Production techniques. The kaolinized raw material or matrix from which kaolin is derived, is normally soft and friable, with little inherent strength. Some kaolins have the appropriate mineral proportions (kaolinite, quartz and feldspar and little or no deleterious accessory minerals), which enables them to be used in the raw state in whiteware ceramic bodies. However, most kaolin undergoes a refining or beneficiation process, which concentrates the kaolinite content and removes undesirable components.

The matrix is either dry-dug or washed out of a production face by high-pressure water monitors (also see

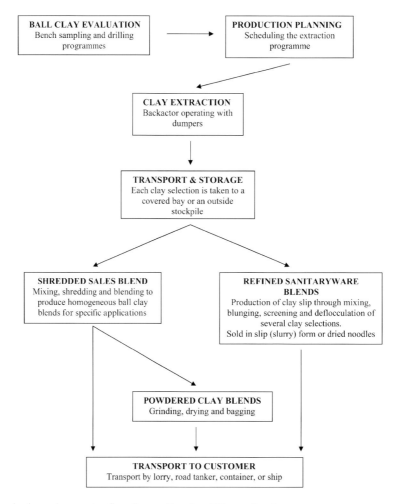

FIG. 14.2. Ball clay production and processing flow diagram (based on UK operations).

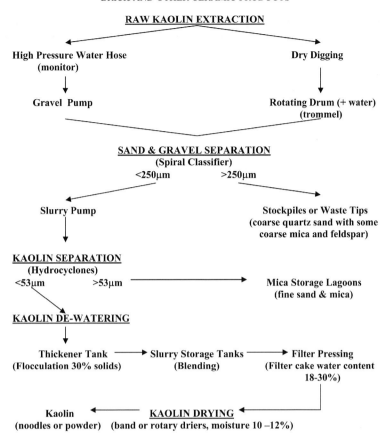

FIG. 14.3. Simplified kaolin production and processing flow diagram.

Chapter 11). Plate 23 shows the hydraulic mining method being utilized at a WBB Devon Clays quarry, south-west Devon, UK. Backactor excavators, face shovels or rotary bucket wheel excavators are used effectively to excavate the matrix in most European deposits. The material is transported by dumper, lorry, railway wagon or conveyor belt to a stockpile or processing site. In the United Kingdom, kaolin is derived from primary deposits developed in the Hercynian granites of southwest England. These deposits are suitable for hydraulic mining and the kaolinized matrix (growan) is broken down or disaggregated by high pressure, automated water monitors. The clay, silt and fine sand fraction flow in suspension to a sump in the base of the pit.

Refining. In most kaolin operations, the first stage of the refining process involves the removal of the coarser material from the kaolin slurry. Vibrating screens, screw classifiers, or bucket wheels are effective in separating the gravel and sand fraction from the silt and clays. Dry dug material is usually fed with water into large rotating drums (or trommels), where the fine fraction goes into suspension. The sand and gravel may then be transferred to an on-site aggregate plant for further processing.

The kaolin slurry passes through a succession of hydocyclones of decreasing diameter, which concentrates the clay fraction (see Plate 24). The underflow from the hydrocyclones (largely consisting of fine quartz and mica) is collected in settling lagoons or mica dams. The whiteness of the clay slurry from the hydrocyclones may be improved by bleaching or using HGMS (cryogenic high-gradient magnetic separation) to separate susceptible Fe containing minerals. The de-watering of the slurry is achieved through thickening, filter pressing (see Plate 25), extrusion of the filter cake through screens and drying to approximately 10% moisture content in a band drier. Most kaolin is sold as a noodled product. A kaolin production and processing flowchart illustrating different options is shown in Figure 14.13. Thurlow (1997) and Bristow et al. (2002) provide detailed explanations of the kaolin extraction and processing in Cornwall.

Selection and quality control. Selective extraction of the kaolin matrix can be attained through face sampling and utilizing closely spaced borehole grids. The hydraulic mining process also lends itself to in-quarry blending, through the merging of clay slurry streams from different production faces. Dry dug material may be selectively stored and then blended before refining. Another important stage of monitoring and blending of the thickened kaolin slurry may also occur in the holding tanks situated before the filter presses. Finally, blending may also occur in the silos holding the noodled product.

The inherent properties of the raw kaolin and (to a certain extent) the refining and beneficiation processes dictate which markets can be supplied from a particular deposit. The most important quality control parameters for ceramic grade kaolins are fired colour, alkali content, particle size and rheological properties.

14.11. Fine ceramic manufacturing processes

The fine ceramic manufacturing process is similar to that used in the heavy ceramic sector in that the same basic steps are followed: the body components are prepared and mixed in the required proportions; the ceramic body is shaped or formed and the resulting 'green' ware is fired. A significant area of difference relates to the nature of the ceramic body prior to shaping or forming. This can often be in the form of a spray dried dust or granulate from which a plastic body is formed or a casting slip for injection into a mould. The firing regime also differs as once-firing and fast-firing are becoming the industry norm in some ceramic sectors.

The ceramic industry has been striving to become more efficient through automation and rationalising the manufacturing processes, and this has been largely achieved in many modern tile plants. These contain long automated, production lines incorporating many modern technological developments such as fast firing roller kilns, in the processing steps. Sanitaryware producers have also been striving to achieve more efficient and cheaper production, and this has been achieved in some plants through the introduction of porous plastic moulds and high-pressure casting. The developments and changes in manufacturing processes often dictate that the ceramic body should perform differently. It is particularly incumbent on clay producers to produce blends that can satisfy the stringent specifications of modern ceramic manufacturers.

The process steps employed by a ceramic manufacturer are to a large extent dependent on the article being produced, and to what degree modern innovations and technology have been introduced into the manufacturing process. Although many processing steps are universally employed in the production of all ceramic ware, some steps are specific to a limited range of articles. Figure 14.4 illustrates the different process routes that can be taken in the production of fine ceramic ware. The major steps in the manufacturing process are body preparation, body forming, drying of the green ware and firing and these are explained below.

14.11.1. Ceramic body preparation

The various body components (clays, flux and filler) are weighed to achieve the correct proportions, and then undergo either blunging or milling. The blunging process involves the mixing of ball clay and kaolin with water to produce a slurry or slip. Deflocculant is added to achieve a clay suspension of greater fluidity. The rheological behaviour of the clay is important, as are any contaminants such as a high soluble salt content, which could adversely affect the density of the slip. The non-clay components (silica and flux) and sometimes all the body components are ground in a ball mill in either a wet or dry grinding process. The grinding media (typically porcelain, flint or steel balls) reduces the particle size of the non-plastic components.

The finely ground or dispersed clay and non-plastic components are mixed in the correct proportions, sieved to remove oversize particles (normally >140 to $>180\,\mu m$), and passed through a magnet to remove contaminant iron minerals. The resultant slip can then follow a number of different process routes that include spray drying, filter pressing and slip casting.

14.11.2. Ceramic article forming

Spray drying is now the most common method of forming dust or granulate for granular pressing. A spray drier is a device in which an atomized suspension of solids in a liquid is dried by hot gases or by direct contact with hot gases. The ceramic body components are mixed, blunged and spray dried to produce a granulate with a moisture content of 2 to 3%. Some manufacturers have their own spray driers, but many purchase the spray dried granulate from a central supplier. Granulate pressing (or isostatic or dust pressing) has become important in some sectors of ceramic manufacturing during the last 20 years (particularly tiles and tableware). The spray dried dust or granulate fills a diaphragm in the form of the required article. Isostatic pressing is achieved through equally distributed pressure exerted by an upper die and simultaneous pressure applied to the base of the diaphragm.

With filter pressing and pug making, the body in slip form is pumped into the filter presses where it is dewatered to approximately 25 to 30%. The resulting cake can be dried and ground, and the resultant dust used for granular pressing. Alternatively the filter cake enters a pug mill to produce a plastic body suitable for forming.

The slip casting process is used to form the most intricate and complex ceramic shapes, including sanitaryware, some tableware and complex refractory and advanced ceramic pieces. The ceramic body is created in slip form for pouring or injecting the clay slurry into a plaster mould. The dewatering of the slip occurs through the sucking action of the mould's fine pore structure. As a result of this process, a semi-dry deposit builds up on the inner surface of the mould to form the cast piece. The de-moulding of sanitaryware is shown in Plate 26.

The factors that influence and control the casting process are the suction pressure, the permeability of the cast layer or deposit and water viscosity. Optimum casting properties can be achieved through the preparation of a high solids content slip with acceptable flow characteristics and careful control of the particle size distribution of the body components.

Most modern manufacturers are anxious to achieve a rapid casting rate, combined with low production losses.

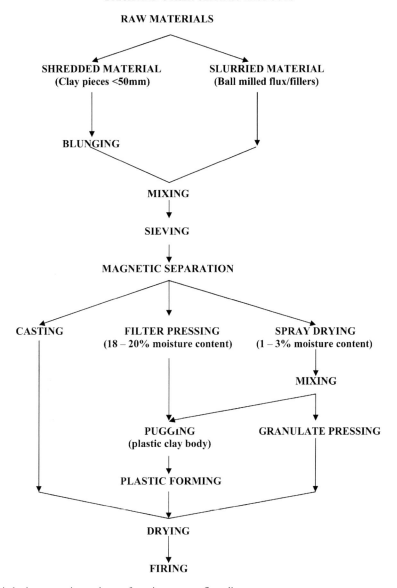

FIG. 14.4. Fine ceramic body preparation and manufacturing process flow diagram.

A low slip residue, easy dispersion, and stable deflocculation and rheological properties can be achieved through careful selection, blending and control of the casting slip components. This helps to ensure a good casting rate, high cast strength and favourable de-moulding properties.

14.11.3. Drying and firing

It is essential to remove the moisture from a body prior to firing. This is achieved by placing the green ceramic ware in a drying oven. Following this, firing converts the relatively weak green ceramic piece into a strong and durable product through the development of a glass. The fired piece will have some or all of the following beneficial properties: high-impact strength, tensile strength, thermal shock resistance, abrasion and chipping resistance, chemical durability and electrical resistance.

There are many variants of the firing process depending on the type of product, its function and body composition. 'Once firing' describes the process where the ceramic body, glaze and decoration are created during a single firing cycle. 'Biscuit firing' is an initial phase of firing to create a porous strong structure, which then undergoes modification during subsequent firing cycles. 'Glost firing' describes the maturation of the glaze on the biscuit piece, and 'Decorating firing' ensures that the decoration is emplaced either onto or into the glaze. Finally 'Fast firing' refers to a rapid firing cycle. Many fine ceramic

TABLE 14.10. *Fine ceramic ware firing cycle*

Firing cycle	Temperature (°C)	Event/process
Heating	Ambient–150	Free water evaporates, shrinkage occurs (ware must start completely dry for fast firing).
	150–300	Binder breaks down in dry pressed ware; temperature increase should be slow and uniform to avoid cracking.
	150–500	Dehydroxylation of clay minerals, generation of steam.
	500–600	α-β quartz transition at 573°C with volume expansion and endothermic reaction.
	600–900	Main oxidation phase.
	200–900	Burning of carbonaceous matter in air. A slow temperature increase from 800 to 900°C (with a soak at 850°C) will ensure complete carbon burnout and avoid specking.
	900+	Sintering and vitrification.
Peak	1160 (Peak)	Peak temperature for earthenware. Ware shrinks due to vitrification. Cristobalite is formed during long periods of high temperature firing.
Cooling	Peak–700	Rapid cooling.
	700–500	Cooling rate is slowed to ensure uniform contraction through α-β quartz transition at 573°C.
	500–300	Rapid cooling.
	300–Ambient	Cooling rate slowed if cristobalite has been formed.

manufacturers are striving to introduce once fired and fast-fired technology into their plants, as a means of reducing costs. However, the body formulation and in particular the ball clay content, play a vital role in determining whether the ceramic piece is suitable for this firing regime. Plate 27 shows a fully loaded trolley of fired sanitaryware emerging from the kiln. Plate 28 shows part of the automated tile manufacturing process.

The firing cycle in the kiln is important as it defines the rate of temperature increase or decrease, peak temperature, periods of static or stable temperature ('soak' period) and the total duration of the piece within the kiln. The stages in a typical firing cycle are summarized in Table 14.10.

Another important factor in the firing process is whether the kiln atmosphere is oxidizing or reducing, as this will directly affect the colour and vitrification of the fired product. In an oxidizing atmosphere, any free iron oxide will form haematite and a red colour is imparted to the product. A reducing atmosphere caused by insufficient oxygen will create fuel rich conditions, and the formation of carbon monoxide instead of carbon dioxide. Any excess iron mineral will form as ilmenite or ferrous silicate and a blue colour results. Porcelain, blue engineering bricks, 'flashed' bricks and special glaze effects can be produced in a reducing atmosphere.

14.12. Future trends in ceramic manufacturing

The last 30 years has seen tremendous improvements in both raw material production and processing and ceramic manufacturing. The twin constraints of tighter specifications and parameters being imposed by ceramic manufacturers and the economic necessity of utilizing as much as possible of a clay deposit, have created a challenging regime for clay producers. Controlled blending of clays to achieve a homogeneous and consistent sales product is now the norm for suppliers in both the fine and structural ceramic sectors (Lawson 2001).

14.12.1. General trends

It is expected that technological developments will enable processes to continue to undergo rapid change. It is likely that the increasing transfer of fine ceramic manufacturing to low production cost countries (especially China and Eastern Europe) will continue, but customers will expect high-quality products as the norm. There will be increasing competition between manufacturers and eroding margins will force innovation and added value products. In addition, the more stringent environmental and ecological legislation that is being imposed, linked to a growing awareness of the concept of sustainability, will impinge on both the raw material producer and ceramic manufacturer.

14.12.2. Processes and materials

Within the structural ceramic sector, it is anticipated that there will be more utilization of alternative materials such as PFA, sawdust, etc. (Šveda 2000) and more interregional blending of clays, especially as some clay selections become scarce. Coupled with these developments it is anticipated that there will be a greater utilization of fast firing roller kilns.

In the fine ceramic sector there will be increasing use of prepared bodies, use of composite flux/fillers and substitution of flint by quartz/calcined quartz. It is also likely that closer scrutiny and control of particle size (packing density) and increased use of pressure casting will be introduced. In sanitaryware, slip casting may be replaced by other forming processes, including injection moulding, isostatic and plastic pressing, whereas tableware will be dry pressed.

The fine ceramic sector is also likely to achieve lower firing temperatures, coupled with more rapid and once firing. There will also be an increase in the production of vitreous bodies, lower expansion bodies (with improved thermal shock resistance) and fashion having a greater influence in dictating market trends.

References

BARLOW, S. G. & MANNING, D. A. C. 1997. The influence of time and temperature on the reactions and transformations of clinochore as a ceramic clay mineral. *British Ceramic Transactions*, **96**, 195–198.

BARLOW, S. G. & MANNING, D. A. C. 1999. The influence of time and temperature on the reactions and transformations of muscovite as a ceramic clay mineral. *British Ceramic Transactions*, **98**, 122–126.

BLOODWORTH, A. J., HIGHLEY, D. E. & MITCHELL, C. J. 1993. *Kaolin, Industrial Minerals Laboratory Manual, British Geological Survey Technical Report WG/93/1*.

BOUGHER, K. 1999. Ceramic casting, challenges for US ball clays. *Industrial Minerals*, April, 77–88.

BRISTOW, C. M. 1968. The derivation of the tertiary sediments in the petrockstowe basin, north devon. *Proceedings of the Ussher Society*, **2**, 29–35.

BRISTOW, C. M. 1988. Ball clays, weathering and climate. In: ZUPAN, A-J. W. & MAYBIN III, A. H. (eds) *Proceedings, 24th Forum on the Geology of Industrial Minerals*, S. Carolina Geological Survey, 25–37.

BRISTOW, C. M., PALMER, Q., WITTE, G. J., BOWDITCH, I. & HOWE, J. H. 2002. The ball and china clay industries of Southwest England in 2000. *Industrial Minerals and Extractive Geology*. Geological Society, London, 17–41.

DUNHAM, A. C. 1992. Developments in industrial mineralogy: I. The mineralogy of brick making. *Proceedings of the Yorkshire Geological Society*, **49**, 95–104.

DUNHAM, A. C., McKNIGHT, A. S. & WARREN, I. 2001. Mineral assemblages formed in Oxford Clay fired under different time-temperature conditions with reference to brick manufacture. *Proceedings of the Yorkshire Geological Society*, **53**, 221–230.

FIEBINGER, W. 1997. Speciality clays for the ceramics industry: Reserves, production, trade, perspectives, cfi/Ber. DKG 74, **9**, 491–496.

GLASSON, N. P. & FORBES, N. R. 2001. Clay systems for improved performance. *cfi/Berichte DKG*, **78**, 22–27.

HIGHLEY, D. E. 1982. *Fireclay*. Mineral Resources Consultative Committee, Mineral Dossier No. 24, HMSO, London.

HIGHLEY, D. E. 1984. China Clay, Mineral Resources Consultative Committee, Mineral Dossier No. 26, HMSO, London.

HILLIER, J. A. & HIGHLEY, D. E. 1997. *United Kingdom Minerals Yearbook 1996*, British Geological Survey.

HILLIER, J. A., CHAPMAN, G. R. & HIGHLEY, D. E. 1999. *United Kingdom Minerals Yearbook 1998*, British Geological Survey.

HILLIER, J. A., CHAPMAN, G. R. & HIGHLEY, D. E. 2001. *Untied Kingdom Minerals Yearbook 2000*, British Geological Survey.

HINKLEY, D. N. 1963. Variability in crystallinity values among kaolin deposits of the coastal plain of georgia and south carolina. *Clays and Clay Minerals*, **11**.

LAWSON, G. B. 2001. Challenges for ceramic raw material suppliers in the 21st century. *cfi/Berichte DKG*, **78**, 30–35.

MANNING, D. A. C. 1995. *Introduction to Industrial Minerals*, Chapman and Hall, London.

McNALLY, G. H. 1998. *Soil and Rock Construction Materials*, Spon, London, 291–310.

MOLDERS, A. & BINDER, D. 1998. Glazed and engobed clay roofing tiles, Ziegel Industrie, 802–809.

PRENTICE, J. E. 1988. Evaluation of brick clay reserves. *Transactions of the Institution of Mining and Metallurgy, Section B Applied Earth Sciences*, **97**, 9–14.

PRENTICE, J. E. 1990. Geology of construction materials. Chapman and Hall, London.

PSYRILLOS, A., MANNING, D. A. C. & BURLEY, S. D. 1998. geochemical constraints on kaolinisation in the St. Austell Granite, Cornwall, England. *Journal of the Geological Society of London*, **155**, 829–840.

PSYRILLOS, A., HOWE, J. H., MANNING, D. A. C. & BURLEY, S. D. 1999. Geological controls on kaolin particle shape and consequences for mineral processing. *Clay Minerals*, **34**, 193–208.

REH, H. 2000. An international trade fair with 60 percent foreign visitors. *cfi/Berichte DKG*, **7**, 19–31.

RIDGWAY, J. M. 1982. Common Clay and Shale, Mineral Resources Consultative Committee, Mineral Dossier No. 22, HMSO, London.

ŠVEDA, M. 2000. The influence of sawdust on the physical properties of a clay body, Ziegel Industrie, 29–35.

THURLOW, C. 1997. *China Clay from Cornwall and Devon—An Illustrated Account of the Modern China Clay Industry*. Cornish Hillside Publications, St. Austell, Cornwall.

Further reading

BATES, R. L. & JACKSON, J. A. 1987. *Glossary of Geology*. American Geological Institute, USA.

DALE, A. J. 1964. *Modern Ceramic Practice*, Maclaren & Sons Ltd, London.

GEORG, H. 1991. Ceramic materials and their uses. cfi/Berichte DKG, **4**, 140–142.

O'BANNON, L. S. 1984. *Dictionary of Ceramic Science and Engineering*, Plenum Press, New York.

RYAN, W. & RADFORD, C. 1997. Whitewares: Production, Testing and Quality Control, The Institute of Materials, London.

15. Cement and related products

15.1. Introduction

15.1.1. What is cement?

This review discusses the important role played by clay minerals in the manufacture and use of cement. The word 'cement' is generic and has a very broad meaning, but we are concerned here with those inorganic cements which find large-scale use in the construction industry.

All types of cement as defined above share certain similarities in the ways they are utilized. Thus, to prepare it for use, the dry cement powder is mixed with water and other chemicals (admixtures), converting it into a pourable liquid, a soft gel, a plastic mass, or a damp granular solid. It is important that the soft mixture can easily be moulded, and that it subsequently sets in a reasonable time. Final curing normally occurs over a longer timescale, to give a product with the required degree of hardness and durability. The mechanism of setting and curing involves hydrolysis and hydration reactions, in which the cement reacts with some of the mixing water to give new chemical compounds which take the form of interlocking crystals. The structure becomes mechanically rigid and strong, due to the interlocking. Cured cement can be used as a construction material in its own right, or as an adhesive which bind aggregate particles together to form concrete or mortar. This is illustrated by two examples, one a simple compound, the other a complex mixture of compounds of variable composition (see the Text Box below).

The chemical structure of cured cement can be compared to that of certain aluminosilicate rocks which occur naturally. For example, hydrated calcium silicate (see Text Box 15.1) in hardened Portland cement has a layered structure which is analogous, at the nanometre scale, to that of phyllosilicate clays. Perhaps this is not surprising, because for economic and practical reasons the most commonly available minerals (i.e. those containing oxides of calcium, aluminium, iron and silicon) have been quarried as raw materials to make cement in the first place. We can view cement technology as a method of treating

---

Cement and cement mixes

Calcium sulphate hemihydrate is one of the most chemically simple examples of cement, as defined above. It is an ionic salt which readily reacts with water under ambient conditions, to form calcium sulphate dihydrate (gypsum). The soft paste, consisting of hemihydrate and water gradually becomes stiff as the recrystallization process removes free water. Simultaneously, an interlocking network of micrometre-size crystals of the dihydrate are formed, and this confers a degree of hardness and strength:

$$2(CaSO_4 \cdot 0.5H_2O) + 3H_2O \rightarrow 2(CaSO_4 \cdot 2H_2O).$$

Anhydrous Portland cement is a more complex mixture, composed mainly of various anhydrous ionic calcium silicates. Portland cement reacts exothermically with water to give a range of crystalline and non-crystalline hydration products, and the whole process can be modified by minor components such as calcium sulphate and aluminium-containing species. When hydration is complete, Portland cement paste is a mixture of ionic salts, including calcium hydroxide (portlandite) and various calcium silicate hydrates (CSH). The silicate anions range from the monomeric orthosilicate (SiO_4^{4-}), though dimeric ($Si_2O_7^{6-}$) and pentameric ($Si_5O_{16}^{12-}$) species, and include small concentrations of higher polymers. Over time, the average degree of polymerization slowly increases. The crystal structure of CSH in cured cement paste is not well defined; indeed it is often referred to as 'cement gel'. It consists of a mass of small interlocking crystals, and regions of amorphous gel, typically of the order of nanometres and micrometres in size. Cement gel is permeated by a system of fine pores, which have diameters also measured in nanometres and micrometres. The forces that hold the crystals together as a coherent mass include hydrogen bonds, ionic bonds, van der Waals forces and forces due to mechanical interlocking.

Portland cement is hydraulic: a compacted mass of the anhydrous material will harden even when fully submerged in water. Not all anhydrous calcium silicates can be described as hydraulic and the cement industry is careful to avoid production of non-hydraulic forms of calcium silicate. Calcium sulphate hemihydrate is not hydraulic because, although it hydrates under water, it fails to harden.

natural rock, so that it can be broken down, moulded into any desired shape, and then re-hardened to give useful artefacts.

However, a major difference between hardened cement paste and natural surface rocks is that hardened cement is less likely to be in chemical equilibrium with the environment. Cement cures slowly; the chemical reactions involved in hydration and recrystallization can continue for many years, and phases which form in the early stages of hardening may not be the most stable phases in the long term. Other slow chemical reactions can occur, either between different components of the concrete itself, or (because cement paste is porous) between the concrete and the external environment. Some of these slow chemical reactions are beneficial, but others can generate internal stresses and lead to degradation of the structure. This can result in a serious loss of mechanical strength which is known by the engineering profession as 'poor durability'. The durability of concrete is of major economic importance, and is the subject of continuing intensive research.

Several textbooks discuss in detail the manufacture, properties and applications of many types of cements in common use, and a selection is listed below:

- *Lea's Chemistry of Cement and Concrete*, 4th Edn. Ed. P.C. Hewlett. Elsevier Butterworth Heinemann, 1998.
- *Cement Chemistry*, 2nd Edn. H.F.W. Taylor. Thomas Telford, 1997.
- *Properties of Concrete*, 4th Edn. A.M. Neville. Longman, 1995.
- *Cement Chemistry and Physics for Civil Engineers*, 2nd English Edn. W. Czernin. Georg Godwin. 1980.

15.1.2. The history and development of modern cement

Over the last 250 years, the development of cement and concrete technology has paralleled economic growth in the industrial nations. Cement and concrete had been widely used by Roman engineers, but their knowledge was based entirely on trial and error, and cement technology suffered a 1300-year decline after the fall of the Roman Empire. By the mid-1700s, cement was expensive and still made on a small scale by burning limestone in brick-lined kilns. The product was variable in quality because the principal raw material varied so much in composition. In particular, the setting time, strength and durability of the resulting mortar and concrete was unpredictable. By 1750 Europe needed new infrastructure for its rapidly expanding manufacturing industry and trading links, and there was an urgent requirement for reliable, cheap cement which hardened rapidly. Smeaton, who was trained in the recently developed principles of systematic scientific research, began to investigate some of the factors which controlled the quality and reliability of cement. He carried out an extensive programme of chemical experiments to discover why some commercial cement was weak and some was strong. Smeaton discovered that the key to producing good quality burnt lime cements was to use an argillaceous limestone of relatively high clay content. He also showed the importance of careful chemical analysis of the limestone, and control of the clay component. His work was vindicated in 1759 when he built the Eddystone lighthouse on exposed rocks 22 km south of Plymouth in the English Channel. The lighthouse was made from granite blocks cemented together with mortar, made by blending a natural pozzolana from Italy with a burnt argillaceous limestone from Aberthaw south Wales. It lasted for 123 years until it was dismantled and rebuilt on Plymouth Hoe, where it stands today. The base platform of Smeaton's tower still survives on the rocks. Smeaton's analysis of the Aberthaw limestone was: carbonate of lime 86.2%, clay 11.2%, water etc. 2.6%.

Other investigators built on Smeaton's pioneering work. In 1796, James Parker discovered that natural clay/limestone nodules, which occurred in abundance in the gravel beds around the Thames estuary, gave remarkably good cement when they were calcined. Parker called his material 'Roman Cement', and it formed the basis of a rapidly growing industry, centred round the coasts of Kent and Essex. However, reserves of the intimately mixed clay/limestone nodules were limited, and many alternative methods of making cement were investigated.

In 1824, Aspdin developed and patented a strong hydraulic cement, which he called 'Portland cement' after Portland stone, a pale grey limestone of exceptional quality from Dorset, which was used in many prestigious building projects of the 18th and 19th centuries. It is interesting to note that Smeaton (1791 and 1793) had earlier written that his own cement 'would equal the best merchantable Portland stone in solidity and durability'. Aspdin's process involved grinding and intimately mixing two minerals, which were limestone-rich and clay-rich respectively. The 'artificial argillaceous limestone' was then calcined to higher temperatures than had previously been attempted, thus forming a primitive cement clinker. Significantly, the two raw materials were widely available, so for the first time cement could be made, in bulk, and at reasonable cost, almost anywhere in the world.

Portland cement was progressively improved in quality and reliability, and by 1860 it had almost displaced the Roman cement. Again, suitable raw materials (clay and chalk) were available in huge deposits in Kent and Essex, so the Thames and Medway valleys became the cradle of the Portland cement industry: UK production was 200 000 tonnes in 1862 and 1 300 000 tonnes in 1898. Portland cement was used to construct the offices, factories, railways and seaports needed for the industrial revolution; concrete rapidly replaced wood as the preferred material of construction in large projects. Detailed histories of this period of cement development are given by Davis (1924) and Francis (1977), and the British Cement Association have published (1999) a booklet: *A brief history of concrete, from 7000 BC to AD 2000*.

Table 15.1, compiled from various sources, shows the growth of the Portland cement industry, in terms of production volume and quality, i.e. strength, over the last 160 years. During that period, the advances in

TABLE 15.1. *Growth of the Portland cement industry from 1843*

Year	World cement production, (tonnes × 10^6)	USA cement production, (tonnes × 10^6)	UK cement production, (tonnes × 10^6)	European cement production, (tonnes × 10^6)	Typical compressive strength,* (MPa)	Typical tensile strength,† (MPa)
1862			0.2			2
1871		(Start of prod'n in USA)			8	
1880		0.00665				
1890		0.0536				
1891						3.6
1898			1.3			
1899		1.6				
1910		13			26	
1924	53.7	25		25	13 (min.)	5.7
1928		30			32	
1975	719		17			
1980					57	
1990	1151	70			60	
1995	1430	72		172	70	
2000	1530		12.5			

* The compressive strength of a block of concrete, made and tested under standard conditions.
† The tensile strength of a bar of neat cured cement paste, made and tested under standard conditions.

production technology have also resulted in significant improvements in fuel efficiency.

There is no doubt that the cement industry is essential to the world economy, as there is no realistic alternative material available for the construction industry. Older materials of construction, such as wood, rammed earth and dimension stone, are still valuable but have a relatively small part to play in major infrastructure projects, in terms of tonnes.

Unfortunately, the production of cement and concrete does exact a considerable ecological price in terms of destruction of the natural landscape, the emission of greenhouse gases, and the generation of undesirable pollutants. Each tonne of Portland cement requires, on average, 1.6 tonnes of quarried minerals, plus 0.1–0.2 tonnes of coal. Table 15.2 shows the major minerals (not including coal) used in 1996 in the USA to make Portland cement. The modern dry process consumes 4.2–4.9 GJt^{-1} (approximately 1500 $kWht^{-1}$), which includes thermal energy, electrical energy used in quarrying and production, and energy used for transportation. During the production of 1 tonne of Portland cement, approximately 1 tonne of CO_2 is liberated into the atmosphere, mainly from the decomposition of limestone and the generation of energy. It is estimated that up to 8% of the global CO_2 emissions, due to human activities, result from the production of concrete.

15.2. Portland cement

15.2.1. Types of clay used in the manufacture of Portland cement

Clays are formed when silicate rocks decompose. The rocks are commonly igneous in nature (e.g. granite), or derived from volcanic ashes. The decomposition process can be complex in nature, sometimes involving several stages of leaching with aqueous fluids, settlement in seas and lakes, and further leaching. Beds of clay used for cement manufacture often contain other fine non-clay minerals, such as silica, iron oxides, titanium dioxide, mica and feldspar. Indeed, some 'clay' beds can have a clay content as low as 20 mass%. Mixtures of different clays are usually present in a given deposit, and interstratified, or mixed-layer, clays are common. The chemical compositions of relatively pure samples of phyllosilicates often found in sedimentary deposits are shown in Table 15.3. Values are not given for those elements present in trace quantities, and it should be noted that, from the cement manufacturer's point of view, it is the chemical composition of the whole deposit which is of importance, rather than the type or composition of the individual clay components.

TABLE 15.2. *Major minerals used in 1996 in the USA to make Portland cement*

Mineral	Tonnes, million
Calcium carbonates	80
Argillaceous limestone	26
Clay, shale etc.	10
Silica	4
Iron minerals	1.5
Gypsum	4
Other	4
Total	**129**

TABLE 15.3. *Typical chemical analyses of clay minerals commonly present in clay sediments*

	Kaolin St Austell UK	Montmorillonite Redhill UK	Montmorillonite Fe-rich* Germany	Illite† Montana USA	Mica‡ Georgia USA
SiO_2	46.2	64.7	52.3	51.5	45.5
Al_2O_3	39.2	18.6	4.43	28.5	36.9
Fe_2O_3	0.23	8.12	22.73	–	–
MgO	–	4.32	2.2	3.0	–
Na_2O	–	3.3	–	–	–
K_2O	–	–	–	9.07	10.17
Combined H_2O	13.8	§	15.75‖	5.5	4.59

Data from Newman 1987.
* Nontronite is montmorillonite with all the Al_2O_3 replaced by Fe_2O_3.
† Illite is also known as hydromica.
‡ Mica is not, strictly, a clay mineral but is often present in clay fractions.
§ Value not given.
‖ Includes free and bound water.

15.2.2. Role of clays in the manufacture of Portland cement

Portland cement is manufactured by calcining an intimate mixture of calcium carbonate, clay and silica. In a typical cement kiln at 1500°C, approximately 20–30% of the total charge melts, and the liquid phase acts as a solvent for converting the raw minerals into cement compounds. The clay fraction plays several important roles in the complex chemical reactions that take place in the calciner.

Firstly, clay provides part of the silica, alumina and iron, which react with the calcium oxide to give the correct blend of anhydrous calcium silicates, aluminates and alumino-ferrites. Secondly, clay is a source of minor elements such as Mg, K, Na, Ti, Fe and F. The minor elements play an essential role in reducing the melting point of calcium silicate, i.e. they act as useful fluxing agents that reduce the thermal energy required. Thirdly, the minor elements act as 'mineralisers', that is they promote the crystallization of certain desirable calcium silicates, By substituting in the crystal lattices, the mineralizing elements stabilize these useful crystal forms of tricalcium silicate and dicalcium silicate, as alite and belite respectively. Without the mineralizers, *pure* tricalcium silicate would convert to dicalcium silicate plus calcium oxide as the clinker cooled below 1250°C. However, in spite of the presence of mineralisers, some dicalcium silicate is unavoidably produced (Table 15.5), and an additional role of foreign ions, especially K, is to stabilize the hydraulic β-form of dicalcium silicate (belite) with respect to the non-hydraulic γ-form. The composition and structure of the various calcium silicates is described in more detail in Section 15.2.3.

The raw materials must be careful blended to achieve the desired chemical composition and final cement quality. The clay component is difficult to characterize by XRD, on account of its fine particle size, poor crystallinity and the occurrence of inter-stratification. Hence XRF is normally used as a routine tool for measuring and controlling the oxide ratios. If the clay contains a high proportion

TABLE 15.4. *Oxide compositions of typical minerals fed to a cement calciner, and of product clinker (examples taken from Lea's Chemistry of Cement and Concrete, 3rd Edition, 1970, reprinted 1988, pages 59,60)*

		Proportion mass%	SiO_2 mass%	Al_2O_3 mass%	Fe_2O_3 mass%	CaO mass%
Required composition of clinker (approx.)			21	6	2.3	65
Mixture (A)	Limestone, high clay	75	15.9	4.7	1.9	41.7
	Limestone, low clay	25	6.9	1.9	0.6	49
	Feed composition		13.7	4.0	1.6	43.5
	Clinker composition		20.5	6.0	2.4	65.5
Mixture (B)	Shale	22.5	37.9	16.5	5.1	5.4
	Limestone, pure	73	1.4	0.5	0.2	53.7
	Sand	4.2	95.0	1.4	1.3	2.7
	Iron oxide	0.3	2.7	6.6	84.0	2.7
	Feed composition		13.6	4.2	1.6	42.7
	Clinker composition		20.3	6.2	2.4	64.1

of kaolin it may be necessary to add extra silica in the form of fine sand. Extra iron oxide may also be added, if there is insufficient in the clay fraction. If coal is used for firing the kiln, coal ash may contribute 2–6 mass% of the final product, so allowances must be made for SiO_2, Al_2O_3, CaO and Fe_2O_3 in the coal ash.

A typical oxide composition of the feed to the calciner is given in Table 15.4. The table also shows how this composition can be attained, either by blending two argillaceous limestones (A) or by blending one argillaceous limestone (calcareous shale) together with pure limestone, sand and iron oxide (B).

Other minerals that may be associated with the clay fraction include mica, feldspar and silica. They have important effects, which may be either beneficial or deleterious, so they must be carefully monitored to ensure that the Portland cement is of consistent quality. It is particularly important to ensure that alkali metal compounds are not present in excess because, being volatile, they might collect in the electrostatic precipitators as 'kiln dust' and require disposal. Also, high levels of alkali metal salts are generally undesirable in concrete; as they may promote the alkali silica reaction (see the Text Box below).

Chlorides may exist in clays of marine origin. Chlorides are volatile and will deposit on cooler surfaces within heat exchangers etc, causing blockages. Normally, chloride is kept below 0.015 mass% of the feedstock (including coal). In any case, chloride should be less than 0.1 mass% in Portland cement, to avoid rapid corrosion of steel reinforcing bars (re-bars) used in concrete structures.

Fluorine is present in many of the kaolinitic clays. Fluoride substitutes for OH^- in the lattice structure, typically at concentrations up to 0.2 mass%. In early cement kilns, fluorides were often added as mineralisers, which promote the formation and crystallisation of desirable alite at the lower kiln temperatures available at that time. In modern kilns, 90% or more of the fluorine in the feed is retained in the clinker so the fluorine content is normally controlled to 0.01–0.15 mass%. Even small levels of fluorine have a significant effect on setting time and rate of strength development. Above 0.15 mass%, fluorine can severely retard strength development.

15.2.3. Composition and hydration reactions of Portland cement

The two main components of Portland cement are both simple ionic orthosilicates, containing SiO_4^{4-} ions, namely tricalcium silicate $(CaO)_3 \cdot SiO_2$ and dicalcium silicate $(CaO)_2 \cdot SiO_2$. They are not pure compounds because the crystal lattices can incorporate various substitution ions, such as K^+ (for Ca^{2+}) and Al^{3+}, Fe^{3+} or Fe^{2+} (for Si^{4+}). These impure forms of tricalcium and dicalcium silicates are known as alite and belite respectively. As an additional complication, both alite and belite exist in several alternative crystal forms, only some of which are hydraulic in that they react rapidly with water to give useful cementitious hydrates which are hard, strong and durable.

As well as alite and belite, two other important components of Portland cement are tricalcium aluminate, $(CaO)_3 \cdot Al_2O_3$, which contains an $Al_6O_{18}^{18-}$ species, and calcium alumino ferrite, $(CaO)_4 \cdot Al_2O_3 \cdot Fe_2O_3$, which has a variable composition. Table 15.5 gives the approximate ranges in chemical composition of each of these principal

---

Alkali metal ions

The origins and forms of alkali metal ions (Na^+ and K^+) present in cement clinkers, and their effects on phase composition, has been reviewed by Taylor (1997).

Although some clay minerals, for example kaolinite and vermiculite, contain low concentrations of alkali metal cations, others notably the illites and hydromicas contain significant amounts. For example, the Westbury (Somerset, England) cement works used the local Upper Jurassic Kimmeridge clay as a source of silica, alumina and iron. Because of concern over ASR, the alkali content was reduced in the mid-1970s by removing much of the kiln dust to landfill, rather than blending it back with the cement clinker. In 1996/7, the clay was replaced by coal fly ash, with lower alkali content. This also had the beneficial effects of reducing the amount of clay quarried, and the amount of kiln dust sent to landfill (Scott & Bristow, 2002).

In concrete, high concentrations of alkali metal ions sometimes have undesirable effects. For example they dissolve in pore water to give highly alkaline solutions which, over a period of months and years, react with certain aggregates containing amorphous silica. The chemical reaction can produce considerable quantities of hydrated calcium/potassium/sodium silicate gel, provided that sufficient alkali metal ions, soluble calcium and reactive silica are present. The gel absorbs excess water and swells, causing the concrete to expand and crack. This is the alkali silica reaction (ASR), also known as alkali aggregate reaction (AAR) or, popularly, as concrete cancer. To limit the formation of swelling gel in concrete exposed to water, many countries specify that Portland cement must contain less than 0.6 mass% as alkali equivalent, and concrete less than 3 kg alkali equivalent per cubic metre (alkali equivalent is defined as the mass total of $Na_2O + 0.658K_2O$).

TABLE 15.5. *Principal constituents of anhydrous Portland cement*

		Alite $(CaO)_3 \cdot SiO_2$	Belite $(CaO)_2 \cdot SiO_2$	Aluminoferrite $(CaO)_4 \cdot Al_2O_3 \cdot Fe_2O_3$	Aluminate $(CaO)_3 \cdot Al_2O_3$
Cement chemists' notation		C_3S	C_2S	C_4AF	C_3A
Oxide analysis of each constituent, mass%	CaO	60–77	55–66	45–51	48–61
	SiO_2	26*	35*	2–5	3–6
	Al_2O_3	1–9	1–5	13–23	22–35
	Fe_2O_3	0.4–4	0.1–2.5	20–30	3–8
	MgO	0.2–2	0.1–1	2–5	0–2
	Na_2O	<1	<1	<1	0.2–4
	K_2O	<1	0.1–3	<1	0.2–3
Abundance of each constituent in Portland cement, mass% by XRD		54–76	5–30	3–8	5–15
Hydration rate		Rapid, 1–100 hours	Slow, 20 hours—years	Very slow, years	Rapid, 1–100 hours
Hydraulicity†		Good	Variable intermediate—good	Slight	Slight

* Average values.
† Contribution to long-term strength of concrete.

components. Table 15.5 also gives the approximate abundance of each component in modern Portland cement, and their individual hydraulic properties.

Cement chemists use a shorthand notation for the chemical species which make up the principal cement compounds, and a partial list is given below:

C(CaO) **S**(SiO_2) **A**(Al_2O_3) **F**(Fe_2O_3) **K**(K_2O) **N**(Na_2O)

It is important to note that the properties of Portland cement, in the field, are critically dependant on the detailed chemical and mineralogical composition. For example, setting and curing times can be controlled by adding calcium sulphate and by changing the particle size distribution. Ordinary Portland cement typically contains 10–13 mass% C_3A and 15 mass% C_2S, but sulphate resisting cement contains less C_3A (e.g. 5 mass%) and more C_2S (e.g. 35 mass%). On curing, this composition generates less of the hydrated calcium aluminate phases, which would normally react with sulphate ions and result in expansion and cracking of the concrete. To ensure that the correct cement is produced and used for any given application, extensive systems for classifying cements have been developed worldwide.

The principal types of Portland cements currently used by the construction industries in Europe and North America are classified according to their properties and mineralogical composition. The European standard, EN 197-1:2000, covers the composition, specifications and conformity criteria of Portland cement and blended Portland cement. There are five main categories, each of which has many possible sub-sections. The corresponding British standards are designated BS EN 197-1, but some of the old BS cement standards have not yet been placed within BS EN 197-1 (Table 15.6). Current information is on the web site of the BCA (British Cement Association): www.bca.org.

The many specifications for different types of Portland-pozzolana cements, CEM II, CEM III, CEM IV and CEM V, reflect their growing importance over the last 50 years as discussed in more detail later.

15.2.4. White Portland cement

White Portland cement is finding increased use in architectural concrete, and applications include both pre-cast and cast-in-place. It is commonly used as a base for making coloured concrete, because coloured pigments

TABLE 15.6. *European, British and US standards for concrete*

Type of cement	EN 197-1	"New" BS EN 197-1	"Old" BS	ASTM
Ordinary Portland	CEM I	CEM I	BS12	Type I
Rapid hardening	CEM I + suffix R	CEM I + suffix R		Type III
Sulphate resisting		BS 4027:1996	BS4027	Type V
Portland-composite	CEM II	CEM II	BS146, BS6588, BS7583	Types IS
Blastfurnace cement	CEM III	CEM III	BS146	P
Pozzolanic cement	CEM IV	CEM IV	BS6610	IP
Composite cement	CEM V	CEM V	(none)	I(PM)

are expensive, and the grey tone of ordinary Portland cement can be detrimental to the final colour.

White Portland cement is made from high-purity raw materials, which are relatively scarce. Therefore, white cement is widely traded internationally, for example 1996 statistics for the USA show that production of white cement was 615 000 tonnes, while imports from Canada, Mexico and Europe totalled 390 000 tonnes. The price of white Portland cement is approximately 2–3 times the cost of ordinary Portland cement. The high cost of white cement is due to the small scale of production, and the relatively expensive raw materials. Also, white cement is often produced at higher temperatures than ordinary Portland cement, for example 1600°C–1650°C compared with 1400°C–1500°C, so energy costs are greater.

The raw materials are white calcium carbonate (chalk, limestone or marble) and white clays, selected for their low iron contents. China clay (kaolin containing muscovite mica, fine silica and feldspar) from the southwest of England was formerly used to make white cement, but in Europe it is now more economic to use white aluminosilicates derived from other deposits such as volcanic ash or diatomaceous earth. Sometimes a blend of purified alumina and finely ground sand is used. The Fe_2O_3 content of clinker must be less than 0.5–0.3 mass%, depending on the degree of whiteness required, and whether the calcining is conducted under reducing or oxidising conditions. Whiteness can be enhanced by converting Fe^{III} to Fe^{II}, followed by rapid cooling. Mn and Cr are even more deleterious to whiteness and must be kept below about 0.03 mass% and 0.003 mass% respectively in the clinker.

Because of their high purities, the raw materials for producing white cement are quite refractory and cryolite, Na_3AlF_6, is sometimes added to increase the proportion of liquid phase, which promotes the formation of C_3S and C_2S and minimizes the presence of free CaO in the clinker. Fluxing is also helped by raising the temperature in the calciner to 1550°C–1650°C and by finely grinding the silica.

Typical composition ranges of white cement are given in Table 15.7.

If the proportion of C_3S is too low, white cement will develop strength more slowly than a normal grey Portland cement. Hamad (1995) gives an example where this is overcome by adding more clay to the initial mix to increase the proportion of C_3A in the clinker, because C_3A enhances early strength development. However, C_3A

TABLE 15.7. *Typical composition ranges of white cement*

Component	Composition range, mass%
C_3S	38–67
C_2S	12–41
C_3A	5–13
C_4AF	0.6–1
Free CaO	2.5 (typical)
K_2O, total	0.11–0.22
Na_2O, total	0.07–0.14

TABLE 15.8. *Effect of added minerals on the composition of white cement clinker*

	Hamad		Blanco-Varela
	Grey	White	
SiO_2	20.3	23.4	23.0
Al_2O_3	6.0	5.2	2.4
Fe_2O_3	3.0	0.25	0.08
CaO	63.5	65.8	69.5
MgO	1.8	0.5	0.8
C_3S	52.2*	48.5*	69.5†
C_2S	18.8*	30.5*	6.8 ± 0.7†
C_3A	10.8*	13.4*	3 ± 1†
C_4AF	9.1*	0.8*	Not reported
Initial set, minute	166–256	129–180	75–160
Final set, minute	250–470	230–245	115–264

* Calculated from chemical composition.
† Determined by XRD.

contributes little to the strength of fully-cured concrete, and a potential problem is that it gives concrete which is liable to suffer from sulphate attack.

Another method of improving the strength development properties of white PC is described by Blanco-Varela *et al.* (1993). They use less kaolin in the initial mixture, so less C_3A is formed, but they use a higher proportion of CaF_2 (5%) and $CaSO_4$ (2%) in the initial mixture to enhance mineralisation and formation of C_3S, (Table 15.8). The feed contained only 3% kaolin and the resulting clinker was composed of only 3% C_3A. However, the clinker also contained a high proportion (69.5%) of C_3S, due to the use of the mixed mineralizer, and this provided rapid setting and good strength development.

15.2.5. Portland—Pozzolana cements

Portland cement is often blended with pozzolanas, which can have several benefits such as reducing costs, reducing energy consumption, utilizing waste materials, and improving the quality of concrete. Pozzolana is a name that was originally given to a fine-grained volcanic ash from Pozzuoli in Italy. Pozzolanas are natural or artificial forms of silica, silicates or aluminosilicates that have the ability to react with calcium hydroxide and water, at ambient temperatures, to give cementitious products. This is known as the pozzolanic reaction. The cementitious products of reaction are hydrated calcium silicates and aluminosilicates. When the pozzolanic reaction is completed, any remaining lime may be slowly neutralised by reaction with carbon dioxide from the atmosphere. This slow conversion of free lime to calcium carbonate contributes to the long-term gain in strength and durability of many lime mortars. Examples of natural pozzolanas include the volcanic ash deposits in Italy and Greece, and the diatomaceous earths in Germany and Denmark. Dehydroxylated clay is an example of an artificial pozzolanas, made by burning and finely grinding clay or shale (see the Text Box on p. 434).

Pozzolanas

Many types of clay can be calcined to give highly reactive pozzalanas. **Metakaolin** is the most common of these because its precursor, kaolin, is widely available commercially in a relatively pure form. When heated to 700–900°C kaolin loses 14 mass% as hydroxyl water, to form metakaolin. The alumina and silica layers become puckered; they lose their long-range order and the powder becomes amorphous with respect to X-ray diffraction. Dehydroxylation causes the clay to become chemically reactive; in particular it is readily attacked by dilute acids and alkalis (but not water) at ambient temperatures. Another important structural change is that much of the aluminium in metakaolin becomes tetrahedrally co-ordinated. This is most evident from Al NMR spectroscopy (Justnes *et al.* 1990; Rocha & Klinowski 1990) and soft X-ray absorption spectroscopy (Roberts *et al.* 1992); indeed the existence of a characteristic tetrahedral AlO_4 resonance is often taken as evidence that metakaolin has been formed. Tetrahedral Al is thought to play an important role in the pozzolanic reactions of metakaolin.

The kinetics of dehydroxylation have been studied (Meinhold *et al.* 1994). Metakaolin can be made by flash calcining at about 1000°C or by prolonged soak calcining at lower temperatures (Dunham 1992). Salvador (1995) reports that the rate of calcining does not influence the pozzolanic properties, providing that dehydroxylation is complete and that the kaolin has not been over-calcined. Above about 800°C kaolin begins to convert to relatively inert ceramic materials such as spinel, silica and mullite.

Pozzolanas can be used in conjunction with Portland cement because calcium hydroxide (portlandite) is one of the hydration products of Portland cement, typically comprising up to 25% of the mass (and volume) of fully cured Portland cement paste. Pozzolanas continuously react with the nascent calcium hydroxide to give new cementitions compounds. This was discovered soon after the development of Portland cement; in the 1870s powdered coal ash and ground granulated slag from blast furnaces were being added to cement. It was found that this technology could reduce costs with little deleterious effect on the properties of cured concrete, but was limited to regions where natural pozzolanas were readily available (e.g. Italy and Greece), or where there was a problem in disposing of ash or blast furnace slag (e.g. UK and Germany).

Later in the 20th century certain general advantages of Portland pozzolana cements were discovered. Stanton showed in 1940 that pozzolanas could have a beneficial effect in preventing deleterious expansion of concrete due to alkali–silica reaction, ASR (Tuthill 1982). A dam was successfully built in California using a known alkali-active aggregate and a blended cement composed of Portland cement and a natural volcanic pozzolana, pumicite. Portland pozzolana blended concretes were reviewed by Malquori (1960).

In the 1960s several dams were built in the Amazon basin, using a blend of Portland cement and locally produced metakaolin. Prior laboratory experiments had shown that the metakaolin could replace up to 20% of the Portland cement and prevent the expansion that would otherwise have occurred, due to the active nature of the local aggregates (Saad *et al.* 1982; Andriolo & Sgaraboza 1985). There was no deleterious effect on the mechanical properties of the concrete. As a result of further research in the 1970s, it was found that highly reactive artificial pozzolanas such as condensed silica fume and metakaolin can significantly improve the durability of concrete, as discussed in Section 15.2.6.

15.2.6. Calcined clays as Portland cement additions

In the 1980s, pure calcined clays were evaluated as high-quality pozzolanas for high-performance (i.e. strong and durable) concrete. Early work in France showed that metakaolin improved the durability of fibreglass reinforced concrete. It reduced the loss in strength and toughness that is normally observed as alkaline solutions in the concrete react with the strands of fibreglass reinforcement. Subsequent work showed that pure metakaolin can usefully replace up to 25 mass% of the cement in Portland cement concrete, and it generally improves durability when the concrete is exposed to aggressive environments, such as salt water, dilute acids and sulphate solutions. It should be noted that this is normally true only if metakaolin is made from a pure feed material, i.e. clay containing at least 85% and preferably 90% kaolin. Metakaolin made from impure clay can be used in low-strength concrete, but it has an unacceptable effect on the mechanical strength of high performance concrete. Table 15.9 gives the chemical analyses of samples of commercially available metakolin, obtained from France, the UK and the USA and used to replace part of the Portland cement in concrete formulations. Metakaolin is X-ray amorphous, but common impurities such as mica, quartz, iron oxide, titania and feldspar are readily detected if present in concentrations of 2% or more. In addition, a pozzolanicity test can be used to assess the quality of metakaolin to be used in concrete. This test involves reacting metakaolin with an excess of calcium hydroxide in aqueous suspension at 80°C for 24 hours, and then determining the unreacted calcium hydroxide by titration.

The use of metakaolin as an addition in concrete has been reviewed (Sabir *et al.* 2001; Jones 2001). Apart from kaolin, other natural clays such as Na- and Ca- montmorillonites, mixed layer mica-smectite, and sepiolite can be converted to pozzolanas by calcining at suitable temperatures (He *et al.* 1995, 1996).

TABLE 15.9. *Chemical analyses of samples of commercially available metakaolin*

	MK (France)	MK (UK)1*	MK (UK) 2†	MK (USA)‡
Composition (mass%)				
SiO_2	55.0	55.4	51.6	51.3–52.6
Al_2O_3	37.5	40.5	41.0	43–45
Fe_2O_3	1.45	0.65	4.8	0.5–1.0
TiO_2	1.45	0.02	0.83	1.75
CaO	0.07	0.01	0.06	0.01–0.1
MgO	0.20	0.12	0.19	0.2–0.25
K_2O	1.17	2.17	0.62	0.1–0.2
Na_2O	0.2	0.13	0.24	0.2–0.3
Loss on ignition	3.0	1.0	1.0	0.7
Pozzolanic reactivity (mg CaO g^{-1} mk)	800		840	

* From a kaolin-rich primary china clay deposit in Devon, England.
† From a secondary ball clay deposit in Devon, England.
‡ Range of secondary kaolins from the USA.

15.3. Lime and hydraulic lime cements

15.3.1. Natural hydraulic lime cements

Hydraulic lime cements were first studied on a scientific basis by Smeaton, as discussed in the Introduction. They are made by calcining a natural argillaceous limestone to 800–1100°C. The rock does not melt and there is no long range diffusion of the elements. Important qualities of the limestone include the clay-limestone ratio, the fineness and intimacy of intermixing of the primary clay and limestone particles, the chemistry of the clay fraction, and the nature of impurities such as iron minerals. The cementitious product is a mixture consisting of various proportions of calcium oxide, calcined clay (for example metakaolin), anhydrous calcium silicates and calcium aluminates.

Deposits of argillaceous limestone suitable for the modern construction industry are rare, so natural hydraulic lime cements are often transported long distances and are expensive. A candidate deposit should be large, be of a suitable mineral composition, and must be consistent in quality. An argillaceous limestone near Grenobles in France is still being exploited after more than 200 years. Hydraulic cement, made by calcining this rock, is fast setting (5–60 minutes) and almost as strong as Portland cement. The compressive strength of concrete containing 400 kg of this cement per cubic metre is 60 MPa. Because of the naturally fine scale of intermixing of the carbonate and clay minerals in this deposit, there is almost complete reaction between the lime and the clay in the calciner, and the product consists of various calcium silicates and virtually no free calcium oxide. This cement is marketed by the Vicat company, under the trade name Vicalpes. Typical ranges for the chemical analysis of Vicalpes cement is given in Table 15.10 (Vicat company literature).

Since about 1990, there has been a revival of interest in lime-based mortars for new construction, as well as for repairing old buildings (Procter 2001). In new buildings, fully-cured lime based mortar is more flexible, is more permeable to water vapour, and has lower adhesive strength than a comparable Portland cement mortar. Therefore mortar is less likely to crack, less likely to trap dampness, and bricks can be more easily recycled when a structure is demolished. When used to repair old buildings, lime based mortar is compatible with the types of mortars and concrete used before 1900; the interface between the repair and the existing structure is less likely to crack and allow the ingress of water.

TABLE 15.10. *Composition range of Vicalpes hydraulic lime cement*

Oxide	Mass%
SiO_2	27–30
Al_2O_3	5–7
Fe_2O_3	3–4
CaO	48–51
MgO	3–4
SO_3	4–6
Loss on ignition	3–6

15.3.2. Lime mortars

Lime mortars are made by calcining relatively pure limestone at temperatures up to 1000°C, and slaking the resulting calcium oxide with water to give calcium hydroxide (CH). Setting and hardening is very slow because it depends mainly on the reaction of calcium hydroxide with atmospheric carbon dioxide. As discussed earlier, pozzolanas (such as volcanic ash from Italy and Greece, diatomaceous earths from Germany and Denmark, or burnt clays) have historically been blended with the lime to increase the rate of setting and curing.

More recently, there has been considerable interest in blending calcined clay with calcium hydroxide. It is possible to adjust the setting time and rate of strength development by using a relatively pure calcined clay, such as metakaolin from china clay, and by correct choice of the clay:calcium hydroxide ratio. Figure 15.1 shows the effect of metakaolin on compressive strength development in two lime cements from the UK, one being mildly hydraulic

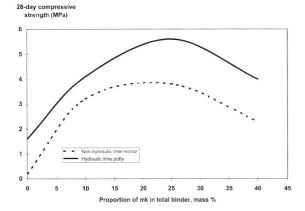

FIG. 15.1. Effect of metakaolin on 28-day strength of lime mortar.

and the other non-hydraulic (i.e. containing very little natural argillaceous material). It is seen that 10 mass% metakaolin in the binder transforms the non-hydraulic lime into a useful hydraulic lime, and increases the rate of strength development in the naturally hydraulic lime.

The pozzolanic chemical reactions between calcined clay and calcium hydroxide are complex. Even in the simplest case where pure metakaolin reacts with pure calcium hydroxide, it is only in recent years, that we have begun to understand the detailed chemistry. This has involved the use of advanced instrumental techniques (such as solid state NMR, analytical electron microscopy and thermal analysis). Experiments using trimethyl silylation, followed by chromatographic separation of polysilanes of different molecular weights, have given some insight into the silicate polymerization reactions.

Dunster et al. (1993) studied the reactions of metakaolin with calcium hydroxide. There is a rapid initial reaction, corresponding to the setting of the mk/CH mixture, followed by a slow reaction, which proceeds over a period of about 30 days. A range of products is formed, including CSH, C_2ASH_8 (gehlenite hydrate, or strätlingite) and hydrated calcium aluminates, C_4AH_{13} and C_3AH_6. Gehlenite hydrate is a well defined crystalline phase, but the composition and structure of CSH is variable.

Metakaolin, in the presence of alkali, provides a source of orthosilicate anions, SiO_4^{4-} and tetrahedrally coordinated aluminate anions, AlO_4^{5-}. It is thought that the main reaction pathway is for the orthosilicate anions to dimerize into $Si_2O_7^{6-}$ units, and these can then join together via SiO_2 or AlO_2^- bridges.

$$(\bullet O \bullet Si_2O_5 \bullet O \bullet AlO_2 \bullet O \bullet Si_2O_5 \bullet O \bullet SiO_2 \bullet O \bullet Si_2O_5 \bullet O \bullet)^{19}$$

Dunster et al. estimated that pure metakaolin reacts with up to 1.6 times its mass of calcium hydroxide. The stoichiometry of the pozzolanic reaction was found to be:

$$AS_2 + 5CH + 5H \rightarrow C_5AS_2H_5$$

where $C_5AS_2H_5$ is an average composition representing a mixture of CSH, C_4AH_{13}, C_3AH_6, and C_2ASH_8. This stoichiometry is similar to that proposed earlier by Murat (1983a, b).

15.4. Other types of cement

15.4.1. 'Inorganic polymer cements' or chemical cements

Pozzolana-lime and the Portland cements have enjoyed a remarkably long history of success, and are likely to be manufactured and used on the large scale for many years to come. However, these cements have well known disadvantages, such as the need for large-scale quarrying, high-energy requirements, and sometimes poor durability. Consequently there have been many attempts to develop cements which are less energy intensive, which utilize waste materials as feedstock, and which are less susceptible to environmental attack. They are based on the pozzolanic reaction between chemically reactive silica or aluminosilicate, and a soluble alkali such as KOH, NaOH, K_2SiO_3, or Na_2SiO_3. The pozzolana can be obtained from sources discussed earlier, such as natural siliceous deposits, fly ash, blast furnace slag, or fumed silica. Alternatively, it can be made by calcining kaolinitic soils or mine wastes containing aluminosilicate clays. Such inorganic polymer cements can be blended with Portland cement to give rapid-setting concrete, which is useful for casting into moulds and for repair work.

These blended cements are often given commercial names, so 'inorganic polymer cement' or 'chemical cement' have been proposed as generic terms. Some examples of inorganic polymer cement are discussed below. It should be noted that much of the information is available in patents applications and from internet web sites, rather than in refereed publications.

15.4.1.1. Geopolymer cement. Geopolymer cements were launched by J. Davidovits in 1978 and he has continued to promote them via the Geopolymer Institute in France (Davidovits 1994). Typically they are chemical cements made from metakaolin, potassium silicate and potassium hydroxide, together with other pozzolanas such as blast furnace slag. They are reported to be used in specialist applications such as rapid repair mortars, fire resistant sandwich panels, insulating boards and high-temperature filters. They have also been used for encapsulating toxic and radioactive waste, on account of their high binding capacity and low permeability.

It is claimed that geopolymer cements require less energy to produce, and that the production process leads to the emission of less CO_2, compared with Portland cement. There is also the possibility that geopolymer cements could be made from a wide range of waste aluminosilicates, including fly ash, clay and albite (Xu et al. 2002).

15.4.1.2. Pyrament concrete. Pyrament was introduced by Lone Star Industries in 1984 as a rapid setting mortar for repairing concrete runways at airports. The

TABLE 15.11. *Typical formulation of Pyrament*

Component	Parts per hundred (mass)
Metakaolin	6.16
Blast furnace slag	3.38
Fly ash	27.05
Silica fume	8.6
Potassium hydroxide	2.81
Portland cement	51.4
Borax	0.61
Citric acid	0.96
Sand	190

TABLE 15.12. *Formulation of typical tectoaluminate cement*

Component	Mass%
Metakaolin	17.5
Calcined siliceous clay (approx. 65 mass% silica)	33.3
Potassium silicate	14.0
Potassium hydroxide	3.5
Blast furnace slag	31.6

concrete was said to be capable of being used by taxiing jumbo jets only six hours after placement.

According to the Lone Star patent, a typical formulation contains a blend of pozzolanas, Portland cement and an alkali metal hydroxide, (Table 15.11).

This formulation was combined with 22 parts of water plus dispersing agent and showed a compressive strength of about 46 MPa after 24 hours curing at 66°C, or 18 MPa after 24 hours curing at ambient temperature.

15.4.1.3. Tectoaluminate cement. This metakaolin-based cement was patented by Holderbank in 1992. Apart from metakaolin, it consists of chemically reactive silica, fly ash activated by calcining with an alkali metal carbonate, a pozzolana such as fly ash or blast furnace slag, and optionally a small proportion of Portland cement.

It is claimed that cured tectoaluminate cement has a three-dimensional zeolitic structure, rather than the one- and two-dimensional structures of the polysilicate anions in cured Portland cement. Hence tectoaluminate cement cures faster and is claimed to be more durable.

One example of a tectoaluminate cement revealed in Holderbank's patent is given in Table 15.12. When 100 parts of this mixture were blended with 24 parts of water, the paste hardened rapidly, giving a compressive strength of 15 MPa after 24 hours and 78 MPa after 28 days.

15.4.1.4. Earth ceramics and calcium silicate bricks. Earth ceramics are made by hydrothermal treatment of a mixture of clay, quartz and calcium hydroxide at 200°C, and have been developed and promoted be the INAX Corporation, Japan. According to INAX, when compared with conventional ceramics, they allow use of lower quality raw materials and require less energy. Comparison of the properties of paving slabs, made from an earth ceramic, with ceramic and cement bricks are shown in the Table 15.13.

It is seen that there would be considerable saving of energy if, in future, earth ceramics were used for sanitaryware (shower trays, toilet bowls, sinks etc.) or to replace other high-quality ceramics.

Earth ceramics appear to be a development of calcium silicate brick technology. Calcium silicate bricks are made by blending quartz sand with approximately 10 mass% of ground quartz flour, 1–5 mass% of calcium hydroxide, plus a little water. The mixture is compressed in a mould and, after demoulding, the bricks are autoclaved in saturated steam at temperatures up to 200°C. The lime reacts with the quartz to give a calcium silicate binding phase with the tobermorite structure. The effect of adding aluminosilicate clays to the mixture is to generate a high proportion of hydrogarnite in the binder.

15.4.2. Clays in calcium aluminate cements

Calcium aluminate cement (CAC) is made by calcining a mixture of bauxite and limestone at 1450°C. The standard cements typically use ferruginous bauxite with up to 20% Fe_2O_3, but purified bauxite can also used to make white CAC. CAC is mainly used for specialist applications where rapid strength gain is required, and also used as a component of refractory concrete.

There is evidence (mainly from the patent literature) that CAC is sometimes blended with various clays, and with Portland cement, in proprietary formulations for certain specialist applications. These include mortars for screeds, renders, grouts and repair patches. The mineral

TABLE 15.13. *Properties of Earth Ceramic*

	Energy of production, (MJm^{-3})	Flexural strength, (MPa)	Compressive strength, (MPa*)	Mohs hardness	Water absorption, (mass%)
Earth Ceramic paver	2.7	6.5	54	3.5	13
Ceramic brick	16	5.3	64	6	9
Cement brick*	2.3	5.6	81	5.5	5
Sanitaryware	40				

* Energy consumed in producing a concrete containing 12 mass% Portland cement.

additions would help to reduce shrinkage, reduce setting times and reduce costs. In refractory mortars, metakaolin can replace part of the CAC and reduce the water demand (US Patent Number 5 976 240). This is said to improve refractory properties, because a reduction in the CAC content means that the concentration of calcium ions (which promote fluxing) in the mortar is also reduced.

CAC is seldom used in large-scale engineering applications, because there is a potential danger that concrete made from CAC can lose strength over a period of months or years. This is due to an internal phase change known as 'conversion', which weakens the cement structure. Concrete made from CAC must be prepared and used under very carefully controlled conditions, to prevent conversion, if long-term strength and durability are important. CAC consists mainly of CA ($CaO \cdot Al_2O_3$) and CA_2 ($CaO \cdot 2Al_2O_3$), together with minor amounts of $C_{12}A_7$. When mixed with water, CAC remains workable for a period of 1–3 hours (i.e. similar to Portland cement). After the cement has set, the hydration product CAH_{10} is formed at ambient temperature, resulting in a very rapid build up of strength. C_2AH_8 plus AH_3 (alumina trihydrate gel) are also formed over a longer period of time, especially if the cement is cured at temperatures above 25°–30°C. Unfortunately, the hexagonal phases CAH_{10} and (especially) C_2AH_8 are not stable and 'convert' to a mixture of cubic C_3AH_6 and gibbsite if the temperature is allowed to rise above about 40°C. C_3AH_6 is denser than its precursors so there is a consequent increase in porosity and decrease in strength. Conversion is normally minimized by controlling the temperature of curing, and by keeping the water/cement ratio below 0.4.

It has been reported that metakaolin can prevent the conversion reaction (Majumdar & Singh 1992, and US Patent 5 624 489). Other pozzolanic additions such as blast furnace slag and micro silica are also effective. These silica-rich additions combine with C_2AH_8 to give C_2ASH_8 (gehlenite hydrate, also known as strätlingite). Gehlenite hydrate is stable and there is no change in density of the solid phases so no porosity is generated. In spite of the potential benefits of using calcined clay with CAC to prevent conversion, this approach is still problematic for major structural applications. Even with pozzolanic additions, there is still a danger of conversion if the temperature is allowed to rise above about 45°C during the first few hours of curing.

15.4.3. Gypsum–Portland–pozzolana cements

Gypsum-based cements are widely used for internal applications such as wall boards, renders and mouldings. External applications are rare because gypsum is soluble in water to the extent of 1.4 kg m^{-3} (as sulphate). When gypsum is exposed to water, the interlocking crystal structure is degraded, and there is a severe loss in strength. There have been many attempts to produce a durable form of gypsum cement because it promises to be a lower cost (and lower energy) alternative to Portland cement. These attempts usually involve blending hemihydrate with another cementitious compound such as Portland cement. The aim is to sheath each gypsum crystal with a water resistant layer—in the case of Portland cement this would be calcium silicate hydrate. However, simple blends of hemihydrate with Portland cement are unsatisfactory because the calcium aluminate component in Portland cement reacts with calcium sulphate to form ettringite. Ettringite is heavily hydrated and has a low density (1730 kg m^{-3}), so if it is formed within concrete that has already hardened it can cause the concrete to swell and eventually crack:

$$CS + C_3A \rightarrow C_6AS_3H_{32}$$
$$[3CaO \cdot Al_2O_3 \cdot (CaO \cdot SO_3)_3 \cdot 32H_2O)].$$

It has been known for many years that the durability of gypsum/Portland cement based mortars is improved by the addition of pozzolanas such as fly ash and blast furnace slag. More recently, it was found that highly reactive pozzolanas such as silica fume, metakaolin and certain natural pozzolanas give even better improvement; gypsum/Portland/reactive pozzolana concrete can retain its strength even after immersing in water for long periods of time.

In patent US 5 958 131 the effect of adding Portland cement and metakaolin to hemihydrate are described. The ingredients were dry-blended, mixed with water and cast into cylindrical moulds. After 24 hours the samples were de-moulded and stored in water for up to 96 days.

The results in Table 15.14 show that addition of Portland cement does not prevent degradation of gypsum mortar in water. By contrast, addition of Portland cement together with metakaolin gives a gypsum mortar which increases in strength over a 96-day period of immersion.

TABLE 15.14. *Effect of Portland cement and metakaolin on the durability of gypsum cement*

				Compressive strength MPa	
Mass hemihydrate	Mass Portland cement	Mass metakaolin	Mass water	28 days immersion	96 days immersion
100	0	0	50	4.7	2.0
90	10	0	50	7.3	5.5
80	20	0	50	8.0	5.5
80	18	2	50	7.7	9.6
80	15	5	50	16	(not meas.)
50	37.5	12.5	50	21.4	22.3

The beneficial effect increases with the Portland cement/hemihydrate ratio and with the metakaolin/Portland cement ratio. This might be because ettringite is formed early in the curing process, rather than after the cement has hardened.

The beneficial effects of metakaolin have also been reported in patent US 6 241 815 B1, which also shows that metakaolin is more effective than silica fume as an addition to gypsum mortar.

In an interesting development, Stav has claimed (patent US 6 197 107) that gypsum mortars of exceptional strength and durability can be made by conducting the curing process under carefully specified conditions. Stav prepared hemihydrate/Portland cement/metakaolin mortars and cured them without allowing any evaporation for a period of up to seven days. He obtained compressive strengths up to 100 MPa, and observed that the matrix appeared to become predominantly amorphous during eight months immersion in water. The intensities of the XRD lines for gypsum and ettringite declined by factors of six and two respectively over this period.

It is too early to predict whether gypsum/Portland cement/metakaolin blends will be commercially successful, but it is evident that there is considerable potential for improving our scientific and technical knowledge in this area. Recent advances in instrumentation have begun to increase rapidly our understanding of the fundamental structure and properties of cementitious materials. The effect this new knowledge will have on the future direction of cement technology is uncertain at present. However, the potential for reducing the high energy cost of cement, by using alternative raw materials such as gypsum and calcined clays, and by incorporating more waste materials into high quality cementitious materials is attractive.

15.5. Use of clay in specialist applications

15.5.1. Lightweight aggregates

Certain clays are used to make lightweight aggregate for concrete. LECA (light weight expanded clay aggregate) is currently manufactured in Norway and Denmark (see http://www.leca.dk). A LECA clay was exploited until recently in Ongar, Essex (southeast England), but is now not commercially available. Deposits of suitable clay are quite rare. Because of high transportation costs the deposit must be near the point of use, and the clay must be free of rock particles and sand. The pellet of clay must begin to melt and form a plastic mass when it is tumbled in a rotary calciner, and at the same temperature a source of gas must be activated. The gas required for foaming is often provided by the decomposition of finely divided ferric iron oxides at 1150°C

$$6\ Fe_2O_3 \rightarrow 4\ Fe_3O_4 + O_2.$$

Other methods of generating gas include the decomposition of carbonate minerals, or the gasification of organic material.

Because clay-based lightweight aggregate is calcined above 1000°C, it is expensive in terms of energy consumption and CO_2 emission. Therefore many alternative, non-clay, materials are used, where locally available. Examples include natural pumice and expanded vermiculite and perlite. Synthetic lightweight aggregate can be manufactured from fly ash (see below), or by crushing aerated (foam) concrete.

There is an extensive technology associated with low density concrete, either made by direct foaming of freshly mixed concrete, or by using lightweight aggregate. They are the subject of various specifications such as ACI 213R-87 and ASTM C 332-87 (Reapproved 1991).

15.5.2. Rheology control

15.5.2.1. Ready mixed concrete. When mortar or concrete has to be poured into a mould, the rheological properties of the slurry is of critical importance in determining the mechanical and visual properties of the final cured product. For example, engineering concrete may be poured into a mould containing a complex network of reinforcing steel bars (rebars). It is essential that the concrete is able to flow rapidly between and round each rebar, thus completely filling the mould. Once in place, the liquid concrete must become more viscous in a relatively short time, so that sand and aggregate do not settle under gravity. This is particularly important in the area beneath a horizontal rebar; if settling does occur here, it could result in regions where the concrete is extremely porous and of very low strength.

Certain clays such as bentonite and attapulgite can provide the required degree of thixotropy to the concrete slurry, although long chain water soluble polymers are also used for this purpose. There is an extensive patent literature on the use of clay minerals to impart thixotropy, but it is not clear whether the technology is widely used at present. In the example US Patent Number 4 673 437, the use of small quantities (about 0.1–0.3 kg m^{-3} concrete) of bentonite or attapulgite to improve the casting behaviour of pre-cast concrete blocks, and the green strength of the blocks. In another application, it is claimed that improved lightweight (aerated or cellular) concrete can be made by using bentonite or attapulgite to stabilize the wet foam structure (US Patent Number 4 900 359). Again, the clay is used at low concentration, up to 0.6 kg m^{-3} concrete.

15.5.2.2. Cement grouts. Two of the major applications of cement based grouts are consolidating soils prior to building work, and constructing underground cut off walls to prevent the migration of undesirable fluids through soil, and these are discussed in more detail below. Other applications include preventing subsidence in existing buildings, filling underground cavities, and in the oil well industry for protecting pipes and repairing sub-sea structures. Bentonite is sometimes used to modify the rheological properties of cement grouts and make them more cohesive. When properly activated (i.e. dispersed), bentonite increases the viscosity of the aqueous phase and prevents the particles of cement from settling under gravity while it is being placed and during setting. This is particularly important in the weaker grouts which have a high water/cement ratio.

In permeation grouting, a relatively weak grout is injected via boreholes into porous soil formations in order to stabilize them when constructing, for example, dams and tunnels. For injection to be successful, the grout should fill all the cracks wider than, say 300 μm, so it is important to carefully control both the particle size of the suspended solids and the slurry rheology. Predispersed sodium bentonite is used to impart the required thixotropy and prevent bleeding, and dispersing agents are added to prevent the grout flocculating.

When grouts are used in large projects, detailed formulations are specifically developed and tailored to the engineering requirements of the project, taking into account the nature of the soil to be grouted and the locally available ingredients. For example in the construction of the Berke dam in Turkey in the early 1990s, a strong grout was used to consolidate the surrounding soil, and Akman & Mutlu (1999) report that the optimum grout formulation (compressive strength within the required range of 10–30 MPa) was:

Portland cement	750 kgm^{-3} grout
Water	600 kgm^{-3} grout
Fine sand	375 kgm^{-3} grout
Na-bentonite	3.75 kgm^{-3} grout
Dispersant	9.0 kgm^{-3} grout

Another example is the Jubilee Line extension of the London underground system. In this project, Haimoni (1999) describes how 11 000 m^3 of permeation grouts were injected under central London to enable shafts and tunnels to be safely constructed in and below the water bearing permeable gravel. Both Portland cement/bentonite grouts and sodium silicate/ester grouts were used in this application. Sometimes the special properties of more exotic chemical grouts, based on soluble silicates or organic resins, are justified on performance criteria, but in general the advantages of Portland cement/bentonite grouts are:

- precise adjustment of rheological properties;
- raw materials are widely available at relatively low cost;
- non-toxic;
- wide range of final cured strength available.

Cement/bentonite-based cut-off walls have been developed since about 1950 to reduce the permeability of undesirable fluids through soil (Jefferis 1997). The cement confers a degree of rigidity (high strength is not normally required for an underground cut off wall) and the bentonite reduces the permeability of pore water.

In a typical example, a landfill site is surrounded by a deep trench, 0.6–0.8 m wide, which is then filled with a cement/bentonite slurry in order to prevent toxic fluids from spreading sideways from the site. A typical formulation for the cement/bentonite slurry is:

Cement slurry (blended Portland/ground granulated blast furnace slag)	100–350 kgm^{-3}
Sodium bentonite	2–6 mass% of the water

The important parameters for a cut off wall are:

- low permeability, typically 1×10^{-9}–1×10^{-8} ms^{-1};
- ability to withstand a confined strain of >5% without cracking, typically under a stress of 50–300 kPa;
- Relatively low minimum unconfined compressive strength—typical values range 50 kPa–2 MPa.

15.5.3. Immobilization of toxic wastes

Toxic pollutants are often removed from the environment by collecting them as a concentrate, which must itself be disposed of safely. This can be achieved either by destroying the concentrated waste, via chemical reactions such as oxidation, or by placing it in secure long-term storage. Particularly dangerous waste is often stabilized in blocks of cement (Conner & Hoeffner 1998), because this reduces future environmental risks arising from physical damage to the storage facility, or leaching of the waste into underground aquifers. Liquid wastes, containing soluble salts or organic compounds, are rendered less mobile if they can be trapped in the highly tortuous pore network of cured cement paste. It has been proposed that retention can be improved if the liquid waste is first absorbed by a suitable clay mineral, since the effect of the fine particles is to increase the tortuousity of the pore network. Much research has been carried out, and many patents granted, but using clay greatly increases the cost of disposal. It is not easy to determine the extent that this technology is used in practice.

Some examples of disposal technology are as follows:

- The plant which treats the effluent from petroleum refineries produces a voluminous sludge containing high levels of hydrocarbons and other toxic organic and metal compounds. It has been suggested (Saikia et al. 2002) that the sludge can be co-calcined with kaolin (30 dry sludge: 70 kaolin) to give a pozzolana which can replace up to 20% Portland cement in concrete.
- Hazardous radioactive isotopes such as ^{137}Cs and ^{90}Sr can be immobilized in concrete made from Portland cement or alkali-activated ground granulated blast furnace slag. There is evidence that leaching is reduced if aluminous pozzolanas such as metakaolin are incorporated in the cement mixture (Guangren et al. 2002). The Al-rich matrix shows higher cation exchange capacity, and improved fixing of Cs$^+$ and Sr^{2+}.
- The effect of (uncalcined) kaolin on the rate of leaching of low-level liquid radioactive waste from concrete has been investigated (Osmanlioglu 2002). It was found that leaching of Cs-135, Cs-134 and Co-60

was significantly reduced as the Portland cement was progressively replaced by kaolin, up to 20 mass% replacement. The optimum replacement level was 5 mass%, when leaching rate is reduced by 50%.

- Conner (1995) reports trials which compared the effects of three additives (activated carbon, organoclay, and rubber particles) on the stabilization of organic compounds in cement. No one additive worked best for all the organic compounds tested. For example organoclay (montmorillionite rendered organophilic by cation exchange with quaternary ammonium ions) was effective in reducing emissions of specific volatile organics, such as 1,1,1-trichloroethane and cyclohexanone, but was ineffective with acetone and methanol.
- A Canadian study (Nehdi & Summer 2003) showed that waste latex paint can be disposed of by blending with concrete, to the extent of 125 kg paint per cubic metre of concrete. Many properties of the concrete were improved by the polymer latex and fine extenders (e.g. titanium dioxide, calcined clay and hydrous clay) present in the paint. In a full-scale trial, a concrete walk way was successfully constructed.

15.6. Recycled clays in concrete

15.6.1. Introduction

There is increasing pressure, based on economic, environmental and energy considerations, to recycle as much as possible. This is particularly applicable to the manufacture and transportation of construction materials, which in the UK is estimated to use 25% of the energy used by industry, while the UK construction industry produces 20% of all UK wastes (Hewlett 2002). Industrial waste includes end-of-life material, such as demolition waste, or by-products such as coal ash and waste from the refining of alumina and ilmenite ($FeTiO_3$). Many industrial waste products contain clay, or were originally made from clay, and some of these can be recycled into cement and concrete. Recycling reduces the amount of quarried rock, and less waste material goes to landfill. Also there can be savings in energy and reduced CO_2 emissions, especially if recycled material replaces limestone.

However, there are well known problems associated with recycling. The construction industry itself tends to have a conservative attitude to change; it has relatively low profit margins, and is sensitive to the effects of economic cycles. Logistical problems can arise simply because an extra (i.e. recycled) ingredient must be added, for example to the cement kiln or concrete batching plant. The extra ingredient requires extra resources in the form of administration, storage, weighing equipment, quality control, etc. The production of concrete is a chemical process, so the chemical composition of all the raw materials must be known and consistent. Recycled waste materials, which are nominally the same but from different sources, can show wide variations in chemistry and mineralogy. This is often a very serious problem in using waste materials in cement and concrete. In addition, many wastes contain potentially toxic components, which might also have a deleterious effect on the properties of concrete, examples being heavy metal ions, toxic organic compounds and chlorides.

It should be noted that, with the exception of fly ash from coal-burning power generating plants and calcium silicate slag from blast furnaces, a relatively small proportion of mineral wastes is recycled into cement and concrete. This is in spite of considerable research and development, directed at overcoming the technical and economic difficulties outlined above. Some examples are given below of the way waste materials which contain clay could be recycled into the production of cement and concrete.

15.6.2. Ash from municipal solid waste (MSW)

The UK produces approximately 450 million tonnes of waste per year, of which 36–40 million tonnes is household solid waste. Over the last 100 years, many publications and patents have discussed methods of using ash from incinerated MSW in cement production. In an interesting recent development, it is proposed that MSW is treated by first removing metal, then drying and shredding (Sikalidis *et al.* 2002). The shredded material is separated into 'heavy' and 'light' fractions. The heavy fraction is mainly stone, brick, ceramics, soil, glass etc., and the light fraction is mainly organic. The heavy fraction is introduced separately into the 1100°C zone of a specially designed cement kiln, which is also fed with the conventional blend of limestone, clay and iron oxides. The light fraction is mixed with petroleum coke and fed to a jet burner for heating the kiln. All the clay minerals present in the various components of MSW (ceramics, bricks, paper, plastics, paint etc.) are recycled into cement production. Adequate cleaning of the flue gases is necessary, but formation of dioxins is minimal because of the high temperature in the kiln.

15.6.3. Ash from industrial processes

Many industrial processes involve the generation of an ash by-product. If clays and calcium minerals are present, they can react together to give chemically reactive cementitious materials in the ash. Three examples are discussed below: coal ash, waste paper ash and oil shale residue. These products are available in large quantities, and have potential for significantly reducing the environmental impact of Portland cement manufacture.

Coal contains considerable quantities of clay, together with other minerals. Spears (2000) gives analytical data for coal used in UK power stations. The non-combustible content is very variable, with a minimum value of 2 mass% and an average of about 16%. It is typically composed of 75 mass% clay, 10% quartz and 15% pyrite plus carbonates. The clays are mainly detrital in origin,

TABLE 15.15. *Chemical composition of ash from typical coals*

	Composition mass%	
Loss on ignition (mainly carbon)	0.3–10	Depends on efficiency of burner
SiO_2	21–56	
CaO	3–30	Can be up to 60 mass%
Fe_2O_3	2–15	Can be up to 20 mass%
Al_2O_3	5–30	
$Na_2O + K_2O$	0.2–1	

and comprise kaolinite, illite, smectite, chlorite and mixed layer minerals. The chemical composition of ash from typical coals is shown in Table 15.15.

Scheetz & Earle (1998) have reviewed the utilization of coal ash, which they describe as a gigascale material, only 27% of which is re-utilized and much of the rest being landfilled. After combustion in power stations, coal forms various types of ash: for example pulverized coal furnaces generate fly ash and bottom ash, while cyclone furnaces generate bottom ash and boiler slag. The quantities of each, generated in the USA in 1999 were 63, 17 and 3 million tonnes respectively (Naik 2002). In addition, 25 million tonnes of an ash/calcium salt mixture was produced by the flue gas desulphurization process. Worldwide it is estimated that 600 million tonnes of coal ash was produced in 2000, compared with 1.4 billion tonnes of Portland cement. Only 10% of this coal ash was used in concrete, so there is potential for much greater use, if the various barriers and problems can be overcome, in particular the highly variable chemical compositions of fly ash from different power stations.

Fly ash is normally in the form of fused glass spheres, 1–100 µm in diameter. It can be used either as a raw material in the production of clinker, or as a pozzolanic material as a partial replacement for cement. As a pozzolana, it can be interground with clinker, or blended with the finished cement. Fly ash can replace up to 50% of the Portland cement in high-quality concrete, but proper quality control is essential. High levels of C, MgO, K_2O, Na_2O and Cl can have deleterious effects on the properties of the concrete. Fly ash should have a high silica content and low calcium content for pozzolanic activity. Concrete containing fly ash gains strength more slowly than conventional concrete, and this is a problem in many applications. There is evidence that the performance of blended Portland/fly ash cement can be enhanced by mixing it with minor proportions of highly reactive pozzolanas, such as metakaolin, thus enabling more fly ash to be recycled (Bai 2002).

Fly ash can also be used as an aggregate, replacing sand in low strength concrete for backfilling etc., or for manufacturing lightweight aggregate. Bottom ash and furnace slag can be used as aggregate in low strength architectural concrete. Artificial aggregate can be made by calcining an intimate 50:50 blend of fly ash and clay and is suitable for making concrete in regions where natural aggregate is scarce (Zakaria & Cabrera 1996).

Cenospheres are hollow glassy particles which occur in fly ash and are separated commercially by a flotation process. Cenospheres can be used to make lightweight concrete. In one example, some of the sand was replaced with cenospheres (25 vol% cenospheres in the concrete). The density of the concrete was reduced from 2300 to 1800 kg m^{-3}, but there were corresponding reductions in compressive strength and fracture toughness (McBride & Shukla 2002).

Another type of ash derives from the combustion of waste sludge generated during the recycling of paper. In Europe and the USA, cellulose fibres are separated and cleaned for recycling, and the ink and mineral fillers constitute a waste sludge. The minerals are mainly kaolin and calcium carbonate, plus some talc. The sludge can be fed into a cement kiln along with the conventional raw materials. Alternatively, it is possible to incinerate the sludge under controlled conditions of combustion to give a pozzolanic ash, which is suitable for blending with Portland cement (Pera & Amrouz 1988; O'Farrell et al. 2002). This is an emerging technology, but is expected to gain in importance as more paper is recycled.

Oil shale contains a significant quantity of organic material in a matrix of fine inorganic minerals. The inorganic component of an American oil shale from Green River has been reported to consist of about 50% carbonate minerals, 15% clay and 12% quartz. The valuable organic substances are decomposed and distilled off by heating to about 800°C. Burnt oil shale is the inorganic residue, and has significant cementitious properties, due to the formation of reactive calcium (alumino)silicates and pozzolanas. It was reported (Burge 1996) that an excellent rapid-hardening concrete was made using a cement blended from 50 parts of burnt oil shale, 33 parts Portland cement, and 12 parts of calcium aluminate, plus aluminium hydroxide and organic additions. The concrete was used in a large trial to repair the runway at Zurich airport in the early 1990s.

15.6.4. Waste bricks

Clay bricks have been crushed and used as an ingredient in concrete since ancient times. Traditionally, clay bricks are ground to a powder and used as a pozzolanic addition to lime or more recently as a partial replacement for Portland cement. Examples are sarooj, from Oman (Hago et al. 2002), homra, which is burnt silt from the river Nile (El-Didamony et al. 2000), and surkhi from India (Sen Gupta 1992). The pozzolanic activity is due to fine amorphous silica and clay minerals, which are present in the raw material used for brick making (Table 15.16.)

Modern bricks, when powdered, vary widely in pozzolanic behaviour and they are generally unsuitable for the cement industry. As discussed by Baronio & Binda (1997), not only can the raw materials contain low proportions of clay mineral, but also modern bricks are often fired above 900°C, which converts any reactive components into relatively inert products such as spinel and mullite. However, recently there has been a re-evaluation of the use

TABLE 15.16. *Compositions of traditional pozzolanic bricks*

	Sarooj (unburnt)	Homra (burnt)	Surkhi (specification for unburnt clay)
SiO_2	33.5	74.8	$\geq 40\%$
Al_2O_3	7.1	14.0	$SiO_2 + Al_2O_3 +$
Fe_2O_3	2.5	5.0	$Fe_2O_3 \geq 70\%$
CaO	22.7	1.25	$\leq 10\%$
MgO	8.4	1.3	$\leq 3\%$
SO_3	1.6	1.8	$\leq 3\%$
LOI	23.6	–	$\leq 10\%$

of factory waste brick as pozzolanic additions, because of the potential economic and environmental benefits. In a study funded by the European Union, Wild *et al.* (1999) and Golaszewski *et al.* (1999) report that current brick production in the European Union is about 30 million tonnes per year, and that the amount of waste varies from factory to factory within the range 2–5%. They found that waste brick from some factories can be ground and used in concrete as a partial replacement for Portland cement, at levels up to 30%. There is a significant reduction in strength during the early stages of curing (up to 90 days), but at longer times strength approaches that of the standard concrete. It is important to note that the waste from some factories has no pozzolanic activity, because it consists mainly of glassy calcium-rich phases. It cannot be used to replace cement without serious reductions in the strength and durability of the concrete. For this reason, it is considered important that the waste from each factory should be considered separately in this application.

From the discussion above, it is evident that bricks obtained from demolition waste are not generally suitable for use as a pozzolanic addition. Instead they should be considered as a potential aggregate in concrete (see Section 15.6.6).

15.6.5. Waste cracking catalyst

Zeolitic cracking catalysts are composed of fine particles of zeolite which are sintered into porous pellets, with diameters typically in the range 50–150 µm. Clays may be used in synthesising the original zeolite, and are often used as the binder in the sintering process. The zeolite pellets are used in fluidized bed and recirculating reactors for cracking crude oils into products of lower molecular weight. During the process, significant erosion occurs, so the waste (or spent) catalyst contains a variable proportion of fine particles. Worldwide, about 400 000 tonnes of waste catalyst is produced each year, and has potential, either to replace some of the fine sand in concrete or as a pozzolanic addition (replacing some of the cement).

Pacewska *et al.* (2002) have shown that spent catalyst (mean particle size 30 µm) can replace 20% of the sand in mortar, and this is beneficial to strength, because of the pozzolanic reactivity of the catalyst. Su *et al.* (2001) found that spent catalyst (particle size 10–130 µm) can replace up to 10% cement in concrete, but at higher replacement levels strength suffered because of the high water absorption of the catalyst, and the resultant poor packing of the mortar mixture. Wu *et al.* (2003) confirmed these conclusions, using fine spent catalyst with particle size in the range 0.5–5 µm. In summary, the high water absorption of waste catalyst is a problem, but waste catalyst can replace some of the sand in concrete, and there is evidence that it can replace some of the cement if it is ground to a mean diameter of 20 µm (Payá *et al.* 1999).

15.6.6. Building rubble as recycled aggregate

Building rubble has long been recycled as aggregate in fresh concrete, especially in low grade concrete. Until recently it has rarely been technically or economically viable to use recycled building rubble in high quality concrete. One problem is the highly variable nature of building rubble, which is composed of different aggregates, sand, bricks, tiles and cement paste, all of which may have originally contained clay minerals.

Another problem arises from the presence of impurities in the rubble (e.g. metal, asphalt, wood, plastic and paper), which may have deleterious effects on the quality of concrete made from rubble. From the calculated lifespan of buildings etc., Kreijer (1981) predicted that the amount of concrete building rubble would increase exponentially during 2000–2020, and that this would present economic and environmental problems, which could only be overcome by more recycling. A NATO Conference was organised to discuss 'Adhesion Problems in the Recycling of Concrete' and the proceedings were edited by Kreijer (1981). More recently, environmental and price pressures, associated with quarrying fresh aggregate and disposing of waste concrete in landfills, have forced industry to re-use more waste concrete (Desmyter *et al.* 1999). Waste concrete is often crushed to rubble and sorted, and used for site levelling and as a sub-base.

The value of many types of rubble is enhanced by improved sorting, including the separation of fines. The fine fraction of building rubble is particularly deleterious in concrete because of its high water absorption. In general, it has been found that carefully crushed and sorted building rubble can replace up to 60% of the aggregate in low-to-medium strength concrete, i.e. concrete with a compressive strength less than 50 MPa (Knights 1999; Chen *et al.* 2003). High strength concrete suffers an unacceptable loss in compressive strength when recycled aggregate is incorporated. Speare & Ben-Othman (1993) examined the durability of concrete, made from recycled concrete aggregate, by exposing it to different aggressive environments. They concluded that durability did not suffer as a result of using recycled aggregate, provided that the concrete was formulated to have a compressive strength greater than about 30 MPa (see Text Box on p. 444).

As discussed above, the fine fraction of building rubble is undesirable in aggregate. However there is evidence that, if it contains a high proportion of cement paste, it can

Concrete recycling

Detailed reviews of recycling in the concrete industry are presented in the book: *Concrete Technology for a Sustainable Development in the 21st Century*, E & FN Spon edited by Gjørv and Sakai (2000). At a conference at the University of Dundee, eighty papers were presented on the same subject, and are available as a volume of proceedings: *Sustainable Concrete Construction*, edited by Dhir, Dyer and Halliday, Thomas Telford (2002). In summary, much work is being carried out with the aim of recycling more building rubble into concrete, and there has been some success so far, but problems of consistency and quality remain.

be ground to a powder which has cementitious properties. Cement paste, even if it is well aged, may not always be fully hydrated and may react with water to give new CSH phases (Katz 2003). Grant & Glasser (2002) suggest that the portlandite in recycled cement paste could be used to activate pozzolanas, and also that the CSH could be partially re-activated by heating to 200–300°C.

15.6.7. Dredging spoils

Dredging spoils are sometimes classified as toxic waste, and must be disposed of accordingly. For example, dredged spoil from the Venice lagoon contains heavy metal ions (mainly lead and zinc) and organic pollutants. Various disposal methods have been evaluated, including melting it by adding 20% waste soda-lime glass as a flux (Bernstein et al. 2002). Dredged spoil from New York harbour is fine (typically 50 mass% < 7 μm), and contains Pb, Cd and PCB's, which are all mainly associated with the clay fraction. It has been suggested (Millrath et al. 2002; Millrath 2003) that the dredged spoil could be used as a filler in concrete, although some form of pre-treatment might be preferable, e.g. washing or calcining. Trials showed that Pb and Cd could be effectively immobilised in concrete, even if the spoil was untreated.

15.6.8. Other waste materials

Numerous other waste products derived from, or containing, clays have been evaluated as feed material to cement kilns, as partial replacement for cement in concrete, or as aggregate. They are relatively minor in terms of tonnes, but may be important locally in terms of environmental benefit, and may contribute in future to reducing the amount of waste material deposited in landfills. Some examples are given below.

- Dealuminized kaolin, which is the silica-rich residue from acid-extracting aluminium sulphate and ferric aluminium sulphate from clay (Mostafa et al. 2001a,b).
- Ash, from burning sewage sludge (Monzó et al. 1999).
- Residues from the refining of crude bauxite and ilmenite ores. Red mud is an iron-rich by-product from the Bayer process, for obtaining pure alumina from bauxite ore. About 35 million tonnes of red mud is produced annually. Although the clay content of red mud is low, it can be used to formulate pozzolanic cements for the construction industry, either without thermal treatment (Gordon et al. 1996), or after activation by calcining in the range 600–800°C (Pera et al. 1997). Various problems, which prevent this being used on the large scale, include the variability of the chemical and mineralogical composition of red mud, and its dark red colour. Ilmenite fines are obtained as a dilute slurry when crude ilmenite ore is concentrated by removing impurities, mostly composed of feldspars, micas, illites and chlorites. The fines are dark in colour, and are usually mixed with condensed fume from the electric arc furnace in which the ilmenite concentrate is reduced to TiO_2. Cyr et al. (2000) have shown that uncalcined ilmenite fines can replace up to 10 mass% Portland cement in concrete, with no loss in strength. The ilmenite is preferably used as the unfiltered slurry, because filtration was found to agglomerate the fine particles, which were then difficult to redisperse fully in concrete.

15.7. Deleterious effects of clays in concrete

As discussed earlier, clays such as montmorillonite, kaolin and metakaolin can be beneficial additions which enhance certain properties of concrete. However, this is with the provisos that the clays are well characterised, are used at precisely controlled addition rates, and that the other components of the concrete (e.g. superplasticiser dose) are optimised to take account of the high surface area of clay minerals. However, clay which occurs as a natural impurity in sand and aggregate is of variable quality and quantity, and therefore can have a deleterious effect on the strength and abrasion resistance of concrete, even if the formulation is modified to take account of its presence.

Many aggregates are washed in order to remove impurity clay minerals. In most cases the washings, which are a dilute suspension of clay, constitute a disposal problem, but occasionally it is possible to recover commercially valuable kaolin as a by product. For example, high-quality white kaolin is recovered by washing the sand deposits at Ione in California. The by-product kaolin is sold as mineral filler to the paper industry, and also converted into metakaolin and used commercially as a pozzolana in concrete.

Clay impurities in aggregate have three main deleterious effects: they modify the rheological properties of wet concrete, leading to higher water/cement ratios, higher superplasticiser demand, and weaker concrete; in cured concrete, aggregated masses of clay can provide regions of porosity and mechanical weakness, especially if they accumulate at the interfacial zone between the cement paste and aggregate; finally, clay can increase the shrinkage which normally occurs while concrete cures and dries, and this can lead to increased crack formation.

Table 15.17 compares the surface areas of typical clays with sand particles of various sizes, and shows the gulf in surface area between fine sand (20 m² kg⁻¹) and the coarsest clay (10 000 m² kg⁻¹).

A methylene blue (MEB) dye test is used to quantify the amount of clay minerals in building sand, and a draft European Standard, prEN 933–9, includes this test. The proposed specification imposes an upper limit to MEB adsorption of 1.0 g dye kg⁻¹ sand. This can be compared to montmorillonite, which typically adsorbs 200 g MEB kg⁻¹, and kaolin, which adsorbs 2 g kg⁻¹.

Ryle & Martin (1999) have evaluated a wide range of natural sands using the proposed Standard, and then measured the 28-day cube compressive strengths of concrete made from each of the sands. The MEB adsorption of the sands varied from zero to 5.2 g dye kg⁻¹ sand, and cube strength varied from 50 MPa (low MEB adsorption) to 20 MPa (MEB adsorption above 4 g kg⁻¹). Note that the highest MEB adsorption corresponds to a montmorilonite/sand mass ratio of about 2.5%; the effect of montmorillonite is to increase the amount of water required to make a workable concrete, and this decreased the cured strength by 60%. A high water/cement ratio also increases the drying shrinkage and hence the probability of crack formation.

Fried et al. (2000) investigated brick houses in Gloucester, UK, built using mortar made from clayey sand. The mortar had crumbled away and daylight was visible through many of the joints. They recovered samples of the 20-year old mortar, dissolved the cement fraction in acid, and tested the remaining sand with MEB. Results showed an MEB absorption of 5.6 g dye kg⁻¹ sand, indicating that the original sand had a high proportion of swelling clay, and this severely reduced the durability of the mortar, due to low strength and high drying shrinkage.

It is important to note that fine inert particles (i.e. < 100 μm in diameter) are not *per se* deleterious to the properties of concrete. They can improve the packing volume and reduce the cement demand for a given concrete strength. Finely ground limestone is routinely added to cement as an inert diluent; it enables concrete of a given strength to be made with a slightly lower Portland cement content, thus reducing the cost and energy content of the concrete. Even montmorillonite does not reduce the strength of concrete, provided that it is of high purity, is well dispersed, and that the water/cement ratio is held constant. For example Fardis et al. (1994) examined the effect of montmorillonite on the hydration of cement, using NMR to monitor the proton spin-lattice relaxation times of the water involved in the hydration process. The small differences in the evolution of relaxation times were attributed to differences in evolution of porosity. They concluded that montmorillonite has no chemical effect on cement hydration, over the 28-day time scale of the experiments, and also that it has no deleterious effect on the compressive and flexural strengths of cement paste, at clay/cement ratios up to 4 mass%. They have supporting evidence from large-scale trials with concrete, in which montmorillonite substituted 1% of the fine aggregates, or 1 and 2% of the cement. In these trials montmorillonite significantly increased the compressive strength of the concrete, but it is important to note that the water/cement ratio was maintained at a constant level, by varying the dose of sulphonated naphthalene superplasticizer in the initial concrete mixture.

References

AKMAN, M. S. & MUTLU, M. 1999. Cement grouting technology for consolidating soils. *Proceedings Specialist techniques and materials for concrete construction, Dundee*, Thomas Telford, London, 69–78.

ANDRIOLO, F. R. & SGARABOZA, B. C. 1985. Use of pozzolan from calcined clay in preventing ASR in Brazil. *Proctology 7th International Conference on ASR*, Noyes, New Jersey, 66–70.

ANON, 2000. *Bentonite Support Fluids in Civil Engineering*, Federation of Piling Specialists, April.

BAI, J. 2002. Metakaolin-pulverised fuel ash-Portland cement binders and their role in mortar and concrete. *Proceedings, Innovations and Developments in Concrete Materials and Construction, Dundee*, Thomas Telford, London, 159–173.

BARONIO, G. & BINDA, L. 1997. Study of the pozzolanicity of some bricks and clays. *Construction and Building Materials*, **11**, 41–46.

BERNSTEIN, A. G. et al. 2002. Inertization of hazardous dredging spoils. *Waste Management*, **22**, 865–869.

BLANCO-VARELA, M. T. et al. 1993. Production and behaviour of a new white sulphate-resistant cement. *Proceedings Concrete 2000 Economic and durable construction through excellence, Dundee*, Spon, London, 1325–1339.

BURGE, T. A. 1996. Highly reactive hydraulic binder. *Proceedings, Concrete in the Service of Mankind, Dundee*, Spon, London, 371–381.

CHEN, H.-J., YEN, T. & CHEN, K.-H. 2003. Use of building rubbles as recycled aggregates. *Cement and Concrete Research*, **33**, 125–132.

TABLE 15.17. *Surface areas of minerals in concrete*

Particle	Typical surface area (m² kg⁻¹)	Diameter of spherical particle with same surface area
10 mm aggregate	0.2	10 mm
1.0 mm sand	2.0	1.0 mm
0.1 mm sand	20	0.1 mm
0.01 mm fines	200	10 μm
kaolin	10,000	0.2 μm
illite	50,000	40 nm
montmorillonite	500,000	4.0 nm

CONNOR, J. R. 1995. Recent findings on immobilization of organics as measured by total constituent analysis. *Waste Management*, **15**, 359–369.

CONNER, J. R. & HOEFFNER, S. L. 1998. A critical review of stabilization/solidification technology. *Critical Reviews in Environmental Science and Technology*, **28** (4), 397–462.

CYR, M., CARLES-GIBERGUES, A. & TAGNIT-HAMOU, A. 2000. Titanium fume and ilmenite fines characterisation for their use in cement-based materials. *Cement and Concrete Research*, **30**, 1097–1104.

DAVIS, A. C. 1924. *A Hundred Years of Portland Cement, 1824–1924*, Concrete Publications.

DAVIDOVITS, J. 1994 High alkali cements for 21st century concretes. *In*: MEHTA, P. K. (ed.) *Proceedings Symposium 'Concrete technology, past, present and future'*, ACI, Vol V, 383–397.

DESMYTER, J., VAN DESSEL, J. & BLOCKMANS, S. 1999. The use of recycled concrete and masonry aggregates in concrete: improving the quality and purity of the aggregates. *Proceedings Exploiting Wastes in Concrete, Dundee*, Thomas Telford, London, 139–149.

DUNHAM, A. C. 1992. Developments in industrial mineralogy: I. The mineralogy of brick-making. *Proceedings Yorkshire Geology Society*, **49**(2), 95–100.

DUNSTER, A. M., PARSONAGE, J. R. & THOMAS, M. J. K. 1993. The pozzolanic reaction of MK and its effects on PC hydration. *Journal of Material Science*, **28**, 1345.

EL-DIDAMONY, H. *et al.* 2000. Homra pozzolanic cement. *Silicates Industrials*, **65**(3–4), 39–43.

FARDIS, M. *et al.* 1994. Effect of clay minerals on the hydration of cement. *Advanced Cement Based Materials*, **1**, 243–247.

FRANCIS, A. J. 1977. *The Cement Industry 1796–1914: a History*, David & Charles, Newton Abbot.

FRIED, A. N., ROBERTS, J. J. & YOOL, A. 2000. The effect of fines content in building sands on the durability of mortar. *Masonry International*, **14**, 51–54.

GOLASZEWSKI, J. *et al.* 1999. The Influence of Ground Brick on the Physical Properties of Mortar and Concrete. *Proceedings, Modern Concrete Materials: Binders, Additions and Admixtures, Dundee*, Thomas Telford, London, 119–130.

GORDON, J. N., PINOCK, W. R. & MOORE, M. M. (1996). A preliminary investigation of strength development in Jamaican red mud composites. *Cement and Concrete Composites*, **18**, 371–379.

GRANT, N. & GLASSER, F. P. 2002. Recycling the cement-rich fraction of waste concrete: a status report. *Proceedings Sustainable Concrete Construction, Dundee*, Thomas Telford, London, 796–804.

GUANGREN, G. *et al.* 2002. Improvement of metakaolin on radioactive Sr and Cs immobilization of alkali-activated slag matrix. *Journal of Hazardous Materials*, **B92**, 289–300.

HAGO, A. W. *et al.* 2002. Effect of varying cement content and curing conditions on the properties of sarooj (artificial pozzolana). *Building and Environment*, **37**, 45–53.

HAIMONI, A. M. 1999. Grouting—materials and techniques on the approach to the millennium. *Proceedings Specialist Techniques and Materials for Concrete Construction, Dundee*, Thomas Telford, London, 11–30.

HAMAD, B. S. 1995. Investigations of chemical and physical properties of white cement concrete. *Advanced Cement Based Materials*, **2**, 161–167.

HE, C. *et al.* 1995. Pozzolanic reactions of six principal clay minerals: activation, reactivity assessments and technological effects. *Cement & Concrete Research*, **25**, 1691–1702.

HE, C. *et al.* 1996. Thermal treatment and pozzolanic activity of sepiolite. *Applied Clay Science*, **10**, 337–349.

HEWLETT, P. C. 2002. Concrete: vade mecum. *Proceedings Concrete Floors and Slabs, Dundee*, Thomas Telford, London, 337–352.

JEFFERIS, S. A. 1997. The origins of the slurry trench cut-off and a review of cement-bentonite cut-off walls in the UK. *Proceedings 1st International Containment Technology Conference and Exhibition, Florida, Feb*.

JONES, T. R. 2001. Metakaolin as a pozzolanic addition to concrete. *In*: BENSTEAD, J. & BARNES, P. (eds) *Structure and Performance of Cements*, 2nd Edn, Spon, London, 372–398.

JUSTNES, H. *et al.* 1990. NMR—a powerful tool in cement and concrete research. *Advances in Cement Research*, **3**, 105–110.

KATZ, A. 2003. Properties of concrete made with recycled aggregate from partially hydrated old concrete. *Cement & Concrete Research*, **33**, 703–711.

KNIGHTS, J. C. 1999. Making the most of marginal aggregates. *Proceedings Utilising Ready-Mixed Concrete and Mortar, Dundee*, Thomas Telford, London, 46–56.

KREIJER, P. C. 1981. Introduction and stating the problem. *In*: KREIJER, P. C. (ed.) *Adhesion Problems in the Recycling of Concrete*, Plenum, New York, 15–34.

MCBRIDE, S. P., SHUKLA, A. & BOSE, A. 2002. Processing and characterisation of a lightweight concrete using cenospheres. *Journal of Materials Science*, **37**, 4217–4225.

MAJUMDAR, A. J. & SINGH, B. 1992. Properties of some blended high-alumina cements. *Cement & Concrete Research*, **22**, 1101–1114.

MALQUORI, G. 1960. Portland-pozzolan cement. *Proceedings of the 4th International Symposium on the Chemistry of Cement*, Washington, 983–999.

MEINHOLD, R. H. *et al.* 1994. A comparison of the kinetics of flash calcination of kaolinite in different calciners, *Transactions of the Institute of Chemical Engineering*, **72**(A), 105–113.

MILLRATH, K. 2003. *Modifying Concrete Matrices with Beneficiated Dredged Material or other Clayey Constituents*, PhD Thesis, Columbia University.

MILLRATH, K. *et al.* 2002. New approach to treating the soft clay/silt fraction of dredged material. *Proceedings of the Dredging 2002, Orlando*, 1–10.

MONZÓ, J., PAYÁ, J. & BORRACHERO, M. V. 1999. Experimental basic aspects for reusing sewage sludge ash (ssa) in concrete production. *Proceedings Exploiting Wastes in Concrete, Dundee*, Thomas Telford, 47–56.

MOSTAFA, N. Y. *et al.* 2001a. Characterisation and evaluation of the pozzolanic activity of Egyptian industrial by-products. I: Silica fume and dealuminated kaolin. *Cement & Concrete Research*, **31**, 467–474.

MOSTAFA, N. Y. *et al.* 2001b. Activity of silica fume and dealuminated kaolin at different temperatures. *Cement & Concrete Research*, **31**, 905–911.

MURAT, M. 1983a. Hydration reaction and hardening of calcined clays: 1- Preliminary investigations on metakaolin. *Cement & Concrete Research*, **13**, 259–266.

MURAT, M. 1983b. Hydration reaction and hardening of calcined clays: 2-Influence of mineral properties on the reactivity of metakaolinite. *Cement & Concrete Research*, **13**, 511–18.

NAIK, T. R. 2002. The role of combustion by-products in sustainable construction materials. *Proceedings Sustainable Concrete Construction, Dundee*, Thomas Telford, London, 117–130.

NEHDI, M. & SUMNER, J. 2003. Recycling waste latex paint in concrete. *Cement & Concrete Research*, **33**, 857–863.

NEWMAN, A. C. D. (ed.) 1987. Chemistry of Clays and Clay Minerals, Mineralogical Monograph Vol. 6, Longman & Mineralogical Society.

O'FARRELL, M. et al. 2002. A new concrete incorporating waste-paper sludge ash (WSA). *Proceedings, Innovations and Developments in Concrete Materials and Construction, Dundee*, Thomas Telford, London, 149–158.

OSMANLIOGLU, A. E. 2002. Immobilization of radioactive waste by cementation with purified kaolin clay. *Waste Management*, **22**, 481–483.

PACEWSKA, B. et al. 2002. Effect of waste aluminosilicate material on cement hydration and properties of cement mortars. *Cement & Concrete Research*, **32**, 1823–1830.

PAYÁ, J. et al. 1999. Fluid catalytic cracking catalyst residue (FC3R). An excellent mineral by-product for improving early-strength development of cement mixtures. *Cement & Concrete Research*, **29**, 1773–1779.

PERA, J., BOUMAZA, R. & AMBROISE, J. 1977. Development of a pozzolanic pigment from red mud. *Cement & Concrete Research*, **27**, 1513–1522.

PERA, J., & AMROUZ, A. 1988. Development of highly reactive metakaolin from paper sludge. *Advanced Cement Based Materials*, **7**, 49–56.

PROCTOR, D. 2001. A case for hydraulic Mortar. *Masonry Construction*, April, 16–22.

ROBERTS, K. J. et al. 1992. Using soft X-ray absorption spectroscopy to examine the structural changes taking place around Si and Al atoms in kaolinite following flash calcination, *Proceedings of the 7th International Conference on X-ray Absorption Fine Structure*, Kobe, August.

ROCHA, J. & KLINOWSKI, J. 1990. ^{29}Si and ^{27}Al magic-angle-spinning NMR studies of the thermal transformation of kaolinite, *Physics & Chemistry of Minerals*, **17**, 179–86.

RYLE, R. & MARTIN, S. J. 1999. The control of harmful fines in concrete sands—a case study. *Proceedings, Utilising Ready-mixed Concrete and Mortar, Dundee*, Thomas Telford, London, 23–32.

SAAD, M. N. A., DE ANDRADE, W. P. & PAULON, V. A. 1982. Properties of mass concrete containing an active pozzolan made from clay. *Concrete International*, July, 59–65.

SABIR, B. B., WILD, S. & BAI, J. 2001. Metakaolin and calcined clays as pozzolans for concrete: a review. *Cement & Concrete Composites*, **23** (6), 441–454.

SAIKIA, N. J. et al. 2002. Cementitious properties of metakaolin-normal Portland cement mixture in the presence of petroleum effluent treatment plant sludge. *Cement & Concrete Research*, **32**, 1717–1724.

SALVADOR, S. 1995. Pozzolanic properties of flash-calcined kaolinite: a comparative study with soak-calcined products. *Cement & Concrete Research*, **25** (1), 102–112.

SCHEETZ, B. E. & EARLE, R. 1998. Utilisation of fly ash. *Current Opinion in Solid State and Materials Science*, **3**, 510–520.

SCOTT, P. W. & BRISTOW, C. M. 2002. (eds) *Industrial Minerals and Extractive Industry Geology*, Geological Society, London, 365–366.

SEN GUPTA, J. 1992. Technology options for the manufacture of calcined clay pozzolana (surkhi). *In*: HILL, N., HOLMES, S. & MATHER, D. (eds.) *Lime and Other Cements*, Intermediate Technology, London, 191–197.

SIKALIDIS, C. A., ZABANIOTOU, A. A. & FAMELLOS, S. P. 2002. Utilisation of municipal solid waste for mortar production. *Resources, Conservation and Recycling*, **36**, 155–167.

SMEATON, J. 1791 and 1793. *A Narrative of the Building and a Description of the Construction of the Eddystone Lighthouse with Stone*, Printed for the author by H. Hughes, London.

SPEARE, P. R. S. & BEN-OTHMAN, B. 1993. Recycled concrete coarse aggregates and their influence on durability. *Proceedings Concrete 2000, Economic and Durable Construction through Excellence, Dundee*, Spon, London, 419–432.

SPEARS, D. A. 2000. Role of clay minerals in UK coal combustion. *Applied Clay Science*, **16**, 87–95.

SU, N. et al. 2001. Reuse of spent catalyst as fine aggregate in cement mortar. *Cement & Concrete Research*, **23**, 111–118.

TAYLOR, H. F. W., 1997. *Cement Chemistry*, 2nd Edn, Thomas Telford, London, 84–87.

TUTHILL, L. H. 1982. Alkali-silica reaction—40 years later. *Concrete International*, (April), 32–36.

WILD, S. et al. 1999. Waste clay brick—a European study of its effectiveness as a cement replacement material. *Proceedings, Exploiting Wastes in Concrete, Dundee*, Thomas Telford, London, 261–273.

WU, J.-H. et al. 2003. The effect of waste oil-cracking catalyst on the compressive strength of cement pastes and mortars. *Cement & Concrete Research*, **33**, 245–253.

XU, H. & VAN DEVENTER, J. S. J. 2002. Geopolymerisation of multiple minerals. *Minerals Engineering*, **15**, 1131–1139.

ZAKARIA, M. & CABRERA, J. G. 1996. Performance and durability of concrete made with demolition waste and artificial fly ash-clay aggregates. *In*: CABRERA, J. G. & WOOLLEY, G. R. (eds.), *Bulk "Inert" Waste: An Opportunity for Use*, Pergamon, New York, 151–158.

Appendix A. Mineralogical and chemical data

This appendix presents data on the clay mineral, non-clay mineral and chemical composition of a small but varied collection ($N = 105$) of clay materials from the British Isles. This collection was assembled and analyzed during the writing of this report, such that the data have not been published previously elsewhere.

Of course, there are many sources of mineralogical and chemical data on clay materials. In the British context these include several well-known publications which have attempted to compile and summarize data from many disparate sources (Perrin 1971; Ridgway 1982; Sellwood & Sladen 1981; Shaw 1981). The monograph compiled by Perrin (1971) was published by the Mineralogical Society of Great Britain and Ireland, and is yet to be superseded in scope; although it should be noted that, at the time of writing, the Mineralogical Society is in the process of preparing a multi-author book review of the knowledge and progress in understanding of the clay mineralogy of the UK stratigraphic column, made since the publication of Perrin's 1971 monograph (Jeans & Merriman 2006).

The publication by Shaw (1981) is an overview of the mineralogy and petrology of the argillaceous rocks of the UK. Shaw's discussion of the non-clay mineralogy of these rocks was severely limited by lack of available data, but trends and patterns in the clay mineral assemblages, based on analyses of clay-sized fractions, throughout the UK stratigraphic column were presented and discussed. The paper by Sellwood & Slade (1981), which appeared in the same thematic volume of the *Quarterly Journal of Engineering Geology*, focused solely on the Mesozoic and Tertiary argillaceous units of the UK. Although ten years had elapsed between the publication of these reviews and that of Perrin the latter works still drew largely on Perrin's compendium and this remains the most extensive and detailed compilation of clay mineral data available in the UK to date. The British Geological Survey (BGS) did compile a limited UK mineralogical mudrock database in the late 1980s/early 1990's which included data from the NIREX (Nuclear Industry Radioactive Waste Executive) candidate sites for the disposal of low-level radioactive waste, but this is no longer publicly available. The publication by Ridgway (1982) contains a compilation of 85 whole-rock, major element, chemical analyses of clay materials from the UK. These chemical analyses may be directly compared with those in this appendix.

Internationally, Weaver's attempts to compile and analyze the trends in Phanerozoic clay mineral assemblages, based mainly on data from the United States, are also well known, beginning with his 1967 publication on 'The significance of clay minerals in sediments'. More recently he has documented and discussed trends and patterns of clay mineral distribution on a global scale in Chapter 9 of his book 'Clays, muds and shales' (Weaver 1989), an account which is remarkable in scope and coverage. Other publications dealing with the stratigraphic distribution of clay minerals include those of Chatzidimitriadis *et al.* (1993) and Viczian (2002), which deal with argillaceous rocks in Greece, and Hungary, respectively. No doubt there are others in existence not referenced here.

All of these aforementioned publications deal primarily with the clay mineralogy of argillaceous rocks as determined on a clay-sized fraction (usually but not always <2 μm) that is physically separated from the bulk material. As pointed out by Shaw (1981) there are much less published data on the non-clay mineral composition (quartz, feldspars, carbonates and total amounts of clay minerals etc.) of mudrocks, and certainly very few attempts to compile or summarise data in any stratigraphic sense. There have, however, been several attempts to determine the average mineralogical composition of shale, as well as the more easily determined average chemical composition. Literature data on these aspects are discussed more fully in Chapter 3 section 6.4. Additionally, a set of almost 100 whole rock mineralogical analyses, mainly from North American shales, can be found in the 'Argillaceous Rock Atlas', compiled by O'Brien & Slatt (1990).

A.1. Purpose and scope

The primary purpose of obtaining and presenting the new data in this appendix is to supplement previous data with further analyses that, above all-else, are internally consistent. Where literature data from diverse and disparate sources are concerned, amongst many other factors, methods and techniques change and vary with time, experience, and purpose. Lack of consistency is, therefore, a perpetual and insoluble problem when data are compiled and compared from different sources. The uncertainties attached to comparisons, and the conclusions that may be drawn from such comparisons, are therefore larger and in many cases unknown. Other problems arising from comparing data from different sources, often obtained using different analytical approaches have been discussed in Chapter 8.

The disadvantage of consistency is that the number of samples that can be compared is inevitably a much smaller number. In this appendix data are presented for 105 samples. The samples range in age from Cambrian to Quaternary, and it must be stressed that such a small set of samples can not be considered as representative in a statistical sense of any of the individual stratigraphic units from which they were collected. Thus it must be stressed that the data are not presented as definitive analyses of clay materials from particular stratigraphic units. They are simply presented as examples of analyses of particular clay materials. Samples from other localities, even the same locality, in the same stratigraphic unit will inevitably be different in detail and in many instances may be wholly different. Nonetheless, certain trends and features in the new data are in general accord with prior knowledge, such as that summarized in the compilations of Perrin (1971), Shaw (1981) and Sellwood & Slade (1981). Some comments are therefore made to put these data into context. In addition, some new features are apparent that have not been established previously and these are also worthy of brief note.

As mentioned above, this appendix also contains data on the 'whole-rock' or 'bulk' mineral composition of British argillaceous units. To our knowledge there are no other compilations of such data in existence at this time. To complement the whole rock mineralogical data, whole rock chemical analyses are also presented. These were obtained on a sub-sample of each specimen. For both whole rock mineralogy and chemistry 'average' compositions have been calculated for all data and these may be usefully compared with other attempts to compute the 'average' composition of shales, as detailed in Chapter 3, section 6.4. Average values have also been calculated for the various stratigraphic groupings into which the data are divided. In most cases, the samples are too disparate in nature for this to be truly meaningful, but in some cases this is useful for comparative purposes.

The samples were donated by various individuals listed in the acknowledgments. Many come from construction sites, but others were also obtained at outcrop, from mineral exploitation pits, and previously assembled collections. On the whole, most contributors donated samples believed to be unaffected by weathering, but some supplied material noted as weathered. Because of the subjective nature of this judgement, as well as it inconsistent assessment, no remarks are recorded on the weathered versus fresh state of the materials. However, the reader may generally assume that the materials are not weathered, unless the presence of some mineral phases clearly suggest otherwise.

A.2. Methods

Mineralogical analyses were made by Dr Stephen Hillier of the Macaulay Institute, Aberdeen. All the analyses have been made using the methods outlined in Hillier (2000) and in more detail in Hillier (2003). The use of one lab, one analyst, and consistent analytical procedure throughout was to ensure internal consistency for comparison of data from a wide range of clay materials distributed throughout the stratigraphic column.

A.2.1. Clay fraction analysis

For clay mineral analyses, samples were dispersed in de-ionised water, and a <2 μm size fraction separated by timed sedimentation according to Stokes' Law. The clay was then deposited on a glass slide using the filter peel transfer procedure and examined in a Siemens D5000 diffractometer in the air-dried state, after glycolation (by vapour pressure method overnight) and after heating to 300°C and 550°C for one hour. For each treatment the sample was scanned from 2–45°2θ, in 0.02° steps, counting for 1 second per step using Co-Kα radiation, selected by a diffracted beam monochromator. Clay minerals were identified by comparison of the response to the various treatments and their effect on peak positions and intensities as detailed in Hillier (2003). Clay minerals identified were quantified using a normalized (sum 100%) reference intensity ratio approach (Hillier 2003) based on factors obtained from calculated X-ray powder diffraction (XRPD) patterns.

A.2.2. Whole rock mineralogical analysis

Each bulk sample was disaggregated and reduced to a fine powder (<2 mm) by hand, and/or using a hammer, depending on hardness. Then a carefully weighed mixture of 2.4 g of each sample and 0.6 g of corundum was wet ground and spray dried to produce a random powder. The 20% corundum addition was used as an internal standard for quantitative phase analysis (QPA). XRPD patterns were recorded from 2–75°2θ, in 0.02° steps, counting for 2 seconds per step, using Co-Kα radiation selected by a diffracted beam monochromator. QPA was performed using an internal standard reference intensity ratio (RIR) method. As detailed in Hillier (2003), a generalized estimate of expanded uncertainty using a coverage factor of 2, i.e. 95% confidence, is given by $\pm X^{0.35}$, where X is the concentration in wt%., e.g. 30 wt% ± 3.3.

A.2.3. Whole rock chemical analysis

Whole rock chemical analysis was made by Mrs. Sieglinde Still and Ms. Michele Heibel at the laboratories of Watts Blake and Berne (WBB) Fuchs Laboratory, Ruppach-Goldhausen, Germany using X-ray fluorescence spectroscopy (XRF). The clays were dried, crushed and mixed thoroughly before being milled to small particle sizes in a vibrating ball mill. The resulting powders were placed in sealed containers prior to approximately 1 g of each being weighed into an alumina crucible. In order to determine values of loss-on-ignition, the crucibles were placed in a furnace at 1000°C for 20 minutes, cooled, and re-weighed immediately afterwards. 0.7000 g of each sample was subsequently weighed into a 95 wt% Pt - 5 wt% Au crucible, to which 3.5000 g of flux was added, giving rise to a flux:fired clay ratio of 5:1. The flux used was Spectroflux 100, Johnson Matthey, Batch No. 40499, of chemical composition $Li_2B_2O_4$. Fusion was carried out in an automatic fusion machine (HD Elektronik, Kleve, Germany), employing a sequence of

3 minutes pre-warming, 4 minutes pre-melting and 6 minutes main melting. Regular agitation of the crucibles was carried out by the machine at 10-second intervals, prior to the melt being poured into pre-heated Pt-Au dishes for casting. The bead-casting cycle was completed by these dishes being cooled in an airstream for 6 minutes. The samples were analysed on a Philips PW2400 Sequential XRF Spectrometer using a multi-element standard-based calibration within Philips SuperQ Software Version 3.0G.

A.3. Notes on data presentation

Data (Table A1) are presented in stratigraphic order but there is not an even distribution of samples and most (about three quarters) are of Mesozoic and Cenozoic age reflecting the distribution of the more significant clay formations in the UK. For each sample 58 numbered columns of data are presented. For identification purposes, columns 1 and 2 contain the stratigraphic unit from which the sample was collected and a National Grid Reference of the locality, as supplied by the originator. Additionally, columns 57 and 58 provide the analysis reference codes. The analytical data are divided into four groups. The first group consists of columns 3–23, which contain data on the whole-rock mineralogical composition of the samples. The second group consists of columns 24–35, which contain data on the clay mineralogy of the clay size fraction (<2 μm). The third group consists of columns 36–45, which contain data on the whole-rock major element chemical composition. And the fourth group consists of columns 46–56, which consists of whole-rock trace element data.

The data are also divided into groups based on stratigraphic age and average values for each data element have been calculated for each of these groupings. As mentioned previously, this is for illustrative purposes only, since some of the groupings are of quite disparate clay materials, whereas other groups are more focused. An overall average composition of all 105 clay materials is also given at the end of the listing.

A.3.1. Whole rock mineralogy

Thirteen different non-clay minerals were identified and are listed in columns 3–15. Columns 16 and 17, give the total dioctahedral layer silicates (Di-layer silicates), and total trioctahedral layer silicates (Tri-layer silicates). These are general groupings, which added together give a total value of layer silicates in the sample. The term layer silicates is used in preference to clay minerals because the analytical method does not distinguish between clay minerals and non clay-sized phyllosilicates such as large grains of muscovite mica or chlorite. However, for most samples it can be assumed that the values in columns 16 and 17 are to a good approximation, a measure of the total clay mineral content in each sample. Columns 19–23 give a further break down of the layer silicate mineralogy of the whole rock. Columns, 19–21 and 23 are all dioctahedral clays, such that their sum is equal to column 16.

Column 20 is an estimate of the total amount of expandable (swells in water) dioctahedral clay. This is calculated as the remaining dioctahedral clay that is unaccounted for after quantifying illite/mica and kaolin separately. For the majority of samples this is an estimate of the mixed-layer illite-smectite content of the whole rock. The only common trioctahedral clay is chlorite so that column 23 is identical to column 17. The main exceptions to this are samples from the Mercia Mudstone, which commonly contain corrensite (trioctahedral chlorite(50)-smectite R1). This mineral is not distinguished from chlorite in the whole rock analysis. Additionally, when present berthierine and vermiculite may also contribute to the measured value for total trioctahedral layer silicate. The reader should therefore check to see if these minerals have been identified in the clay fraction in order to judge if the usual equivalence between total trioctahedral clay (column 17) and chlorite (column 23) in the whole rock is a valid assumption.

A.3.2. Clay fraction mineralogy

The main clay minerals identified in the samples are illite (mica), mixed-layer illite-smectite, kaolin, and chlorite. Other clay minerals are much rarer. Precise identification of the mixed-layer illite-smectite identified has not been attempted, but based on the overall character of the XRPD patterns and peak positions an estimate of its expandability has been made as listed in column 35. Expandability may be equated with the percentage of smectite layers in the mixed-layer clay. Some of the diffraction data also suggest that it may be a gross oversimplification to identify some of the expandable clay minerals as mixed-layer illite-smectite and that other layer types, such as vermiculite may be present. Nonetheless, for the purposes of this appendix if the clay mineral collapsed to 10Å after heating it is labeled mixed-layer illite-smectite. In addition, to an approximate measure of expandability, column 34 indicates whether or not a peak at 17Å was observed, following treatment of the sample with ethylene glycol. When the clay mineral has been identified as illite-smectite, the presence of this peak is a reliable indicator of a so-called randomly interstratified variety of mixed-layer illite-smectite. Frequently, this clay mineral is incorrectly identified as pure smectite, which also has a peak at 17Å. It was also noted that some chlorites where partially susceptible to heating at 300°C. This is an unusually low temperature for chlorite to be labile, and this behavior could indicate the presence of berthierine. This is indicated by a # symbol in the berthierine column (column 33).

A.3.3. Whole rock chemical data

The whole rock chemical data require little explanation. There are only three significant points to note. Firstly, the data are presented normalized to a 100 wt% minus Loss On Ignition (LOI) basis. This was deemed the best way to present the data for direct comparison with the whole rock

TABLE A.1. *Mineralogical and chemical composition of a collection of clay materials from the British Isles*

Stratigraphic Unit	Nat.Grid Ref.	Whole-rock mineralogy wt.%															Whole rock-clay mineralogy wt.%								Clay fraction (<2μm) clay mineralogy wt.%										
1	2	Quartz 3	K-feldspar 4	Plagioclase 5	Calcite 6	Dolomite 7	Siderite 8	Aragonite 9	Pyrite 10	Hematite 11	Goethite 12	Jarosite 13	Anatase 14	Amphibole 15	Di-layer silicate 16	Tri-layer silicate 17	Total 18	Illite/mica 19	Exp. di–layer silicate 20	Kaolin 21	Chlorite 22	Paragonite 23	Illite/Mica 24	Illite-smectite 25	Kaolin 26	Chlorite 27	Smectite 28	Vermiculite 29	Corrensite 30	Kaolin/Smectite 31	Paragonite 32	Berthierine 33	17Ang. peak 34	%EXP I-S 35
Quaternary																																			
Sandy alluvial clay R. Thames	SU 76 82	81.3	5.8	2.0											12.5		101.6	11.3	1.2				4	83	10	3							#	60%
Alluvial clay R. Thames	SU 84 84	27.7	2.8	0.9	36.1			4.2			4.3				16.5	8.7	92.5	14.7		1.8			4	89	7								#	70%
Alluvial clay R. Avon	ST 752 644	32.7	6.5	1.6	19.7										30.4	14.1	99.6	14.1	11.6	4.7			12	76	10	2							#	55%
Glacio-lacustrine clay	SN 1761 4285	11.0	2.6	2.6	3.3										60.7	14.1	94.3	27.5	21.3	11.9	8.7		44	30	20	6								20%
Glacio-lacustrine clay	NJ 968 267	11.4	7.2	7.5										1.7	41.8	23.6	93.2	22.3	14.7	4.8	23.6		69		16	3									
Brick Earth	TQ 5857 1895	52.6	7.0	5.1											32.0		96.7	6.4	21.6	4.0			12	77	11								#	70%
Brick Earth	TQ 087 780	64.5	9.2	5.9											14.0	4.0	97.6	5.0	6.0	3.0	4.0		14	74	12									50%
Skipsea Till	TA 228 438	33.9	3.6	3.1	18.3	1.7									26.9		87.5	14.4	3.2	9.3			26	41	29	4								20%
Boulder Clay, Dublin	O 1750 3715	37.6	2.9	4.2	35.5	1.3			0.8						13.0	2.0	97.3	7.6	4.4	1.0			70	8	6	16								10%
Boulder Clay, Dublin	O 1825 3045	33.9			45.4										13.4	2.8	99.5	8.2	5.2		2.0		32	63		5								20%
Boulder Clay, Dublin	O 1609 3804	40.1	3.6	5.2	17.3	1.4					1.5				20.0	6.9	94.5	11.7	4.3	4.0	2.8		48	33	6	12		1						15%
Average (N = 11)		**38.8**	**4.7**	**3.7**	**16.0**	**0.4**		**0.4**	**0.1**		**0.5**			**0.2**	**25.6**	**5.6**	**95.8**	**10.7**	**10.8**	**4.2**	**5.6**		**30.5**	**52.2**	**11.5**	**4.6**		**1.2**						
Eocene/Oligocene																																			
Ball clay, Abbrook Member	SX 861 750	18.1	3.4	2.0											70.0		93.5	38.1	2.1	29.8			26	27	47									20%
Ball clay, Abbrook Member	SX 861 750	10.6	2.8	1.6							1.0				77.9		92.9	28.8	0.5	48.6			16	22	62									20%
Ball clay, Abbrook Member	SX 861 750	38.4	4.8	1.4							1.0				47.9		93.5	13.7	3.7	30.5			16	21	63									20%
Ball clay, Courtmoor Member	SS 517 118	12.3	2.4	1.0							1.3				70.7		87.4	17.4	25.8	27.5			25	23	52									20%
Ball clay, Southacre member	SX 853 755	5.1	2.4	1.3											86.4		96.5	22.6	13.7	50.1			21	16	63									20%
Ball clay, Stover Member	SX 852 755	1.0													96.3		98.1	3.1	4.2	89.0			2	1	97									20%
Ball clay, Westbeare Member	SS 514 120	33.7	2.4	1.0			0.8								60.4		98.5	16.1	18.2	26.1			21	28	51	1								20%
Ball clay, Westbeare Member	SS 514 120	24.4	4.2	1.2											68.5		99.2	15.2	19.4	33.9	2.0		17	26	57									20%
Ball clay, Westbeare Member	SS 514 120	13.8	3.5	1.3					1.2						75.3		95.1	8.0	34.5	32.8	2.8		11	46	42	12								20%
Average N = 9		**17.5**	**2.9**	**1.2**			**0.1**		**0.2**		**0.4**				**72.6**		**95.0**	**18.1**	**13.6**	**40.9**	**6.9**		**17.2**	**23.3**	**59.3**	**4.6**								
Eocene																																			
London Clay	TQ 05 73	42.3	2.5	1.2			0.4				0.4				41.8	6.0	94.6	18.0	4.9	18.9	6.0		25	35	30	10								50%
London Clay	TQ 05 73	43.2	2.2	1.2			0.4				0.6				40.2	6.8	94.6	18.3	3.4	18.5	6.8		23	38	28	11							##	50%
London Clay	SU 803 696	65.8	4.9	1.8											23.0	4.4	99.9	5.6	13.3	4.1	4.4		10	75	15								#	55%
London Clay	SU 803 696	35.2	5.4	1.6											39.0	7.6	88.8	9.0	25.6	4.4	7.6		14	73	13								#	70%
London Clay	TQ 312 809	21.4	5.3	0.7							2.0				60.0	8.1	99.8	15.2	36.2	8.6	8.1		19	68	12	1							#	75%
London Clay	TQ 312 809	17.2	4.1	0.6							3.0				62.2	12.3	102.8	22.3	31.0	8.9	12.3		17	67	14	2							#	75%
London Clay	SU 4448 1643	25.5	6.7	0.7					1.2						44.4	11.2	96.7	13.7	24.6	6.1	11.2		20	59	18	3							#	85%
Average N = 7		**35.8**	**4.4**	**1.1**			**0.1**		**0.2**		**0.9**				**44.4**	**8.1**	**96.7**	**14.6**	**19.9**	**9.9**	**8.1**		**18.3**	**59.3**	**18.6**	**3.9**								
Palaeocene																																			
Reading Formation	SZ 4438 0856	23.1	2.5	1.6							1.3				72.7		99.9	21.5	42.0	9.2			27	53	20						23		#	60%
Reading Formation	SU 4458 1697	50.4	2.7										1.2		41.3		96.9		37.8	3.5			7	56	14								#	60%
Reading Formation	SZ 3054 8524	20.0	3.7	1.6							5.1				60.7	4.4	95.5	28.5	24.8	7.4	4.4		42	37	17	4								20%
Reading Formation	SZ 3054 8524	31.4	1.8	1.0							10.8				50.9		95.9	19.5	24.6	6.8			19	65	16								#	75%
Reading Formation	SU 4581 6970	36.6	2.8										0.9		58.5		98.8	4.9	48.8	4.8			9	54	16						21		#	70%
Reading Formation	SU 825 795	41.7	3.8	1.3							2.8		0.4		51.8		101.8	8.9	38.0	4.9			12	66	10						12		#	70%
Average N = 6		**33.9**	**2.9**	**0.9**							**3.3**		**0.4**		**56.0**	**0.7**	**98.1**	**13.9**	**36.0**	**6.1**	**0.7**		**19.3**	**55.2**	**15.5**	**0.7**					**9.3**			

APPENDIX A

Stratigraphic Unit	Nat.Grid Ref.	Whole-rock chemical composition: Major elements wt.%										Whole-rock trace element composition ppm									Analysis reference	
1	2	36	37	38	39	40	41	42	43	44	45	46	47	48	49	50	51	52	53	54	55	56	57	58
		SiO_2	TiO_2	Al_2O_3	Fe_2O_3	CaO	MgO	K_2O	Na_2O	P_2O_5	LOI	Mn	V	Cr	Co	Ni	Cu	Zn	As	Zr	Ba	Pb		
Quaternary																								
Sandy alluvial clay R. Thames	SU 76 82	85.88	0.41	4.29	2.73	0.72	0.25	0.92	0.26	0.14	4.2	357	48	236	11	727	17	249		366	242	57	cwp-97	680030
Alluvial clay R. Thames	SU 84 84	45.38	0.26	4.66	3.20	21.63	0.46	0.66	0.15	0.23	23.2	727	59	183	11	478	99		108	142	177		cwp-46	679979
Alluvial clay R. Avon	ST 752 644	55.72	0.61	10.55	5.46	9.78	0.94	2.52	0.33	0.46	13.4	987	89	161	16	375	39	176	26	271	440	27	cwp-108	680041
Glacio-lacustrine clay	SN 1761 4285	49.88	0.95	24.03	6.80	2.63	1.67	3.76	0.52	0.14	9.5	582	175	127	18	65	46	62	77	155	551	41	cwp-28	679961
Glacio-lacustrine clay	NJ 968 267	52.01	0.93	19.47	8.69	1.50	3.35	4.00	1.31	0.18	8.3	849	144	153	22	249	52	139	44	181	630		cwp-58	679991
Brick Earth	TQ 5857 1895	75.58	0.77	11.39	2.89	0.60	0.94	2.48	0.82	0.09	4.2	480	97	233	12	747	11	96		466	432	81	cwp-29	679962
Brick Earth	TQ 087 780	80.85	0.76	7.88	1.90	0.46	0.51	2.08	0.77	0.23	4.4	371	70	212	9	472		65	30	565	421	14	cwp-95	680028
Skipsea Till	TA 228 438	55.84	0.58	11.20	4.30	10.14	1.56	1.82	0.71	0.11	13.6	240	89	203	14	751	1		14	224	344	7	cwp-30	679963
Boulder Clay, Dublin	O 1750 3715	51.22	0.35	6.38	3.11	17.87	1.29	1.31	0.65	0.11	17.5	1223	63	171	11	487	39	86	52	139	262	38	cwp-100	680033
Boulder Clay, Dublin	O 1825 3045	46.22	0.28	5.50	2.38	21.80	0.64	0.97	0.47	0.12	21.5	570	63	128	9	368	37	76	62	105	185	38	cwp-101	680034
Boulder Clay, Dublin	O 1609 3804	60.60	0.50	8.81	4.27	9.74	1.61	1.74	0.82	0.13	11.6	914	81	168	14	472	42	81	23	176	350	25	cwp-105	680038
Average (N = 11)		**59.9**	**0.6**	**10.4**	**4.2**	**8.8**	**1.2**	**2.0**	**0.6**	**0.2**	**11.9**	**664**	**89**	**180**	**13**	**472**	**35**	**93**	**40**	**254**	**367**	**30**		
Eocene/Oligocene																								
Ball clay, Abbrook Member	SX 861 750	55.83	1.40	27.40	1.07	0.25	0.34	3.07	0.43	0.06	10.0	33	172	122	7	79	22	125	66	283	586	2	cwp-50	679983
Ball clay, Abbrook Member	SX 861 750	51.54	1.05	31.94	1.07	0.20	0.29	2.48	0.34	0.07	10.9		145	113	6		11	22	28	160	479	10	cwp-51	679984
Ball clay, Abbrook Member	SX 861 750	68.45	1.41	20.12	0.65	0.15	0.24	1.67	0.24	0.06	6.9		97	137	6	280	42	123		286	387	87	cwp-52	679985
Ball clay, Courtmoor Member	SS 517 118	53.10	1.21	29.57	1.31	0.16	0.48	2.80	0.41	0.09	10.7		165	139	8	81	10		86	344	683		cwp-56	679989
Ball clay, Southacre member	SX 853 755	42.26	0.67	30.48	1.17	0.21	0.33	2.28	0.24	0.06	22.2		119	127	8	20	3	192		131	380	68	cwp-48	679981
Ball clay, Stover Member	SX 852 755	44.47	0.42	37.25	0.92	0.14	0.14	0.48	0.00	0.06	16.0	68	37	20	7	1	34	26	86	68	194		cwp-49	679982
Ball clay, Westbeare Member	SS 514 120	64.53	1.57	21.92	1.14	0.12	0.35	2.10	0.45	0.07	7.6	33	118	136	6	161	14	15	10	241	505	29	cwp-53	679986
Ball clay, Westbeare Member	SS 514 120	59.96	1.46	25.54	0.66	0.13	0.36	2.16	0.34	0.08	9.2		138	123	6	99		10		146	545	27	cwp-54	679987
Ball clay, Westbeare Member	SS 514 120	53.35	1.61	28.84	1.08	0.13	0.40	2.32	0.43	0.13	11.6	23	167	163	9	115	9	16		275	621	41	cwp-55	679988
Average N = 9		**54.8**	**1.2**	**28.1**	**1.0**	**0.2**	**0.3**	**2.2**	**0.3**	**0.1**	**11.7**	**18**	**129**	**120**	**7**	**93**	**16**	**59**	**31**	**215**	**487**	**29**		
Eocene																								
London Clay	TQ 05 73	66.95	1.46	15.70	6.23	0.24	0.77	2.18	0.28	0.09	5.9	278	124	146	18	305	43	104	39	276	407	40	cwp-7	679940
London Clay	TQ 05 73	67.29	1.48	15.51	6.25	0.28	0.79	2.13	0.28	0.11	5.6	335	146	168	18	478	46	138	440	286	516	202	cwp-8	679941
London Clay	SU 803 696	79.64	0.71	8.55	2.36	0.52	0.55	1.87	0.41	0.06	5.1	138	80	263	10	907	16	54	54	297	310		cwp-41	679974
London Clay	SU 803 696	67.83	0.85	12.90	6.68	0.71	1.38	2.82	0.46	0.10	6.1	170	127	254	19	503		149		305	429	74	cwp-42	679975
London Clay	TQ 312 809	60.32	0.96	16.83	4.77	1.41	2.62	3.35	0.32	0.20	9.1	195	183	198	14	136	31	218		195	350	64	cwp-83	680016
London Clay	TQ 312 809	54.62	0.95	17.39	7.88	1.90	2.92	3.31	0.26	0.18	10.4	167	200	172	21	225	62	15	20	184	343	26	cwp-84	680017
London Clay	SU 4448 1643	58.68	0.91	14.83	6.74	3.82	2.36	3.04	0.17	0.14	9.1	412	171	196	19	328	32	102	22	227	341	36	cwp-13	679946
Average N = 7		**65.0**	**1.0**	**14.5**	**5.8**	**1.3**	**1.6**	**2.7**	**0.3**	**0.1**	**7.3**	**242**	**144**	**196**	**17**	**412**	**33**	**111**	**82**	**253**	**385**	**63**		
Palaeocene																								
Reading Formation	SZ 4438 0856	59.40	0.91	22.38	5.12	0.62	0.85	3.05	0.29	0.07	7.2	93	161	145	15	165	158	64	12	196	477	31	cwp-14	679947
Reading Formation	SU 4458 1697	56.24	1.01	22.30	8.86	0.49	1.12	3.47	0.28	0.13	5.9	349	162	186	23	161	36	147	7	170	486	34	cwp-15	679948
Reading Formation	SZ 3054 8524	71.53	1.58	12.88	4.41	0.38	0.52	2.88	0.06	0.10	7.6	60	116	140	14	313	27	22	8	286	173	32	cwp-16	679949
Reading Formation	SZ 3054 8524	62.26	0.94	16.34	10.54	0.54	0.70	2.17	0.19	0.07	6.1	109	123	218	26	229	38	87	21	227	345	36	cwp-17	679950
Reading Formation	SU 4581 6970	64.33	1.51	15.58	8.64	0.52	0.69	1.30	0.07	0.05	7.2	77	155	139	22	186	23	22	14	252	232	32	cwp-35	679968
Reading Formation	SU 825 795	68.78	1.01	15.93	3.24	0.48	0.84	1.86	0.12	0.05	7.6		118	166	12	412	32	34	88	274	268		cwp-57	679990
Average N = 6		**63.8**	**1.2**	**17.6**	**6.8**	**0.5**	**0.8**	**2.1**	**0.2**	**0.1**	**6.9**	**115**	**139**	**166**	**19**	**245**	**49**	**62**	**25**	**234**	**330**	**28**		

APPENDIX A

Stratigraphic Unit	Nat.Grid Ref.	Whole-rock mineralogy wt.%															Whole rock-clay mineralogy wt.%							Clay fraction (<2μm) clay mineralogy wt.%										
		Quartz	K-feldspar	Plagioclase	Calcite	Dolomite	Siderite	Aragonite	Pyrite	Hematite	Goethite	Jarosite	Anatase	Amphibole	Di-layer silicate	Tri-layer silicate	Total	Illite/mica	Exp. di-layer silicate	Kaolin	Chlorite	Paragonite	Illite/Mica	Illite-smectite	Kaolin	Chlorite	Smectite	Vermiculite	Corrensite	Kaolin/Smectite	Paragonite	Berthierine	17 Ang. peak	%EXP I-S
1	2	3	4	5	6	7	8	9	10	11	12	13	14	15	16	17	18	19	20	21	22	23	24	25	26	27	28	29	30	31	32	33	34	35
Cretaceous																																			
Gault Formation	TQ 093 139	17.6	2.8	1.3						0.8					72.6	5.5	100.6	26.4	30.7	15.5	5.5		18	58	22			2						50%
Gault Formation	TQ 093 139	17.4	2.5	1.1						1.0					73.2	5.4	100.6	21.6	37.3	14.3	5.4		19	63	16			2						50%
Gault Formation	TL 1887 3463	6.5	1.5	0.4	43.9										43.0		95.3	12.9	21.6	12.3			12	65	23								#	60%
Gault Formation	SU 505 925	23.5	5.6	1.2	1.2						3.5				57.8	4.7	97.5	19.0	26.5	12.3	4.7		10	67	23								#	55%
Gault Formation	TL-450595	10.2	1.8	0.7	27.7				0.7						54.2	7.4	102.7	17.6	25.1	11.5	7.4		7	60	25	2							#	55%
Gault Formation	SU 804 407	27.6	6.6	1.4											53.6		89.2	26.0	34.1	7.0			7	66	27	1							#	70%
Gault Formation	SP 93750 24865	18.5	3.5	1.2					0.9						62.4	6.4	92.9	26.0	27.0	9.4	6.4		7	82	10								#	75%
Gault Formation	SU 755 288	23.3	3.8	1.0	0.7				0.8						68.3		97.9	16.1	39.1	13.1			13	71	14			2					#	60%
Gault Formation	SP 4940 2284	12.5	3.1	1.0	33.7				0.8						43.7		94.8	8.2	26.4	9.1			8	84	8								#	75%
Gault Formation	TL 5188 2346	12.8	1.9	0.7	33.1				0.8						46.9		96.2	9.4	27.5	10.0			15	65	20								#	70%
Gault Formation	TL 5188 2346	6.2	0.4	2.0	41.2			2.6							44.8		97.2	4.8	30.6	9.4			8	72	20								#	55%
Average N = 11		**16.0**	**3.0**	**1.1**	**16.5**			**0.2**	**0.4**		**0.5**				**56.4**	**2.7**	**96.8**	**15.9**	**29.6**	**10.9**	**2.7**		**11.8**	**68.5**	**18.9**	**0.3**		**0.5**						
Weald Clay	TQ 112 372	37.5	3.0	1.3											39.0	14.9	95.7	16.3	5.0	17.7	14.9		24	38	19	19							#	50%
Weald Clay	TQ 112 372	43.8	3.2	1.3											38.0	13.3	99.6	18.6	2.0	17.4	13.3		25	29	24	22							#	50%
Weald Clay	SY965815	37.1	0.7	0.5	0.6	1.5			1.3		4.9				47.8	4.6	96.7	10.6	32.5	4.7	4.6		14	67	19								#	75%
Speeton Clay	TA 1500 7571	18.1	1.4	0.7	5.3	0.4									62.6	5.7	96.6	24.6	26.3	11.7	5.7		17	56	23	4								50%
Average N = 4		**34.1**	**2.1**	**1.0**	**1.5**	**0.4**			**0.3**		**1.2**				**46.9**	**9.6**	**97.2**	**17.5**	**16.5**	**12.9**	**9.6**		**20.0**	**47.5**	**21.3**	**11.3**								
Jurassic																																		
Kimmeridge Clay	SY 92 78	34.3	3.6	1.4	2.6				2.5						44.2		88.6	25.0	10.0	9.2			14	69	17									20%
Kimmeridge Mudstone	SY 92 78	13.7	1.6	0.6	38.6		2.8		2.0				0.8		38.4		99.7	12.5	15.1	10.8			13	60	19									20%
Kimmeridge Clay	SY700736	17.6	4.9	1.2	6.4				4.4		2.0		0.7		54.3	4.0	97.3	26.2	16.5	11.6	4.0		17	57	23	3								20%
Oxford Clay	SP 369 074	12.5	2.6	1.3	27.1	2.0		4.5	2.1				0.6		42.9	5.4	95.2	22.9	13.3	6.7	5.4		26	50	21	3								20%
Oxford Clay	SP 369 074	11.9	2.4	1.1	27.1	1.5			1.3						44.6	5.1	95.0	16.5	21.0	7.1	5.1		21	46	28	5								20%
Oxford Clay	TL 504525 246467	21.1	2.6	1.1	3.7			6.0	2.5						56.8	6.9	100.7	21.3	27.3	8.2	6.9		19	63	16	2								40%
Blisworth clay, Bathonian	SP 964 662	8.2													90.1		99.7		83.4	6.7			3	15	30					52				
Blisworth clay, Bathonian	SP 964 662	6.3	0.4		0.6					9.8					81.5		98.7		69.9	11.6			5	59	18	6				14				
Blisworth clay, Bathonian	SP 964 662	6.9					2.9			2.9			0.6		91.2		101.6		82.2	9.7			9	67	15	2				9				
Lias Whitby Mudstone	NZ 8317 1599	17.5	1.9	1.7					0.9						58.9	12.6	95.3	32.0	9.0	17.9	12.6		17	64	13	6								25%
Lias Dyrham	SO 835 149	21.2	5.0	3.0									1.8		50.4	14.0	96.5	9.7	24.6	16.1	14.0		14	57	22	7							#	55%
Lias Redcar Mudstone	NZ 9778 0223	16.5	1.8	0.7	4.6				3.2						70.5		97.3	22.8	21.6	22.8			17	54	21	8							#	25%
Lias Redcar Mudstone	NZ 9644 0113	20.5	1.7	1.5	2.3	2.5			2.0						55.0	10.6	96.1	20.3	13.2	21.5	10.6		16	63	17	4								25%
Lias Charmouth Mudstone Fmt.	SO 9136 3075	9.7		1.4	23.8	3.1	1.8		0.6						52.1		92.5	22.7	35.8	6.7			16	61	16	4							#	55%
Lias Charmouth Mudstone Fmt.	SP 369 074	27.1	4.4	4.1					2.1						44.1	12.7	96.0	12.7	10.1	10.1	12.7		14	45	35	6								55%
Lias Belemnite Marl	SY 3816 9264	9.6	2.1	1.7	43.5	4.8			2.6						28.6	7.1	100.1	12.8	11.1	4.7	7.1		24	65	5	6								25%
Lias Black Vein Marl	SY 3739 9288	6.0	2.1	1.5	22.6				6.3						46.5	6.5	90.5	14.5	14.6	6.5			13	68	14	5								75%
Lias Lower Lias Clay	SO 946 272	17.4	3.1	1.1	22.0	2.0	0.8		0.8						40.9	8.5	97.2	14.6	18.3	8.0	8.5		22	50	22	6								50%
Lias Lower Lias Clay	SO 946 272	17.0	2.2	2.2	26.0	0.8	0.4		0.7						35.4	8.0	94.0	15.4	10.8	8.0	8.0		20	60	10	3								20%
Lias Lower Lias	ST 750 647	9.1	3.3	1.1	47.5										30.9	6.1	98.0	18.8	5.5	3.6	6.1		27	60	10	3								20%
Lias Lower Lias	ST 750 647	10.6	4.5	1.6	43.8										33.0	5.4	98.9	23.1	7.9	2.0	5.4		30	57	12	1								20%
Lias Blue, Angulata Zone	SY 3315 9138	10.0	0.8	0.8	35.3	3.2	1.0		3.1						34.4	11.8	102.6	11.2	19.5	3.7	11.8		22	70	4	1								55%
Lias Angulata Clays	SK 490615 351788	9.2	1.5	1.0	9.6	1.7			3.1						57.7	7.2	91.0	18.3	32.1	7.2	7.2		15	58	20	7							#	30%
Average N = 24		**14.5**	**2.4**	**1.3**	**16.9**	**0.8**	**0.6**	**0.5**	**1.7**	**0.6**	**0.1**		**0.1**		**51.4**	**5.7**	**96.7**	**15.7**	**25.7**	**10.0**	**5.7**		**17.4**	**57.0**	**18.5**	**3.9**			**3.3**					
Triassic																																		
Mercia Mdst.	SO 852 709	26.9	3.6	3.3	1.7	8.9				1.4					34.5	17.7	98.0	28.4	6.1	9.2	17.7		51	13	16				20					<5%
Mercia Mdst.	SO 852 709	26.4	3.3	2.7	1.6	10.4				1.4					30.7	20.2	96.7	23.0	7.7	10.8	20.2		49	17	14				20					<5%
Mercia Mdst. Cr. Bish. Fmt.	SK 472519 320612	18.2	6.2	3.1		11.2									47.0	12.9	89.3	38.0	9.0	12.9			77	13	10				4					<5%
Mercia Mdst. Cr. Bish. Fmt.	SK 4792 3430	35.1	6.7	4.2		19.8									26.0	7.5	99.3	25.0	1.0		7.5		77	6	13									<5%
Mercia Mdst. Edwal. Fmt.	SK 4462 3305	18.6	5.5	1.4	6.5	4.1				1.6					35.7	27.2	100.4	19.5	16.2	6.2	27.2		34	7	11	13			48					<5%
Mercia Mdst. Gunth. Fmt.	SK 4590 3472	26.6	10.2	2.5	2.5	9.7				1.2					28.1	22.7	102.4	21.9	6.2		22.7		46	11	8	11			35					<5%
Mercia Mdst. Radcl. Fmt.	SK 4479 3179	18.9	11.2	1.1		30.4				1.1					21.9	10.3	94.9	20.2	1.7		10.3		52	32	8	8			8					<5%
Mercia Mdst. Sneit. Fmt.	SK 4391 3097	20.1	11.5	2.5						1.3					46.5	14.4	98.4	41.4	5.1	3.7	14.4		70	23		7									<5%
Average N = 8		**23.9**	**7.3**	**2.5**	**1.5**	**11.8**				**1.0**					**33.8**	**16.6**	**98.4**	**27.2**	**6.6**	**5.7**	**16.6**		**57.0**	**15.3**	**10.9**	**3.9**			**16.9**					

454

APPENDIX A

Stratigraphic Unit	Nat.Grid Ref.	Whole-rock chemical composition: Major elements wt.%												Whole-rock trace element composition ppm											Analysis reference	
1	2	SiO_2	TiO_2	Al_2O_3	Fe_2O_3	CaO	MgO	K_2O	Na_2O	P_2O_5	LOI	Mn	V	Cr	Co	Ni	Cu	Zn	As	Zr	Ba	Pb		
		36	37	38	39	40	41	42	43	44	45	46	47	48	49	50	51	52	53	54	55	56	57	58
Cretaceous																									
Gault Formation	TQ 093 139	56.31	0.90	22.63	6.02	0.30	1.13	3.52	0.18	0.18	8.7	60	210	158	18	191	79	97	98	253	400		cwp-9	679942
Gault Formation	TQ 093 139	57.44	0.94	23.24	3.73	1.13	1.14	3.64	0.18	0.20	8.2	26	223	182	13	91	70	98	29	248	424	21	cwp-10	679943
Gault Formation	TL 1887 3463	31.36	0.53	13.14	4.28	22.24	1.39	2.07	0.15	0.08	24.6	1173	95	81	13	145	37	72	4	174	217	25	cwp-34	679967
Gault Formation	SU 505 925	57.87	0.87	18.18	6.04	1.60	1.28	2.98	0.35	0.09	10.6	292	168	163	17	341	36	111	53	243	349	45	cwp-59	679992
Gault Formation	TL450595	39.80	0.68	15.75	4.20	15.43	1.43	2.60	0.20	0.10	19.7	441	102	105	13	83	19	80	28	160	264	35	cwp-82	680015
Gault Formation	SU 804 407	60.16	0.75	14.88	5.02	4.18	1.48	3.06	0.33	0.09	9.9	91	106	158	15	235		118	97	355	356		cwp-94	680027
Gault Formation	SP 93750 24865	55.14	0.83	16.81	3.63	6.04	1.86	3.01	0.22	0.11	12.2	195	152	158	13	217	22	170		224	336	99	cwp-98	680031
Gault Formation	SU 755 288	57.88	0.92	19.11	6.87	1.08	1.34	2.97	0.16	0.08	9.4	193	177	244	19	273	39	111	30	239	378	38	cwp-18	679951
Gault Formation	SP 4940 2284	53.89	0.80	15.89	5.48	6.51	1.76	2.66	0.20	0.06	12.5	194	129	151	17	697	35	129	396	238	385	170	cwp-19	679952
Gault Formation	TL 5188 2346	40.13	0.65	13.56	3.74	17.73	1.40	2.13	0.19	0.09	20.3	364	95	110	12	249	58	69	9	203	244	20	cwp-20	679953
Gault Formation	TL 5188 2346	32.86	0.54	12.65	4.26	22.13	1.39	1.97	0.17	0.07	23.7	1466	106	85	13	150	44	74	22	169	209	23	cwp-21	679954
Average N = 11		**49.4**	**0.8**	**16.9**	**4.8**	**8.9**	**1.4**	**2.8**	**0.2**	**0.1**	**14.5**	**409**	**142**	**143**	**15**	**243**	**40**	**103**	**70**	**228**	**324**	**44**		
Weald Clay	TQ 112 372	67.10	1.42	15.77	6.31	0.27	0.78	2.13	0.29	0.09	5.7	302	141	147	18	307	37	104	25	277	437	34	cwp-11	679944
Weald Clay	TQ 112 372	69.20	1.49	14.65	5.88	0.25	0.74	2.11	0.28	0.07	5.2	282	135	146	17	337	33	90	29	272	419	34	cwp-12	679945
Weald Clay	SY965815	64.15	0.83	15.03	8.17	1.18	0.67	1.60	0.13	0.09	8.0	528	125	264	21	613			29	280	259		cwp-45	679978
Speeton Clay	TA 1500 7571	55.01	0.87	16.66	3.17	3.42	1.23	3.05	0.78	0.06	11.2	160	187	264	12	66	35	102	21	216	357	23	cwp-93	680026
Average N = 4		**63.9**	**1.2**	**16.6**	**5.9**	**1.3**	**0.9**	**2.2**	**0.4**	**0.1**	**7.5**	**318**	**147**	**175**	**17**	**331**	**26**	**74**	**26**	**261**	**368**	**23**		
Jurassic																									
Kimmeridge Clay	SY 92 78	64.54	0.94	15.85	2.77	2.32	1.09	2.89	0.45	0.12	8.9	156	131	181	11	178	92	147	9	245	314	8	cwp-43	679976
Kimmeridge Mudstone	SY 92 78	30.69	0.54	11.32	6.82	18.92	1.26	1.26	0.34	0.53	28.2	270	124	117	18	149	74	35	31	114	173		cwp-44	679977
Oxford Clay	SY700736	51.66	0.87	18.37	4.00	5.46	1.51	3.36	0.24	0.31	14.1	290	144	161	13	147	74	60	70	162	315		cwp-47	679980
Oxford Clay	SP 369 074	40.99	0.71	14.85	3.40	16.98	1.51	2.68	0.26	0.12	18.4	440	140	135	12	65	38			129	249	25	cwp-3	679936
Oxford Clay	SP 369 074	40.63	0.73	14.86	3.73	16.87	1.51	2.70	0.24	0.12	18.5	323	130	184	12	60	48	74	116	136	255		cwp-4	679937
Oxford Clay	TL 504525 246467	51.31	0.92	17.50	3.42	4.87	1.37	2.91	0.30	0.14	17.1	91	119	196	12	234	29	86	12	183	327	19	cwp-99	680032
Blisworth clay, Bathonian	SP 964 662	52.17	1.18	23.24	3.02	1.56	2.69	0.93	0.13	0.05	11.9	62	258	117	22	60	66	51	7	259	248	31	cwp-73	680006
Blisworth clay, Bathonian	SP 964 662	48.46	1.12	24.03	11.43	0.96	1.17	2.12	0.11	0.12	10.4	118	315	208	27	71	59	51	20	229	255	34	cwp-74	680007
Blisworth clay, Bathonian	SP 964 662	50.85	1.07	24.60	7.85	1.10	1.23	2.32	0.11	0.07	10.7	106	223	163	21	60	47	65	8	203	206	30	cwp-75	680008
Lias Whitby Mudstone	NZ 8317 1599	52.16	0.96	20.94	7.40	0.63	1.55	3.25	0.60	0.13	12.3	66	153	152	19	224	26	73	69	168	430	8	cwp-91	680024
Lias Dyrham	NZ 835 149	54.44	0.91	17.87	4.20	2.88	1.42	2.68	0.71	0.20	14.6	255	126	175	14	129	10		37	161	370		cwp-89	680022
Lias Redcar Mudstone	NZ 9778 0223	50.68	0.99	22.04	6.90	2.71	1.63	3.11	0.53	0.07	11.2	195	294	163	18	91	9	145	26	189	328	33	cwp-90	680023
Lias Redcar Mudstone	NZ 9644 0113	53.83	0.96	18.92	8.03	2.43	2.00	3.08	0.67	0.16	9.7	564	185	195	21	316	25		85	234	336	27	cwp-92	680025
Lias Charmouth Mudstone Fmt.	SO 9136 3075	37.97	0.68	16.01	5.02	14.20	2.34	2.25	0.31	0.09	21.0	471	141	114	14	117	64	131	4	144	245	35	cwp-22	679955
Lias Charmouth Mudstone Fmt.	SP 1795 3699	59.25	0.68	17.26	6.53	1.69	1.56	2.79	0.69	0.16	9.0	429	131	164	18	283	31	118	21	203	398	4	cwp-85	680018
Lias Belemnite Marl	SY 3816 9264	28.43	0.47	9.73	4.73	25.10	2.69	2.03	0.50	0.09	26.1	514	169	85	13	168		55		233	241	39	cwp-87	680020
Lias Black Vein Marl	SY 3739 9288	32.31	0.59	13.67	7.90	13.42	1.59	2.26	0.35	0.19	27.5	453	308	200	15	282		31		82	246	91	cwp-88	680021
Lias Lower Lias Marl	SO 946 272	47.03	0.74	14.46	4.23	11.82	2.19	2.57	0.64	0.12	16.1	359	129	133	13	40	38	23	23	218	297		cwp-5	679938
Lias Lower Lias Clay	SO 946 272	44.26	0.69	12.93	4.69	14.41	2.19	2.29	0.57	0.10	17.8	691	104	160	13	243	46	79	16	229	317	35	cwp-6	679939
Lias Lower Lias	ST 750 647	31.20	0.48	10.32	4.25	23.74	1.30	2.57	0.32	0.12	25.5	704	102	145	13	140	53	108	18	151	275	30	cwp-106	680039
Lias Lower Lias	ST 750 647	34.19	0.49	10.87	4.59	21.90	1.46	2.80	0.33	0.13	23.0	731	93	118	16	111	59	104	18	162	637	27	cwp-107	680040
Lias Blue, Angulata Zone	SY 3315 9138	28.55	0.43	9.54	5.98	19.69	2.05	2.22	0.26	0.15	30.8	280	268	118	11	152	169			114	2112	53	cwp-86	680019
Lias Angulata Clays	SK 490615 351788	45.88	0.80	20.04	7.48	6.37	2.35	3.10	0.85	0.10	12.9	639	143	138	20	135	35	85	9	121	327	37	cwp-36	679969
Average N = 24		**44.8**	**0.8**	**16.7**	**5.6**	**10.0**	**1.6**	**2.5**	**0.4**	**0.1**	**17.2**	**357**	**171**	**157**	**16**	**150**	**52**	**66**	**26**	**177**	**387**	**25**		
Triassic																									
Mercia Mdst.	SO 852 709	56.12	0.69	13.76	4.99	3.74	5.66	4.10	0.57	0.15	10.0	641	88	139	15	352	28	121	22	196	494	33	cwp-1	679934
Mercia Mdst.	SO 852 709	55.65	0.69	13.63	4.82	4.08	5.96	4.09	0.56	0.15	10.2	674	90	131	14	341	126	115	10	190	502	34	cwp-2	679935
Mercia Mdst. Cr. Bish. Fmt.	SK 472519 320612	52.22	0.72	15.81	6.93	3.46	5.14	4.71	0.79	0.14	9.9	316	111	128	18	316	36	22	19	163	399	27	cwp-33	679966
Mercia Mdst. Cr. Bish. Fmt.	SK 4792 3430	58.16	0.58	10.00	2.34	6.01	6.21	3.07	0.74	0.12	12.5	500	100	213	18	551	17	63	51	252	706	48	cwp-27	679960
Mercia Mdst. Edwal. Fmt.	SK 4462 3305	51.01	0.69	13.10	6.14	5.18	9.43	3.19	0.25	0.18	10.6	707	81	147	10	340	36	83	12	264	652	27	cwp-24	679957
Mercia Mdst. Gunth. Fmt.	SK 4590 3472	56.64	0.72	12.49	5.05	4.15	6.86	3.80	0.36	0.17	9.5	333	89	160	15	475	32	66	26	355	497	33	cwp-23	679956
Mercia Mdst. Radcl. Fmt.	SK 4479 3179	41.33	0.50	11.20	4.55	9.69	9.02	4.31	0.16	0.13	18.8	2457	62	96	10	221	54	84	44	105	336	31	cwp-25	679958
Mercia Mdst. Snett. Fmt.	SK 4391 3097	59.56	0.88	18.27	6.47	0.68	2.51	6.34	0.50	0.18	4.4	389	139	152	18	318	39	110	14	245	559	34	cwp-26	679959
Average N = 8		**53.8**	**0.7**	**13.5**	**5.2**	**4.6**	**6.3**	**4.2**	**0.5**	**0.2**	**10.7**	**752**	**95**	**146**	**15**	**364**	**46**	**96**	**25**	**221**	**518**	**33**		

APPENDIX A

Stratigraphic Unit	Nat.Grid Ref.	Whole-rock mineralogy wt.%															Whole rock-clay mineralogy wt.%								Clay fraction (<2μm) clay mineralogy wt.%										
1	2	Quartz 3	K-feldspar 4	Plagioclase 5	Calcite 6	Dolomite 7	Siderite 8	Aragonite 9	Pyrite 10	Hematite 11	Goethite 12	Jarosite 13	Anatase 14	Amphibole 15	Di-layer silicate 16	Tri-layer silicate 17	Total 18	Illite/mica 19	Exp. di—layer silicate 20	Kaolin 21	Chlorite 22	Paragonite 23	Illite/Mica 24	Illite-smectite 25	Kaolin 26	Chlorite 27	Smectite 28	Vermiculite 29	Corrensite 30	Kaolin/Smectite 31	Paragonite 32	Berthierine 33	17Äng. peak 34	% EXP I-S 35
Carboniferous																																		
Etruria Marl	SP 435470 288500	20.7	3.8	0.9											74.2		99.6	74.2					28	70	2									15%
Middle–Lower Etruria Measures	SJ 832 460	21.6		0.4			10.7								61.6		96.1	9.5	23.6	28.5	4.7		23	37	40									20%
Middle–Lower Etruria Measures	SJ 832 460	37.3	2.2	0.7			0.4								51.8		99.6	7.5	20.6	23.7	4.7		20	47	33									20%
Westphalian B/C	NZ 324 483	19.8	1.4	1.5						3.5	1.8		1.1		70.3	4.7	98.8	65.0	0.6	4.7	10.2		33	50	10	7								15%
Westphalian B	SK 370 442	27.4	2.2								3.7		1.0		57.0	10.2	97.8	37.2	7.3	12.5	10.2		27	52	10	11								15%
Westphalian B	SD 606 045	35.0	2.1	5.9			11.0								34.5	7.7	98.2	23.5	5.1	5.9	7.7		32	44	16	8								15%
Westphalian A	SS 900 060	20.3	1.4	1.1									1.4		59.7	14.7	98.6	51.0	0.0	8.7	14.7		33	43	16	8								10%
Accrington Mudstone	SD 757 305	20.6	1.7	4.6					0.6				1.4		53.2	18.0	99.5	45.0	0.0	7.8	18.0		29	49	12	10								10%
Westphalian,Carb. Limestone Group	NT 269 599	9.3	1.1				8.8						0.7		75.9		96.4	10.0	24.3	41.6			6	36	58									15%
Namurian	SD 577 648	19.9	3.5	4.0			2.1						0.8		51.1	14.1	95.5	15.4	14.1	21.6	14.1		9	55	26	10								25%
Average N = 10		23.2	1.9	1.9			3.3		0.1	0.4	0.6		0.6		58.9	6.9	97.8	33.8	9.6	15.5	6.9		24.0	48.3	22.3	5.4								
Devonian																																		
Stromness Flags (Devonian)	HY 245 077	16.6	28.6	4.9	2.8	6.4			1.4						37.3	14.1	98.0	37.3	8.5	26.2			92	8										<5%
Caithness Flagstones, Middle ORS	ND 257 743	10.3	12.2	10.9	2.4	2.6			1.7						31.6	26.2	97.9	23.1	1.8	15.5			37	9	11	43								<5%
Caithness Flagstones, Middle ORS	ND 263 353	31.6	21.0	12.4	4.5										13.8	15.5	98.8	12.0		1.8	15.5		41	3		56								<5%
Middle ORS	NH 865 754	25.4	9.0	12.5	4.2										27.8	21.8	100.7	21.1		6.7	21.8		15	32		53								20%
Lower ORS	SO 973 227	23.1		4.9											49.5	18.8	96.3	49.5			18.8		71	9		20								<5%
Lower ORS	ST 308 915	21.7	1.4	3.3						2.9					52.4	18.3	100.0	34.1	18.3		18.3		44	31		25								10%
Meadfoot Beds, Lower Devonian	SW 78 62	31.8	2.5	1.0											42.5	21.9	99.7	40.5	2.0		21.9		50			43	7						#	100%
Devonian	SX 518 546	21.6	1.7	3.3						1.5					54.9	13.2	96.2	54.9			13.2		75			25								
Average (N = 8)		22.8	9.6	6.7	1.7	1.1			0.4	0.6					38.7	17.0	98.5	34.1	4.7		17.0		53.1	11.5	1.4	33.1	0.9							
Lower Palaeozoic																																		
Llandovery K-bentonite	NT 1956 1584	18.9	2.5	8.0	0.7										66.2	5.1	101.4	66.2			5.1		11	88								1		5%
Silurian	SJ 265 100	31.3	2.6	5.1											48.9	12.1	100.0	16.3	32.6		12.1		26	56	9	9						#		15%
Ordovician slate	NS 9613 1697	26.8	2.9	13.7											30.7	23.9	98.0	30.7			23.9		47	1		52								<5%
Ordovician, Caradoc mudstone	SH 3645 3482	25.1	0.6	4.0					1.2						45.4	20.8	97.1	45.4			20.8		77	9		14						#		<5%
Ordovician	SH 490 615	19.4	0.7	2.6											60.5	18.8	102.0	18.5	42.0		18.8		20	72		8						#		20%
Ordovician	SM 948 336	35.3	2.5	0.9					0.8						37.4	20.9	97.8	31.4			20.9	6.0	59	9		13						#		
M.Cambrian K-bentonite	SM 7861 2425	31.5													50.6	16.9	99.0	50.6			16.9		78		9	13					28			<5%
Cambrian	SH 617 643	30.7	4.0	13.0						5.3					31.5	10.5	95.0	26.5			10.5	5.0	51			16					33			
Average N = 8		27.4	2.0	5.9	0.1				0.3	0.7					46.4	16.1	98.8	35.7	9.3		16.1	1.4	46.1	29.4	1.1	15.6					7.6	0.3		
Average, all data, N = 105		23.9	3.7	2.4	7.5	1.3	0.5	0.2	0.5	0.2	0.6	0.0	0.1	0.0	48.7	7.5	97.1	20.8	17.5	10.2	7.5	0.1	27.0	44.9	17.6	7.1	0.1	0.2	1.3	1.2	0.6	0.0		0.4

APPENDIX A

Stratigraphic Unit	Nat.Grid Ref.	Whole-rock chemical composition: Major elements wt.%										Whole-rock trace element composition ppm											Analysis reference	
1	2	36	37	38	39	40	41	42	43	44	45	46	47	48	49	50	51	52	53	54	55	56	57	58
		SiO_2	TiO_2	Al_2O_3	Fe_2O_3	CaO	MgO	K_2O	Na_2O	P_2O_5	LOI	Mn	V	Cr	Co	Ni	Cu	Zn	As	Zr	Ba	Pb		
Carboniferous																								
Etruria Marl	SP 435470 288500	55.79	0.82	16.75	14.39	0.39	1.55	4.44	0.17	0.20	5.4	174	125	140	33	201	129	65	33	279	323	34	cwp-104	680037
Middle–Lower Etruria Measures	SJ 832 460	52.13	1.11	20.50	10.95	0.43	0.99	1.55	0.10	0.07	11.7	3749	178	115	26	140	84	84	8	171	404	31	cwp-102	680035
Middle–Lower Etruria Measures	SJ 832 460	63.58	1.24	16.77	8.94	0.25	0.62	1.41	0.07	0.10	6.8	945	154	160	23	278	45	76	10	263	589	30	cwp-103	680036
Westphalian B/C	NZ 324 483	56.64	0.77	23.13	3.57	0.34	1.51	5.01	0.41	0.08	8.4	300	147	162	12	119	56	123	14	138	685	33	cwp-70	680003
Westphalian B	SK 370 442	59.01	1.02	20.99	5.16	0.31	1.62	3.76	0.25	0.12	7.6	603	145	130	16	195	58	109	30	158	505	37	cwp-69	680002
Westphalian B	SD 606 045	61.43	0.91	14.18	9.51	0.60	1.40	2.32	1.02	0.16	8.1	1568	92	174	24	602	44	96	16	297	455	36	cwp-72	680005
Westphalian A	SS 900 060	58.03	0.93	24.56	2.55	0.23	1.60	4.49	0.73	0.12	6.6	110	159	139	11	124		151		135	766	41	cwp-66	679999
Accrington Mudstone	SD 757 305	58.75	0.92	21.80	4.65	0.36	1.99	4.07	0.92	0.13	6.3	226	146	131	15	177	97	48	27	156	575		cwp-67	680000
Westphalian,Carb. Limestone Group	NT 269 599	43.15	0.81	23.83	7.16	0.59	0.70	1.77	0.04	0.23	21.5	1336	154	156	19	187	66	48	45	102	274	37	cwp-71	680004
Namurian	SD 577 648	55.66	0.91	20.95	6.31	0.50	1.69	2.64	0.77	0.13	10.3	466	107	177	18	324	50	80	21	204	408	35	cwp-68	680001
Average N = 10		**56.4**	**0.9**	**20.3**	**7.3**	**0.4**	**1.4**	**3.1**	**0.4**	**0.1**	**9.2**	**948**	**141**	**148**	**20**	**235**	**63**	**83**	**20**	**190**	**498**	**31**		
Devonian																								
Stromness flags (Devonian)	HY 245 077	58.91	0.69	18.14	2.19	2.37	1.31	9.73	0.78	0.08	5.6	208	127	195	10	390	32	27	73	131	500	45	cwp-76	680009
Caithness Flagstones, Middle ORS	ND 257 743	52.87	0.73	17.66	8.25	2.61	3.97	5.64	1.76	0.14	6.2	588	138	151	22	301	65	83	51	136	605	43	cwp-77	680010
Caithness Flagstones, Middle ORS	ND 263 553	59.16	0.69	12.91	4.06	4.63	3.16	4.99	1.81	0.16	8.2	566	113	195	13	499	55	58	60	309	496	43	cwp-79	680012
Middle ORS	NH 865 754	59.68	0.75	15.63	6.09	2.58	3.70	3.79	1.93	0.16	5.5	440	94	161	18	362	58	101	47	193	844	40	cwp-78	680011
Lower ORS	SO 973 227	57.97	0.95	19.78	7.40	0.40	2.73	4.50	0.82	0.16	5.1	454	138	168	20	335	22	107	32	188	516	46	cwp-80	680013
Lower ORS	ST 308 915	56.25	0.82	19.19	8.69	0.45	3.19	4.25	0.81	0.12	6.0	672	134	173	22	346	31	250	26	165	656	74	cwp-81	680014
Meadfoot Beds, Lower Devonian	SW 78 62	61.14	0.93	19.47	5.70	0.33	2.46	4.23	0.44	0.17	4.9	852	153	204	17	272	108		131	211	574		cwp-96	680029
Devonian	SX 518 546	56.93	0.85	23.04	5.02	0.27	2.63	5.34	0.80	0.12	4.8	727	141	133	16	38	10	120	44	149	626		cwp-65	679998
Average (N = 8)		**57.9**	**0.8**	**18.2**	**5.9**	**1.7**	**2.9**	**5.3**	**1.1**	**0.1**	**5.8**	**563**	**130**	**173**	**17**	**318**	**47**	**93**	**58**	**185**	**602**	**36**		
Lower Palaeozoic																								
Llandovery K-bentonite	NT 1956 1584	58.78	0.81	21.25	3.52	0.99	2.18	5.25	1.40	0.19	5.4	942	44	45	12	136	24	70	2	360	991	32	cwp-37	679970
Silurian	SJ 265 100	63.28	1.00	18.64	5.49	0.29	1.49	3.47	0.88	0.07	5.3	128	122	166	17	181				305	512	1	cwp-64	679997
Ordovician slate	NS 9613 1697	59.14	0.92	17.92	6.84	0.27	4.35	3.46	1.95	0.19	4.8	170	147	220	18	176	11	64	43	258	597	22	cwp-38	679971
Ordovician, Caradoc mudstone	SH 3645 3482	58.62	0.91	20.68	6.45	0.22	1.98	4.63	0.65	0.10	5.6	269	139	123	19	59	69	14	104	149	524	36	cwp-39	679972
Ordovician	SH 490 615	53.61	0.87	25.14	5.85	0.40	1.84	3.37	0.88	0.13	7.6	1826	123	114	17	93		24	50	82	635		cwp-62	679995
Ordovician	SM 948 336	63.19	1.01	20.28	4.32	0.20	1.45	3.16	0.78	0.21	5.0	1759	129	118	14	161		243		180	997	47	cwp-63	679996
M.Cambrian K-bentonite	SM 7861 2425	57.92	0.82	22.13	3.50	0.60	1.42	5.27	0.66	0.13	6.5	8662	78	107	13	122	77	273	17	218	878	125	cwp-40	679973
Cambrian	SH 617 643	62.00	0.78	18.58	7.45	0.63	1.87	3.07	2.04	0.10	3.2	1293	102	71	20	105	13	103	30	141	475	36	cwp-61	679994
Average N = 8		**59.6**	**0.9**	**20.6**	**5.4**	**0.4**	**2.1**	**4.0**	**1.2**	**0.1**	**5.4**	**1881**	**110**	**120**	**16**	**129**	**24**	**99**	**31**	**212**	**701**	**38**		
Average, all data, N = 105		**54.8**	**0.9**	**17.5**	**5.2**	**4.8**	**1.8**	**3.0**	**0.5**	**0.1**	**11.3**	**561**	**135**	**155**	**15**	**254**	**41**	**84**	**38**	**213**	**444**	**33**		

mineralogical data. Secondly, some trace elements were below the limit of detection in some samples. A blank cell in the table indicates such data. Thirdly, organic matter and sulphur are other chemical components that will be present in some samples in significant amounts but these were not analysed.

A.4. Comments on the mineralogical data

The following comments on the mineralogical compositions of the clay materials analyzed for this appendix are made simply in relation to existing data, mainly from Perrin (1971). The aim is to attempt to put the new data into context with existing data and to remark upon features which are not known to have been noted previously, or that require emphasis. In keeping with Perrin (1971) and in geological tradition this starts with the oldest clay materials.

A.4.1. Lower Palaeozoic

A total of eight Lower Palaeozoic samples were analyzed, which may be compared with 46 reported by Perrin (1971). In the clay fraction, these samples consist of assemblages of 'well crystallized' illite and chlorite, with very little if any trace of expandable mixed-layer illite–smectite, and only rare occurrences of kaolin. Perrin (1971) noted the exception of bentonites to this general trend since these may contain larger amounts of more highly expandable mixed-layer illite–smectite and this is borne out by the present data. Additionally, the present data indicated the occurrence of the sodium mica called paragonite in two samples, and berthierine which was positively identified in one sample and suspected in several. In the whole rock none of the analyzed samples contained any significant amount of any carbonates, but the generality of this observation is unknown. It is also notable that the Lower Palaeozoic samples are the only set of samples where, on average, plagioclase is more abundant than K-feldspar.

A.4.2. Devonian

Devonian samples show a similar pattern to the Lower Palaeozoic samples, in that illite and chlorite are dominant, kaolin rare and mixed-layer illite–smectite moderate in abundance but generally of low expandability. One sample was unusual in consisting of an assemblage of extremely well crystallized, illite and chlorite together with minor but pure smectite. In the bulk analyses carbonates are not uncommon, with both calcite and dolomite present, and some samples have unusually high contents of K-feldspar.

A.4.3. Carboniferous

The most notable feature of the Carboniferous samples is that relatively high concentrations of kaolinite are observed. This is a well-known feature (Perrin 1971), indeed some types of Carboniferous mudrocks may be almost pure kaolin and have been commercially exploited as a result. Compared to older rocks, concentrations of expandable mixed-layer illite–smectite are also generally elevated, but highly variable. In the bulk analyses the frequent occurrence of the iron carbonate siderite is a notable feature.

A.4.4. Permo-Trias

All the materials analyzed are Mercia Mudstone, formerly Keuper Marl. Although sample coverage of the Permo-Trias is clearly unbalanced in a geological sense, economically and from an engineering perspective the Mercia Mudstone is the most important clay material of this age. The most notable feature of this group of samples is the common presence of abundant chlorite and corrensite. Other notable features of the clay mineralogy include the complete absence of kaolinite and the low concentrations of mixed-layer illite–smectite, which is generally detectable but of very low expandability. Perrin (1971) also noted the rare (absent?) occurrence of kaolin in Mercia Mudstone. In the bulk analyses substantial amounts of dolomite are very common and the abundance of potassium feldspar is frequently high.

A.4.5. Jurassic

Samples from the Jurassic are very varied. However, as far as the clay mineralogy is concerned it is notable that the appearance of a peak at about 17Å in diffraction patterns of glycolated samples becomes common in Jurassic samples. This peak is due to the presence of random mixed-layer illite–smectite. Previously, especially in the older literature, this peak has often been taken to indicate the presence of pure smectite. Although pure smectite may be encountered, the results of the current analyses indicate that this is rarely the case. Thus many of the clays minerals identified as smectites in some of the older literature are probably more accurately identified as random mixed-layer illite–smectites. Data in this appendix shows when a 17Å peak was observed, and also gives an estimate of the expandability (% smectite layers) of the mixed-layer clay. Small amounts of chlorite were detected in all the Liassic samples. Previously Perrin (1971) noted that the occurrence of chlorite tended to become sporadic above the Lias. The new data presented here, generally concur with this observation. Three samples all from the same locality in the Blisworth clay contained appreciable amounts of mixed-layer kaolinite-smectite. This is a clay mineral that has not previously been reported in the published literature from any geological clay formation in the UK. As far as the bulk analyses are concerned, traces of dolomite are notable in many Liassic samples, siderite also is a common carbonate detected in minor amounts throughout the Jurassic set, but the main carbonate is calcite, which can be abundant.

Jurassic samples are also conspicuous for the consistent, but not ubiquitous presence, of minor amounts of pyrite.

A.4.6. Cretaceous

Most of the Cretaceous samples examined for this report are from the Gault Formation. Compared to other samples, on average, the Gault contains more mixed-layer illite-smectite in the clay fractions than any other stratigraphic grouping analyzed for this report. Some samples also contained traces of a clay mineral behaving like vermiculite, and this feature was also noted in Perrin (1971). Analysis of the whole rock indicates that some Gault samples are also exceedingly clay-rich $\geq 70\%$, although like some Jurassic samples substantial amounts of calcite may 'dilute' the silicate component.

A.4.7. Palaeocene

All Paleocene samples are from the Reading Formation. The most distinctive feature here was the recognition of several samples that contain mixed-layer kaolinite-smectite. To our knowledge this clay mineral has not been reported previously from this formation. Kaolinite-smectite is, however, well known in the early Eocene Argiles Plastiques formation from the Paris Basin, France, which is of a very similar age, for example see Amouric & Olives (1998).

A.4.8. Eocene/Oligocene

The Eocene samples analyzed are all London Clay, but in addition a set of Devon Ball Clays of Eocene/Oligocene age were also analyzed. The London Clay samples are generally rich in mixed-layer illite-smectite, but also may contain realtively large amounts of kaolinite. Chlorite is also a consistent component. Some London Clay samples contained minor amounts of carbonates, but in common with the other Tertiary samples analyzed carbonates do not appear to figure as much as they do in Mesozoic samples. Another notable feature is the potassium feldspar content, which appears to be higher than average. As expected the Ball Clays were all dominated by kaolinite, but most also contain substantial amounts of illite and illitic mixed-layer illite–smectite. In contrast to the London Clay samples, chlorite was not found in any of the Ball clays.

A.4.9. Quaternary

The Quaternary samples are an extremely varied set and have extremely varied compositions. Undoubtedly, this is a reflection of their diverse origins as glacial, fluvi-glacial, alluvial and aeolian materials. One notable feature is the generally elevated feldspar contents, with respect to both potassium and plagioclase feldspar. Calcite is also a major constituent of some samples. Both of the Brick Earth samples have high quartz contents.

Acknowledgements. G. West, P. Hobbs, R. Merriman, S. Kemp, C. and M. Adams, K. Privett, G. Witte, M. Czerewko, J. Cripps, R. Price, A. Merrit, M. Crilly, P. Sherwood, P. Horner, M. Lawler, P. Halsam-Brunt, G. M. Reeves and D. Jordan are gratefully acknowledged for provision of samples.

References

AMOURIC, M. & OLIVES, J. 1998. Transformation mechanisms and interstratification in conversion of smectite to kaolinite: an HRTEM study. *Clays and Clay Minerals*, **46**, 521–527.

CHATZIDIMITRIADIS, E., TSIRAMBIDES, A. & THEODORIKAS, S. 1993. Clay mineral abundance of Greek shales and slates through geologic time. *Neueo Jahrbuch für Mineralogie–Monatshefle*, **7**, 297–311.

HILLIER, S. 2000. Accurate quantitative analysis of clay and other minerals in sandstones by XRD: comparison of a Rietveld and a reference intensity ratio (RIR) method and the importance of sample preparation. *Clay Minerals*, **35**, 291–302.

HILLER, S. 2003. Quantitative analysis of clay and other minerals in sandstones by X-ray powder diffraction (XRPD). *International Association of Sedimentologists, Special Publication*, **34**, 213–251.

JEANS, C. V. & MERRIMAN, R. J. (eds) 2006. Clay minerals in onshore and offshore strata of the British Isles. *Clay Minerals* (special issue), **41**. Mineralogical Society, London. 600pp.

O'BRIEN, N. R. & SLATT, R. M. 1990. Argillaceous Rock Atlas. Springer Verlag, New York, 141pp.

PERRIN, R. M. S. 1971. *The Clay Mineralogy of British Sediments*. Mineralogical Society, London. 247pp.

RIDGWAY, J. M. 1982. *Common Clay and Shale*. Mincral Resources Consultative Committee. HMSO. London. Mineral Dossier 22.

SELLWOOD, B. W. & SLADEN, C. P. 1981. Mesozoic and Tertiary argillaceous units, distribution and composition. *Quarterly Journal of Engineering Geology*, **14**, 263–276.

SHAW, H. F. 1981. Mineralogy and petrology of the argillaceous sedimentary rocks of the UK. *Quarterly Journal of Engineering Geology*, **14**, 277–290.

VICZIAN, I. 2002. Typical clay mineral associations from geological formations in Hungary—A review of recent investigations. *Geologica Carpathica*, **53**, 65–69.

WEAVER, C. E. 1967. The significance of clay minerals in sediments. In: NAGY, B. & COLOMBO, U. (eds), *Fundamental Aspects of Petroleum Geochemistry*. Elsevier, Amsterdam, 37–76.

WEAVER, C. E. 1989. *Clays Muds and Shales*. Developments in sedimentology, 44. Elsevier, Amsterdam. 820pp.

Appendix B. Properties data

This appendix presents compiled data for various UK clay and mudrock formations arranged in order of increasing age. The data are mainly those presented by Cripps & Taylor (1981, 1986, 1987), in British Geological Survey (BGS) reports on the engineering geology of particular areas relating to particular deposits, and from other published sources.

The data are intended to serve as a general guide as to the typical properties and variability of the materials concerned. The tables also provide a reference source for further information about the formations. The values quoted are intended as guidance only and should not be used as a substitute for investigations into the properties of the materials. It is the intention that the data will enable investigations to be more focussed and to assist with the interpretation of test data. The properties covered in the tables are ones that are commonly quoted in site investigation reports and publications, including consistency, density and strength. Stiffness, consolidation, permeability and compaction parameters are also quoted for some formations but many parameters mentioned in Chapter 4 are not included. In most cases this is because few values of these parameters are quoted in the literature. Chapter 4 includes background and discussion of the parameters quoted in the tables and also of some additional parameters.

The data presented by Cripps & Taylor (1981, 1986, 1987) were compiled from a great number of published sources available at that time, which were the result of a comprehensive literature review. These data are presented as a mean and range of values, where the range does not include what appear to be extreme values. Although such values may be an accurate representation of the material tested, they do not reflect typical behaviour. A less extensive literature review has been performed to complete the present data tables. Instead data from several sources are presented for comparison. The BGS reports, however, are the result of comprehensive studies involving the selection of data from what are judged to be the most reliable and comprehensive site investigations that, in some cases, have been supplemented with additional testing. Values taken from BGS Reports are usually quoted as a median or sometimes a range of medians, with the first and third percentiles. This is so that the extreme values do not give rise to a biased view of the materials. This latter method has a sounder scientific basis than the more subjective one used by Cripps & Taylor (1981, 1986, 1987) who quote average and range values but exclude extreme values. In many cases, percentiles other than the first and third quartiles and the number of samples tested, together with the values for some other parameters such as the grading and amounts of sulphate present, are also given in the BGS Reports. Besides the numerical data themselves, many of the original data sources provided further information about the tests and materials tested. Salient points are included in the discussion in Chapter 4, but where an important decision is to be based on the values of parameters, the reader would be advised to refer to the original data sources.

As explained in Chapter 4, variation in the values of geotechnical parameters arises as a consequence of differences of grain size, mineralogy and structure of the materials. In turn these depend on the environment and processes of sedimentation, the loading history, including present depth of burial, and action of weathering processes subsequent to deposition as well as being a consequence of the effects of disturbance during sampling and the methods and conditions of testing, particularly the moisture content and stress conditions. Although data have been selected from sources judged to be reliable, unfortunately not all authors provide full details of the methods and conditions of testing. For example, consolidation and deformation parameters are dependent on the stress range used for determinations but these may not be quoted. In view of the variation of consolidation parameters with stress level, most of the BGS Reports quote the coefficient of consolidation and volume compressibility in terms of defined classes. The values quoted in the tables are the upper and lower limits of these classes or, where appropriate, several classes. Not all authors are explicit about the grade of weathering of the material tested. Most of the data have been derived in the course of site investigations, although some originate from research projects or, as noted above for BGS Reports, specific programmes of testing. Therefore it can be expected that where not otherwise indicated, the conditions are those that, would apply for standard site investigations in which a range of weathered and unweathered materials derived form depths of less than about 20 m will have been tested. Where possible, the weathering grade is specified and the data being assembled into those for unweathered (zone I, fresh and zone II, slightly weathered) and weathered (zone III, moderately weathered and zone IV, completely weathered) materials.

Differences in the composition, loading history and weathering give rise to variation of geotechnical parameters with depth and location. In some cases, for instance where deposits are laid in thin laminae of

TABLE B1. *Engineering properties of tills from England, Wales and Scotland*

Material						Till (Boulder Clay)							
Place	NE England		Leeds	Deeside	Wrexham	N Midlands England	Nottingham	Birmingham West	Taff Valley, South Wales	SW Essex	Glasgow		
Authors	Jackson & Lawrence (1990)		Lake et al. (1992)	Culshaw & Crummy (1988)	Waine et al. (1990)	Waine et al. (1991)	Forster (1992)	Forster (1991d)	Fookes et al. 1975	Culshaw & Crummy (1991)	McKinlay (1969)		
Weathering grade	UW	W									UW		W
Moisture content w%	17 [1–38]	21 [8–40]	16 [13–19]	14 [2–43]	19 [14–25]	15 [13–18]	14 [12–17]	14 [11–17]	13 [3–21]	21 (17–25)	13 [9–16]		20 [10–30]
Liquid limit w_L %	40 [5–61]	45 [27–63]	34 [29–39]	33 [15–66]	32 [25–41]	32 [28–7]	32 [28–41]	28 [24–35]	24 [14–35]	42 (33–59)	31 [32–42]		32 [23–42]
Plastic limit w_P %	17 [9–30]	20 [13–32]	17 [16–19]	17 [10–31]	17 [15–20]	17 [15–19]	16 [14–19]	15 [13–18]	14 [8–20]	21 (18–23)	17 [12–27]		14 [7–22]
Plasticity index I_p %	23 [0–45]	25 [10–43]	16 [13–21]	15 [2–39]	16 [11–20]	15 [12–19]	17 [14–22]	14 [10–18]	10	21 (15–33)	14 [7–21]		18 [11–26]
Clay fraction %			15 [9–19]	9 [0–34]		27			21 inc silt		18		9
Bulk density γ_b Mg/m³	2.12 [1.51–2.21]	2.03 [1.52–2.27]	2.10 [2.03–2.17]	2.14 [1.63–2.37]	2.16 [2.08–2.23]	2.16 [2.07–2.23]	2.12 [2.08–2.18]	2.19 [2.11–2.05]		2.00 (1.93–2.06)	2.24 [2.08–2.40]		2.08 [1.84–2.32]
Dry density γ_d Mg/m³	1.87 [1.45–2.21]	1.76 [1.67–1.86]	1.92 [1.80–1.98]	[1.49–2.37]	[1.68–1.86]	1.85 [1.75–1.95]		1.94 [1.84–2.05]		1.64 (1.57–1.76)			
Undrained cohesion c_u kPa	132 [max 445]	108 [max 379]	93 [64–147]	63 [max 175]	130 (90–190)	85 (48–135)	193 (58–250)	80 (40–125)	116 [36–221]	70 (54–108)	s = 150 [20–100] #		s = 70 [20–60] #
Undrained friction angle ϕ_u °	1 [0–40]	2 [0–32]	5 [0–9]	9 [0–33]	0	0 (0–5)	9 (6–12)	4 (0–12)		0 (0–13)			
Effective shear strength c', ϕ'°	* [0.05–1.5]				$\phi' = 31.5°$	$c' = 8$ $\phi' = 29$			$c' = 0$ $= 35$		$c' = 10–24$ kN/m² $\phi = 28–32°$		
Coef vol change m_v, m²/MN	* [0.05–1.5]	* [0.05–1.5]	* 0.1–0.3 [0.1–0.3]	* 0.1–0.3	* 0.1–0.3	* 0.5–1 (0.05–1)	* 0.1–0.3 (0.1–1.5)	* 0.05–1 (<0.05–0.3)					
Coeff consolidation c_v, m²/year	* [0.1–100]	* [0.1–10]	* 1–10 [1–10]	* 1–10	* 1–10	* 1–10 [1–10]	* 1–10 [1–10]	* 1–10 (1–100)					
Compaction max dry density Mg/m³	1.86 [1.54–2.08]	1.77 [1.53–2.01]		2.02 [1.74–2.21]	2.00	2.03							
Compaction opt moisture content %	14 [8–56]	15 [10–22]		10.3 [6–21]	10	9			16				

Average and range of values 24 [20–29]. Median and 1st and 3rd quartiles 23 (18–24) where known.
* Data grouped into classes — median and 1st & 3rd quartile or average and range classes indicated.
UW, unweathered values; W, weathered values; # Shear strength values calculated from $s = [c_u + K_o \gamma h \tan \phi_u][1 + \sin \phi_u]$ where $K_o = 1.5$ and $h = $ depth.

TABLE B2. *Engineering properties of various tills and other drift deposits from England*

Formation	Cromer Till	Contorted Drift	Chalky Boulder Clay	Hunstanton Till	Hessle Till	Withernsea Till	Skipsea Till	Basement Till	Lower Boulder Clay	Upper Boulder Clay
Place		North Norfolk				East Yorkshire			Teesside	
Author					Bell (2002)					
Weathering grade					Weathered					
Moisture content w %	13 [12–16]	16 [13–19]	24 [22–25]	18 [17–19]	23 [12–19]	12 [12–19]	16 [14–18]	17 [14–20]	14 [5–21]	13 [10–17]
Liquid limit w_L %	35 [27–40]	25 [19–29]	37 [32–45]	37 [34–40]	47 [38–53]	34 [22–39]	30 [20–36]	36 [28–42]	33 [22–49]	31 [27–38]
Plastic limit w_P %	17 [14–20]	14 [9–18]	20 [18–21]	18 [15–23]	22 [20–26]	18 [15–21]	16 [14–19]	20 [16–23]	15 [11–20]	13 [9–16]
Plasticity index I_P %	19 [13–24]	11 [8–13]	18 [14–26]	20 [15–23]	25 [17–32]	17 [12–20]	14 [9–18]	19 [12–22]	19 [10–34]	18 [13–23]
Clay fraction %	[18–30]	[12–20]	[32–55]	[15–30]	[6–37]	[10–27]	[15–35]	[22–40]	21 [16–30]	24 [17–38]
Dry density γ_d Mg/m^3									1.89 [1.77–1.99]	1.83 [1.63–1.99]
Uniaxial compressive Strength σ_c kPa	176 [154–224]	160 [124–180]	110 [104–120]	158 [152–184]	106 [96–138]	160 [140–172]	186 [182–194]	186 [163–212]		
Undrained cohesion c_u kPa	35 [26–48]	26 [20–46]	27 [16–49]	29 [22–43]	35 [22–98]	30 [17–62]	29 [17–50]	38 [22–59]	266 [63–494]	162 [38–352]
Undrained angle of friction ϕ_u °	4 [2–6]	6 [3–10]	1 [0–3]	5 [3–9]	7 [5–8]	9 [5–19]	12 [10–21]	9 [6–17]	[0–4]	[0–4]
Effective cohesion c' kPa	14 [12–19]	11 [6–20]	11 [7–16]	12 [8–18]	T 26 [10–80] S 20 [16–30]	S 23 [17–42] T 26 [21–38]	S 28 [22–25] T 27 [25–45]	S 34 [19–42] T 29 [23–47]	48 [35–72]	44 [30–67]
Effective angle of friction ϕ' °	29 [26–32]	30 [27–33]	24 [21–28]	29 [26–34]	S 25 [13–24] T 24 [16–25]	S 25 [16–34] T 24 [20–30]	S 30 [24–36] T 26 [20–38]	S 29 [20–36] T 24 [20–34]	36 [32–39]	34 [31–37]
Residual friction angle ϕ_r' °					20 [13–23] $c_r' = 1$ kPa	21 [18–27] $c_r' = 1$ kPa	25 [19–35] $c_r' = 1$ kPa	23 [18–30] $c_r' = 1$ kPa		
Coef vol comp m_v, m^2/MN	[0.13–0.16]	[0.09–0.18]	[0.15–0.24]	[0.12–0.19]					[0.12–0.14]	[0.11–0.14]
Coeff consol c_v, m^2/year	[1.3–2.8]	[0.94–3.81]	[0.84–3.01]	[1.25–3.3]					[2.6–4.8]	[3.3–5.4]

Average and range of values 24 [20–29] where known. S, shear box test; T, triaxial test.

TABLE B3. *Engineering properties of various clay containing superficial deposits from England and Wales*

Material	Alluvium								Laminated Clay			Marine Deposits
Place	SW England (Exeter)	NE England	N England (Leeds)	Birmingham West	NW England (Deeside)	N Wales (Wrexham)		N Midlands England	N. England (Skipton)	NE England	N Wales Wrexham	Teesside	SW England (Exeter)
						Clay & silt	Clay & peat						
Authors	Forster (1991a)	Jackson & Lawrence (1990)	Lake et al. (1992)	Forster (1991d)	Culshaw & Crummy (1988)	Waine et al. (1990)		Waine et al. (1991)	Threadgold & Weeks (1975)	Jackson & Lawrence (1990)	Waine et al. (1990)	Bell (2000)	Forster (1991a)
Moisture content w%	25 (21–30)	31 [19–80]	26 [19–33]	22 (16–32)	22 [11–40]	22 (17–26)	51 (41–73)	29 (20–39)	27 [26–28]	26 [14–37]	23 (20–26)	[22–35]	36 (29–54)
Liquid limit w_L%	42 (33–54)	52 [23–113]	38 [33–46]	36 (28–44)	30 [13–48]	35 (29–34)	60 (47–73)	39 (32–50)	45 [40–50]	48 [28–72]	44 (39–49)	[29–78]	54 (40–70)
Plastic limit w_p%	19 (16–23)	35 [20–89]	23 [20–27]	19 (15–23)	17 [11–23]	21 (19–24)	32 (28–44)	21 (18–25)	20 [17–25]	20 [15–27]	22 (19–24)	[18–31]	26 (23–32)
Plasticity index I_p %	23 (16–33)	16 [2–25]	15 [12–21]	18 (12–22)	15 [5–25]	15 (13–21)	25 (17–34)	18 (13–22)	22 [20–28]	28 [10–45]	22 (19–25)	[16–49]	28 (18–40)
Clay fraction %	12 (7–16)		22	18	[1–35]	10							
Bulk density γ_b Mg/m³	1.95 (1.87–2.01)	1.94 [1.79–1.99]	1.95 [1.85–2.10]	2.01 (1.86–2.16)	2.04 [1.80–2.32]	1.99 (1.83–2.06)	1.70 (1.48–1.77)	1.89 (1.76–2.02)		1.98 [1.77–2.21]	2.06 (1.99–2.10)	[1.19]	1.82 (1.60–1.95)
Dry density γ_d Mg/m³	1.61 (1.50–1.72)		1.415 [1.36–1.57]	1.57 (1.31–1.84)			1.17 (1.28–1.49)	1.43 (1.28–1.49)		1.59 [1.58–1.61]	1.68 (1.56–1.75)	[1.45]	1.37 (1.18–1.44)
Undrained cohesion c_u kPa	55 (33–86)	41 [10–130]	40 [21–65]	24 (13–67)	64 [0–140]	30 (11–35)	18 (13–22)	26 (20–46)	[40–100]	71 (20–134)	80 (50–120)	60 [27–102]	22 (18–40)
Undrained angle of friction ϕ_u °		1 [0–8]	0 [0–6]	0 (0–5)	9 [0–32]	0	0	0		1.5 [0–32]	0		2 (1–5)
Effective cohesion c' kPa									25			0	
Effective friction angle ϕ' °									28		22 (20–23)	14	
Coef vol change m_v, m²/MN	* 0.1–0.3 (0.1–1.5)	* [0.05–0.3]	* [0.1–0.3]	* 0.05–0.1 (<0.05–0.3)	* [0.05–1.5]	* 0.1–0.3	* 0.3–1.5	* 0.1–0.3	0.25 [0.1–0.45]	* [0.05–0.3]	* 0.1–0.3	H 1.51–2.03 V 2.21–2.96	* 0.1–1.5
Coef consolidation c_v m²/year	* 1–10 (0.1–10)		* [1–100]	* 1–10 (<0.1–10)	* [0.1–100]	* 1–10			2 [2–3]	* [1–100]	* 1–10	H 7.73–8.00 V 1.05–1.09	* 0.1–10
Max dry density γ_d Mg/m³		1.57 [1.39–1.74]								1.67			
Optimum moisture content w_{opt}%		20 [14–25]								16			

Average and range of values 24 [20–29]. Median and 1st and 3rd quartiles 23 (18–24) where known. H, horizontal sample; V, vertical sample.
* Data grouped into classes – median and 1st & 3rd quartile or average and range classes indicated.

TABLE B4. *Engineering properties of Tertiary clay formations from southern England*

Formation	Ball Clays	Barton Clay		Bracklesham Beds			London Clay Formation			Claygate Beds		Woolwich and Reading Beds	Lambeth Group
Place	SW England	S England					SE England		London	SE England		London	SE England
Authors	Best & Fookes (1970)	Cripps & Taylor (1986)					Culshaw & Crummy (1991)		Forster (1997)	Culshaw & Crummy 1991		Cripps & Taylor (1986)	Entwisle et al. (2005)
Weathering grade		F, S	M,H		F, S	M, H	UW	W		UW	W	UW	
Moisture content w%	[18–31]	23 [17–32]		[19–26]	[19–28]	[23–49]	30 (27–32)	28 (25–30)	[10–42]	27 (25–29)	28 (26–31)	22 [15–27]	21 (17–25)
Liquid limit w_L, %	[61–92]	59 [45–82]		[52–68]	70 [50–105]	80 [66–100]	77 (70–84)	78 (69–86)	[30–103]	58 (40–72)	70 (60–80)	52 [42–67]	52 (40–63)
Plastic limit w_p, %	[25–35]	24 [21–29]		[15–22]	28 [24–35]	28 [22–34]	26 (23–29)	27 (24–29)	[19–66]	23 (23–26)	25 (21–27.5)	24 [15–30]	21 (18–25)
Plasticity index I_p %	[31–57]	35 [21–55]	41		47 [41–65]	44 [36–55]	51 (44–57)	51 (43–58)	[43–58]	34 (20–46)	47 (37–54)	[20–37]	31 (23–39)
Clay fraction %	[71–94]	42 [35–70]			[48–61]	[48–61]	55 (47–63)	(57)	[40–70]	8 (3–15)			10 (0–20)
Bulk density γ_b, Mg/m³	[1.92–2.18]			2.07	[1.92–2.04]	[1.70–2.00]	1.91 (1.88–1.95)	1.95 (1.92–1.98)	[1.83–2.35]	1.97 (1.92–2.00)	1.95 (1.90–1.98)		2.06 (1.99–2.13)
Dry density γ_d Mg/m³							1.48 (1.43–1.54)	1.55 (1.50–1.60)	[1.23–2.13]	1.55 (1.50–1.62)	1.54 (1.46–1.58)		1.70 (1.61–1.79)
Shear strength s_u, kPa	[79–1245]	122–150 [50–350]	40 [20–210]	143	100–400 [80–800]	100–175 [40–190]	98 [72–137]	145 (98–203)	[25–525]	90 (58–128)	75.5 (55–102.5)	400 [84–814]	(100–465)
Effective cohesion c' kPa	[34–62]	[18–24]	[7–11]	24 [0–55]	[17–252]	[12–18]			[0–14]				7 (0–24)
Effective friction angle ϕ' °	[20–26]	[27–39]	[18–24]	25 [18–32]	[20–29]	[17–23]			[20–35]				27 (20–33)
Residual friction angle ϕ_r' °	[10–15]		15		[9.4–17]	14 [10.5–22]	(14–16)	9–17	[14–16]				27 (18–34)
Coef vol change m_v, m_s m²/MN				m_t [0.07–0.5] m_s [0.2–0.3]	m_t [0.002–0.01] m_s [0.003–0.04]	m_t [0.05–0.18]	* <0.05– (<0.05–0.1)	* 0.1–0.3 (0.05–0.3)	[0.03–1.18]	* 0.05–0.1 (<0.05–0.3)		m_t 0.04 [0.002–0.1]	* 0.1 (0.07–0.12)
Coeff consolidation c_v m²/year					[0.3–60]	[0.2–2.0]	(0.5–2)	* (7.5–60)	[0.2–60]	* 1–10 (1–100)		[0.75–95]	* 5 (3–11)
Modulus of elasticity, E_{sec} (MPa)					[25–141]	[10–35]							
Permeability $k \times 10^{-10}$ m/s					Lab 2 In situ [3–300]			300				[0.3–30]	
Max dry denty γ_d Mg/m³							1.60 (1.58–1.81)	(1.56)		1.72 (1.67–1.74)			1.89 (1.78–1.98)
Opt moisture cont w_{opt} %							23 (18–24)	(24)		17 (14–20)			12 (11–15)

Average and range of values 24 [20–29]. Median and 1st and 3rd quartiles 23 (18–24) where known.
* Data grouped into classes – median and 1st & 3rd quartile classes indicated.
Weathering grades: F, fresh; S, slightly weathered; M, moderately weathered; H, highly weathered; UW, unweathered; W, weathered.

TABLE B5. *Engineering properties of various Cretaceous age over-consolidated clay formations of from southern England*

Formation		Gault Clay				Fullers Earth Clay	Atherfield Clay	Weald Clay			Wadhurst Clay	Speeton Clay
Place	S England	Central and E England	SE England	S & SE England	S & E England	Central England			SE England				E England
Authors		Forster *et al.* (1995)		Cripps & Taylor (1987)		Forster (1991*b*)			Cripps & Taylor (1987)				Bell (1994)
Weathering grade				F, S	M, H		F, S	S, M	F, S	S, M			
Moisture content w%	(23)	(30, 28)	(26, 27)	22 [18–30]	35 [30–42]	29 (26–34)	46	[19–36]	5 [25–34]	[42–82]	[11–32]		18.5 [14–22]
Liquid limit w_L %	(50)	(80, 67)	(68, 74)	75 (68–80) [60–120]	[70–92]	77 (67–87)		[35–85]	[36–55] [25–30]	[26–31]			56 [39–70]
Plastic limit w_P %	(24)	(27–30)	(23, 30)	28 (23–32) [18–36]	[25–34]	28 (26–30)		[18–32]	[25–30] [11–28]	[26–40]			22 [15–29]
Plasticity index I_p %	(22)	(43, 41)	(41, 44)	43 (41–48) [27–55]	[44–80]	49 (42–58)		[17–55]	[20–32]	[45–74]			34 [24–42]
Clay fraction %				[15–65] [38–60]	[50–62]		69	[52–59]					58 [41–71]
Bulk density γ_b Mg/m³		(1.89, 1.94)	(1.90, 2.01)	1.94 (1.89–2.01) [1.96–2.13]	1.99 [1.93–2.01]	1.90 (1.84–1.97)	1.77		2.41				1.95 [1.84–2.20]
Dry density γ_d Mg/m³		(1.62, 1.47)	(1.50, 1.53)	1.51 (1.46–1.57)		1.48 (1.40–1.56)							1.65 [1.56–1.71]
Shear strength s_u kPa	105	(97, 104)	(83, 104)	105 (85–148) [56–1280]	[17–76]	92 (65–145)					[40–220]		Intact 84 [65–108] Rem 59 [38–88]
Effective cohesion c' kPa				13 (21–26) [25–124]	[10–14]			[0–21]		13			33 [30–40]
Effective friction angle ϕ' °				24 (1–24) [19–53]	[23–25]			24		[13–18]	[12–14]		30 [24–53]
Residual friction angle ϕ_r' °				10 (7–25) [12–19]	[13–14]			[12–16]		[10–20]			
Coef vol compressibility m_v m²/MN				[0.04–0.1] [0.005–0.08]	[0.075–0.12]	* 0.05–0.1 (<0.05–0.3)	0.007		0.002				0.07 [0.05–0.09]
Coeff consolidation c_v m²/year				0.4–10.4 [0.09–1.4]	[1–3.6]	* 1–10 (1–100)							0.2 [0.12–0.32]
Permeability $k \times 10^{-12}$ m/s				Lab (16–112)	2	In situ [0.2–0.7]	0.004					Lab 200	Oedmtr 440

Average and range of values 24 [20–29]. Median and 1st and 3rd quartiles 23 (18–24) where known.
* Data grouped into classes — median and 1st & 3rd quartile classes indicated.
Weathering grades: F, fresh; S, slightly weathered; M, moderately weathered; H, highly weathered; UW, unweathered; W, weathered.

APPENDIX B

TABLE B6. *Engineering properties of various Jurassic age over-consolidated clay formations from southern and eastern England*

Formation	Ampthill Clay Fm		Kimmeridge Clay				Oxford Clay					West Walton Fm	Fullers Earth Clay	
					Upper	Middle	Middle	Lower	Lower	Lower				
Place	S Central England	England	S Central England	England	S Central England		S Central & E England			S Central England		SW England		
Authors	Forster (1991b)	Cripps & Taylor (1987)	Forster (1991b)	Cripps & Taylor (1987)	Forster (1991b)		Cripps & Taylor (1987)			Forster (1991b)		Cripps & Taylor (1987)		
Weathering grades	F, S	F, S	F, S	F, S			F, S	F, S	M, H			F, S	M, H	
Moisture content w %	28 (24–32)	[23–88]	23 (21–29)	21 [18–22]	22 (20–26)	23 (21–26)	25 [22–28]	21 [15–32]	29 [17–36]	25 (22–29)	25 (22–29)	33	[17–40]	
Liquid limit w_L %	61 (53–81)	[79]	51 (42–59)	79 [70–81]	49 (35–80)	67 (54–75)	58 [53–63]	60 [45–70]		52 (47–61)	59 (52–65)	100	[40–95]	
Plastic limit w_P %	25 (22–28)	[30]	22 (19–25)	27 [24–66]	24 (21–25)	21 (19–25)	[21–33]	33 [18–47]		25 (22–28)	25 (22–28)	61	[20–51]	
Plasticity index I_p %	35 (31–53)	[49]	29 (16–42)	53 [24–59]	26 (23–31)	44 (30–54)	[31–39]	32 [12–50]		26 (22–34)	33 (28–39)	[20–39]		
Clay fraction %		[64]		[57]	(3–57)		64 [60–68]	52 [30–70]		33 [17–36]	47 (45–53)	[38–68]	[38–65]	
Bulk density γ_b Mg/m³	2.00 (1.93–2.05)	[2.10]	2.06 (1.95–2.14)	[2.08]	2.11 (2.02–2.15)	2.01 (1.95–2.07)	[2.03]	2.03 [1.84–2.05]	1.71–2.10	1.94 (1.84–2.03)	2.02 (1.92–2.08)	1.86		
Dry density γ_d Mg/m³	1.55 (1.55–1.57)		1.59 (1.46–1.70)		1.57 (1.52–1.61)	1.64 (1.54–1.71)					1.56 (1.43–1.67)			
Shear strength s_u or undrained cohesion c_u kPa	80 (64–120)		140 (90–240)	130–470 [70–500]	270 (190–390)	106 (70–190)	110–360 [45–510]	360–1100 [96–1300]	52–93	100 (60–155)	145 (90–225)		[10–120]	
Undrained friction angle ϕ_u °	0 (0–3)		0	#	0	0	#	#	#	0	0			
Effective cohesion c' kPa		[23–48]		[14–67]			10	10–480	0–20			0.005	0	
Effective friction angle ϕ' °		[17–32]		[14–23]			31	23–40	21.5–28				17	
Residual friction angle ϕ_r' °		[10–14]		[10–18]			15	18 [12.5–18.5]		13–18.5			12	
Coef vol comprs. m_v m²/MN	* 0.1–0.3 (0.05–0.3)		* 0.1–0.3 (0.05–0.3)	[0.002–0.22]	* 0.1–0.3 (0.05–0.3)	* 0.1–0.3 (0.05–0.3)	[0.05–0.12]	[0.003–0.12]		* 0.1–0.3 (0.05–0.3)	* 0.1–0.3 (0.05–0.3)			
Coeff consolidation c_v m²/year	* 0.01–1 (0.01–1)		* 1–10 (0.01–1)	[0.25–2.7]	* 0.1–1 (0.01–1)	* 0.1–1 (0.01–1)	[0.4–1.27]	0.93 [0.5–>40]		0.1–1 (0.1–10)	* 0.1–1 (0.1–1)			
Modulus of elasticity, E_{sec} (MPa)							40 [10–30]	50–135 [45–230]						
Permeability k m/s				10^{-10}–10^{-12}						5×10^{-10}		2×10^{-10}		

Average and range of values 24 [20–29]. Median and 1st and 3rd quartiles 23 (18–24) where known.
Weathering grades: F, fresh; S, slightly weathered; M, moderately weathered; H, highly weathered; UW, unweathered; W, weathered; #, shear strength value shown, s_u ($\phi_u = 0$).
* Data grouped into classes – median and 1st & 3rd quartile classes indicated.

TABLE B7. Engineering properties of various Lower Jurassic (Lias) over-consolidated clays from England

Formation	Lower Lias clay			Lias				Blue Lias	Charmouth Mst	Redcar Mst	Scunthorpe Mst	Dyrham Fm	Whitby Mst
				Lower			Upper			Lower Lias		Middle Lias	Upper Lias
Place	SW England			SW, Mid & E England				SW & Mid England	SW, Mid & E England	NE England	E Mids & E England	SW & E Mids England	SW, Mid, E & NE England
Authors	Coulthard & Bell (1993)			Cripps & Taylor (1987)						Hobbs et al. (2005)			
Weathering grade	UW	W	F, S	M, H	F, S	M, H							
Moisture content w %	20 [16–26]	24 [19–44]	18 [10–22]	20–32 [25–47]	13–18 [12–27]	20–38 [20–38]		23 (19–26)	21 (18–25)	18 (15–20)	17 (13–21)	23 (19–26)	21 (18–24)
Liquid limit w_L %	59 [27–32]	62 [56–75]	59 [53–63]	59–62 [56–75]	54 [39–72]	60 [56–72]		51 (45–58)	52 (46–59)	51 (44–54)	46 (40–53)	46 (42–51)	53 (49–58)
Plastic limit w_P %	27 [25–29]	30 [26–34]	27 [25–29]	23–25 [22–40]	26 [18–45]	28 [28–36]		23 (20–26)	24 (21–26)	23 (20–25)	22 (9–25)	23 (21–26)	24 (22–26)
Plasticity index I_P %	32 [28–36]	32 [26–42]	32	36 [33–39]	33 [19–40]	32 [24–42]		29 (24–34)	28 (23–34)	27 (23–31)	24 (20–30)	23 (18–28)	29 (29–34)
Clay fraction %	34–52	5–48		53 [50–56]	38 [18–65]	48 [28–68]		42 (37–50)	37 (28–50)		41 (32–0)	22 (16–27)	40 (32–48)
Bulk density γ_b Mg/m³			[1.94–2.00]	[1.96–2.05]	[1.87–2.09]	[1.79–1.96]		2.00 (1.94–2.08)	2.05 (1.98–2.12)		2.05 (1.99–2.13)	2.03 (1.97–2.10)	2.05 (1.99–2.13)
Dry density γ_d Mg/m³								1.64 (1.59–1.70)	1.65 (1.54–1.74)		1.74 (1.66–1.87)	1.61 (1.54–1.73)	1.71 (1.64–1.76)
Shear strength s_u Undrained cohesion c_u kPa	s_u 111 [83–146]	s_u 100 [63–203]	s_u [95–179] 12.5 MN/m²#	s_u [28–57]	s_u [30–1200]	s_u 30–150 [20–180]		c_u 139 (88–250)	c_u 125 (84–187)		c_u 127 (91–215)	c_u 113 (72–191)	c_u 123 (93–164)
Effective cohesion c' kPa	20			[0–5]	[1–17]	[18–25]		43 (11–4)	12 (5–25)		26 (17–50)	19 (12–36)	15 (0–25)
Effective friction angle ϕ' °	32			[24–27]				46 (28–28)	28 (32–32)		23 (20–28)	33 (28–36)	26 (24–31)
Residual friction angle ϕ_r °	29			[12.5–16]		[5–13.5]							
Coeff vol compressibility m_v m²/MN				0.2	[0.002–0.003]	[0.42–0.67]							
Coeff consolidation c_v m²/yr					100 [1000–10000]								
Modulus of elasticity E_{sec} MPa			[22–52]MPa 35.5 GNm²#	10 GN/m²#	[22–52]MPa 35.5 GNm²#	[10] GN/m²#							
Permeability k m/s × 10⁻¹⁰					[0.01–4000]	[3–12]							

Average and range of values 24 [20–29]. Median and 1st and 3rd quartiles 23 (18–24) where known. # Samples from deep boreholes.
Weathering grades: F, fresh; S, slightly weathered; M, moderately weathered; H, highly weathered; UW, unweathered; W, weathered; #, shear strength value, s_u ($\phi_u = 0$) or undrained cohesion value shown.

TABLE B8. *Engineering properties of Mercia Mudstone (Triassic) from England*

Formation				Mercia Mudstone					
Place	NW England	Midlands	England	SW England			Midland & E England		NE England	
Authors				Hobbs et al. (2002)						
Weathering grade	M -H	S-H, R	S -H	S -H	M -H	S-H, R	S-H, R	S -H	S -H	S -H
Moisture content %	20 (15–24)	20 (16–24)	18 (15–22)	19 (15.5–25)	23 (20.5–26)	20 (17–26)	19 (15–22)	18 (15–23)	26 (21–30)	12 (8–15)
Liquid limit %	44 (34–52)	32 (29–36)	32 (29–36)	37 (33–49)	37 (35–40)	40 (34–47)	32 (29–32)	34 (31–38)	52 (46–60.5)	30 (28–33)
Plastic limit %	23 (19–28)	20 (18–23)	18 (17–21)	24 (21–29)	22 (17–24)	23 (20–27)	20 (18–22)	19 (17–22)	28 (26–32)	19 (18–20)
Plasticity index %	19 (14–24)	13 (10–15)	13 (11–17)	15 (12–19)	16 (13–20)	17 (14–20)	12 (10–16)	14 (12–18)	24 (20–29)	11 (10–13)
Bulk density Mg/m³	2.01 (1.96–2.06] H	2.01 (1.92–2.07) M,H,R	2.04 (1.98–2.10)	2.27 (2.03–2.42) F-H	2.07 (2.02–2.10)	2.12	2.01 (1.92–2.08) M,H,R	2.03 (1.95–2.10) S-M	1.96 (1.87–1.99) M	2.13
Shear Strength kPa	121 (79–184) H	82 (52–127)	155 (94–220)	103 (77–125)	126 (82–175)	129 (89–174) M,H	94 (64–137) M.H.R	90 (60–153)	89 (65–103)	103
Uniaxial Comp Strength MPa	0.61 (0.34–1.54)	0.3 (0.1–0.7) S,M		8.1 (3.9–12.0) F-M				3.61 (2.39–7.13) F-M		
Coef vol compressibility* m²/MN				0.19 (0.15–0.24) Stress range 50–100kPa 0.12 (0.09–0.16) Stress range 100–200kPa 0.08 (0.07–0.11) Stress range 200–400kPa						
Coef consolidation* m²/yr				6.5 (1.3–19.6)⁺ Stress range 50–100kPa 6.1 (1.1–18.7)⁺ Stress range 100–200kPa 5.5 (1.2–21.4)⁺ Stress range 200–400kPa						
Moduli of deformation MPa				E_{vi}' (96) E_v' (328) E_{hi}' (70) E_h' (34) G_t (25) G_{ur} (125)						

Median and 1st and 3rd quartiles 23 (18–24) where known. ⁺Coefficient of consolidation values shown for 10th and 90th percentiles.
Weathering grades: F, fresh; S, slightly weathered; M, moderately weathered; H, highly weathered; R, reworked; S-H, R – slightly, moderately and highly weathered and reworked. Grade as shown for column, unless indicated for cell.
Deformation parameters: E_{vi}' Initial Young's modulus (effective stress), vertical samples, stress range 0–3MPa; h, horizontal; G_{ur}, shear modulus for unloading and reloading cycle.
* Consolidation data apply to all areas.

TABLE B9. *Engineering properties of Mercia Mudstone and Aylesbeare Mudstone of Triassic age from England*

Formation	Mercia Mudstone							Aylesbeare Mudstone
Place Authors	England Cripps & Taylor (1987)		England Chandler & Forster (2001)		Forster (1991c)	Midlands, England Forster (1992)	Waine et al. (1991)	Atkinson et al. (2003)	SW England Forster (1991a)
Weathering grade	F, S	M, H	F, S	M, H				M, H	
Moisture content w %	[5–15]	[12–35]	[5–15]	[12–35]	19 (15–24)	19 (15–23)	15 (13–16)	24 [12–33]	16 (15–19)
Liquid limit w_L %	34 [25–35]	[17–60]	[35]	[25–60]	34 (31–38)	35 (30–40)	30 (27–31)	32 [27–34]	39 (34–45)
Plastic limit w_p %	23 [17–25]	[17–33]	[17–25]	[17–27]	19 (17–22)	20 (17–23)	16 (16–19)	22 [14–23]	22 (20–24)
Plasticity index I_p %	11 [10–15]	[10–35]	[10–15]	[17–33]	14 (11–18)	15 (11–30)	13 (12–15)	10 [5–13]	17 (14–22)
Clay fraction %	[10–35]	[10–50]	[10–35]	[10–50]		15 (0–24)		3 [1–8]	
Aggregation ratio			[2.5–10]	<2.5				23 [8–36]	
Bulk density γ_b Mg/m³	[2.24–2.48]	[1.84–2.32]	[2.24–2.50]	[1.83–2.34]	2.06 (1.96–2.15)	2.04 (1.93–2.13)	2.15 (2.09–2.16)		
Dry density γ_d Mg/m³					1.70 (1.96–2.15)	1.77 (1.65–1.87)	1.85 (1.79–1.87)		
Undrained cohesion c_u kPa					108 (61–166)	129 (69–204)	119 (52–196)		126 (85–170)
Undrained angle of friction ϕ_u °					10 (0–20)	0 (0–7)	0 (0–22)		11 (6–19)
Uniaxial compressive strength σ_c MPa	[1.5–2.8]	[0.6–1.5]							
Effective cohesion c' kPa	[14–80]	[0–383]	>25	<20					
Effective friction angle ϕ' °	[26–40]	[16–48]	>40	[25–42]					
Residual friction angle ϕ_r' °	[23–32]	[18–29]	[23–32]	[22–29]					
Coef vol compressIty m_v m²/MN	[0.004–0.31]	[0.03–0.4]				* 0.1–0.3 (0.1–1.5)			* 0.5–1 (<0.05–0.3)
Coef consolidation c_v, m²/yr	9–3000	In situ 2–2000				* 1–10 (1–10)			* 0.1–1 (0.1–10)
Deformation modulus E_{sec} MPa		50–1000							
Permeability k m/s $\times 10^{-10}$									
Maximum dry density γ_d Mg/m³					1.75 (1.62–1.92)				
Optimum moisture content w_{opt} %					16 (12–20)				

Average and range of values 24 [20–29]. Median and 1st and 3rd quartiles 23 (18–24) where known.
Weathering grades: F, fresh; S, slightly weathered; M, moderately weathered; H, highly weathered; UW, unweathered; W, weathered.
* Data grouped into classes – median and 1st & 3rd quartile classes indicated.
* Apparent cohesion (higher value for $\sigma_3' = 1.9$ MPa)

TABLE B10. *Engineering properties of some mudrock formations of Permian age from England*

Formation	Whipton Formation	Monkerton Formation	Edlington Formation	Cadeby Formation (Mudstone)
Place	SW England		E Midlands, England	
Authors	Forster (1991a)		Forster (1992)	
Moisture content w%	13 (12–13)	22 (22–25)	19 (17–21)	14 (12–18)
Liquid limit w_L%	30 (27–32)	39 (22–40)	48 (37–55)	48 (37–51)
Plastic limit w_p %	13	22 (16–24)	20 (17–30)	21 (20–23)
Plasticity index I_p %	18 (17–21)	16 (6–17)	29 (18–32)	25 (21–31)
Bulk density γ_b Mg/m³		2.02–2.06	2.06 (2.00–2.11)	2.14 (1.93–2.13)
Dry density γ_d Mg/m³		1.62–1.70	1.82 (1.74–1.84)	
Undrained cohesion c_u kPa			96 (59–136)	148 (128–194)
Undrained angle of friction ϕ_u°			0 (0–11)	
Coeff vol compress. m_v m²/MN		* 0.3–1.5	* 0.1–0.3	* 0.1–0.3 (0.1–0.3)
Coeff consolidation c_v m²/yr		* 1–10	* 1–10	* 1–10 (1–10)

Median and 1st and 3rd quartiles 23 (18–24) where known.
* Data grouped into classes – median and 1st & 3rd quartile classes indicated.

differing composition, the properties will vary on a very small-scale and the material may also display anisotropy in that the properties depend on orientation. In other cases the deposit may be variable on a regional scale. For instance the amount of clay-sized material present in London Clay increases to the east and upwards though the deposit. In the case of Lias Clay the materials differ depending whether deposition took place in rapidly subsiding basinal areas or the intervening shallow-water shelf areas. Gault Clay varies for a similar reason and Mercia Mudstone, which was deposited in a series of ephemeral water bodies, mudflats and alluvial plains, is a highly variable deposit. Chapter 4 includes discussion of these effects.

Geotechnical parameters vary for a number of reasons connected with the material itself and also as a consequence of the sampling and testing conditions. It is possible that further testing may reveal behaviour that has not been previously observed. It needs to be emphasized that the data presented in this Appendix are intended for guidance purposes only and do not reduce the need for specific investigations into the behaviour of the materials. However the data may assist with the design of investigations and with the interpretation of test results.

It is strongly emphasized that all the information given in this Appendix must be regarded as providing only a general appreciation of clay properties and should not be relied upon for specific test values for particular clays.

CAUTION
Use of Formation-specific Data for Clay Deposits

In this Working Party report, various example data and 'typical values' of properties are included for particular types of material and named geological formations. These are given strictly for guidance purposes. As the performance of clay materials and clay deposits is likely to be highly variable and dependent not only on material properties, but on mass-specific and site- or source-specific conditions, great care must be taken in using these values. Assessments of particular clay materials should be based on property data obtained for representative samples from actual sites or sources and must not rely on the example data or 'typical values' given in this report.

TABLE B11. *Engineering properties of mudrocks of Carboniferous age from England*

Formation	Coal Measures							Etruria Marl			Crackington Formation
	Mudstone		Carbonaceous mudstone (shale)			Seatearth						
Place Authors			England Cripps & Taylor (1981)						Midlands, England Forster (1991d)			SW England Forster (1991a)
Weathering grade	F, S	M, H	F, S	M, H	F, S	M, H		M	H	
Moisture content w%	8 [5–8]	[5–8]	9 [9–14]	[9–14]	34	11	9–22 17–44	9 (8–10)	12 (9–15)	23 (16–30)
Liquid limit w_L%	42	[39–49]	[44–51]	[42–45]	[18–21]	[30–34]	[35–52] [43–79]	28	34 (30–39)	44 (35–55)
Plastic limit w_p%	29	30	26	[27–32]	16	[20–23]	[8–32]	18	18 (17–20)	22 (20–25)
Plasticity index I_p%	13	[9–19]	[15–23]	16		[12–15]	[12–25]	10	15 (12–19)	24 (15–29)
Clay fraction %		[24–53]		[37–87]		[33–77]				
Bulk density γ_b Mg/m³	2.22	[2.19–2.27]	2.18	2.22		[2.10–2.21]		2.29 (2.25–2.34)	2.18 (2.11–2.25)	
Dry density γ_d Mg/m³	2.05	2.08	[1.97–2.00]	2.03		[1.89–1.98]		2.10		
Undrained cohesion c_u kPa		[15–335]	[15–335]	[15–335]			[120–620] [40–240]	141 (100–180)	112 (69–174)	63 (58–70)
Undrained angle of friction ϕ_u°		0		0			0	9 (0–15)	0 (0–13)	0
Uniaxial compressive strength σ_c MPa	[58] [76–100]		[20] [2–50]	[15–78]	[5–30]					
Effective cohesion c' kPa	f [0–131]	[0–15]	f [0–15]	[0–49]	[0–16]	[24–162]				
Effective friction angle ϕ'°	f [45.5]	21	f [37–39]	[29–39]	[32–37]	[31–39]				
Residual friction angle ϕ_r°		12–14	i 12	16	i [2–32]	[13–26]	[8.5–17.5]			
Apparent cohesion c_a MPa	i [2–13]		i 32							
Apparent angle of friction ϕ_a°	i [28–39]				i [12–39]					
Coef vol compressity m_v m²/MN		0.1	f 25	0.08						
Coef consolidation c_v m²/yr		10		12						
Deformation modulus E_{sec} MPa	i [16000–25000] f [10–200]	[3000–35000]	i [10000–50000]	[10000–49000]	i [2–32] f [32–138]	i [8000–10000]				

Average and range of values 24 [20–29]. Median and 1st and 3rd quartiles 23 (18–24) where known.
Weathering grades: F, fresh; S, slightly weathered; M, moderately weathered; H, highly weathered; UW, unweathered; W, weathered; f, fissured sample; i, intact sample.

TABLE B12. *Engineering properties of Carboniferous Coal Measures age*

Formation	Mudstone and siltstones	Mudstone	Millstone Grit and Coal Measures Mudstones		Productive Coal Measures		Mudstone	Mudstones	Shales	Lower Coal Measures	Middle Coal Measures	Mudstones
Place	NE England	NE England (Tow Law)	NW England		NE England		N Wales			Midlands, England		
Authors	Jackson & Lawrence (1990)	Bell et al. (1996)	Lake et al. (1992)		Culshaw & Crummy (1988)		Bell et al. (1996)	Waine et al. (1991)		Forster (1992)		Forster (1991d)
Weathering grade		UW	UW	W	UW	W	M-H*	CW*	H	H			
Moisture content w%	15 [4–25]	7 [6–10.5]	9 [8–12]	17 [13–22]	[2–8]	18 [10–34]	6	14	14 (11–18)	16 (11–21)	17 (13–21)	16 (12–21)	15 (11–20)
Liquid limit w_L%	43 [33–53]	30 [27–34]	31 [29–33]	42 [36–52]		[30–47]	31	40–41	39 (34–46)	48 (38–48)	31 (25–45)	43 (37–52)	41 (37–48)
Plastic limit w_P%	18 [16–24]	13 [12–16]	21 [19–22]	23 [21–26]		[16–23]	15	25	20 (17–25)	21 (18–23)	21 (19–25)	20 (17–23)	20 (18–23)
Plasticity index I_p%	24 [17–35]	16 [15–18]	10 [8–12]	19 [14–25]	20	[14–24]	15	15–16	17 (14–22)	27 (16–35)	12 (6–21)	23 (20–28)	22 (18–27)
Clay fraction %			32	29 [18–40]			43	49–53					
Bulk density γ_b Mg/m³	2.12 [1.81–2.48]	2.51	2.15 [2.08–2.30]	2.05 [1.95–2.14]		[2.00–2.16]	2.04	2.11–2.12	2.08 (2.00–2.23)	2.15 (2.02–2.22)	2.04 (1.98–2.13)	2.11 (2.03–2.19)	2.07 (1.99–2.19)
Dry density γ_d Mg/m³	1.72 [1.47–1.89]	2.34			2.29		1.92	1.84–1.86			1.74 (1.63–1.81)	1.71 (1.59–1.89)	1.75 (1.67–1.91)
Undrained cohesion c_u kPa		[26–45]	105 [76–110]	76 [53–110]	100	[72–170]			80 (48–122)	95 (48–160)	155 (110–188)	141 (94–252)	107 (58–141)
Undrained angle of friction ϕ_u°			20 [12–23]	8 [0–15]	20	0			2 (0–7)	0 (0–9)	0 (0–16)	0	4 (0–10)
Uniaxial compressive strength σ_c MPa	129 [93–469]			7 [1–13]	[16.5–18.6]							8 (5–12)	
Apparent cohesion c_a MPa	7 [0–34]												
Apparent angle of friction ϕ_a°													
Coef vol compressity m_v m²/MN	[0.05–0.1]									0.1–0.30			0.05–0.1
Coef consolidation c_v m²/yr	[1–100]									1–10			>100
Maximum dry density γ_d Mg/m³	1.96 [1.92–1.99]												
Optimum moisture content w_{opt}%	9 [8–10]												

Average and range of values 24 [20–29]. Median and 1st and 3rd quartiles 23 (18–24) where known. * Values for 3 samples tested.
Weathering grades: F, fresh; S, slightly weathered; M, moderately weathered; H, highly weathered; CW, completely weathered; UW, unweathered; W, weathered.

References

ATKINSON, J. H., FOOKES, P. G., MIGLIO, B. F. & PETTIFER, G. S. 2003. Destructuring and disaggregation of Mercia Mudstone during full-face tunnelling. *Quarterly Journal of Engineering Geology and Hydrogeology*, **36**, 293–303.

BELL, F. G. 1994. The Speeton Caly of North Yorkshire: An investigation of its geotechnical properties. *Engineering Geology*, **36**, 257–266.

BELL, F. G. 2000. *Engineering Properties of Soils and Rocks*, 4th ed. Blackwell, Oxford.

BELL, F. G. 2002. The geotechnical properties of some till deposits occurring along the coastal areas of eastern England. *Engineering Geology*, **63**, 49–68.

BELL, F. G., ENTWISLE, D. C. & CULSHAW, M. G. 1997. A geotechnical survey of some British Coal Measures mudstones, with particular emphasis on durability. *Engineering Geology*, **46**, 115–129.

BEST, R. & FOOKES, P. G. 1970. Some geotechncail and sedimentary aspects a ball-clays from Devon. *Quarterly Journal of Engineering Geology*, **3**, 207–238.

CHANDLER, R. J. & FORSTER, A. 2001. *Engineering Geology in Mercia Mudstone*, Construction Industry Research and Information Association (CIRIA), London.

COULTHARD, J. M. & BELL, F. G. 1993. The engineering geology of the Lower Lias Clay at Brockley, Gloucestershire, UK. *Geotechnical and Geological Engineering*, **11**, 185–201.

CRIPPS, J. C. & TAYLOR, R. K. 1981. The engineering properties of mudrocks. *Quarterly Journal of Engineering Geology*, **14**, 325–346.

CRIPPS, J. C. & TAYLOR, R. K. 1986. Engineering characteristics of British over-consolidated clays and mudrocks I. Tertiary deposits. *Engineering Geology*, **22**, 349–376.

CRIPPS, J. C. & TAYLOR, R. K. 1987. Engineering characteristics of British over-consolidated clays and mudrocks II. Mesozioc deposits. *Engineering Geology*, **23**, 213–253.

CULSHAW, M. G. & CRUMMY, J. A. 1988. *The Engineering Geology of the Deeside Area*. Technical Report WN/88/7. British Geological Survey, Keyworth, Nottingham.

CULSHAW, M. G. & CRUMMY, J. A. 1991. *South West Essex – M25 Corridor: Engineering Geology*. Technical Report WN/90/2. British Geological Survey, Keyworth, Nottingham.

ENTWISLE, D. C., HOBBS, P. R. N., NORTHMORE, K. J., JONES, L. D., ELLISON, R. A., CRIPPS, A. C., SKIPPER, J. S., SELF, S. J. & MEAKIN, J. L. 2005. *The Engineering Geology of UK Rocks and Soils: The Lambeth Group*. Report IR/05/006. British Geological Survey, Keyworth, Nottingham.

FOOKES, P. G., HINCH, L. W., HUXLEY, M. A. & SIMONS, N. E. 1975. Some soil properties in glacial terrain – the Taff Valley, South Wales. *The Engineering Behaviour of Glacial Materials*. Proceedings of the Symposium held at the University of Birmingham, April 1975, 100–123. The Midland soil Mechanics and foundation Engineering Society, Birmingham.

FORSTER, A. 1991*a*. *The Engineering Geology of the Exeter District*, 1:50,000 Geological Map Sheet 325. Technical Report WN/91/16. British Geological Survey, Keyworth, Nottingham.

FORSTER, A. 1991*b*. *The Engineering Geology of the Area around Thame, Oxfordshire*, 1:50,000 Geological Map Sheet 237. Technical Report WN/91/4. British Geological Survey, Keyworth, Nottingham.

FORSTER, A. 1991*c*. *The Engineering Geology of Coventry*, 1:50,000 Geological Map Sheet 325. Technical Report WN/91/16. British Geological Survey, Keyworth, Nottingham.

FORSTER, A. 1991*d*. *The Engineering Geology of Birmingham West* [The Black Country]. Technical Report WN/91/15. British Geological Survey, Keyworth, Nottingham.

FORSTER, A. 1992. *The Engineering Geology of the Area around Nottingham*, 1:50,000 Geological Map Sheet 126. Technical Report WN/92/7. British Geological Survey, Keyworth, Nottingham.

FORSTER, A. 1997. *The Engineering Geology of the London Area*, 1:50,000 Geological Map Sheet 256, 257, 270 and 271. Technical Report WN/97/27. British Geological Survey, Keyworth, Nottingham.

FORSTER, A., HOBBS, P. R. N., CRIPPS, A. C., ENTWISLE, D. C., FENWICK, S. M. M., HALLAM, J. R., JONES, L. D., SELF, S. J. & MEAKIN, J. L. 1995. *The Engineering Geology of British Rocks and Soils: Gault Clay*. Technical Report WN/94/31. British Geological Survey, Keyworth, Nottingham.

HOBBS, P. R. N., HALLAM, J. R., FORSTER, A., ENTWISLE, D. C., JONES, L. D., CRIPPS, A. C., NORTHMORE, K. J., SELF, S. J. & MEAKIN, J. L. 2002. *The Engineering Geology of British Rocks and Soils: Mudstones of the Mercia Mudstone Group*. Research Report RR/01/02. British Geological Survey, Keyworth, Nottingham.

HOBBS, P. R. N., ENTWISLE, D. C., NORTHMORE, K. J., SUMBLER, M. G., JONES, L. D., KEMP, S., SELF, S., BUTCHER, A. S., RAINES, M. G., BARRON, M. & MEAKIN, J. L. 2005. *The Engineering Geology of UK Rocks and Soils: The Lias Group*, Report IR/05/008. British Geological Survey, Keyworth, Nottingham.

JACKSON, I. & LAWRENCE, D. J. D. 1990. *Geology and Land-use Planning: Morpeth-Bedlington- Ashington*, Technical Report WN/90/14. British Geological Survey, Keyworth, Nottingham.

LAKE, R. D., NORTHMORE, K. J., DEAN, M. T. & TRAGHEIM, D. G. 1992. *Leeds: A Geological Background for Planning and Development*. Technical Report WN/91/1. British Geological Survey, Keyworth, Nottingham.

MCKINLAY, D. G. 1969. Engineering properties of boulder clays. Engineering Group Regional Meeting, University of Strathclyde, September 1969. Geological Society, London.

THEADGOLD, L. & WEEKS, A. C. 1975. Deep laminated clay deposits in the Skipton area. *The Engineering Behaviour of Glacial Materials*. Proceedings of the Symposium held at the University of Birmingham, April 1975, 203–208. The Midland Soil Mechanics and foundation Engineering Society, Birmingham.

WAINE, P. J., CULSHAW, M. G. & HALLAM, J. R. 1990. *The Engineering Geology of the Wrexham Area*. Technical Report WN/90/10. British Geological Survey, Keyworth, Nottingham.

WAINE, P. J., HALLAM, J. R. & CULSHAW, M. G. 1991. *The Engineering Geology of the Stoke on Trent Area*. Technical Report WN/90/11. British Geological Survey, Keyworth, Nottingham.

Appendix C. Part 1: Some World and European clay deposits

This table should be read in conjunction with the text in Chapter 5, particularly Sections 5.3 and 5.5. Commercially exploited minerals are capitalized.

Period (Sub-era)	Epoch	Age	Age (Ma)	Formation/Environment	Location	PRODUCT/Mineral
QUATERNARY	Holocene					
	Pleistocene					
	Late Pliocene				Greece, Milos Island	BENTONITE
	Early to Late Pliocene			Barreiras Fm, Sedimentary -deltaic deposits	Brazil, Amazon Basin	CHINA CLAY
	Early Pliocene			Marine mudrocks with interlayered ash beds	Japan	BENTONITE
	Late Miocene				Spain	BENTONITE
	Late Miocene				Germany	BENTONITE
	Late Miocene			Marine clays and sands	Italy, Sardinia	BENTONITE
	Late Miocene				Australia, Cape York	CHINA CLAY
	Middle Miocene			Fort Preston Fm.	USA, Florida	CHINA CLAY
	Middle Miocene				Czechoslovakia, Karlov and Vary areas	CHINA CLAY
	Middle Miocene				New Zealand, Maungaparerua	CHINA CLAY
	Early to Middle Miocene				Indonesia, Belitung and Banka islands	CHINA CLAY
	Oligocene			Sedimentary deposits from residual source areas	India, Kerala and Gujarat	BALL CLAY
	Oligocene		35	Westerwald Fm, Sedimentary Lacustrine Clay	Germany	BALL CLAY
(TERTIARY)	Eocene			Wilcox Fm., sedimentary deposits on eastern edge of Mississippi Basin	USA, Tennessee and Kentucky	BALL CLAY
	Eocene			Huber Fm	USA, South Carolina and Georgia.	CHINA CLAY
	Eocene				Mexico, Tlaxcala and Morelos region	FULLER'S EARTH
	Eocene			Weathering of granite	India, Kerala and Jaipur	CHINA CLAY
	Palaeocene to Eocene			Hydrothermal alteration of granites	Ukraine	CHINA CLAY
	Palaeocene to Eocene			Hydrothermal alteration of granites	UK, Cornwall	CHINA CLAY
	Palaeocene			Residual and sedimentary	USA, California	Kaolinite
	Palaeocene			Porter's Creek Fm	USA, Texas,	FULLER'S EARTH
	Palaeocene		65	Buffaloe Creek Fm	USA, South Carolina and Georgia.	CHINA CLAY

Period (Sub-era)	Epoch	Age	Age (Ma)	Formation/Environment	Location	PRODUCT/Mineral
CRETACEOUS	Late				Morocco, Trebia Bouhoua and Providencia areas	BENTONITE
	Late			Shales	Brazil, Campos Basin	Kaolinite and Smectite
	Late			Shales	Cameroon, Douala Basin	Smectite
	Late			Ewtaw and Ripley Fm	USA, Texas and Alabama	SOUTHERN (Ca) BENTONITE
	Late			Mowry and Belle Fourche Shales	USA, Wyoming, Montana and South Dakota	WYOMING (Na) BENTONITE
	Late				Spain, Guadalajara and Valencia regions	CHINA CLAY
	Late				Canada, Alberta and Manitoba	(Ca) BENTONITE
	Early		145	Dakota Group shallow marine mudstones	USA,	Common Clay, illite plus kaolinite
JURASSIC	Late			Claystones	Africa, Congo	Common Clay
	Late			Clays	Switzerland, Molasse basin	Common Clay
	Late			Clays	France, Paris basin	Common Clay
	Late			East Berlin Fm	USA, north-east	Common Clay
	Late			Morrison Fm	USA, central and west	Common Clay
	Early		205	Shales	Morocco, Guerief Basin	Common Clay
TRIASSIC	Middle		250		Germany, Hirscau and Schnaitenbach areas	CHINA CLAY
PERMIAN	Late			Dewey lake Fm	USA	Common Clay
	Late			Rustler Fm	USA	Common Clay
	Late			Salado Fm	USA	Common Clay
	Early			Castille Fm	USA	Common Clay
	Early			Bell Canyon Fm	USA	Common Clay
	Early		290	Road Canyon Fm	USA	Common Clay
CARBONIFEROUS	Late	Stephanian		Atoka Fm	USA	Common Clay
	Late	Westphalian		Bloyd Shale Fm	USA	Common Clay
	Early to Late	Namurian		Hale Fm	USA	Common Clay
	Early to Late	Namurian		Imo Fm	USA	Common Clay
	Early to Late	Namurian		Springer Shale	USA, Oklahoma	Common Clay, illite and smectite
	Early to Late	Namurian		Pitkin Fm	USA	Common Clay
	Early to Late	Namurian		Fayetteville Fm	USA	Common Clay
	Early			Besa River Black Shales	Canada, British Columbia	Common Clay
	Early			Shales	USA, Montana	Common Clay
	Early			Tesnus Fm shales	USA, southern	Common Clay
	Early		365	Stanley Fm shales	USA, southern	Common Clay

Period (Sub-era)	Epoch	Age	Age (Ma)	Formation/Environment	Location	PRODUCT/Mineral
DEVONIAN	Late	Frasnian		Black Shales	USA, eastern	Common Clay
	Late			Marine shales	South America, Parano Basin	Common Clay
	Late		415	Shales	North-west Africa, Tindouf and Polignac Basins	Common Clay
SILURIAN	Late	Pridoli		Wills Creek Shales	USA, West Virginia	Common Clay
	Late	Pridoli		Clinton Group	USA, New York to Alabama	Common Clay
	Early	Wenlock	445	Crab Orchard Shales	USA, Kentucky	Common Clay
ORDOVICIAN	Late				Scandinavia	Potassium Bentonites
	Late	Ashgill		Cincinnatian Shales	USA, Indiana and Ohio	Common Clay, illite and chlorite
	Late	Caradoc		Decorah Shales	USA, Minnesota	Common Clay
	Late	Caradoc	495	Glenwood Shales	USA, Minnesota	Common Clay
CAMBRIAN	Middle			Shales	Norway and Sweden	Common Clay, mainly illite
	Late	Merioneth		Conasauga Fm	USA, Tennessee	Common Clay, mainly illite
	Late	Merioneth		Appalachian Basin	USA, Eastern	Common Clay, mainly illite
	Late	Merioneth		Francona Fm	USA, Wisconsin	Common Clay, mainly illite
	Middle	St. David's	550	Marine Shales	Israel	Common Clay, mainly illite

Appendix C. Part 2: British clay deposits

This table should be read in conjunction with the text in Chapter 6.

Period (Sub-era)	Epoch	Age	Age (Ma)	Formation (Member)	Location	Former names
QUATERNARY	Holocene	(Flandrian)	<10Ka	*Alluvium*	Valley bottoms, flood zones, UK	
	Pleistocene		<2Ma	*Head*	Slopes & valleys, UK	
	Pleistocene	(Anglian - Devensian)	<2Ma	*Till (undifferentiated)*	N. & Mid England, Wales, Scotland	Boulder clay
	Pleistocene	(Anglian - Devensian)	<2Ma	*Glaciolacustrine*	In association with till	
	Pleistocene		<2Ma	*Clay-with-flints*	Chalk upland, S. England	
	Pleistocene	(Devensian)	0.013–0.12	*Brickearth*[1]	Hants, Kent, Essex, E. Anglia, Devon	
	Pleistocene	(Devensian)	0.013–0.12	Withernsea Till	S. Yorks	
	Pleistocene	(Devensian)	0.013–0.12	Skipsea Till	S. Yorks	
	Pleistocene	(Wolstonian)	0.13–0.2	Basement Till	S. Yorks	
	Pleistocene	(Wolstonian)	0.13–0.2	Welton Till	Lincs	
	Pleistocene	(Wolstonian)	0.13–0.2	Fremington Till	Devon	
	Pleistocene	(Anglian)	0.25–0.3	Cromer Till	Norfolk	
	Pleistocene	(Anglian)	0.25–0.3	North Sea Drift F	Norfolk	
	Pleistocene	(Baventian)	1.6	(Easton Bavents Clay M)	Suffolk	
	Oligo/Mio		5.3–33.7	*Ball clays*[2]	W. Dorset, N. Devon, S. Devon	
	Oligo/Mio		5.3–33.7	*China clays*[2]	Cornwall	
(TERTIARY)	L. Eocene	Luteln/Bartonian	34–37	(Bembridge Marl M)	Isle of Wight (IOW)	
	L. Eocene	Luteln/Bartonian	37–41	Barton Clay F	Hants	
	E. Eocene		36–55	Bagshot F[1]	E. Hants	Bagshot Beds
	E. Eocene	Ypresian	49–55	London Clay F[1]	London, Essex, Wilts, Hants, IOW	
	E. Eocene	Ypresian	49–55	Harwich F	E. Anglia	
	Palaeocene		55–56	Reading F[1]	E. London, Wilts, Hants, Berks, IOW	Reading Beds
	Palaeocene		55–56	Woolwich F	E. London, Wilts, Hants, IOW	Woolwich Beds
	Palaeocene		56	Ormesby Clay F	Norfolk, Suffolk	
CRETACEOUS	Early	Albian	99–112	Gault F[1]	E. Mids, IOW, Dorset, Hants, Kent	
	Early	Barrem–Albian		Roach F	Lincs	Sutterby Marl + Fulletby Beds
	Early	Hauteriv/Barremian		Tealby Clay F	Lincs	
	Early	Ryazanin-Hauterivian		Claxby Ironstone F	Lincs	
	Early	U. Aptian	29	(Marehill Clay M)	Hants	
	Early	Aptian/Barremian	120	Atherfield Clay F	Kent, Surrey, Sussex, Dorset, IOW	
	Early	Hauteriv/Barremian	121–132	Weald Clay F[1]	W. Sussex, Kent, S.E. Surrey	Wealden Marl
	Early	Hauteriv/Barremian	121–132	Wessex F	IOW	
	Early	Valanginian	132–137	(Grinstead Clay M)	W. Sussex, Kent, S.E. Surrey	Hastings Beds
	Early	Valanginian	132–137	Wadhurst Clay F[1]	W. Sussex, Kent, S.E. Surrey	Hastings Beds
	Early	Valanginian	132–137	(Fairlight Clay M)[1]	W. Sussex, Kent, S.E. Surrey	Hastings Beds
	Early	Ryazanian/Albian	99–142	Speeton Clay F	Yorks	

APPENDIX C

Period (Sub-era)	Epoch	Age	Age (Ma)	Formation (Member)	Location	Former names
JURASSIC	Late	Kimmeridgian	151–154	Kimmeridge Clay F	Dorset to N. Yorks	Oaktree
	Late	Oxfordian	154–159	Ampthill Clay F	Beds	
	Late	Oxfordian		West Walton F	Oxon, N. Lincs	included in Oxford Clay
	Mid-Late	Oxfordian/Callovian	154–164	Oxford Clay F[1]	Dorset to N. Yorks	Fen Clay F, Clunch Clay
	Middle	Callovian	163	(Kellaways Clay M)	Dorset to East Midlands	
	Middle	Bathonian	164–169	(Frome Clay M)	Dorset, Wilts	Part of Fuller's Earth
	Middle	Bathonian	164–169	Forest Marble F	Dorset, Bucks	
	Middle	Bathonian	164–169	Blisworth Clay F	Beds, N. Lincs	Great Oolite Clay
	Middle	Bathonian		Rutland F	Bucks, N. Lincs	Upper Estuarine Series
	Middle	Bathonian	164–169	Hampen F	Gloucs, Oxon	Hampen Marly F
	Middle	Bathonian		Sharp's Hill F	Oxon, Bucks	
	Middle	Bathonian	164–169	Fuller's Earth F	Dorset, Gloucs	
	Middle	Aalenian		Birdlip Limestone F	N. Gloucs	Snowshill Clay, Naunton Clay
	Early	Toarcian	182–190	Whitby Mudstone F	Mendip to Yorks	Upper Lias (Clays/Clay F)
	Early	Toarcian		Bridport Sand F	Dorset	Midford Sands, Cotswold Sands
	Early	Toarcian	180–190	Cleveland Ironstone F	N. Yorks	Kettleness Beds
	Early	Pliensbachian		Dyrham F	Gloucs to E. Midlands	Middle Lias silts and clays
	Early	Hettangian/Pliensb.	190–206	Redcar Mudstone F	N. Yorks	Lower Lias
	Early	Sinem/Pliensb.	190–202	Charmouth Mudstone F	Dorset to Yorks	Blue Marl, Lower Lias clays
	Early	Rhaet/Sinemurian	195–210	Scunthorpe Mudstone F	E. Mids to Yorks	Part of Lower Lias
	Early	Rhaet/Sinemurian	202–206	Blue Lias F	Dorset to E. Midlands	Hydraulic Limestones
TRIASSIC	Late	Rhaetian	206–210	Lilstock F	Dorset, S.Wales	Cotham Beds
	Late	Rhaetian	206–210	Westbury F	Dorset, S.Wales	Westbury Beds
	Late	Scyth./Rhaet.	206–221	Blue Anchor F	Devon, S. Wales, Midlands	Tea Green Marl
	Late	Scyth./Rhaet.	206–221	Twyning Mudstone F	Worcs	
	Late	Scyth./Rhaet.	206–221	Eldersfield Mudstone F[*1]	Worcs	
	Late	Scyth./Rhaet.	206–221	Tarporley Siltstone F[1]	Devon, S. Wales, Yorks, Lancs	
	Late	Scyth./Rhaet.	206–221	Cropwell Bishop F[1]	Notts	Keuper Marl
	Late	Scyth./Rhaet.	206–221	Edwalton F[1]	Notts	Keuper Marl
	Late	Scyth./Rhaet.	206–221	Gunthorpe F[1]	Notts	Keuper Marl
	Late	Scyth./Rhaet.	206–221	Sneinton F[1]	Notts	Keuper Marl
	Late	Scyth.	210–221	Exmouth Mst. & Sandst F	E. Devon	Exmouth F (inter alia)
	Late	Scyth.	210–221	Littleham Mudstone F	E. Devon	Littleham Beds (inter alia)
	Late	Carnian/Norian	210–227	Brooks Mill Mudstone F	Cheshire	Keuper Marl (upper)
	Middle	Ladinian	227–234	Wych Mudstone F	Cheshire, Staffs	
	Middle	Anisian/Ladinian	227–242	Byley Mudstone F	Staffs	Keuper Marl (middle)
	Middle	Anisian/Ladinian	227–242	Breckells & Kirkham Mst. F	W. Lancs	
	Middle	Anisian	234–242	Bollin Mudstone F	Staffs	Keuper Marl (lower)
PERMIAN	Late			Edlington F	Yorks	Middle Marl, Permian Middle Marl
	Late			Eden Shales F	Cumbria	

APPENDIX C

Period (Sub-era)	Epoch	Age	Age (Ma)	Formation (Member)	Location	Former names
CARBONIFEROUS	Late	Stephanian	310–315	Tile Hill Mudstone F	Warks	Tile Hill Marl G
	Late	Westph D/Steph.	310–315	Salop F	Staffs/Salop	Keele F (part), Enville F. (part)
	Late	Westphalian D	300–315	Halesowen F	Staffs	
	Late	Bolsovian, Westph C	300–315	Etruria F[1]	Staffs	Etruria Marl (inter alia)
	Late	Westphalian	300–315	Ruabon Marl F	Flints, Denbighsh	
	Late	Duckmant. Westph B		Upper Coal Measures F[1]	N & W England, S. Wales, S. Scotland	
	Late	Duckmant. Westph B		Middle Coal Measures F[1]	N & W England, S. Wales, S. Scotland	
	Late	Duckmant. Westph B		Lower Coal Measures F[1]	N & W England, S. Wales, S. Scotland	
	Late	Duckmant. Westph B		Lawmuir F[1]	Scotland (Central Coalfield)	
	Late	Namurian	315–325	Crossdale Mudstone F	Yorks, Lancs	Crossdale Shales
	Early-late	Namurian	315–325	Kilbirnie Mudstone F	Scotland	Limestone Coal Group
	Early-late	Namurian	315–325	Passage F[1]	Scotland (Central, Fife, Lothian)	
	Early-late	Namurian	315–325	Marros Mudstone F	Wales, Bristol	Middle Shales G (Millstone Grit)
	Early-late	Namurian	315–325	Kenfig F	Wales, Bristol	Plastic Clay Beds (Millstone Grit)
	Early	Namurian/Visean		Accrington Mudstone F[1]	Lancs, Yorks	
	Early	Namurian/Visean		Bowland Shale F	E. Midlands, Warks, Lancs, Wales	Edale Sh, Bowland Sh, Holywell Sh
	Early	Dinantian	325–363	Laggan Cottage Mst. F	N.E. Arran, Scotland	
	Early	Dinantian	325–363	Hodder Mudstone F	Lancs	Worston Shales
	Early	Dinantian	325–363	Crackington F[1]	Devon	Culm Measures
	Early	Tournas/Chadian		Cwmyniscoy Mudstone F	Glamorgan	
	Early	Tournas/Chadian		Tongwynlais F	Glamorgan, Bristol	
	Early	Chadian/Holkerian		Hodder Mudstone F (Morte Slates M)		
DEVONIAN	Late	Frasnian	370	Red Marls F	S. Wales	
	Early	Gedinnian (Lochkov.)	400	Raglan Mudstone F	Gwent, Wales	Raglan Marl
SILURIAN	Late	Pridoli	417–419	Temeside Shale F	Wales, Salop	
	Late	Pridoli	417–419	(Upper Ludlow Shales M)	Wales, Salop	
	Late	Ludlow	419–423	(Lower Ludlow Shales M)	Wales, Salop	
	Late	Ludlow	419–423	Wenlock Shale F	Wales, Salop	
	Early	Wenlock	423–428	(Purple Shales M)	Wales, Salop	
	Early	Llandovery	428–443	Birkhill Shale F	S. Uplands, Scotland	
	Early	Llandovery	428–443	Browgill F	Wales	
	Early	Llandovery	428–443	(Stockdale Shales M)	Wales	
	Early	Llandovery	428–443	Pistyll Gwyn Mudstones F	Powys, Wales	Pistyll Gwyn Mudstones M
	Early	Llandovery	428–443	Skelgill F	Wales	V2 Beds
	Early	Llandovery	428–443	Laundry Mudstone F	Wales	
	Early	Llandovery	428–443	Maxwellton Mudstone F	S.W. Scotland	
ORDOVIC.	Early	Ashgill	443–449	Grugan Mudstone F	N.W. Wales	
	Middle	Llandeilo	458–464	Auchensoul Bridge Mst. F	S.W. Scotland	Rastrites Maximus Mudstones
	Middle	Llanvirn	464–470	Superstes Mudstone F	S.W. Scotland	Crugan Mudstones
	Early	Tremadoc	485–495	Shineton Shales F	Wales	

Key:
[1], Brick clays; [2], porcelain clays.
* Subdivided in E. Midlands, Cheshire, Lancashire.
F, Formation; M, Member; chlor, chlorite; kaol, kaolinite; smec, smectite; Mst, mudstone; E, Early; L, Late.

Appendix D. Test methods for clay materials

The following Tables indicate the Clauses or Sections in the appropriate references in which details of the procedures outlined in each Section of Chapter 9 are to be found. The Notes for all Tables are listed at the end.

TABLE D1. *Samples and laboratory practice*

	Procedures in:	
Procedure	BS 1377:1990 [Part] *Note (1)*	M S L T [Volume] *Note (2)*
Preparation of disturbed samples	[1] 7	[1] 1.5 [2] 9.5
Preparation of undisturbed samples	[1] 8	[2] 9.1 to 9.4
Environmental requirements	[1] 6	[1] 1.6
Calibration of measuring instruments	[1] 4.4.1 to 4.4.4	[1] 1.7.3 [2] 8.4.1 to 8.4.5 [3] 17.2
Laboratory reference standards	[1] 4.3	[1] 1.7.2 [2] 8.4.7
Checks on test equipment	[1] 4.4.5	[1] 1.7.4 [2] 8.4.6 [3] 17.3
Precision for reporting test results		[1] Table 1.8 [2] Table 8.1

TABLE D2. *Classification test procedures*

	Procedures in:		
Test procedure	BS 1377:1990 [Part]	M S L T [Volume]	ASTM *Note (3)*
Moisture content	[2] 3.2	[1] 2.5.2	D 2216
Liquid and plastic limits	[2] 4.3, 5.3	[1] 2.6.4, 2.6.8	D 4318
Shrinkage limit — definitive	[2] 6.3	[1] 2.7.2	
— ASTM	[2] 6.4	[1] 2.7.3	
Linear shrinkage	[2] 6.5	[1] 2.7.4	
Density — linear measurement	[2] 7.2	[1] 3.5.2, 3.5.3	
— immersion	[2] 7.3	[1] 3.5.5	
— water displacement	[2] 7.4	[1] 3.5.4	
Particle density — small pyknometer	[2] 8.3	[1] 3.6.2	D 854
— gas jar	[2] 8.2	[1] 3.6.4	
Particle size — pipette	[2] 9.4	[1] 4.8.2	D 422
— hydrometer	[2] 9.5	[1] 4.8.3	
Soil suction			See Note (4)

TABLE D3. *Chemical test procedures*

Test procedure		BS 1377:1990 [Part]	M S L T [Volume]	Other
			Procedures in:	
pH value	Indicator papers		[1] 5.5.1	
	Electrometric	[3] 9.5	[1] 5.5.2	
	Colorimetric		[1] 5.5.3	
	Lovibond comparator		[1] 5.5.4	
Sulphate content	Total, acid extraction	[3] 5.2, 5.5	[1] 5.6.2, -5	Note (5)
	Water soluble (gravimetric)	[3] 5.3, 5.5	[1] 5.6.3, -5	
	Water soluble (ion exchange)	[3] 5.3, 5.6	[1] 5.6.6, -7	
	In ground water	[3] 5.4, 5.5, 5.6	[1] 5.6.4, 5.6.6	
Organic content	Peroxide oxidation		[1] 5.7.3	
	Dichromate oxidtion	[3] 3.4	[1] 5.7.2	
Carbonate content	Rapid titration	[3] 6.3	[1] 5.8.2	
	Gravimetric	[3] 6.4	[1] 5.8.3	
	Calcimeter		[1] 5.8.4, -5	
Chloride content	Water soluble	[3] 7.2	[1] 5.9.3	
	Acid soluble	[3] 7.3	[1] 5.9.5	
	Mohr's method		[1] 5.9.4	Note (6)
Total dissolved solids		[3] 8	[1] 5.10.2	
Loss on ignition		[3] 4	[1] 5.10.3	
Concentration of certain salts			[1] 5.10.4	Manufacturer's instructions

TABLE D4. *Compaction-related test procedures*

	Procedures in:		
Test procedure	BS 1377:1990 [Part]	M S L T [Volume]	ASTM
Moisture/Density relationship –			
'light' compaction (1 litre mould)	[4] 3.3	[1] 6.5.3	D 698
(CBR mould)	[4] 3.4	[1] 6.5.5	D 698
'heavy' compaction (1 litre mould)	[4] 3.5	[1] 6.5.4	D 1557
(CBR mould)	[4] 3.6	[1] 6.5.5	D 1557
Moisture Condition Value	[4] 5.4	[1] 6.6.3	
MCV/Moisture Content relationship	[4] 5.5	[1] 6.6.4	
Rapid assessment of suitability	[4] 5.6	[1] 6.6.5	
California Bearing Ratio	[4] 7	[2] 11.7.2	D 1883

TABLE D5. *Dispersability and durability test procedures*

	Procedures in:			
Test procedure	BS 1377:1990 [Part]	M S L T [Volume]	ASTM	Other
Pinhole test	[5] 6.2	[2] 10.8.2	D 4647	Note (7)
Crumb test	[5] 6.3	[2] 10.8.3		Note (7)
Cylinder test		[2] 10.8.6		Note (8)
Dispersion test	[5] 6.4	[2] 10.8.4	D 4221	Note (7)

TABLE D6. *Consolidation test procedures*

	Procedures in:		
Test procedure	BS 1377:1990 **[Part]**	M S L T **[Volume]**	ASTM
One-dimensional (oedometer) consolidation (BS)	[5] 3	[2] 14.5	
(ASTM)		[2] 14.5.8	D 2435
Constant rate of strain consolidation		[3] [25.4 *]	D 4186
Consolidation in hydraulic cell	[6] 3	[3] 22.1 to 6	
One-way vertical drainage	[6] 3.5	[3] 22.6.2	
Two-way vertical drainage	[6] 3.6	[3] 22.6.3	
Drainage radially outwards	[6] 3.7	[3] 22.6.4	
Drainage radially inwards	[6] 3.8	[3] 22.6.5	
Isotropic consolidation in triaxial cell	[6] 5	[3] 20.2	

* Vol 3, first edition, out of print.

TABLE D7. *Permeability test procedures*

	Procedures in:	
Test procedure	BS 1377:1990 **[Part]**	M S L T **[Volume]**
Falling head		[2] 10.7
Hydraulic consolidation cell methods	[6] 4	[3] 22.7
Triaxial cell method	[6] 6	[3] 20.4

TABLE D8. *Shear strength (total stress) test procedures*

	Procedures in:		
Test procedure	BS 1377:1990 **[Part]**	M S L T **[Volume]**	ASTM
Laboratory vane	[7] 3	[2] 12.8	D 4648
Unconfined compression (autographic apparatus)	[7] 7.3	[2] 13.5.2	
Unconfined compression (load frame)	[7] 7.2	[2] 13.5.1	D 2166
Undrained triaxial compression	[7] 8	[2] 13.6	D 2850
Direct shear (small and large shearbox)	[7] 4 & 5	[2] 12.7.1-.4	D 3080
Residual shear strength (small and large shearbox)	[7] 4 & 5	[2] 12.7.5	
Ring shear	[7] 6	[2] 12.9	

TABLE D9. *Shear strength (effective stress) test procedures*

Triaxial test procedure	Procedures in:		
	BS 1377:1990 **[Part]**	MSLT **[Volume]**	ASTM
Saturation stage	[8] 5	[3] 18.6.1	D 4767 (8.2)
Consolidation stage	[8] 6	[3] 18.6.2	D 4767 (8.3)
Consolidated Undrained test	[8] 7	[3] 18.6.3	D 4767 (8.4.2)
Consolidated Drained test	[8] 8	[3] 18.6.4	

TABLE D10. *'Special' test procedures*

Type of test	Procedures in:	
	MSLT **[Volume]**	AGSG *Note (9)*
Continuous loading consolidation		9.6
Stress paths	[3] 21.2	9.8.7
SHANSEP		9.8.8
Coefficient B-bar	[3] 21.4	9.8.4
Initial effective stress		9.7.3
Triaxial extension		9.8.3
Anisotropic consolidation		9.8.5
Stress ratio K_o		9.8.6
Resonant column		9.10.3
Simple shear		9.10.4
Cyclic triaxial		9.10.5

Notes

(1) BS 1377:1990: *British Standard methods of test for soils for civil engineering purposes.* British Standards Institution, London.
(Number in square parentheses and bold indicates Part No.)

(2) M S L T: Head, K. H. *Manual of Soil Laboratory Testing* (2nd edn), Vol. 1, 1992; Vol. 2, 1994; Vol. 3, 1998. J. Wiley, Chichester. Vol. 1, 3rd edn, was due to be published (2006) by Whittles Publishing, Dunbeath, Caithness.
(Number in square parentheses and bold indicates Vol. No.)

(3) ASTM: Annual Book of ASTM Standards 2005, Section 4 Construction, Volume 04.08 *Soil and Rock (I): D 420-D 5611*. American Society for Testing and Materials, West Conshohocken, Pennsylvania, USA.

(4) Procedures are not yet included in these publications, but are described in: Crilly, M. S. & Chandler, R. J. *A method of determining the state of desiccation in clay soils.* BRE Information Paper IP4/93. February 1993. Building Research Establishment, Watford.

(5) Building Research Establishment Special Digest 1, *Concrete in Aggressive Ground, Part 1: Assessing the Aggressive Chemical Environment*. Revised edition, 2005. Construction Research Communications, London.
Reid, J. M., Czerewko, M. A. & Cripps, J. C. 2001. *Sulfate specification for structural backfills*. TRL report 447, Transport Research Laboratory, Crowthorne.

(6) Bowley, M. J. 1995. *Sulphate and acid attack on concrete in the ground: recommended procedures for soil analysis*. Building Research Establishment Report. BRE, Garston, Watford, Herts.

(7) Sherard, A. L., Dunnigan, L. P. & Decker, R. S. 1976. 'Identification and nature of dispersive soils'. *Journal of Geotechnical Engineering Division, ASCE*, Paper 12052, April 1976.
Sherard, A. L., Dunnigan, L. P., Decker, R. S. & Steele, E. F. 1976. 'Pinhole test for identifying dispersive soils'. *Journal of Geotechnical Engineering Division, ASCE*, Paper 11846, January 1976.
Sherard, A. L., Ryker, N. L. & Decker, R. S. 1972. 'Piping of earth dams of dispersive clay'. *Proceedings of the ASCE Specialty Conference. The performance of earth and earth-supported structures, Vol. 1, pp 602–611.*

(8) Atkinson, J. H., Charles, J. A. & Mnach, H. K. 1990. Examination of erosion resistance of clays in embankment dams. *Quarterly Journal of Engineering Geology*, **23**, 103–108.

(9) *AGS Guide to the Selection of Geotechnical Soil Laboratory Testing*. February 1998. Association of Geotechnical & Geoenvironmental Specialists, Beckenham, Kent.

Glossary

A

Ablation till	Glacial deposit, or boulder clay, containing a wide variety of grain sizes, but not necessarily containing clay minerals, which was transported in or on an ice sheet that has subsequently melted; thus tends to be normally consolidated.
Acidolysis	Term used for a form of weathering, particularly in the French literature, that is governed by the presence of organic matter in soil.
Active clay	See Activity.
Activity	An indication of the proportion of clay in a soil that has swelling properties. Numerically activity is defined as the ratio of the plasticity index to the clay size (<0.002 mm) material, expressed as a percentage of the dry weight. Normal clays have activities in the range 0.75 to 1.20, inactive clays have activities less than 0.75 and active clays have activities greater than 1.20.
Activity chart	A graph of plasticity index (y-axis) plotted against clay content (x-axis). A line on the chart at $45°$ passing through the origin denotes an activity of 1.
Adobe	Unfired earth bricks.
Advection	Movement (of contaminants) due to movements of water.
Aeolian	Processes or deposits formed by wind action, e.g. loess.
Aeolian clays	Clays formed by aeolian processes.
Aerobic	In the presence of oxygen. Oxygenated environments.
Aggradation	A clay mineral transformation process or processes responsible for reconstituting parts of a clay minerals structure previously degraded by other processes.
Aggregate	Crushed rock or natural fragmentary rock material, including sand and gravel, used for unbound compacted fills or sub-bases or in the manufacture of bound materials such as concrete or tarmacadam.
Air entry point	The minimum water content at which the soil remains saturated under atmospheric conditions.
Air void content	The ratio, expressed as a percentage, of the volume of air in a soil (usually compacted material) to the total volume of the soil.
A-line	A line on the plasticity chart that separates inorganic clays (which plot above the A-line) from micaceous silts and organic clays (which plot below the A-line).
Allochthonous	A term applied to rock fragments or other sediment/rock forming materials that have been transported to a site of deposition, usually in a different environment (e.g. block of limestone transported into a muddy sedimentary environment).
Allogenic	Derived minerals or particles in a sediment.
Allophane	Non-crystalline or poorly crystallised clay mineral formed in some soils, particularly andosols, by rapid weathering of volcanic glass.
Alluvial	Description of conditions in which suspended detrital matter is transported and deposited by rivers and floods.
Alluvial Clays	Clays laid down under alluvial conditions.

Alluvium	A general term for unconsolidated detrital material, such as clay, silt, sand and gravel, deposited by rivers and streams as sorted or semi-sorted sediment in the stream-bed or on the flood plain.
Alteration	Changes to the chemistry and/or physical character of minerals due to earth processes, for example hydrothermal action, but not including weathering.
Amorphous	In relation to minerals, amorphous refers to those minerals (solids) that do not have a definite crystal structure (e.g. glasses).
Anaerobic	In the absence of oxygen. Oxygen deficient environments.
Anchizone	A zone or grade of low-grade metamorphism, usually defined by specific values of illite crystallinity and which lies between the lower grade diagenetic zone and the higher-grade epizone.
Andesite	A fine-grained intermediate, volcanic, igneous rock.
Andosols	Porous soils of low bulk density formed by rapid weathering of volcanic ash and containing complexes of humus and allophane.
Angle of friction	One of the parameters used to express the shear strength of rock and soil defined as the slope of the Mohr envelope. Geomaterials may display both friction and cohesion, where in many cases friction increases with the applied normal stress, cohesion is a constant equal to the value of shear stress for zero normal stress.
Anion	an atom or group of atoms with one or more additional electrons to produce a net negative charge e.g OH^-, SO_4^{2-}.
Anisotropic	Term describing material for which the physical properties depend upon the direction relative to some defined axes, such as foliation, bedding or cleavage.
Apparent angle of friction	The slope of a tangent to a curved Mohr envelope, where friction is not constant for different values of normal stress, as is frequently the case with rocks.
Apparent cohesion	The value of shear stress for zero normal stress arising from triaxial tests on rock in which the Mohr envelope is a curve, so that friction is not constant for different values of normal stress, and a best fit tangent is drawn.
Apparent friction	The friction mobilized on a shear surface or within a material which resists relative movement or deformation. Numerically it is equal to the tangent of the angle of friction multiplied by the stress acting normally to the shear surface or potential shear surface.
Apparent viscosity	A property displayed by certain fluids in which the viscosity increases if the material is left in a stationary condition.
Aquifer	A water-bearing stratum of soil or rock that is relatively permeable to water.
Aquitard	A stratum of soil or rock that is relatively impermeable to water.
Argillaceous	Pertaining to sediments or sedimentary rocks composed predominantly of silt and clay grade particles and, typically, significant amounts of clay minerals in their composition.
Ash	See Volcanic ash and Pyroclastic.
Atomic number	Number of protons in an atomic structure.
Attapulgite	See Palygorskite.
Atterberg Limits	Consistency criteria for comparing water content and condition of clay material: comprise liquid, plastic & shrinkage limits, which are determined by carrying out standard tests on the material.
Attrition Mill	A machine in which materials are pulverised between counter-rotating toothed metal discs.
Auger	Device for sinking boreholes which is turned in the ground and withdrawn with material clinging to it or contained within it or which forces cuttings to rise to the surface on spirals.
Authigenic	Minerals formed or developed *in situ*/in place in sediments.
Autochthonous	In place, i.e. not transported.

Available lime	The amount of pure lime (calcium oxide) in an additive. For quicklime this is likely to be almost 100%, whereas for hydrated lime it would be lower.

B

BP	Before present (years).
Backscatter electrons	High-energy electrons produced by elastic collisions of an incident electron beam with a solid (mineral) specimen (e.g. in scanning electron microscopy).
Banksman	The trained operative working with an excavator or crane to guide the driver in blind spots, also acting as a watchman for the geologist (or other individual) working in a trial pit.
Basalt	A fine grained, volcanic, mafic (basic) igneous rock.
Batholith	A large intrusive mass of igneous rock (most commonly consisting of granitic rock).
Bauge	See Cob.
Bed	A unit of deposition in a sedimentary rock, usually separated from units laid immediately beneath and above by bedding planes or bedding surfaces.
Bedding	a rock surface parallel to the surface of deposition which may or may not have a physical expression.
Bedrock	The unweathered rock below any layers of soil and/or superficial deposits.
Beidellite	A clay mineral member of the smectite group.
Belt press	Apparatus in which water is squeezed from wet spoil by being pressed between filter belts passing around sets of rollers.
Bentonite	A rock composed primarily of a mineral of the smectite group, such as montmorillonite. The term has no genetic connotations, but is most often applied to deposits formed by the alteration of air-fall volcanic ash in a marine environment. Bentonites are mined as industrial raw materials, with many uses, for example as a low permeability clay in civil engineering, drilling and water retaining. They occur naturally in two main forms: sodium or calcium bentonite depending on the exchangeable cation. Calcium bentonite may be converted to sodium bentonite artificially by cation exchange.
Berthierine	An iron-rich 7Å clay mineral that may be formed authigenically in sedimentary environments and is a member of the kaolin-serpentine group of clay minerals.
Bingham fluid	A fluid whose shearing resistance increases linearly with shear rate, from an initial non-zero value (the yield strength) at zero shear rate.
Bingham solid	A solid which displays a yield stress that, once exceeded, will flow where the rate of shear is proportional to the shear stress.
Biogenic	Material formed by processes associated with living organisms.
Biota	The group of living organisms that may inhabit sediment or soil.
Biotite	A common variety of mica; black mica.
Bioturbation	The mixing of sediment by the activities of the sediment biota, such as feeding and burrowing.
Birefringence colours	Range of colours observed in many minerals when examined as standard thin-sections in a petrological microscope under crossed polars.
Blastfurnace slag	Rock-like by-product of the iron-making process in a blastfurnace, which may be crystalline when air-cooled or glassy when quenched rapidly. Crushed air-cooled slag is used extensively as an aggregate, whereas the glassy variety is often ground to a powder and used as a cement-replacement material.
Bleed	The separation of water from a slurry, usually appearing on the surface as the solid material settles.
Bloating	The distortion of fired pieces caused by gas entrapment.
Blunging	The process of mixing clays and other ceramic materials in a liquid, usually water, to form a slurry or slip.

Boehmite	An aluminium oxy-hydroxide mineral (AlO(OH)), associated with bauxite.
Borrow pit	An area where material has been excavated or 'borrowed' for use as fill at another location, usually for a specific construction project. Thus they have a limited lifetime, which may assist in obtaining planning consents in areas where these would not normally be given. Once formed, they may also be used for the disposal of (usually inert) unacceptable, waste or surplus earthworks materials. Examples include obtaining clays for use as bulk fill, and sands and gravels (with or without processing) for capping layer material.
Bottom dump bucket	Bucket for an excavator from which material is discharged from the rear or bottom, i.e. from the opposite end from which it entered. This avoids having to rotate the bucket to empty it.
'Boulder Clay'	More correctly known as Till or Glacial Till. Material deposited by glacial processes, which usually has a wide range of grain sizes and may contain little clay (but much silt and fine sand) and frequently a number of much larger fragments. It is not uncommon for there to be a proportion of boulders, sometimes up to metres in diameter. Sometimes boulders are completely absent.
Bragg's Law	Defines the conditions for constructive X-ray diffraction by the wavelength of the incident X-rays (λ), the interatomic distance between the planes of atoms (d) and the angle of incidence of the X-ray beam to the planes of atoms (θ) For constructive X-ray diffraction to occur the following equation must be satisfied: $2d\sin\theta = n\lambda$, where n is an integer.
Brecciated	Fragmented condition of a rock, typically caused by tectonic or other movements; glacial brecciation also occurs.
Brickearth	Re-worked loess.
Bronsted and Lewis Acids	Bronsted Acids act as acids by donating protons, whereas Lewis Acids act as acids by accepting electrons.
Brownfield site	Any land that has been previously developed or requires work done to it to bring it into use. Site or land that has been developed previously, usually for an industrial purpose.
Bucket, bowl or hopper capacity	Measured with the bucket held upright as 'struck' i.e. approximately the volume of the bucket full with water or 'heaped' i.e. with material in the bucket and free standing above the material in the bucket. These can be quantified in accordance with national and international standards.
Bulk density	The mass of the material, including any water it contains, divided by the volume this mass occupies.
Bulk modulus	The change in volume of a material, divided by the increase in stress to cause that change.
Bund	Small embankment or a fill area that has an environmental function, e.g. reduction in noise and visual impact or to retain water.
Burial diagenesis	The diagenesis that occurs as a result of burial of sediment in a sedimentary basin and consequent changes in temperature, pressure, and pore fluid composition.

C

Calcareous	Containing readily detectable amounts of calcium-bearing minerals, such as calcite (calcium carbonate).
Calcination	The process of subjecting a material to a high temperature, but without fusion, which results in the expulsion of volatiles and changes the physical and chemical properties (e.g. heating limestone to create lime).
Calcrete	A variety of duricrust formed by precipitation of calcium carbonate.
Caldera	A large crater, usually formed as a result of the collapse of a volcano following a major volcanic eruption.
Capping	A relatively high strength and high stiffness layer placed and compacted on weaker fills to act as a foundation for the construction layers (e.g. highway pavement or rail ballast) above, or a low-permeability layer placed on the top of a fill.
Carbonaceous	Containing readily detectable amounts of natural organic matter such as coal, or other more finely dispersed sedimentary organic matter.

Carbonate	Mineral or minerals containing carbonate (CO_3) in their structure, e.g. calcite, dolomite and siderite. Rocks comprising >50% carbonate minerals.
Cardhouse	A term used to refer to an arrangement of platy clay particles, by analogy to a structure built with playing cards
Casting Slip	Ceramic body mixed with a deflocculant to form a slurry.
Catena	An association of soils that can be related in a regular fashion to changes in a set of soil forming conditions, most often topography and hydrology.
Cathodoluminescence	Visible light emitted when electron beam interacts with a solid (mineral) specimen.
Cation	An atom or group of atoms which is deficient in one or more electrons to produce a net positive charge e.g. Al^{3+}, $(NH_4)^+$
Celadonite	Soft green grey mineral of the mica group, with a similar structure to glauconite.
Cement	(*Geology*) Minerals that form after the deposition of sediments and cause grains to be bound together. (*Engineering*) General term for binding material, but especially for the binder in concrete and mortar. See Portland cement.
Chalk	Fine grained white biogenic limestone. Originally used in a stratigraphic sense (Chalk with a capital C) as the equivalent to the Upper Cretaceous in parts of NW Europe.
Chamosite	An iron-rich chlorite.
Chamotte	A calcined plastic refractory clay (or grog) which is used as a non-plastic component in refractory compositions.
Chert	A rock that mainly consists of microcrystalline and/or cryptocrystalline silica, often forming in bands or layers within limestone. Flint is a form of chert found in Cretaceous Chalk.
China clay	See Kaolin.
Chlorite	A general name for a group of phyllosilicate minerals that have an approximate 14Å basal spacing, are usually non-expandable in water or organic solvents and can be rich in iron and/or magnesium. Chlorites are one of the important groups of clay minerals. See also Swelling chlorite.
Clast	Fragments (rock and/or mineral grains) in a rock or soil deposit that were derived from older rocks.
Clastic	A form of sediment or sedimentary rock composed of fragments of other rocks and minerals derived from weathering and erosion processes.
Clay	(*General*) A naturally occurring material composed primarily of fine-grained minerals, which is generally plastic at appropriate water contents and will harden when dried or fired. Although clay materials usually contain significant amounts of clay minerals, other materials may also be present. Associated phases in clay commonly include quartz, feldspars and organic matter. (*Engineering*) Naturally derived material (from eroded and weathered older rock) with a grain size (by definition) of less than 0.002 mm (or 2 μm). However, non-engineers sometimes use grain size limits of 4 μm or 1 mm to define the upper limit for clay grade materials. (*Soil mechanics*) Very fine-grained soil of colloid size, consisting mainly of clay minerals. It is a plastic cohesive soil which shrinks on drying, expands on wetting, and when compressed gives up water. Under the electron microscope, clay crystals have been seen to have a platy shape in which, for most clays, the ratio of length to thickness is about 10:1; for Wyoming bentonite it is about 250:1 (like mica). Clays are described for engineering purposes by their consistency limits. A 'very soft' clay is one with an unconfined compressive strength of less than 35 kPa; 'soft' clay has 35–70 kPa; 'medium' or 'firm' has 70–140 kPa; 'stiff' clay has 140-280 kPa. (*Stratigraphy*) A fine-textured, sedimentary, or residual, deposit, comprising clay minerals and other clay-sized constituents. (*Ceramics*) Clay for use in the manufacture of pottery, bricks and other ceramics must be fine-grained and sufficiently plastic to be moulded when wet; it must retain its shape when dried, and sinter together, forming a hard coherent mass without losing its original shape, when heated to a sufficiently high temperature.
Clay 'dabbs' or dabbin	The regional term given to the earth buildings of the Solway Plain area of NW England and SW Scotland.

Clay grade	See Clay (Engineering).
Clay mineral	An aluminosilicate mineral group, which forms on average about 45% of all minerals in sedimentary materials. The most common clay minerals are kaolinite, montmorillonite and illite. Their phyllosilicate crystal structure is similar to that of mica, i.e. sheeted layer structures with strong intra- and intersheet bonding but weak interlayer bonding. The type of clay that is dominant in sedimentary material depends, during early stages of weathering, on the mineral composition of the parent rock; but in later stages, it depends completely upon the climate. Some clay materials are composed of a single clay mineral, but in many there is a mixture of minerals. In addition to the clay minerals, most clay materials contain varying amounts of non-clay minerals such as calcite, feldspar, pyrite and quartz. Also, many clays contain organic material and water soluble salts. Some clay materials may contain clay minerals that are non crystalline (e.g. allophane). There are examples in which allophane is a major component of the clay material.
Claystone	Term sometimes used for an argillaceous sedimentary rock of clay grade (clay sized particles) mainly comprising clay minerals.
Clob	See Cob.
Clom	See Cob.
Cob	Unfired clay material or 'earth' used for building. Regional earthen materials or earthen building techniques are known, for example, as *cob* in Devon, *clom* in South Wales, *clob* in Cornwall and *witchert* in Buckinghamshire, *bauge* in France, *tapis* in Spain and *zabour* (or *zabur*) in North Yemen.
Cohesion	(*General*) A term often used to express a form of attraction between soil particles, thus giving them strength. However, such strength is normally due to pore water suctions in incompletely saturated soils, rather than significant attraction between particles. (*Soil mechanics*) The value of shear strength of a soil at zero normal stress.
Cohesive soils	Soils possessing strength mainly owing to the surface tension of capillary water; usually containing over about 35% of silt and clay sized (<0.06 mm) material.
Collapsible soils	Soils that will undergo a large reduction in volume in response to small disturbances or changes in moisture content.
Colloid	Ultra-fine particles usually in the size range of 1to 1000 nm, which when dispersed in an aqueous medium represents a state intermediate between a suspension and a true solution.
Colluvium	A fine sediment of variable texture formed by run-off and creep on slopes and frequently found on or at the toe of hill slopes, where it may overlie a residual soil.
Comminution	The breakdown of a solid material into finer particles, by mechanical means. In wet comminution, the solid material is broken down while suspended in liquid, as a slurry.
Common clay and shale	Primarily an economic term for clays and shales that are used for relatively low value purposes and commodities, such as brick making.
Compaction	(*Geology*) This term is used to refer to the physical processes that form a dense coherent rock from loose unconsolidated sediment. (*Soil mechanics*) The process of expelling air from a soil in a fill by the application of a dynamic loading (e.g. by rolling or vibration). The process involves particular construction plant and an understanding of the grading and moisture content of the soil. cf Consolidation.
Compressed blocks	Earth blocks manufactured from dry earth without fibre, compacted in a machine, using the same principle as rammed earth.
Compressibility	The change in volume of geomaterials caused by increases in stress. Numerically it is the inverse of bulk modulus.
Compression index	The reduction in void ratio for unit change in the logarithm of the increase in effective stress bringing about that change.
Concretion	A zone of enhanced cementation within a sedimentary rock, usually because of the precipitation of cementing material around a nucleus such as a fossil.

Conditioning	Modifying the properties of excavated material (in tunnelling machines) by injection of materials such as clays, polymers and foams. Modifying the properties of fill in advance of compaction or soil stabilisation.
Consistency index	The difference between the water content and the liquid limit of a soil divided by its plasticity index.
Consolidation	(*Geology*) The process by which a sediment becomes a rock. (*Soil mechanics*) The process by which a soil increases in density as a result of water being squeezed out by the imposition of a constant or increasing pressure. cf Compaction.
Continental margin	The zone at the edges of continents where continental crust meets oceanic crust, including the continental rise, slope and shelf.
Cookeite	A rare dioctahedral chlorite containing lithium, formed by hydrothermal alterations.
Coordination	Tetrahedral and Octahedral: Shape of the coordination polyhedra formed by nearest neighbour atoms in a crystal structure. When a cation such as silicon is surrounded by four oxygens, the structure is termed a tetrahedra. When a cation such as aluminium is surrounded by six oxygens, the resulting polyhedra has eight sides and is termed an octahedron.
Cornubia	Alternative name for southwest England.
Corrensite	A mixed-layer or interstratified clay mineral consisting of a regular interstratification of chlorite and smectite layers.
Cracking	Thermal decomposition of a substance into fractions of lower molecular weight.
Cratonic	Geologically very old and large, stable areas of the Earth's continental crust. 'Cratons' are found in many continents, generally in the interiors, and are more or less synonymous with 'shields'.
Creep	A mode of deformation in materials brought about by the application of a constant stress. Down-slope movement of superficial materials on hill sides.
Critical point drying	Drying of a sample under effective zero surface tension. This is usually achieved by miscibly displacing the formation fluids sequentially with ethanol, amyl acetate and finally pressurised liquid CO_2. The temperature is then raised above the critical point of CO_2 above which the gas cannot exist in a liquid state and pressure restored to atmospheric whilst keeping temperature above the critical point. Consequently the sample is dried without the development of disruptive gas–liquid–solid interfaces.
Critical state	State of soil after failure has occurred and the particles are undergoing turbulent flow. At the critical state, there are direct proportion relationships between shear stress and normal stress and between the volume of the soil and the logarithm of the normal stress.
Critical state line	The line that represents the state of soil after failure has occurred and the particles are undergoing turbulent flow. See also Critical state.
Cross polarized light	The use in a petrological microscope of an analyzing polariser mounted with its direction of polarization at right angles to the light source polariser. Such a configuration provides conditions in which the birefringence colours of certain minerals illuminated by transmitted light can be viewed, which can be used to assist with the identification of minerals.
Crown	Top or highest part of the arch of a tunnel.[opposite: Invert]
Crystalline	Solid structure that shows a regular atomic lattice arrangement.
Crystallinity	Estimation of the degree of crystalline ordering of a crystal structure, usually based on empirical measures such as the sharpness of X-ray reflections (q.v).
Cut-off	An underground barrier to the movement of fluid or gas.

D

d-spacing	spacing between inter-atomic planes usually measured in Angstroms (Å) where $1 \text{ Å} = 10^{-10}$ m.
Dabbs or dabbin	See Clay dabbs or dabbin.
Defects	(Australia & North America) See Discontinuity.

Deflocculant	Substance used to disperse agglomerations of clay minerals (or bodies), for example, to achieve a fluid slip at a high density in the ceramic industry.
Deltaic	Formed in a delta, a build-up of sediment that may form where a river enters the sea or a lake.
Density	The mass of unit volume of a substance. In the SI system, density is thus expressed as kg/m^3 (or its equivalents). See also Bulk density.
Detrital	Description of rock and/or mineral grains, transported as detritus and deposited as sediments, that were derived from pre-existing rocks either within or outside the area of deposition. Any particle derived from pre-existing rocks by processes of weathering and/or erosion. (See Clastic, q.v.)
Detritus	Collective term for loose rock and mineral fragmental material, such as sand, silt and clay, derived from older rocks by mechanical means, mainly abrasion and disintegration by erosion and weathering.
Diagenesis	All the chemical, physical, and biologic changes undergone by a sediment and any of its components after its initial deposition, and during and after initial transformation into rock, excluding weathering and metamorphism.
Dialysis	A process of separating compounds in solution or suspension through a semi-permeable membrane.
Diametral	In the direction of a diameter. In the splitting of a sample cylinder, for example, a 'diametral' load is applied along the side of the cylinder, thus parallel to the diameter of the cylinder.
Diamicton	Glacial (or other) sediment containing wide range of particle types and sizes. cf Polymict and Polymictic.
Diaphragm wall	A concrete wall constructed underground in a trench to form foundations, a retaining wall or a cut-off wall.
Diaspore	$Al_2O_3.H_2O$, a mineral used in refractories or as an abrasive. An aluminium oxy-hydroxide mineral, common in some bauxites and in some hydrothermal altered aluminous rocks.
Dickite	A clay mineral of the kaolin group of identical composition to kaolinite, but with a different crystal structure.
Diffusion	Movement (of contaminants) from areas of high concentration to areas of low concentration.
Diffusive	by the process of diffusion.
Dioctahedral & trioctahedral sheet structures	in sheet silicate structures the octahedral layers are of two types. In dioctahedral structures 2/3 of the octahedral (six fold co-ordinated sites) are occupied by trivalent cations (Al^{3+}, Fe^{3+}) whilst in trioctahedral structures all sites (3/3) are occupied by divalent cations (Mg^{2+}, Fe^{2+}).
Directional drilling	Sub-horizontal drilling using steerable bits to provide inclined and horizontal bores.
Discontinuity	A break in the continuum of a soil or rock, arising from depositional changes, structural dislocations and weathering effects, but generally considered to be a mechanical break; includes bedding planes, joints, fissures and faults. In Australia and North America, commonly termed 'defects'.
Dispersion	The condition of a slurry in which the individual clay particles do not aggregate into flocs.
Dispersive	A property of a soil where, owing to the presence of water, the soil particles are prone to disaggregate.
Dispersive soil	A soil which, on immersion in static water, will go into suspension.
Dissolution	The process by which a substance is dissolved, usually in water.
Distal	The sedimentary deposits formed furthest from the source of the sediment.
Disturbed sample	A sample of soil in which the structural integrity has been disrupted or which has been deformed compared with the natural state.
Docking	The wetting of bricks during the construction process to prevent the rapid uptake of water from the mortar in brickwork.

Domain	Well ordered blocks of a crystal structure.
Donbassite	A dioctahedral chlorite of extremely high aluminous composition in which both octahedral sheets are dioctahedral. (qv).
Drag box	Steel framed and sided box, that can be placed in a deep trench to enable operatives to work in the trench in safety. The steel sides of the box provide safety against any instability of the sides of the trench. As work is progressed, the box is pulled forward into a newly excavated section of trench, usually by an excavator.
Drawdown	Reduction in water level or pressure, usually as a result of pumping or lowering the level of water in a reservoir. In water retaining structures, it can lead to instability as water remaining in the fill is subject to high positive pore water pressures.
Drift deposits	Sedimentary materials, usually of glacial origin, overlying the older bedrock formations.
Dry Strength	(*Ceramics*) The ability to maintain shape after drying. (*Soil mechanics*) Strength of a clay material after oven drying at 105°C.
Drying Shrinkage	Loss of volume during drying.
Durability	The resistance of materials to change brought about by weathering action or other processes of degradation. See also Slake durability.
Duricrust	Hardened ground surface horizons formed in soil profiles or unconsolidated sediments by precipitation of various compounds from moisture evaporation, which may subsequently become buried by sedimentary processes. Common duricrust examples are 'ferricrete' formed by the precipitation of iron oxide, 'calcrete' by calcium carbonate and 'silcrete' by silica.

E

Earthenware	A glazed or unglazed, non-vitreous, opaque ceramic whiteware, which has a water absorption of greater than 3%.
Earthmoving	The various operations involved in the excavation, loading, transportation, modification, placement and compaction of materials in or on the ground.
Earthworks	The structures that are formed as the result of earthmoving.
Effective stress	The difference between the total stress and pore water pressure, where total stress is the force per unit area acting at a particular point. At depth in the ground the total stress will be due to the weight of the overlying ground and any forces imposed on the ground.
Efflorescence	A formation of powdery or crystalline salt on the surface of materials such as brickwork caused by the migration and precipitation of soluble salts from the interior of the body.
Eh	Redox (reduction-oxidation) potential measured relative to a standard hydrogen half cell in which oxidation is defined as a loss of electrons from an atom, with a positive redox potential, and reduction, the addition of electrons to an atom, with a negative redox potential.
Elasticity	Deformation in which the strain is proportional to the stress (Hooke's Law) and in which there is complete recovery of strain on removal of the stress.
Eluviation	The process of leaching in a soil, which mainly removes iron and calcium
Embankment	A raised structure built of soil or rock usually to impound water or to carry a road, railway or canal.
Engineering soil	Gravels, sands, silts, clays, peats and all other loose material, including topsoil down to bedrock. In the UK, for example, London clay and Weald clay are 'engineering soils', but rock materials to geologists.
Engineer's strain	See Shear strain.
Engobe	A slip coating applied to ceramic bodies to mask the colour and texture of the body, and to impart colour and opacity.
Environmental bund	A raised structure of soil or rock used as a visual and noise screen. Differs from an embankment in that the engineering requirements may be less onerous as the structure does not carry any load other than its own self-weight.

Epizone	The highest zone or grade of low-grade metamorpishm, transitional between the lower grade anchizone, and the higher grade greenschist facies of classic regional metamorphism. The boundaries of the epizone can be defined in terms of illite crystallinity.
Ethylene glycol	1,2-dihydroxyethane $(CH_2OH)_2$, used in anti-freeze and coolant substances, that causes characteristic expansion of swelling clays by a fixed amount that can be measured by X-ray diffraction analysis to help define the nature of the swelling clays.
Eustatic	Pertaining to worldwide changes in sea level.
Evaporite	A sedimentary rock or mineral originating principally from direct precipitation from seawater and other saline waters. Common examples are gypsum and halite.

F

Facies	A particular set of pressure and temperature conditions for metamorphic rocks, or a particular set of depositional conditions for sedimentary rocks.
Fast firing	Process in which ceramic ware is subjected to a rapid firing cycle, normally in small cross-section kilns.
Feldspar	A very important group of silicate rock forming minerals, further subdivided into potassium feldspars (K-feldspars) and plagioclase feldspars.
Felsic	Description of a rock rich in feldspars, usually in association with quartz.
Fermentation	A microbially mediated process in which organic matter is incompletely decomposed to carbon dioxide and water, thus resulting in the production of volatile organic compounds, such as alcohol or methane.
Ferric illite	A type of illite rich in ferric iron which may form authigenically at surface temperatures.
Ferricrete	A variety of duricrust formed by the precipitation of iron oxide.
Ferromagnesian	Minerals rich in iron and magnesium. Syn. Mafic.
Filter cake	A thin layer of highly impermeable clay, which can form on the surface of more permeable soils/minerals by deposition from a slurry filtering into the ground, or on to fine meshes used in filtration processes.
Fireclay	A clay material that has refractory properties. In the UK, many seatearths (underclays) in the Coal Measures are fireclays.
Fissility	The ability of a rock to be broken along closely spaced parallel planes.
Fissure	A type of joint seen in certain clays. In rocks, a fissure is an open planar discontinuity.
Flashing	Increasing kiln temperature to stabilise reduced phases (procedure for producing blue bricks).
Flint	A variety of chert formed in the Cretaceous Chalk of Northern Europe.
Floc	A mass of aggregated clay particles bound to each other by their own electrical attraction within a natural electrolyte fluid. An assembly of a number of clay particles bonded together to form a larger 'particle'; hence flocculation, flocculated.
Flocculation	The process of aggregation of clay (colloidal) particles either through chemical or biological means, which results in them becoming capable of settling out of suspension and becoming deposited due to increased size and increased settling velocities.
Fluviatile	Deposited by or pertaining to rivers.
Fluxing	The fusion or melting of a substance resulting from the combined influence of chemical reaction and heat.
Fly ash	Fine ash material derived from the flues of power stations burning pulverized coal. Pulverized-fuel ash (PFA) is a fly ash.
Fly rock	Rock fragments that are caused to travel a relatively large distance in the air as the result of blasting. This is a potential safety hazard and can lead to death and injury.
Foliation	The planar arrangement of textural or structural features in any type of rock, for example, 'schistosity' in metamorphic rocks (see Schist).

Fool's gold	See Pyrite.
Fourier transform	A mathematical relationship between the energy in a transient and that in a continuous energy spectrum of adjacent component frequencies. In infrared spectroscopy, the principle is used to enhance spectra.
Fracture	A general name for a structural discontinuity in rocks and rock masses, such as a joint or bedding plane.
Framework silicate	Silicate structures composed of silica tetrahedra that are linked together by the sharing of the four oxygens of each tetrahedron with neighbouring tetrahedra to form a framework structure.
Freeze drying	Drying by the removal of fluid from frozen solids under very low pressure to limit disruption of the fabric/texture of the sample.
Fresh	Description of rock in an unweathered condition, so that material shows no evidence of the action of weathering processes.
Friction	Force developed between two contacting bodies, or within a body, that resists relative movement or deformation.
Fuller's earth	**1.** A clay deposit dominated by the clay mineral montmorillonite, historically used for de-greasing fleeces ('fulling'). See also Bentonite. **2.** A stratigraphic name for a division in the Middle Jurassic of Britain (Fuller's Earth clay), which contains substantial beds of Fuller's earth.
Fundamental particles	Individual sheet silicate (or phyllosilicate) particles that are multiples of 10 Å thick.

G

Gap grading	An aggregate or sediment particle size distribution in which one particular size is absent from the range. For example, some tills are gap graded in that they consist of clay, silt, and gravel sized material, but not sand.
Gel	A substance with properties intermediate between the liquid and solid states. See also Thixotropy.
Gel strength	The resistance of a gel to static shearing forces.
Gelifluction	See Solifluction.
Geogrid	Polymer perforated sheet with high tensile strength. See Reinforced clay.
Geomaterials	Natural geological materials or materials manufactured therefrom, usually used for construction purposes.
Geomembrane	A solid sheet of polymer material for use as an impermeable barrier in geotechnical construction.
Geophysical surveying	The study of the variations in the natural or induced physical properties of the Earth's crust with the object of gaining knowledge about the subsurface, such as the location and structure of a mineral deposit.
Geotectonics	Regional scale tectonic processes and products.
Geotextile	A woven or non-woven (heat-bonded or needle-punched) sheet material made from polymer or sometimes natural fibres for use in geotechnical construction.
Geosynthetic	A general name for a Geotextile or Geomembrane.
Gibbsite	The most common aluminium hydroxide mineral, often found in highly weathered tropical soils.
Glacial clay	Clay deposited by glacial processes.
Glacial drift	See Till.
Glacial till	See Till.
Glacio-marine	Sediments of glacial origin deposited in a marine environment.
Glauconite	An iron and potassium-rich clay mineral belonging to the illite-mica family that forms most often authigenically in shallow marine sedimentary environments.

Glaucony	A facies that contains glauconitic clay minerals, these may range from glauconitic smectite to glauconite *sensu stricto*.
Glaciogenic	Resulting from action of glaciers.
Glaciotectonic	Process of deformation of earth or rock materials by glacial forces.
Glycerol	1,2,3 trihydroxypropane $(CH_2OH)_2\ CH(OH)$ causes characteristic expansion of swelling clays by a fixed amount that can be measured by X-ray diffraction analysis to help define the nature of the swelling clays.
Gneiss	A banded high-grade metamorphic rock.
Grading	The particle size distribution of an aggregate or sediment, which may be well graded (a spread of sizes), uniformly graded (a predominance of one size) or gap graded (a spread of sizes but with an absence of one size). By contrast, in geotechnical engineering, 'well graded' is used to used to describe a predominance of one particle size, as in 'well sorted' to a geologist. See also Sorting.
Gravel	Deposit or aggregate comprising particles between 2 and 60 mm in size.
Green strength	(*General*) The strength required to maintain shape in the wet or plastic state. (*Ceramics*) Strength of a clay material rendered plastic by addition of water at a specific moisture content.
Greenfield site	An area previously undeveloped and therefore undisturbed by construction, mining or waste disposal.
Gres Porcellanato	Unglazed, vitreous firing, white bodied floor tile.
Gritstone	Term sometimes used for a (usually) coarse sandstone with angular particles. Previously used in the UK as a group term for aggregates comprising any clastic rock with sand-sized particles.
Grog	A ground mixture of firebrick, pottery etc, which is added in small amounts to some ceramic bodies to control thermal expansion.
Groundmass	The finer-grained material constituting the main body of a rock, in which is set relatively larger crystals or particles. See also Porphyritic.
Grout	Fluid material that is injected into the ground to fill voids or fractures, and then solidifies to increase the strength or reduce the permeability of the ground.

H

Habit	General shape of mineral crystals and crystalline aggregates that reflect their crystal structure and the conditions of formation.
Halloysite	A mineral of the kaolin group that contains interlayer water. See also Hydrated halloysite and Metahalloysite.
Halmyrolysis	Submarine weathering.
Hammer mill	A crushing device in which materials are reduced in size by the impact of hammers revolving in a vertical plane within a steel casing; used to crush ores and other large solid masses.
Hard pan	Strongly cemented material in unconsolidated sediments, often just below the ground surface; usually formed by precipitation of mineral cements from groundwater. In North America, the term is used for cemented glacial deposits.
Head	Superficial granular deposits on valley sides and floors formed by flow of soil and rock fragments from higher ground, usually under conditions of high water content.
Header	A brick placed in a course or layer of brickwork with the short side facing out.
Hercynian	A late Palaeozoic European orogenic event (tectonic plate collision).
Hemi-pelagic	Sediment that is partly pelagic (qv) and partly terrigenous in origin.
Hoffman Kiln	An efficient periodic multi-chamber kiln in which the chambers are so constructed that combustion gases and cooling air may be used to dry and preheat the ware and subsequently fire the ware with a minimum loss of heat.

Hooke's Law	See Elasticity.
Horizonation	The organisation of a soil profile into distinctive soil horizons.
Hydrated halloysite	The variety of the clay mineral halloysite containing all four molecules of water of hydration.
Hydraulic conductivity	A measure of the ease with which fluids may pass through a material.
Hydraulic trap (containment)	A method of reducing or preventing leachate flow out of a landfill facility by inducing inward flow of water from the surrounding ground.
Hydrocyclone	Apparatus for separating solid suspended matter from liquids, which works as a centrifuge with the spinning motion of the fluid induced by the way in which the fluid flows through the apparatus
Hydrograph	A plot of the rate of water flow past a given point with time.
Hydrolysis	A decomposition reaction involving water. Commonly applied to the reaction between silicate minerals and water or an aqueous solution. Chemical interaction resulting in the dissociation of water.
Hydrothermal alteration	Changes to the chemical composition and physical structure of rocks caused by the passage of hot water solutions from deeper in the crust. Alteration is a sub-surface process, whereas weathering is a surface process.
Hypersaline	salinity substantially greater than seawater.
Hysteresis	The retardation of an effect behind the cause of the effect. In stress-strain curves for materials, hysteresis is indicated when the unloading cycle gives a curve that does not coincide with the original loading curve.

I

ICP-AES	Inductively coupled plasma—atomic emission spectroscopy. A method of inorganic chemical analysis that involves the formation of an ionic plasma by the high temperature (~8000 K) ignition of a liquid sample as an aerosol in a gas plasma and the resultant optical emission spectrum is then analysed to determine elemental compositions.
ICP-MS	Inductively coupled plasma—mass spectroscopy. A method of inorganic chemical analysis that involves the formation of an ionic plasma by the high temperature ignition of a liquid sample which is then introduced into a mass spectrometer for analysis of the mass spectrum of the ionised sample.
Illite	A 2:1 clay mineral group common in sedimentary rocks and characterized by 10Å X-ray diffraction spacing. The aluminium-silicate layers are separated by potassium ions and do not expand and contract with moisture changes. Illite develops from the alteration of micas and alkali-feldspars in alkaline weathering conditions.
Illite crystallinity	An empirical measure of diagenetic and low-grade metamorphism, usually expressed as the Kubler index, the peak width at half peak height of the 001 illite peak on an X-ray powder diffraction pattern of the clay sized-fraction.
Inactive clay	See Activity.
Induration	Natural process of hardening of sediments into rock by pressure, heat, or cementation.
Infra-red radiation	Electromagnetic radiation with wavelengths beyond the visible range, in the range of 700 nm to 1 mm.
Inheritance	One of the three possible origins for clay minerals, inherited clay minerals are allochthonous (i.e. the constituents have been transported to the location of deposition). See also Neoformation and Transformation.
Intercrystalline	Relating to the regions, usually pore spaces, between crystal grains. In soils, changes to the amount of intercrystalline water may cause the soil volume to change.
Interstratified clay mineral	See Mixed-layer clay mineral.

Intracrystalline	Relating to the interior of crystals. In swelling clays, changes in intracrystalline cations and the addition or removal of intracrystalline water molecules cause the crystal to change volume.
Intrusives	Bodies of igneous rock which have intruded themselves into pre-existing rocks, often along some line of weakness, examples include dykes and sills.
Invert	The lowest part of the inner surface of a tunnel. [opposite: Crown]
Ion	Atom with a positive or negative charge.
Ion beam milling	Process whereby a sample extracted from a standard thin section (q.v) is gently eroded by a stream of ions or atoms in a vacuum to produce specimens for transmission electron microscopy analysis.
Ion exchange	The reversible exchange of ions, either cations or anions, in a crystal structure without changes to the morphology of the structure.
Iron pyrites	See Pyrite.
Isomorph	Solid phases with different chemical compositions but the same crystal structure.
Isotropic	Having the same properties in all directions. Cf Anisotropic.

J

Jaw Crusher	A rock crushing machine; either a single toggle machine, in which a hinged jaw moves toward and away from a fixed jaw in a rapidly alternating fashion, or a double toggle machine in which both jaws move.
Joint	A fracture or discontinuity in a rock mass along which there has not been relative displacement in the plane of the structure.

K

Kaolin	(*Mineralogy*) A group of minerals that is the most common of the clay minerals with the two layer 1:1 type of structure and are low plasticity clay minerals, including kaolinite, dickite, nacrite and halloysite. (*Geology*) A deposit or rock consisting of kaolin minerals, most commonly kaolinite (syn. china clay).
Kaolinite	The most common mineral of the kaolin group.
Kaolinised	Description of a rock changed to one containing mainly clay minerals of the kaolin group, often by weathering or hydrothermal processes.
Karst	Features including rock mass cavities and also the topography created in limestone terrain by the solution effects of percolating water and subterranean streams. Named after the Karst region of former Yugoslavia.
Kibbler	A crushing or shredding machine consisting of contra-rotating toothed or spiked drums.
Kubler index	An empirical but formal measure of illite crystallinity (qv).

L

Lacustrine	Description of deposits formed in a lake.
Lagoonal	Description of deposits formed in a lagoon.
Laminated	Exhibiting thin (typically less than about 10 mm) distinct layers of sediment, often of differing grain-size or colour.
Laterite	A red tropical residual soil material, commonly formed from the weathering of basic igneous rocks, that typically consists mainly of kaolinite and is impregnated with, cemented by or partly replaced by hydrated oxides of iron and aluminium. Originally described in India.
Lateritic	Description of rock and soil materials in the process of becoming laterite.
Latocrete	Red hardened crust that sometimes forms on the dried surface of laterite.
Leachate	Liquid waste derived from landfill material
Lenticular	A lens-like geological layer of limited lateral extent.

Leucocratic	Light coloured igneous rocks that are rich in felsic minerals.
Lewis and Bronsted Acids	Lewis Acids act as acids by accepting electrons, whereas Bronsted Acids act as acids by donating protons.
Liquefaction	The ability of some soils to flow in response to a slight disturbance, for example by an earthquake.
Liquid limit	The percentage water content at which a clay changes from being a plastic solid to being a viscous liquid, as defined by a standard test.
Liquidity index	An indication of the *in situ* plasticity condition of clay .The relative natural water content of a clay on a scale from 0 to 1 between the plastic limit and the liquid limit. It is equal to the water content minus the plastic limit divided by the plasticity index.
Lithogenetic	Relating to the processes of formation of geological materials/sedimentary rocks.
Lithology	Visible characteristics of rock, including mineral composition, grain-size and structure; the type of rock. Usually applied to sedimentary rocks in outcrop and hand specimen.
Lithorelict	Surviving portion(s) of parent rock in a gradually weathered rock mass that is(are) weaker than the original material.
Lodgement Till	Type of glacial deposit (or Till, or boulder clay), containing a wide variety of grain sizes but not necessarily containing clay minerals, deposited beneath an ice sheet that has subsequently melted; thus tends to be overconsolidated.
Loess	Sediment formed of predominately silt sized material deposited by the action of wind.

M

Ma	Mega annum, i.e. a million (1 000 000) years.
Made ground	Ground made by infilling natural depressions or artificial pits using reworked natural soil or man-made materials. See also Soil (Geotechnical).
Mafic	Rich in magnesium and iron minerals. Syn. Ferromagnesian.
Matric potential/suction	That part of water potential due to the interaction of water with colloids and to capillary forces (surface tension); often important in influencing suction (negative pore pressure) in unsaturated clay soils.
Matrix	Finer material between framework grains, e.g. clay matrix between sand sized framework grains in a sediment.
Maximum dry density	The highest density attained in a standard test to determine the compaction properties of an engineering soil, where dry density is the dry weight of the material divided by the volume it occupies.
Metabentonite	A bentonite that has altered by diagenesis or low-grade metamorphism, such that it consists predominantly of mixed-layer illite-smectite rather than smectite.
Metahalloysite	The variety of the clay mineral halloysite having no water of hydration.
Metamorphism	The natural process of alteration of rocks largely in the solid state by great heat and/or pressure.
Metastability	State of apparent stability, often because of the slowness with which equilibrium is attained. In clay materials, the property of collapse due to flooding, whilst under constant stress.
Mica	A group of phyllosilicates, (or sheet silicates), characterized by non-exchangeable interlayer cations, including muscovite and biotite. Some common clay mineral micas include illite and glauconite.
Microgranite	A medium-grained, porphyritic, acid igneous rock containing phenocrysts of quartz and feldspar. The term 'quartz porphyry' has been used previously.
Microgravimetric	Measurement or analysis of small changes in mass.
Micro-organisms	Very small animals and plants; strictly, those that can only be seen using a microscope.
Microtunnelling	Pipe jacking using a tunnelling machine remotely controlled from the surface; in the UK this refers only to tunnels of diameter less than 1.0 m.

Milliequivalents	Unit of measurement for ion exchange related to the equivalent weight of an ion i.e its molecular weight divided by its valence, and expressed as multiples of that equivalent weight divided by 1000.
Mineral	A naturally occurring crystalline solid with a definable chemical composition. Also, a geological material with an economic value.
Mixed-layer clay mineral	A clay mineral, composed of different clay mineral sheet silicate units occurring in a stacking sequence to form a coherent crystal structure (e.g. illite-smectite) with characteristics that may be assigned to more than one type of discrete clay mineral. Syn. Interstratified clay mineral.
Modulus	The ratio of the stress imposed on a body to the amount of deformation thus caused.
Modulus of rigidity	The ratio of the shear stress to the shear strain thus caused.
Modulus of rupture	A measurement of tensile strength used mainly in the ceramics industry in which a prismatic sample of clay is supported at two places on one side and failed by loading the midpoint on the opposite side.
Mohr's circles	A two dimensional graphical construction that relates the principal stresses acting in a body to the shear and normal stresses thus induced, where the shear and normal stresses are plotted at the same scales on the vertical and horizontal axes of the graph. If the Mohr's circles represent the conditions for failure, then the common tangent drawn for a series of circles defines the failure condition for the material, where the slope of the line is the value of angle of friction and the intercept on the vertical axis is the value of cohesion. For many soils, the Mohr envelope is a straight line, whereas for rocks it is liable to be curved.
Moisture content	In most applications, the mass of water in a material expressed as a percentage of the mass of the dry solids (dried at 105°C to constant mass), or, in some applications, as the mass of wet soil.
Monodisperse	System of colloidal dispersion having particles all of effectively the same size.
Montmorillonite	The most common clay mineral of the smectite group.
Mud	Soil or sediment consisting largely of clay and silt-sized particles, i.e. grain size less than 0.06 mm. Popularly, mud is found in a liquid or semi-liquid state.
Mudrock	Sediment whose modal grain size is in the mud grade.
Mudstone	A general term used to refer to an indurated sedimentary rock composed primarily of clay and silt sized minerals (see Mud), and often more specifically to one that that does not show obvious fissility for which the term shale (qv) may sometimes be more appropriate.
Munsell colour system	A system of colour notation devised by Albert Munsell, which describes colour according to three attributes: hue, value (lightness or darkness) and chroma (strength or saturation). The Munsell scale of chromaticity values gives approximately equal magnitude changes in visual hue. A range of Munsell colour charts is available for use with soils and other geomaterials.
Muscovite	A common variety of mica; white mica.

N

Nacrite	A rare kaolin group mineral, identical in composition to kaolinite, but with a different crystal structure.
Neoformation	Literally 'new formed', one of the three possible origins for clay minerals. See also Inheritance and Transformation.
Nodules	Spherical or ovoid concretions in rocks composed of minerals that precipitated out after deposition.
Non-marine	Sediments deposited in lakes, or rivers, or other continental environments, i.e. not marine environments.
Nontronite	A clay mineral member of the smectite group.
Normal compression line	The linear relationship between the void ratio and the logarithm of effective stress for a normally consolidated soil.

Normal stress	The force per unit area acting at right angles to a plane.
Normally consolidated	A deposit, usually a clay, that has not borne an overburden greater than that at the present day. Many lacustrine and alluvial clays are in this condition, because the overburden on them has undergone increase but has never been reduced by, for example, the removal of part of the overburden by erosion.

O

Octahedral coordination	See Coordination.
Odinite	An iron-rich 7Å clay mineral, a member of the serpentine-kaolin group. Odinite appears to contain significant amounts of both ferrous and ferric iron and have an octahedral occupancy that is intermediate between dioctahedral and trioctahedral.
Oedometer	Laboratory apparatus in which the one-dimensional consolidation characteristics of a small soil specimen can be obtained over a range of applied static (loading) pressures.
Ooidal	Description of a sediment consisting of sand-sized spherical accretionary particles (ooids) cemented together. Also see Oolitic.
Ooids	Spherical or subspherical grains consisting of concentric lamellae around a nucleus; usually composed of carbonates but can also be of Fe minerals such as berthierine (qv). Syn. Ooliths.
Ooliths	See Ooids.
Oolitic	Texture of sediment or sedimentary rock comprising ooids or ooliths, hence 'oolitic limestone' (or sometimes 'oolite').
Optimum moisture content	The moisture content measured in a standard engineering soil (qv) compaction test that provides the maximum dry density.
Organic maturity	Estimates of the changes in the character and composition of organic matter during burial diagenesis (qv) usually measured by changes in organic chemistry, optical reflectivity and coloration.
Overburden	In a general geological sense, the unconsolidated sediments which overlie the solid rock (or 'bedrock'). In mineral extraction or quarrying, the unconsolidated or consolidated material that has to be removed to provide access to the mineral.
Overburden pressure	Pressure or vertical stress at a particular point in the ground resulting from the mass of the material above that point.
Overconsolidated	State of compaction of clay material in which the present overburden pressure is less than the maximum effective stress borne by the material during its history. The moisture content is less than the value for an equivalent deposit consolidated from slurry up to the same pressure.
Overconsolidation ratio	The ratio of the maximum overburden pressure on a deposit at some time in the past to the present value. In normally consolidated (qv) materials this ratio will be unity, whereas in overconsolidated materials it will be greater than this.
Overcut	The excavation of the ground to a diameter slightly greater than the pipes in pipe jacking or microtunnelling, to allow the pipes to be pushed easily through the ground.
Oxisols	A highly weathered group of soils found mostly in the tropics and noted for their high content of oxides.
Oxic/Anoxic	Related to general oxidising/reducing conditions.

P

Palaeoclimate	Former climate in geological past.
Palaeosoil	A fossil soil, i.e. no longer actively forming, usually buried below the present ground surface. Such a soil may be of Recent origin or very ancient.
Palygorskite	A magnesium rich clay mineral, usually occurring in fibrous form, common in some evaporitic environments. Syn. Attapulgite.
Particle size distribution	The ranges of sizes of particles in a disaggregated material, such as sediment, soil or aggregate, usually expressed as a histogram or a frequency diagram. Syn. Grading.

Peak strength	The maximum stress that can be borne by a soil or rock in an undisturbed condition. Typically fracturing or a large increase in the rate of deformation occurs once the peak strength has been attained, but the materials may not lose all its load carrying capacity. In many clays, continued shearing after the peak condition will lead to a reduction in strength to the ultimate value (qv), and with further shearing, the strength will be reduced to the residual value (qv).
Ped	A naturally-formed unit or mass of soil, such as an aggregation of soil particles.
Pedogenic	Soil forming process.
Pelagic	A term applied to sediments accumulated in the deep sea and oceans.
Penetrometer	Instrument for assessing the in situ strength of soil by measuring the resistance to the controlled penetration by a probe, usually cone-shaped. A laboratory type of penetrometer is also used for the standard determination of liquid limit.
Periglacial	A term used for the region adjacent to a ice sheet, including the climate, topography and natural processes that occur in that region.
Perimarine	Adjacent or marginal to a marine environment.
Permeability	A measure of the ease with which a fluid (or gas) is able to pass through a material.
Permeameter	Laboratory apparatus for measuring the permeability of soils.
Periglacial processes	Mechanisms associated with active freeze–thaw cycles in the area around an ice sheet, or in areas of similar climate where permanent frozen ground develops, giving rise to a characteriztic suite of effects on materials and deposits.
Petrofabrics	Description of rocks/soils by their textures and granular or crystal inter-relationships.
Petrography	Systematic description and classification of rocks.
Petrology	General term for the overall study of rocks through their mineralogy, geochemistry, textures, field relationships and especially their formation.
Petrological microscope	Microscope principally used for the analysis of thin sections of rock mounted on glass slides and examined in transmitted plane polarised or crossed polarized light.
pF	A unit for expressing the magnitude of suction. It is the logarithm to the base 10 of the suction in centimetres of water. For example, a suction of 10 cm water is pF 1.0 and a suction of 100 cm water is pF 2.0. Oven dryness is pF 7.0.
pH	Measurement of acidity/alkalinity by the activity of hydrogen ($pH = -\log[H^+]$), on a scale of 1–14, with 7 being neutral (as in pure water), values <7 being acid and those >7 being alkaline.
Phenocryst	In igneous rocks, a relatively large well-developed crystal found within a finer-grained groundmass; phenocrysts and groundmass together constituting a 'porphyritic' texture.
Phyllite	A term applied to argillaceous rock that has been altered by low-grade metamorphism, and which is coarser grained and less perfectly cleaved than slate, but finer grained and better cleaved than schist.
Phyllosilicates	A group of silicate minerals composed of tetrahedral and octahedral sheets joined together to form T:O (1:1) or T:O:T (2:1) layers, between which there may also be interlayers of hydrated cations and water, where required to compensate excess negative charge arising from substitutions of cations by other cations of lower valence. Also known as sheet (qv) or layer silicates.
Phytoplankton	Planktonic (qv) plants.
Piezometer	An instrument for measuring water or gas pressure or level in the ground.
Pipe jacking	A method of constructing tunnels by pushing a string of pipes through the ground behind a tunnelling shield within which the ground is excavated.
Pit	See Quarry (qv).
Pitchers	Recycled waste fired product from ceramic production, which is often ground and utilised in ceramic bodies, glazes and colouring compounds (similar to Grog).

Pitchstone	A volcanic glass with dull resinous lustre.
Placer deposits	A mass of sand, gravel or similar material resulting from the weathering, erosion and sedimentation of rock and enriched in mineral particles, such as gold, platinum and tin, that have been derived from rocks, or veins formed within the rock mass, and concentrated usually as a result of their high specific gravity.
Plane polarised light	In a petrological microscope, thin-sections of rock samples are viewed by transmitted polarized light. Plane polarised light refers to the condition in which the thin-section is viewed *without* a second polariser set with its polarization direction at right angles to the light source polariser being used. The colour and range of colours displayed by minerals thus viewed is of assistance in their identification. cf Cross polarized light.
Plankton	Animals and plants that live in the surface layers of seas, rivers, ponds and lakes; hence 'planktonic'.
Plasticity	(*Clay mineralogy*) A property that is greatly affected by the mineral and chemical composition of the material. For example, some species of chlorite and mica can remain non-plastic even after grinding to flakes where more than 70% of the material is less than 2 μm esd (equivalent spherical diameter). In contrast, other chlorites and micas become plastic upon grinding to flakes where only 3% of the material is less than 2 μm esd. Plasticity may be affected also by the aggregate nature of the particles in the material. (*Soil mechanics*) Property of a clay material which permits it to undergo permanent deformation under stress in any direction without rupturing up to an elastic yield point and to retain its shape after the stress is removed.
Plasticity chart	A graph of plasticity index (*y*-axis) plotted against liquid limit (*x*-axis) used in the classification of fine soils. See Atterberg limits.
Plasticity index	The difference (in percentage points) between the liquid and plastic limits of a soil.
Plastic limit	The percentage water content at which a clay changes from being a brittle solid to being a plastic solid, as defined by a standard test.
Plastic viscosity	The rate of increase of shear resistance with shear rate for a Bingham fluid.
Plate tectonics	A unifying interpretation of the Earth's major geological structures, plates of the lithosphere (crust), and their relative movements, boundaries and effects.
Plutonic	Adjective to describe large igneous intrusions ('plutons') which have cooled at great depth in the Earth's crust.
Point counting	Microscopic analysis of thin-sections (qv) of rocks involving the systematic counting of identified grains /crystals that intersect the microscope cross-hairs as the microscope stage is moved in fixed lateral steps. Quantitative analysis of thin-sections usually requires the counting of hundreds (or even thousands) of grains/crystals.
Poisson's ratio	The ratio of the strain in the direction of an applied stress to the strain at right angles to this.
Polydisperse	System of colloidal dispersion having particles of varying size.
Polygenetic	Of multiple origins.
Polymictic	Description of detrital rock or sediment consisting of fragments or grains of many different rock and/or mineral materials.
Poorly graded	See Grading.
Pore pressure	The pressure of the air or, more usually the water, occupying the pore space in a soil or rock. It is equal to the total stress minus the effective stress.
Porosity	(*Hydrogeology*) The pore space within a rock that may be occupied by a fluid. (*Soil mechanics*) The ratio between the volume of voids in a soil to its total volume.
Porphyritic	An igneous rock texture, comprising relatively large well-developed crystals (phenocrysts) set in a finer-grained groundmass.
Portland cement	Hydraulic cement invented in England in 1824 and named for its supposed resemblance, when set, to Portland stone. Now the most commonly used manufactured cement worldwide. It is manufactured by fusing limestone and clay or shale, which is then ground to a powder

	with a small proportion of gypsum. Unhydrated Portland cement principally consists of a mixture of tricalcium silicate (50–70%), dicalcium silicate (15–30%), tetracalcium alumino-ferrite (5–15%) and tricalcium aluminate (5–10%). On hydration, complex calcium-silicate-hydrates are formed, together with calcium hydroxide (portlandite) and other minor phases.
Pozzolanas	Natural or artificial forms of silica, silicates or aluminosilicates that have the ability to react with calcium hydroxide and water, at ambient temperatures, to give calcium-silicate cementitious products.
Precipitation	(*Chemical reactions*) The formation of an insoluble solid, such as a mineral, by a reaction which occurs in solution. (*Meteorology*). The amount of rain (or snow or hail) falling on an area per unit time.
Preconsolidation pressure	The maximum value of consolidation (qv) pressure borne by a sediment since its deposition. In overconsolidated (qv) clays, this will be greater than the present value of overburden pressure.
Pressure	Force per unit area. Syn. Stress.
Principal stress	Any system of shear and normal stresses acting on a body can be represented by three orthogonal principal (normal) stresses. These are respectively the major, intermediate and minor principal stresses in terms of magnitude.
Pseudomorph	A mineral that mimics the habit (qv) characteristic of another mineral often as a result of it replacing that mineral
Puddle clay	Natural clay of high plasticity that is worked with sufficient water to produce a homogeneous material of low permeability. Traditionally used for lining canals and as the core of some types of dam. Originally worked by feet. Occasionally less correctly termed 'puddled' clay (qv).
Puddled clay	See Puddle clay.
Pugging	The creation of an extruded plastic ceramic body.
Pulverized-fuel ash (PFA)	See Fly ash.
Punned	Description of soil, hard-core and similar material that has been compacted (or rammed) by repeated blows from a heavy-ended tool (a punner), to densify the material in order to create a stronger, denser layer of low permeability.
Pyrite	A mineral comprising iron sulphide, FeS_2; formerly known as 'iron pyrites' and popularly known as 'fool's gold'.
Pyritous	Containing pyrite (qv).
Pyroclastic	Description of fragmental material, such as volcanic ash, that has been blown into the atmosphere from a volcano by explosive activity.
Pyrolysis	Chemical decomposition by heating that forms the basis for the analysis of organic compounds by the thermal degradation of complex organic structures to simpler more readily analysable products.
Pyrophyllite	An aluminous phyllosilicate found in some metamorphic or hydrothermally altered rocks.

Q

Quarry	A surface excavation or series of excavations made in order to extract minerals. Non-rock (e.g. sand and gravel) quarries may locally also be known as pits. In the UK, the term 'quarry' also has a meaning under the Quarries Regulations, which include workings for minerals in disused tips but exclude mineral workings for construction works or mines and excavations for roads and railways. See also Borrow pit.
Quartz	The most common silica (SiO_2) mineral, widely distributed in a great variety of rock types.
Quartz Porphyry	A medium-grained, porphyritic, acid igneous rock containing phenocrysts of quartz and feldspar. The term 'microgranite' (qv) is preferred for modern usage.
Quick clay	A clay that will lose shear strength and may liquefy if disturbed.

R

Reaction Interval	The temperature band within which a reaction, change of state or the formation of new minerals occurs.
Reconstituted soil	Soil material that has been removed from its original position in the ground and then re-compacted in the laboratory, usually prior to testing.
Recrystallisation	The process by which the crystals in rocks are reformed, usually into larger entities in the solid state.
Rectorite	A mixed-layer clay mineral consisting of a regular 1:1 interstratification of smectite and illite/mica layers.
Red-black toposequence	A type of soil association or catena common in the tropics where red soils are found in dryer upland areas and black soils at lower wetter elevations and in depressions. The red soils are usually kaolinitic and the black soils smectitic.
Refractories	Refractory materials, typically used for lining furnaces etc.
Refractoriness	The property of a material to withstand high temperatures, the environment, and conditions of use without change in its physical or chemical identity.
Refractory	The property of being resistant to high temperature and changes in temperature.
Regolith	The surface layer of fragmented or unconsolidated soil or rocky materials.
Residual Soil	Surface soil formed *in situ* by degradation of parent material by weathering processes.
Reinforced clay	Clay masses in which geogrids (polymer perforated sheets with high tensile strength), geotextiles or other materials are used to enhance their strength.
Residual shear strength	The shear resistance that occurs on a shear surface within a clay material when it has been subjected to a large displacement. The value is a constant that depends on the mineralogy and grain size of the material.
Rheology	The study of the flow properties of materials.
Riffling	Reduction in quantity of a sample of material by dividing it into two approximately equal proportions by passing the material through a sample splitter (riffle box) of the appropriate size.
Road base	The main structural element in a road pavement. See also Sub-base.
Roseki	A type of hydrothermal clay mineral deposit, characterised by abundant pyrophyllite as well as other aluminous phases such as diaspore and kaolinite.

S

Sabkha	Shallow ephemeral water bodies in an arid environment. Commonly intertidal and supratidal flats in an arid environment that are the sites for evaporite salt formation.
Sand	Sediment, deposit, aggregate or soil with particles having a grain size between 0.06 and 2 mm (various other size definitions in use).
Sanitary ware	Shower trays, toilet bowls, sinks etc. and other high quality ceramics
Saponite	A trioctahedral smectite, generally rich in magnesium.
Saprolite	Weathered rock with changed chemical and mineral composition and lower strength that is still recognisable as rock, for example by the visible preservation of rock structures such as joint or bedding planes.
Saturation	The degree to which all the pore space in a material is occupied by water.
Schist	A type of metamorphic rock characterised by a distinct metamorphic particle alignment ('schistosity'). See also Foliation.
Seam	(*Mining*) A relatively thin and laterally extensive stratum in a sequence, usually of an economic mineral, for example coal. (*Geology*) A layer that is different from the rest of the sedimentary sequence, for example a clay seam in a limestone formation.
Seatearth (or Seat earth)	Former soil horizon, usually fine-grained and containing the remains of plant roots, that underlies a carbonaceous deposit (peat or coal). A British term that is more or less equivalent to the American term 'underclay'. (qv)

Secant pile wall	A continuous wall formed underground by overlapping circular-section piles.
Secondary electrons	Low energy electrons produced by inelastic collisions between an electron beam and the surface of a solid specimen.
Sediment	Solid material, organic or inorganic in origin, that has been transported and deposited by wind, water, or ice. It may consist of fragmented rock material, products derived from chemical action or from the secretions of organisms. Loose sediment, such as gravel, sand or mud, may become consolidated and/or cemented to form coherent sedimentary rock.
Sedimentary rock	A rock formed from the consolidation of detrital sediment (qv) that has accumulated in layers (including sandstones and shales), or a chemical rock formed by precipitation (including some limestones and evaporites), or an organic rock consisting largely of the remains of plants and animals (including some limestones and coal).
Sedimentation	The processes by which sediment accumulates to form a deposit.
Sensitive tint plate	A section of a mineral (e.g. quartz) cut to a standard thickness to produce a relative retardation of 560 nm and used as an accessory plate for the analysis of thin sections in the petrological microscope (qv).
Sensitivity	The ratio of the undrained strength of undisturbed soil to the strength of reconstituted soil at the same moisture content.
Sepiolite	A fibrous magnesium-rich clay mineral with a modulated layer structure.
Sericite	A white fine-grained potassium mica, close to muscovite in composition, occurring as an alteration product of various aluminosilicate minerals. The term 'sericite' tends to be used imprecisely for any fine-grained micaceous alteration or weathering products.
Serpentine	A sub-group of the 7Å phyllosilicates, usually trioctahedral.
Sesquioxides	Oxides containing two kinds of atom or radical in the proportion 2:3; the prefix 'sesqui' means 1½.
Shale	A fine-grained (grain size <0.06 mm) sedimentary rock that exhibits fissility (i.e. tendency to split into thin sheets). Certain mudstones (qv) develop into shales as a consequence of weathering action.
Shear box	A device for holding a soil specimen for the determination of its shear strength.
Shear strain	A measure of the distortion a body suffers as a result of the imposition of stress. The so-called 'engineer's strain' is the linear distance a point in the body moves compared with its original position divided by the thickness of the body undergoing distortion, measured at right angles to the plane of the surface over which the shear stress (qv) acts.
Shear stress	The stress causing distortion of a body. A stress caused by a force acting parallel to a surface.
Shear surface	A discontinuity along which there has been shear displacement, rendering the discontinuity of lower strength than previously
Sheet silicate(s)	See Phyllosilicates (qv).
Shelf	The continental shelf, the region of shallow seas surrounding continents that is distinguished from the deeper ocean basins.
Shield	See Cratonic. Syn. Craton.
Shrinkage limit	The moisture content below which there is no further decrease in volume of a soil with decreasing moisture content.
Silcrete	A variety of duricrust formed by the precipitation of silica.
Siliceous	Rich in silica, SiO_2.
Silt	**1.** A detrital particle, finer than very fine sand and coarser than clay, in the range of 0.002 to 0.06 mm (various other size definitions are in use). **2.** An uncemented clastic sediment of silt-sized rock or mineral particles. **3.** Fine earth material in suspension in water.
Sinter/Sintering	The bonding of powdered materials by solid-state reactions at temperatures lower than those required for the formation of a liquid phase.
Slake durability	A measure of the resistance of material to degradation by slaking.

Slaking	A degradation process in geomaterials caused by alternating cycles of wetting and drying.
Slate	A low-grade metamorphic argillaceous rock, characterised by the presence of a 'slaty' cleavage, which allows it to be split into large thin sheets.
Slip	A suspension or slurry of finely divided ceramic materials in a liquid.
Slurry wall	Concrete wall constructed *in situ* using mud, usually bentonite, slurry to support the sides during excavation. On the completion of excavation, concrete may be introduced at the base of the excavation and the slurry is displaced upwards to form a subsurface concrete wall. In some cases, a low permeability boundary is formed by backfilling the trench with bentonite. (cf Diaphragm wall).
Smectite	A group of high plasticity clay minerals, with a 2:1 sheet silicate structure, characterised by a layer charge of less than 0.6 per formula unit, the presence of interlayer cations and the ability to absorb water, as well as other liquid substances, between the layers and swell. The members of the group are distinguished by chemical composition, magnitude and location of layer charge, and the nature of the octahedral sheet, which may be dioctahedral or trioctahedral. Montmorillonite is a variety of smectite; bentonites and some Fullers' earths are materials that are rich in montmorillonite.
Soak Period	A period of static temperature during the firing cycle (often at peak temperature).
Soffit	Underside of any deck, beam, arch or similar structural unit.
Soil	(*General*) The unconsolidated materials at the surface of the Earth that support plant and animal life. (*Geotechnical*) An assemblage of discrete particles in the form of a deposit, usually of mineral composition but sometimes of organic origin, which can be separated by gentle mechanical means and which includes variable amounts of water and air (and sometimes other gases). A soil commonly consists of a naturally occurring deposit forming part of the earth's crust, but the term is also applied to 'made ground' consisting of replaced natural soil or man-made materials exhibiting similar behaviour (e.g. crushed rock, crushed blastfurnace slag, fly ash). See also Engineering soil.
Soil modification	General term for improving the properties of soil. With lime, the 'modification' is usually the immediate improvement in workability, placeability, compactability and strength of soil, whereas '*stabilisation*' is additional long-term improvement in the strength and durability of the soil.
Soil profile	A sequence of layers or horizons that reflect the development from the parent material and composition of a soil.
Soil stabilisation	See Soil modification.
Sol	See Thixotropy.
Solifluction	The process of slow flow of soil from high to low ground, whilst oversaturated with water. Examples of soliflucted deposits are found on many slopes in clay deposits in south eastern England, which were formed when periglacial conditions existed. Solifluction in periglacial conditions is specifically termed 'gelifluction'.
Solifluction shears	Shear surfaces resulting from the slow down-slope movement of unconsolidated materials or debris containing some fines as a result of alternate freeze and thaw of the water in these materials. May also include shear planes from episodes of more rapid downslope failure or flow.
Sorting	The geological processes by which fragmented materials are separated into fractions of different sizes. Well sorted materials contain fragments of a similar size, unsorted materials consist of fragments of a wide range of sizes and poorly sorted materials contain a range of sizes but with an absence or predominance of one particular size. See also Grading.
Souring	A weathering process in which claystone, slate or shale becomes soft and malleable, losing their hard brittle nature.
Spray drying	**1.** A method of drying involving spraying a sample as an aqueous suspension into a heated chamber so that the solid dries in the form of spherical spray droplets, producing a fine granulated powder. **2.** The creation of an atomised slip that dries on contact with hot gases or a hot liquid to produce a fine granulate or dust.

Stevensite	A trioctahedral smectite.
Stereoscope	An optically simple device holding lenses and sometimes mirrors for the inspection of pairs of aerial photographs or images; the image viewed in stereo allows the viewer to gain an appreciation of the relief of the land.
Stiffness	The ability of a material to resist deformation.
Stoneware	A dense, vitreous or semi-vitreous ceramic ware (produced from semi-refractory plastic clay).
Stop chocks or blocks	Low barriers, kerbs or bunds to stop or inhibit the tyres of vehicles from passing a given point without acting as a potentially damaging barrier to the body of the vehicle.
Strain	A measure of the linear, volumetric or shear deformation that results from the application of a stress.
Stress	Force per unit area. Syn. Pressure.
Stress relief	Reduction in overburden stress due to the erosion of overlying materials.
Sticky point	The minimum moisture content for clay to stick to metal tools.
Stretcher	A brick placed in a course or layer of brickwork with the long side facing out.
Sub-base	In a pavement, especially a road, a layer of usually granular material below the main structural element in a pavement (road base).
Suction	The stress-free negative pore water pressure of an element of soil. Suction can be expressed in any units of pressure, but the pF scale is commonly used.
Sudoite	A species of dioctahedral chlorite.
Sulphate reduction	The chemical and biochemical processes by which sulphate is reduced to sulphide.
Surfactant	A material which reduces the surface tension of water.
Swelling	An increase in volume of a geomaterial.
Swelling chlorite	A variety of the clay mineral chlorite that exhibits interlayer swelling.
Swelling index	The increase in void ratio for unit change in the logarithm of the decrease in effective stress bringing about that change.
Swelling limit	The boundary between structured and normal shrinkage, where during normal shrinkage volume change is proportional to the water loss, whereas during structured drying it is higher than this and during normal drying it is lower than this.

T

Tactoid	A unit of organisation of clay particles.
Talc	A trioctahedral phyllosilicate mineral rich in magnesium.
Tapis	See Cob.
Talluvium	Coarse hillslope deposit comprising talus (scree) and finer materials.
Tectonic	Relating to the stresses and deformation effects, including folding and faulting, in the Earth's crustal plates.
Tensile strength	Maximum tensile stress that can be borne by a material before fracturing or a large increase in the rate of strain occurs. May be determined directly by applying a tensile force to a specimen, or indirectly by loading a prismatic sample that is supported at two points or by splitting a cylindrical sample by the application of a diametral compressive load.
Terrigenous	Derived from land.
Tephra	Air-fall debris from a volcano.
Terra cotta	Red or brown, unglazed earthenware, sometimes used for decorative building components.
Tetrahedral coordination	See Coordination.

Thermal expansion	Reversible or permanent change in the dimensions of a body as a result of temperature increase.
Thin-section	A rock or soil cut as a thin-section, usually 0.03 mm thick, cemented onto a glass slide for examination in a petrological microscope (qv).
Thixotropy	(*Physics*) Rheological property in which the viscosity or state of the material is influenced by applied stress. (*Ceramics*) The tendency of some casting slips to thicken when left standing or be fluid when agitated. (*Soil mechanics*) The property of some slurries to alternate between a gel (solid) when left undisturbed and a sol (viscous fluid) when agitated.
Till	Term used for the deposits left by glaciers; generally synonymous with the looser terms Boulder clay and Glacial drift. Syn. Glacial till.
Tip	An accumulation or deposit of mineral or waste materials. In quarries, this can be in a solid or liquid state, and may include overburden dumps, spoil heaps, stockpiles and lagoons. In the UK these are covered by the Quarries Regulations. Elsewhere tips are generally temporary or permanent disposal areas for unsuitable, surplus or waste materials and may be covered by Waste Management Regulations.
Titrometric	Quantitative chemical analysis method involving the reaction of a measured volume of one liquid added to a fixed volume of another liquid, to determine the volume of the added liquid required to complete the reaction.
Tonstein	Literally 'clay stone' in German. In English usage a clay formed by the alteration of volcanic ash in a fresh water environment and consisting predominantly of kaolinite.
Topsoil	The uppermost part of a soil layer that, by its humus content, supports vegetation. In the UK, the topsoil is typically approximately the uppermost 150 mm of soil, but in wet tropical climates it may be a metre or more in thickness.
Total stress	The force per unit area acting at a particular point. At depth in the ground the total stress is the mass of the overlying ground and any forces imposed on the ground per unit area. It is equal to the effective stress plus the pore water pressure.
Trafficability	Ability of vehicles (e.g. construction plant) to traverse the ground.
Transformation	One of the three possible origins for clay minerals whereby some structural element of a precursor mineral is retained. See also Inheritance and Neoformation.
Treated clay	Clay that has been mixed with either lime or cement or both to improve its physical characteristics.
Tremie (pipe)	A pipe used to deliver concrete or grout to the base of a trench or pile bore, to prevent segregation of the concrete or contamination by slurry. See Diaphragm wall and Slurry wall.
Triaxial test	A strength test for soil or rock in which an axial load is applied to a specimen, which is also subjected to a lateral pressure or confining pressure at right angles to this.
Trioctahedral	The composition and structure of the octahedral sheet of a clay mineral if all three octahedral sites are occupied by cations.
Trophic	Pertaining to nutrition.
Trophic level	The level at which energy in the form of food is transferred from one organism to another in the food chain.
Tropical weathering	Weathering by the combined effects of heat and moisture under tropical climate conditions.

U

Ultimate strength	The maximum stress that can be borne by a soil in its critical state before a large increase in the rate of strain occurs. In overconsolidated (qv) clays the ultimate strength will be less than the peak strength, and with further shearing the shearing resistance will decrease to the residual value.
Ultrabasic	Adjective describing a group of igneous rocks, relatively low in silica and extremely rich in ferromagnesian minerals such as olivine and pyroxene. Syn. Ultramafic.
Ultramafic	See Ultrabasic.

Ultraviolet radiation	Electro-magnetic radiation below the visible light spectrum i.e. wavelengths less than 360 nm.
Unconfined compressive strength	The uniaxial compressive load required to cause failure of an unconfined specimen of soil (or other material) divided by the cross-sectional area of the specimen at right angles to the axis of loading. Units are force per unit area, which in the SI system is kN/m^2 (or its equivalents).
Unconfined tensile strength	The maximum tensile stress that can be borne by a material before it fractures or undergoes a rapid increase in strain rate, where the stress is applied in one direction only. May be determined directly by applying a tensile force to a specimen or indirectly by loading a prismatic sample that is supported at two points or by splitting a cylindrical sample by the application of a diametral compressive load.
Underclay	A type of clay normally found under a coal bed or peat and often with fossilised rootlets or other evidence that it formed as a soil. See also Seatearth.
Undisturbed sample	A sample taken in such a way that the chemical and physical condition of the sample is not changed.
Undrained shear strength	The maximum stress that can be borne by a material before it fractures or undergoes a rapid increase in strain rate where the test takes place under undrained or total stress conditions, i.e. pore pressure changes due to the application of stress are not allowed to dissipate during the test.
Uniaxial compressive strength	The maximum compressive stress that can be borne by a material before it fractures or undergoes a rapid increase in strain rate, where the stress is applied in one direction only.
Uniaxial tensile strength	The maximum tensile stress that can be borne by a material before it fractures or undergoes a rapid increase in strain rate, where the stress is applied in one direction only. May be determined directly by applying a tensile force to a specimen or indirectly by loading a prismatic sample that is supported at two points or by splitting a cylindrical sample by the application of a diametral compressive load.
Uniformly graded	See Grading.

V

Valence	Term used to define oxidation state of chemical elements and ions.
Varve	A layer of sediment deposited in one year, typically exhibiting grain size variations with season (coarser in summer and finer in winter); hence 'varved clays'. Almost exclusively used for sediments deposited in glacial melt-water lakes.
Varved clays	See Varve.
Veins	Fractures or fissures in rock or soil materials that are infilled with minerals.
Verdine	A group of green marine clays, rich in iron, typically characterised by the presence of the mineral odinite (qv).
Vermiculite	A swelling clay mineral with a 2:1 sheet silicate structure, and a layer charge of between 0.6–0.9 per formula unit that differentiates them from smectites.
Vertisol	A type of soil, formally defined, but generally having a high content of smectite which imparts shrink-swell characteristics to soil, leading for example to the development of deep vertical cracks during dry periods.
Vitrification	(*Physics*) Process of altering a material into a glassy state. (*Ceramics*) Gradual melting, in which increasing quantities of liquid are produced as the temperature is raised.
Vitrification Range	The temperature interval between the first appearance of glass and the melting point.
Vitrified	Converted into glass or a glassy state.
Void ratio	In a soil or rock, the volume of voids divided by the volume of dry solids in the sample.
Volcanic	Description of igneous rocks and pyroclastic deposits that are formed by extrusion from a volcano.
Volcanic ash	See Pyroclastic.
Volcanogenic	Formed by processes associated with volcanic processes.

W

Water absorption
: A measure of the amount (by mass) of water that a body or substance will absorb, or assimilate, under standard conditions, expressed as a percentage of the dry weight of the body or substance.

Water potential
: A measure of the free energy of water in a solution, as in a soil sample, and hence of its tendency to move by diffusion, osmosis or as a vapour.

Weathering
: The physical and chemical processes by which rocks are broken down and decomposed by external agencies such as water, wind, ice, temperature changes and the actions of plants and bacteria.

Well graded
: See Grading.

Whiteware
: A general term for a ceramic body that fires to a white or ivory colour.

Witchert
: See Cob.

Workability
: The combination of properties which contribute to the ease with which plastic materials can be mixed, handled, transported and placed.

X

XRD
: See X-ray diffraction.

XRF
: See X-ray fluorescence.

X-ray reflections
: Constructive X-ray diffraction when Bragg's Law (q.v.- see X-ray diffraction) is satisfied.

X-rays
: Electromagnetic radiation of wavelengths in the range of 0.01–20 nm (analytical X-rays in the range 0.02–2 nm) generated when an atom is bombarded by electrons—each element produces a characteristic X-ray spectrum.

X-ray diffraction (XRD)
: Method for the analysis of crystalline solids involving the diffraction of a monochromatic X-ray beam in which the regular array of atoms in the crystal structure act as planes of a diffraction grating to the incident X-rays. See also Bragg's Law.

X-ray powder diffractometer
: (*Position sensitive*) In a position sensitive X-ray diffractometer the powdered sample is mounted in the path of the fixed incident beam of monochromatic X-rays and rotated in a horizontal plane. The diffracted X-rays are detected by a fixed array of detectors arranged in an arc. (*Scanning*) In a scanning diffractometer the powdered sample is mounted in the path of the incident monochromatic X-rays and is rotated in a plane about a horizontal axis to change the angle of incidence of the fixed X-ray beam (θ) whilst the X-ray detector rotates at twice the rate of the sample.

X-ray fluorescence (XRF)
: A technique of chemical analysis in which a sample is irradiated with X-rays. This produces secondary fluorescent X-rays that are characteristic of the elements present. Analysis of the fluorescent spectrum enables quantitative identification of certain of the chemical elements present in the sample.

Y

Yield
: The point that defines the end of elastic deformation such that the stress–stain relationship is no longer linear.

Young's modulus
: The slope of the stress-strain graph for elastic deformation.

Z

Zabour (or Zabur)
: See Cob.

Zooplankton
: Planktonic animals.

Index

Note: **bold** page numbers indicate tables; *italic* page numbers indicate figures; 'g' after a page number indicates a Glossary entry (e.g. 123g).

A-line, 92, *229*, 485g
ablation till, 485g
ACEC Classification, 296
acidity, 22, *22*, 23
  dispersive soils, 114
  in hydothermal alteration, 56–7, *56*
  tests, 235, **236**
acidolysis, 34, 485g
activity charts, *94*, *123*, 485g
activity (soil), 92, 94, **94**, *94*, 230, **230**, 485g
adobe, 390, *390*, **391**, 394–5, 485g
advection, 95, 364, 485g
aeolian clays, 485g
aeolian transport, 44–5, *47*
aerial photography, 181–2, *181–2*
aggradation, 37, 485g
aggregates, 485g
  lightweight, 439
aggregation, 41, 81, 116 (*see also* flocculation)
Aggressive Chemical Environment for Concrete (ACEC) Classification, 296
air entry point, 91–2, **91**, 485g
air void content, 100, 238, 293–4, *295*, 337, *338*, 485g (*see also* compaction; voids ratio)
allophane, 4, 24, **74**, **94**, 118, 485g
alluvial clays, 39, 120, 141, **464**, 485g (*see also* sedimentary processes)
alluvium, 486g
alteration, 486g (*see also* hydrothermal alteration)
aluminium hydroxides, 24 (*see also* diaspore; gibbsite)
aluminium interlayered clays, 34
aluminium oxides, 24
aluminium-silicon oxyhydroxides, 24
aluminium sulphate, 114
alunite, 66
Amazon (river), 38, 39, *40*
Ampthill Clay, 125–6, 165, **467**
analcite, 25
analytical methods, 199–222, 235–6, **236–7**
anatase, 64
anchizone, 51, 486g
Ancholme Group, 162
andesite, 486g
andosols, **116**, 117, 118, 140, 486g
angle of friction, 486g
anhydrite, 24, 160, 163, 216
anion, 486g
anion exchange capacity, 218
anisotropic consolidation test, 272
apparent angle of friction, 486g
apparent cohesion, 112, 262, 486g
apparent friction, 486g
apparent viscocity, 88, 351, 486g

aquifer, 486g
aquitard, 486g
aragonite, 23
archaeological sites, mineral workings, 289
argillite, **60**, 81
ash, volcanic, 46, 48, 62–3, 64
ash waste, recycling, 441–2
ASTM standards
  concrete, **432**
  mudrock description, 84
  test procedures, **481–4**
Atherfield Clay Formation, 166, **466**
atomic emission analysis, 215
atomic number, 486g
attapulgite *see* palygorskite
Atterberg limits, 89, 92, 486g (*see also* liquid limit; plastic limit; shrinkage limit)
attrition mill, 486g
auger, 486g
autographic unconfined compression test, 259, *259*
available lime, 487g
Aylesbeare Mudstone, **470**

B-bar test, 271, *272*, *273*
backscatter electrons, 487g
bacteria, 25, 49
ball clays, 16, 170
  ceramics, 98, 403, **404**, 405
    fine ceramics, 413, 417
    structural ceramics, 408
  deposits, 121, 149, 150, 417, **475**
  engineering properties, **465**
  extraction and processing, 340, 418–20, *420*
  formation, 63–4
  production statistics, **7**
banksman, 487g
barite (barium sulphate), 24, 216
Barton Clay, 121, 168, **465**
basalt, 487g
batholith, 487g
bauge *see* cob building
bauxite, 24, 116, 437, 444
bed (deposition), 487g
bedding, 487g
bedrock, 5, 29, 141, 154, 487g
beidellite, 17, 57, 58, 487g
belt press, 487g
Benton Shale, 64
bentonite (*see also* montmorillonite)
  absorption of pollutants, 377
  composition, 17
  concrete additive, 439
  definition, 487g

deposits, 150, **475**
formation, 46, 48, 64–5, 67
grouting, 375–7, 439–40
in radioactive waste storage, 368–9, 370–3, *371*
slurries, 347–62, *348*, *353*, **356**, *356–7*, *363–4*, **382**
swelling properties, 65
bentonite-cement mixes *see* cement-bentonite mixes
bentonite enhanced sands (BES) *see* sand-bentonite mixes
berthierine, 15, 46, 487g
BES *see* sand-bentonite mixes
Bingham fluid, 88, 351, 487g
Bingham solid, 487g
biogenic material *see* organic matter
biological assays, 220
biological processes, 25–6, 41, 49, 155
biota, 487g (*see also* organic matter)
biotite, 59, 487g
bioturbation, 487g
birefringence colours, 487g
black clay soils, 116, **116**, 118
black shales, 43, 60
blastfurnace slag, 487g
bleed, 487g
blending, 419
Blisworth Clay Formation, 164, 458
bloating, 411, 487g
Blue Lias Formation, 126, 127, 162, *162*, **468**
blunging, 422, 487g
boehmite, 24, 488g
boreholes
geophysical investigations, 187–8, 191–2, *191–2*
sealing, 381
borrow pit, 488g
bottom dump bucket, 488g
boulder clay, 43–4, 119, 141, 171, **462–3**, 488g
bowl capacity, 488g
Bowland Shale Formation, 158
Bracklesham Beds, 121, **465**
Bragg's law, 488g
brick clays, 98, 161, 165, 166, 407–8
Brickearth, 172, 488g
bricks, 406
calcium silicate, 437
earth, 390
engineering properties, 411–12
ideal particle size, *408*
production, 409–11
statistics, *7*
recycling of clay, 442–3, **443**
Bronsted Acid, 488g
brownfield site, 488g
BS 1377:1990, 91, 109, 223, **236–7**, **481–4**
BS 5930:1999
classification and description of clays, 76, **77**, **78**, 80–1
classification and description of mudrocks, 81, **81**, 84, **84**
definition of clay, 4–5
field investigations logging, 193
BS EN 197-1, 432
bucket capacity, 488g
building *see* construction; earthen architecture
building regulations, earth buildings, 392, *393*
building rubble, recycling of clay, 443–4
bulk density, 231, **334**, 488g
bulk modulus, 105–6, 488g
bunds, 284–5, 488g
burial diagenesis, 488g
buried structures, corrosion, 296–7

cable tool boring, 191
CAC (calcium aluminate cement), 437–8
Cadeby Formation, **471**
calcareous microfossils, 66
calcination, 488g
calcined clays, 434, 435–6
calcite, 23, 53, 62, 66, 155
Calcite Compensation Depth, 66
calcium aluminate cement (CAC), 437–8
calcium carbonate (calcite), 23, 53, 62, 66, 155
calcium silicate bricks, 437
calcium sulphates, 114, 216, 432, 438 (*see also* anhydrite; gypsum)
calcrete, 488g
caldera, 488g
calibration, test equipment, 225–6
California bearing ratio (CBR), 101, 295
tests, 238, 242, *242–4*
Cambrian deposits, 50
World and European, 141–2, **477**
canals
embankments, 276–7, *277*, 298
puddle clay liners, 377, 382–3
capping
contaminated land, 377
definition, 488g
highways, 306
landfill covers, 364–7, *364–5*
carbonate apatite group, 25
carbonates (*see also* aragonite; calcite; dolomites; siderite; vaterite)
analytical methods, 215, **236**
definition, 489g
diagenetic reactions, 53
formation, 49
mineralogy, 23
in shales, 62
Carboniferous deposits
United Kingdom, 129–31, **456**, **457**, 458, **472–3**, **480**
stratigraphy, 157–9, *157*
World and European, 143–4, **476**
cardhouse, 489g
carnallite, 25
Cassagrande soil classification chart, *93*
casting slips, 413, 422–3, 489g
catalysts, cracking, 443
catalytic behaviour, organic matter, 26
catena, 489g
cathodoluminescence, 489g
cation, 489g
cation exchange capacity, 114
CBR *see* California bearing ratio (CBR)
CD (consolidated drained compression) test, 267–8, **267**, *269*
CDM (Construction (Design and Management)) regulations, 330
celadonite, 16, 419, 489g
celestene (celestite), 24, 216
celestite, 24, 216
cement-bentonite mixes
grouting, 376, 439–40
slurries, 348–52, *350*
soft piles and cut-off walls, 373–5
cementation shales, 81–2, 83
cements, 427–39, **429**, 489g
CEN prENV 1997-2, 224
cenospheres, 442
Cenozoic deposits, *145*, 146–7 (*see also specific periods*)
ceramic clays, 98–9, 407–8, 413, 417–22, *420–1*

additives and impurities, **99**, 402–3, **406**
  moisture content, **87**, 98–9
ceramic properties, **406**
ceramics, 401–25
  clay minerals, 404–5
  composition, 402–4, **404**
  fine ceramic products, 412–24
  historical development, **402–3**
  structural products, 405–12, **409**
chalk, 81, 489g
Chalk Marl, 159
chamber kilns, 411
chamosite, 15, 46, 53, 489g
chamotte, 489g
Charmouth Mudstone Formation, *163*, 164, **468**
chemical analysis, 214–18, 235–6, **236–7**, **482**
  data, 449–59, **452–7**
chemical cements, 436–7
chemical treatment, engineering fills, 304–7
chert, 23, 53, 157, 489g (*see also* flint clays)
china clay *see* kaolin
chlorides
  analytical methods, **237**
  in cement, 431
chlorite-smectites, 20, 53, 57
chlorites (*see also* chamosite)
  ceramics, 405, **406**
  definition, 489g
  distribution, *37*
  formation, 34, 37–8, 53, 57
  mineralogy, 17, 19, *19*
  properties, 21, **74**, **94**
clamp firing, 411
classification
  clay minerals, **15**
  clays, 75–8, **76**, *76*, **77**
  earthworks materials, 330, **331**, 337
  mudrocks, 60, **60**, 81–4, **81**, *82*, **83**, 85, **86**
  permeability, **95**
  plasticity, 92
  soils, *6*, 75–80, **77**, *79*, 227–35
  test procedures, **481**
clast, 489g
clay-cements, 376
clay cycle, 29, *30*
clay dabbs, 395, 489g
clay fraction, 234
clay lump, 394–5
clay minerals, 13–23, *14–21*, **15** (*see also* illite; kaolinite; montmorillonite)
  ceramics, 404–5, **406**
  definition, 490g
  distribution, *37*
  formation, 30–2, *30–3* (*see also* inheritance (clay formation); neoformation; transformation (clay formation))
  properties, 20–3, **74**
clay rich soils, 58–9 (*see also* vertisols)
clay-with-flints, 172
Clay Working Party, 7–10
Claygate Beds, 121, **465**
claystones, 81, **81**, 82, *82*, 490g
climate, role in clay formation, 33–4, *33*, *35*, 36, 139 (*see also* rainfall; wind)
climatic changes, 147
clom, 389–90, *389*, **391**, 392, 394, *394*, 395
coal ash, recycling of clay, 441–2, **442**
Coal Measures, 63, 129–31, 157, 158, **472**

cob building, 389–90, *389*, **391**, 392, 394, *394*, 395, 490g
coccoliths, 66
coefficient of consolidation, **109**
coefficient of earth pressure at rest, 107, 272
coefficient of volume compressibility, **109**
cohesion, 78, 110, 112, 262, 490g (*see also* shear strength)
cohesive soils, 490g
collapsible soils, 114–15, 490g
colloid, 490g
colloidal suspensions, 88
colluvium, 490g
colour, 24, **99**, 184, 403
comminution, 490g
common clay and shale, 59–63
  ceramics, 407
  composition, **61**, **62**
  definition, 490g
  deposits, 147, 149, **476–7**
  production statistics, **7**
compaction, 100–1 (*see also* consolidation)
  clay formation, 54–5, *54*, *55*
  definition, 490g
  earthworks, *292*, 332, 333–5, 337–9
  effect on permeability, 365, *366*
  engineering fills, 291–2, *291*, 293–4
  mechanical improvement, 302, 304
  related to moisture content, 100–1, *101*, 238, 239–42, *239*, *240*
  sand-bentonite mixes, *368*
  tests, 238–40, *239*, **240**, **482**
compaction moulds, **239**, 240
compaction plant, *318–19*, 324–6, **329**, 333–5
compaction shales, 81–2, 83
composition
  analytical methods, 199–222, 235–6
  ceramics, 402–4, **404**
  data, 449–59, **452–7**
  mineralogy, 13–27
  mudrocks and shales, 60–3, **61**, **62**, *62*, 82
  river suspended sediment, 39, *41*
Comprehensive Soil Classification System, 78
compressed blocks, 390–1, *390*, **391**, 490g
compressibility, 104–9, **107**, **109**, 490g (*see also* deformation behaviour)
compression index, 92, **108**, 117, 490g
compression tests, 259–60, 263–8
compressive strength, *108*, 109
  bricks, 412
  mudrocks, 82, **84**, 111–12, **113**
  puddle clay, 379–381
concrete (*see also* cements)
  clay additives, 439
  clay impurities, 444–5, **445**
  recycled clays in, 441–4
  reinforced, corrosion, 296
  standards, **432**
concretion, 490g
conditioning, soil, 358, 491g
cone penetration testing, 192–3
consistency index, 92, **94**, 491g
consolidated drained compression (CD) test, 267–8, **267**, *270*
consolidated undrained compression (CU) test, 266–7, *268–70*
consolidation, 104–7, **107**, **109** (*see also* compaction; compressive strength; over-consolidation)
  definition, 491g
  lateritic soil, *117*
  tests, 243–55, *247–9*, *250–2*, **251**, **253**, *261*, **483**

constant rate of strain test, 247–8, *250*
construction (*see also* earthworks; highways; pile construction)
    earthmoving, 317–18, **328**
    soil problems, 112–17
Construction (Design and Management) Regulations 1994, 330
containment (impermeable barriers), 362–81
contaminated land treatment, 377
continental margin, 491g
    sedimentary processes, 39–41, **41**, *41*, *42*
cookeite, 57, 491g
coordination (crystal structure), 491g
core drilling, 192, *192*
Cornish china clay, 66–7, *67* (*see also* kaolin)
cornubia, 491g
corrensite, 45, 491g
corrosion, buried structures, 296–7
cracking (decomposition), 491g
    catalysts, 443
Crackington Formation, **472**
creep, 491g
Cretaceous deposits
    United Kingdom, 124–5, **454**, **455**, 459, **466**, **478**
        stratigraphy, *162*, 166–7
    World and European, 145–6, **476**
critical moisture content, 99
critical point drying, 209, 491g
critical state, 491g
critical state line, 491g
critical state strength, 109
Cropwell Bishop Formation, *160*
cross polarized light, 491g
crown (tunnel), 491g
crumb test, 243
crushing, 408
crystal structures, 14–25, *14–21*
crystallinity, 491g
crystallinity indices, 203–4, **204**, *204*
CU (consolidated undrained compression) test, 266–7, *268–70*
cut-off walls, 373–5, 440, 491g
cuttings (infrastructure), 276
cyclic trixial test, 273
cylinder test, 243

d-spacing, 491g
dabbs, clay, 395, 489g
dams *see* earth dams
Darcy's Law, 94–5
data precision, laboratory tests, **206**, 226–7, **227**
data sources, exploration, 179–81
deep sea clays, 44, 65–6, *65*
definitions of clay, 1–5, 489g
deflocculants, 422, 492g
deflocculation, 114
deformation behaviour, 90–1, 102–3, *102*, 111–12 (*see also* elastic properties)
deltaic deposits, 39, 141, 492g
density, 87, **87**, 492g (*see also* bulk density; dry density)
    tests, 231–3, *232*
deposition (sedimentation), 38–45
deposits (*see also* distribution of clays)
    United Kingdom, 153–75, **478–80**
    World and European, 139–51, **475–7**
description
    clays, 80–1
    mudrocks, 84–5

desert regions, clay formation, 44, *47*
desiccation cracks, 184
detrital mineralogy, 492g
detritus, 492g
Devonian deposits
    United Kingdom, **456**, **457**, 458, **480**
    World and European, 143, **477**
dewatering, highway construction, 309–10
diagenesis, 23, 48–56, *49*, 492g
dialysis, 492g
diamicton, 492g
diaphragm wall construction, 352, 354–5, *355*, 373–5, **382**, 492g
diaspore, 24, 66, 492g
diatoms, 53, 66
dickite, 15, 66, 492g
differential scanning calorimetry (DSC), 208, *208*
differential thermal analysis (DTA), 207–8, *208*
diffusion, 95, 492g
dioctahedral sheet structures, 492g
dipmeter tools, 187
direct shear test, 260–2
directional drilling, 492g
discontinuity, 492g
dispersibility, 114, 381
    tests, 242–3, **245**, **482**
dispersion, 492g
dispersion method, 243
dispersive clays, 88, 114, 243, 283–4, 492g
dispersivity *see* dispersibility
distal deposits, 492g
distribution of clays, 5, *8*, *37*
disturbed soils, 110, 111, 115, 492g
docking, 412, 492g
dolomites, 23, 53, 62
domain (crystal), 493g
donbassite, 57, 493g
double hydrometer test, 243
downhole geophysical logging, 187–8
drag box, 493g
drainage
    earthworks, 288, 340
    role in clay formation, 34
drawdown, 493g
dredging spoils, recycling of clay, 444
drift deposits, 76, 154, 172, **464**, 493g (*see also* glacial tills)
drilling
    geophysical investigations, 191–2, *191–2*
    using bentonite slurries, 358, 361
dry density, 231
    compaction properties, 238, 291–2, *291*, *292*, 293
    engineering fills, 294–5, *294–5*
dry strength, 78, 493g
drying process, ceramics, 410, 423
drying shrinkage, 493g
DSC (differential scanning calorimetry), 208, *208*
DTA (differential thermal analysis), 207–8, *208*
durability
    definition, 493g
    earth buildings, 392
    mudrocks, 82–4, **83**
        tests, **482**
duricrust, 493g
dust, mineral workings, 288
dynamic probing, field investigations, 191
Dyrham Formation, 126, 127, *162*, 164, **468**

earth bricks, 390
earth buildings, 387–400, *388*, **398**, 399
earth ceramics, 437, **437**
earth dams, 279–84, *282*, *283*, 310, **378**, *378*
   failures, 298–9
   puddle clay cores, 377, 378–81
earth-pressure-balance tunnelling machine (EPBTM), 358
earthen architecture, 387–400, *388*, **398**, 399
earthenware, 413, 493g
earthmoving, 290–1, 315–45, **341**, 493g
earthmoving plant, *316–17*, 320–4, **328**, 343–4 (*see also* trafficability)
earthworks, 275–313 (*see also* earthmoving; mineral workings)
   compaction, *292*, 332, 333–5, 337–9
   definition, 493g
   design, 276–86
   engineering fills, 290–307, 330, **331**, 337
   failure, 297–9, **298**, 339–40
   planning, 286–90
Edlington Formation, 127, **471**
EDS (energy dispersive spectrometry), 215
Edwalton Formation, 160, 161
effective stress, 95–6, 97–8, 493g
effective stress shear strength, 104, 107, 110, 133
   tests, 263–8, *265*, **266**, *266*, *268–9*, 271, **483**
efflorescence, 412, 493g
Eh (redox potential), 493g
elastic properties, 88–94, *89* (*see also* deformation behaviour; stress-strain relationships)
elasticity, 493g
Eldersfield Mudstone, 160, 161
electrical methods, geophysical mapping, 186, *186*, 187–8
electrical porcelain, 413
electron microprobe analysis (EMPA), 216–17
electron microscopy, 209–14
eluviation, 493g
embankments (*see also* bunds; earth dams; engineering fills)
   definition, 493g
   design, 336
   historical development, *281*
   infrastructure, 276–9, *277–9*, 280, 295–6, 297–8, 300–1
   stability, 96, 277–9, *281*, 297–8, 339–40
EMPA (electron microprobe analysis), 216–17
energy dispersive spectrometry (EDS), 215
engineered earth, 392–3
engineering applications, classification of clays, 76–8, **77**
engineering fills
   classification, 330, **331**, 337
   design, 276–85
   maintenance, 297–307, **302–3**
   material characteristics, 290–8, *291–2*, *294–5*, 337
engineering properties, **73**, 101–12, 461–74, **462–73**
engineering soil, 493g
engobe, 493g
environmental bunds, 284–5, 493g
environmental impacts (*see also* landscaping)
   mineral workings, 183, 287–9
Eocene deposits, **452**, *453*, 459
EPBTM (earth-pressure-balance tunnelling machine), 358
epizone, 51, 494g
erodibility, puddle clay, 381
erosion processes (sedimentation), 38–45
ethylene glycerol, 202, 203
ethylene glycol, 494g
Etruria Formation, 129, 158–9, **472**
ettringite group, 24, 114, 438, 439

evaporite, 494g
evolved gas detection and analysis, 209
excavatability, 330, 332, **332**, **333**
excavation *see* earthmoving; earthworks
excavations, trial, 190
excavators, 419
expansive clay soils, 112–14
exploration, 177–97
   geophysical, 185–8, *186*
   intrusive, 188–97
   non-intrusive, 179–85, *180*
extraction *see* mineral workings
extrusion process, 409, 410

fabric structures, 42–3, *45*, *46*
   analytical methods, 209–14
facies, 494g
failure, earthworks, 297–9, **298**, 339–40 (*see also* slope stability)
falling head test, 255–6, *255*
fast firing, 411, 423–4, 494g
feldspars, 23, 49, 54, 62, 458, 459, 494g
felsic rocks, 494g
   clay formation, 34–5, *36*, 57
fermentation, 494g
ferric illite, 45, 494g
ferricrete, 494g
field investigations, 189–96
field mapping, 183–4, *184*
fills (earthworks) *see* engineering fills
filter cake, 349–50, *350*, 351, 494g
filters, in earthworks, 283–4, 298
fine ceramics, 412–13
   clays for, 413, *420–1*
   manufacturing processes, **414–16**, 422–5, *423*, **424**
fire resistance, earth buildings, 392
fireclays, 16, 63, 98, 159, 407–8, 494g
firing, ceramics, 410–11, **411**, 423–4, **424**
fissility, 43, 82, 85, 494g
fissures, in clay, 95, 107, 109–10, 113, 133, 494g
flashing, 494g
Fletton process, 410
flint, 494g
flint clays, 16, 64, 418
floc, 494g
flocculation, 41, 41–2, 43, *44*, 75, 348, 494g
floor tiles, 412–13
flow behaviour, 87–8
fluidity, 88
fluxing, 494g
fly ash, 494g
   recycling of clay, 442
fly rock, 494g
foliation, 494g
foraminifera, 66
Forest Marble Formation, 164
forming processes, ceramics, 409, 422–3
fossils, in clay formation, 53, 66
foundations (*see also* substructure fills)
   embankments, 339
Fourier transform, 495g
Fourier Transform infra-red (FTIR) spectrometry, 206
fracture, 495g
framework silicate, 495g
freeze drying, 209, 495g
friction, 495g
Frome Clay, 164

FTIR (Fourier Transform infra-red) spectrometry, 206
Fullers earth, 17, 65, 150, 495g
Fullers Earth Clay, 164, **466**, **467**, **475**
fundamental particles, 495g
fungi, 25

Ganges, 38
gap grading, 495g
Gault Clay, 124–5, 167, *167*, **466**
GCL (geosynthetic clay liners), 364, *365*, 367–8, *369*
gel strength, 495g
gels, 87, 495g
geochemical analysis, 214–18
geogrid, 302, 304, *304*, 495g
geological classification, 75–6
geomaterials, 495g
geomembranes, 364, *365*, 367–8, *369*, 495g
geophysical investigations, 185–8, *186*
geophysical surveying, 495g
geopolymer cement, 436
geosynthetic clay liners (GCL), 364, *365*, 367–8, *369*
geosynthetics (clay reinforcement), 302, 304, *304*, 495g
geotechnical properties, 101–12
geotectonics, 29–30, 495g
geotextiles, 302, 304, *304*, 495g (*see also* geomembranes)
gibbsite, 24, 34, 495g
glacial activity, 147, *148*
glacial clay, 495g
glacial tills, 43–4, 119, 141, 171, **462–3**
glauconite, 16, 18, 45, 495g
glaucony, 45–6, 496g
glycerol, 202, 203, 496g
gneiss, 496g
goethite, 24
grading, 496g
  clay materials, 78–80, *235*
  slurries, *357*
grading curves, *76*, 80
grain size *see* particle size
granulate pressing, 422
grass, fill stabilization, 304
gravel, 496g
Great Oolite Group, 162, 164
green strength, 496g
greenfield site, 495g
gres porcellanato, 340, 412, 496g
griegite, 24
gritstone, 496g
grog, 496g
ground stresses, 95–6, *95*
groundmass, 496g
groundwater, in mineral workings, 193–4, 288–9, 335–6
grout, 496g
grouting, 349–50, 352, 375–7, 439–40
gypsum, 24, 114, 155, 163, 216
gypsum-Portland-pozzolana cements, 438–9, **438**

habit (crystalline), 496g
haematite *see* hematite
Halesowen Formation, 158, 159
halide minerals, 25
halite, 25, 159, 160
halloysite
  ceramics, 405, **406**
  definition, 496g
  formation, 32, 57, 66, 67

mineralogy, 15, *17*
soil properties, 117, 118
water absorption, 21, 26
halmyrolysis, 46, 496g
hammer mill, 496g
hard pan, 496g
haul roads, trafficability, 83, 332–3, **334**
head deposits, 172, 496g
header (brickwork), 496g
health and safety (*see also* slope stability)
  earthmoving sites, 300, 343–4
  field investigations, 190
  lime and cement stabilization, 307
  mineral workings, 286, 287–9
Health and Safety at Work Act 1974, 343
heave, 113–14 (*see also* swelling properties)
heavy clay products, 405–12, **409**
hematite, 24
hemipelagic muds, 40
high-alumina cement, 437–8
highways
  embankments, 278–9, *279*
  Leighton–Linslade Bypass, 340–3, *342*
  Newbury Bypass, 308–10, *310*
  substructure capping, 306
Hoffman kilns, 411, 496g
hopper capacity, 488g
horizonation, 497g
Hughley Shales, 157
hydrated halloysite, 497g
hydration, 26
hydraulic conductivity *see* permeability
hydraulic consolidation cell
  consolidation test, 249–53, *250–2*, **251**, **253**
  permeability test, 256–7, *256*
hydraulic lime cements, 435–6, **435**
hydraulic mining, 420–1
hydraulic trap, 364, 497g
hydrocyclone, 497g
hydrograph, 497g
hydrolysis, 497g
hydrometer test, 234, *234*
hydrothermal alteration, 55–7, *56*, 67, *67*, 497g
hydrothermal clays, 140
hydrothermal kaolin deposits, 66–7, *67*, 149, **149**
hypersaline, 497g
hysteresis, 497g

ice, role in clay formation, 43–4
ICOLD Register of Dams, 281, 282
ICP-AES analysis, 215, 497g
ICP-MS analysis, 215, 497g
igneous activity (*see also* volcanic ash)
  hydrothermal alteration, 55–7, *66*, 67, *67*
igneous rocks, clay formation, 34–5, *36*, 57, 66
illite (*see also* ferric illite; glauconite)
  ceramics, 405, **406**
  definition, 497g
  distribution, *37*
  formation, 34–5, 38, 49, 50–2
  ion exchange, 21
  mineralogy, 15–16
  properties, **74**, *93*, **94**, *94*
  water absorption, 21
illite crystallinity, 51, *52*, 203–4, **204**, *204*, 497g
illite-smectites, 20, *20*, 50–2, 57

illitization, 50–2, *51*, *52*, 55
image analysis (microscopy), 212–13
imogolite, 24
improved clay, 301–7
impurities
   in ceramic clays, **99**, **406**
      clay, in concrete, 444–5, **445**
incinerator ash, recycling of clay, 441
inductively coupled plasma atomic emission spectrometry (ICP-AES), 215
inductively coupled plasma mass spectrometry (ICP-MS), 215
induration, 60, 497g
infrared (IR) spectroscopy, 206, *207*
infrared radiation, 497g
inheritance (clay formation), 30–2, 33–4, 497g
initial effective stress test, 272
inorganic geochemical analysis, 214–18
inorganic polymer cements, 436–7
instability *see* slope stability
intercrystalline swelling, 21, 113
International Commission on Large Dams, Register of Dams (ICOLD), 281, 282
interparticle swelling, 21, 113
interstitial water analysis, 216
interstratified clay minerals *see* mixed layer clay minerals
intracrystalline swelling, 113
intrusives, 498g
invert (tunnel), 498g
ion, 498g
ion beam milling, 498g
ion exchange, 498g
ion exchange capacity, 217–18
ion exchange properties, 21–3, *21–2*, **23**, 24, 65, 114
IR (infra-red) spectroscopy, 206, *207*
iron carbonate *see* siderite
iron hydroxides, 24
iron minerals, 413 (*see also* goethite; hematite; pyrite; siderite)
iron oxides, 24, 403, 424, 431, 439
iron pyrites *see* pyrite
iron sulphides, 24, **24**, 49, 366 (*see also* pyrite)
ISO 14688:1996, 5
isomorph, 498g

jarosite, 24, 163
jasper, 23
jaw crusher, 498g
joints, in clays and mudrocks, 107, 109–10, 133, 498g
Jurassic deposits
   United Kingdom, 125–7, **126**, **132**, **454**, **455**, 458–9, **467–8**, **479**
      stratigraphy, 161–5, *162*
   World and European, 144, **476**

K-bentonites, 65
kaolin, 161 (*see also* dickite; halloysite; kaolinite; nacrite)
   ceramics, 404, **404**
      fine ceramics, 417–18
   definition, 498g
   deposits, 66–7, *67*, 149–50, **149**, 417–18, **475**
   extraction and processing, 420–2, *421*
   formation, 53, 59, 63–4, 66–7, *67*
   ion exchange properties, 21
   mineralogy, 15
   production statistics, **7**
kaolinite
   in ball clay, 64
   ceramics, 98, 404–5, **406**

   definition, 498g
   distribution, *37*
   formation, 34, 37, 48, 53, 57
   mineralogy, 15, *16*
   properties, **74**, *93*, **94**
   water absorption, 21
karst, 498g
kerogen, 25, 219
kerolite, 45
Keuper Marl *see* Mercia Mudstone
kibbler, 498g
kilns, 411
Kimmeridge Clay, 126, 165, 431, **467**
Kubler index, 51, 203–4, **204**, 498g

laboratory tests
   compositional and textural analysis, 199–222
   properties, 223–74, **481–4**
lacustrine deposits, 65, 141, 498g
lagoons
   mine and quarry, 286
   sealing, 381
lake deposits, *47*, 65, 141
Lambeth Group, 121, 168, *169*, 172, **465**
laminated clays, 120, **464**
lamination (sedimentation), 42–3
landfill liners and covers, 362, 364–8, *364–5*, 440 (*see also* waste disposal / reuse)
landscaping, earthworks, 285, *285*, 309
laterite, 498g
lateritic soil, 116–17, **116**, *117*, 498g
latocrete, 498g
layer silicates, 13–15, *14*, *15*
leachate, landfill liners, 365–7, 498g
leaching, clay formation, 34
leaching techniques, analytical methods, 216, **216**
LECA (light weight expanded clay aggregate), 439
Leighton–Linslade Bypass, 340–3, *342*
lepidocrosite, 24
Lewis Acid, 499g
Lias clay, 125, 126–7, 162, *162*–4, **468**
light weight expanded clay aggregate (LECA), 439
lime and cement stabilization, 114, 296–7, 304–7, *305*, 326
lime cements, 435–6
lime mortars, 435–6
limestones, 428, 435
limonite, 24
linear shrinkage, 231
liners
   canals, 377, 382–3
   landfill, 362, 364–8, *364–5*, 440
liquefaction, 499g
liquid limit, 89, 91, 100, 499g
   tests, 228, *229*
liquidity index, 82, 92, 499g
   tests, 229
lithology, 499g
lithorelict, 499g
load factors, **334**
lodgement till, 499g
loess, 44, 115, 499g
logging
   boreholes, 193
   geophysical, 187–8
   trial pits, 189

London Clay, 50, 121, 123–4, 459
  engineering properties, **465**
London Clay Formation, 169–70, *170*
Ludlow Shales, 157

mackinawite, 24
made ground, 499g
mafic rocks, 499g
  clay formation, 34–5, *36*, 57
maghemite, 24
magnesium sulphate, 24, 216
maintenance, engineering fills, 297–307, **302–3**
mapping surveys, 183–8, *186*
marcasite, 24, 296
marine environments (*see also* continental margin; deep sea clays)
  clay deposits, 65–6, *65*, **464**
  clay formation, 39–41, *43*, 45–8, 65–6
marls, 60, 159
mass spectrometry, 215
matric potential, 499g
matric suction, 97–8, 499g
matrix (sediments), 499g
maximum dry density, 238, 499g
MCV *see* Moisture Condition Value (MCV)
MEB (methylene blue) dye test, 217, 445
mechanical improvement, engineering fills, 302, 304
Mercia Mudstone
  formation, 45, 53, 62
  mineralogy, 458
  properties, 127–9, 132, **469**
  stratigraphy, 159–61
Mesozoic deposits, 144–6, *145* (*see also specific periods*)
metabentonites, 65, 499g
metahalloysite, 15, 116, 499g
metakaolin, 434, **434**, 435–6, *436*
metamorphism, 29, 30, *30*, 48, 51, 499g
metastability, 172, 499g
methylene blue (MEB) dye test, 217, 445
micas, 15–16, 32, 35, 203, 499g
micro-organisms, 25–6, 49, 209, 220, 499g
microfossils, 53, 66
microgranite, 499g
microscopy, 209–13
microstructure
  clay minerals, 14–23, *14–21*
  formation, 42–3
microtunnelling, 358, *359*, 499g
milliequivalents, 500g
milling, 422
Millstone Grit, 129, 157, 158
mineral, 500g
mineral rights, 287
mineral stability diagrams, 34, *36*
mineral workings
  earthmoving, 317–20, **328**, 340, **341**
  end uses, 188
  environmental impacts, 183, 287–9
  filling by grouts, 376
  fine ceramic clays, 419
  planning, 286–91
  quarry tips, 285–6, 317–20, 330
  site restoration, 289, 319
  structural clays, 408
  surface and groundwater, 193–4, 288–9, 335–6
mineralogy, 13–27
  analytical methods, 199–209
  data, 449–59, **452–7**

Miocene, USA, 45
mixed layer clay minerals, 19–20, 32, 34, **202**, 203, **203**, 500g
  (*see also* chlorite-smectites; illite-smectites)
modulus of rigidity, 500g
modulus of rupture, 111, 500g
Mohr circles, 104, *105*, 106, 500g
Mohr-Coulomb criterion, 104, *105*
Moisture Condition Value (MCV), 101, 295, 337
  tests, 238, 240–2
moisture content, 87–8, **87**, **91** (*see also* pore pressures)
  bricks, 412
  ceramic clays, **87**, 98–9
  in compaction, 100–1, *101*, 238, 239–42, *239*, *240*
  definition, 500g
  effect on plasticity, 88–9, *89*, 91–2, *91*
  effect on strength, 110–11, *112*, 114–15, 116, *117*
  engineering fills, 291–2, *291–2*, 294–5, *294–5*
  related to suction, 97–8, *98*, 113–14
  related to weathering, 133
  tests, 227–8, *241–2*
Monkerton Formation, **471**
montmorillonite
  ceramics, 405, **406**
  definition, 500g
  formation, 57, 64
  infra-red spectra, *207*
  mineralogy, 17, 347
  properties, 17, **74**, *93*, **94**, *94*, 100, 113
mud, 500g
mud and stud, 395
mud deposits, continental shelves, 41, *41*, *42*
mudrocks
  classification, 60, **60**, 81–4, **81**, *82*, **83**, 85, **86**
  composition, 60–3, **61**, *62*, 82
  definition, 500g
  description, 84–5
  durability, 82–4, **83**
  fabric and microstructures, 42–3, *45*, *46*
  porosity, 54–5, *54*, *55*
  strength, 82, **84**, 111–12, **113**, *113*
muds, drilling, 358, 361
mudstone, 59–60, **60**, **81**, *82*, 500g
mudwalls, 395
Munsell colour system, 500g
muscovite, 16, *17*, 500g

nacrite, 15, 66, 500g
natural moisture content *see* moisture content
negative pore pressure *see* suction, soils
neoformation, 30–2, 34, 45, *47*, 500g
Newbury Bypass, 308–10, *310*
nodules, 500g
noise barriers, 284, *284*
noise, mineral workings, 288
nontronite, 17, 45, 500g
normal compression line, 500g
normal stress, 501g
normally consolidated clays, 106–7, 501g

oceans *see* marine environments
odinite, 46, 53, 501g
oedometer, 501g
oedometer test, 245–8
oil shale, recycling of clay, 442
Oligocene deposits, **452**, **453**, 459
ooids, 501g

opal, 23, 53, 66
optical microscopy, 209
optimum moisture content, 238, 501g
Ordovician deposits
    United Kingdom, **480**
    World and European, 142–3, **477**
organic-clay complexes, 26
organic matter, 25–6, 48–9, 66, 155 (*see also* biological processes; micro-organisms)
    analytical methods, 218–20, **236**
    effect on soil stabilization, 297
organic maturity, 501g
organic petrography, 219–20, **220**
Ormesby Clay Formation, 168
Orville Dam, California, 310
over-compaction, 295, 339
over-consolidation, 106, 108, 133, 153–4, 297–8, 501g
over-consolidation ratio, 501g
overburden, 501g
overburden pressure, 501g
overcut, 501g
overstressing (compaction), 295, 339
Oxford Clay, 98, 125–6, 164–5, **467**
oxisols, 58–9, 501g

Palaeocene deposits, **452**, **453**, 459
palaeoclimate, 501g
palaeogeography, 141–7
palaeosoil, 501g
Palaeozoic deposits (*see also specific periods*)
    United Kingdom, **456**, **457**, 458
    World and European, 141–4, *142*
palygorskite (attapulgite)
    applications, 347
    concrete additive, 439
    definition, 501g
    formation, 45, 57, 139
    mineralogy, 20, *21*, 201
    plasticity, **94**
    water absorption, 21, 26
paper waste, recycling, 442
paragonite, 57
particle density, tests, 233, **233**
particle size
    for bricks and tiles, *408*
    classification of clays, *6*, 78–80, 81, 82, *82*
    effect on settling, 41–2
    river suspended sediment, *40*, *41*
    sedimentary rocks, *13*
    tests, **226**, 233–4
particle size distribution *see* grading
peak shear strength, 109, 262
peak strength, 502g
ped, 502g
pedological classification, 78
pelagic sediments, 44, 66, 502g
Penarth Group, 161
penetrometers, 228, 502g
Pennines-style dam, 283, 377, 377–81, **378**, *378*
periglacial processes, 502g
perimarine deposits, 65, 502g
permeability, 43, 54–5, *55*, 94–5, 100
    classification, **95**
    definition, 502g
    effect of compaction, 365, *366*
    effect of pollutants, 367
    puddle clay, 380
    tests, 255–8, *255–7*, **483**

permeameter, 502g
Permian deposits
    United Kingdom, **471**, **479**
    World and European, **476**
Permo-Triassic deposits
    United Kingdom, 127–9, 458
        stratigraphy, 159–61, *160*
    World and European, 143–4
Peterborough Member, 164–5, 410
petrofabric analysis, 209–14, 502g
petrography, 502g
petrological microscope, 502g
petrology, 502g
pF *see* suction
Pfefferkorn test, 99
pH *see* acidity
phenocryst, 502g
phlogopite, 59
phosphate minerals, 25
phyllite, 502g
phyllosilicates, 13–15, *14*, *15*, 502g
phytoplankton, 502g
piezometers, groundwater measurement, 194, 502g
pile construction, using bentonite slurries, 352–5, *353*, 373
pinhole method, 243, **245**, *245*
pipe jacking, 358, *359*, 502g
pipes (ceramic), 406
pipette test, 234, *234*
pitchers, 502g
pitchstone, 503g
placer deposits, 503g
plane polarised light, 503g
plankton, 503g
planning applications, mineral workings, 286–7
plastic clay, 362–82
plastic limit, 89, 91, 92, 100, 503g
    tests, 228, 229–30, 295
plastic viscosity, 503g
plasticity, 26, 75, 88–94, *89*, *91*, **94**
    definition, 503g
    effect of lime, 305, *305*
plasticity charts, *79*, 92, *93*, *122*, 229, 503g
plasticity index, 91, 92, 99, 503g (*see also* activity charts)
    tests, 228, 229, 230
plate tectonics, 29–30, *31*, *140*, 503g
point counting, 503g
Point Load Test, 112
Poisson's ratio, 503g
pollutants
    absorption by bentonite materials, 377, 440
    effect on clay permeability, 367
pollution, from mineral workings, 288–9
polymer cements, 436–7
polymer-sand-bentonite mixes, 349–50
ponds, sealing, 381
porcelain, 413
pore blocking, 349
pore pressures, *95*, *96*, 97–8
    in compression, 104, 111, 112 (*see also* effective stress)
    definition, 503g
    measurement, 192, 248, 252, *252*, 253
    slope stability, 288, 294, 298, 339–40
porosity
    ceramics products, 411–12, *412*, 413
    definition, 503g
    mudrocks, 54–5, *54*, *55*

porphyry copper deposits, 67
Portland cement, 427–34, **429**, **430**, **432**, **433**, 503–4g
Portland-pozzolana cements, 433–4 (*see also* gypsum-Portland-pozzolana cements)
potassium
    effect on clay firing, 410
    illitization, 50, 204
    shale content, 63
    slurry additive, 358
potassium feldspar, 458, 459
pozzolanas, 433–4, 504g
Pre-Carboniferous deposits, 156–7
precipitation, 504g
precision, laboratory tests, **206**, 226–7, **227**
preconsolidation pressure, 254, 504g
pressure, 504g
principal stress, 504g
production statistics, 7, 401
prospecting *see* exploration
protozoa, 25
pseudomorph, 504g
puddle clay, 283, 377–81, **379**, **380**, 382–3, 504g
pugging, 504g
pulverized fuel ash, 442
punned clay, 283, 504g
pyknometer method, 233
Pyrament concrete, 436–7, **437**
pyrite, 24
    in ceramics, 401, 413
    definition, 504g
    deleterious chemical effects, 296–7, 336
    formation, 49, 163
    geochemical analysis, 216, 235
pyrolysis, 504g
pyrophyllite, 53, 57, 66, 504g
pyrrhotite, 24

quantitative mineralogical analysis, 204–6
quarries *see* mineral workings
Quarries Regulations 1999, 286, 301, 330
quartz
    in ceramics, 403, **404**, 410
    definition, 504g
    formation, 49, 53, *53*
    mineralogy, 23
    mudrock composition, 61–2, 82
quartz porphyry, 504g
Quaternary deposits
    United Kingdom, 119–20, **452**, **453**, 459, **465**, **478**
        stratigraphy, 170–3, *170*
    World and European, 139, **475**
quick clays, 115, 504g

radioactive waste storage, 368–73, **370**, *371*
radioactivity logging, 187
radiolaria, 53, 66
railways, embankments, 276–8, *278*, *280*, 298, 300–1
rainfall, role in clay formation, 35, *36*
rammed clay, 283
rammed earth construction, 389, **391**, 398
rapid firing kilns, 411
reaction interval, 505g
Reading Formation, 121, 168, *169*
ready mixed concrete, 439
Recent deposits, United Kingdom, 119–20
reconstituted soil, 505g

recrystallisation, 505g
rectorite, 57, 505g
recycled clays, 441–4
red-black toposequence, 34, 59, *59*, 505g
red clays, deep sea, 44, 65–6, *65*
red tropical soils, 116–17, **116**, *117*
Redcar Mudstone, **468**
reference intensity ratio (RIR) method, 205
refining, 419–20, 421
refractories, 505g
refractoriness, 505g
refractory properties, 63, 98, 505g (*see also* fireclays)
regolith, 505g
reinforced clay (geogrid), 302, 304, *304*, 505g
relative consistency, 229
relict moss movements, *185*
remote sensing, 181–2, *181–2*
repair
    earth buildings, 396
    earthworks, 300, 301
reservoirs (*see also* earth dams)
    sealing, 381
Reservoirs Act (1975), 282, 299
residual clays *see* tropical residual soils
residual shear strength, 110, 133, *133*, 505g
    tests, 262–3, *263*
residual soil, 505g
resistivity logs, 187
resonant column test, 272
restoration, sites of mineral workings, 289, 319
rheological blocking, 349
rheological properties, 88, 439–40
rheology, 505g
Rietveld methods, 205
riffling, 505g
ring shear test, 262–3, *264*
RIR (reference intensity ratio) method, 205
rivers, sedimentary processes, *38*, *39*, *40*, **41**, *41*, 141 (*see also* alluvial clays)
road base, 505g
roads *see* haul roads; highways
Rock-Eval pyrolysis, 218–19, *219*, *220*
rock type, role in clay formation, 34–5
roller-hearth kilns, 411
rolling resistance, **334**
Roseki, 66, 505g
Rowe cell *see* hydraulic consolidation cell
rupture, modulus, 111
Russian Platform, 50
rutile, 413

sabkha, 505g
safety *see* health and safety; slope stability
samples
    analysis, 199, *200*
    field investigations, 194–6, *196*
    particle size, **226**
    preparation, 215, 223–5, *224–6*, **481**
sampling programmes (exploration), 188–9
sand, 505g
sand-bentonite mixes, 349, 349–50
    compaction characteristics, *368*
    landfill liners, 367
    in radioactive waste storage, 372
    shrinkage, *369*

sanitaryware, 413, **416**, 505g
saponite, 17, 45, 57, 505g
saprolite, 505g
saturation, soil, 87, **87**, 96–8, 505g (*see also* air entry point; moisture content)
scanning electron microscopy (SEM), 209–12, **210**, *211–13*
schist, 505g
Scunthorpe Mudstone, **468**
sea basins *see* marine environments
seam (layer), 505g
seatearth, 16, 63, 505g
seawater, bentonite compatibility, 348
secant pile walls, 373, *373*, 506g
secondary electrons, 506g
sediment, 506g
sedimentary processes, 36–48, *40–3*, **41**, *47*, 66, 141
sedimentary rocks, 13, *13*, 59, 506g (*see also* mudrocks)
sedimentation, 506g
seepage control, in earthworks, 284, 298
seismic refraction methods, 186–7, *187*
selenite, 164–5
SEM (scanning electron microscopy), 209–12, **210**, *211–13*
sensitive tint plate, 506g
sensitivity, 111, **111**, 506g
separation plants, bentonite slurries, 362, *363*
sepiolite
　definition, 506g
　formation, 45, 57
　mineralogy, 20, *21*
　water absorption, 21, 26
sericite, 16, 405, 506g
serpentine, 15, 46, 57, 202–3, 506g
sesquioxides, 506g
settlement (*see also* consolidation)
　clay fills, 292–3, 299
settling limit, **91**
settling (sedimentation), 41–2
shales, 81–2, *82*, 506g (*see also* common clay and shale; mudrocks)
shear box, 506g
shear planes, 293
shear strain, 506g
shear strength, 103–4, *108* (*see also* effective stress shear strength; residual shear strength; undrained shear strength)
　engineering fills, 293, *294*
　mudrocks, 112, *113*
　parameters, 109–11, **110**
　puddle clay, 380, **380**
　related to weathering, *131*
　sensitivity classes, **111**
　tests, 258–68, 273, **483**
shear stress, 90–1, 102–3, *102*, 109, 506g
shear surface, 506g
shearbox test, 260–2, *261*, *263*
sheet silicates, 13–15, *14*, *15*
shelf (continental) *see* continental margin
shredding, 419
shrinkage, 75, 112–14, *230*
　ceramic clays, 99
　sand-bentonite mixes, *369*
　tests, 230–1, *231*
shrinkage limit, 89, 91, 506g
shrinkage ratio, 230
siderite
　in ceramics, 401
　formation, 49, 53
　mineralogy, 23
　mudrock composition, 62, 166
silcrete, 506g
silica, diagenesis, 53, *53*
siliceous microfossils, 53, 66
silicification, 23
silt ponds, 286
silts, 506g
　distinguishing from clay soils, 76, **78**
siltstones, **60**, 81, **81**, 82
Silurian deposits
　United Kingdom, 157, **480**
　World and European, 143, **477**
simple shear test, 273
sintering, 410, 506g
size reduction, 408, 422
Skipsea Till Formation, *171*
slake durability, 506g
Slake Durability Index, 83
slaking, 82, 83, 507g
slates, 48, 64, 143, 156, 168, 507g
slip, 507g
slip casting, 413, 422–3
slope stability
　durability of mudrocks, 83
　earthworks, 287, 330, 339–40, 343
　embankments, 277–9
　engineering fills, 294
　failure, *178*, 288, 297–8, **298**
slurries, 88, 347–62, *350*, *356–7*
slurry wall, 507g
smectites (*see also* beidellite; bentonite; chlorite-smectites; illite-smectites; montmorillonite)
　definition, 507g
　deposits, *37*, 38, 150
　formation, 34, 38, 45, 46, 48, 57
　illitization, 50–1, *51*
　ion exchange properties, 21
　mineralogy, 16–17, *18*
　water absorption, 20, 26
soak period, 424, 507g
sodium, ion exchange, 65, 114
sodium sulphate, 24, 216
soffit, 507g
soft mud process, 409
soil activity, 92, 94, **94**, *94*, 230, **230**
soil chlorites, 34
soil conditioning, 358
soil modification, 507g (*see also* soils, stabilization)
soil profiles, 33, 507g
Soil Taxonomy system, 78
soils
　classification, 6, 75–80, **77**, **79**, 227–35
　definition, 507g
　description, 80–1
　formation, 33–6, *33*, 58–9
　grouting, 349–50, 375–6
　laboratory tests, 223–74
　properties, 73–81, 87–118, **94**
　stabilization, 114, 115, 296–7, 301–7, **302**, **303**, 326
solid (bedrock) deposits, 76, 154
solid waste, recycling of clay, 441
solifluction, 172, 184, 507g
solifluction shears, 507g
sorting (geological), 507g
sounding rig, *193*

souring, 408, 507g
Specification for Highway and Bridge Works, 330, **331**, 333, 337
specimens *see* samples
Speeton Clay, 124, 166–7, **466**
spray drying, 422, 507g
stability *see* slope stability
stabilization (soil), 114, 115, 296–7, 301–7, **302**, **303**, 326
standards (*see also* BS 1377:1990; BS 5930:1999; ISO 14688:1996)
 concrete, **432**
 earth building, 393
 exploration, 178, 179
 puddle clay, 382–3
 test methods, **481–4**
stereoscope, 508g
stevensite, 45, 508g
sticky point, 91, **91**, 508g
stiffness, 101–2, 508g
stiffness index, 99
stoneware, 413, **415**, 508g
stop chocks/blocks, 508g
strain, 508g
stratigraphy, British clays, *9*, 153–74, **155**
strength parameters, 101–2, 103–4, *103–4*, 109–12 (*see also* compressive strength; shear strength)
stress, 508g
stress path tests, 270–2, *270*, *270*
stress relief, 508g
stress–strain relationships, 88, *89*, 90–1, 101–2, 112
stresses, ground, 95–6, *95*
stretcher (brickwork), 508g
structural clay products, 405–12, **409**
structural clays, 407–8, **407**
structural damage
 buried structures, 296–7
 problem soils, 112–17
structural performance, earth buildings, 392
submerged density, 231–2
substructure fills, 285, 295–6, 299
suction, soils, 96, 97–8, **98**, *98*, 113–14, 508g (*see also* pore pressures)
 tests, 234–5
sudoite, 57, 508g
sulphate minerals, 24 (*see also* barite; celestite)
sulphate reduction, 508g
sulphur species (*see also* aluminium sulphate; calcium sulphates; iron sulphides; magnesium sulphate; sodium sulphate)
 analytical methods, 216, **236**
 detrimental effects, 235, 296–7
superficial deposits, 76, 154, 172, **464** (*see also* glacial tills)
surface areas
 measurement, 213–14, **214**, *214*
 minerals in concrete, **445**
surface waters (*see also* rivers)
 impact of mineral workings, 288–9, 335–6
surfactant, 508g
surveying, 181–8, *182*, *184* (*see also* exploration)
suspensions, 87–8, **87**
swelling chlorites, 21, 508g
swelling index, 108, 508g
swelling limit, 91–2, **91**, 508g
swelling properties, 75, 107–9, 112–14, 508g (*see also* activity (soil))
 bentonites, 65
 clay minerals, 20–1
sylvite, 25

tableware, 413
tailings lagoons, 286
talc, 57, 508g
talluvium, 508g
tanking systems, 381
tapis *see* cob building
Tarporley Siltstones Formation, 161
tectoaluminate cement, 437, **437**
tectonic plates, 29–30, *31*, *140*, 508g
Tees Laminated Clay, 120
TEM (transmission electron microscopy), 212
tensile strength, 111, 112, 508g
tephra, 508g
terminology, 184 (*see also* definitions of clay; glossary)
terra cotta, 508g
terrigenous clays, 59–63
Tertiary deposits, 121–4, 167–70, *168* (*see also specific periods*)
 United Kingdom, **465**, **478**
 World and European, **475**
test methods
 bentonite slurries, 350–2, **356**
 compositional and textural analysis, 199–222
 properties, 223–74, **481–4**
 puddle clay, 377
test specimens *see* samples
textural fabric, 42–3, *45*, *46*
 analysis, 199–222
thaumasite, 24
thaumasite attack, 163, 296
thermal analysis, 207–9
thermal expansion, 509g
thermal gravimetric analysis, 208–9, *208*
thermal performance, earth buildings, 392
thermal properties, 99–100
thin-section, 509g
thixotropy, 88, 348, 439, 509g
tile clays, 98, 407–8
tiles, 406–7, *408*, 412–13, **414**
tills, 43–4, 119, 141, 171, **462–3**, 509g
tips, quarry waste, 285–6, 317–20, 330, 509g
titanium minerals, 413 (*see also* anatase)
titanium oxide, in kaolin, 64
TOC (total organic carbon) analysis, 218, *218*
tonsteins, 48, 64–5, 509g
topography, role in clay formation, 34
topsoil, 509g
tosudite, 57
total dissolved solids, analytical methods, **237**
total organic carbon (TOC) analyses, 218, *218*
total stress, 509g
toxic waste, immobilization, 440–1 (*see also* radioactive waste storage)
traffic, quarry, 288
trafficability, earthmoving plant, 295, 332–3, **334**, 509g
transformation (clay formation), 30–2, 34, 37, 509g
transmission electron microscopy (TEM), 212
transport processes (sedimentation), 38–45
treated clay, 509g
tremie (pipe), 509g
trenches, 189, 343, 354
Tresca criterion, 103, *105*
trial pits, 189–90, *190*, *196*
Triassic deposits
 United Kingdom, **454**, **455**, **469–70**, **479**
 World and European, **476**

triaxial cell
    compression test, 259–60, *260*
    consolidation test, 253–4, *254*
    effective stress tests, 265, *266*, *268–9*
    permeability test, 257–8, *257*
triaxial test, 509g
trioctahedral sheet structures, 492g, 509g
troilite, 24
trophic level, 509g
tropical areas, clay formation, 34, *35*, 46, 58, 59, *59*
tropical residual soils, 113, 115–17, **116**, 140
tropical weathering, 509g
tunnel kilns, 411
tunnelling, 326–7
    durability of mudrocks, 83
    using bentonite slurries, 355–8, *356*, 360–1

ultimate strength, 509g
ultraviolet radiation, 510g
unconfined compressive strength, 111, 111–12, 259, 510g
unconfined tensile strength, 510g
under-compaction, 339
underclays, 16, 63, 510g
undisturbed sample, 510g
undrained shear strength, 92, 110–11, 133, 510g
uniaxial compressive strength, 82, *82*, 111–12, 510g
uniaxial tensile strength, 111, 510g
Unified Soil Classification System, **79**
United Kingdom
    clay and mudrock properties, 117–38, **462–73**
    clay deposits, **478–80**
    clay stratigraphy, *9*, 153–75, *156–7*, *160*, *162*, *168*, *170*
    geographical distribution of clays, *8*
unsaturated clays, 96–8
Upnor Formation, 168

valence, 510g
Van Krevelen diagram, 219, *220*
vane shear test, 258–9, **258**, *258*
varved clays, 120, 510g
vaterite, 23
veins, 510g
verdine, 510g
Verdine facies, 46
vermiculite
    definition, 510g
    formation, 32, 34, 37, 59
    identification, 202
    mineralogy, 17, *19*
    water absorption, 20, 26
vertisols, 58, *58*, 116, 118, 510g
viscosity, 88, *351*
visual barriers, 284, *284*
vitreous hotelware, 413
vitrification, 410, 412, 510g
vitrification range, 510g
vivianite, 25
voids ratio (*see also* air void content)
    definition, 510g
    related to soil strength, 109, 117
    tests, 233, 246–7, *249*, 252, 254

volcanic ash, 46, 48, 62–3, 64
volcanic glass, in mudrocks, 62–3

Wadhurst Clay Formation, 166, **466**
wall tiles, 412
Warwickshire Group, 158–9
wash boring, 192
waste disposal / reuse (*see also* landfill liners and covers; radioactive waste storage)
    bentonite slurries, 361–2, *364*
    earthmoving, 330
    mineral workings, 287–8
    recycled clays in concrete, 441–4
    toxic waste immobilization, 440–1
waste tips, quarries, 285–6, 317–20, 330
water absorption / adsoption, 20–1, 26, 412, 511g (*see also* moisture content)
water potential, 511g
watercourses *see* surface waters
wattle and daub, 391, 393
wavelength dispersive spectrometry (WDS), 215
Weald Clay, 124, 166, **466**
weathering
    clay formation, 32–6, 155
    definition, 511g
    mudrocks, 82–3, 85, **86**
    related to moisture content, 133
    related to shear strength, *131*
Weaver Index, 203–4, **204**
Wenlock Shales, 157
West Walton Formation, 165, **467**
Westbury Formation, 161
Whipton Formation, **471**
Whitby Mudstone, 162, 164, **468**
white Portland cement, 432–3, **433**
whiteware products *see* fine ceramics
wildlife habitats, impact of mineral workings, 289
wind, in clay formation, 44–5, *47*
window sampling, field investigations, 191
witchert *see* cob building
Woolwich Formation, 121, 168
workability, 511g

X-ray diffraction, 511g
X-ray fluorescence, 511g
X-ray fluorescence spectrometry, 214
X-ray goniometry, 213
X-ray powder diffraction (XRD) analysis, 199–206, *201–3*, **202**, *205*
X-ray powder diffractometer, 511g
X-ray reflections, 511g
X-rays, 511g

yield, 511g
yield stress, 88
Young's modulus, 511g (*see also* stress-strain relationships)

zeolites, 24–5, **25**, 57
    cracking catalysts, 443
zooplankton, 511g